The Behaviour, Population Biology and Physiology of the Petrels

The Behaviour, Population Biology and Physiology of the Petrels

JOHN WARHAM

Zoology Department
University of Canterbury
Christchurch, New Zealand

ACADEMIC PRESS
Harcourt Brace & Company, Publishers

London · San Diego · New York · Boston · Sydney · Tokyo

ACADEMIC PRESS LIMITED
24–28 Oval Road
LONDON NW1 7DX

U.S. Edition Published by
ACADEMIC PRESS INC.
San Diego, CA 92101

This book is printed on acid free paper

A catalogue record for this book is available from the British Library

ISBN 0-12-735-4158

Typeset by Paston Press Ltd, Loddon, Norfolk, UK
Printed and bound in Great Britain by the University Press, Cambridge

Contents

Preface

When beginning this companion volume to *The Petrels* back in 1990 I would have been alarmed to know that 5 years' steady work would be needed to review topics ignored or only touched upon in the earlier book. In practice, the amount of information proved to be very extensive and scattered, added to which some aspects, such as the study of tubenoses at sea, have been intensively researched recently and it has been difficult to keep abreast of developments.

On account of page limitations the text is quite condensed, so that some topics may seem to be treated rather summarily and discussion curtailed, while other, perhaps lesser-known, aspects are given more prominence. However, I have tried to ensure that ample referencing points the reader to the appropriate literature in such instances.

I have been greatly helped by others who have provided data, tape-recordings, and preprints of work in the press, and to many workers and their publishers who have permitted the use of their illustrations: these are acknowledged in the text. Colleagues in the Zoology Department have continued their support and I am also indebted to our University Librarians for help over many years. A particular debt is owed to colleagues who have commented on parts of the text. These include D. G. Ainley, W. R. P. Bourne, V. Bretagnolle, J. C. Coulson, M. E. Forster, R. Gales, M. P. Harris, J. A. L. Hector, M. J. Imber, J. Jacob, G. R. Martin, D. B. Peakall, L. McPherson, K. A. Nagy, D. T. Parkin, A. M. Paterson, C. J. Pennycuick, P. A. Prince, D. R. Thompson, B. M. Wenzel and G. C. Whittow. Errors are my own, and I hope not to have passed any on—'from one naturalist to another as an heir-loom'—to use F. W. Hutton's words from long ago!

My final gratitude goes to my wife Pat, who can now look forward to a husband free to tackle rather more mundane tasks, such as painting the house, and even find time for a holiday.

References to the earlier volume take the form *The Petrels*, p. 00. Statistical means are ±1 SD unless otherwise stated, and nomenclature remains that of the checklist in *The Petrels*, pp. 424–433, except that I have used *Pseudobulweria* for *aterrima, rostrata* and *macgillivrayi*.

John Warham
Christchurch

CHAPTER 1
Petrel Populations

Contents

I Introduction

Judging by their worldwide distribution, numbers and biomass, the Procellariiformes are the most successful of all seabirds. Petrel populations vary in size from the very small to the very large.

Biggest aggregations occur at high latitudes, particularly in the subantarctic. Although some northern hemisphere birds such as the storm petrels *Oceanodroma leucorhoa* and *O. furcata* reach large numbers, and *Fulmarus glacialis* about 4.3 million pairs, most of the rather few species there have relatively small populations, perhaps due to past and present competition from the auks. For example, Lensink (1984) estimated that the three breeding Alaskan petrels totalled about 11 million birds, the 12 species of auks amounting to some 35.6 million birds. It seems more likely that their competition has held down the northern petrel radiation rather than a lack of protection from predators, since burrowing auks do well.

Some estimates of world populations of petrel species have been given, mostly of the easily counted surface nesters. For instance, almost all Laysan Albatrosses *Diomedea immutabilis* breed on islands of the Hawaiian Leeward Chain which, according to Fefer *et al.* (1984), support 759 000 breeding birds plus an estimated 1 771 000 non-breeders, that is, a total of about 2.5 million albatrosses (Table 1.1).

Other estimates of world populations include those for *Macronectes giganteus* of some 209 000 birds (76 200 breeders; 26 700 non-breeders, 87 100 prebreeders and 19 000 chicks, and for *M. halli* of 47 200 birds (17 200 breeders; 6000 non-breeders; 19 700 prebreeders and 4300 chicks)—Hunter (1985). The Tasmanian Muttonbird *Puffinus tenuirostris* was calculated to have a total breeding stock of about 23 million birds (Skira *et al.*, 1985), to which must be added an undetermined number of non-breeding, mainly 1–7 year-olds.

The biggest populations, running into many millions of pairs, are of small animals, the great albatrosses numbering only a few thousands of pairs; for example *D. exulans c.* 36 000 breeding pairs and *D. epomophora c.* 16 000 breeding pairs (Robertson, 1975a).

On the basis of its wide distribution in all the oceans, Wilson's Storm Petrel *Oceanites oceanicus* has been mooted as the world's most abundant seabird. This seems unlikely, and being a talus and crevice nester, is one of the most difficult birds to census—the numbers of prions and diving petrels probably far outnumber that species, though less wide-ranging at sea.

At most breeding places the total biomasses of these smaller birds may also exceed those of the larger species, but few data are available. Nor does this generalization always hold; for example, the 390 000 breeding pairs of Black-browed Albatrosses *D. melanophrys* (3.7 kg) at the Falkland Islands would be approximately matched by

Table 1.1 Estimated numbers of breeding and non-breeding seabirds in the Northwestern Hawaiian Islands. From Fefer *et al.* (1984)

	Estimated % of non-breeders	Estimated number of breeders	Estimated number of non-breeders	Estimated number of birds
Black-footed Albatross	50	98 820	98 820	197 640
Laysan Albatross	70	759 140	1 771 330	2 530 470
Bonin Petrel	50	662 500	662 500	1 325 000
Bulwer's Petrel	50	206 250	206 250	412 500
Wedge-tailed Shearwater	66	522 800	1 014 850	1 537 650
Christmas Shearwater	56	5920	7530	13 450
Sooty Storm-petrel	31	15 000	6740	21 740
Red-tailed Tropicbird	50	22 470	22 470	44 940
Masked Booby	27	4740	1750	6490
Brown Booby	30	950	410	1360
Red-footed Booby	45	10 220	8360	18 580
Great Frigatebird	69	19 700	43 850	63 550
Sooty Tern	63	2 661 000	4 530 890	7 191 890
Gray-backed Tern	50	102 000	102 000	204 000
Blue-gray Noddy	50	8000	8000	16 000
Brown Noddy	45	185 600	151 850	337 450
Black Noddy	63	32 550	55 420	87 970
White Tern	63	29 860	50 840	80 700
TOTAL		5 347 520	8 743 860	14 091 380

the mass of about 11 million pairs of prions (130 g), considerably more than those islands support.

II Colony sizes

Sklepkovych and Montevecchi (1989) estimated 3.36 million pairs of *O. leucorhoa* breeding on 523-ha Baccalieu Island, Newfoundland, where 62% of burrows held an adult, a chick, or an egg. The next largest colony of this species is at Daikokujima Island, Japan, with about 2 million birds (Hasegawa, 1984). In the southern hemisphere some 3–5 million pairs of Antarctic Prions *Pachyptila desolata* together with about 1 million Thin-billed Prions *P. belcheri* breed in various parts of the Kerguelen Archipelago (Jouventin, 1994a), as do 2–5 million pairs of South Georgian Diving Petrels *Pelecanoides georgicus* and perhaps 1–3 million pairs of *P. urinatrix* (Table 1.2). In the 98 ha of forest on South East Island there are an estimated 1.3 million burrows, 63% of them housing *Pelagodroma marina* (West & Nilsson, 1995). At The Snares Islands, some 2.75 million pairs of *Puffinus griseus* may breed (Warham & Wilson, 1982). Although here, as elsewhere, nest density varies with vegetation cover and substrate, and the whole island can be regarded as one big colony, doubtless many subcolonies are contained therein, as with *Calonectris diomedea* (Thibault, 1994).

Congregations of larger petrels may also reach impressive sizes. There are some 2 million pairs of Shoemakers *Procellaria aequinoctialis* at South Georgia (Croxall *et al.*, 1984b) and about 166 000 breeding pairs of *D. melanophrys* on Steeple Jason, Falkland Islands (Thompson & Rothery, 1991).

There are few data on the variations in colony sizes. Skira *et al.* (1985) showed that Tasmanian colonies of *Puffinus tenuirostris* are quite compact. Their size distributions were: covering <1 ha, 62 colonies; 1–10 ha, 76 colonies; 11–100 ha, 25 colonies and >100 ha, 4 colonies. The largest colony, on Babel Island, covers 380 ha and had an estimated 2.86 million burrows.

A *Estimating the sizes of petrel populations*

Estimation of population sizes for group-breeding species such as petrels would appear to be easy when the colonies are discrete and data are needed solely on breeding pairs, but if an assessment of a total population at a particular colony is required, then the non-breeders and chicks must be counted, and as the younger birds probably do not visit the colony at all, estimates for these are difficult to acquire. Failed breeders too may often be at sea, and some birds will probably be on 'sabbatical leave' (*The Petrels*, p. 223). The time for making a census will vary with the objective: if to obtain annual production then counts of chicks at or approaching fledging will be needed; if for checks on year-to-year variation, then fairly crude estimates may suffice. Recently, at-sea counts have been used to estimate population sizes of several tubenoses by Spear *et al.* (1995).

1 *Surface nesters*

Albatrosses and giant petrels lend themselves readily to direct counting of individual birds or indirectly from photographs, particularly if taken from the air.

Table 1.2 Numbers of breeding pairs of seabirds at the French subantarctic islands. From Jouventin (1994a)

Species	Saint-Paul & Amsterdam	Crozet	Kerguelen
A. patagonica	—	455 000	120–140 000
P. papua	—	9000	10–15 000
E. chrysolophus	—	2–3 × 10^6	1.5–2 × 10^6
E. moseleyi	55 000	—	—
E. chrysocome	—	120–150 000	150–200 000
Diomedea exulans	10	1960	800–900
D. melanophrys	—	980	3300
D. chrysostoma	—	5940	7860
D. chlororhynchos	37 000	7030	50
D. cauta	—	5	—
Phoebetria fusca	240	2620	3
P. palpebrata	—	2280	4–5000
Macronectes halli	—	1017	1000–1500
M. giganteus	—	1313	3–5
Daption capense	—	200–300	1000–2000
Pachyptila belcheri	—	10–20	500 000–1 × 10^6
P. desolata	—	100–200	3–5 × 10^6
P. salvini	150–200	6–8 × 10^6	—
P. turtur	5–10	20–30 000	1000–2000
Halobaena caerulea	—	40–60 000	1–2 × 10^6
Pterodroma macroptera	—	60–100 000	100–200 000
P. lessonii	—	100–200	10–30 000
P. mollis	10–50	30–50 000	—
Lugensa brevirostris	—	40–60 000	50–100 000
Procellaria aequinoctialis	—	20–30 000	30–60 000
P. cinerea	5–10	2–5000	10–20 000
Puffinus carneipes	400–600	—	—
P. assimilis	10–20	—	—
Oceanites oceanicus	—	10–20 000	400–800 000
Fregetta tropica	—	5–6000	5–10 000
F. grallaria	10–20	—	—
Garrodia nereis	—	500–1000	1000–2000
Pelecanoides georgicus	—	2–3 × 10^6	2–5 × 10^6
P. urinatrix	—	1–2 × 10^6	1–3 × 10^6
P. albiventer	—	815	20–50
P. verrucosus	—	—	10–12 000
Anas eatoni	—	600–700	15–20 000
Chionis minor	—	2–3000	3–5000
C. lonnbergi	16	500–600	500–1000
L. dominicanus	—	600–800	3000–5000
S. virgata	—	150–200	1000–2000
S. vittata	200	100–120	1000–2000
S. fuscata	1	—	—
Number of breeding species	14	37	35

However, this method is less suitable with surface-nesting species in the tropics as the birds tend to shelter beneath coral heads or vegetation.

Counting birds on eggs raises the problem of birds in nests and apparently breeding that actually lack eggs and will never lay that season. Often such birds can be detected and eliminated from the true breeders by their postures and behaviour, though some errors are likely, none the less. With ground counts of petrels on cliffs or mountain sides there is usually 'dead ground' where pairs are known to nest; in that event, without aerial or boat surveillance, estimates may have to be made on the basis of vocal and aerial activity in relation to similar but visible areas nearby, for example Warham and Bennington (1983). Nest densities were discussed in *The Petrels*, pp. 215–216.

2 *Burrow nesters*

Most estimates are based on counts of burrow entrances within marked quadrats to get mean burrow densities for the different vegetation types followed by the integration over the burrowed areas, these latter determined by ground or aerial surveys and vegetation maps. Techniques have been discussed by Wormell (1976), Hunter *et al.* (1982) and others. The line transect method has also been used. Naarding (1980) used 100-m^2 transects. A 100-m line was stretched across the colony and a PVC tube, 1-m long, threaded on to the line through a small hole drilled at the 0.5 m mark. While moving the tube along the line, every burrow within its extremities was scored.

It is usually easy to get raw data of this kind from randomly sited quadrats, but much more difficult to determine burrow occupancy with statistical accuracy. Often the presence or absence of a nesting adult, an egg or a chick is hard to detect—the burrow is too long, too twisted, is that of another kind of petrel, a non-breeder or prospector, of a rabbit or an auk, two nests may have one entrance and one nest more than one entrance, and so on. If only short or straight-tunnelled nests are selected, for example for using fibre-optics or video probes (Dyer & Hill, 1991, 1992), then a bias may be introduced, perhaps favouring young birds. For such reasons, reliable burrow-occupancy figures are seldom available—see also *The Petrels*, p. 209. Dyer and Hill (1992) by direct examination at two islands, found that only 43 and 56% of *Puffinus pacificus* burrows held birds on eggs during the incubation stage. Later they developed a new approach for comparisons between seasons (Dyer & Hill, 1995).

Hunter *et al.* (1982) used tape playback of the appropriate species to establish burrow occupancy, judging from the sizes of the entrances which species' tape to play. *Pachyptila desolata* and *Halobaena caerulea*, whose similar-sized entrances were indistinguishable, responded to each other's calls. Where these occurred together, checks were made when *H. caerulea* was incubating and later when the prion was incubating.

Tape playback may be useful and save time but, at least for some petrels at some stages of the season, calls of both sexes are needed as only males respond to males, females to females, for example with *Puffinus puffinus* (Warham, pers. obs.; Brooke, 1990, p. 189), but James and Robertson (1985e) got consistent results with this species. Otherwise, the effectiveness of playback for establishing burrow occupancy has not been determined and the responses need study, for example do chicks react to calls of either sex independent of their nutritional state?, etc.

Wormell (1976) and Thompson and Thompson (1980) calculated the surface areas of shearwater 'greens' (Section XI below) from aerial photographs, correcting for altitude. From sample ground plots, they calculated burrow densities correcting these for slope to give burrow density per unit planar area and hence total burrow numbers.

Other estimates have involved modified Lincoln Index techniques based on recaptures of marked birds. Orians (1958) attempted this from ringing and recoveries of *P. puffinus* marked as nestlings and as adults. However, as Orians suspected, petrels do not 'randomize'. He also thought that another source of bias was that young, non-breeding birds spent more time on the surface and were thus more often captured, and Harris (1966b) showed that adults of known age caught on the surface were younger than those incubating below it, that is, many surface birds were non-breeders. Furthermore, Perrins *et al.* (1973) concluded that established nesters that had already been handled learnt to avoid subsequent capture.

Non-breeding birds from other colonies may be quite plentiful at times, are often well represented in mist-net catches of storm petrels, and most have feathered brood patches (Furness & Baillie, 1981). The presence of such wandering birds is another factor precluding simple capture–recapture population estimation (Love, 1978).

3 *Crevice nesters*

Petrels nesting beneath boulders, in scree and talus breakdowns may be impossible to count with any degree of precision. Here, little can be done except to get some idea of numbers using mark–recapture techniques, quantifying responses from playbacks in the prelaying or incubation stages, or from other indices of abundance.

Copestake *et al.* (1988) mist-netted *Oceanites oceanicus* at various stages of the laying cycle, identified breeding females by the state of the cloaca, and estimated their numbers by capture–recapture and hence the number of breeding males and also the number of birds of unknown status without breeding partners.

III Coloniality

A *Introduction*

Although petrels are often solitary at sea, when on land they are seldom, if ever, so dispersed as to be beyond the sight or hearing of conspecifics. The most dispersed are probably some storm petrels, such as *Fregetta tropica* and *Garrodia nereis*.

There are few records of single pairs nesting successfully. The Wandering Albatross with a chick at Heard Island (Johnstone, 1982) and *Fulmarus glacialis* on Surtsey may be exceptions. Another example may be the single pair of *Puffinus puffinus* that had a downy chick at Penikese Island, Massachusetts in 1973 (Bierregaard *et al.*, 1975).

As most young petrels are philopatric (*The Petrels*, p. 228) and, especially after they have bred, very seldom shift to other colonies, new settlers are presumably non-breeders or older prebreeders. As a colony grows, the mean age of its members may be expected to increase. There may well be a minimum size for a colony to succeed and this could be important with endangered species. As nesting islands are often in short supply, philopatry helps reduce the risks of birds getting 'lost' or scattered and tends to keep colony numbers above any possible failure level (Fisher, 1975b, p. 296).

Discussions on coloniality in seabirds (Lack, 1968; Wittenberger & Hunt, 1985; Siegel-Causey & Kharitonov, 1990) mostly focus on diurnally active, surface-nesting species but most petrels are nocturnally active when on land. Why are they colonial?

One possibility is that colonies act as information centres. For example, the predawn mass departures of burrowing species could well enable inexperienced birds to follow older ones to distant, perhaps ephemeral, food concentrations.

That colonial breeding reduces predation seems unlikely except as regards the 'swamping' effect, where the petrels are too numerous for predators to affect productivity, and those with extended laying periods, and therefore at greater risk from natural predators, often nest when these are missing; for example *Pterodroma macroptera* and *Puffinus assimilis* breed in the southern winter when skuas are mostly absent. Undoubtedly, predation may have considerable impact at times, particularly on small birds such as *Oceanodroma leucorhoa* and prions, but generally such losses to natural predators while ashore seem relatively unimportant.

There is some anecdotal evidence for a 'swamping' effect. At the Snares Islands many *Puffinus griseus* come ashore in broad daylight despite a significant population of breeding and non-breeding *Catharacta skua lonnbergii* (Warham & Wilson, 1982). The number of skua-killed birds left on the ground is minute in relation to the huge shearwater population. The skuas appear to be overwhelmed by the masses of birds alighting near them: perhaps they have easier prey available—penguin eggs and the smaller, less obstreperous prions and gadfly petrels.

That coloniality is forced on petrels because of a shortage of breeding space within range of other necessary resources seems generally unlikely (Fisher, 1954), particularly at large islands such as South Georgia, Gough, Kerguelen or the Aucklands. Sometimes saturation may indeed be achieved. The shearwaters *Puffinus gravis* at Nightingale Island and *P. griseus* at the Snares Islands seem to occupy all the available ground not already taken up by other species, with many eggs laid on the surface (*The Petrels*, p. 300).

What may prove to be a key factor in colonial breeding is mutual stimulation, as shown by the attractiveness of taped playbacks (see Chapter 12.VII.E.1).

Whatever the stimuli involved, they may at times be so effective that first-time breeders persist in attempting to find space in crowded colonies and apparently ignoring suitable habitat nearby. Rowan (1965) reported that despite acute congestion of *P. gravis* at Nightingale Island there was ample space available at Inaccessible Island a mere 32 km away where there was a sparse population with few, if any, surface eggs laid. Rowan's question as to why such areas remain unexploited, resulting in an annual wastage of reproductive potential, remains unanswered.

Not all petrels are so rigidly conservative, for example *Fulmarus glacialis*, and *Oceanodroma leucorhoa* and *O. furcata* re-established thriving colonies on some Alaskan islands where they had previously been exterminated by foxes (Lensink, 1984) and see Section VII below.

B *Colony characteristics*

Petrels aggregate to nest in places adjacent to the sea—on islands, headlands and coastal mountains where the birds have a view of the surrounding ocean. In the rather few instances of inland breeding, such as some *F. glacialis* in northern Britain and Greenland today (see Fig. 9.7) and some *Procellaria* and *Pterodroma* formerly in inland New Zealand, the sea has generally been but a few minutes flying time away, though not visible from the nest site. The major exceptions are the Snow and Antarctic Petrels *Pagodroma nivea* and *Thalassoica antarctica* which may nest far from open water (*The Petrels*, p. 64).

A prime requirement is that the site be isolated from predatory land mammals. In the northern hemisphere this means that species such as fulmars use cliffs inaccessible to foxes and hazardous to others, such as mustelids. Some birds none the less get taken, for example where colonies of *O. furcata* coexist with predatory River Otters *Lutra canadensis* (Speich & Pitman, 1984).

A further essential need is that the colony must be within feeding range of adequate food resources. As most petrels can fly long distances when off their nests, and they and their chicks survive long fasts, most are not restricted to inshore feeding.

In the Southern Ocean suitable islands seem to be in short supply, particularly in the Pacific Ocean sector. Those within range of adequate food may carry huge populations of many species. The prize must go to the Crozets with about 25 million breeding seabirds of 37 species, 26 of which are Procellariiformes (Table 1.2), a richness ascribed partly to a varied and complex oceanography.

Suitable breeding areas must also have the right substrates—broken talus breakdowns for crevice nesters, deep soils for the big burrows of *Procellaria* petrels, ledges for sooty albatrosses, and relatively flat ground with exposed ridges for take-offs in calms for the great albatrosses *Diomedea exulans* and *D. epomophora*, etc. Bare rocky islands support fewer species but even mollymawks like *D. cauta salvini* and *D. c. eremita* nest successfully on rocky stacks with little nesting material except scraps of vegetation, tiny pockets of soil, old bones and a few stones. Where there is no vegetation stones alone suffice, for example with *Daption capense*; or even frozen stomach oil—*Pagodroma nivea*.

C *Colony development*

Young petrels are less philopatric than old ones and in their early years may turn up in colonies very distant from their own. A few even settle there as did the Laysan Albatrosses described by Fisher and Fisher (1969). Harris (1972) analysed recoveries of Manx Shearwaters at distant colonies; many were young birds and others, ages unknown, he suspected were non-breeders. They could have been attracted by the night-time din ashore and/or have joined offshore evening rafts and flown in with these after dark.

The parties of shearwaters and storm petrels that call and mill around at night over headlands where they may land but not nest, for example of *Puffinus puffinus*

(Coward, 1895) and *P. griseus* (pers. obs.), presumably also consist of young birds during this exploratory phase. Very few of about 10 000 *O. leucorhoa* mist-netted at Kent Island have ever been seen again (Huntington, pers. comm.) and he concluded that most were probably prebreeding visitors from elsewhere. And although most such birds evidently return eventually to their natal colonies, it is from their ranks that the vanguard of any expansion seems likely to be drawn.

Some such displaced birds may become imprinted to their new location like the lone Short-tailed Albatross *Diomedea albatrus* born on Torishima Island in March 1964 that appeared on Midway Atoll each winter breeding season from 1972 to 1983 without ever attracting a partner. Had it done so there is little to suggest that it could not have bred and perhaps started a new colony many kilometres from its birthplace.

The growth of some new petrel colonies has been documented, for example that of the small colony of *D. epomophora sanfordi* in New Zealand (Fig. 1.1), this being unusual in that the age and status of most of the colonists have been known. Fisher (1954) found that small colonies of *F. glacialis* produced relatively fewer eggs than did large ones but thought that much of the improved success with increasing colony size could have been attributable to the bigger colonies having more old and experienced birds.

Storey and Lien (1985) described the development of the *P. puffinus* colony at Middle Lawn Island, Newfoundland. In 1974 some were heard on land, 2 years later flocks were noted offshore and some birds found onshore were evidently not breeding. Burrows were first seen in 1977 when three eggs were laid and a chick hatched. Thereafter growth was rapid and by 1981 there were 221 burrows and a population estimated at at least 350 birds. Many of the early nests were in rock crevices and only when these had been occupied were burrows dug. These were grouped in distinct subcolonies as more recent arrivals excavated between existing burrows, suggesting the influence of mutual stimulation. That many of the shearwater colonists were young birds born in West Wales was shown by recoveries of 10 bearing rings. Five were yearlings, one was 2 years old, three were 5 years old and one 6 years old, only the last four having reached the normal age for first breeding.

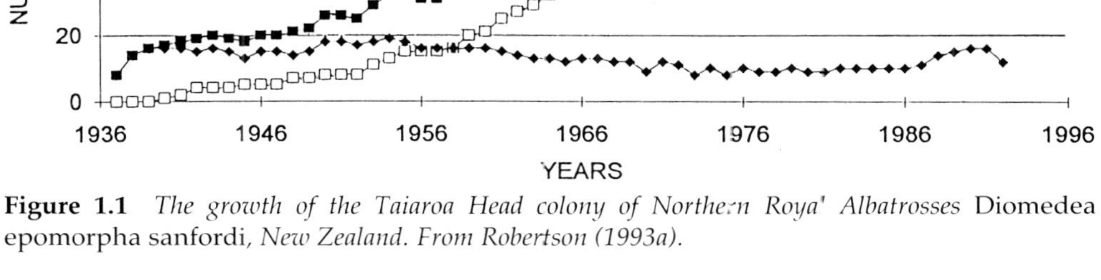

Figure 1.1 *The growth of the Taiaroa Head colony of Northern Royal Albatrosses* Diomedea epomorpha sanfordi, *New Zealand. From Robertson (1993a).*

IV Population structure

A *Introduction*

Petrel populations have a high proportion of birds of quite advanced age owing to reproduction being delayed until as late as a mean of 13 years in the larger albatrosses (Table 1.3). Their low productivity rate is balanced by a high adult survival rate; that is, they are all K-'strategists'. Birds that survive to breed have high expectations of further life. Because the young birds remain at sea for 1–7 years, the proportions of these age classes in a population cannot be determined—most can neither be recovered nor recognized as members of a particular age group. Those that do not come ashore to be identified are usually assumed to have died, although some still alive will have shifted permanently elsewhere unbeknown to the demographer.

This section of the population can be regarded as a 'strategic reserve' helping ensure the survival of the species in the face of natural catastrophes such as the volcanic eruptions of 1902, 1939 and 1941 at Torishima Island, or unnatural ones like the slaughter of *D. immutabilis* by the plume hunters early in the 20th century. Although their natal colonies may be shattered, these younger birds may form a nucleus from which a population can rebuild, as has *Calonectris diomedea borealis* at Grand Salvage Island.

Some of the long-term studies of petrel populations have involved repeated handling of the study birds, so that their component compositions may be atypical of undisturbed populations. Examples are the 27-year study of the Fisher Island population of *P. tenuirostris* (Wooller *et al.*, 1989) and the 34-year study of *F. glacialis* at Eynhallow in the Orkney islands (Ollason & Dunnet, 1988). It is suspected that recruitment of birds born at these colonies was affected by handling, some young birds, discouraged from staying, nesting elsewhere.

Intermittent breeding among petrels was once considered to be restricted to the 'great' albatrosses, which, if successful, breed every other year. However, substantial proportions of the breeders in some populations miss a year now and then, for example *C. diomedea* (*The Petrels*, p. 223) and *Pagodroma nivea* (Chastel *et al.*, 1993). In the Basque population of the storm petrel *Hydrobates pelaguicus*, Hemery *et al.* (1986) reported that about 31% of mature adults did not breed in successive years. Wooller *et al.* (1990) found that 10–11% of *Puffinus tenuirostris* that had formerly bred were absent each year and 15–18% were present without being known to produce an egg. Both absentee and non-laying rates decreased with older birds. Such intermittency, which may vary from year to year, obviously complicates calculations of mortality rates and population structures.

To categorize the members of a colony or population depends on long-term examinations of birds marked as chicks and without loss of marks. Few such colonies exist and the best information tends to derive from work on albatrosses and fulmars—their diurnal activities and erect postures help individual recognition at a distance. Burrowing birds generally need handling to check identities.

Major year-to-year changes might arise, for example when some category suffers heavy mortality for some reason—perhaps when a mammalian predator has gained access to the colony or when breeders are killed at their nests by tidal waves or sandstorms, so that prebreeders, then absent, become a higher proportion of the whole.

B *Cory's Shearwater at Grand Salvage Island*

Changes to the population of *C. d. borealis* at this Atlantic island have been described by Mougin *et al.* (1987). The birds had been harvested since at least the start of the 19th century and poaching continued even after the birds gained legal protection. After the 1975 and 1976 debacles only about 5000 breeders were left. Fortunately these, and the reserve of prebreeders then at sea, took advantage of the guarding of the island from 1977. This led to a dramatic turn-round in the fortunes of the colony. There was a marked shift in the proportions of breeders, advanced prebreeders (4–8-year-olds) and immature prebreeders (1–3-year-olds). In 10 years, according to these researchers, the proportion of breeding birds has increased steadily (Fig. 1.2A), and, in the next 10 years, is predicted to stabilize (Fig. 1.2B).

C *The Sooty Albatross at Possession Island*

Phoebetria fusca lays biennially if successful and, on average, does not do so until 11.9 + 1.5 year ($n = 22$) at Ile de la Possession, Crozet Islands (Jouventin & Weimerskirch, 1984). Figure 1.3A shows their calculated breakdown of the community structure in October, the month of egg laying. On shore 29.3% of the population (B) consists of couples that have laid, 6.3% (C) are potential reproducers, birds that have yet to find their partners or pairs yet to lay, and 6.2% (D) are advanced prebreeders regularly visiting the colony. At sea are (A) 21.8% of birds that bred successfully the year before, (E) 33.1% of prebreeders and (F), 3.2% of birds aged 14+ years that are not breeding. Later, in February (Fig. 1.3B), the birds ashore consist only of 17.2% of the population that is raising chicks plus 4% (D) of advanced prebreeders. All the rest, including a large proportion (B) of birds that have lost egg or chick, are now at sea.

D *Non-breeders and prebreeders*

Assessment of the non-breeders (those not currently breeding but physiologically capable of doing so) in a population is difficult, not only because many prebreeders are at sea and unaccounted for, but because it is seldom that all the original chicks, even in a restricted area, were marked, and so separable from immigrants or visitors. Even marked birds that shift to breed elsewhere will often never be reported.

Prebreeders (birds that have never attained sexual maturity) may be recognizable by their behaviour and by the timing of their visits to the colonies. Serventy (1967) tabled occurrences of known-age *Puffinus tenuirostris* (*The Petrels*, fig. 12.8), but pointed out that as search-efforts varied over time his figures did not show the precise frequencies of the visitors to the colony. Usually the younger prebreeding petrels return later than the non-breeders. Many prebreeders have no attachment to particular sites but wander around, calling vigorously and displaying with other

Table 1.3 Ages at first breeding, adult survival rates and expectations of further life for petrels

Species	Age at first breeding (years) Min.	Mean	s_x	e_x	Locality	Reference
Diomedea exulans	7(F)	9.6(F)	0.97	33	Crozet Is.	Weimerskirch (1992)
	7(M)	10.4(M)				
D. epomophora	6	8.5–10.6	0.94		New Zealand	Robertson (1993a)
	8		0.97	36	New Zealand	Richdale (1952)
D. irrorata	3	3–11	0.96	25	Galapagos	Harris (1979)
D. immutabilis	5(F)	8.9(F)	0.95	19	Midway I.	van Ryzin & Fisher (1976); Fisher (1975b)
	6(M)	8.4(M)				
D. melanophrys	6	10	0.88	7.8	Kerguelen	Jouventin & Weimerskirch (1988); Prince *et al.* (1994)
	8	10	0.93	13	S. Georgia	
D. chrysostoma	10	12	0.95	19.5	S. Georgia	Prince *et al.* (1994)
D. chlororhynchos	5	8.9	0.91	10.6	Amsterdam I.	Weimerskirch *et al.* (1987)
D. bulleri	—	—	0.89	8.5+	Snares Is.	Richdale & Warham (1973)
Phoebetria fusca	7	12.2	0.95	19.5	Crozet Is.	Weimerskirch *et al.* (1987)
P. palpebrata	9	12.0	0.97	33	Crozet Is.	Weimerskirch *et al.* (1987)
Macronectes giganteus	6	—	0.90	9.5	S. Georgia	Hunter (1984b)
	5	8.3	—	—	S. Orkney Is.	Conroy (1972)
	6	11.4	—	—	Macquarie I.	Woehler & Johnstone (1988)
M. halli	6	9.7	—	—	Macquarie I.	Woehler & Johnstone (1988)
			0.90	9.5	S. Georgia	Hunter (1984b)

Fulmarus glacialis	6	10.3(F) 8.4(M)	0.97	30+	Orkney Is.	Ollason & Dunnet (1978, 1988)
F. glacialoides			0.90–0.95	12.8	Terre Adelie	Mougin (1975); Guillotin & Jouventin (1980)
Daption capense	3	6	0.93–0.95	17.7	S. Orkney Is.	Beck (1969); Hudson (1966)
Pagodroma nivea	5	10.6(F) 9.1(M)	0.93	16.2	Terre Adelie	Chastel *et al.* (1993)
Bulweria bulwerii	—	—	0.95	18.3	Salvage Is.	Mougin (1989)
Pterodroma phaeopygia	6	—	0.80–0.93	—	Hawaii	Simons (1984)
Pachyptila desolata	3	5–6	—	—	S. Georgia	Croxall (1982a)
P. turtur		4–5	0.84	5.9	New Zealand	Richdale (1965b)
Calonectris diomedea	7	9	0.95	22	Salvage Is.	Mougin *et al.* (1987)
Puffinus griseus	—	6	0.93	13.5	New Zealand	Richdale (1965b)
P. tenuirostris	—	7.6	0.80–0.94	4–16	Bass Strait	Bradley *et al.* (1989)
P. puffinus	—	6–7	0.91	9–10	Wales	Brooke (1990)
Oceanites oceanicus	—	—	0.91	10.4	S. Orkney Is.	Beck & Brown (1972)
Hydrobates pelagicus	3	4–5	0.88	7	Wales	Scott *in* Cramp & Simmons (1977)
Oceanodroma leucorhoa	4	5	0.86	—	E. Canada	Huntington (pers. comm.)
Pelecanoides urinatrix	2	2+	0.75	3.5	New Zealand	Richdale (1965b)

M, male; F, female

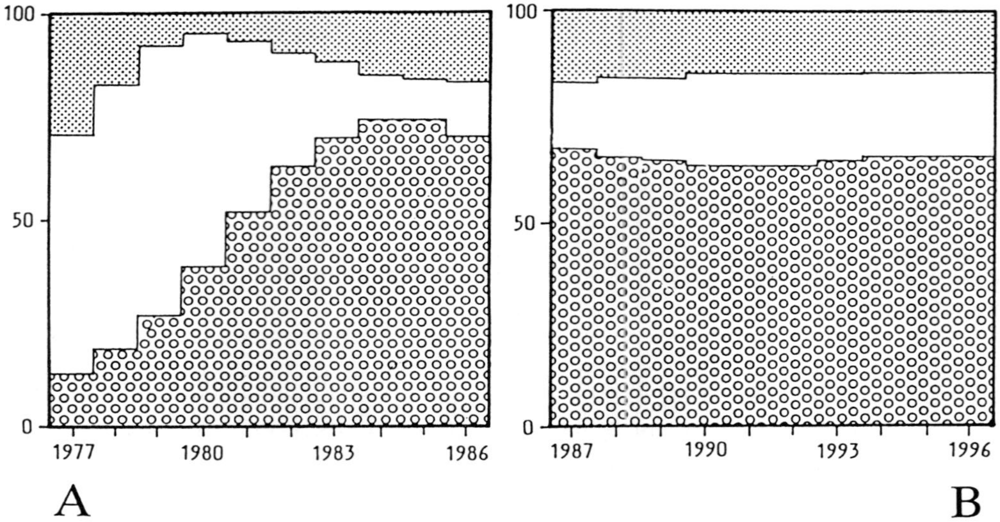

Figure 1.2 *Per cent changes in the structure of a population of Cory's Shearwaters at Grand Salvage Island. Breeding adults white; 4–8-year-olds stippled; 1–3-year-olds circles. (A) Between 1977 and 1986. (B) Expected changes during the next 10 years. From Mougin* et al. *(1987).*

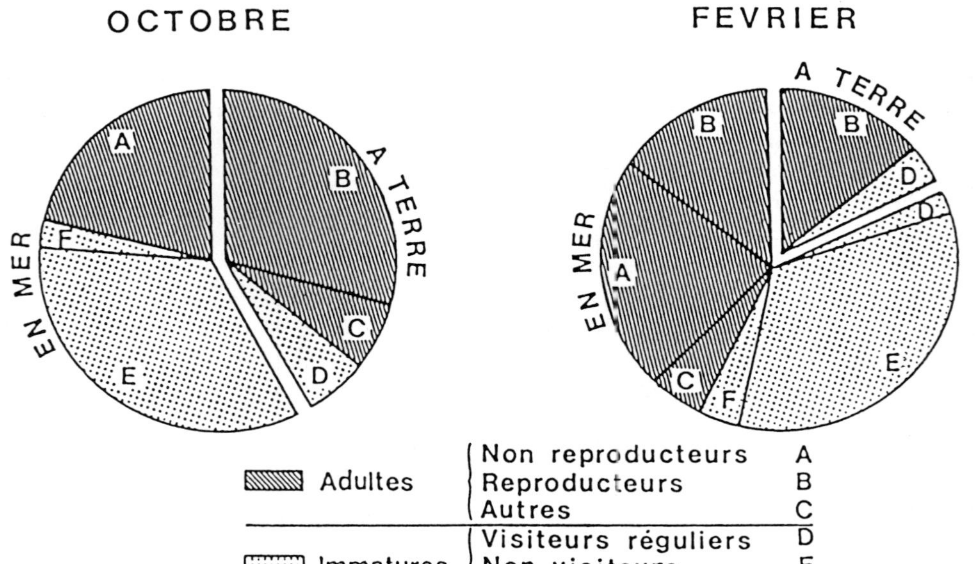

Figure 1.3 *Structure of the albatross* Phoebetria fusca *population at the Crozet Islands in October and February on land and at sea. A, Non-breeders; B, breeders; C, others; D, regular visitors; E, non-visitors; F, occasional visitors. From Jouventin and Weimerskirch (1984).*

prebreeders or with breeders lacking partners. With nocturnally active species such birds are responsible for much of the calling after dark, and, if commonly calling in flight, non-breeders tend to do so vigorously as they overfly their colony whereas established nesters tend to be more cryptic, and are often quite silent until entering their nests. Prebreeders and non-breeders may dominate the ground surface after dark while the breeders are engaged below it. This predominance may be shown from mist-net catches. Absence of a brood patch is not diagnostic of non-breeding, of course, as advanced prebreeders also develop them, though not swollen cloacae as with breeding females. Non-breeders often have both nest sites and partners and so may be unrecognizable as never having bred until it is clear that they have failed to produce an egg.

In some studies the proportions of birds that have previously bred but fail to do so the following season, despite coming ashore (i.e. not having taken 'sabbatical' leave), is high. Hunter (1984b) found that from 27 to 57% of breeding giant petrels that were colour-banded in one season were ashore without laying during the next 2 years. Brooke (1990, p. 184) found that about 20% of breeding *P. puffinus* miss breeding each year.

Richdale was the first to categorize petrels by status. He estimated that in the 1953/54 season, out of 512 *P. griseus*, 29% were breeding, 26% were 'keeping company', 28% were found alone in a burrow without an egg and 17% were birds caught on the surface evidently of unknown status; that is, no less than 71% of birds ashore were non-breeders or prebreeders, a very high figure except for a young, growing colony (Richdale, 1963, p. 68). Equivalent figures for *Pelecanoides urinatrix* were breeders 73%, 'keeping company' 9%, alone in nest 4%, caught on surface 7.5%, known to have been missed 7% (Richdale, 1965a, p. 41). As he pointed out, this high proportion of breeders is partly a consequence of some *P. urinatrix* breeding in their second year, so that a greater proportion of the prebreeders would be expected ashore than with a species with a long period of immaturity. Richdale also noted that birds that lost their partner quickly obtained a new one. Another complicating factor here may have been that some birds relayed after losing their egg as established by Astheimer and Grau (1990).

From a 5-year study of Alaskan *F. glacialis* Hatch (1989) found that non-breeders comprised 30–35% of the population in the spring, 17.3% of which had nests but no egg; the others—'prebreeders'—lacked nests and moved around a lot. For *Puffinus tenuirostris*, Wooller *et al.* (1989) also determined the proportions of past breeders still alive but absent or that failed to lay (see Table 1.6).

Woodward (1972) counted albatrosses on Kure Atoll and in 1969 the breeding *D. nigripes* were marked with coloured streamers facilitating their identification when off their nests. Birds without nests comprised 0–55% of the breeding population of 235–335 *D. nigripes* pairs, according to the time of the year. Among 309 of these unattached birds 12.6% were non-breeders, 34.3% were prebreeders born on Kure, 3.9% had been banded elsewhere, 6.8% of unknown status had been banded on Kure, and the remaining 42.4% were unmarked birds of unknown origin and status, thought to be from another island. If so, 53% of those lacking nests were immigrants. A rather similar situation was found with the *D. immutabilis* population of 800–1600 pairs that bred annually.

The very high proportions of immigrants lacking nests in this albatross population would be expected of an expanding one. Woodward attributed this to two main developments—to bulldozing activities that had increased the available nesting habitat, and to disturbance on the Midway Island colonies where attempts were being made to eradicate albatrosses from military runways. Many *D. immutabilis* born on Midway were recovered on Kure, some staying to breed.

V Age structures of petrel populations

There have been numerous attempts to categorize particular petrel populations by age class, based on analyses of birds marked as chicks, but because of the absence of true figures from a petrel's early years at sea, life tables such as those of Westerskov (1963), Tickell (1968), Fisher (1975a) and Mougin (1989) involve assumptions that may be invalid. These include the proposition that once breeding has begun the annual mortality in a stable population is constant. This results in tables in which a proportion of birds is theoretically still alive at 80 years old, for example Tickell (1968) and Table 1.4.

Table 1.4 Life table for the Sooty Albatross *Phoebetria fusca* at Possession Island, The Crozets. From Weimerskirch (1982)

Age (yr)	No. of Birds	% in the population	% surviving to this age
0	650 eggs laid	—	—
0.5	234	6.35	36.0
1	147	3.99	22.6
2	141	3.83	21.7
3	135	3.66	20.8
4	130	3.53	20.0
5	125	3.39	19.2
6	120	3.25	18.4
7	115	3.12	17.7
8	110	2.99	17.0
9	106	2.88	16.3
10	102	2.77	15.7
11	98	2.65	15.1
12	94	2.54	14.5
15	83	2.25	13.1
20	68	1.85	10.6
25	55	1.49	8.6
30	45	1.22	7.0
35	37	1.00	5.7
40	30	0.81	4.6
45	24	0.65	3.8
50	20	0.54	3.1
60	13	0.35	2.1
70	9	0.24	1.4
80	6	0.16	0.9

Table 1.5 Age structure of a population of banded Laysan Albatrosses *D. immutabilis* at Midway Atoll in 1972–73. From Fisher (1975a)

Recaptures	Recaptured at minimum ages (years)						
Total no.*	10–14	15–19	20–24	25–29	30–34	35–39	40+
14 420	7979	790	30	13	23	18	11

*Birds fewer than 10 years of age at time of recapture are excluded.

Other workers have drawn up tables of the proportions of birds of known breeding experience, ignoring those lost at sea before first return to their natal colony. Fisher's figures for *D. immutabilis* (Table 1.5) were based on 14 420 recaptures. However, as Fisher pointed out, his population was increasing rapidly, so that his estimated age structure may have been skewed towards the younger age classes.

In a 'model' to simulate the population composition and factors regulating the population of the Hawaiian Petrel *Pterodroma phaeopygia*, Simons (1984) calculated that, at stability, about 52% of the population would be immature (<6 years old), 19% 6–11 years, 12% 12–17 years, 8% 18–23 years, 5% 24–29 years and 3% 30–35 years old. His variables included breeding frequency, yearly adult mortality, yearly fecundity, breeding success and immature survival.

A *Age at first return to the colony*

Larger species with extended longevity also stay at sea longer before making their first landfalls, have longer prebreeding periods and more delayed sexual maturity than the smaller species and, in a stable population must breed repeatedly just to replace themselves. However, the diving petrel *Pelecanoides urinatrix*, though not the smallest, appears to make its first landfall at an earlier age than the rest: some return as yearlings and these breed the next season (Richdale 1965a, p. 43). In contrast both sexes of *Pagodroma nivea* returned at 8.1 ± 3.0 year (Chastel *et al.*, 1993). Medium-sized *Puffinus puffinus* and *P. tenuirostris* first appear as 2-year-olds (Serventy, 1967; Perrins *et al.*, 1973), whereas, at The Crozets, first visits of 187 *D. exulans* were at an average age of 7.8 years (4–14 years) according to Weimerskirch and Jouventin (1987). A few *D. immutabilis* reappear as 2-year-olds (*The Petrels*, p. 254).

B *Age at first breeding*

Petrels breed for the first time when aged from 2 years in *Pelecanoides urinatrix* (Richdale, 1965a) to 12.1 (9–16 years) for male *D. exulans* and 11.2 (7–16 years) for females (Weimerskirch & Jouventin, 1987) to a mean of 13 years for *D. chrysostoma* (Croxall, 1982a). Most medium-sized species take 5–7 years before producing their first eggs (Table 1.3). With some of the larger species there are significant differences between the sexes as with *D. exulans* and *D. immutabilis*, but in others no sexual dimorphism in this respect has been demonstrated, for example in *Puffinus tenuirostris* (Wooller *et al.*, 1988).

Although smaller petrels do tend to breed for the first time when they are younger than do the larger ones, the trend is not constant. For example, among the albatrosses, *Phoebetria* tends to delay breeding more than does the much larger *D. exulans* (Fig. 1.4). With *D. immutabilis* more males of 9 years or less bred over 9 seasons (Fig. 1.5), but by age 10 years and thereafter the number of female breeding recruits was greater in each age class. This figure also shows how the increase in 10-year-olds in 1966 carried over to the 11- and 12-year-olds in 1967 and 1968. Likewise, the fall in 10-year-olds in 1968 was reflected in that of the 11- and 12-year-old classes of 1969 and 1970.

The mean age at first breeding may also vary from colony to colony and from time to time. Weimerskirch and Jouventin (1987) found that recruitment into the Wandering Albatross population at The Crozets took place at a successively younger age during their 10-year study (Fig. 1.6), an effect, not a consequence of sampling variation. This fall in the average age of first breeders corresponded with a decline in the Crozet population. A similar situation was found at South Georgia where the population has also fallen (Croxall *et al.*, 1990a) as has the time between first returning and first breeding (Pickering, 1989). By reducing the age at first breeding some compensation for increased mortality occurred, as has been documented for other vertebrates. On the other hand, figures given by Brooke (1990, p. 170) suggest that Skokholm Island *Puffinus puffinus* in the 1970s started to breed about 1 year later than during the previous decade. Brooke suggested that increased competition for burrows excluded 5-year-olds from nesting.

Figures for the ages at first breeding of burrowing petrels tend to be higher than

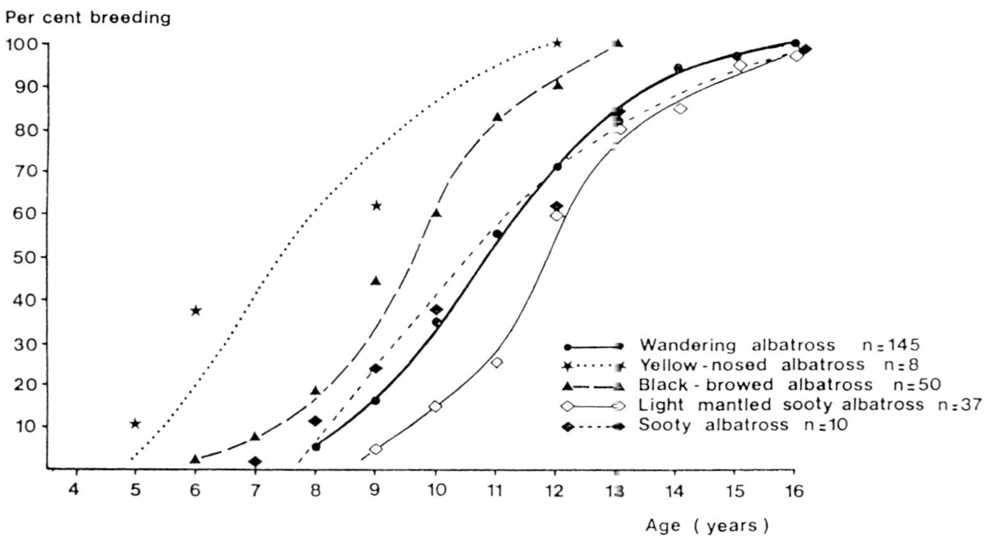

Figure 1.4 *Proportions of five albatross species at the Crozet Islands breeding for the first time in each age class. From Jouventin and Weimerskirch (1988).*

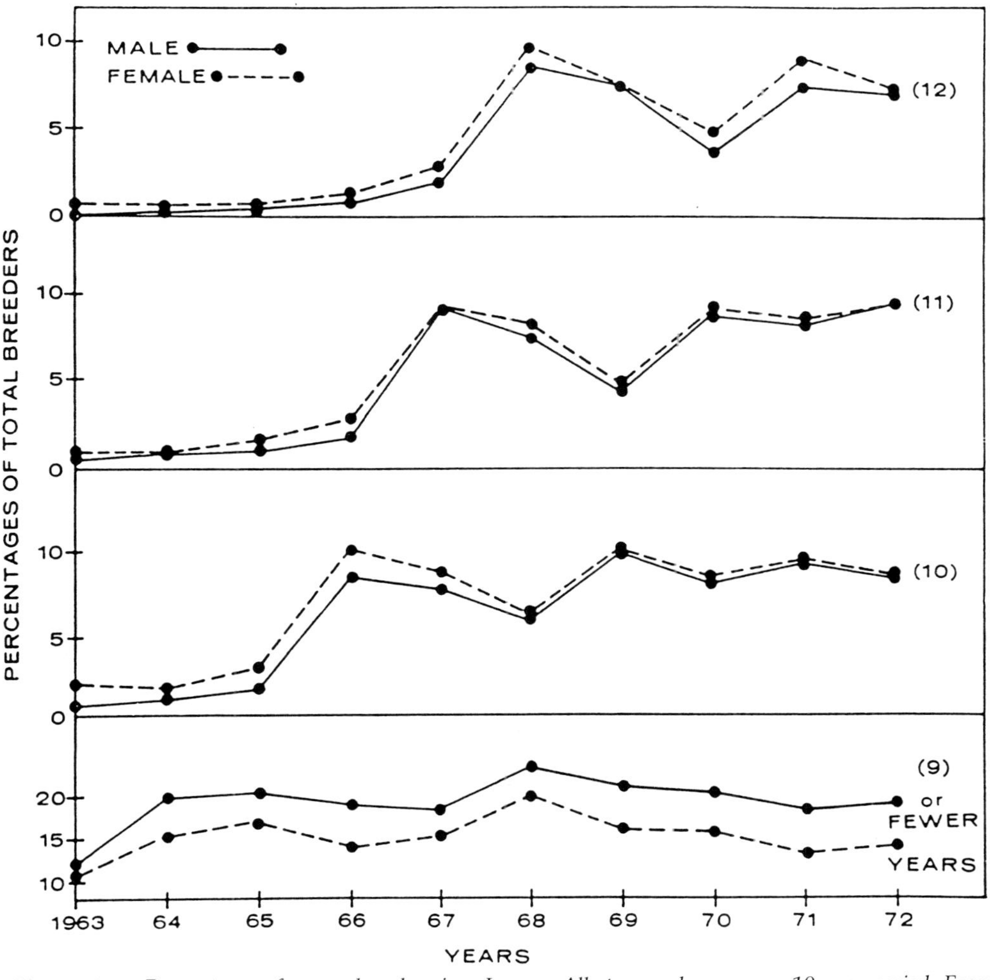

Figure 1.5 *Percentages of young breeders in a Laysan Albatross colony over a 10-year period. From Fisher (1976).*

the true means because it is seldom that the field worker can be sure that a bird first found with an egg had not previously laid elsewhere.

C *Fecundity*

The fecundity of petrels, as measured by the number of chicks fledged annually, obviously varies from year to year because of environmental factors, because of changes of mates, etc. Richdale (1965a, p. 44) reported that all of 17 *Pelecanoides urinatrix* in one sample of mainly 2- and 3-year-olds fledged a chick—a very atypical pattern. Hatch (1987a), over a 9-year study of *F. glacialis*, had a more typical result: his birds fledged 0.41 chicks annually on average over that period.

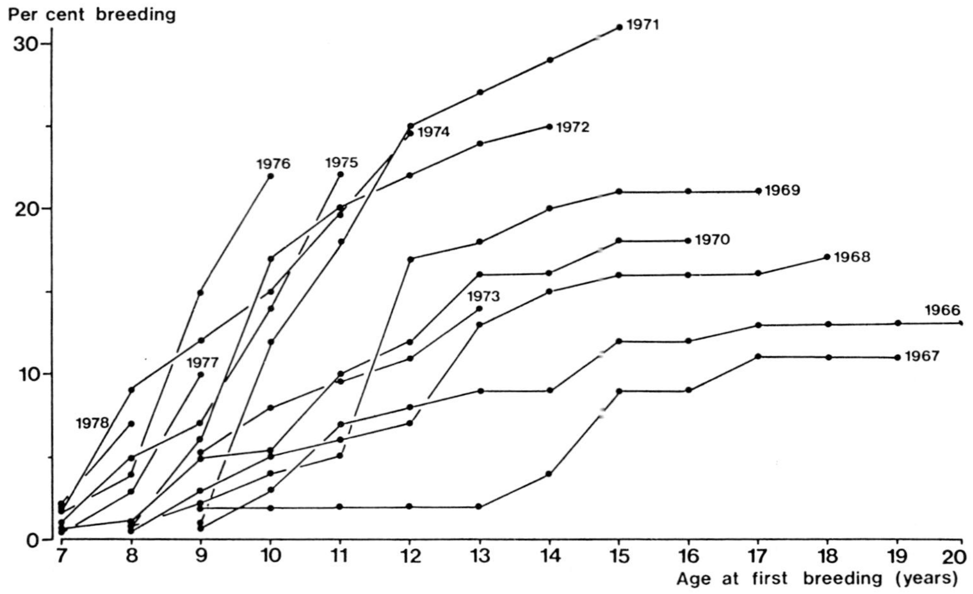

Figure 1.6 *Proportions of Crozet Island Wandering Albatrosses breeding for the first time at different ages in each cohort of fledglings banded from 1966 to 1976. From Weimerskirch and Jouventin (1987).*

Wooller *et al.* (1989, 1990) and Bradley *et al.* (1989) showed that those *Puffinus tenuirostris* that started breeding very young were relatively unsuccessful at first compared with those that deferred breeding more. Table 1.6 also shows changes in mean fecundity with breeding experience, increasing from 0.43 chicks per year for first attempts to about 0.7 at about the tenth attempt and with an apparent decline towards the end of their breeding lives. Their total production varied with their mean lifespans (Fig. 1.7A); those that started early produced more young in their lifetimes. Birds that produced high numbers of flying chicks also produced more that returned themselves to breed (Fig. 1.7B). Overall, 71% of all the birds that had completed their reproductive lives produced no offspring that returned to their natal island. Thus only a small part of the breeding stock was responsible for most of the next generation. How many bred elsewhere is not known.

Jouventin and Weimerskirch (1988) determined population variables for five albatrosses. Pairs of *D. melanophrys* and *D. chrysostoma* fledged a chick about every second year, those of *D. exulans* one chick about every third year, and the sooty albatrosses about one every fourth year.

D *Longevity and survival*

That some petrels are long-lived has been known for many years. For example, Harting (1887) reported the capture of an 'immense' albatross (doubtless *D. exulans*) that had been marked about 38 years previously.

Petrels not only live longer than most other seabirds of similar sizes, but larger

Table 1.6 Mean annual mortality, breeding success, percentage attendance and percentage laying of Short-tailed Shearwaters according to breeding experience. From Wooller *et al.* (1989)

Years since first bred	Mean (±SE) annual % mortality	% of eggs laid that resulted in young fledged	% of breeding birds known to be alive but not present at colony	% of breeding birds present at colony that did not lay	Sample size
0	12.8±1.7	43	—	—	470
1	9.6±1.6	49	15	25	345
2	8.3±1.6	59	17	24	297
3	5.0±1.3	60	17	20	268
4	5.1±1.4	66	11	19	264
5	7.8±1.8	60	15	16	233
6	6.0±1.7	69	9	21	228
7	9.1±2.2	69	11	16	201
8	8.3±2.3	74	11	19	172
9	8.0±2.4	71	10	12	154
10	14.9±3.5	79	11	18	126
11–12	11.1±3.2	72	11	13	195
13–15	10.8±3.9	69	8	16	212
16–19	10.1±4.8	65	9	24	168
20–27	20.6±10.0	56	10	10	134

species tend to survive better than small ones (Table 1.3). Using mean values for 16 species, Croxall and Gaston (1988) found that:

$$y = 0.79 + 0.019 \log_e x,\ r^2 = 0.27,\ P < 0.05$$

where y is mean annual adult survival rate and x is mean adult mass (g).

Although there are no direct data for survival during their early years at sea, various estimates suggest that, as expected, a higher mortality rate obtains then than later. In samples totalling 3821 prebreeding *D. exulans* born over 6 years at South Georgia, 49% were estimated by Croxall *et al.* (1990a) to have survived to age 5 years—45% being known males, 48% being known females (Table 1.7). The 3% differences for the sexes was not significant, although female adults survived significantly poorer by about 2% than males. Weimerskirch *et al.* (1987, p. 1051) calculated that at the Crozet Islands this same albatross during its prebreeding years (1–9) survived at an annual rate of 0.31 which rose to 0.97 when breeding. Another analysis by Weimerskirch and Jouventin (1987) of data for annual survival from fledging to 5-year-old for 9 yearly cohorts (1966–1974) averaged 0.72, but between 5 and 11 years rose to 0.92. Survival of breeders was significantly higher during 1977–1983 than during 1968–1976, probably owing to mortality in fishing nets during the latter period. Throughout, the annual survival of females (0.90) was significantly lower than that of males (0.94), as at South Georgia. Crozet Island breeders >28 years old survived less well (0.93) than 16–22-year-olds at 0.97 ($P < 0.01$, n=256) (Weimerskirch, 1992).

Calculating survival rates for biennial breeders is complicated because not all

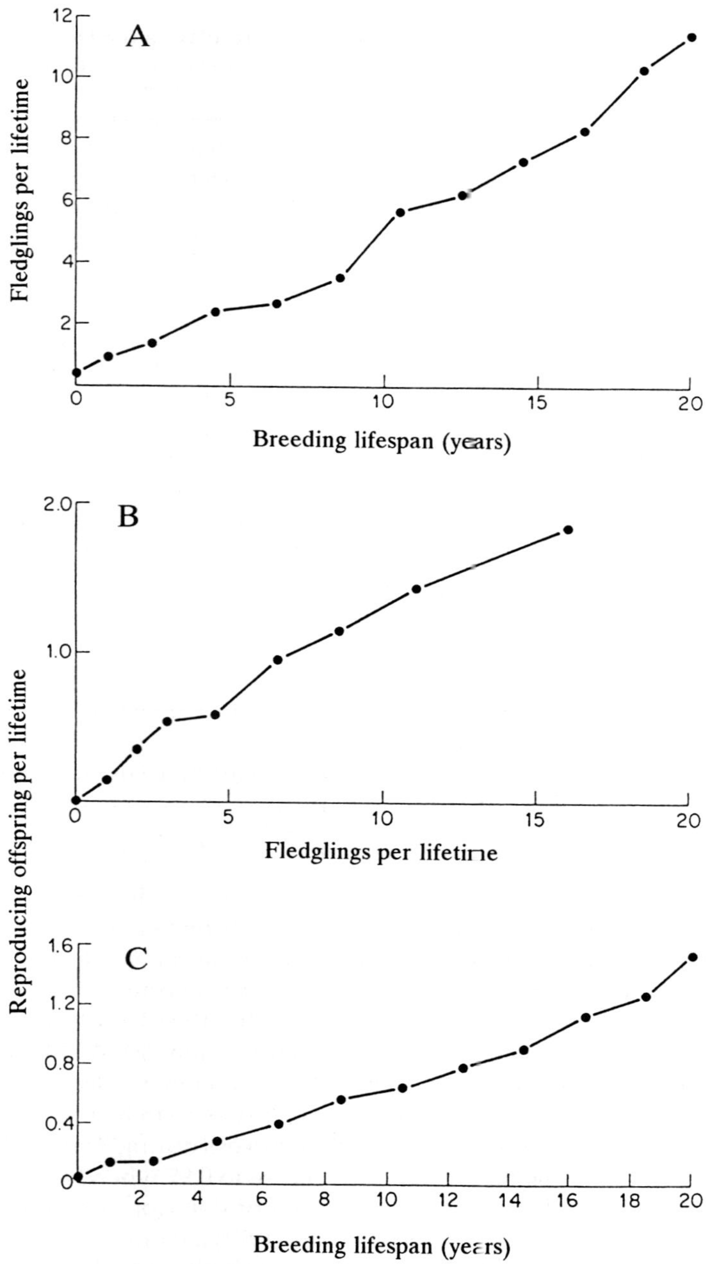

Figure 1.7 *For Short-tailed Shearwaters known to have completed their breeding careers: (A) The relationship between the mean duration of a bird's completed lifespan, from the time it first bred until final disappearance, and the mean number of free-flying young that it produced during that period. (B) The relationship between the mean number of free-flying young produced in a lifetime and the mean number of those young that returned to the colony to breed. (C) The relationship between the mean duration of a bird's completed lifespan and the mean number of reproducing offspring produced during that period. From Wooller* et al. *(1989).*

Table 1.7 Estimated survival rates of juvenile Wandering Albatrosses at South Georgia. From Croxall *et al.* (1990a)

		Estimated percentage survival to age 5 years (SE)*					
		Known males		Known females		All birds	
Cohort	Birds ringed	Crude	Adjusted	Crude	Adjusted	Crude	Adjusted
1972	368	39.6	43.5 (6.6)	35.3	41.5 (5.7)	38.9	45.3 (4.6)
1973	75	24.0	24.0 (7.0)	45.3	60.0 (16.8)	36.0	42.8 (8.6)
1975	854	40.3	41.8 (2.6)	45.7	46.9 (2.5)	45.5	47.2 (1.8)
1976	847	38.0	39.2 (2.6)	38.4	40.2 (2.6)	39.8	42.5 (2.0)
1977	806	46.9	48.4 (2.7)	51.3	52.6 (2.6)	50.4	52.5 (2.0)
1978	871	50.3	52.4 (2.6)	49.3	53.1 (2.7)	50.6	53.9 (1.9)
Mean†	—	—	44.6 (1.1)	—	47.7 (1.3)	—	48.9 (1.0)

*Crude estimates based on birds known to be alive. Adjusted value obtained by capture–recapture analysis.
†Weighted mean of the individual survival estimates with weights inversely proportional to the estimated sampling variances.

successful pairs miss one year. For example, in large samples of *D. chrysostoma* 1% bred the next year, 61% 2 years later, and 14% bred 3 years later. Likewise 58% of failed breeders tried again the next year, 22% not until 2 years later and 8% not until 3 years had elapsed (Rothery & Prince, 1990). These authors developed a model for calculating survival in such circumstances which was further developed by McDonald and Caswell (1993). Annual survival rates for adult mollymawks over 14 years are shown in Fig. 1.8.

Significant differences in adult survival rates have been demonstrated for some petrels nesting in different locations. For the albatross *D. melanophrys* breeding at Kerguelen the mean annual rate was 0.881, those breeding at South Georgia surviving better, annual rate 0.920 (Jouventin & Weimerskirch, 1988). These authors ascribed the difference to an increased mortality at the hands of Indian Ocean fishermen from which the population at South Georgia was largely exempt.

For the albatross *D. immutabilis*, Fisher (1975b) reported that mean annual survival in the first 3 years of reproduction was high, and the rates increased after the 14th year of life (the sixth and seventh of breeding). After 20 years of age there was some evidence of a decrease in survival. Fisher found that annual survival for breeding males averaged 94.7%, of females 94.6%. Nor were there significant differences in the survival of adult male and female *P. puffinus* (Brooke, 1990, p. 186), *P. tenuirostris* or *F. glacialis* (Dunnet & Ollason, 1978; Bradley *et al.*, 1989).

There are few data for the survival of the smaller petrels. Richdale's figures for *Pelecanoides urinatrix* varied considerably between four seasons, but his preferred annual mortality rate of 18.2% breeding birds for two normal years gives a low value of 0.82 for the survival rate and a mean expectation of further life of about 5 years (Richdale, 1965a, p. 48). These are the lowest values recorded for a tubenose and tie in with this bird's ability to breed when much younger than is known of any other petrel. Inshore feeding when breeding may also be a significant factor here.

Bradley *et al.* (1989) detailed statistical problems in estimating age-dependent survival and analysed rigorously selected samples of data for *Puffinus tenuirostris* breeding from their first to their 27th nesting attempts. They found that birds laying for the first time experienced relatively high rates of mortality which fell in subsequent years but after 10 years of annual breeding began to rise again (Fig. 1.9), suggesting an old-age effect.

A senility effect had been mooted earlier for this species by Serventy (1967), and operates in other seabird populations like those of the Kittiwake *Rissa tridactyla* (Coulson & Wooller, 1976). In his 31-year study of *Oceanodroma leucorhoa*, Huntington (pers. comm.) found that survival (73%) was below average after the first year of breeding, slightly higher (82%) after the second year, but very constant at between

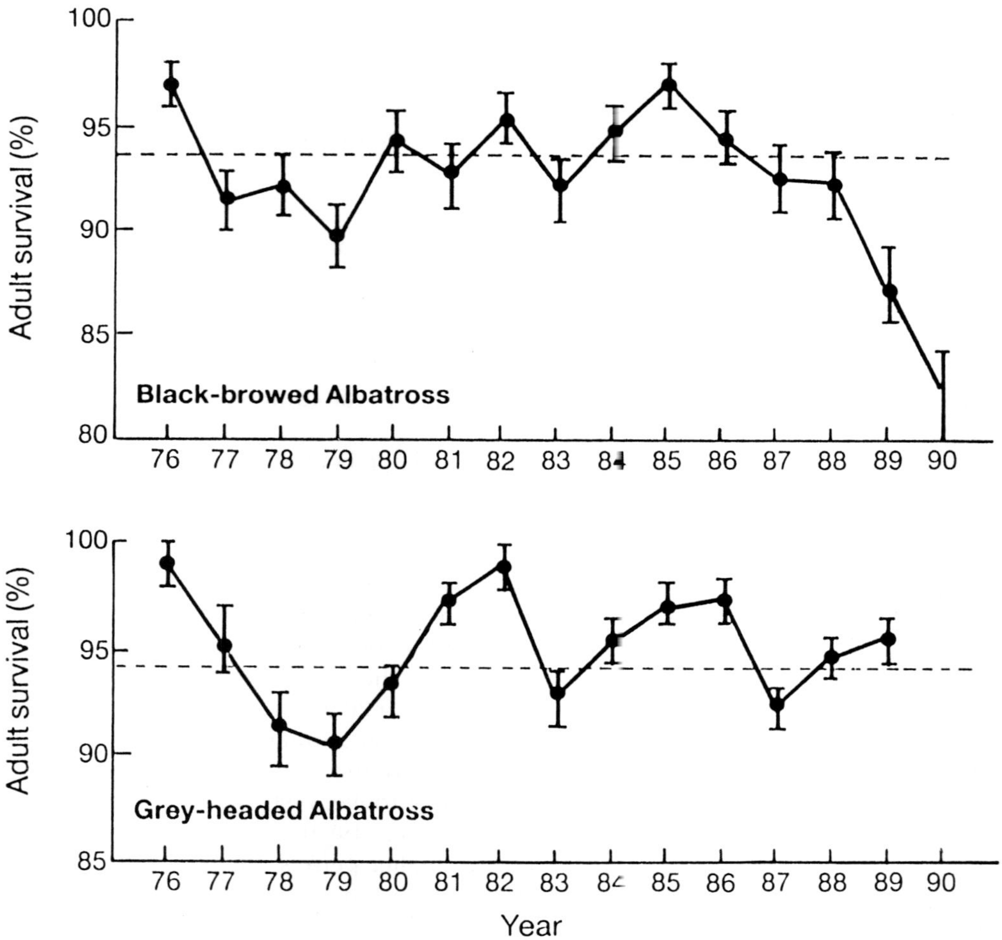

Figure 1.8 *Average survival rates (±SE) for Black-browed and Grey-headed Albatrosses. Dotted lines show arithmetic mean survival 1976–1988 inclusive. From Prince* et al. *(1994).*

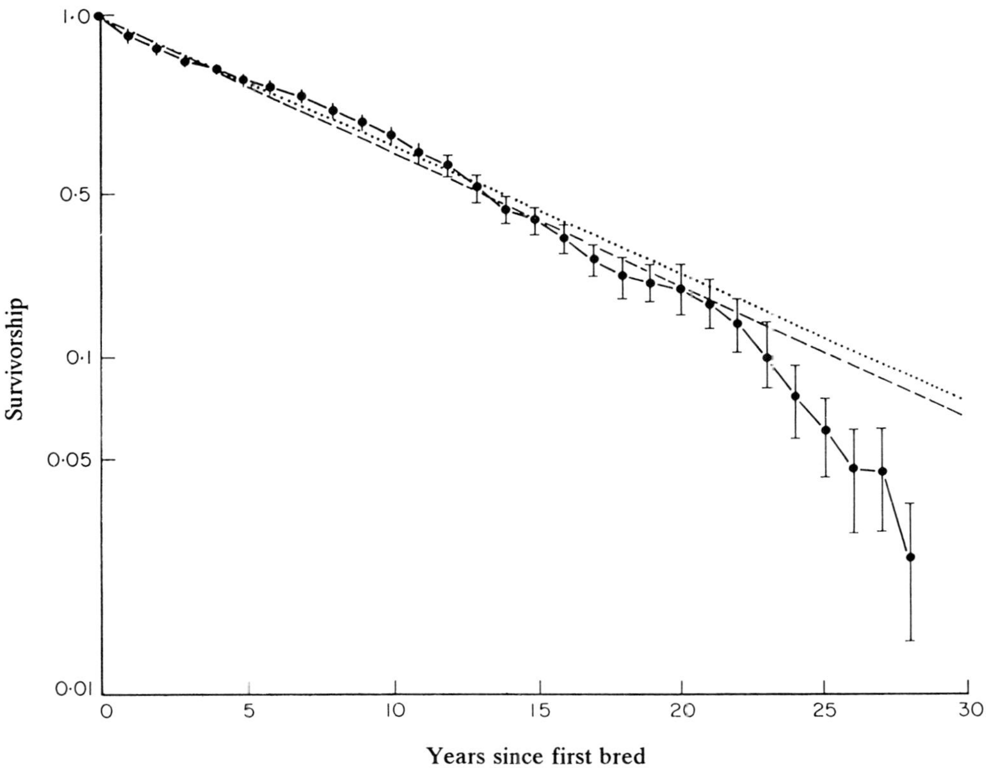

Figure 1.9 *The empirical survivorship* (l_x) *and theoretical survivorship* ($l_{(x)}$) *curves for Short-tailed Shearwaters, in relation to the years since they bred for the first time. The theoretical functions are for the exponential (––––) and Weibull (. . . .) distributions fitted to the data by the algorithm of Aitken and Clayton (1980). From Bradley* et al. *(1989).*

85 and 91% from the third to 17th year. There was a marked decline among birds that had bred for more than 25 years and these birds were less successful. A similar effect is indicated for *D. epomorphora sanfordi*. Robertson (1993a) found an apparent decline in survival for birds >25 years old, particularly among females; that is, for 20–24-year-old males (females) $s_x = 0.986$ (0.972), for 25–41-year-old males (females) 0.943 (0.894).

Unfortunately, the delayed maturity of tubenoses combined with their long lives means that even where the initially marked cohorts are of substantial size and no ring loss has occurred, after 30 years or more so few still remain that random errors may mask senility effects. Richdale and Warham (1973), for example, noted a decline in the survival rate of *D. bulleri* 22+ years after ringing, but ascribed this to the difficulties of tracking down the few birds still alive and to decreased search effort. In fact, later recoveries did not correct the steepening of their curve, but too few birds were still alive to justify assuming that the increased mortality was real. Ollason and Dunnet (1988) reported that analyses of the survival of cohorts with histories of breeding ranging from 2.5 to 19.5 years, failed to reveal any evidence of increasing mortality with age, although the 95% confidence limits inevitably widened for old birds. In any event, in their final years the few very old birds probably contribute little to the population's production.

Even if there is a consistent decrease in the survival of old birds, their potential longevity is considerable. The greatest longevities are to be expected from large birds like albatrosses, and the oldest known wild bird is still the female *D. epomophora*, in 1993 aged at least 61–62 years (Robertson, 1993a). Fisher (1975a) gave particulars of 15 *D. immutabilis* aged 35–42 years and Sagar and Warham (1993) a *D. bulleri* at least 51 years old. Among smaller, nocturnally active species one *Hydrobates pelagicus* storm petrel was 20 years old (Cramp & Simmons, 1977, p. 164), a male *O. leucorhoa* was recovered when 29 years old, a Bulwer's Petrel *Bulweria bulwerii* at 22 years (Clapp *et al.*, 1982), and a *P. puffinus* 40+ years old (Mead & Clark, 1990).

Long-lived birds in a stable population must breed repeatedly to replace themselves. This system confers resilience on the species. The delayed breeding that is a concomitant of longevity insures against extinction by a catastrophic density-independent event, while a failure of a population to breed in a particular year has a much reduced long-term effect than for a short-lived species with little opportunity to recoup before death intervenes.

VI Mortality

Few adults or prebreeders die on the colonies: most die at sea. Direct data on such losses are rare but *Pelecanoides urinatrix* is recorded as being eaten by 'snapper' with up to three birds in one fish. Because of their philopatry and colony tenacity the deaths of established breeders are inferred by their absence from the breeding territories. Many causes of mortality for birds of the various genera are given in *The Petrels*, chapters 2–10.

On land, the loss of adults occurs mainly during the breeding season when substantial losses of eggs and chicks may also take place. These latter tend to be greater with young, inexperienced birds, as has been shown for *D. immutabilis* by Fisher (1975b), for *D. exulans* by Du Bost and Segonzac (1976) and for *C. diomedea* by Mougin *et al.* (1987), among others. Losses of eggs and chicks vary greatly between pairs, colonies, habitats and seasons, for example with *D. irrorata* (Table 1.8).

A *Loss of eggs*

As soon as they are laid, eggs begin to be lost through a wide variety of causes. Precise figures are difficult to obtain. It is impracticable to monitor adequate samples of nests continuously and frequent direct inspections may increase egg loss. In most studies there is a category 'egg disappeared, cause unknown'.

A major cause of egg loss is abandonment by one or both parents. Most of this occurs early in incubation. Abandonment during the first 10 days accounted for over 50% of the egg mortality in nine species of Procellariidae studied by Mougin (1975, p. 134), and a similar pattern has been found with *F. glacialis*, *Puffinus tenuirostris* and

Table 1.8 Mortality of petrel eggs and chicks

Species	% eggs lost (*n*)	% chicks lost (*n*)	Reference
Diomedea exulans	15.6 (45)	15.8 (38)	Mougin (1970c)
	27.8 (385)	18.7 (278)	Tickell (1968)
	28.2 (262)	10.6 (188)	Du Bost & Segonzac (1976)
D. irrorata	65.2 (262)	49.5 (91)	Harris (1973)
	48.5 (134)	49.3 (69)	Harris (1973)
	51.0 (316)	38.1 (155)	Harris (1973)
	44.2 (258)	52.1 (144)	Harris (1973)
D. immutabilis	45 (350)	9 (193)	Fisher (1975b)
D. bulleri	22 (90)	27 (51)	Sagar & Warham (in press)
Phoebetria palpebrata	29.4 (34)	30.0 (10)	Mougin (1970a)
P. fusca	29.9 (784)	49.0 (549)	Jouventin & Weimerskirch (1984)
Macronectes giganteus	14 (100)	9.4 (86)	Hunter (1984b)
	39.7 (174)	31.4 (105)	Conroy (1972)
M. halli	63 (147)	22 (54)	Mougin (1975)
Fulmarus glacialoides	14 (200)	12 (171)	Norman *et al.* (1992)
F. glacialis	45.6 (103)	16.1 (56)	Mougin (1967)
	37 (1778)	19 (1196)	Hatch (1993)
Pagodroma nivea	49 (75)	55 (18)	Brown (1966)
	36.7 (1555)	16.7 (1555)	Chastel *et al.* (1993)
Pachyptila salvini	20.8 (72)	31.6 (57)	Derenne & Mougin (1976)
Calonectris diomedea	32 (many)	14.6 (many)	Mougin *et al.* (1987)
Procellaria aequinoctialis	21.4 (14)	45.5 (5)	Mougin (1970b)
Oceanites oceanicus	64.6 (82)	72 (22)	Beck & Brown (1972)
Hydrobates pelagicus	38 (214)	33 (133)	Cramp & Simmons (1977)
Oceanodroma castro	51 (154)	33 (76)	Allan (1962)
	57 (268)	40 (107)	Harris (1969a)
O. monorhis	49 (many)	20 (many)	Lee & Pyong-Oh Won (1988)
O. furcata	27 (85)	6 (62)	Hatch (1986)

others (Dunnet *et al.*, 1963; Serventy & Curry, 1984). Much of this neglect may arise from the inability of new breeders to lay down adequate reserves or perhaps to environmental deterioration. For example, irregularity in the return of storm petrels to their nests because of contrary winds, leading to egg neglect, has been suggested by Boersma and Wheelwright (1979) and Ainley *et al.* (1990). Selected data for egg loss, chosen from good samples, are given in Table 1.8.

Egg losses often vary greatly from one season to another. Thus out of 4691 eggs of *D. immutabilis* laid over eight seasons, losses varied from 3.4 to 16% with a mean of 7.2% (Fisher, 1971, p. 37). This author also listed causes of loss during 1962–1965 as: desertions 20–30%; storms 0–5%; researcher activity 1–3%; direct effects of man 2–5%; and interference from other albatrosses 1–2%. The causes of desertions were: nests too close together 1–3%; egg broken by parents 2–4%; egg missing, cause unknown 5–8%; failure of living mate to return on time 10–20%; death of mate 1–3%; interference by second pair trying to take over the nest 1–4%; confusion of pairs from adjacent nests 0–1%; storms 0–38%; unknown 10–30% and handling of birds by researcher 2–4%. Fisher (1971, p. 39) illustrated the impact of bad weather on egg loss and in relation to the biological activity of the colony as a whole (Fig. 1.10).

In recent years the Royal Albatrosses on the Forty-Fours Islets have been forced to nest on rocky substrates owing to the loss of vegetation. Considerable egg mortality from cracking ensues and helps account for 53% of the lay being lost in 1993 (Robertson, 1994). Sometimes egg loss is total. In 1983 only about 15% of the Waved Albatrosses *D. irrorata* on Hood Island attempted to breed and torrential rain associated with El Nino caused all to desert (Rechten, 1985).

B *Loss of chicks*

Many chicks die at hatching, for example in *Diomedea epomophora* (Sorensen, 1950), *Daption capense* (Pinder, 1966), and *Diomedea immutabilis* (Fisher, 1975b). Tiny chicks expire for unknown reasons; some are trampled on by clumsy parents. Another peak in losses occurs around fledging when naive birds are taken by predators, but the precise extent of this is mostly unknown and clearly hard to quantify.

The figures in Table 1.8 are very heterogeneous, and it is difficult to draw general conclusions from them. Clearly there is much variation from place to place, for example in the figures for *Diomedea exulans*. Any trends such as differences between surface and underground nesters are not apparent and may require much longer study for elucidation.

C *Adult losses on land*

1 *Losses from environmental factors*

Unseasonal weather conditions comprise a major source of deaths of petrels on land, mainly to chicks and to eggs. Important among these are snow blockage of nest sites and the freezing of eggs and chicks of birds like *T. antarctica, Pagodroma nivea* and *Oceanites oceanicus* (Roberts, 1940; Pryor, 1964; Brown, 1966; Beck & Brown, 1972). Flooding in heavy rain is probably the most widespread and important natural hazard for burrowing species, but may also affect surface nesters like *Diomedea immutabilis* (Fisher, 1971). The mortality mainly hits the chicks as these seem more susceptible to chilling than the eggs (*The Petrels*, p. 252). Royal Albatrosses on Middle Sister Island, Chathams Group have been blown off their nests and their eggs lost during violent storms. Conversely, three 26–29-year-old females of this same species at Taiaroa Heads, New Zealand died of heat exhaustion following atypical periods of high temperatures, low humidity and little wind (Robertson, 1993a), and dehydration is a common cause of chick loss for *D. immutabilis* (Sileo *et al.*, 1990a).

Other density-independent sources of deaths include tidal waves, volcanic eruptions, sandstorms, and El Nino perturbations. For example, Duffy (1990) pointed to unusually high numbers of beach-washed *Puffinus tenuirostris, F. glacialis, Lugensa brevirostris, Pachyptila turtur* and *Halobaena caerulea* in Australasia and South Africa as possible effects of El Nino Southern Oscillation (ENSO). These may reach to

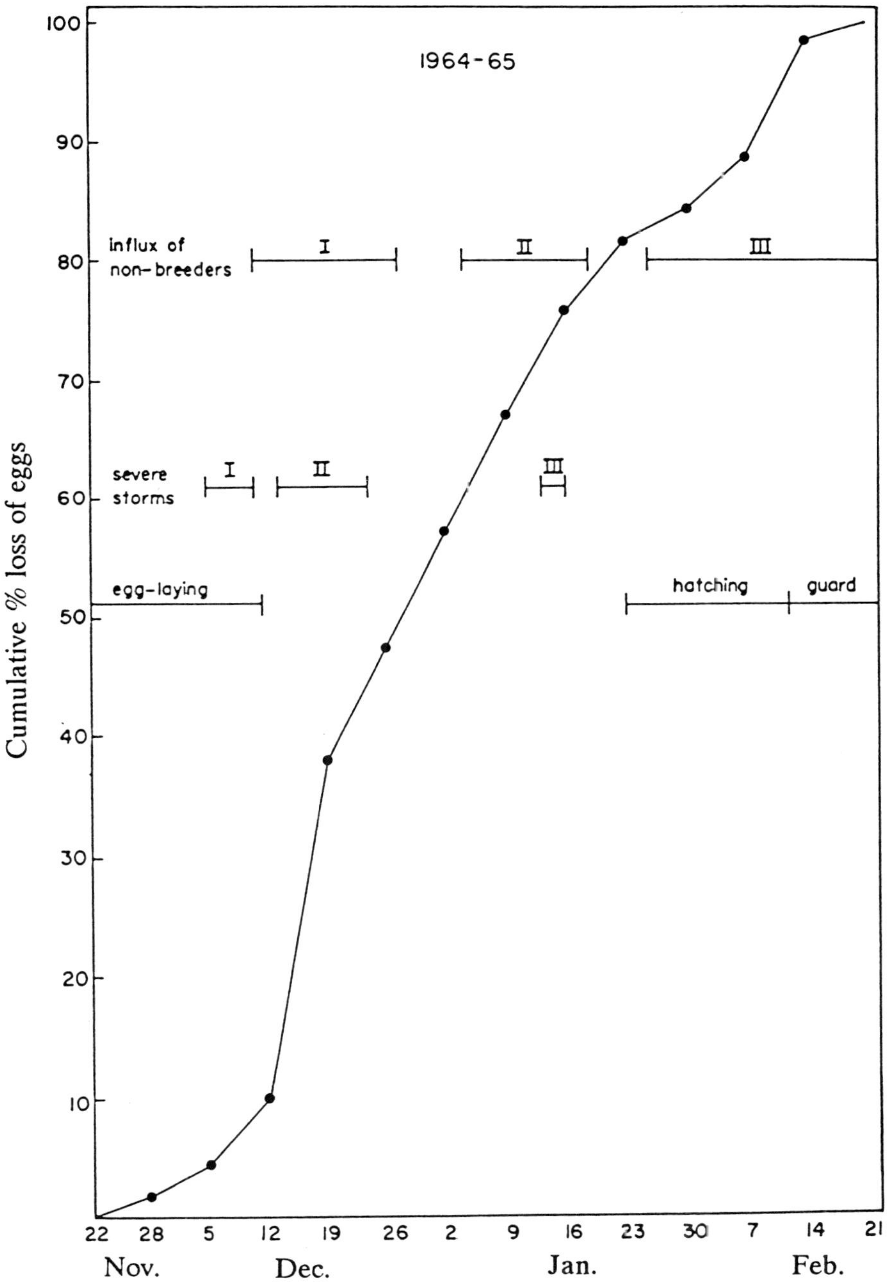

Figure 1.10 *The cumulative impact of Laysan Albatross egg and chick losses in relation to the weather and colony activity in 1964–65. Based on 350 fertile eggs. From Fisher (1971, p. 39).*

Antarctica where the poor breeding of Snow Petrels was ascribed to ENSO events a year previously (Chastel *et al.*, 1993).

Losses due to tidal waves, for example to *Puffinus nativitatis*, have usually been inferred rather than documented, but Munro (1947) described the destruction of a colony of *P. pacificus* by tidal waves in Hawaii. An eruption in 1952 at San Benedicto Island, Mexico, exterminated all the seabirds there including any *P. auricularis* (Brattstrom, 1963), but whether that species has re-established itself from a 'strategic reserve' of prebreeders, is not clear (Jehl, 1982). Landslides at colonies of *Pagodroma nivea* in Heimefrontfjella, West Antarctica, were believed to have destroyed many nests in the 1963–64 summer (Bowra *et al.*, 1966). At Laysan Island, Willett (1919) reported the mass smothering of nesting Sooty Storm Petrels *Oceanodroma tristrami* by wind-blown sand. Chicks of beach-nesting *D. nigripes* are likewise lost by sand burial during gales (Rice & Kenyon, 1962).

There may be natural hazards at the breeding places. These include blowholes as are found at the Albatross Island colony of *D. cauta* (Macdonald & Green, 1963) and at the Snares Islands where *Puffinus griseus* is the main victim. Birds that fall into these pits and can neither fly out vertically nor scale the surrounding cliffs succumb to predators or starve to death. Even aerially competent petrels may be trapped. Gun pits on South Channel Fort Island in Victoria, Australia claimed the lives of many White-faced Storm Petrels. This ceased when the pits were covered with wire netting but resumed when it was stolen (Gillham & Thomson, 1961).

A special kind of an environmental hazard is provided by the mosquito *Aedes taeniorhynchus* which took blood meals from *D. irrorata* by day and by night. These insects abounded after heavy rains so that where the insects were at high densities at least half the eggs were deserted. These then became encased in mud or were eaten by the mockingbird *Nesomimus macdonaldi* (Anderson & Fortner, 1988).

The accidental entrapment and deaths of *Pterodroma leucoptera* by sticky *Pisonia* seeds has been referred to (*The Petrels*, p. 107), but more bizarre are those of adult *O. leucorhoa* whose plumage is contaminated in their burrows by the slime from slugs that hide there. Unable to fly, the birds are eaten by gulls (Campbell & Stirling, 1968).

Pressure on space by conspecifics appears to be causing density-dependent production losses in biennially-breeding Chatham Island *D. epomophora sanfordi*. Because of the high egg losses already noted, in the following years large numbers of birds return to renest and the colonies are excessively crowded. In the 1973/74 season 80% of the breeders were trying to nest and productivity has declined greatly since then (Robertson, 1991).

2 *'Wrecks'*

Petrels usually predominate among beach-stranded seabirds, often driven to land by strong onshore winds. These 'wrecks' are particularly common along the western coasts of South America, South Africa, Australia and New Zealand where the enormous fetch of the westerlies provides a huge catchment area and tubenoses are the predominant birds.

The potential of such events in providing information on what species occur locally was realized early by Alexander (1916) and Whitlock (1927) in Western

Australia and by Falla (1922) in New Zealand, and many museum specimens have been collected during beachcombing expeditions. Nowadays beach surveys are regularly organized—see for example Ryan *et al.* (1989) and Powlesland (1986).

Strandings may occur over a wide area. The 1952 wreck of *O. leucorhoa* documented by Boyd (1954), probably involved the whole Atlantic population, and while most casualties were in Britain and Ireland, birds turned up in France, Holland, Germany and even Switzerland. The principal cause of the mortality was deemed to be persistent strong wind and the emaciated state of the victims with no subdermal fat, little tissue glycogen and empty stomachs.

Storm-driven birds may end up far inland. Among many examples are the *P. macroptera* 200 km from the sea in Western Australia, *Procellaria aequinoctialis* 250 km up the Amazon and *Puffinus griseus* 480 km inland in North Carolina. In eastern North America such birds have been seen flying around lakes and rivers—up to hundreds of *O. leucorhoa* at a time. In one North Carolina wreck thousands of these petrels dead or dying formed 'a blanket from the water's edge into the grass' and fishermen carried baskets of the birds home to eat (Pearson, 1899).

A major irruption during 1984 into South African waters of *Lugensa brevirostris, Halobaena caerulea* and *Pachyptila belcheri,* birds seldom seen locally, was analysed by Ryan *et al.* (1989). Although the strandings were related to passages of cold fronts with strong onshore winds, the birds had appeared at sea at least 2 weeks earlier and these authors believed that the wind was only a proximate factor in their deaths. They speculated that the irruption of these more southerly species into local seas resulted from a major physical or biological perturbation in the Southern Ocean. The wrecks occurred a year after the 1983 ENSO and there were similar wrecks in Australasia (Powlesland, 1986). It was suggested that the ultimate cause was ENSO-induced changes to oceanic circulation leading the birds to shift to seas to which they were ill adapted, the winds taking a toll of already weakened birds.

Heavy mortalities also accompany the migrations of *Puffinus griseus* and *P. tenuirostris*. These take place both in home waters and in their contranuptial quarters. Beach-washed birds of both species peak on New Zealand coasts in November–December and in May. The early birds seem most likely to be returned prebreeders; the May ones are fledglings often still with down on their napes. They appear to have fledged underweight and lack the strength to cope with strong onshore winds (Stonehouse, 1964). Wrecks of *P. tenuirostris* occur mainly in eastern Australia at irregular intervals. The biggest events are of birds back in Australian waters after battling the South-East Trades and most are first-year birds. They may be so ravenous for food as to try to steal fishermen's baits. Serventy *et al.* (1971, p. 35) believed that they die because of the unreliable plankton of the Tasman Sea which, in some years, results in a scarcity of *Nyctiphanes australis* on which the shearwaters depend. The inexperienced birds perish in great numbers and are washed ashore from Queensland to Victoria and even as far as Stewart Island, south of New Zealand.

Wrecks on Japanese shores mainly involve *P. tenuirostris* and *O. leucorhoa* and locally breeding *Calonectris leucomelas*. Samples from the occasional mass deaths of *P. tenuirostris* that occur in May and June proved to be mostly young birds (Oka & Maruyama, 1986). These workers suggested that such birds had begun their migration with inadequate fat stores and that the incidence of 'wrecks' in Japan might correlate with fluctuations of *N. australis* in Australian seas. The birds' physical condition was studied by Oka & Maruyama (1985) and their ages by skull ossification (Sugimori *et al.*, 1985). Nishigai *et al.* (1981) determined from pathological and haematological examination that the young birds died of anaemia and malnutrition.

3 *Predation*

Natural predators of petrels on land have doubtless had important effects on the past evolutionary history of the group, particularly in directing them into ecological niches where such pressure is relaxed, such as nocturnality, burrowing, and use of island sanctuaries.

a. Avian predators. Today natural predators are mainly avian, particularly gulls and skuas. Gulls may have large appetites. One *Larus marinus* swallowed five adult *O. leucorhoa* in 20 min (Gross, 1935). Such predation is readily detected as gulls typically turn a petrel's skin inside out. Few reports give quantitative information, and what is available is difficult to interpret.

Gynn (1984) counted *P. puffinus* corpses on Skokholm between 1957 and 1983. There were more dead birds when many pairs of *L. marinus* bred than when few did so. In 1957 2465 corpses were collected and 27 pairs of these gulls nested, in 1983 when 11 pairs bred 1373 corpses were picked up while in 1968 the counts were 841 dead birds and three pairs of gulls. Many of the victims were prebreeders or chicks, but Brooke (1990, p. 48) estimated that, over the whole period, under 10% of the total annual mortality was ascribable to gulls and he did not believe that this level of predation threatened the population which may indeed be increasing.

Watanuki (1986) estimated that *L. schistisagus* took 13.2% of the annual population of adult *O. leucorhoa* at Daikoku Island, Japan. The incidence of this predation was high when the energy needs of the gulls and their chicks were at a peak (Fig. 1.11). Watanuki thought that the reduction in petrel activity ashore on moonlight nights was primarily in response to the risk of predation, and Bretagnolle (1990b) confirmed that five other tubenoses behaved similarly, the non-breeders especially (cf. *The Petrels*, p. 253).

In the tropics, gulls and skuas are generally absent and predation by other birds slight. There is little information on the role of frigate-birds here but Shallenberger (1973, p. 87) described *Fregata minor* at Midway Atoll forcing *P. pacificus* that flew overhead before dark to disgorge and elsewhere some *P. lherminieri* have been reported as killed by frigate-birds.

In high latitudes the very long, light nights add to the risks for petrels, especially for non-breeders, although such conditions may aid food getting at sea. And, as most gulls and skuas breed in the summer, winter-nesting petrels such as *Pterodroma macroptera* and *Puffinus assimilis* largely escape their attention. Other winter breeders, such as *Pterodroma hypoleuca*, experience no such predation either, as neither gulls nor skuas share their breeding places. Why this bird should still be entirely nocturnal on land is unclear but we know nothing of past predators that might have enforced such behaviour.

Much skua predation seems to occur of birds on the surface at night and of birds near their nests by day, for example *Pagodroma nivea*. The techniques used by skuas

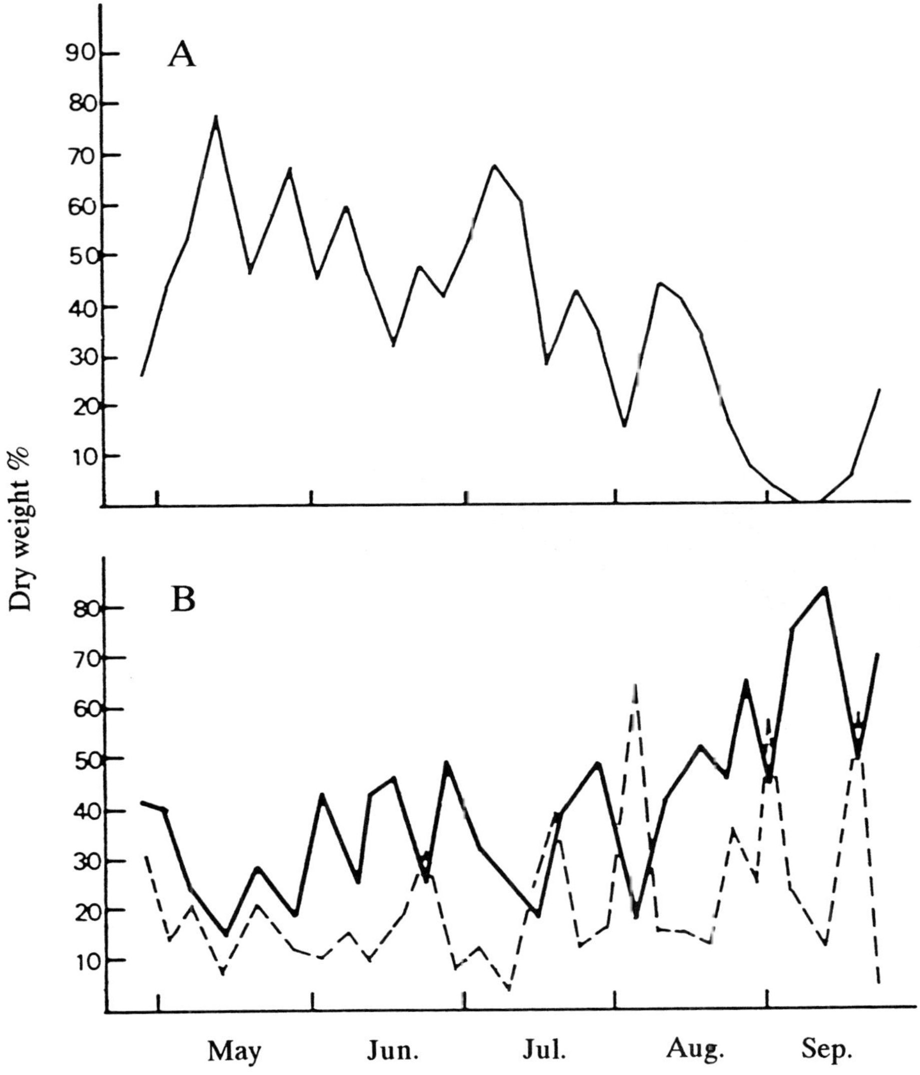

Figure 1.11 *Seasonal changes in the diet of Slaty-backed Gulls* Larus schistisagus *as shown by dry-weight percentages. (A) Adult Leach's Storm Petrels. (B) Fish (dark line) and other foods (broken line). From Watanuki (1986).*

to take subantarctic petrels were described by Sinclair (1980). In the dry valleys of Antarctica bones of skua-killed Snow Petrels have piled up over many years in the middens of *Catharacta maccormicki* (Fig. 1.12).

From analyses of pellets, Jones (1980), Schramm (1983) and Ryan and Moloney (1991a) determined the prey of *C. s. lonnbergi* at Macquarie and Marion Islands and Tristan da Cunha. The skuas fed mainly within their territories and took the most abundant small petrels there, for example *Pachyptila desolata* at Macquarie and

Figure 1.12 *McCormick's Skua chick on nest surrounded by Snow Petrel bones. Photo: W.Richter.*

P. salvini at Marion Island. Large petrels were thought to be under-represented in the pellets. For example, the Shoemaker *Procellaria aequinoctialis* was abundant at Marion Island yet evidently was seldom taken; presumably the prions provided abundant and easier prey. These big birds may be eaten by skuas at times, for example on Antipodes Island where a short study found that they comprised 21% of the skua's petrel diet (Moors, 1980) and where Warham and Bell (1979) described daytime attacks by as many as eight skuas in consort.

Southern Skuas excavate shallow petrel burrows with their bills, trenching back from the entrances to reach the chicks or adults of birds like *Pachyptila desolata* and *Pelecanoides georgicus*. Small species are also taken by day on the wing, the straight flight of diving petrels making their capture easy. The flight of prions is usually too erratic for most skuas, but some are caught, and in the Antarctic *Oceanites oceanicus* is taken on the wing (Burton, 1968; Beck & Brown, 1972).

Other natural avian predators, whose pressures may be locally important, include sheathbills *Chionis* spp., while owls *Asio flammeus*, crows *Corvus* spp. and Bald Eagles *Haliaeetus leucocephalus* prey on *Oceanodroma leucorhoa* (Abe *et al.*, 1972; DeGange & Nelson, 1982; Holt, 1987). *Asio galapagoensis* also kills significant numbers of *O. castro*, *O. tethys* and *Puffinus lherminieri* at the Galapagos (Harris, 1969a,b). Harris saw this owl hunting on the wing after dark but also taking the diurnally active *O. tethys* by waiting near a nesting crack and jumping feet first onto the emerging petrel. At these same islands the gadfly petrel *Pterodroma phaeopygia*, nesting in rocky recesses and shallow caves, is dragged out and killed by a buzzard *Buteo galapagoensis* (Cruz &

Cruz, 1987b). Manx Shearwaters are also taken by *B. buteo* (Warman & Warman, 1985), while at the Falkland Islands caracaras *Phalcoboenus australis* get some *Pachyptila belcheri* (Strange, 1980).

Tristan Thrushes *Nesocichla eremita* are reported as killing adult *Fregetta grallaria* (Ryan & Moloney, 1991b); even more enterprising is the mockingbird *Nesomimus macdonaldi*, which not only attacks unattended *Diomedea irrorata* eggs and kills the small nestlings, but also takes food directly from the parents' bills as they disgorge to their chicks (Brosset, 1963, p. 87).

The Weka *Gallirallus australis*, a flightless rail, can be an aggressive predator, has been implicated in the deaths of small petrels at Macquarie Island and was largely responsible for the decline of *Pterodroma cookii* and *P. inexpectata* on Codfish Island, New Zealand (Blackburn, 1968) before being removed. On Laysan Island, Bristle-thighed Curlews *Numenius tahitiensis* use stones to crack open uncovered (but not necessarily abandoned) eggs of *D. immutabilis* and *D. nigripes* (Fig. 1.13).

b. Non-avian predators. The only invertebrates regularly implicated on preying on petrels are crabs. Murphy (1924a) described how sand crabs *Ocypode cursor* hunted young *Pelagodroma marina* in their burrows at night and Harris (1969a, p. 131) reported *Grapsus grapsus* preying upon nestling *O. castro*.

Some petrels lose chicks or eggs to reptiles. Snakes take eggs and small chicks of *Puffinus tenuirostris*, *C. leucomelas* and *Pelagodroma marina* (Jones, 1937; Serventy *et al.*, 1971, p. 130; Yoshida, 1981). According to Penny (1974, p. 122) the skink *Mabuya wrightii* ate about 10% of the eggs of *Puffinus pacificus* on Cousin Island, Seychelles, in 1970. The Tuatara *Sphenodon punctatus* also eats young prions (Harper, 1976; Newman, 1987), and nesting Green Turtles *Chelonia mydas* sometimes dig up burrows of Wedge-tailed Shearwaters.

Apart from foxes, there are a few instances of native mammals eating petrels in northern latitudes. The best documented is the case of the River Otters *Lutra canadensis* that prey upon the storm petrels *O. furcata* and *O. leucorhoa* at Fish Island, Alaska (Quinlan, 1983). The colonies at 60°N are the most northerly in the Pacific with continuous daylight during most of the nesting season. Quinlan found that otter predation on adult birds was high, much greater than that from raptors and crows. The eggs were not eaten but most predation occurred during egg laying and incubation, mainly of birds in burrows, few of those in crevice sites. It was estimated that only 24% of 204 eggs of *O. furcata* produced chicks whereas, in an otter-free exclosure, 68% of 25 eggs gave rise to fledged chicks. Once a nest in a high-density area had been detected others were quickly found—a high cost for colonial breeding.

Unusual predators are the Red Deer *Cervus elephas* that kill Manx Shearwaters for the calcium in their bones, a habit first noted by Wormell (1969). On the Scottish island of Rùm (Rhum), deer attack fledglings caught on the surface at daybreak. They bite off their victim's head and chew only the bones of wings and legs (Furness, 1988a). Furness estimated that only about 4% of all the fledglings were taken. Such activity has not been reported elsewhere, for example at South Georgia or Kerguelen where introduced Reindeer *Rangifer tarandus* share habitats with many kinds of petrel.

c. Alien predators. Tubenosed birds are particularly vulnerable to attack by land mammals introduced from mainland ecosystems. The birds mostly have evolved no effective direct protection from such attack apart from oil spitting and few burrowing species do that effectively. Petrels tend to be noisy at night and their body odour may guide nocturnal hunters such as cats to occupied nests. Particularly at risk are the prebreeders as these display and call from the surface—in a burrow a bird is reasonably protected if the tunnel is too small to admit the carnivore. But burrows

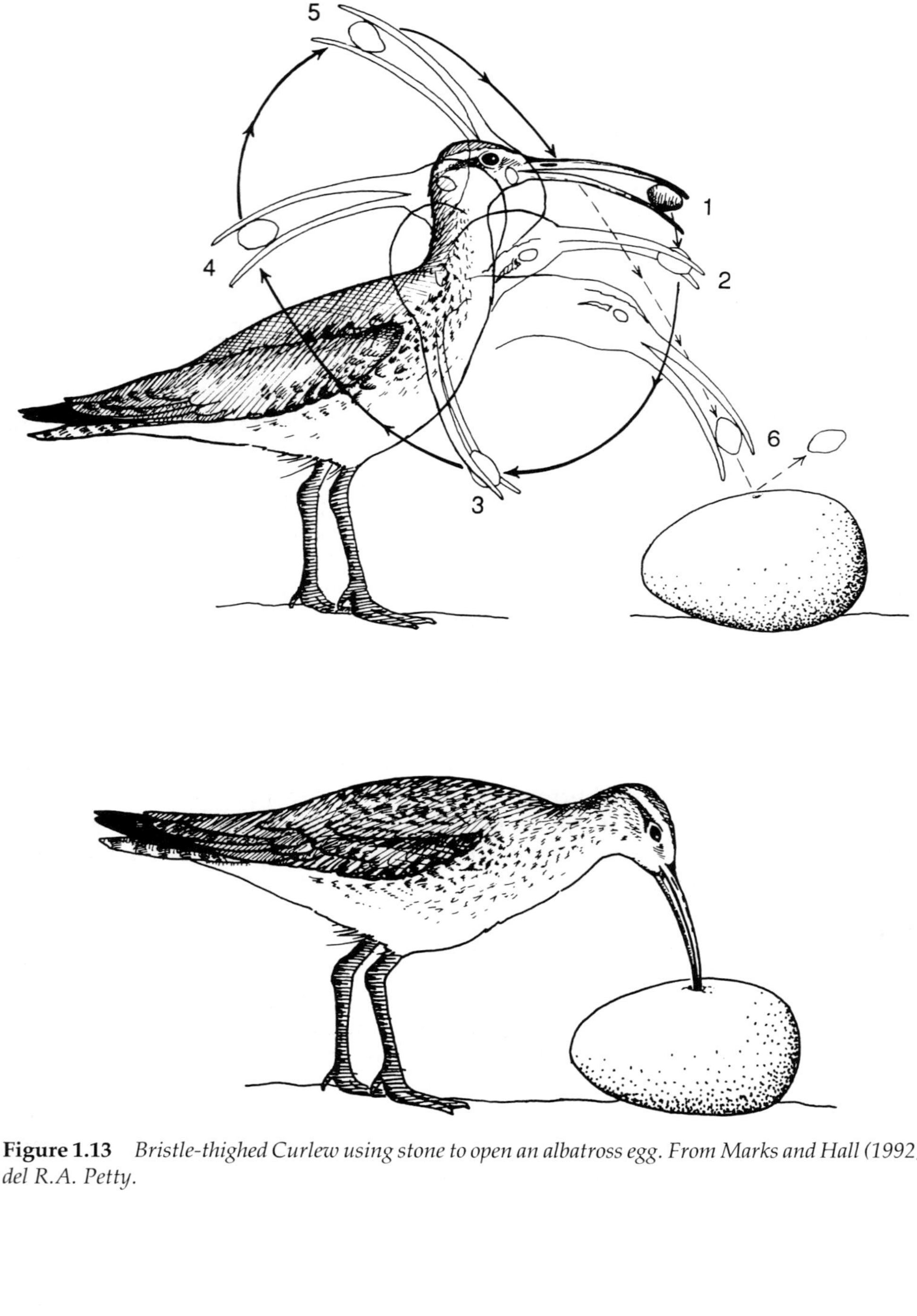

Figure 1.13 *Bristle-thighed Curlew using stone to open an albatross egg. From Marks and Hall (1992), del R.A. Petty.*

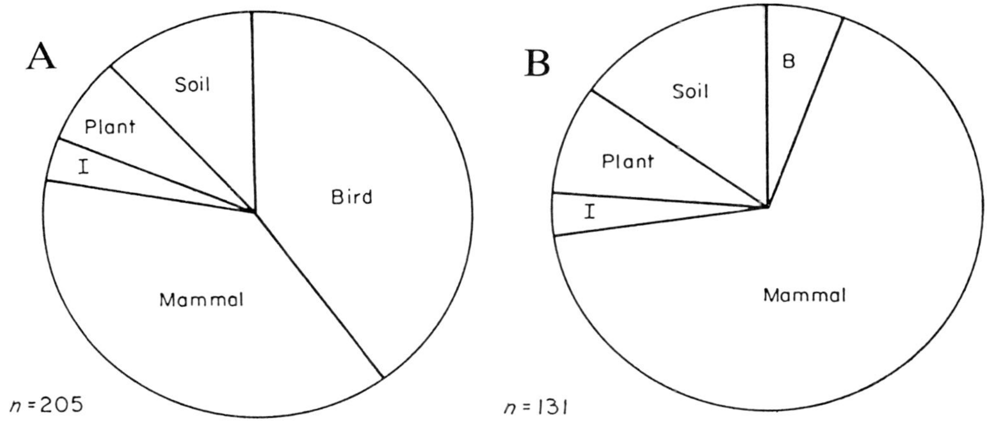

Figure 1.14 *Results of analyses of fox scats from Cape Woolamai from September 1965 to August 1968. Scats have been separated into two samples based on whether Short-tailed Shearwaters were (A) present or (B) absent from the nesting colonies. The samples have been subdivided into the five major foods. I, insects; B, birds. From Norman (1971).*

that may provide protection from gulls and skuas give little protection from rats or mustelids small enough to enter the tunnels, where the petrel is effectively trapped. Large, surface-nesting albatrosses and giant petrels can defend themselves and seem to be immune from any pest except dogs. A notable exception is the strange case of *D. immutabilis* and *Rattus exulans* on Kure Atoll (*The Petrels*, p. 46).

General reviews of the effects of alien predators have been provided by Moors and Atkinson (1984) and Johnstone (1985). Control and elimination of mammalian predators is discussed in Chapter 12.VII.B.

In the Aleutian Islands introduced Arctic Foxes *Alopex lagopus* drastically reduced populations of petrels and other marine birds with large numbers of storm petrel remains being found in fox dens (Sekora *et al.* ,1979). Norman (1971) investigated the predation by European Foxes *Vulpes vulpes* at colonies of *P. tenuirostris*, one at Cape Woolamai with rabbits present, the other at rabbit-free Benison Island. At Woolamai the major victims were rabbits (Fig. 1.14). At the other location counts of shearwater corpses peaked in October and fell in February–March. Norman concluded that with rabbits present the effects on the birds were relatively unimportant, but where these were absent fox predation was substantial, the main victims being the noisy, surface-active prebreeders and non-breeders.

There are many more instances of damage by cats than by dogs. Some of the better documented accounts include those of Imber (1987) for *Procellaria parkinsoni* and of van Aarde (1980) and van Rensburg and Bester (1988) for various species at Marion Island. Jones (1977, 1980) provided data on cat kills at Macquarie Island where, after the rabbit, *Pterodroma lessonii* and *Pachyptila desolata* were the most frequent items in both scats and stomach contents. The incidence of cat predation on Macquarie Island petrels was also discussed by Brothers (1984) who estimated that they might kill 26 000 prions annually.

Cats take adults, nestlings and eggs. When in their burrows big aggressive birds like *Procellaria aequinoctialis* can probably keep cats at bay, but unguarded chicks are very vulnerable and the big burrows allow easy access. As hungry chicks tend to call after dark, this too gives a further clue to their whereabouts. All young petrels are particularly endangered during their exploratory emergences late in their development to exercise their wings.

Introduced mustelids, for example Stoats *Mustela erminea* and mongooses *Herpestes auropunctatus* in New Zealand and Hawaii respectively, presumably speeded the extinction of birds like *Pterodroma inexpectata* from the South Island and *P. phaeopygia* from some of the Hawaiian islands.

Pigs can devastate petrel colonies by rooting out burrows and devouring their contents. Howell and Webb (1990), for example, found that all the nests of the rare *Puffinus auricularis* on the Revillagigedo Islands, Mexico, had been dug out, with shearwater remains littering the destroyed burrows.

Except perhaps for the water rat *Hydromys chrysogaster* of Bass Strait, all the rats (*Rattus rattus, R. norvegicus, R. exulans*) found in petrel colonies have got there with the help of man. Their impacts on seabirds were reviewed by Norman (1975) and Atkinson (1985). Norman pointed out that most accounts of rat predation on petrels were circumstantial, with few local studies of rat foods having been made. In his work on a colony of *P. tenuirostris* Norman (1970c) found that *R. rattus* was mainly vegetarian and ate only unattended eggs and dead nestlings.

While many earlier accounts of the impacts of rodents on petrels lack firm data, recent studies have established that rats may be very significant predators of petrel eggs and chicks. For example, field experiments by Woodward (1972) and Grant *et al.* (1981) revealed that *Pterodroma hypoleuca* lost many eggs to *R. rattus*. Tompkins (1985) described methods of distinguishing predation by this animal on *P. phaeopygia* from that due to dogs, cats, pigs, owls and hawks. The rats were the major causes of egg and chick mortality.

4 *Disease*

Disease in petrels is well known only in *Fulmarus glacialis* and *Puffinus puffinus*. Ornithosis in the Northern Fulmar was studied in some detail because it caused human fatalities. The effects on the fulmar populations have not been assessed. Fulmars certainly die from the disease but there seems to be no evidence of associated mass mortalities.

Puffinosis in Manx Shearwaters has been described from the West Wales islands (*The Petrels*, pp. 169 & 170) and Brooke (1990, pp. 144–167) provides a detailed account of experiments to help understand its causes. It mostly affects nestlings during the desertion period and Brooke found that about 15% of those with blistered feet died before they flew.

Ornithosis also occurs in *P. tenuirostris* fledglings but consequent mortalities on the nesting islands have not been seen during the postnuptial migrations (Mykytowycz *et al.*, 1955). Miles and Shrivastav (1951) suggested that this virus could not be highly infective in man. However, while working with the 1954 strain, Mykytowycz himself had a moderately severe attack, with symptoms of pneumonia.

The 'limy bird disease' of *P. tenuirostris* is a non-infectious condition resulting in the blockage of the lower alimentary tract with sodium urate concretions that causes heavy chick loss in some seasons. The disease was studied by Mykytowycz (1963), Munday (1966) and Munday *et al.* (1971), the associated gastrointestinal flora by Mushin and Ashburner (1952). The cause of the blockages remains obscure.

Avian pox was reported in chicks of *D. melanophrys* at the Falkland Islands by Thompson (1989), and was accompanied by considerable mortality, as has also happened with *D. immutabilis* in the North Pacific.

VII Population changes

The numbers of petrels on land vary during the year, tending to peak just before egg laying with many non-breeders ashore. The numbers of those species having compressed laying periods may show a cyclic pattern reflecting on- and off-duty stints, and this may carry over into the early chick stage. Minima are reached during the contranuptial season when only biennial breeders and sedentary ones like *Macronectes* are regularly ashore.

On a broader scale, seabird breeding locations, populations and their marine habitats are not static, and Myers (1979) pointed out that new colonies formed by young birds may represent a continual adjustment to changing conditions like shifts in sea surface temperatures, with older colonies that are less well placed in the new environment gradually fading away.

Following population changes requires comparable counts at the same time over several years. For example, the Snares Island *D. bulleri* population in 1969 was about 4750 pairs (Warham & Bennington, 1983) as against 8460 pairs in 1992 (Sagar *et al.*, 1994), but two counts 23 years apart are quite inadequate to determine population trends.

Most recent changes have been decreases due to human intervention, for example to habitat destruction, the importation of alien mammals and commercial fishing (e.g. Croxall *et al.*, 1990a). Changes to southern species were reviewed by Jouventin and Weimerskirch (1991).

A *Population increases*

Changes in the numbers of species with small populations may be difficult to identify. For example, *Pterodroma hasitata* is being seen with increasing frequency off southeastern USA, with 161 off Capes' Lookout and Hatteras on 7 October 1980 (Le Grand, 1981). Whether these sightings reflect a real population increase on the little-known Caribbean nesting grounds, are due to a shift in migratory patterns in response to changed oceanographic conditions, or simply result from more people looking for seabirds at sea, is unclear.

1 *The case of the Northern Fulmar*

The most dramatic population increase of any petrel and perhaps of any bird, has been that of *F. glacialis*. Like most increases, this has been accompanied by an expansion of breeding range. Being a large, surface-nesting and diurnally active species, its change of status has been well recorded and was outlined in *The Petrels*, pp. 65, 71–72.

The increase in the large-billed race *F. g. auduboni* was noted at the beginning of this century and Harvie-Brown (1912a,b) published a survey of its distribution and spread. Fisher (1952a, 1966) mapped the colonies and worked out their histories in great detail and traced the expansion from Iceland in about 1640, thence to The Faroes about 1839, to the British Isles in 1878 (except for St Kilda where it had long been established). Fisher (1952a,b, 1966) showed that the mean annual increase for all colonies except St Kilda varied from 6.4 to 16.5%. According to Cramp *et al.* (1974) the overall increase in the British Isles between 1929 and 1949 was about 420%, that between 1949 and 1969/70 about 280%. They calculated that, excluding St Kilda, the British fulmar population was still growing at almost 7% per annum compounded. The expansion continues, with small colonies becoming established in France (Brittany 1960; Normandy 1971, Picardy 1979, Pas de Calais 1983). All these colonies are small and confined to cliffs. Mercier (1987) reviewed the increases in occupied sites in Picardy between 1972 and 1986 and described the breeding cycles there. Small numbers have also bred in Heligoland since 1972.

The hypothesis that expansion of the fulmar's range and its use of non-traditional sites is partly induced by a shortage of natural ones is not supported by the results of a recent study. Olsthoorn and Nelson (1990) evaluated the quality of fulmar nesting sites at a colony on the Aberdeenshire cliffs and considered that there were plenty of suitable ledges still unoccupied.

In 1970, the Northern Fulmar became the first bird to become established on Surtsey, the volcanic island off Iceland that was born in 1963, and in Iceland there are fulmar colonies 50 km from the sea (see Fig. 9.7). This ability to breed far inland is shared by the southern fulmars *Pagodroma nivea* and *Thalassoica antarctica* which succeed under much more severe conditions. Why does the Southern Fulmar *F. glacialoides* not do that too?

Salomonsen (1935, pp. 251–252) pointed out that a prerequisite for a fulmar 'explosion' would be a reduction in the species' attachment to its nesting places and suggested that some individual birds experienced a genetic mutational change affecting their philopatric behaviour. Fisher (1952a, 1966) ascribed the increase to the appearance of a new source of food—offal from Arctic whalers and, later, from fishing trawlers. However, Salomonsen (1965) believed that the initial expansion occurred independently of human activity as he could find no evidence for any increase in fishing or whaling in Icelandic waters when the expansion is deemed to have begun. He proposed that the shift of genotype better adapted the subspecies to warmer southern waters, and particularly to feeding on the macroplankton of offshore seas in winter, a hitherto unexploited niche for the bird.

Brown (1970) concluded that availability of fish offal does not control the fulmar's distribution in the western Atlantic even of birds belonging to the eastern Atlantic expansionary population allegedly linked to the offal supply. He postulated that changing oceanographic conditions will be found to be responsible for the population increase. Further light may be thrown on the matter as light-phase birds now breed on islets off Newfoundland (Nettleship & Montgomerie, 1974; Montevecchi *et al.*, 1978; Cairns *et al.*, 1986); others prospecting around St Pierre et Michelon (Desbrosse & Etcheberry, 1989).

The role of filter-feeding by fulmars (*The Petrels*, p. 49) has yet to be examined.

The expansion of the boreal population of the Northern Fulmar shows that, despite a long period of immaturity, a clutch size of one egg, no relaying (i.e. no ability to increase production within the same year with a sudden improvement in conditions) and some tendency to take 'sabbatical', non-breeding years (Ollason & Dunnet, 1983), the potential for a rapid build up is high once some key restraint is relaxed.

A fulmar expansion may also be taking place in the North Pacific. At least four colonies there may be new, and these are growing, but figures for the large colonies are as yet inconclusive (Hatch, 1993).

2 *The case of the Laysan Albatross*

The range expansion of this albatross is less dramatic and more recent than that of the Northern Fulmar. The data have been analysed by Gould and Hobbs (1993) who estimated annual growth rates at the various islands, for example at Midway Atoll these were 7.1% for *D. immutabilis* (and 1.1% for *D. nigripes*). The change may represent a rebound of a depressed population and a re-establishment of a former range.

In recent years, the US Navy has taken good care of the Midway birds, the other island populations have recovered, and the species has established (or re-established) a foothold on the main Hawaiian Islands, particularly on Kauai, since 1970. A number of breeding attempts at various sites have been made—see also Chapter 12.V. The expansion has carried some birds across the North Pacific to western Mexico where small colonies have been founded (Pitman, 1985; Dunlap, 1988; Oberbauer *et al.*, 1989; Howell & Webb, 1990; McDermond & Morgan, 1993).

That this expansion has been eastwards is not unexpected as the epicentres of all the North Pacific albatrosses tend to shift eastwards during the non-breeding season (see Fig. 2.16), with some birds reaching coastal waters from British Columbia to Mexico, and it is perhaps significant that *D. immutabilis* has not re-established colonies to the south and southwest of the Hawaiian Chain, for example on Johnston and Wake Islands and the Marianas. The colony on Guadalupe Island, Mexico started just after the 1982/83 ENSO. Gallo-Reynoso and Figueroa-Carranza (pers. comm.) suggested that the colonizers might be experienced birds forced to forage far from their normal range and, having discovered the island, stayed.

Increased sightings of this albatross in the northern Gulf of California and in inland California and Arizona (Newcomer & Silber, 1989) support the idea of birds exploring for new territories and there seems to be no obvious reason why *D. immutabilis* should not build up a self-sustaining population on islands off Mexico and California, as suggested by Pitman (1985).

3 *Some lesser expansions*

A bird that is extending its range, this time in a southerly direction, is *Pterodroma nigripennis*. Since 1980 it has bred on Mangere Island in the Chathams, and recently has been increasingly in evidence flying around offshore islands and northern headlands. As far as is known, these exploratory birds do not originate from the colonies on the Chatham or Three Kings Islands, but may have come from the Kermadecs Group, where another gadfly petrel, *P. neglecta*, and *P. nigripennis* are now abundant on Macauley Island with about 300 000 pairs nesting on Curtis Island (Tennyson & Taylor, 1990a).

Two petrels that illustrate how a species can rebound once a restraint is lifted are Buller's and Cory's Shearwaters. *Puffinus bulleri* has only ever been known to nest on the two Poor Knights Islands off northern New Zealand—Tawhiti (132 ha) and Aorangi (68 ha). Harper (1983) detailed how at the latter island feral pigs ate out the vegetation and routed out the several species of resident petrels so that none could be found in 1924 (Falla, 1924). On the larger island *P. bulleri* flourished, free from introduced pests. After the shooting of the last pig in 1936 the petrels reappeared on Aorangi, presumably from Tawhiti, and by 1981 there were approximately 200 000 birds on Aorangi. So far there is little sign of this species expanding its breeding range.

Population changes of Cory's Shearwater *Calonectris diomedea borealis* breeding on Grand Salvage Island have been followed in detail (Zino, 1985; Mougin *et al.*, 1987; Mougin & Roux, 1988). This provides another example of a petrel's ability to spring back from a very low point, and where the long prebreeding period (7–13 years) may have proved valuable—keeping a substantial part of the population beyond the reach of vandals. It is also an instance of population increase without range expansion.

Other extensions of range that may or may not reflect increases in populations, involve the establishment of new, small colonies far from the customary breeding places. Examples are the Manx Shearwaters already referred to (Section III.C above) that have begun to nest in the Western Atlantic at Middle Lawn Island and are prospecting Colombier Island (Mactavish, 1992), and the Shy Albatross at Penguin Island, The Crozets, far from their nearest colony in Bass Strait (Jouventin, 1990). No doubt such colonizations have occurred many times during the course of petrel evolution and until quite recently would have gone unrecorded, particularly if they involved small, nocturnally active kinds.

B *Population declines*

Clear instances of petrel populations in natural decline and uninfluenced by human activity are hard to find outside the fossil record. Some species seem always to have been rare in historical times, for example *Pterodroma macgillivrayi*, or to be in small numbers, for example *Pachyptila crassirostris*. The latter may perhaps be restricted by the availability of crevice-nesting habitat; the former could well have once been eaten by tribesmen, although there seems to be no evidence for this, so that the species may be heading for extinction.

VIII Population regulation

As with other animals, petrel populations may be regulated through some essential need being in short supply at some critical phase in the annual cycle, so as to impose

Table 1.9 Estimated breeding populations (pairs) of seabirds at South Georgia. From Croxall and Prince (1980) and Prince and Croxall (1983)

King Penguin	*A. patagonica*	22 000
Chinstrap Penguin	*P. antarctica*	2000+
Gentoo Penguin	*P. papua*	*c.* 100 000
Rockhopper Penguin	*E. chrysocome*	10–15
Macaroni Penguin	*E. chrysolophus*	5 million+
Wandering Albatross	*Diomedea exulans*	4300
Black-browed Albatross	*D. melanophrys*	60 000
Grey-headed Albatross	*D. chrysostoma*	60 000
Light-mantled Sooty Albatross	*Phoebetria palpebrata*	5000
Southern Giant Petrel	*Macronectes giganteus*	5000–6000
Northern Giant Petrel	*M. halli*	2500
Cape Pigeon	*Daption capense*	20 000
Snow Petrel	*Pagodroma nivea*	3000
Antarctic Prion	*Pachyptila desolata*	22 million
Fairy Prion	*P. turtur*	small nos
Blue Petrel	*Halobaena caerulea*	70 000+
Shoemaker	*Procellaria aequinoctialis*	2 million
Wilson's Storm Petrel	*Oceanites oceanicus*	*c.* 600 000
Black-bellied Storm Petrel	*Fregetta tropica*	10 000
Grey-backed Storm Petrel	*Garrodia nereis*	Scarce
South Georgian Diving Petrel	*Pelecanoides georgicus*	2 million
Common Diving Petrel	*P. urinatrix*	3. 8 million
Blue-eyed Shag	*P. atriceps*	2000–5000
Brown Skua	*C. lonnbergi*	2000–5000
Dominican Gull	*L. dominicanus*	500–1000
Antarctic Tern	*S. vittata*	1000+

a density-dependent mortality, cessation of breeding or recruitment, or through predation or parasitism. The possible mechanisms have been reviewed by Croxall and Rothery (1991) and Wooller *et al.* (1992) for seabirds in general but the complex of regulatory factors is unknown for any tubenose population at present; their importance no doubt varies from species to species, from time to time, from population to population and from one colony to another. And, as many populations have been drastically reduced by man, factors that formerly limited their growths should now be relaxed and the potential for recovery with protection high.

In some instances, shortage of nesting habitat may be limiting further growth. For example, the *Phoebetria* albatrosses choose grassy ledges for their nests, a habitat that is not at all abundant in the subantarctic where they breed, so that it is not surprising to find that they are among the least plentiful members of their communities (Table 1.9). Here man's influences are negligible; the birds have not been eaten since the sealing days and at their breeding stations they appear to experience little, if any, disturbance. A small-scale example is the decline in success of biennially breeding *D. epomophora* at the Forty-Fours following habitat degeneration and the resulting overcrowding with nests only 1–1.5 m apart (Robertson, 1994). Because of the high egg losses, large numbers of birds try to renest in the following years and in the

1973/74 season 80% of the breeders were attempting to breed; productivity has declined greatly since then (Robertson, 1991). Chastel *et al.* (1993) found that some nest sites of *Pagodroma nivea* were very productive, others, partly through snow blockage, fledged few chicks.

Shortage of space, however, has more usually been invoked as regulating populations of burrow-nesting species. Rowan (1952) suggested that there was insufficient room for more burrows of *Puffinus gravis* at Nightingale Island where many eggs were laid on the surface and it was thought that all the burrows were occupied. This may also be a factor at the Snares Islands *P. griseus* colony where all suitable ground is burrowed and many surface eggs are laid (Warham & Wilson, 1982). Although Rowan believed that *P. gravis* did not defend its nest site except from within, later work has shown that vigorous fighting over burrows is common at high nest densities and that securing a site may not be easy, for example with *P. puffinus* (Brooke, 1990, p. 90), but experimental proof of nest-site limitation is lacking.

Alternate use of scarce nest sites by two species that use them at different times should reduce competition for them, for example the winter breeding/summer breeding *Pterodroma macroptera/Puffinus carneipes, Pterodroma hypoleuca/Puffinus pacificus* and *Procellaria cinerea/P. aequinoctialis*. However, because of their long breeding periods, well-grown chicks of the one species tend to be still in occupation when the other arrives (Warham, 1956; Warham & Bell, 1979) and the benefits, if any, are unclear.

Brooke (1990, p. 91) pointed out that where competition for nests exists, as he believed holds for *Puffinus puffinus*, a density-dependent control could be exerted by delaying the age of first breeding of those lacking a nest, hence reducing the recruitment rate. Likewise, if a population is falling, a reduction in the age of first breeding would help to compensate for the decline—as may be occurring with *D. exulans* in response to 'by-catch' losses at sea (Weimerskirch & Jouventin, 1987; Croxall *et al.*, 1990a). Brooke also speculated on the role of nest quality as a regulatory factor. In large populations relatively more pairs would be forced to use marginally satisfactory sites, so that fecundity as a whole would decline and so would recruitment.

Food shortage at a critical time—during the prelaying exodus, chick rearing or postnuptial moult, for example—has often been mooted as the main factor maintaining populations within current limits. Non-availablility of food might arise through changes in oceanographic conditions whereby a food resource retreated into deeper water or otherwise shifted out of range. Food could also become denied to a petrel by intra- or interspecific competition for it, or by the birds themselves over-cropping their prey populations (e.g. Ashmole, 1963). In these circumstances only the best-adapted would feed well enough to breed and the proportion of non-breeders and the mortality rates would increase.

Whether competition for food is usually of great significance and whether any petrels deplete their food stocks is questionable. Certainly, where concentrations of food occur, there are often concentrations of petrels and other predators, with consequent intra- and interspecific competition, the most visible being the scavenging around fishing boats. Furthermore, segregation of feeding zones between the sexes, age classes, and populations of some species, and between species, has been suggested for southern albatrosses by Weimerskirch *et al.* (1985, 1988) and would support the idea of competitive exclusion (Cairns, 1989). Particularly when breeders feed inshore, the size of the food resource may limit colony size. However, the significance of competition at sea for population regulation for petrels in general is unclear. That many search pelagic seas and are widely dispersed suggests little competition. It may well prove that it is the ability of the birds to find widely scattered foods, liable to change at short notice with local changes in oceanography, that is limiting.

Brooke (1990, p. 172) believed that his study population of *P. puffinus* has increased in the last 100 years but he thought that because young parents were as good at feeding their chicks as older ones, it seemed unlikely that there was intraspecific competition for food.

Predation on land by skuas, gulls and the like seems to have but a minor role in depressing petrel populations under natural conditions. Although at high petrel densities there may be more predators, their populations are often checked by the absence of petrels during the non-breeding season—usually the winter months when alternative foods are scarce—so that the 'carrying capacity' of the predator's habitat is limited accordingly.

The hypothesis that petrels control their own numbers in a Wynne-Edwardsian fashion, that is, that 'sabbatical years' are taken to help the success of their kin, has proved controversial. It seems more likely that those that do not nest, either as prebreeders or as past breeders skipping a year, do so because they are unable to breed then. They may have failed to put on the weight needed to sustain them during the long fasts required by the tubenose breeding system, a failure that could have been caused by many things that make food harder to find, such as persistent and unseasonal gales driving the birds far from preferred foraging areas.

The mutual stimulatory effects of calling and display activity may have a role in population regulation, but have yet to be measured (see Chapter 4.VI.E). For example, in a declining population with nests getting further apart, the reduced stimulation could conceivably delay gonad maturation and reduce fecundity, with the lower levels of calling deterring young birds from being recruited. Or the effects might be in the opposite direction, for example with high levels of the night-time shearwater chorus deterring landfalls by the younger immatures and delaying their onset of breeding, whereas at quieter colonies recruitment might begin at an earlier age.

IX Petrel communities on land

A *Introduction*

Where essential needs on land and sea are available, breeding stations may support many kinds of petrel, together with other seabirds and their associated predators. The combinations of seabirds encountered today, however, often only represent a sample of those present before human influences were felt. Finding examples of breeding stations free from those influences is difficult; even apparently remote islands may prove not to be in their pristine state, for example Henderson Island in the Central Pacific whose birds experienced heavy human predation in the past (Steadman & Olson, 1985).

Under undisturbed conditions the natural partitioning of breeding space may be readily apparent, the various spatial niches being occupied by particular species. Associations of breeding birds often include a range of sizes reflecting the ordinal adaptive radiation. Typically there are one or more large surface-nesting species and, below ground, large and small shearwaters, gadfly petrels and probably one or two species of storm petrel. In the southern hemisphere, particularly on islands where the birds can forage to the convergences, the communities may be more diverse being composed of a range of surface nesters—a great albatross (usually *D. exulans*), several smaller *Diomedea* and a *Phoebetria*, one or both *Macronectes*—a ledge nester (usually *Daption capense*), and a variety of burrowing nesters including several shearwaters and *Pterodroma*, even more small species—prions *Pachyptila*, storm petrels like *Fregetta tropica* and *Garrodia nereis* and a diving petrel (usually *Pelecanoides urinatrix*).

That large numbers of tubenoses can breed below ground while others nest on the surface allows full use of the limited terrain within range of adequate food, for example at the Bounty Islands *Diomedea cauta* and the penguin *Eudyptes sclateri* occupy surface sites, the prion *Pachyptila crassirostris* rock crevices below them.

There tends to be little interspecific contact between these various tubenoses, with the bigger kinds ignoring the smaller unless their 'individual distances' or burrows are violated. Burrowing forms tend to reduce competition by occupying different physical niches and substrates (Table 1.10) and having different phenologies (Fig. 1.15). Species occupying the same terrain may compete, a larger bird taking over the nest of a smaller one by scraping out its eggs or chick, as *Puffinus pacificus* may dispose *Pterodroma hypoleuca* and *Fulmarus glacialoides Thalassoica antarctica* (Orton, 1968).

B *Some major communities*

The most populous petrel islands are The Crozets in the southern Indian Ocean. This archipelago provides breeding places for 37 seabirds, 26 of which are tubenoses (see Table 1.2). The data here illustrate the effects on communities of alien mammal introductions. Ile aux Cochons lacks any gadfly or storm petrel: the birds suffer from cats (Derenne & Mougin, 1976). In contrast, Ile de l'Est is in a pristine state and supports a great range of small- and medium-sized species.

To some extent the number of species breeding on an island will depend on the land area available—a larger island will probably have a greater variation in terrain. The Crozets cover some 500 km^2 but Kerguelen, with some 300 islands and islets totalling some 7000 km^2, supports only 38 species, and with an almost identical composition to that of The Crozets. Total numbers breeding at these two places also appear to be rather similar. Why the overall population does not exceed that of The Crozets is not clear.

At higher southern latitudes the species diversity declines and gadfly petrels disappear: the situation at South Georgia, an island free of mammalian predators, is

Table 1.10 Breeding habitat and dispersion of South Georgia seabirds. From Croxall and Prince (1980)

Species	Breeding habitat	Breeding dispersion	Inter-nest distance or density
A. patagonica	Flat beaches	Usually large colonies	1 m
P. antarctica	Beaches, slopes	Medium colonies	0.75 m
P. papua	Flat beaches, tussock	Small colonies	1 m
E. chrysolophus	Steep coastal slopes	Very large colonies	0.5 m
Diomedea exulans	Tussock flats	Loose aggregations	10–20 m
D. chrysostoma	Tussock slopes	Medium colonies	1–2 m
D. melanophrys	Steep tussock slopes	Medium colonies	1–2 m
Phoebetria palpebrata	Tussock cliffs	Solitary–small groups	5–10 m
Macronectes giganteus	Tussock flats	Loose aggregations	5–10m
M. halli	Tussock flats (often coastal)	Loose aggregations	5–10 m
Daption capense	Ledges of coastal cliffs	Small groups	—
Pagodroma nivea	Crevices of high (300 m above sea level) inland cliffs	Small groups	—
Pachyptila desolata	Tussock flats, slopes	Dense colonies	1400
Halobaena caerulea	Tussock flats, slopes	Locally dense colonies	720
Procellaria aequinoctialis	Tussock ridges, hills, slopes	Locally dense colonies	40
Oceanites oceanicus	Coarse scree, rubble, cliff crevices	Medium colonies	—
Pelecanoides georgicus	Fine, high scree (100–250 m above sea level)	Small colonies	200
P. urinatrix	Very steep coastal tussock slopes	Local medium colonies	—

For burrow-dwelling species values are breeding densities (occupied burrows per 1000 m^2) for optimum habitats.

shown in Table 1.9. Here petrel numbers and biomass are high, but only 25 seabirds nest, 16 of them being procellariiforms.

In the North Atlantic there are few islands, even remote ones like St Kilda, without a history of human occupation and their present range of seabirds may not reflect that of prehuman times. Presumably populations of albatrosses like *D. anglica* once shared suitable islands with shearwaters and storm petrels.

Much the same is true of most petrel breeding stations in the North Pacific. Here, typified by the much modified, but still very important, islands of the Hawaiian Leeward Chain, we find communities with two albatrosses *D. immutabilis* and *D. nigripes*, a shearwater *Puffinus nativitatis* and a gadfly petrel *B. bulwerii* all laying on the surface and a small range of underground nesters—a large shearwater

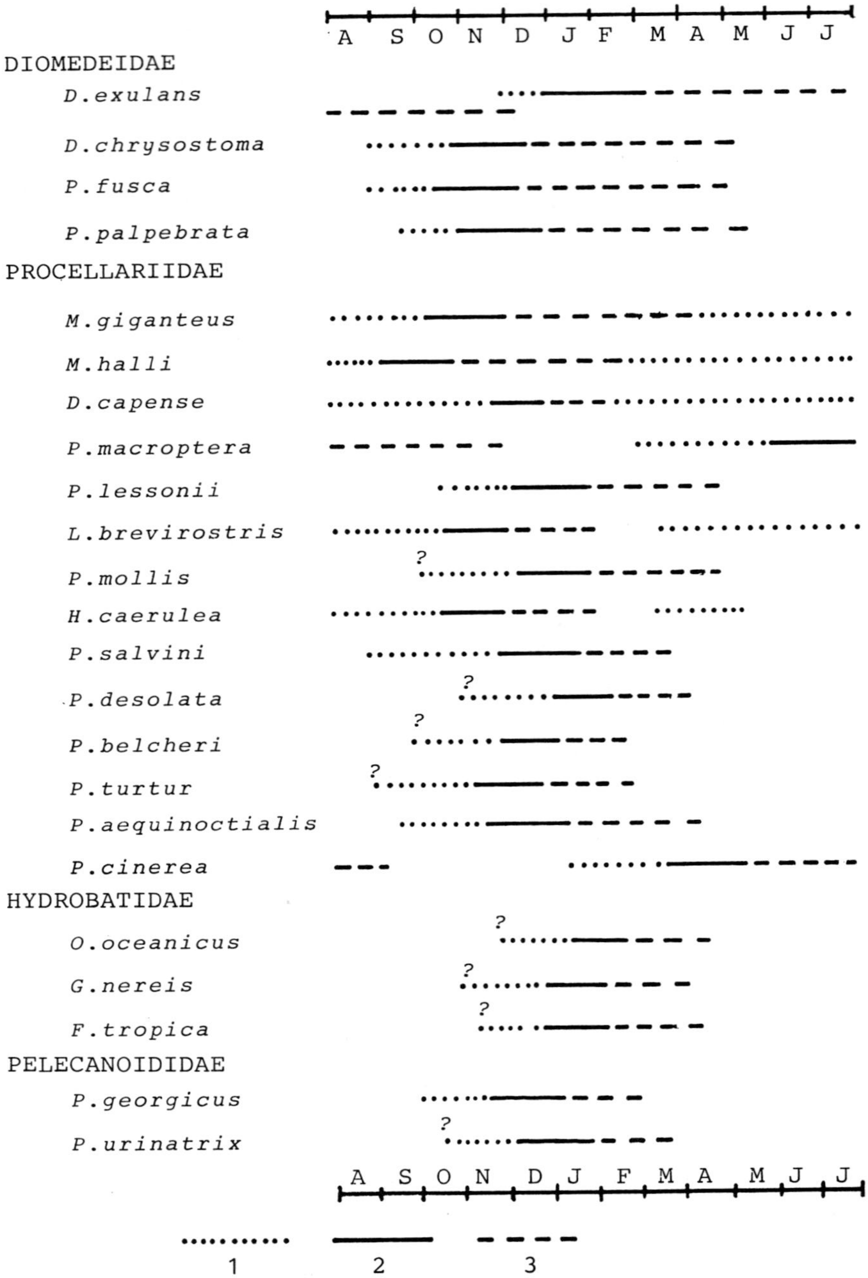

Figure 1.15 *Reproductive cycles of petrels at the Crozet Islands. 1, Adults ashore; 2, incubation; 3, chick rearing. From Jouventin* et al. *(1982).*

P. pacificus, a small gadfly petrel *Pterodroma hypoleuca* and a storm petrel *O. tristrami*. These birds share their habitats, but hardly interact with, a range of terns, boobies and tropic birds.

This string of islands and atolls is the most important breeding station in the North Pacific supporting among 14 million seabirds over 6 million petrels (see Table 1.1). The species composition tends to be rather similar from island to island, but the proportions of the components vary with the nesting niches available. Thus Nihoa, although quite small, is high and rocky with abundant sites for crevice and cavity nesters and supports the world's largest populations of *B. bulwerii* and *O. tristrami*, whereas the 13 low, sandy islets of French Frigate Shoals provide level sand for burrowing species, while Lisianski is a large flat sand and coral island supporting probably the world's biggest colony of *P. hypoleuca*.

Table 1.1 shows that although the Sooty Tern *S. furcata* is the most abundant species, its biomass totals only about 1.4×10^6 kg, whereas *D. immutabilis* equals *c.* 7.8×10^6 kg so that tubenoses dominate in terms of biomass and therefore in their effects on oceanic resources. As elsewhere, the breeding cycles of the seabirds are staggered and conflicts between potential competitors reduced.

Tropical petrel communities consist of rather few species with smallish populations, at least as compared with many in middle and high latitudes. Presumably this is partly due to a poorer and less seasonal food supply in surrounding seas. For example, the seabird fauna of the Gambier Islands (23°S; 135°W) consists of five tubenoses (*Pterodroma arminjoniana, Puffinus nativitatis, P pacificus, P. lherminieri* and *Nesofregetta fuliginosa*); two tropic birds (*Phaethon rubricauda* and *P. lepturus*); and a booby (*Sula leucogaster*) (Lacan & Mougin, 1974). Habitat variety may also be limited and many sites are low, sandy cays where an overlay of hard coral breccia hinders burrowing. High cliffs may be available, for example in Fiji and Samoa. Some, such as *Nesofregetta* and *P. pacificus*, may burrow but most tropical petrels lay on the surface. These include three *Pterodroma*, aerial predators being absent.

The species' compositions of the avifaunas of several islands where tubenoses predominate have been examined, for example by Barrat and Mougin (1974) and Hatch and Hatch (1990a,b).

Even when examination is restricted to the petrel faunas of islands in pristine condition, there are unaccountable absences. For example, why does *Procellaria aequinoctialis* not breed on Nightingale Island when it does so on nearby Inaccessible and why no *Calonectris diomedea* on Little Salvage Island when it does so on Grand Salvage, and so on? It seems remarkable that philopatry should be so rigid that 'spill-over' populations have not developed long ago. Again, the Snares Islands, although only 330 ha and heavily used by nine species of petrels, still have apparently suitable habitat for storm petrels, yet have none. Some such anomalies may have arisen through unknown historical factors, for example past disruptions from changing sea levels and vulcanism.

C *Factors affecting distributions on land*

The distribution of petrel colonies at any island tends to be governed by each species' nesting requirements, for example by substrate preference, availability of take-off

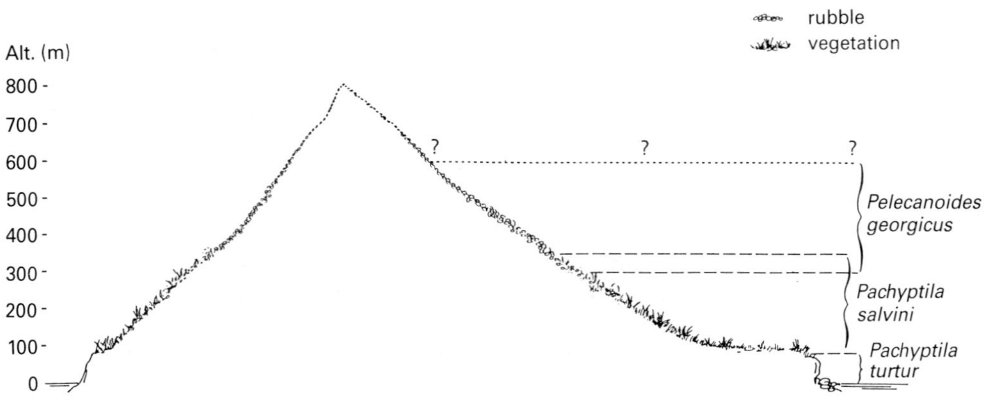

Figure 1.16 *Zonation of nesting habitats of three petrels on Pig Island, The Crozets. From Derenne and Mougin (1976).*

points, shelter from sun, wind and rain, etc., so that quite clear zonation patterns may develop. When similar-sized species share a habitat, competition may lead to niche compression, whereas the same species elsewhere, nesting alone, may occupy a much broader niche.

Mammalian predation may also simplify zonal patterns. Figure 1.16 shows these for three small species on Pig Island, Crozet, where cats take a heavy toll. Here *Pachyptila turtur* breeds solely at low altitudes in cliffs above the sea, *P. salvini* at medium altitudes in heavy soils covered in vegetation, and *Pelecanoides georgicus* higher up in stony rubble above the vegetation line and in a zone relatively free of cats and skuas (Derenne and Mougin, 1976). Yet at pest-free Ile de l'Est in the same group, the last two petrels cohabit at all altitudes. At The Crozets and elsewhere where many species share a breeding place, temporal and reproductive segregation may reduce competitive exclusion (Fig. 1.15).

X Habitat modification

A *Introduction*

Surface nesters may lay on bare ground, but the only species to burrow below such ground routinely is *P. georgicus*. Otherwise, when ashore petrels influence both the vegetation and the soil. They are important in transferring nutrients and biological materials from the sea to the land in the form of faeces, egg shells, dead eggs and the bodies of adults and chicks that die there. They also affect plant growth and soil characteristics by trampling and their cropping of vegetation for nest-making and burrow-blocking. The plant communities are often isolated and with relatively few species and so are particularly vulnerable to external disturbance (Elton, 1958, pp. 77–81; Holdgate & Wace, 1961).

Petrels will dig under almost any kind of vegetative cover providing that the soil allows this. If not, they may lay on the surface sheltered by the foliage, but sometimes too dense a ground cover deters both burrowers and surface nesters. They may use otherwise unsuitable surroundings when these have changed since they first bred. For example, some *D. immutabilis* nest in thickets of *Casuarina* on Midway Island, presumably retaining nest sites established when the ground was uncluttered by trees.

These influences may be beneficial or detrimental to the habitat, according to the circumstances. Freshly ejected faeces kill seedlings so that the community around the nests tends to be of quite a different composition from that beyond the birds' influences. Prostrate and coprophilic plants often dominate and may become very lush and green, standing out from the surrounding vegetation.

The enrichment of soils below bird cliffs by faeces has long been noted and the plant associations examined, for example by Summerhayes and Elton (1923), but studies of the relationships of petrels *per se* to vegetation, soils, and of the effects of environmental factors, including those of introduced mammals, were pioneered by Gillham (1956a,b) working in west Wales. She later extended her studies to Australasia (Gillham, 1957, 1960a,b,c, 1961a,b,c, 1962; Gillham & Thomson, 1961) and to Macquarie Island (Gillham 1961c). Further work in southern Australia, often on the same sites as examined by Gillham, was done by Norman (1967, 1970a). Vegetation associated with ground-nesting *F. glacialis* was examined by Hepburn and Randall (1975).

The most detailed examinations of the effects of subantarctic petrel populations have been carried out at Marion Island by Smith (1976,1977,1978,1979). Also, at that island Williams (1978), Williams *et al.* (1978) and Burger *et al.* (1978) attempted to quantify the minerals and energy transferred to the terrestrial ecosystem by cat-killed carcasses, and of guano. Siegfried *et al.* (1978) examined the contributions of albatross and giant petrel eggs to that system and Siegfried (1982) reviewed the roles of the birds there.

B *Effects on plants*

1 *Physical damage*

a. Surface-nesting species. Smith (1990) followed the revegetation at and below a former south Shetlands colony of *Macronectes giganteus* over 20 years. On the stony ground once occupied by the birds the fructose lichen *Usnea antarctica* has flourished while at the bird's previous launching site the moss *Polytrichum alpestre* became dominant at the expense of lichens epiphytic on other mosses. Joly *et al.* (1987) analysed the effects of trampling and nest-building on the vegetation around nests of *D. exulans* at Possession Island, The Crozets.

On flat terrain mollymawk nest columns may stand on bare rock, most of the soil between them having been washed out to sea. Hall and Williams (1981) recorded that at Marion Island, where concentrations of *Diomedea* and *Phoebetria* albatrosses bred on steep slopes, the combination of vegetation cropping and the impacts of landings on saturated ground caused local slumping because the soil often lay beyond the angle of repose, only bound in place by the vegetation. Similarly, at the

Snares Islands, some pedestal nests of *D. bulleri* based on *Poa* stools sit on strata that are inclined towards the sea. These stools are very easily toppled through undermining by *Pterodroma inexpectata* which may have several nests below a single plant.

b. Burrowing species. Burrowing petrels tend to line their nests with surface litter, cropping grasses and herbs sometimes at appreciable distances from their burrows. Shearwaters tend to snip off any seedlings near their burrows so that regeneration occurs only where the young plants are protected, for example by fallen logs. Where the birds nest at high densities beneath a forest canopy they may completely suppress a ground understorey, the bare soil between the burrows looking as if swept by a vacuum cleaner (Fig. 1.17).

Maesako (1985) considered mainly the physical impacts of *C. leucomelas* on the *Machilus thunbergii* forest and its associated understorey. He found that the numbers of species and individual plants declined with increasing nest density, trampling and cropping resulting in a depauperate vegetative community and a simplified stratification of the forest. Seeding *M. thunbergii* grew only among moist *Carex* patches devoid of burrows.

Burrowing in peat opens up the surface layers which retain much of the rain and tends to dry out the soil. Here long-rooted plants are favoured. Burrowing also produces spoil that may be thrown 1–2 m from the entrance and up to 1 m^3 of soil may be removed from a burrow (Hall and Williams, 1981, p. 19). Some plants are

Figure 1.17 *Ground cleaned by Sooty Shearwaters at the Snares Islands. Note the leaning* Olearia *trees undermined by the burrowing. Photo: Author.*

destroyed in the process, some are buried and some root systems exposed and disturbed.

In the southern hemisphere many petrels breed among tussock grasses, for example *Poa* spp. Soil eroding from around the plants aggravated by the birds' activities leaves the tussock supported on peaty 'stools' (Fig. 1.18). Walkways ramifying among the tussocks are kept open by trampling. On Antipodes Island, the pillars are so crowded together and so honeycombed below with burrows that birds moving on the surface are completely hidden, the maze of narrow passages used by petrels and parrots forming a distinct 'superterranean' zone (Warham & Bell, 1979).

Most tussock grasses are tough and little affected by normal trampling, but at the height of the nesting season the highways between them may be worn bare. This was detailed by Gillham (1962) for the Bass Strait colonies of *Puffinus tenuirostris* among *Poa poiformis* and she compared this pattern with that of *Armeria maritima*-dominant vegetation on *P. puffinus* colonies of west Wales where burrows penetrate the stabilizing mounds and tracks ramify between them (Gillham, 1956a). She described the 'bird evading' plants of the southern Victorian colonies of *P. tenuirostris* which take advantage of the stored nutrients during the winter when the petrels are at sea, only to succumb later to bird pressure and die out during dry summers.

Because of their large numbers and concentration at the few places available for take-offs, burrowing petrels can affect big changes in the vegetation below their launching sites. Thousands may use a single boulder or tussock stool in an hour, and this continues throughout a long breeding season. Many crash to the ground below. Tussocks may be worn down to their stumps, killed and eventually eroded away. In Bass Strait succulents like *Carpobrotus rossi* below launching rocks are often badly damaged by *P. tenuirostris*. At the Snares Islands a swathe about 1.5 m wide was cut through a dense belt of *Hebe elliptica* by the strikes of *P. griseus*. This is a very tough bush, and few petrels succeed in burrowing below it, but the persistent bombardments of falling bodies wiped the plants away leaving a mud smear ready to be washed to sea in the next big rain (Warham, pers. obs.).

Intense burrowing in forests on peaty soils may cause tree fall in high winds. On the 'mutton-bird' islands of New Zealand the trees of surface-rooted *Olearia lyallii* and *O. angustifolia* are very susceptible to undermining by *P. griseus* and *Pterodroma inexpectata* whose nests honeycomb the ground. In gales the branches sway and the ground heaves as the roots rock to and fro. Many trees become semiprostrate (Fig. 1.17) and the fall of one often brings down others so that open areas are created (Fineran, 1973; Johnson, 1982).

In some situations bared soil may become a seed bed and be revegetated during the birds' absence. Norman (1970a) described how damaged *Poa poiformis* tussock recovered following rains and a simplified ecosystem was maintained with the invasion of ephemerals like *P. annua* and *Senecio lautus*.

2 *Manuring: effects on plants*

Petrel faeces and other biological products are rich in N, P, K and Ca. These may either stimulate growth in some plants, or depress it in others. Guano-scorched plants, particularly seedlings, often die. Soil pH is usually changed.

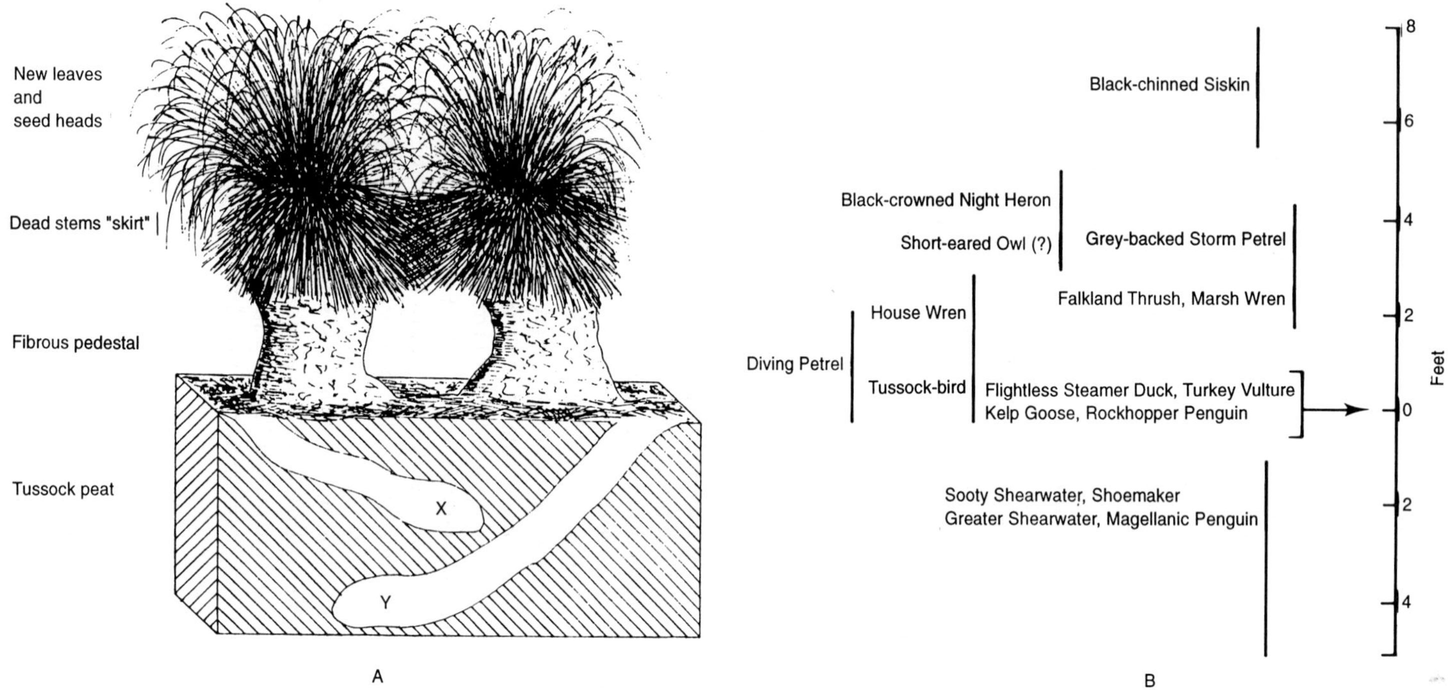

Figure 1.18 *Tussock grass on the Falkland Islands showing typical community of birds using this habitat for breeding. A. Vertical profile with burrows of (X) Sooty Shearwater and (Y) Magellanic Penguin. B. Vertical zonation of nest sites in tussock meadows. From Woods (1970).*

Table 1.11 Percentage presence of plants growing on, or in vicinity of, Wandering Albatross nests, July 1965 (100 nests examined). From Huntley (1971)

Plants growing on nest	%	Plants 'introduced' to vicinity of nest	%
Montia fontana	26	*Poa cookii*	82
Poa cookii	17	*Montia fontana*	71
Callitriche antarctica	14	*Callitriche antarctica*	55
Cotula plumosa	8	*Cotula plumosa*	20
Ranunculus biternatus	3	*Poa annua*	12
Poa annua	3	*Ranunculus biternatus*	8
Uncinia dikei	1		

Some estimates of the annual production of guano have been made for larger species at Marion Island (Burger *et al.*, 1978). The contributions of four albatrosses and two giant petrels were quite minor in comparison with that of the penguins, but no figure for the outputs of the many burrowing petrels was given. Fugler (1985) established the guano production by chicks of *Halobaena caerulea, Pterodroma macroptera, P. mollis, Procellaria cinerea* and *P. aequinoctialis* and determined its chemical composition. The large *Procellaria* produced about 1.8 g (dry wt) guano day^{-1}, the smallest, *H. caerulea* 0.6 g day^{-1}. The chemical composition of these faeces was essentially similar to that of surface-nesting seabirds nearby.

Owing to this manuring a range of coprophilous plants tends to dominate the flora of a petrel colony. Van Zinderen Bakker (1971) noted how the distribution of *Poa cookii* at Marion Island coincided with old and new colonies of *Macronectes*, especially downhill from them; in effect by nitrifying the soil and stimulating the growth of the tussock, the birds were creating a more favourable habitat on the grey lava. Similarly, stimulation by run-off from mollymawk colonies often leads to increased growth and stool formation of tussocks which tends to improve conditions for burrowing petrels.

Detailed analyses of the effects of manuring on the plants and soils around nests of *D. exulans* in a Crozet Island bog were given by Joly *et al.* (1987). They describe the pedological changes and the perturbation in the plant communities downhill from unused nests. All the bryophytes had disappeared and *Callitriche antarctica, Montia fontana* and *Agrostis magellanica* were dominant in turn, according to distance from the nest. The open zone around the nest was first colonized by the foliose alga *Prasiola crispa*. These authors also found that while the number of plant species fell in the enriched soil, the number of soil invertebrates increased. On Prince Edward Island, Huntley (1971) found that *A. magellanica* growing within the influence of *D. exulans* had bigger leaves than when it grew without manuring. Where the albatross nests were most dense the mire vegetation was replaced by a rich green association dominated by *P. cookii, C. antarctica* and *M. fontana* as a result of nitrification (Table 1.11). On nearby Marion Island the vegetation around nests of *D. exulans* and both *Macronectes* spp. is dominated by dark-green *P. cookii* which contrasts with the drab, yellow-green surroundings (Smith, 1976).

On Western Australian islands guano deposition stimulates the growth of the succulent *Zygophyllum fruticulosum* which often forms a thick trailing mat over low bushes on colonies of *Puffinus pacificus* and *Pelagodroma marina*, and birds may die when trapped in its wiry stems (Gillham, 1961a). On these islands a range of annuals springs up on 'doormat' areas of unoccupied burrows, as also occurs on Manx Shearwater colonies in Wales. The plants are often the same in both places, for example *Poa annua, Stellaria media, Cerastium viscosum* and *Sonchus oleraceus* (Gillham, 1961a). She also noted several algae, for example the diatom *Navicula mutica*, common in shearwater burrows, with the lichen *Cladonia pyxidata* a characteristic species of burrow mouths (Gillham, 1956b).

On some of the Bass Strait islands the storm petrel *P. marina* burrows at high densities, commonly 2–4 m^2. Gillham (1963) scored the occurrence of the associated plant species. The birds nest under a mat of soft-leaved, nitrophilous *Poa poiformis* ('silver tussock') whereas, for some unknown reason, when associated with *Puffinus tenuirostris* this same species is stiff-leaved. Many storm petrels also nest under sheets of the nitrogen-loving *Carpobrotus rossi* and *Tetragonia implexicoma*, at times simply burrowing into masses of this latter herb and, in effect, laying on the surface.

Animal–soil–nutrient–plant relationships at subantarctic Marion Island have been examined by Smith (1976, 1977, 1978, 1979). Inland slopes are carpeted with the fern *Blechnum penna-marina*, which is replaced by luxuriant *Poa cookii*-dominated communities where petrels burrow. *Acaena magellanica* is also abundant here. The soils and the plants in these communities contain higher N, P, and other minerals than do these same plants in non-tussock areas. In subpolar soils the vegetation around albatross and giant petrel colonies is more luxuriant than the same plants achieve elsewhere.

Johnson (1975) looked at the colonization of old nests of *D. bulleri* at Little Solander Island, and considered that as the nest disintegrates there seems to be fortuitous race between scrub, fern or grass to become the sole occupier of the site. At subantarctic Antipodes Island, a groundsel, *Senecio radiolatus antipodus*, has been considered an associate of manured soils, but Godley (1989) found that, while it was often common and luxuriant around the deserted nests of *D. exulans* and *Macronectes halli*, it also occurred on open, apparently unmanured ground as well, although these could have been on pockets of manured soil from extinct colonies.

On the Rockefeller and Alexandra Mountains of inland Antarctica, *P. crispa* flourishes around meltwater below colonies of *Thalassoica antarctica* and *Pagodroma nivea* (Broady, 1989; Broady *et al.*, 1989) and at Signy Island very high algal counts were found below nesting *Daption capense* and *P. nivea* (Broady, 1979). Close by such nests, however, few algae occurred, presumably because of the toxicity of the guano at high concentrations (see also Ryan & Watkins, 1989). On nunataks in east Antarctica, these researchers found that plant cover was directly related to the influence of breeding *P. nivea*, with soil concentrations of N, P and K being greater in the colonies than elsewhere. A major plant associated with the manuring was the lichen *Umbilicaria decussata*.

On Skokholm, west Wales, Gillham (1956a) reported that the soil above the burrows of rabbits, puffins (*Fratercula arctica*) and *Puffinus puffinus* was dominated by *Armeria maritima*, but if the latter was absent *Rumex acetosa* and *R. acetosella* were plentiful. Gilham traced the course of even quite tortuous burrows by the strip of stimulated *Rumex* overhead or by the stunted nature of other plants. Part of the effect was evidently caused by the ability of the long *Rumex* roots to reach moisture when the roof of the burrow had dried out. Similarly, on some New Zealand colonies, *Disphyma australe* survives around burrows because it is rooted at a distance from the nests.

C *Effects on soils*

The nutrient input often improves soil fertility and burrowing helps soil turnover and aeration and, with the burying of plant litter, tends to improve soil structure and its suitability to support vegetation. In sandy soils the spoil often accumulates before burrow entrances to form bared 'doormats'. Wind and water erosion often start at the entrances and atmospheric conditions at the roots may be changed. Extensive burrowing at the edges of cliffs with friable substrates may accelerate erosion as at some colonies of *P. pacificus* at Norfolk Island.

Petrels have some direct effects on soils, apart from digging and burrowing. Fineran (1973) described how the constant trampling and scratching of *P. griseus* on New Zealand islands produced a thin layer of powdered peat which, in wet weather, becomes rolled into small pellets by the webbed feet so that the surface acquires a granular appearance.

1 *The chemical compositions of soils*

From her field experiments Gillham (1956b) worked out the roles of the various components of guano (nitrates, phosphates, exchangeable calcium, carbonates, acidity and the composition of organic matter) on soils of the west Wales islands.

In the breeding season the nitrate in the soil on the floor of a *P. puffinus* burrow under *Armeria* was about twice that at the burrow entrance and four times that in the *Armeria* hummock forming the roof. When the birds left for the winter the nitrate diffused away and formerly bare areas developed a robust growth of *Poa annua* and other coprophiles (Gillham, 1956b).

Analyses of the acid peat blanket burrowed by petrels on the Snares Islands were given by Flint and Fineran (1969) who noted the low values of C/N (50–60%) presumed to be due to fertilization by birds, and Smith (1976, 1977) analysed the soils on Marion Island. Other analyses of similar petrel-manured substrates were presented by Kennedy (1978) and Okazaki *et al.* (1993).

At South Georgia levels of Na, K, Ca, Mg, P and N in leaves of *P. foliosa* growing among burrows of the prion *Pachyptila desolata* were considerably higher than those from unburrowed ground or from the margins of seal wallows (Walton and Smith, 1979). They suggested that the high values came from aerial defecation by the birds at night. Engelskjon (1986) recorded very high P levels in soils from around *Pagodroma nivea* and *T. antarctica* colonies in inland Antarctica and Ryan and Watkins (1989) give more detailed analyses of ornithogenic soils from the same area.

In the opposite hemisphere at Torishima Island the nitrogen content of the volcanic ash soils in the *D. albatrus* colony was found to have high levels of nitrogen and high ratios of ^{15}N to ^{14}N, apparently due to evaporation of ammonia during which fractionation occurs (Mizutani *et al.*, 1986). They suggested that such high ratios could be used to identify sites of abandoned colonies.

2 *The chemical composition of petrels*

Although under natural conditions a major means by which dead birds become incorporated in soils and vegetation is predation by skuas and gulls, introduced predators may have a larger effect. At Marion Island, Williams (1978) estimated that the *c.* 2100 cats consumed about 400 000 burrowing petrels annually amounting to some 35 t dry weight including 8 t of feathers. Williams analysed the mineral contents of the principal species involved (Table 1.12). The contribution of the feathers is noteworthy.

3 *Effects on soil moisture and burrow microclimates*

Gillham (1956a) found that there were much greater fluctuations in moisture content in burrowed than in unburrowed soil. Burrowing generally makes dry soils drier and wet ones wetter (Gillham, 1957). In hot weather the soil over burrows may dry out rapidly, in wet weather they aid the percolation of water to lower levels and, particularly in peaty or clayey soils, flooded burrows may remain waterlogged when the ground elsewhere has dried out.

The air inside burrows is usually about 10% more humid than that outside and burrowed soils tend to have more varied temperatures than those that are undisturbed. This may lead to drought at the nest entrance, an effect hastened by wind and insolation—depending on the exposure of the opening. Light penetration down the burrow affects the direction of growth of plant rhizomes, for example of *Carex arenaria*.

Another consequence, probably ascribable to variation in soil moisture near the burrow, is that burrowing may lead to the development of both moisture-loving and xerophytic vegetation. Particularly in abandoned burrows, Gillham (1956a) found that the accumulation of moist air suited the development of bryophytes and other small plants. Variations in water content and temperature were greatest in porous mineral soils and here xeromorphic plants above the burrows tended to be minute, hairy rosettes, closely pressed to the surface and suffering little water loss.

XI Shearwater 'greens'

In some areas where shearwaters nest that are subject to regular rains, wind and frosts, the ground has a characteristically hummocky appearance (Fig. 1.19). This is caused by burrowing, the piling up of ejecta and its erosion, as well as the life-form of the overlying vegetation. This is often cushion-shaped. On the west Wales shear-

Table 1.12 Concentrations (ppm) of selected elements in petrels at Marion Island, (A) in the bodies of 12 species, (B) in the feathers of 11 species. From Williams (1978)

Species	Micro-elements				Macro-elements						
	Cu	Cd	Zn	Mn	Sr	Mg	K	Na	Ca	P	N
A											
Pterodroma macroptera	25.02	18.85	106.45	3.52	147.36	890	6790	4930	42 500	14 980	65 480
Lugensa brevirostris	10.29	5.00	84.80	2.83	68.63	1060	5540	4750	37 860	15 560	108 530
Pterodroma mollis	9.49	1.61	98.55	4.88	33.91	910	5740	5730	26 800	8430	96 400
Pachyptila salvini	15.36	11.50	108.85	3.88	93.12	1140	7640	6640	24 000	7890	82 520
P. turtur	10.98	13.00	108.68	3.51	84.00	1110	8070	6880	51 720	10 000	97 090
Halobaena caerulea	7.96	4.60	71.50	2.57	48.98	700	4750	2040	16 070	10 760	126 070
Pelecanoides urinatrix	9.64	8.32	94.46	3.32	86.31	940	7200	4850	58 590	11 860	97 090
P. georgicus	9.58	19.75	97.54	3.64	103.91	1080	7990	5680	42 920	—	98 120
Procellaria aequinoctialis	6.00	16.00	128.00	5.00	83.00	570	4990	4100	37 800	21 000	84 060
P. cinerea	4.78	10.25	71.75	2.22	69.49	810	5240	4110	15 210	21 500	68 760
Fregetta tropica	15.60	0.01	105.00	2.80	30.09	1010	—	4800	28 450	18 050	65 710
Garrodia nereis	17.78	0.01	94.35	4.18	43.93	1460	1050	1050	28 700	15 340	81 760
B											
Pterodroma macroptera	49.07	0.01	115.85	2.64	9.65	436	96	1058	1155	1492	107 142
Lugensa brevirostris	14.17	0.16	88.24	1.07	17.38	1069	454	5164	1540	3166	90 877
Pterodroma mollis	14.42	0.27	103.99	4.06	16.15	303	367	1318	3188	2531	142 971
Pachyptila salvini	15.41	0.01	110.00	5.15	16.05	524	422	2813	850	2526	121 578
P. turtur	17.02	0.15	112.55	2.64	9.65	274	1118	1866	1099	2805	145 572
Halobaena caerulea	16.64	3.44	364.00	6.00	15.72	752	592	3304	4238	5285	102 666
Pelecanoides urinatrix	22.84	0.02	104.06	2.03	16.50	934	964	5076	850	1725	97 742
P. georgicus	19.07	0.44	111.88	4.71	17.67	374	947	4079	8269	2496	102 551
Procellaria cinerea	17.00	0.01	90.00	5.00	10.00	660	390	3900	850	22 500	140 000
Fregetta tropica	9.88	8.88	92.50	4.05	55.50	922	6234	5401	1160	37 500	159 905
Garrodia nereis	11.06	7.92	73.92	9.31	89.06	949	6519	4209	—	—	—

Figure 1.19 *Shearwater 'green' on Rùm. Photo. Peter Wormell.*

water islands this hummocky habitat is the creation of shearwaters, puffins and rabbits, the dominant plant being Sea Thrift *Armeria maritima*. On the nitrogen-rich soils the plants are a brighter green than on unburrowed ground. The result is a belt of 'greens', often extending inland from the cliff edges.

On the Scottish island of Rùm 'greens' occur mostly above the 580 m contour and are created by *Puffinus puffinus* digging in infertile soils derived from ultrabasic rock (Wormell, 1976). The soil is enriched by the deposition of marine material by the petrels, and it and the resulting vegetation support a range of beetles, Diptera and earthworms and a pyralid moth *Catopria furcatellus* restricted to this habitat (Wormell, 1976). Applications of NPK fertilizer to unburrowed and almost bare ground by Ferreira and Wormell (1971) showed that the basically infertile soils could support a herb-rich *Agrotis/Festuca* grassland similar to that created by the birds.

Such greenswards may be found elsewhere, for example on shallow soils at Campbell Island where non-breeding mollymawks congregate to display in 'clubs' on small plateaux along the clifftops (*The Petrels*, cover photo).

XII Effects of farming practices

Farming and other human activities on land used by petrels is usually detrimental to them; sometimes it may help. On Motuara Island, New Zealand, burrows of *P. griseus* are restricted to ground that was formerly tilled when the island was farmed, and on Heron Island, Queensland, Hill and Barnes (1989) found that burrow densities of *P. pacificus* were highest in areas with piles of debris from *Pisonia* trees felled to make way for buildings.

A *Grazing animals*

The effects of petrels on habitats are often amplified by the parallel influences of grazing mammals, particularly by rabbits, deer, goats and sheep. The usual result is increased degradation of the plant and soil cover. Sometimes such animals help petrels by opening up heavy vegetation to allow burrowing, for example tracks made by grazing Cape Barren Geese *Cereopsis novae-hollandiae* allow *Pelagodroma marina* access to ground from which they would otherwise be excluded by dense carpets of *Tetragonia* (Gillham, 1963).

Rabbits and goats were often taken to islands as possible food sources. The results have sometimes been catastrophic. Rabbit grazing on Laysan Island led to its almost complete devegetation and then the rabbits evidently became extinct. The remnant *D. immutabilis* albatrosses re-established themselves following the regeneration of the plant cover (Bailey, 1956). Gillham and Thomson (1961) traced the combined effects of rabbit grazing, trampling by man, wind, and flooding in high tides, in the degradation of a colony of *P. marina* in Port Phillip Bay, Victoria. They also followed the progress of the colonization of a small islet nearby that was not subjected to man-made disturbance.

On the other hand, the removal of sheep from Campbell Island made no apparent change to the population of *D. epomophora* there, although a reversion to a natural ungrazed vegetation association has occurred (Dilks & Wilson, 1979).

Gillham (1956a) showed that high-density burrowing by rabbits at Skokholm had detrimental effects on the soils and vegetation and, combined with the activities of the petrels, could initiate soil erosion. She also assessed the effects of rabbit grazing on seabird islands off the coast of Victoria (Gillham, 1961b). Norman (1967, 1970a) re-examined one of these where the rabbits had almost been exterminated by myxomatosis. Their denudation of the vegetation and their burrowing near a large colony of *Puffinus tenuirostris* had caused extensive soil loss. Sheet erosion appeared to have centred on an old abandoned shearwater colony, remnants of whose burrows were present in the sand, and erosion gullies, caused by rain, had led to outwashes and spills. But between 1965 and 1968 a previously eroded 6.5-ha area of blown sand was revegetated by a succession of plant communities. Meanwhile the petrels spread out from their stronghold in the *Poa poiformis* tussocks into the adjacent vegetated ground, the probable cycle of events and return to a stable state being shown in Fig. 1.20.

Domestic cattle have often been blamed for losses of petrel nests by burrow breakthrough. However, the trampling tends to consolidate loose soils and thereby lessen erosion. Sheep have also been considered deleterious, but experiments on *Puffinus tenuirostris* colonies revealed that pairs nesting where sheep were excluded neither produced more young than those in grazed areas, nor did the sheep prevent the expansion of existing colonies or maintain a turf impenetrable to burrowing birds (Norman, 1970b).

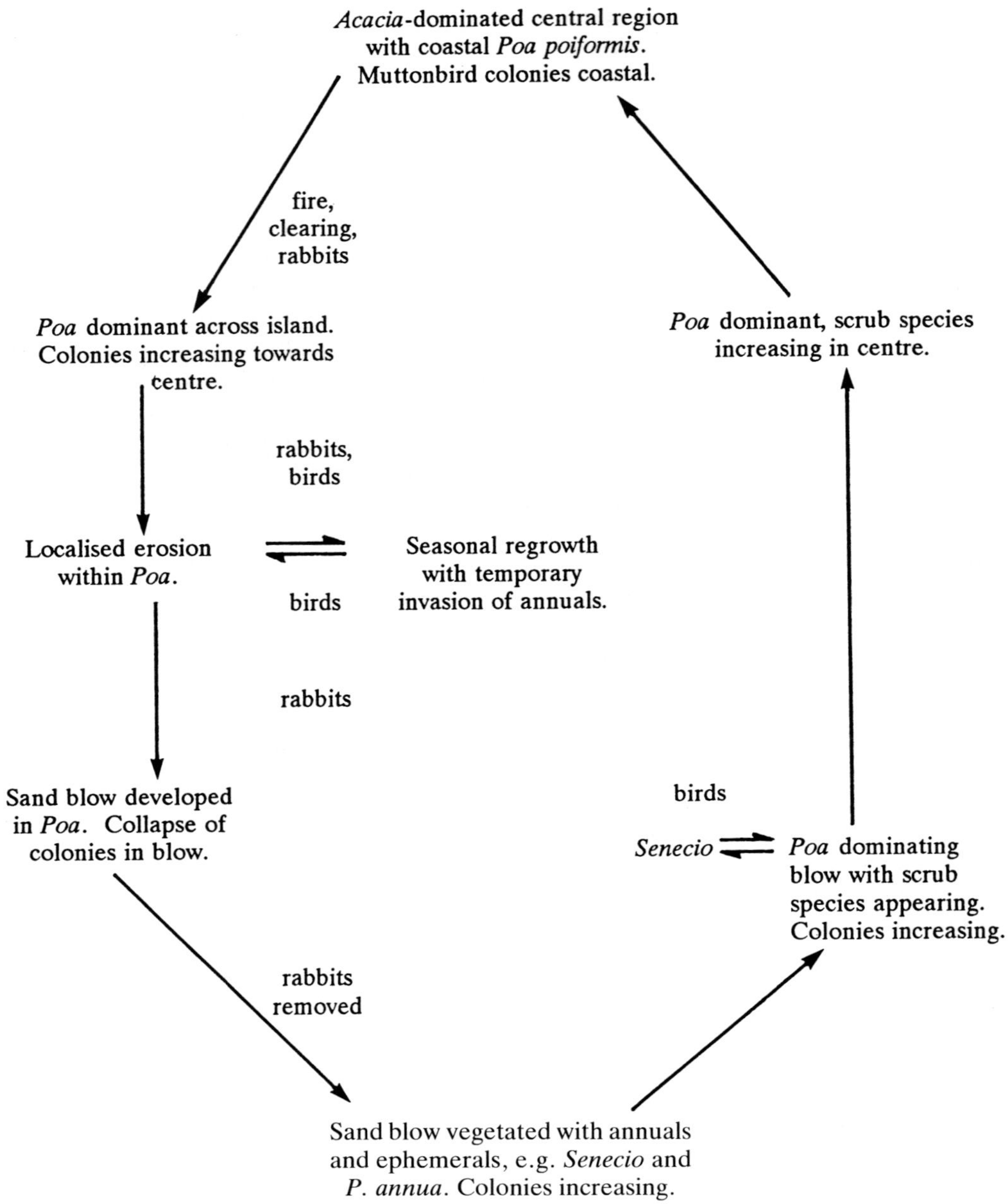

Figure 1.20 *Schematic representation of changes in the vegetation of Rabbit Island, Wilson's Promontory, Victoria in relation to an area of sand blow and the distribution of the Short-tailed Shearwater colonies. From Norman (1970a).*

B *Fire*

Fire has often been used on grassy islands to convert petrel-burrowed ground into habitat more suitable for grazing stock. Some surface erosion usually follows a fire and may cause the loss of the colony while fires during the breeding season kill birds and fires in peat may be serious. One on Inaccessible Island, Tristan da Cunha, is said to have burnt for a month (Hagen, 1952).

XIII Seed dispersal

Petrels have the potential to influence their ecosystems by importing seeds and other propagules, a topic briefly reviewed by Godley (1989).

Falla (1960) discussed the (circumstantial) evidence for seed dispersal by petrels. He cited instances where they nested at high densities in vegetation that produced adhesive seeds and drew attention to the frequency with which fledglings left with down still attached to their feathers, one young *P. tenuirostris* with a heavy ruff of down being recovered alive at least 1600 km from its nest. On subantarctic islands the barbed fruits of *Acaena* spp. seem particularly well adapted for dispersal by this means. Such seeds have been seen embedded in the plumage of *H. caerulea, Pelagodroma marina* and *D. exulans* (e.g. photo in Huntley, 1971), and *Acaena* is commonly abundant around petrel colonies like those of *Pachyptila desolata* at Macquarie Island (pers. obs.). Falla (1960) emphasized that many millions of petrels leave their nests annually to disperse widely. As they have been doing this for many thousands of years and probably on an even larger scale before their population declines, the potential of petrels as seed carriers and of influencing plant distributions seems considerable.

CHAPTER 2
Petrels at Sea—Distribution, Dispersal and Migration

Contents

I Introduction

Omitting those species such as *Puffinus assimilis* and *Pelagodroma marina* that nest in both hemispheres, about 70% of the 103 species are southern birds. If the comparison were by numbers or biomass the difference in favour of the southern hemisphere would be still greater due to the enormous contribution of prions, shearwaters, diving and gadfly petrels in the south. The large populations in the Southern Ocean arise because the shortness of their food webs allows a higher proportion of the total productivity to reach upper trophic levels than obtains in most other seas. Most southern seabirds have been isolated from terrestrial mammals, whereas man's activities have been a major factor in the declines of many populations in the north in recent times, although the fossil record shows that some northern petrels became extinct before man evolved.

Allowing for those that nest in several oceans, such as *Bulweria bulwerii, Oceanodroma leucorhoa* and *Fulmarus glacialis* (Atlantic and Pacific), *Puffinus pacificus* (Indian and Pacific), *Pterodroma macroptera* (Indian, Atlantic and Pacific) and *Puffinus griseus* (Pacific and Atlantic), the largest ocean, the Pacific, has the most resident species, about 62% of the total, the Southern Ocean (seas >50°S) and the Atlantic, about 13% each, the Indian Ocean 11%, and the Mediterranean about 1% of the whole. Some 21% of the species nest in the tropics and about 5% in the polar regions.

First reports of the relationships of petrels with the marine environment were in the accounts of early voyagers such as Solander, the Forsters and Peale, who recorded sightings and often collected specimens. Until recently, studies of oceanic distributions were mainly of academic interest, but the increasing exploitation of marine resources such as oil and fisheries, and concerns about pollution, have led to a dramatic increase in work on the distributions, abundances and roles of seabirds in marine ecosystems. Few such studies relate solely to tubenoses but in most of them they feature prominently. The recent literature is massive and space permits review of only some of the more revealing examples. For general reviews see Bourne (1963, 1980), Ashmole (1971), Brown (1980), Eakin *et al.* (1986) and Hunt (1990, 1991a).

Atlases of seabird distributions in various areas include those by Watson *et al.* (1971), Gould *et al.* (1982), Powers (1983), Brown (1986), Pitman (1986), Harrison *et al.* (1989), Webb *et al.* (1990), Morgan *et al.* (1991) and Tickell (1993).

Much information is also available from reports of species seen on ocean transects, a pioneer of which was Wynne-Edwards's (1935) paper on seabirds of the North Atlantic. Those of Rankin and Duffey (1948) during wartime over the same seas, complemented the earlier work, since when many other transects with much information on petrels have been published.

II Recording petrels at sea

Most early voyagers who logged their sightings of seabirds also gave their noon positions and often sea-surface temperatures (SSTs), but their identifications are often unclear. Even today, a fair proportion of petrels seen from a ship or fishing boat is unidentifiable because too far off or bad weather conditions hinder sustained examination. Consequently, many reports include categories like 'unidentified all dark shearwater' and so on (see, e.g. the problem of separating *Puffinus griseus* and *P. tenuirostris*; Briggs *et al.*, 1987, p. 22).

A major difficulty concerns the varying conspicuousness of different species: single birds of camouflaged kinds such as prions, and birds sitting on the sea are easily missed; albatrosses and giant petrels visible from afar particularly if on the wing, etc. King (1970, p. 84) examined the effect of wind on sea counts, finding that calm seas favoured sighting small, low-flying birds such as *O. leucorhoa* and *B. bulwerii*, whereas medium to high winds favoured albatrosses and shearwaters such as *P. griseus*, and as the higher the wind the higher such birds swung into the sky, the greater the distance at which they were visible.

Breeding, non-breeding and immature tubenoses cannot usually be separated at sea. Exceptions include some albatrosses whose young birds have duller bills (mollymawks) and darker plumage than adults—(*Diomedea albatrus*, *D. irrorata* and *Macronectes*). Adult *D. exulans antipodensis* (and presumably *D. e. amsterdamensis*) have usually been scored as juveniles because of their dark plumages (e.g. Wood, 1992). Nor can many subspecies be identified at sea. Exceptions include Spectacled Shoemakers *Procellaria aequinoctialis conspicillata* (all from Inaccessible Island) and the pale-eyed *D. melanophrys impavida* (all from Campbell Island).

Until recently, observers conducted their sightings differently so that comparisons between different censuses and transects were impossible. Standardizations of counting practices, for example by the Royal Naval Bird Watching Society, helped. Griffiths (1981) highlighted biases in different methods, Powers (1982) the relationships between two of these, and Tasker *et al.* (1984) recommended three methods for different purposes. The latter's suggestions have been widely followed, for example

10-min counts of all birds seen within a defined band width, usually 300 m, this being determined with the aid of a rangefinder (Cline *et al.* 1969; Heinemann, 1981; Zink, 1981): see also Gould and Forsell (1989), Spear *et al.* (1992a) and Spear and Ainley (in prep.). Examples of censuses based on 10-min counts include those of Powers (1983), Brown (1986), Stahl (1987), Ryan and Cooper (1989) and Morgan *et al.* (1991).

However, 10-min counts have been criticized by van Franeker (1994), who found them giving densities about twice as high as from the instantaneous 'snapshot' counts of Tasker *et al.* (1984) Furthermore, where birds are infrequent and short counts are impracticable, 30-min ones have been used (e.g. Ainley *et al.*, 1984; Ribic & Ainley, 1988/89).

Transect data are not continuous, being broken by night-time gaps or poor viewing conditions. Another major difficulty with data collected from ships is the varying responses of seabirds to the vessels, some unaffected, some being repelled and others attracted, biases discussed by Griffiths (1981) and Duffy and Schneider (1984), among others. In Antarctic seas biases are given by huge numbers of petrels crowding to roost on icebergs and similar distortions arise around fishing fleets, for example at least 1000 birds were attracted to each stern trawler in the Benguela Current (Abrams, 1985a).

Aerial counts have also been used, for example by Rankin and Duffey (1948), these having the advantage that they can provide estimates over large areas if the flight-paths are appropriately designed. Aerial surveys have been used with *F. glacialis* by Forsell and Gould (1981), Bourne (1982a) and Webb *et al.* (1990) and this and other petrels off California by Briggs *et al.* (1987), and for Alaskan species by Harrison (1982). Hunt *et al.* (1981) compared the results of shipborne and aerial counts. Gaston and Smith (1984) and Briggs *et al.* (1985a,b) discuss the problems arising with aerial censussing.

Results are often given in birds km^{-2}. Ainley *et al.* (1984) even estimated the numbers and biomass of seabirds in the whole of the Ross Sea in mid-summer—3000 *Macronectes giganteus*, 2900 *F. glacialoides*, 5.04×10^6 *Thalassoica antarctica*, 1.61×10^6 *Pagodroma nivea* and 0.36×10^6 *Oceanites oceanicus*, a total of 7.03×10^6 tubenoses making 72% of the total birds present but a mere 17.9% of the total biomass due to the weighting of penguins.

Some workers, such as Kuroda (1988, 1991), divided their sea areas into cells. 'Marsden Squares'—areas bounded by 10° latitudes and longitudes—have been used, notably by Jespersen (1930) and Fisher (1952a, p. 298). Figure 2.1 is an example of the mapping of *Puffinus griseus*, which revealed that only relatively small numbers penetrate the North Sea, mainly by way of the northern approaches.

III Factors affecting distributions at sea

The occurrence patterns of petrels at sea depend on the locations of their breeding places, the habitats to which they are adapted, their foods and food availability. A feature that sets petrels apart from all other volant birds is their considerable ability to fast so that breeding birds can search widely for trophic resources. The smaller species have less freedom, at least when nesting, and their distributions then lie closer to home.

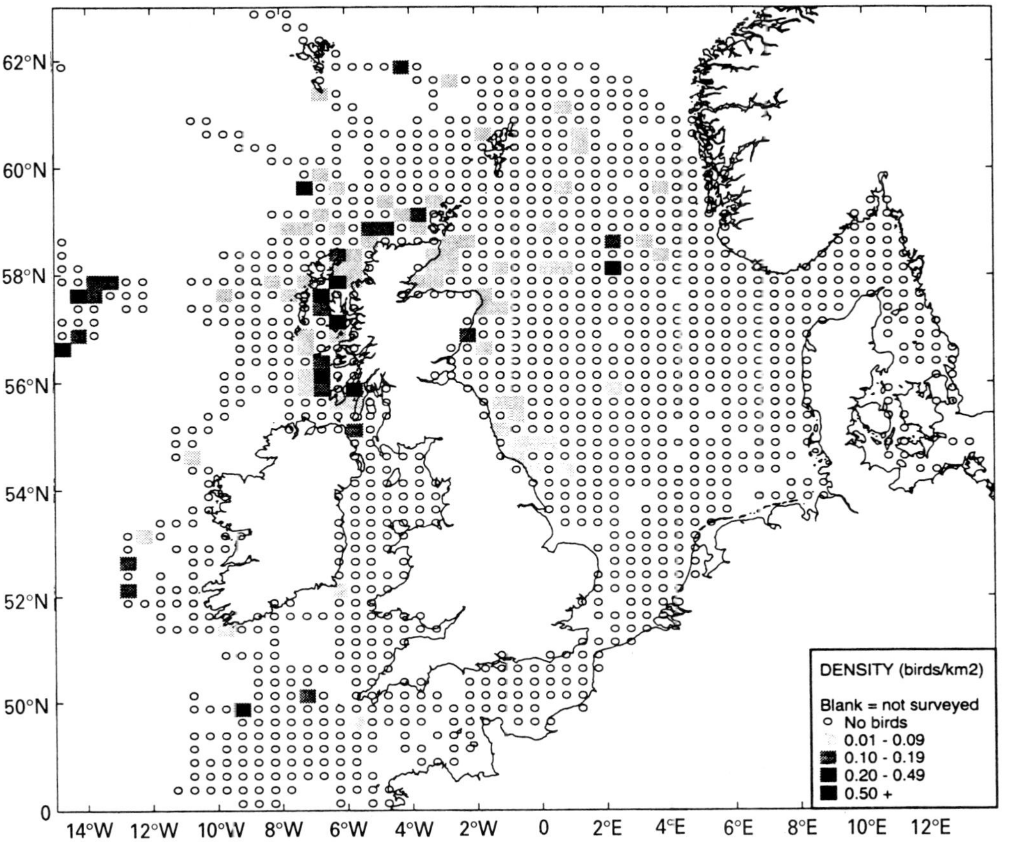

Figure 2.1 *Distribution of Sooty Shearwaters in the North Sea, July to October. Shading gives the numbers of birds seen per kilometre travelled; solid lines designate four zone boundaries, these being, from north to south, northern North Sea, northern Scotland, eastern and southern Scotland, and England. From Stone* et al. *(1995).*

A complex of other factors (Bourne, 1981) determines where petrels occur at sea, factors that complicate the unravelling of their importance to particular species in particular places. These factors include the status of the bird, that is whether it is breeding, a non-breeder or a recently fledged bird, these often being found in different areas at a particular time of the year. Petrels may also search further afield during the incubation stage than when rearing chicks, see Section 2.IX below. In the African sector of the Southern Ocean, Griffiths' analysis (1983) showed that both species' richness and total numbers of birds were markedly reduced during the winter months, particularly at higher latitudes, presumably as a result of the seasonal shift north with the extension of the sea ice (Figs. 2.2A & B). Numbers at sea also vary according to the stage of the breeding cycle, for example with more at sea during the prelaying exodus than when incubation has begun, and so on.

Season also exerts a major effect, particularly in open waters of higher latitudes because the wintertime descent of the zooplankton reduces the food there and many

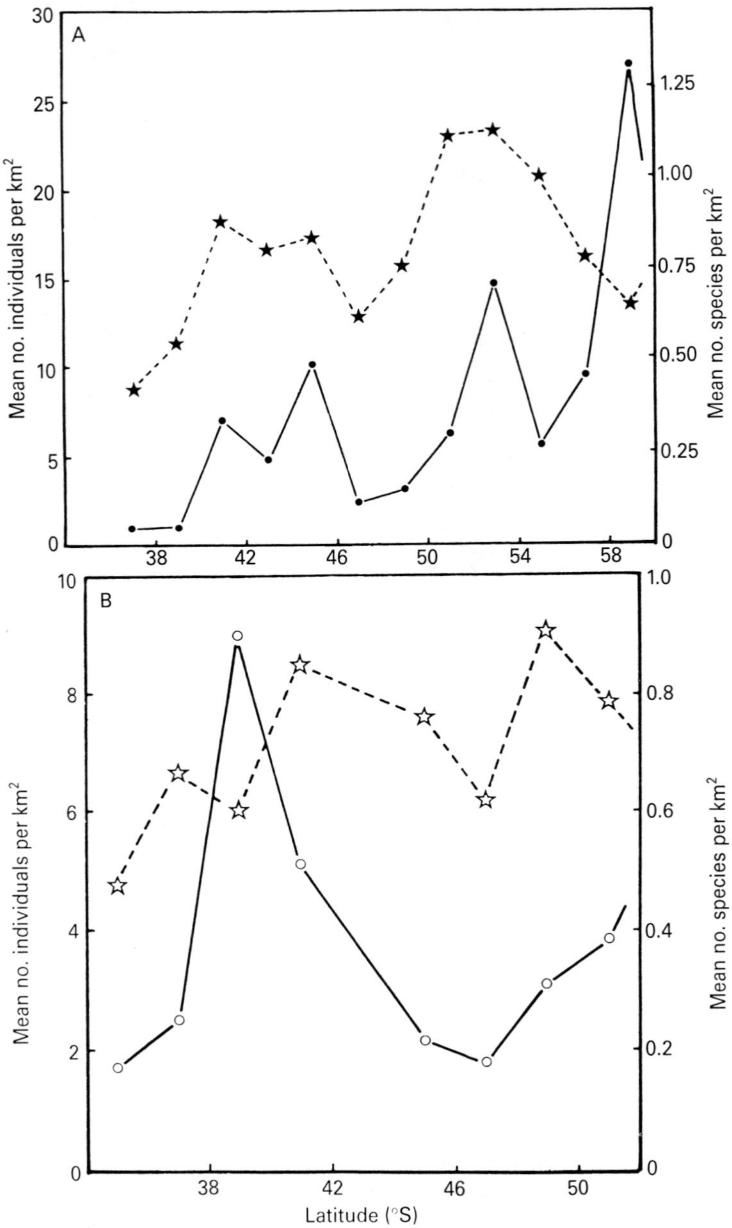

Figure 2.2 *Total numbers of birds (circles) and species richness (stars), of seabirds between Antarctica and South Africa, (A) in summer, (B) in winter. Data are for 32 spp. of tubenoses and seven of Charadriiformes. From Griffiths (1983).*

petrels migrate. Others disperse so that spacial segregation may increase in winter and spring as suggested by Stahl *et al.* (1985) and Fig. 2.6. Some factors are man-made—fishing fleets and sewage outfalls that provide offal for scavengers, for example, and these too may be seasonal.

As Bourne (1963) pointed out, apart from the atmosphere, the sea appears to be the oldest and is still the largest and simplest habitat available to living things, and because of the ease of dispersal through seawater the marine fauna is still rather simple compared with similar areas of land. It supports only about 250 species of birds, the land about 8500. The Procellariiformes comprise a very important part of this marine biotope, whether considered on the basis of number of species, of birds or of biomass.

Nevertheless, as Brown (1980a) emphasized, the ocean is far more than a flat, wet surface separating birds from their food, and, even in the short term, the marine habitat is neither uniform nor static. Seabird distributions and their habitat features must be examined at appropriate scales of space and time (Hunt and Schneider, 1987): at the level of biogeographical regions—'mega-scale', at 'macro-scale' level with features 1000—3000 km in size, 'meso-scale' of 100—1000 km, for example within a feature like the Benguela Current, at 'coarse scale', 1–100 km in size, or 'fine scale' patchiness of areas only metres to hundreds of metres across. For example, in a study of patchiness of *P. griseus, D. melanophrys, D. cauta* and *Procellaria aequinoctialis* in the Benguela Current, patch scale varied from 7 to 23 km and bird abundance was not independent of the orientation of the transect through the current (Schneider and Duffy, 1985).

It was in Murphy's classic *Oceanic Birds of South America* (1936) that the principles of oceanic biogeography were first applied significantly to marine ornithology. However, the relative importance of physical and chemical variables and food are seldom clear, factors may operate simultaneously but on different scales, and species transgress temperature or other boundaries when migrating or hungry. The various hypotheses anent the importance of physical and biological factors were reviewed by Ainley *et al.* (1994). Their work in Antarctic seas supports their view that a petrel's distribution is tied in with the physical characteristics of the habitat and prey availability rather than prey preferences (e.g. Ainley *et al.*, 1984).

A *Latitude and longitude*

The zonal nature of seabird distributions was appreciated by scientists such as Gould (1865) and Hutton (1865) who recorded sightings of tubenoses during voyages to and from Australia. A major early study of the link between latitude and distributions was that of Dixon (1933) who analysed his records of the great albatrosses, and of *D. chlororhynchos, D. chrysostoma, D. melanophrys, Macronectes, Procellaria cinerea, P. aequinoctialis, Pterodroma lessonii, F. glacialoides, Daption capense* and prions when covering 250 000 miles under sail in the Southern Ocean. He tabulated the seasonal distributions of 35 295 birds within 10 latitudinal zones between 20°S and 60°S. More recent studies, mainly of petrels, that emphasize latitudinal zonations include those of Bierman and Voous (1950), Gould (1983),

Harper *et al.* (1990) and Bretagnolle and Thomas (1990) in the Southern Ocean, and Shuntov (1974) for the whole of the Pacific and Indian Oceans.

Jouventin *et al.* (1982) analysed sightings of tubenoses during transects across the south Indian and Southern Oceans in summer. These revealed some clear segregations, for example the distribution of *Diomedea chlororhynchos* being to the north of its breeding place, The Crozets, with *D. chrysostoma* only to the south of there. The two *Phoebetria* spp. were less clearly delimited (Fig. 2.3A) but *Pterodroma mollis* and *Lugensa (P.) brevirostris* were (Fig. 2.3B), while *D. exulans* and *Procellaria aequinoctialis* ranged widely across the convergences. Stahl (1987) examined later sightings based on 10-min counts. The segregations of the two albatrosses and the two gadfly petrels was not quite as clear cut as indicated by the earlier study. He found that within the area surveyed the immatures of all three mollymawks had narrower and more northerly ranges than those of the adult birds, with many along the Subtropical Convergence and with quite a concentration of *D. melanophrys, D. chrysostoma* and *Phoebetria fusca* half way between Crozet and Kerguelen.

The effects of latitude on day lengths and petrel distributions, for example long nights favouring nocturnal feeders such as *D. immutabilis* and *Pterodroma inexpectata,* short ones diurnal feeders such as *D. nigripes,* have not been examined. High latitudes favour daytime feeders in summer but restrict them in winter, and this may be one reason why some high-latitude species shift to lower latitudes during these shorter days, for example *F. glacialis* (Section XI.C below), but others, such as *Pagodroma nivea,* may stay in preferred habitats (Ainley *et al.,* 1992).

B *Sea-surface temperature (SST)*

Early students of marine birds realized that particular petrels and seabirds were associated with cool, temperate or warm seas and SST has been used to define zones of surface water and their associated seabird faunas, for example by Murphy (1936), Kuroda (1960a, 1991), Ashmole (1971), Serventy *et al.* (1971) and Shuntov (1974). Kuroda (1991) plotted the temperature preferences on a monthly basis of 36 tubenoses of the North Pacific. He found that *Pterodroma inexpectata* and *Puffinus tenuirostris* tolerate SSTs to 2°C and 3°C respectively, *P. griseus* to 4°C, *P. carneipes* and *Pterodroma solandri* to 7°C, *Puffinus bulleri* and *Pterodroma neglecta* to 9°C. Warm-water species included *P. pycrofti* to 24°C, *Pterodroma externa* (17–20°C), *P. longirostris* to 18°C, *P. nigripennis* to 20°C, and *P. cookii* to 16°C. In Chilean waters in winter Jehl (1973b) correlated seabird occurrences with SST from transect data and, further north, he did this for birds off western Central America in April (Jehl, 1974a). He noted two distinct assemblages there then: *Puffinus creatopus, Halocyptena microsoma, P. opisthomelas, P. auricularis,* and probably *Oceanodroma melania* were wholly or mainly restricted to waters above 27°C, *P. pacificus, P. lherminieri, Sula sula, S. dactylatra, Procellaria parkinsoni* and *O. tethys* to water exceeding 27°C. In each group there was one small and one large shearwater. Unseasonal changes in SST may also be accompanied by unusual visitors as when mass immigrations of *Puffinus opisthomelas* off California coincided with invasions of exceptionally warm water (Helbig, 1983). In the Southern Ocean Ainley *et al.* (1994) found a stratification of petrel species outwards from the pack ice partly as a function of SST.

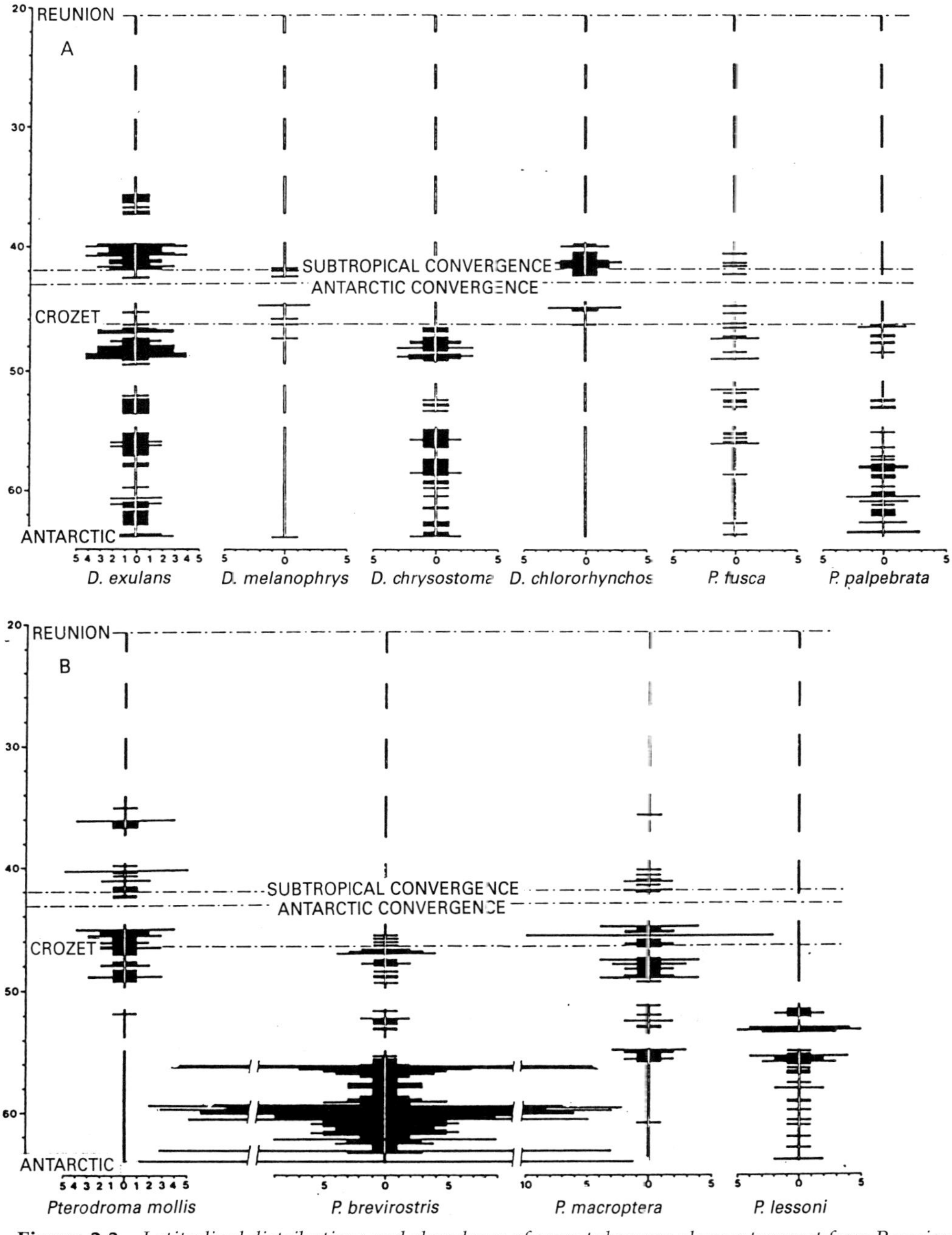

Figure 2.3 *Latitudinal distributions and abundance of some tubenoses along a transect from Reunion to Crozet towards the Antarctic coast in February, (A) for six albatrosses, (B) for four gadfly petrels. The convergences plot close together as they were crossed obliquely: gaps reflect night-time travel. Bars for* Lugensa brevirostris *broken to accommodate high numbers. From Jouventin* et al. *(1982).*

Possible changes in the distributions of shearwaters in the Atlantic in response to changes in SST were discussed by Brown (1991), and for the Pacific research on the effects of changes in atmospheric circulation associated with changes in surface and subsurface currents and upwellings—El Nino Southern Oscillation (ENSO)—on the breeding of seabirds see Ainley *et al.* (1988), Ainley *et al.* (1990), Duffy (1990) and Ribic *et al.* (1992).

ENSO of 1982–84 was quite severe. ENSO brings higher SSTs, deeper thermoclines, higher sea levels and heavier rainfall. Some birds such as *Pterodroma alba* and *Puffinus pacificus* at Christmas Island and *O. tethys* and *D. irrorata* at the Galapagos deserted their nests or did not try to breed during that event. *Pterodroma phaeopygia* at the Galapagos lost many nestings from flooding and the chicks grew more slowly in 1983 (Cruz & Cruz, 1985), which they ascribed to poorer feeding conditions at sea.

When a petrel fails to appear at a regular nesting place during or after an ENSO event we do not know whether the birds have simply shifted their feeding grounds too far from the nest to allow breeding there, have remained in familiar waters without finding enough food to put on the necessary weight to breed, or have died unrecorded at sea. However, Ribic *et al.* (1992) found that during ENSO 1986–87 the common petrel species stayed in the waters of suitable characteristics and whether they bred must have depended on the resources available there.

Where numbers quickly bounce back to pre-ENSO levels, the inference is that the birds survived, but are these the same birds that formerly bred or drawn from the 'strategic reserve' of prebreeders? From transects in Galapagos waters during and after ENSO 1982/84 Duffy and Merlen (1986) found *Puffinus lherminieri* and *O. tethys* abundant at sea, whereas *O. castro* and *Pterodroma phaeopygia* were more plentiful there during the event. They concluded that the shearwaters emigrated and then returned but speculated that the gadfly petrels might have fed closer to the islands during ENSO and thus gave high counts, or that many died then, depressing their numbers in later counts.

C *Currents and fronts*

Murphy (1936) highlighted the importance of the circulation of air and seawater to the distribution of seabirds. For example, he listed the petrels and other birds endemic to the Humboldt (Peru) Current: these include *Oceanodroma tethys, Oceanites gracilis* and *Pelecanoides garnotii* and others, mainly of southern origins, that congregate there seasonally after nesting, for example *Daption capense, F. glacialoides, M. giganteus* and *Puffinus griseus,* all attracted to the richness of the food made available by this current.

Upwellings of nutrient and plankton-rich water attract high densities of petrels, whether offshore or above seamounts and other subsurface irregularities, and where drifting organisms are concentrated along lines of convergence. For example, high plankton densities are found along the northern edge of the South Equatorial Current where *P. l. pycrofti* is a plentiful planktivore (Spear & Ainley, in prep.). Veit and Hunt (1992) found that the influence of the Weddell Sea–Scotia confluence was important to tubenoses nesting on the South Orkney Islands nearby, but its attractiveness varied considerably from species to species.

Mixing may result from tidal streaming. For example, the seas over the Newfoundland Banks are frequently mixed by strong tidal flows over shallows and shoals and are consequently highly productive. Cool-water petrels such as *Puffinus griseus*, *P. gravis*, *Oceanites oceanicus* and *Oceanodroma leucorhoa* are abundant there in the northern and eastern parts, while warm-water ones like *Calonectris diomedea* and *P. lherminieri* use the southwestern area and the Mid-Atlantic Bight within the influence of the Gulf Stream.

There are also well-developed divisions of the surface waters due to rather abrupt changes in SST and circulation—the convergences and divergences. The degree to which the distributions are restricted by the convergences varies with their intensity and from place to place, but is often greatest in the Southern Ocean.

D *Chemical composition*

Levels of salinity and nutrients vary in different parts of the oceans but few attempts have been made to link the distributions of petrels to particular water types. Pocklington (1979), however, did examine relationships between some petrel distributions and SST and salinity in the Indian Ocean. Figure 2.4A shows how two closely related species *B. bulwerii* and *B. fallax* favoured different water types, *B. bulwerii* being recorded only over 'intermediate salinity' (IS) water, *B. fallax* only over the more saline part of this water and neither favouring the low temperature (LT) water used by the three gadfly petrels (Fig. 2.4B). The environmental preferences of the five or six storm petrels of that ocean were much less clear, with *Oceanites*

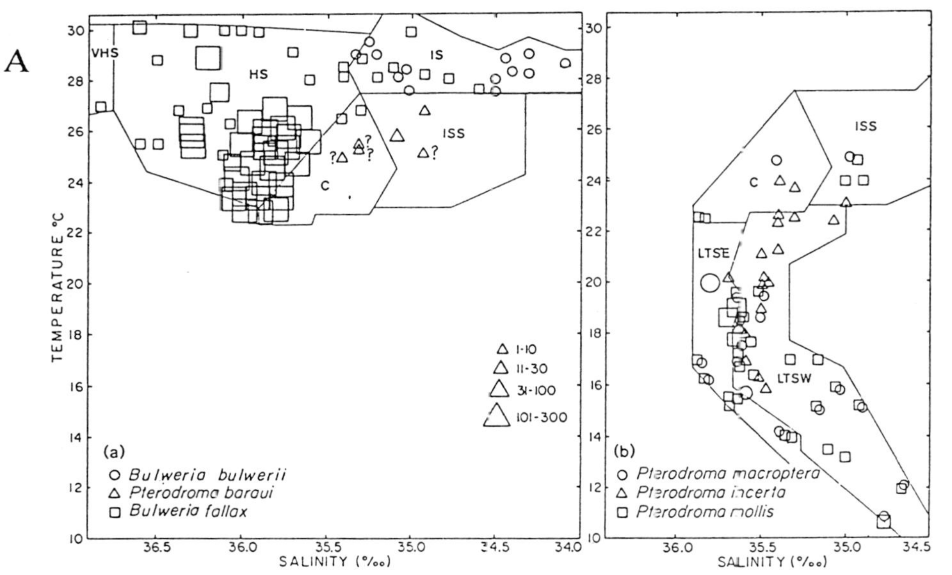

Figure 2.4 *Distributions of some petrels in the Indian Ocean in relation to SST and salinity of (A) two warm-water* Bulweria *spp. and* Pterodroma baraui*, and (B) of cool-water gadfly petrels. From Pocklington (1979).*

oceanicus occurring in all Pocklington's water types but with *Oceanodroma monorhis* choosing warmer seas than *O. matsudairae*.

Brown *et al.* (1975) also used SST and salinity values in analyses of seabird distributions of the Chilean fiords where both can change sharply over very short distances and ranged between 2.9°C—15.7°C and 0.4—33.6‰. Here, among tubenoses he found that *Diomedea exulans/epomophora, Pelecanoides* sp. and *Puffinus griseus* preferred high salinity water, temperature being less important. *D. melanophrys* had a broader salinity preference whereas *Puffinus creatopus* and *Procellaria aequinoctialis* were most plentiful over warm, highly saline waters, but absent over cold, equally saline seas. It appeared that temperature was the key factor. Wahl *et al.* (1989) developed Brown's approach in a detailed analysis of summertime seabird distributions in the North Pacific by water type and oceanographic region. They found that many species used a broad range of water types with *Puffinus griseus, O. furcata* and *O. leucorhoa* in all zones. During these censuses boundaries between regions were seldom sharp except for the subarctic one, *Pterodroma externa* and *P. cookii* being very concentrated on its colder, low salinity side.

Other studies that include SST and salinity data are those of Gould (1983) and Blaber (1986). Ainley *et al.* (1993, 1994) found that SST was much more important than salinity in ice-influenced waters of the Southern Ocean.

E *Water depths*

Petrels forage over shallow coastal waters, over seamounts, continental shelves and in deep, blue water beyond the shelves. There are often higher densities of birds along the shelf breaks.

In the Ross Sea, Ainley *et al.* (1984) also found bird densities in summer much higher over the continental slope compared with shallower seas inshore or in deep water further out. The petrels dominant over the shelf break were *T. antarctica* and *Pagodroma nivea*; *Phoebetria palpebrata, Pterodroma inexpectata* and *Pachyptila desolata* were only seen offshore of this region; inshore, in shallower seas, the common petrel was *O. oceanicus*.

Briggs *et al.* (1987) used water depths as one variable in their examination by principal components analysis of the seabirds off the California coast where the varied currents, seafloor topography, SST patterns and upwellings produce a complicated hydrological picture (Fig. 2.5). These workers found that the density of Sooty Shearwaters varied inversely with water depth and distance from the shore (Component I); the Principal Component (PC) space of Northern Fulmars and Black-footed Albatrosses was particularly correlated with latitude and SST (Component II), and the densities of Leach's Storm Petrels and Buller's Shearwaters varied as distance from the shore and water depth increased (Component III). In contrast, the density of Pale-footed Shearwaters varied inversely on latitude with SST reflecting their preference for warmer, southern waters. Figure 2.5A shows that *D. nigripes* prefers coastal Californian waters, its numbers peaking from May through July (15 000–75 000 birds in early summer), being more abundant to the north. Briggs and his fellow workers found these birds more plentiful over the continental slope than over the shelf or deep waters with a northwards withdrawal from April on. The

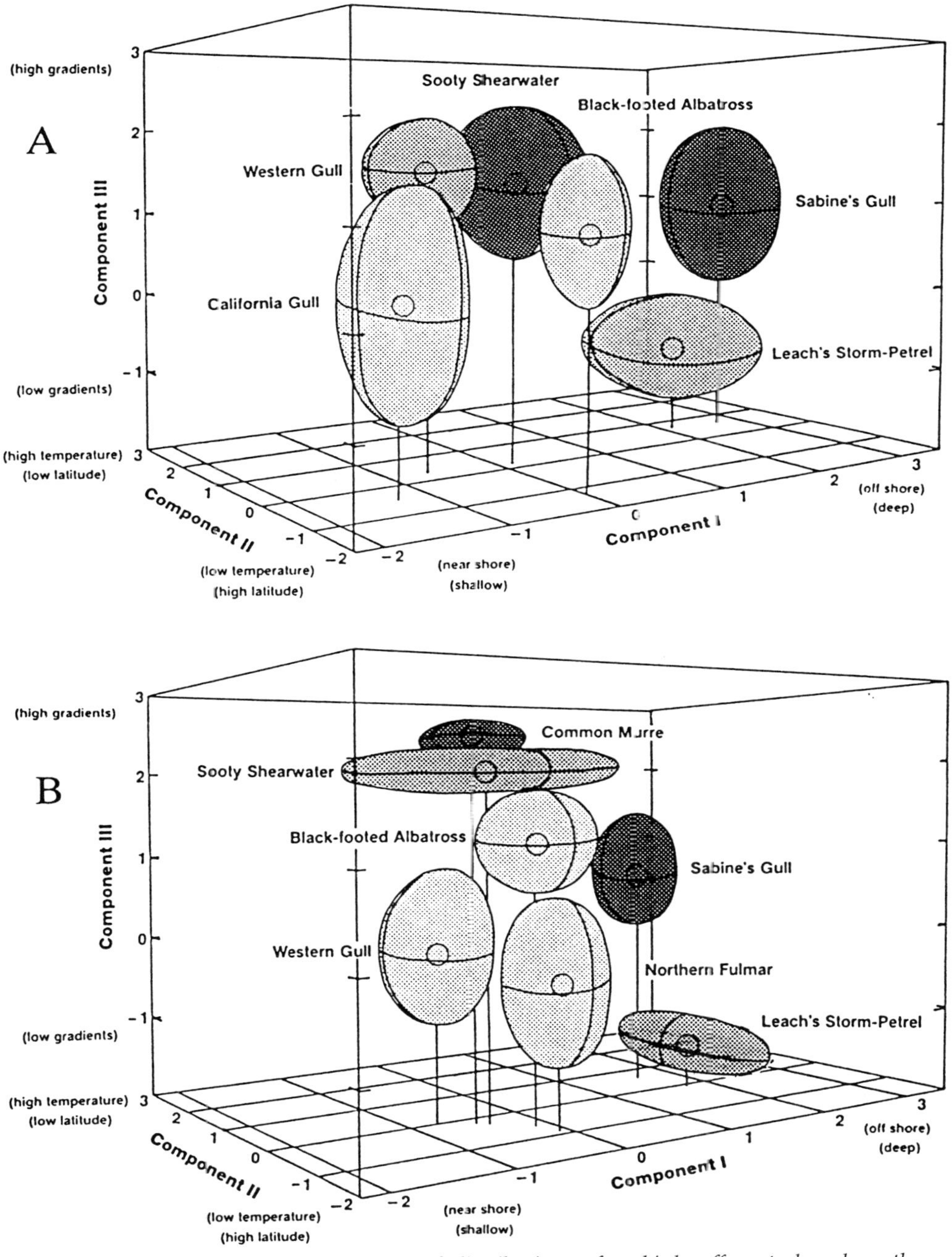

Figure 2.5 *Principal component analysis of distributions of seabirds off central and northern California, (A) in August 1981, (B) in June 1982. Component I represents the covariance between water depth and distance offshore; Component II the inverse relationship between SST and latitude; Component III the variance of the SST gradients. Mean scores and 95% confidence limits are represented by central spheres and ellipses in three dimensions. From Briggs* et al. *(1987).*

concentrations of birds were greater over areas with upwellings and warm California Current eddies that also attracted important trawl fisheries with offal for scavenging albatrosses.

Other studies examining oceanic distributions of petrels in relation to water depths include a cluster analysis by Stahl *et al.* (1985) of foraging zones in the Crozet Island sector of the Southern Ocean (Fig. 2.6) and the patterns of distribution of four albatrosses over the New South Wales continental shelf by Wood (1993).

F *Wind and weather*

The most obvious instances of a link between wind zones and petrel distributions are provided by the albatrosses which need winds to allow progress by slope-soaring and without which they rest on the sea to await a breeze. Presumably their ability to fast then stands them in good stead, for even in the middle of the 'Roaring Forties' there are occasional calms and flat seas, often with fog. There are few detailed observations on petrel behaviour under these conditions, but Manikowski (1971) found a significant relationship between wind speed and the percentage of *F. glacialis* sitting on the sea in the shelf slope area of the Labrador Current with more swimming birds when the force fell below 4° Beaufort. Haney (1987a) noted that most *Pterodroma hasitata* seen sitting on the sea were experiencing winds of ≤4 knots, suggesting that flight occurs at higher wind velocities.

The lack of wind in the tropics has been used to explain the rarity with which the southern albatrosses and giant petrels reach northern waters, but other major barriers may be their difficulty of finding food and of thermoregulating effectively since they are equipped for life where wind-chill is high. Similarly, slope soaring is also impossible over ice shelves and broken pack (see Section V below)—a possible reason for the scarcity of *F. glacialis* among large ice fields (Manikowski, 1971). This author also proposed that these and other seabirds move round atmospheric pressure systems in the North Atlantic to stay within conditions favouring feeding. However, satellite weather photographs do not support this idea according to Bourne (1981), who thought that too many variables were involved to form any clear conclusions. Blomqvist and Peterz (1984) also related occurrences of cyclones to clockwise movements of *F. glacialis* and *Puffinus griseus* off southern Sweden.

Ainley and Boekelheide (1984, fig. 7) compared average wind speeds relative to

---

Figure 2.6 *Distributions of Crozet Islands seabirds in February and September. In February 80% of species were feeding chicks, in September most species were not breeding. Breeding species, filled circles, non-breeding ones, open circles. The 23 petrels plotted were* Diomedea exulans *(WDA);* D. melanophrys *(BBA);* D. chrysostoma *(GHA);* D. chlororhynchos *(YNA);* Phoebetria fusca *(SOA);* P. palpebrata *(LMA);* Macronectes giganteus *(SGP);* M. halli *(NGP);* Daption capense *(CAP);* Pachyptila belcheri *(THP);* P. salvini *(SAP);* P. turtur *(FAP);* Halobaena caerulea *(BLP);* Pterodroma macroptera *(GWP);* P. lessonii *(WHP);* Lugensa brevirostris *(KRP);* P. mollis *(SFP);* Procellaria aequinoctialis *(WCP);* P. cinerea *(GRP);* Fregetta tropica *(BLS);* Garrodia nereis *(GBS);* Pelecanoides urinatrix *(CMD);* P. georgicus *(SGD). From Stahl* et al. *(1985).*

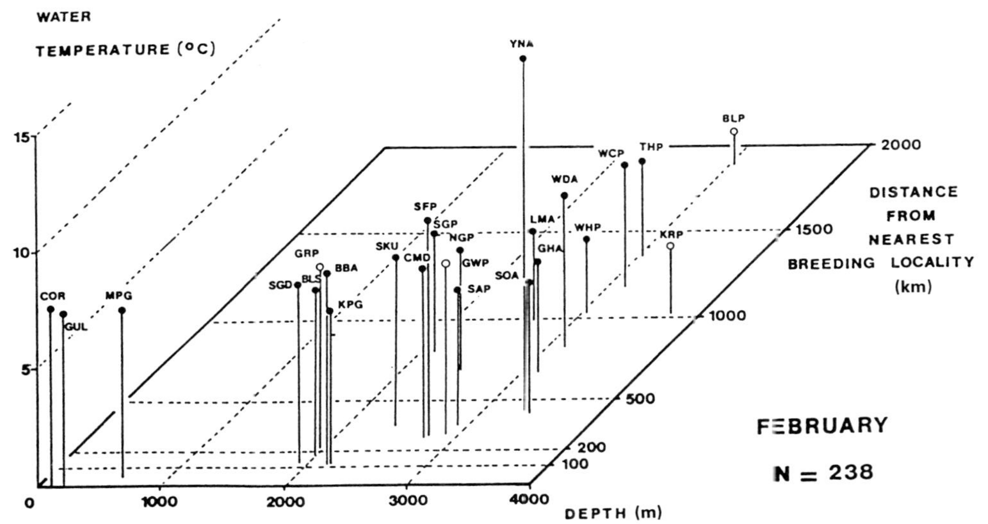
WATER
TEMPERATURE (°C)
15
10
5
0
1000
2000
3000
4000
DEPTH (m)
100
200
500
1000
1500
2000
DISTANCE
FROM
NEAREST
BREEDING LOCALITY
(km)
FEBRUARY
N = 238
YNA
BLP
WCP
THP
WDA
SFP
SGP
LMA
WHP
NGP
KRP
GRP
SKU
CMD
GHA
BBA
GWP
SOA
SGD
BLS
SAP
COR
MPG
KPG
GUL

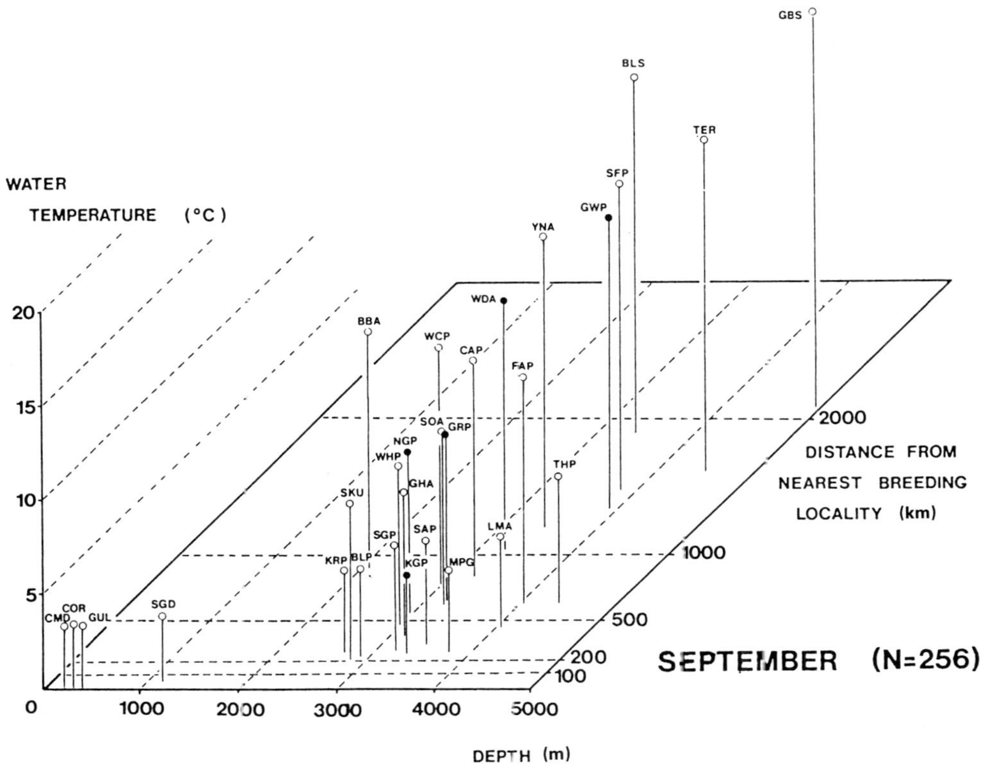
WATER
TEMPERATURE (°C)
20
15
10
5
0
1000
2000
3000
4000
5000
DEPTH (m)
100
200
500
1000
2000
DISTANCE FROM
NEAREST BREEDING
LOCALITY (km)
SEPTEMBER (N=256)
GBS
BLS
TER
SFP
GWP
YNA
WDA
BBA
WCP
CAP
FAP
SOA
GRP
NGP
WHP
THP
GHA
SKU
LMA
SGP
SAP
KRP
BLP
KGP
MPG
CMD
COR
GUL
SGD

1°C intervals of SST during transects through the tropics, subantarctic and Antarctic seas. They found that compared with their respective average values, deviations from the mean in southern seas still left 8–15 knots of wind, but only 2–5 knots in the tropics, suggesting that flight there may be more energetically expensive.

The importance of other factors such as water clarity has yet to be evaluated and just how, if at all, petrels use and respond to clues such as olfactory gradients is inadequately researched.

IV Coarse-scale distributions

Murphy (1936, pp. 65–81, 1964) used oceanographic concepts based on the SST and salinity discontinuities of water masses to describe three concentric life-zones in the Southern Ocean, the Antarctic Zone from Antarctica north to the Antarctic Convergence (Polar Front), the Subantarctic Zone between that convergence and the subtropical one, and the Subtropical Zone extending into lower latitudes from the Subtropical Convergence. These various broad-scale zones have characteristic seabird faunas, with some species being restricted to particular zones, for example *T. antarctica* and *Pagodroma nivea* to the Antarctic Zone, and so on.

Life-zones were used to classify seabirds and petrels by many later workers such as Holgersen (1957), Carrick and Ingham (1967, 1970), Ashmole (1971), Serventy *et al.* (1971), Watson *et al.* (1971) and Shuntov (1974).

Although the boundaries of life-zones can be sharp, at times they are quite diffuse (hence e.g. Polar Front) and some petrels transcend them, notably birds such as *Pterodroma inexpectata, Puffinus griseus, P. tenuirostris* and *Phoebetria palpebrata* which nest in the Subantarctic Zone but also feed in the Antarctic one (see Fig. 2.22).

Other examples of 'coarse-scale' studies are those of Wood (1990a,b,c; 1992), who examined the distributions of tubenoses off New South Wales over 2 years from 23 monthly transects extending seawards to 89 km. Of the 4710 albatrosses seen, 54% were *D. melanophrys*, 34% *D. chlororhynchos*, 6% each of *D. exulans* and *D. cauta*, these all peaking in winter and spring. Most were over the continental slope, but *D. exulans* preferred the lower slope, depth 1500–4200 m. The bulk of the 18 612 shearwaters—*Puffinus carneipes, P. pacificus* and *P. tenuirostris*—were in pelagic waters but the 375 *P. gavia/P. huttoni* were inshore. The figures for *P. pacificus*, the most abundant shearwater, showed a decline or absence from April to October coinciding with their migration, and a peak in November coinciding with the prelaying exodus.

V Ice zones

Frozen or partly frozen seas comprise a special habitat. In the north polar region the only tubenose affected is *F. glacialis*. Fisher (1952a, pp. 305–317) during his 'beating the bounds' of the Northern Fulmar's range, reviewed all the sightings in the pack- and ice-bound seas. Even in May and June these birds are at their breeding colonies in northeast Greenland and Franz Josef Land with the ice front still hundreds of miles away, and Fisher considered that even if leads in the ice were lacking, the birds could still reach the sea. His most northerly record was of two at 86°35′N on 6

September. The ice is then at its furthest retreat but in October the sea quickly starts to freeze and by November and December the petrels have made their greatest withdrawal from polar waters. Brown and Nettleship (1981) pointed out that in Arctic Canada and east Greenland, fulmar colonies are all within range of wide areas of open water even when much of the sea's surface has continuous pack-ice cover.

Pack-ice of medium density supports high densities of petrels in the seas around Antarctica, but the range of species is small. The southern pack-ice is a bigger zone than the northern one and forms the major feeding zone for *T. antarctica* and *Pagodroma nivea*. It lacks shearwaters and albatrosses presumably because of the difficulty of flight and take-off in calm waters—even loose pack dampens waves. The avifauna was described by Murphy (1936), Routh (1949), Erickson *et al.* (1972), Zink (1978, 1981), Ainley and Jacobs (1981), Ainley and Boekelheide (1984), Griffiths (1983), Montague (1988) and Plotz *et al.* (1991) and Ainley *et al.* (1984, 1994), among others.

Ainley *et al.* (1994) used a multidisciplinary approach to follow bird distributions in the Weddell Sea ice in autumn, winter and spring, simultaneous collection of a wide range of environmental data allowing subsequent cluster and other analyses. They looked at the densities in various types of pack-ice, in a near-to-ice open water zone, and in a far-from-ice region.

The pack-ice had permanent residents in *P. nivea, T. antarctica* and *F. glacialoides*, these core 'ice-birds' being supplemented in autumn and spring by smaller numbers of other species. The findings were cross-checked against a springtime study in the Ross Sea (Ainley *et al.*, 1984) where the 'ice-birds' were again *P. nivea* and *T. antarctica*, but with *Oceanites oceanicus* a prominent member.

VI Fine-scale distributions

From earlier, broad-scale studies, attention has extended to the examination of mesoscale features revealed by satellite imagery such as ephemeral upwellings, vortices, plumes, filaments and wind-streaks, for example off California (Briggs *et al.*, 1987). Such phenomena often bring nutrient- and food-rich water to the surface and attract foraging petrels. Some of these features can be seen from the deck of a ship when 'tide-lines' on the surface are marked by flotsam and feeding petrels, or the birds alone may signal an anomaly, as when 1000+ *O. oceanicus* were feeding along a line of scum off Arabia where the SST dropped from 30°C to 21°C. A parallel example was furnished by Pitman and Ballance (1990) who encountered hundreds of *Oceanodroma leucorhoa* concentrated where the salinity abruptly fell from 34‰ to 33‰ and convergent flow was visible with a streak of foam about 1-km long. Brown (1988b) showed that in summer the storm petrels *O. leucorhoa* and *Oceanites oceanicus* off Nova Scotia are concentrated over eddies, 'boils', streaks and other small turbulences partly formed by tidal streams crossing shallow reefs. Sampling revealed that plankton densities along the streaks were significantly higher than outside them. The storm petrels fed at the streaks, not in the adjacent water.

Fine-scale phenomena may also be long lasting, as, for example, zones of mixing and vortices around seamounts, headlands and islands. Increased activity around islands may not always be merely that of birds moving to and from their feeding

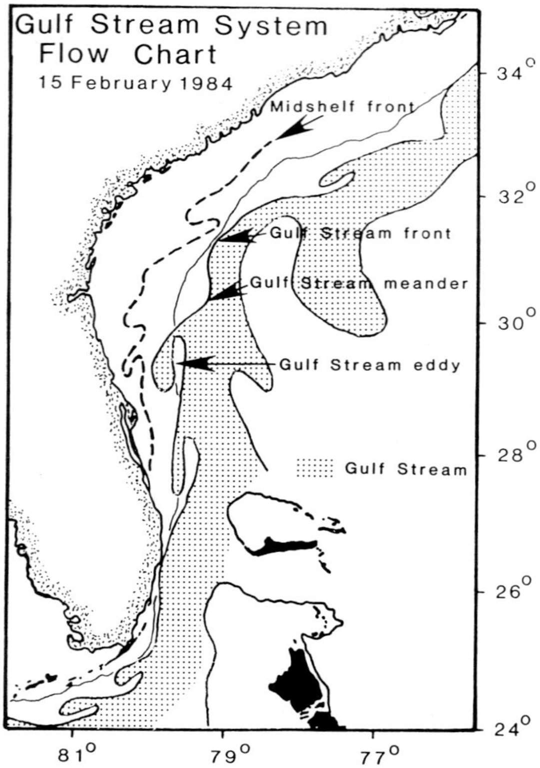

Figure 2.7 *Gulf Stream meanders and eddies off Florida and the Carolinas as revealed by satellite imagery. From Haney (1989a).*

grounds, nor result in a food-depleted 'halo' as postulated by Ashmole (1963); there may be fine-scale effects because of nutrient run-off of faeces of the birds themselves. For example, downstream of the Snares Islands, which lie in a prevailing westerly current, and where perhaps 6 million birds nest on 330 ha, one would expect there to be a lens of enriched water downstream supporting high levels of planktonic activity and providing food 'on the doorstep' for birds such as diving petrels and prions. As some species are ashore almost throughout the year, the leaching of nutrients seems likely to be rather continuous and partly dependent on rainfall.

Fine-scale distributions of *Oceanodroma castro, C. diomedea* and *Pterodroma hasitata* at fronts and eddies generated by the Gulf Stream off the southeastern United States (Fig. 2.7) were investigated by Haney (1985, 1987b), Haney and McGillivary (1985) and Haney (1987a). The birds were counted during transects across the outer continental shelf where the warm, oceanic Gulf Stream water intermingles with cooler shelf water so that productivity is enhanced and organisms normally at greater depths tend to be trapped at shear fronts and thus available to the birds. A range of physical variables was recorded simultaneously to characterize the marine conditions. Food availability was not monitored but direct observations showed, for example, that Cory's Shearwaters swam around pecking at the surface and dipping their heads below water.

The occurrence of *P. hasitata* off the eastern United States peaked where upwellings linked to Gulf Stream meanders and eddies occurred along the frontal boundary (Fig. 2.8), particularly at the older and smaller eddies. Further east the petrels were seen only over seamounts and submarine ridges where the current created topographical turbulence.

Haney (1989a) used infrared radiometry to follow SST changes during a census of *P. hasitata* over four types of marine habitat—resident shelf waters, warm filaments from the Gulf Stream, the Gulf Stream itself, and a cold core. He found that the birds did not use all these habitats equally, nor did they use the different habitats in proportion to the area available. They preferred the cold ones.

Remote sensing techniques allow the instantaneous recording of sea-surface conditions so that changes can be followed on the 'meso' scale and seabird distributions can be checked by shipboard or aerial surveys as done by Briggs *et al.* (1987). The implications and applications of these techniques to the study of seabirds at sea are discussed by Haney (1989a,b) and methods for analysing the data by Haney and Solow (1992).

VII Introduction: movements and migrations

The extent of petrel movements varies from limited circulations offshore, to the commuting of breeding birds between nesting and breeding grounds, long-range dispersal flights typical of fledglings, irregular irruptions and 'wrecks', to the regular shifts of all or a major part of a population from one area to another and its return; that is, true migrations.

During their first phase of pelagic life the young birds explore and learn the characteristics and resources of the marine environment, the more remote bounds of which they may perhaps never need to exploit again, as their range from the natal island contracts and the bird's knowledge of food-finding increases. Exploratory movements during which birds visit colonies other than their own are discussed in Chapter 4.V.

The nature of the various movements is being unravelled from recoveries of marked birds of known age, from direct observations by seafarers and plotting of individual birds by radiotelemetry. It may soon be possible to follow migrating flocks with the help of satellite photography. Away from the breeding places, recoveries of marked birds tend to be few and, while most deaths presumably take place at sea, band returns are biased by most being from inhabited coasts: few petrels are taken at sea except from fishing hooks and nets. Likewise those that tend to feed offshore, such as *D. melanophrys, D. chlororhynchos, C. diomedea* and giant petrels, tend to be recaptured; more pelagic species, such as the *Phoebetria* albatrosses, are seldom recovered.

Age and status are usually indeterminable in a free-flying bird, but occasionally this may be possible, for example with immature mollymawks. In the hand, fledglings may still carry down on their napes, their feathers fresh, their skulls and bills soft and their claws unworn, criteria useful in sorting out beach-wrecked birds.

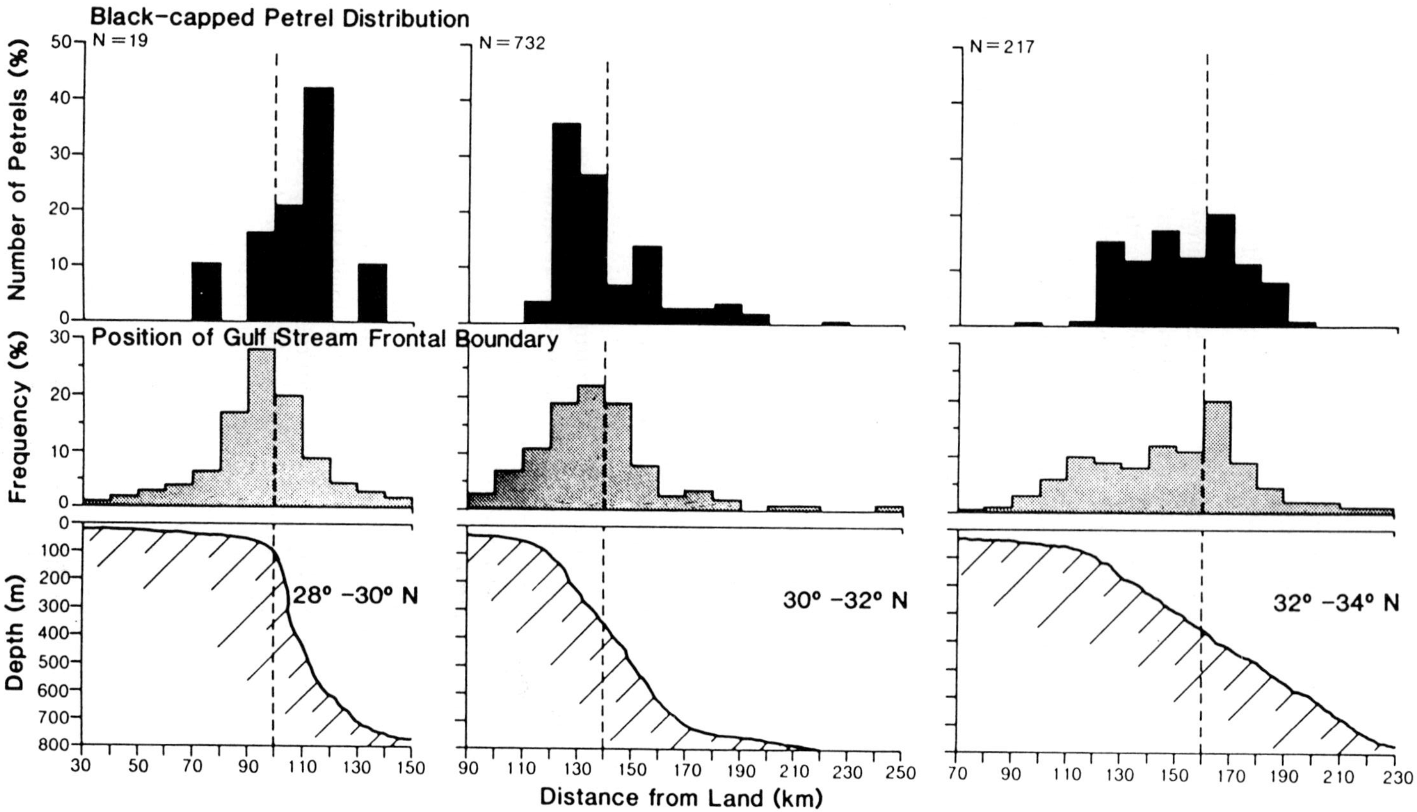

Figure 2.8 *Cross-shelf distribution of the Black-capped Petrel in relation to water depth, distance from land and position of the Gulf Stream front at three levels of latitude off the southeastern United States. N, number of birds, vertical dotted lines mean positions of the front. From Haney (1987a).*

VIII Local circulatory movements

Watchers often see groups of petrels heading in one direction and others of the same species flying in the opposite one. In some cases it is clear that the same birds are involved, flying circulatory courses. These movements may suggest premigratory restlessness, for example with *Puffinus griseus* and *P. gravis* (Collins, 1884), but often no such shift is pending and the reasons for the flights are obscure. Bourne (1982a) examined movements of seabirds off eastern Scotland in relation to wind strength and direction. He concluded that foraging *F. glacialis* from the northern colonies can be drifted downwind south into the North Sea and return in a concentrated movement up the east coast of Britain. No doubt many similar local movements occur in all the oceans.

Similar movements may also take place during migrations. Marchant (1977) watched streams of *P. tenuirostris* off New South Wales returning from their annual foray to the North Pacific. From the time that the main southward passage began he saw a persistent northwards flow in which the individual birds and flocks maintained their directions and integrity in the face of the southwards stream (Fig. 2.9). The birds were so tightly packed that he was surprised that no head on collisions occurred. Observations further out showed that this was not part of a wider circulatory movement. These counterflow movements were not seen in some later years, suggesting fluctuations in marine conditions.

Another example is provided by the flights of the Levantine Shearwater *P. puffinus yelkuoan*, the 'damned souls', that circulate through the Bosphorus and round the Black Sea (Robel & Konigstedt 1976) (Fig. 2.10).

IX Movements between breeding and feeding places

The distances to which petrels fly when breeding have usually been inferred by calculating ranges possible from flight speeds over the known periods of absence, that is, during the prelaying exodus, incubation and chick rearing, ranges during the last phase being judged to be shorter than during incubation. Even when a petrel is feeding in seas hard by a nesting place it is a large assumption that that particular bird breeds there.

Some past deductions of ranges have proved inaccurate. Lockley (1953) plotted the inward- and outward-bound streams of Manx Shearwaters passing headlands and islands and analysed recoveries of ringed birds mostly caught by Bay of Biscay fishermen, 850 km away. He proposed that the birds that followed the movements of the sardine shoals there were breeders from the Scillies and Wales. However, Harris (1966a) found that 87% of the recoveries were of 1–4-year-olds, too young to be nesting and only one was known to have had a chick. He concluded that the birds could not reach the Bay, feed, and return in the 1.7 days between chick meals, and that those feeding on the sardines were sexually immature non-breeders. Brooke (1990), re-examining the recoveries, now of 339 birds, deduced that breeders from west Wales were feeding no further than 400 km away, even during their 6–7 day off-duty stints during incubation, whereas females during their prelaying exodus might reach the main sardine nursery area in the southeast corner of the Bay.

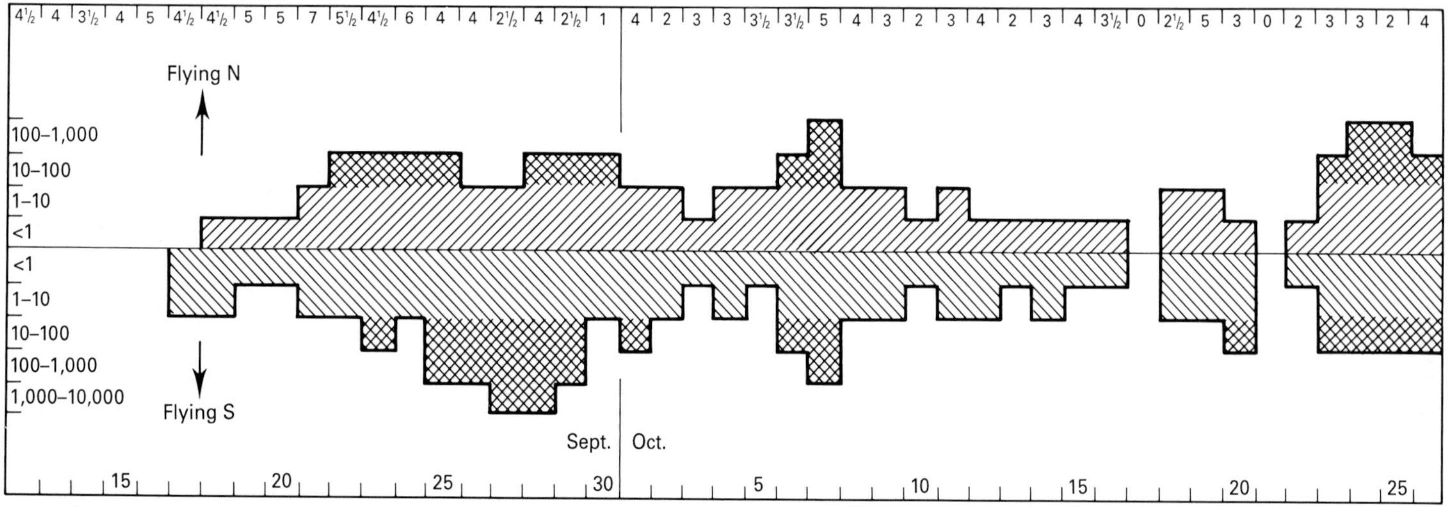

Figure 2.9 *Daily passage of* Puffinus tenuirostris *off New South Wales during late September and early October 1974, showing average numbers passing per minute by order of magnitude. Hours of watching per day along top, significant passages cross-hatched. Were the southbound birds newly arriving migrants, the others already returned, and heading for feeding grounds? From Marchant (1977).*

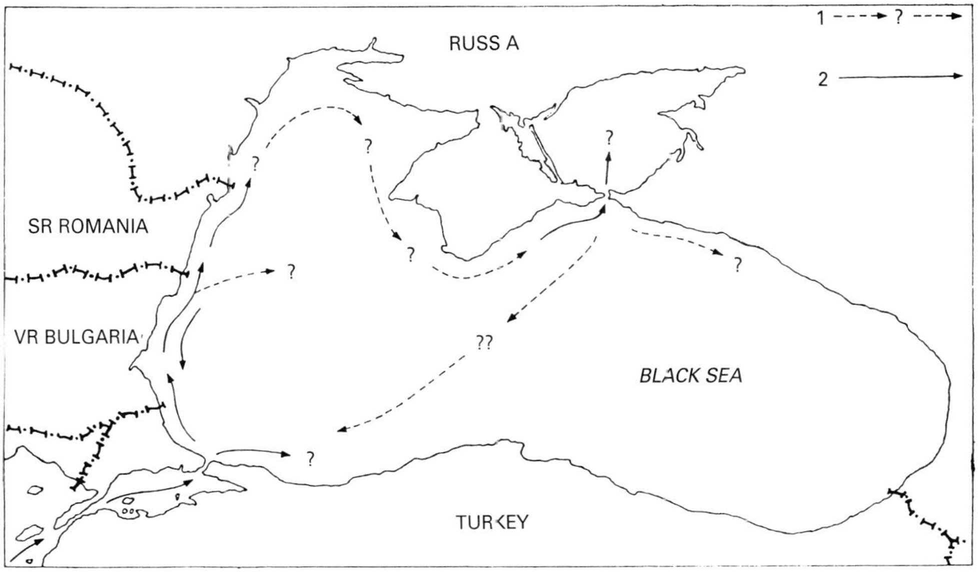

Figure 2.10 *Circulatory patterns of* Puffinus puffinus yelkouan *in the Black Sea during May to July: hypothetical routes shown by dashed lines. From Robel and Konigstedt (1976).*

Some feeding ranges have been deduced from diets when composed of foods with restricted distributions; in one instance by the presence in the stomach of identifiable pumice whose nearest source at sea was known (Sutherland, 1965). Skira (1986) deduced from their diet that Tasmanian *P. tenuirostris* fed locally when breeding, citing the absence of *Euphausia superba* in their stomachs as evidence that they were not travelling to Antarctic waters. Similarly Montague *et al.* (1986) deduced from their dependence on *Nyctiphanes australis* that the Phillip Island birds ranged only within Bass Strait or over the continental shelf.

Dye marking of birds has helped to elucidate positions at sea. This has been used with albatrosses, for example by Tickell, 1968. Weimerskirch *et al.* (1988) marked breeding *D. exulans, D. melanophrys* and *D. chrysostoma* at Kerguelen with various dyes; with the last species, birds at two colonies were differently marked. The 3989 birds with chicks yielded 828 sightings. In general, most were of birds over the shelves around Kerguelen, Heard and McDonald Islands at water depths <500 m. With *D. melanophrys*, there was some suggestion that birds from different colonies fed in different places. Most from Ile de Croy were seen over the northern half of the Kerguelen Shelf or on a smaller shelf to the west and those from the mainland colony over the south and eastern parts of the shelf and to the southeast as far as Heard Island (Fig. 2.11). Birds from both colonies intermingled over the southeastern section of the shelf. Unfortunately, where the birds were most abundant was where the trawler fleets for notothenid fish were most active, and Black-brows, in particular, follow fishing boats. The few resightings of *D. chrysostoma* by the Weimerskirch team suggest that these range much further afield when raising chicks than *D. melanophrys*—over Antarctic as well as subantarctic seas as far as 1850 km from their nests.

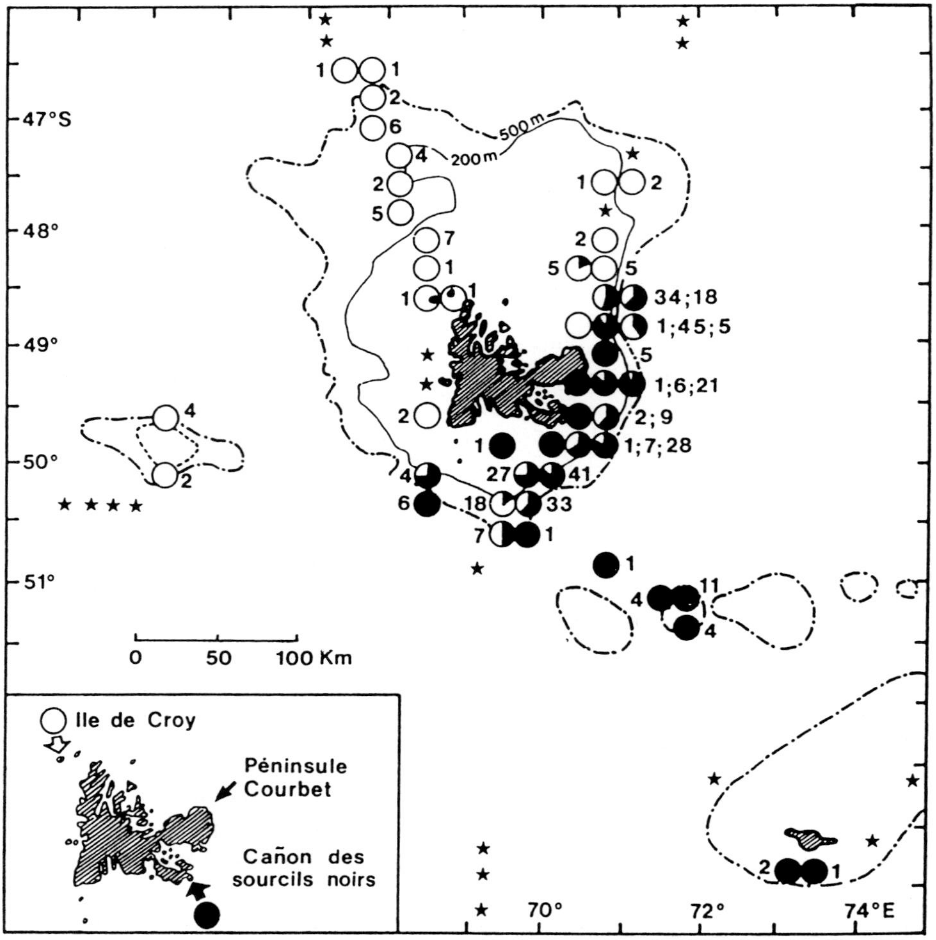

Figure 2.11 *Sightings of dyed breeding* Diomedea melanophrys *from two colonies at Kerguelen around the Kerguelen–Heard Islands shelves. The amounts of black and white in a circle indicate the proportions of birds from Ile de Croy (white) and from Canon des Sorcils Noir (black) in 15′ latitude × 20′ longitude squares. The total number of dyed birds seen in each square is given beside each circle; the stars indicate no dyed birds. From Weimerskirch* et al. *(1988).*

The first successful tracking by satellite seems to have been that of Southern Giant Petrels in 1985 (Strickwerda *et al.*, 1986). In preliminary trials one breeding bird travelled over 2000 km off the Antarctic Peninsula in 30 days before transmission ceased. The transmitter packages used also recorded external temperatures and activity but none of the experimental birds returned to their colony.

After several false starts in various parts of the world using albatrosses (these being large enough to carry the transmitters), excellent results are now being achieved. In due course further miniaturization will allow movements of smaller species such as the transequatorial migrants to be monitored, but if good, statistically satisfactory samples are needed, the cost seems likely to be substantial. Furthermore, the batteries now used are exhausted after 30–50 days and further development awaits more powerful sources or efficient solar power generation. For technical data see Prince *et al.* (1992) and Weimerskirch *et al.* (1992).

Two *D. exulans*, breeding birds of either sex, were tracked from South Georgia while rearing chicks. The female made three foraging trips totalling 13 951 km, the male two, covering at least 9280 km (Prince *et al.*, 1992). The female searched seas mainly to the north and on her longest trip of 6479 km spent at least 2 days of an 8-day flight about 240 km off the Brazilian coast. The male tended to hunt further south and on one trip spent 3 days in Drake Passage (Fig. 2.12). The minimum average flight speed of the male (20 km h^{-1}) was less than that of his partner (35 km h^{-1}), perhaps because the male spent more time on the water. However, he reached 88 km h^{-1} during two successive fixes; she managed only 79 km h^{-1} under these conditions. Most distance was covered by day, 58% for the female, 77% for the male and they flew longer on nights when the moon was new than when it was full. These ranges make quite feasible the suggestion that breeding birds from the New Zealand subantarctic islands forage in eastern Australian waters.

The most detailed study to date is of birds from the Crozet Islands (Jouventin and Weimerskirch, 1990a; Weimerskirch *et al.*, 1992, 1993). These workers tested the accuracy of the satellite fixes on an incubating bird, finding that 73% of the locations given were within 1 km of the nest and only 6% were more than 3 km out. They also checked the lengths of the foraging trips by birds with and without transmitters and showed that there was no overt difference in behaviour.

During incubation, off-duty Wandering Albatrosses foraged in long loops of 2000–15 000 km, travelling mainly by day, some reaching polar seas. So too did five telemetred *Phoebetria palpebrata* from Macquarie Island, moving to individual foraging areas in pelagic seas not far from the Antarctic coast (Weimerskirch and Robertson, 1994). During the guard stage, when they alternate frequently on the nest, *D. exulans* fed over the continental shelf, while in the postguard stage they mixed long pelagic trips with shorter inshore ones. Flight paths varied with wind speed and direction, the birds generally keeping the wind on their quarter (Fig. 2.13).

Weimerskirch *et al.* (1993) found that the foraging ranges of *D. exulans* contracted as hatching approached, shortened still more during the guard stage, but thereafter extended outward again (see also Walker *et al.* (1995)). Weimerskirch *et al.* (1994a) and Chaurand and Weimerskirch (1994) also measured the times that parents were absent and their consequent changes in mass, as well as the mass of the meals delivered to the chicks for six species—*Pelecanoides urinatrix*, *Halobaena caerulea*, *Pachyptila belcheri*, *D. chlororhynchos*, *D. melanophrys* and *D. exulans*. With all except *P. urinatrix* and *D. melanophrys*, both sexes undertook both long and short trips to get food. The two burrowing species did this on alternate occasions, the others took more short forays than long ones. The parent birds stored energy during these long trips, energy they used to power the short ones foraging for their chick when they lost energy and their mass fell. This system was seen as a compromise between the demands of the chick and the need for the adults to forage without losing condition, and further experiments indicate that parents can adjust their foraging according to their body condition so as to avoid increased mortality (Weimerskirch *et al.*, 1995). The two petrels not using this system were feeding in neritic waters, the others were pelagic.

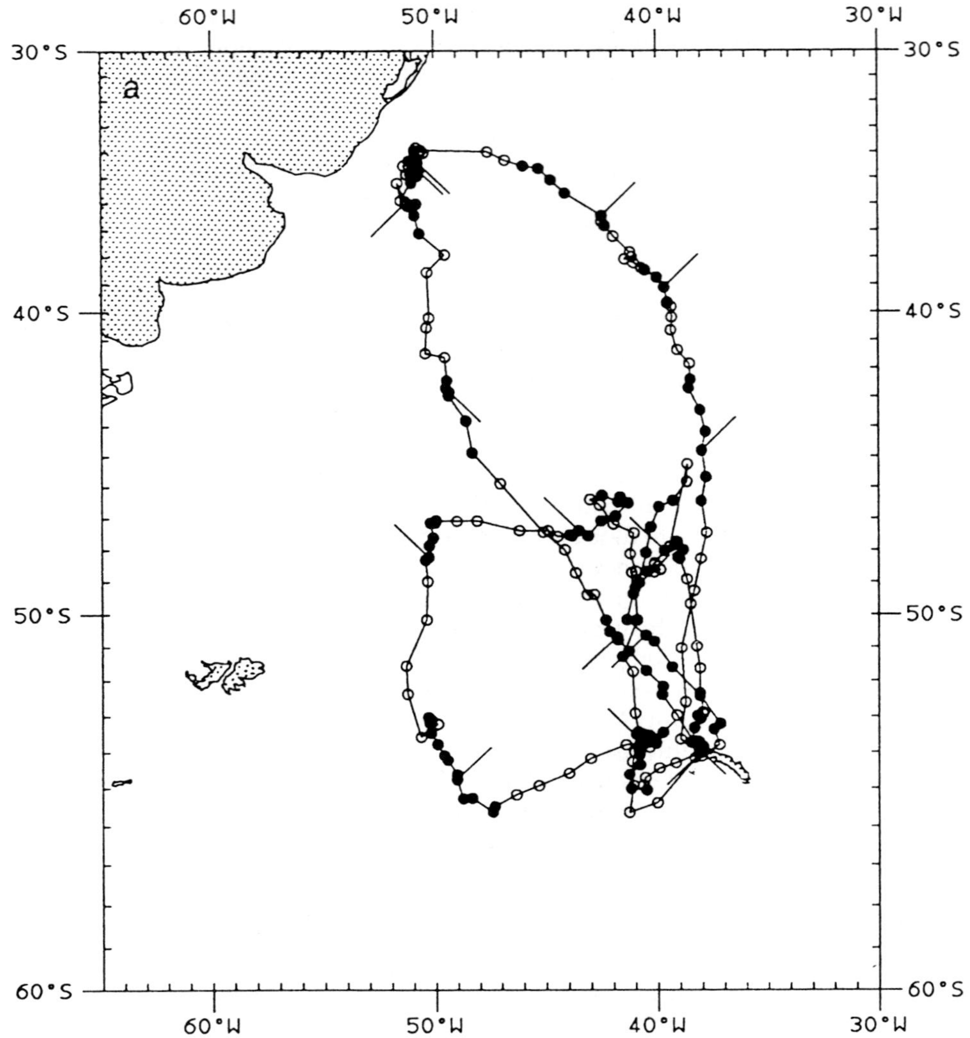

Figure 2.12 *Tracks followed by a female* Diomedea exulans *tending a chick at South Georgia on three successive foraging trips. Open circles indicate daytime records, filled circles night-time ones, / means date change. The trip to the Brazilian coast took her to a major area used by long-line tuna fishermen. From Prince* et al. *(1992).*

X Dispersal

Birds that shift seasonally but to no regularly occupied location may be said to disperse, although the distinction from migration may be blurred. Those not tied to returning to their breeding colony include some younger non-breeders, failed breeders, and those taking 'sabbatical years' off (*The Petrels*, p. 123). Some such may go to preferred feeding grounds on a regular basis; that is, they migrate. Another category is that of the contranuptial movements of sedentary species. Some established breeders may visit their colonies outside the breeding season. These include both *Macronectes*, some gadfly petrels such as *Pterodroma macroptera*, and prions (*The Petrels*, p. 128).

Dispersal movements are most easily seen with southern species which have all the Southern Ocean available to them—see Weimerskirch *et al.* (1985). On their first flight petrels may leave in batches triggered by wind gusts (*The Petrels*, p. 376). Whether any congregate offshore before final departure, as does *F. glacialis*, is unrecorded. Weimerskirch *et al.* (1985) reported two *Daption capense* chicks from the same Antarctic colony recaptured within 4 days of one another at the same place in New Zealand. This could mean that immatures keep together in flocks and/or that each colony has its particular wintering area.

The giant petrels provide good examples of apparently passive downwind dispersal of fledglings, because large numbers have been ringed and recovery rates for these big, obvious birds are relatively high. However, downwind flight is unusual among petrels, and Spear and Ainley (in prep.) report that cross-wind movement is the norm, upwind travel occasional.

Most giant petrels turn up on, or close to, land and are found near populated coasts, particularly around Australasia which lies athwart the westerlies. Sightings

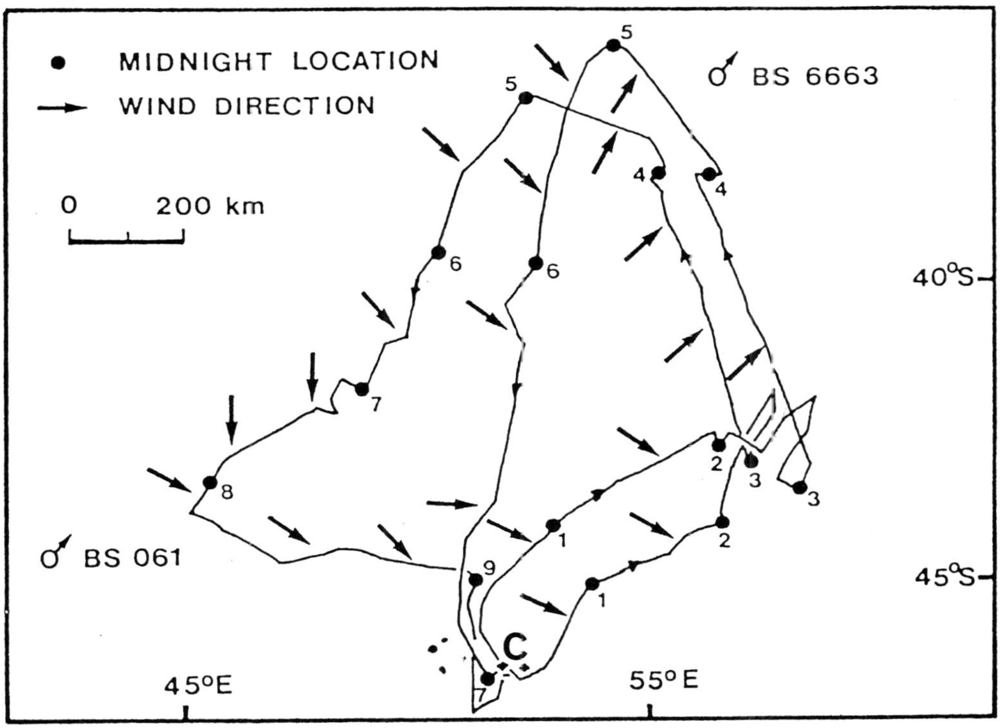

Figure 2.13 *Tracks of two male* Diomedea exulans *from The Crozets (C) at the same time, showing wind directions and strengths in knots (numbers). The birds tended to fly with the wind on the quarter. From Weimerskirch* et al. *(1993).*

of both species show that they are pelagic as well; perhaps, as Hunter (1984c) postulated, older birds prefer deep water, younger ones neritic seas, and the latter thus tend to end up more often in beach wrecks: certainly most of those found dead are birds of the year. And, as Ingham (1959) pointed out, simple downwind drift does not control all the fledglings for some of those from Macquarie Island escape the westerlies, reaching low latitudes and dying on subtropical beaches.

Analyses of the recoveries of banded South Georgian birds were made by Hunter (1984c), of those from Macquarie Island by Woehler and Johnstone (1988), and from Crozet and Kerguelen by Voisin (1990). The recoveries support the view that most young birds travel eastwards on leaving their nests. Furthermore, their mean ages tend to increase the further to the east their recoveries (Conroy, 1972; Hunter, 1984c; Weimerskirch *et al.*, 1985; Voisin, 1990). Robertson and Kinsky (1972) demonstrated a similar effect with young *Diomedea epomophora*. Thus Woehler and Johnstone (1988) showed that *M. halli* fledgelings from Macquarie Island tend to reach New Zealand beaches in March, South American shores in May, South Africa in June, but not until July the Australian mainland, inferring that they have flown around the world to reach there, as Dixon (1933) long ago suggested was true of southern albatrosses (see also Robertson and Kinsky, 1972). Such circumstantial evidence awaits confirmation from satellite telemetry.

Woehler and Johnstone (1988) found no banded *M. halli* south of 50°S which supports the views of Bourne and Warham (1966) and Johnstone (1974), that in the summer this bird mostly forages north of the Antarctic Convergence, *M. giganteus* south of it. Both are more numerous in Australasian seas in the austral winter and Hunter (1984c) suggested that these form a major wintering zone for both species. Another factor may be that juveniles from Crozet, Marion and Kerguelen Islands can reach these seas quite quickly before experiencing the heavy initial mortality (Voisin, 1990). Recaptures of immature *M. giganteus* in New Zealand show that some remain in coastal waters for several months and even in a particular area during two successive years (Kinsky, 1958).

Shaughnessy and Voisin (1981) found *M. halli* plentiful in Nambibian waters from mid-August to January, that is, during their breeding season—being attracted by the pupping of the fur seals. The nearest colonies are on Gough and the Prince Edward Islands, all 3000 km away, but whether the scavenging birds were non-breeders or nesters stocking up during their off-duty spells, was not determined.

Recoveries of older birds suggest a contraction of range towards the natal island with age (Hunter, 1984c; Woehler and Johnstone, 1988; Voisin, 1990). Some older birds have been found on islands remote from their own. Three *M. halli* born on Kerguelen and The Crozets caught as non-breeders at South Georgia may have just been passing through during their prebreeding wanderings (Hunter, 1984c), but the female *M. giganteus* banded as a chick at Macquarie Island on 26 January 1961 and recaptured at Cape Crozier, Antarctica on 7 February 1967 could have shifted colonies, and at least one breeder from Macquarie Island has nested later at Cape Crozier (Woehler and Johnstone, 1988).

Some very rapid movements take place. The fledgling travelling from Kerguelen to Chile in 3 months (*c.* 14 000 km directly or *c.* 21 000 downwind) may have flown against the westerlies. One *M. giganteus* from the Antarctic Peninsula reached Fiji when only 6 months old and another from South Georgia evidently covered at least

14 000 km to New Guinea in 10–12 weeks. Others reach low latitudes in cool waters, as with those found off Peru in the Humboldt Current.

Other instances of dispersals of young birds are less easily separated from the movements of adults which may shift seasonally to preferred feeding areas. For example, Jouventin *et al.* (1982) suggested that in the Indian Ocean at least, immature *D. exulans* fed in different zones from the adults, at least during the breeding season. Whether, in their prebreeding years, birds return regularly to the same areas at particular times of the year—that is, migrate—is unclear. Another complication is that meteorological and other conditions vary from season to season: a lull in the westerlies when a batch of chicks makes its first flight may mean fewer recoveries downwind so that the pattern of dispersal may vary between years.

Banding of nestling *D. melanophrys* at South Georgia and the Falkland Islands by Tickell (1966) showed that most moved north, 92% of the recoveries of South Georgian birds being from South Africa with small numbers reaching Australasia, while the Falkland Island birds ended up on South American beaches, the majority far to the north. However, most recoveries came from fishermen and so revealed the whereabouts of the fishing fleets rather than the birds' 'natural' dispersal pattern. Periods of high mortality occur on New Zealand beaches from April to September (Powlesland, 1985). The first part of this was thought to be mostly fledgling *D. m. impavida* from Campbell Island. These disperse northwards to southeast Australian waters and the Pacific, the later high casualty rates being maintained by a wave of eastwards-drifted birds from colonies in the south Indian Ocean.

The oceanic distribution of the diving petrels *Pelecanoides urinatrix* and *P. georgicus* when not breeding is virtually unknown and it has been generally assumed that they disperse around and to the north of their nesting colonies in the southern winter, but, at least in summer, birds have been seen, if not identified to species, far from any nesting place, for example towards the Ross Sea as far as 64°S (Ainley *et al.*, 1984). These must have been non-breeders. Bourne (pers. comm.) saw a large eastward movement of *P. georgicus* west of South Georgia in the spring, apparently on a return flight.

XI Migration

A *Introduction*

Regular seasonal shifts of a population from breeding to contranuptial areas and back again are common among tubenoses. Other regular movements cover only relatively short distances, for example, the wintertime northwards shift of most Southern Ocean species such as *D. exulans* as documented by Shuntov (1974), Jouventin *et al.* (1982), and others. Even ice-loving *Pagodroma nivea* and *T. antarctica* must move north with the pack-ice periphery where there is some open water.

In many cases the migrants completely desert their nesting islands after breeding, but no migration is entirely complete and for all well-studied species there are reports or specimens of birds that have remained in local seas at either end of the migration.

Migration routes are not precisely known for any petrel. Complicating factors are that different age-classes may follow different paths which in any event may not be fixed but vary according to conditions at the time, for example a migrant stream may be wind deflected, while the numbers passing any place may vary from year to year in response to variations in food distributions or upwellings, etc. Sometimes such variations may lead to veritable invasions of birds seldom seen before, for example Newell (1968).

The migrations of Procellariiformes can be categorized according to the kinds of movements made (Serventy, 1953; Kuroda, 1957a). The biggest group comprises the transequatorial migrants. Four species shift from the northern to the southern hemisphere, but 17 move in the opposite direction—at least eight gadfly petrels, seven shearwaters and two storm petrels.

How these navigate over wide stretches of apparently featureless sea is unclear and there have been no major experimental investigations of navigation by tubenoses since the homing trials with *Puffinus puffinus* by Lack and Lockley (1938) and Matthews (1953, 1964)—reviewed by Brooke (1990, pp. 66–70),—with *D. immutabilis* (Kenyon & Rice, 1958), and with *Oceanodroma leucorhoa* by Billings (1968) (Fig. 2.14). Recently homing of *C. diomedea* has been studied using direction recorders (Dall' Antonia *et al.*, 1995). None of these tests involved crossing the Equator and transequatorial migrants face many navigational problems. They encounter changing star patterns, sun movements and magnetic fields as they skim rapidly across the latitudes. Some of Matthews's birds indicated their ability to navigate by day over unknown terrain using a sun-compass and internal clock, but seemed unable to do so after dark. Do the migrating shearwater flocks continue their flights on dark overcast nights?

Whether tubenoses can orientate to a magnetic field is unknown: their brains appear not to have been examined for the presence of magnetite cells; but the ability to detect the dip of the geomagnetic field, for example, has been mooted as a navigational aid for birds and one that could perhaps be inherited genetically.

Another category is of birds that cross the latitudes, but only within their own hemisphere. These latitudinal migrants include birds that shift southwards in the contranuptial season, such as *F. glacialis*, *O. leucorhoa* and *O. furcata* on the Pacific coasts of North America, and north, such as many Southern Ocean species.

B *Longitudinal migrants*

Birds that move to east or west, crossing the longitudes include Atlantic *F. glacialis*, all the North Pacific albatrosses, *O. monorhis*, *O. matsudairae*, *Pterodroma rostrata* and *Puffinus huttoni*. Likewise, most of the population of *P. puffinus yelkouan* leaves the Balearic Islands to pass through the Straits of Gibraltar in May and June to feed in the Bay of Biscay and off Brittany, returning to the Mediterranean in September and November (Le Mao and Yesou, 1993). This bird is one of a group whose movements may involve dispersion and longitudinal or transequatorial migrations. Bourne (1982b) pointed out that *P. p. mauretanicus*, *P. p. yelkouan*, and *P. gavia* are mainly coastal shearwaters that move north and east or west chasing shoaling fish while they moult, whereas *P. puffinus* and *P. huttoni* take similar prey in the South Atlantic and off northwestern Australia respectively.

With most of these petrels there is also a latitudinal component to their migration,

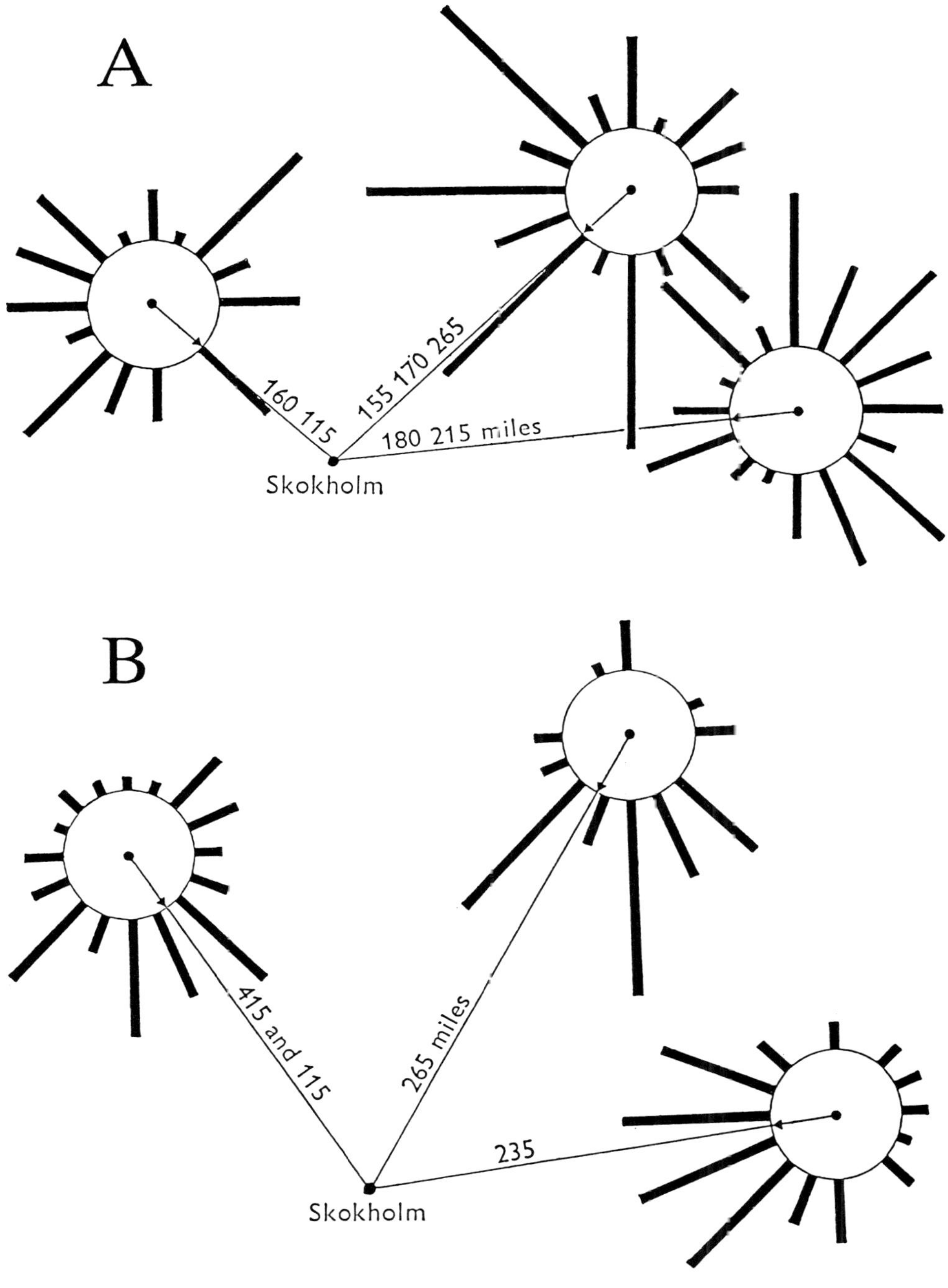

Figure 2.14 *Orientation of released Manx Shearwaters, vanishing bearings gathered into compass points; shortest spoke = 1 bird. A, disorientated scatter when set free under heavy cloud; B, homewards orientation when released in sunny conditions. From Matthews (1968).*

for example with the 300 000–2 × 10^6 *O. leucorhoa* that winter in the Bay of Biscay from the North American colonies (Hemery and Jouanin, 1988). Similarly, *P. huttoni* travels northwards from New Zealand's South Island and across the Tasman Sea towards Australia. Specimens and sightings off northern and northwestern Australia support the hypothesis of a circumnavigation around that continent (Warham, 1981). The birds return to New Zealand in October and November when in some years many are beach-washed along the west coast of the North Island. The virtual absence of dead *P. huttoni* there between April and August, when many dead *P. gavia* are found, together with the numbers off northwest Australia (Halse & Halse, 1988), suggest that *P. huttoni* is a total migrant. Most breeding *P. gavia*, however, seem to stay around New Zealand throughout the year (Wragg, unpubl.), although some, possibly younger birds, also visit Australian seas.

One of the best documented longitudinal movements is the east–west migration of Wandering Albatrosses from their bases in the Indian and South Atlantic Oceans to and from coastal New South Wales (*The Petrels*, p. 43). Many are ringed and of known age and provenance. However, not all the breeders make this regular journey: where do the others go during their off-duty years? And do all fly back home from NSW upwind or downwind through Drake Passage? Some certainly do go against the wind as evidenced by those tracked by transmitters from NSW around southern Australia and into the Indian Ocean between 30° and 45°S, including one that reached its nest on The Crozets just 103 days after the transmitter was attached, having travelled about 20 000 km (Nicholls *et al.* 1995) (Fig. 2.15C). Four birds were followed. One remained offshore for a week before its battery was exhausted. Some foraged in the Great Australian Bight (Fig. 2.15 A, B) and one went as far as west of Marion Island (Fig. 2.15C) and back towards the Crozets before contact was lost. The plots suggest that the birds track round the edge of the rotating high and low pressure cells to keep the wind to one flank and so progress against the prevailing westerlies.

The aggregations off eastern Australia are not solely of adult birds, some immature birds in dark plumage also go there regularly and with repeated capture over the years their plumage development has been recorded.

In the North Pacific the centres of the populations of *D. immutabilis* and *D. nigripes* also shift laterally, and the movements have been plotted from visual evidence and from the recoveries of marked birds by Fisher and Fisher (1972), Sanger (1974a,b), Shuntov (1974), Kuroda (1988) and Robbins and Rice (1974). These last authors based their findings on 324 recoveries of *D. immutabilis* from 165 890 birds banded from 1937 to the end of 1969 and of 399 *D. nigripes* recovered from 48 727 banded from 1940 to the end of 1969. As usual, many pelagic reports came from fishing boats but that bias affects *D. nigripes* much more than *D. immutabilis*. The birds were banded in the Hawaiian Leeward Islands and included many of known age.

Figure 2.16 shows the ranges of adult birds (6 or more years old) on a bimonthly basis. The migration of older non-breeders takes place in April and May, that of the breeders in June and July, but few birds were recovered then, perhaps, as these authors surmised, because they fly quickly to their summering seas and don't get caught by fishermen. In the contranuptial season the adult Laysans range mainly north of 40°N and then overlap the Black-foots. At the gradual southward shift starting in October their distributions tend to become even more distinct: the

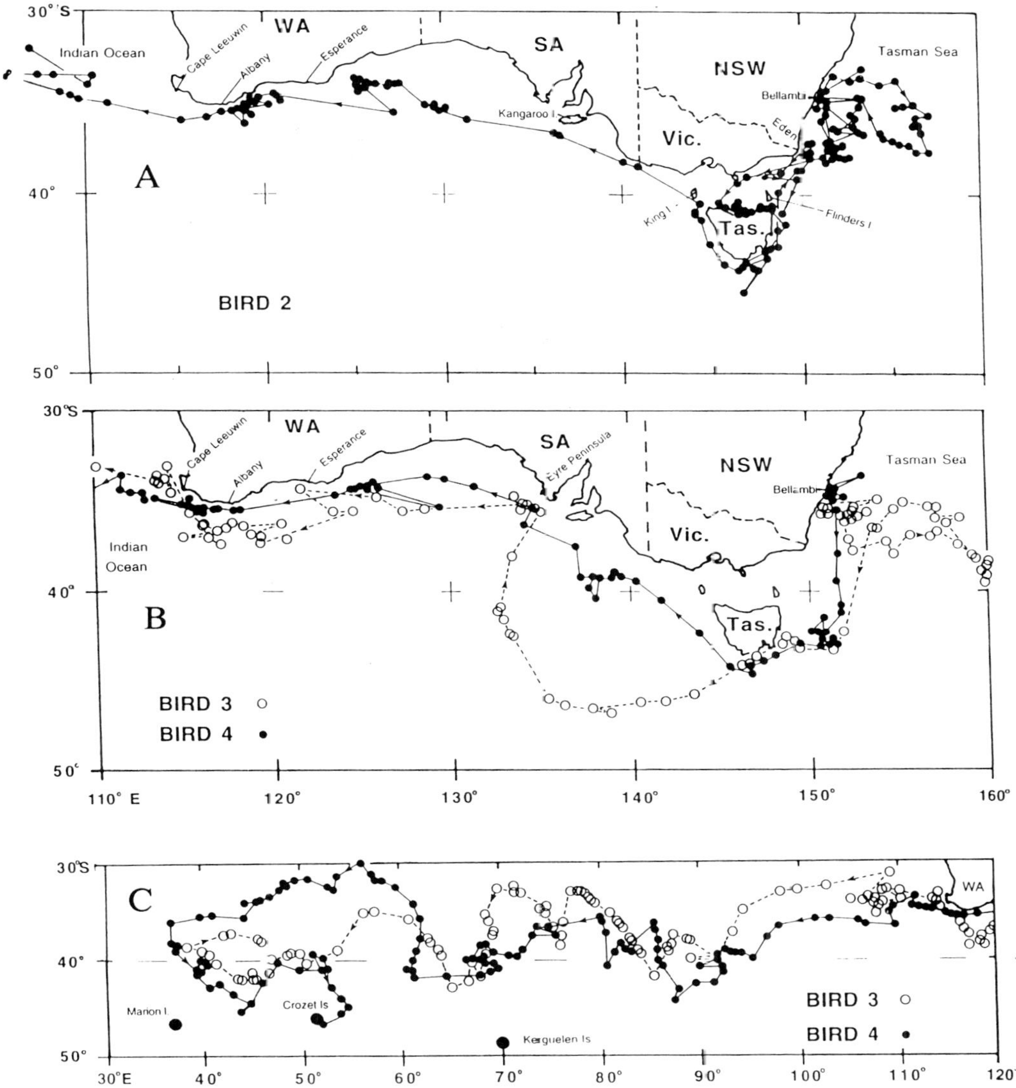

Figure 2.15 *Satellite tracking of* Diomedea exulans *from New South Wales. (A) A female that flew into the Tasman Sea, then to northeast Tasmania and back to NSW before heading round southern Australia to the Indian Ocean. (B) Flights of bird 3 (a male?) and 4, a female, into the Tasman Sea and back, thence past southern Australia to Cape Leeuwin, Western Australia. (C) Flights of birds 3 and 4 from there across the Indian Ocean. Contact with 3 was lost when north of The Crozets and with 4 when at its nest on Possession Island. From Nicholls* et al. *(1995).*

Laysans west of 180°E and north of 30°N, the Black-foots to the east of this meridian and south of 30°N. Their ranges again tend to coincide in February when both are foraging frequently and over shorter ranges while rearing chicks. Sanger (1974b) thought that it was quite possible for breeding birds to forage off western North

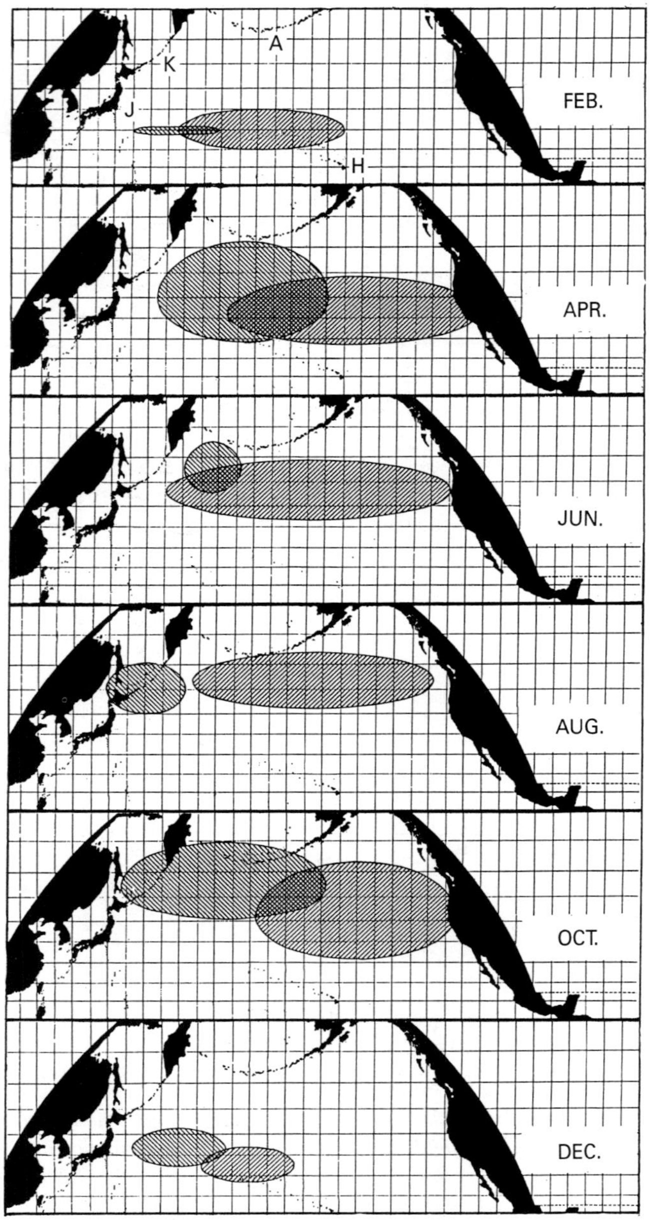

Figure 2.16 *Comparative bimonthly distributions of adult* Diomedea immutabilis *(left) and* D. nigripes *banded at Midway Atoll and other islands of the Hawaiian Leeward Chain. Shaded areas enclose ±1* SD *from the mean co-ordinates of all pelagic band recovery locations. A, Aleutian Is.; K, Kurile Is.; J, Japan; H, Hawaiian Is. From Robbins and Rice (1974).*

America and get back to their nests within 2–3 weeks, although most probably only went 2400–3200 km.

Robbins and Rice's (1974) data for younger birds were less extensive but showed that their winter dispersion tended to be much wider than those of the adults. Young birds of both species during their first summer and autumn at sea remain mostly south of the adults and Laysans in their first winter are closer to Japan than the old birds. Both species move east and west during their first year but by the end of their first winter the Laysans have moved west towards Japan while the young Black-foots shift to the east. In their next four summers the Laysans gradually shift from Japanese seas to the adult's contranuptial area and by their third winter at sea some visit their natal islands in March. Young Black-foots in their later prebreeding are in the eastern Pacific and closer to the coasts of North America than the adults.

Fisher and Fisher (1972) related the concentrations of Laysans in summer off eastern Japan to highly productive waters where the Oyashio and Kuroshio Currents converge and they also believed that the western Aleutians were a major feeding ground in the summer, another region where turbulence generates high planktonic activity. Sanger (1974b) also pointed out that the abundance of *D. nigripes* off western North America correlated with seasonal changes in the upwelling patterns along that coast and its preference there for cooler SSTs, the region delimited by the 14°C and 22°C surface isotherms being much larger in the eastern Pacific than in the west. The distribution of commercial fisheries is one confounding problem.

C *Latitudinal migrants*

An example of a petrel that migrates seasonally across the latitudes without crossing the Equator is *Puffinus opisthomelas*. This breeds off Baja California in the boreal winter and its postnuptial migration is mainly northwards along the California coast where the birds usually reach Monterey Bay from mid-October to late November. A few travel as far as British Columbia during years when SSTs off California and Oregon are unusually warm.

Another petrel shifting northwards after nesting is the Bonin Petrel *Pterodroma hypoleuca*, a winter breeder. It leaves its nesting places in May and June to feed east of Honshu but has gone by the end of August. During this northern migration it selects waters with SSTs 16–29°C (Tanaka and Kaneko, 1983).

The Kerguelen Petrel *Lugensa brevirostris* may not follow the usual pattern of Southern Ocean species in moving north in the winter. Lambert (1984) suggested that after breeding the birds move down to the edge of the pack ice. Whitehouse and Veit (1994) found it off the Antarctic Peninsula in winter and Ainley *et al.* (1994) reported its year-round presence in open water of the Weddell Sea.

The Northern Fulmar *F. glacialis* is another species some of whose populations move south during the boreal winter. Analysis of 300 recoveries of British ringed birds by Macdonald (1977) showed that the fledglings disperse widely into the Atlantic Ocean and European Arctic seas and in their second year reach the rich feeding grounds off Newfoundland and to the south and west of Greenland, and not until their fourth year does their range contract towards the breeding colonies. All the British birds in the western Atlantic appear to be prebreeders; the breeders must remain much closer to their colonies (*The Petrels*, p. 238). Fisher (1952a, p. 447) proposed that prebreeding European fulmars congregated in 'nursery areas'—that is, the Newfoundland Banks and Varanger Fjord—but Macdonald's data, admittedly biased by the many returns from fishing boats, did not confirm this: it remains unproven.

In the eastern Pacific Northern Fulmars shift south and appear in large numbers off California from October through to March or April. The populations here reach a peak in December and January (Fig. 2.17), with perhaps 225 000–360 000 birds. A low in mid-winter of 35 000–95 000 birds indicates movements through Californian waters of birds wintering off Mexico (Briggs *et al.*, 1987).

D *Transequatorial migrations*

The switch from one hemisphere to another allows migrants to enjoy perpetual summer—feeding on seasonal food flushes when nesting and concentrating in the opposite hemisphere after breeding where upwellings and fronts provide food to support the large numbers without interspecific competition normally becoming excessive. For a summary of the migrations of *B. bulwerii, P. gravis, P. griseus* and *C. diomedea* in the Atlantic (Bourne, 1995).

1 *North to South*

a. Puffinus puffinus *Manx Shearwater.* Although something like 262 603 Manx Shearwaters had been ringed by the end of 1989 and over 3600 recoveries made, the fine details of their migratory pathways have yet to be revealed, partly because very few birds are recovered in mid-ocean and partly because most that are found are on beaches in well-populated areas. Thomson (1965) and Brooke (1990) attempted to unravel these patterns. A major segment of the population spends the northern winter off the coasts of Argentina, Brazil and Uruguay. This includes many fledglings in their first autumn and 1-year-olds (some of which spend the northern summer there as well), together with older birds which alternate between their northern breeding places and the South American wintering one.

The outward and return routes appear to differ, those leaving Europe passing close to the bulge of Africa before cutting diagonally across the central Atlantic to South America helped by the northeast Trades, and perhaps returning on pathways more to the west and thence north through the central North Atlantic to the Welsh, Irish and Scottish colonies. The evidence for these return routes could be much stronger, however. In addition, there is a significant presence nowadays of birds off the New England and eastern Canadian coasts. Some are 1-year-olds spending the northern summer there and sharing the seas with masses of *P. gravis* and *P. griseus.*

b. Calonectris diomedea *Cory's Shearwater.* Like the Manx, Cory's Shearwater also uses the offshore waters of the South American Atlantic coast as a contranuptial

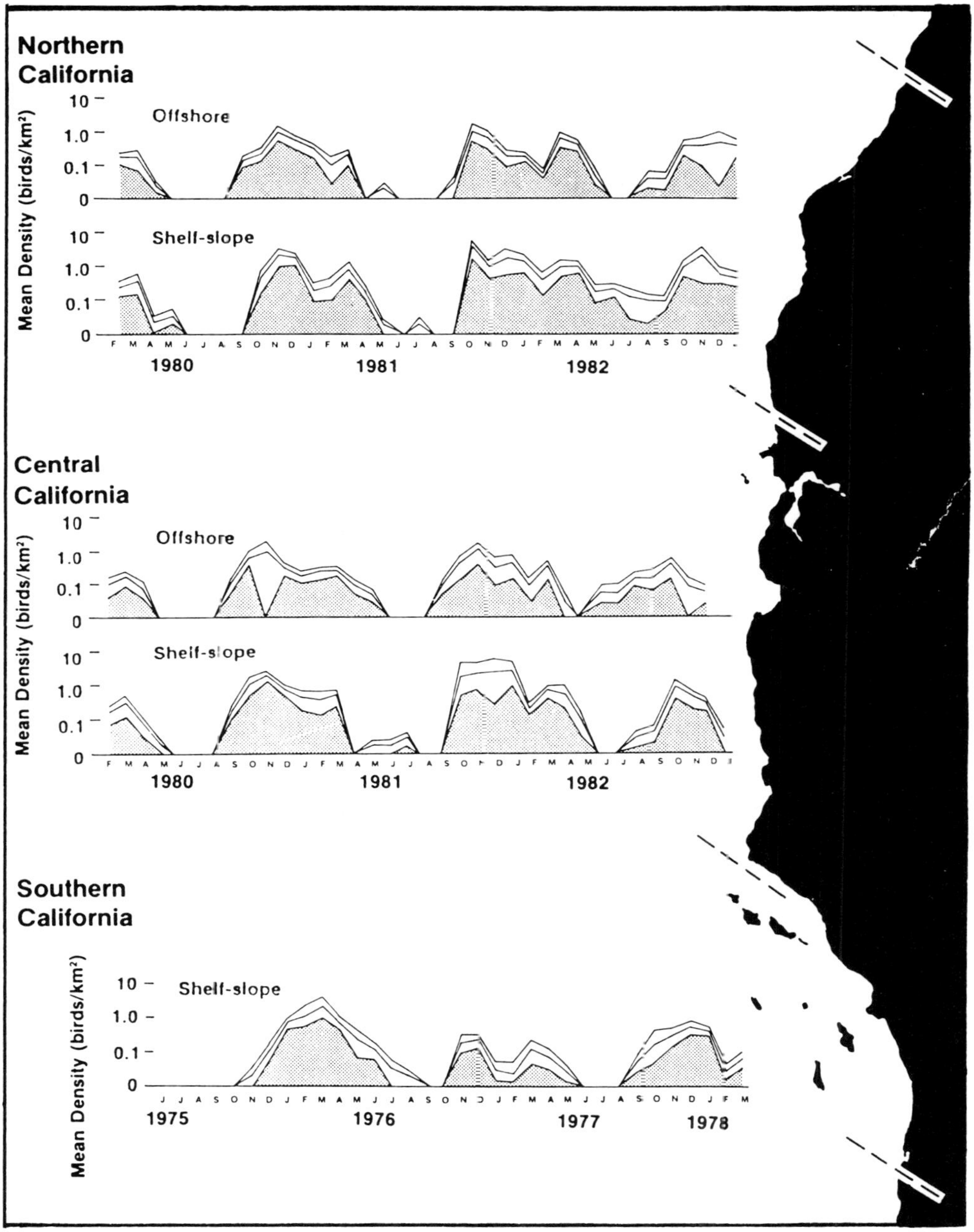

Figure 2.17 *Comparison of monthly mean densities (birds km^{-2}) of* Fulmarus glacialis *in three regions off California. Three curves give mean densities ±1 SE, shaded values lying >1 SE below the mean. After winters of high abundance (1976 and 1981) some birds summered off California. From Briggs* et al. *(1987).*

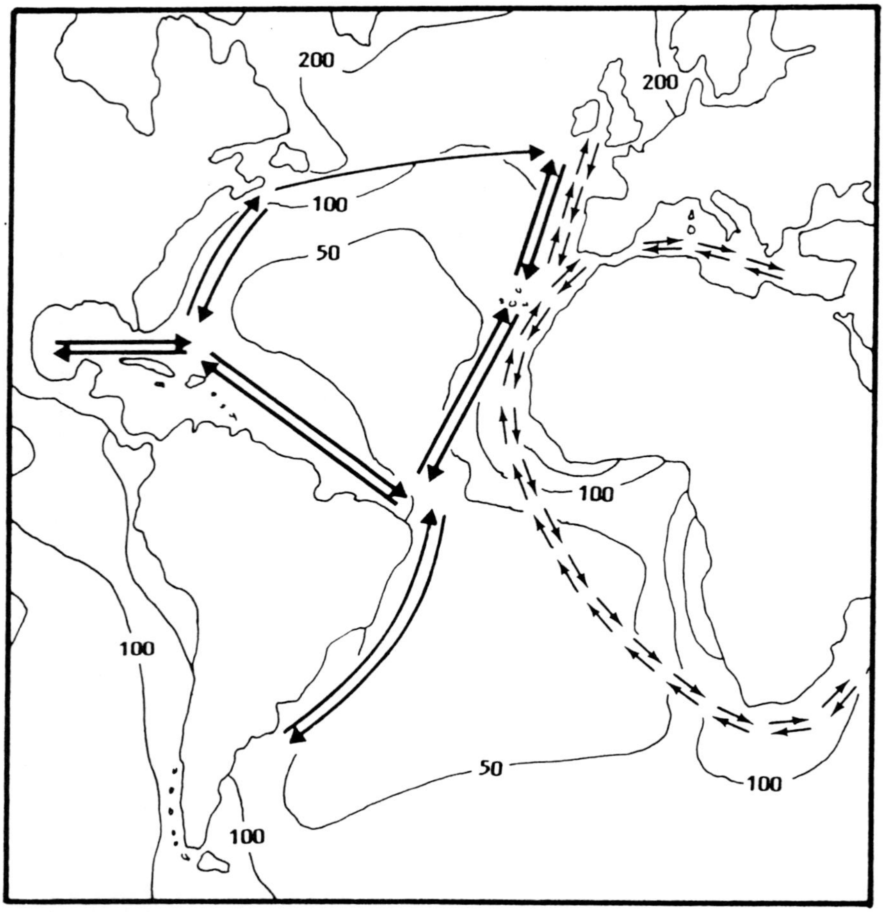

Figure 2.18 *Possible routes taken by the two migratory subspecies of* Calonectris diomedea *as deduced by Mougin* et al. *(1988). Thick arrows,* C. d. borealis; *thin arrows* C. d. diomedea.

feeding ground. Mougin *et al.* (1988) concluded that the two migratory subspecies had separate wintering grounds, the typical race leaving the Mediterranean to winter off southern Africa with some penetration along the Agulhas Current into the Indian Ocean, while *C. d. borealis* headed for South America to feed off the coasts of Brazil south to Argentina (Fig. 2.18). The adults of both were thought to retrace their steps in returning to the breeding places in March. Immatures of both are also found in the southern regions, but Mougin *et al.* (1988) suggested that those of *borealis* head north at the start of the southern winter, spend some time in warm waters off Central America, moving still further north to coastal waters off eastern USA and Canada before finally returning to the breeding places either via the North Atlantic and Europe or retracing their flight to South America and thence northeastwards across the central Atlantic. Immature *C. d. diomedea* retrace their flights back to the Mediterranean although some spend the boreal summer off the North Atlantic coasts of Europe (Yesou, 1982).

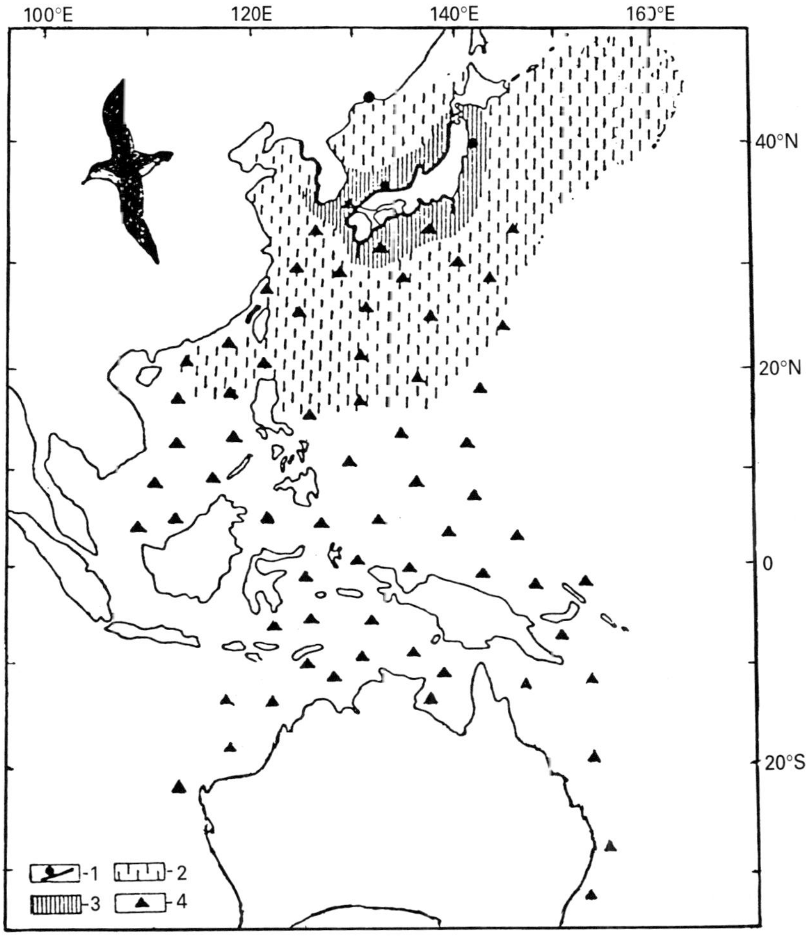

Figure 2.19 *Migration of* Calonectris leucomelas. *1, Breeding places; 2, boreal summertime distribution; 3, main concentration areas in summer; 4, postnuptial distribution. Adapted after Shuntov (1974).*

c. Calonectris leucomelas *Streaked Shearwater.* The Streaked Shearwater also moves south after breeding, the bulk of the population evidently wintering in Philippine and Indonesian waters. Some pass into the Indian Ocean and reach Sri Lanka, but more appear to feed off the northwest coast of Western Australia and others off Queensland and fairly regularly as far south as Sydney (Fig. 2.19).

d. Oceanodroma *storm petrels.* Two summer-breeding Japanese storm petrels also shift south for the northern winter and like *C. leucomelas* some enter the Indian Ocean. These are *Oceanodroma monorhis* and *O. matsudairae*. Both are adapted to warm seas and both have been identified over the whole of the central Indian Ocean, *O. matsudairae* as far west as Somalia, *O. monorhis* as far as the Arabian coast (Bailey *et al.*, 1968). These authors suggested that *O. monorhis* might reach the Indian Ocean via the Straits of Malacca, *O. matsudairae* by a more southerly route through the Indonesian chain, an idea borne out by later sightings in the Timor Sea, off northwest Australia, and by specimens from Christmas Island.

The storm petrel *Hydrobates pelagicus* is a distinct transequatorial migrant with birds from the Scottish colonies having been recovered in South African waters and the routes down the western seaboards of Iberia and Africa in both directions being supported by many sightings (van Oordt & Kruijt, 1953; Cramp & Simmons, 1977). The bulk of the population appears to spend the boreal winter in the cool Benguela Current down to 38°S. Because specimens collected in mid-winter off West Africa were all juveniles, it has been suggested that birds of the year winter north of the Equator off West Africa rather than continuing south with the older birds.

2 *South to north migrations*

a. Oceanites oceanicus *Wilson's Storm Petrel.* From its circumpolar breeding stations this bird performs its transequatorial migrations through all of the world's oceans, being the only petrel to do that. It spends the austral winter in temperate and subtropical seas, not in Arctic ones, as might perhaps be expected of an Antarctic breeder. The ability of a sparrow-sized seabird to undertake such vast journeys on an annual basis is the more remarkable as its low-aspect ratio wings limit its ability to travel by dynamic soaring, although it uses slope-soaring extensively. It may also be wind assisted for parts of its journey. Presumably, that the bird can find surface plankton almost continuously makes this migration possible, but a relatively slow crossing of the wide, food-poor tropical seas would seem particularly dangerous. The pathways used appear to be quite complex and few birds have been ringed. The birds are uncommon among beach-washed debris, their movements having been deduced mainly from specimens and sightings at sea. In the Atlantic the migration was described by Murphy (1918), for the Atlantic and Pacific by Murphy (1936), for Australian seas by Serventy (1952) and Wood (1990c) and worldwide by Roberts (1940).

It has been suggested that the younger birds winter at lower latitudes than the older ones. Lee (1987) has shown that the species is plentiful over the continental shelf off North Carolina in the austral summer, for example 351 were counted on an offshore trip on 20 December, 1986 under normal weather conditions. At that date breeders are back at their nests: it seems unlikely that these 'summering' birds were breeders on a 'sabbatical year' off.

The ramifications of this storm petrel's movements will be difficult to untangle. There is not only the question of where the different age classes go but also whether there is any sexual differentiation, as observations of 63 birds collected at the Marshall Islands suggest (Huber, 1971). All proved to be adult but those taken in April were predominantly female, May birds mainly male, suggesting different timetables for the sexes.

One consequence of this migration pattern is that it removes the birds to warmer climes posing fewer thermoregulatory problems and from seas subject to violent winter storms that might blow them far down wind and prevent feeding. Present knowledge of the movements in each ocean is summarized below.

i. Atlantic Ocean. In April the vanguard moves rapidly northwards and by the third week reaches the warm coastal waters of the USA evidently keeping to the west side of the ocean with some birds entering the Sargasso Sea. By June they are abundant along the whole of the Atlantic coast of North America with thousands to be seen off vantage points like Cape Cod. The main concentrations are off Maine and the shelf break off these northeast coasts as far as Nova Scotia. Some may cross the North Atlantic to the Iberian coast, joining birds from a smaller movement that appears to travel up the eastern Atlantic and concentrate in the Bay of Biscay.

Murphy (1918) judged that all 49 birds collected off Bahia between April 15 and May 1, 1916 were but 2 months out of the nest and with others he determined as juveniles, whereas those taken far offshore were adults, suggesting that these might migrate pelagically, the young offshore.

The return begins in September, the movement being mainly back down off South America and by November some birds are back in west Antarctic waters north of the pack-ice. Bourne (1963) pointed out that as *Oceanites oceanicus* withdraws towards the south its place is taken in central Atlantic waters by *Oceanodroma leucorhoa* and in offshore West African seas by *H. pelagicus*.

ii. Indian Ocean. Birds fly north offshore past the coast of Western Australia in April and May but the movement appears to be broadbased, the species being seen off Amsterdam Island from February to April. Many winter in the Somali Current during the Southwest Monsoon, especially along the edge of the shelf. Bailey (1968) thought that the main migration was through the central region of the ocean and arrivals in Sri Lankan seas are mainly of single birds, not of flocks (Phillips, 1955). He found many wintering off Colombo, 51 birds being released from a ship on a single night after being dazzled by its lights. The petrels depart from Sri Lanka during the first 2 weeks of November after collecting in very large flocks, sitting on the water in calm weather, and they concentrate similarly off the Arabian coast in September before the southwards movement (Phillips, 1955; Bailey, 1968).

iii. Pacific Ocean. The main body here appears to move less far northwards than it does in the Atlantic Ocean. In the eastern Pacific many travel north in April and May. Pitman (1986) saw very few during his 4333 h of marine bird observations in the eastern tropical Pacific and the birds are rare in Hawaiian seas so the movements are evidently mainly west of 160°W. Huber (1971) recorded a northward movement past the Marshall Islands and a similar passage has been seen off New Caledonia. The return route seems unclear but large numbers are in the Humboldt Current in August and Crossin (1974) considered that fat, healthy birds collected near the Marshall Islands in October and November with somewhat enlarged gonads were on their southward migration.

In the southwest, hundreds are seen in April flying off New South Wales in some years but Wood (1990c) saw most over deep water there and speculated that the main northwards movement on the west Pacific is in the pelagic zone. While many appear to winter off northern Australia and New Guinea (Serventy, 1952; Coates,

1985), there appear to be no big concentrations like those in the North Atlantic and Indian Ocean. Some go further and presumably the few that reach Japanese seas travel in the west Pacific too.

b. Pelagodroma marina *White-faced Storm Petrel.* This is also found in all the oceans, but only the race *P. m. dulciae* is a clear transequatorial migrant, the southwest Pacific form, *P. m. maoriana*, only marginally so, and the Atlantic forms not. A specimen of *P. m. maoriana* was taken in the Gulf of Guayaquil at 2°35′S, 81°20′W, and the birds well known as visitors in the boreal winter to the eastern tropical Pacific (Loomis, 1918; Crossin, 1974), have been thought to belong to this race, a view supported by the recovery of a banded Chatham Island's bird at 9°52′S, 88°42′W (Imber, 1984). He pointed out that this supports Crossin's suggestion that the southwest Pacific birds move to the Humboldt Current and then fly north and west of the Galapagos. Pitman (1986) found them along the Equatorial Front with concentrations around the Galapagos and between there and Ecuador. A more direct return route may be taken to New Zealand, as many are back in home waters within a month of their peak numbers in Galapagos seas (Imber, 1984).

Some from eastern Australia perhaps go less far. Imber (1984) documented sightings of *P. marina* mainly in July in an area north of New Zealand towards Norfolk, the Kermadec and Fiji Islands; others are in the Coral Sea in May, so that part of the population may winter along the Subtropical Convergence here.

Western Australian birds winter in the northern Indian Ocean and Arabian Sea between May and September, being found well offshore, often in small flocks (Voous, 1965; Bailey, 1966). Birds banded on Mud Island, Victoria, have also been recovered to the west suggesting that Bass Strait populations may also go to the Indian Ocean but, in contrast, an adult *P. m. dulciae* was recovered far to the east on a New Zealand beach in May, 1983 (Imber, 1984).

The movements of the Atlantic populations—*P. m. hypoleuca* (Salvage Is.); *P. m. eadesi* (Cape Verdes) and *P. m. marina* (Tristan da Cunha and Gough Is.) are little known but they appear to disperse rather than migrate after breeding. More data are needed. Some cross the Atlantic towards eastern USA where they are found 70–650 km from the coast over the continental slope, between August and October, and mostly north or west of the Gulf Stream (Watson *et al.*, 1986).

c. The southern migratory shearwaters. Six species of *Puffinus* and one *Procellaria* cross the Equator, some to the far north almost to the pack-ice. It was formerly thought that *Puffinus tenuirostris* followed circular or looping pathways (e.g. Serventy, 1953), but more recent studies indicate that broad-front, more direct routes are taken, earlier results being biased by the greater likelihood of recoveries and sightings around the peripheries of the oceans, most recoveries being of birds of the year.

i. Procellaria parkinsoni *Parkinson's Petrel.* This is the only *Procellaria* to make regular transequatorial migrations. It was first found in the northern hemisphere during the Californian Academy of Science Expedition of 1905–1906 to the Galapagos (Loomis, 1918) and was subsequently recorded widely in the eastern tropical Pacific (ETP) and off the coasts of southern Mexico, Central America and Ecuador between *c.* 14°N and 5°S and out along the Equatorial Front to about 100°W (Jehl, 1974a; Pitman, 1986). Pitman and Ballance (1992) found it plentiful in areas of upwelling associated with the Gulf of Tehuantepec and Panama throughout the year except for January.

ii. Puffinus creatopus *Pale-footed Shearwater.* These leave their breeding stations on the Chilean islands in March and April to spend the austral winter off western North America, occasionally reaching the Bering and Alaskan Seas. Their numbers reached about 130 000 birds off central California with about 400 000 off the southern sector. They return to the nesting grounds in November and December. In the central Pacific they occur as far as 160°W (Pitman, 1986), so that they are not confined solely to offshore seas.

iii. Puffinus carneipes *Flesh-footed Shearwater.* Considerable numbers spend the austral winter in the Arabian Sea and Gulf of Oman (Bailey, 1966). The birds presumably mostly come from the Western Australian colonies; at least one of the typical race has been identified from Colombo. The birds leave their breeding places in late April and early May to appear in the Arabian Sea by late May, perhaps by tracing a great circle route across the southern sector of the ocean. Whether those that nest in the Recherche Archipelago off southern Australia join this movement is not known.

However, there have been many recoveries of adults banded at Lord Howe Island (ssp. *hullianus*), showing their migration to the northwest Pacific, most leaving the island in May, but some are in Korean waters as early as late March and in the Sea of Japan, the Okhotsk Sea and offshore Japan in June. According to Shuntov (1974, p. 183), in June the birds are concentrated in the northwest of the Sea of Japan where there are flushes of *Cololabis* fish and squid. In June through to August they are found both over the shelf and in deep waters, both here and in the Okhotsk Sea, where the species is common in the northern summer. Smaller numbers also occur off British Columbia in April and May and again in August and September, suggesting a separate route through the central Pacific to and from the breeding places, perhaps mainly of New Zealand birds.

iv. Puffinus pacificus *Wedge-tailed Shearwater.* The movements of the many populations in the Indian and Pacific Oceans on both sides of the Equator are poorly understood. Light-morph birds breed mostly in the northern hemisphere and with few sightings south of the Equator, so that long-distance migration appears to be mainly restricted to southern, dark-plumaged birds although both morphs are present, for example, in the ETP (Pitman, 1986).

Recoveries of birds banded off NSW include seven from the Philippines during the contranuptial period, including one found dead 17.5 years after banding as an adult 6500 km to the south. Some eastern Australian birds perhaps travel further and are among the dark-phase ones plotted by King (1974) in the central Pacific during June and July. Jenkins (1979) reported that *P. pacificus* seems to leave and return about the same time at the islands of the southwest Pacific, and he noted that their return in force at the end of October tallied well with King's observation of the large decrease in the eastern Pacific at the end of September.

Jenkins also pointed out that the NSW *P. pacificus* return about 6 weeks before those of the southwest Pacific colonies, and that they must do so along a different route, for the species is absent from those islands' waters in August when the Australian birds would be passing through.

Light-phase birds from the Hawaiian Islands evidently move south to the eastern Pacific in the northern winter, perhaps out along the Equatorial Countercurrent, returning via the North Equatorial Current (King, 1974).

v. Puffinus bulleri *Buller's Shearwater.* This species nests in the austral summer. It is an easy bird to recognize at sea so most sightings there can be relied upon. In the austral winter its main habitat is in subarctic seas and a notable feature is the marked increase in sightings there in recent years almost certainly due to more observers but also to a build up of numbers at the breeding ground.

The changing distributions around New Zealand were outlined by Jenkins (1988). The birds return to northern New Zealand seas in September and the young birds probably leave in May. This may account for the large rafts seen off the nesting islands, near which there was a decline in birds in April, suggesting a departure of breeding birds in that month. In June, July and August they are virtually absent from local waters. A sighting of 15 on 31 August at 32°S, 175°W by Chapman (1981) may have been part of the van of the returning stream.

In the northeastern Pacific the distribution of the birds and their timing have been well documented by Wahl (1985). In June he already found small flocks resting on the surface in the mid-Pacific between 35–45°N and 160–180°E. Guzman and Myres (1983) reported 100 birds only 140 km west of the British Columbian coast on 29 August and 200 at 50°N, 145°W on 27 August, and the species is now a regular autumn migrant off British Columbia (Campbell *et al.*, 1990).

Throughout August, September and October considerable numbers feed offshore from California to British Columbia. Wahl suggested that a flock of about 1300 that joined other shearwaters in August 1976 off the Channel Islands, southern California, might have been the van of the southern leg of the migration of the breeding birds, but later sightings must have mostly been of non-breeders. Peak numbers off California were in August and September (Briggs *et al.*, 1987). Here they feed over upwellings along the shelf break but become concentrated more inshore when the upwelling slackens and Briggs *et al.* suggested that numbers are lowest in coastal waters there during years of warm SSTs.

In the northwest Pacific Buller's Shearwater was unrecorded before 1951 when the first specimen was collected near the Kurile Islands. Sleptsov (1960) found the species fairly widespread between 150 and 160°E and 30 and 46°N. Nakamura and Hasegawa (1979) reviewed the position and reproduced a photograph showing some 50 birds of a flock at about 44°N, 173°E in July 1978. It seems that in the austral winter this shearwater is widespread off northern Japan particularly around and south of the Kuriles to about 30°N, but also as far east as 151°W.

The status of the birds seen off Peru and Central Chile in February and March is unclear: Blake (1977) regarded them as only casual visitors, but Ainley (pers. comm.) saw large flocks of moulters there in May 1978.

The routes taken across the Pacific are unknown. Probably the movements are on broad fronts in both directions, an easterly one taking the birds to the northwest Pacific and a central one to the eastern areas.

vi. Puffinus gravis *Greater Shearwater.* There are many accounts of this species off the eastern seaboard of North America, in the North Atlantic, North Sea and off Greenland (Rees, 1963; Grafe, 1973; Cramp & Simmons, 1977; Brown *et al.*, 1981; Lee, 1986) but the migratory pathways are still not completely defined, although recent information seems to support the pattern mooted by Voous and Wattel (1963).

The birds leave the southern islands from about April and May with fledgelings following from May until August (Rowan, 1952). Eakin *et al.* (1986) believed that they encountered part of the northern movement in early May during a transit from Buenos Aires to the northern edge of the Antarctic Convergence, when many were flying north. The species has also been seen between 10 and 25°S and 20 and 35°W flying north or northwest towards the Brazilian coast in May and June with others further east heading northwest (Barritt, 1992). These and other sightings indicate that these shearwaters fly northwest past the horn of Brazil and the West Indies, where mass mortalities have been reported on Surinam and Trinidad (Collins & Tikasingh, 1974; Mees, 1976), and then turn up off the Atlantic North American coast in May and June. By then they are widespread in the western North Atlantic and in July and August are as far as 66°N in Davis and Denmark Straits and have also spread east to European coasts south to the Iberian Peninsula.

Very large numbers winter and moult off the Grand Banks, Newfoundland, Nova Scotia and Greenland. An estimated 200 000 over George's Bank as late as 11 November, many satiated and sleeping, counted by Powers and Van Os (1979) must have been non-breeders. These authors suggested that the birds were flocking ready for migrating south. Bourne (pers. comm.) points out that the many along the shelf break in the Bay of Biscay long after breeding has begun must be non-breeders too, as must those moulting off Tierra del Fuego in January. Bourne also reports them plentiful and moulting in Falklands seas in March to May. Probably some age classes do not leave the South Atlantic.

Return routes are less clear. Some fly south in the western side of the ocean but those seen apparently on their way south in mid-Atlantic in September (Bourne, 1970, 1995), in Azores seas, and at 34°N, 31°W may have been moving on a broad front as suggested by Bourne. Large numbers seen off Rio de la Plata in mid- to late November by Brenning and Mahnke (1971) may have been part of the late returning stream.

vii. Puffinus griseus *Sooty Shearwater.* The Sooty Shearwater migrates to the North Atlantic and North Pacific Oceans (Figs. 2.20, 2.21). Few have been ringed, most being adults of unknown status from the Snares Islands, and all 10 birds recovered from there and elsewhere in Australasia have been in Japan, seas south of the Aleutian Islands, and off North America.

Successful breeders leave nesting areas in mid- to late April, fledgelings up to a month later. Whether all start migrating immediately is not clear, but during April birds stream northwards past the South Island of New Zealand and, according to Shuntov (1974, p. 79) so do others far to the east. King (1970, p. 60) recorded the

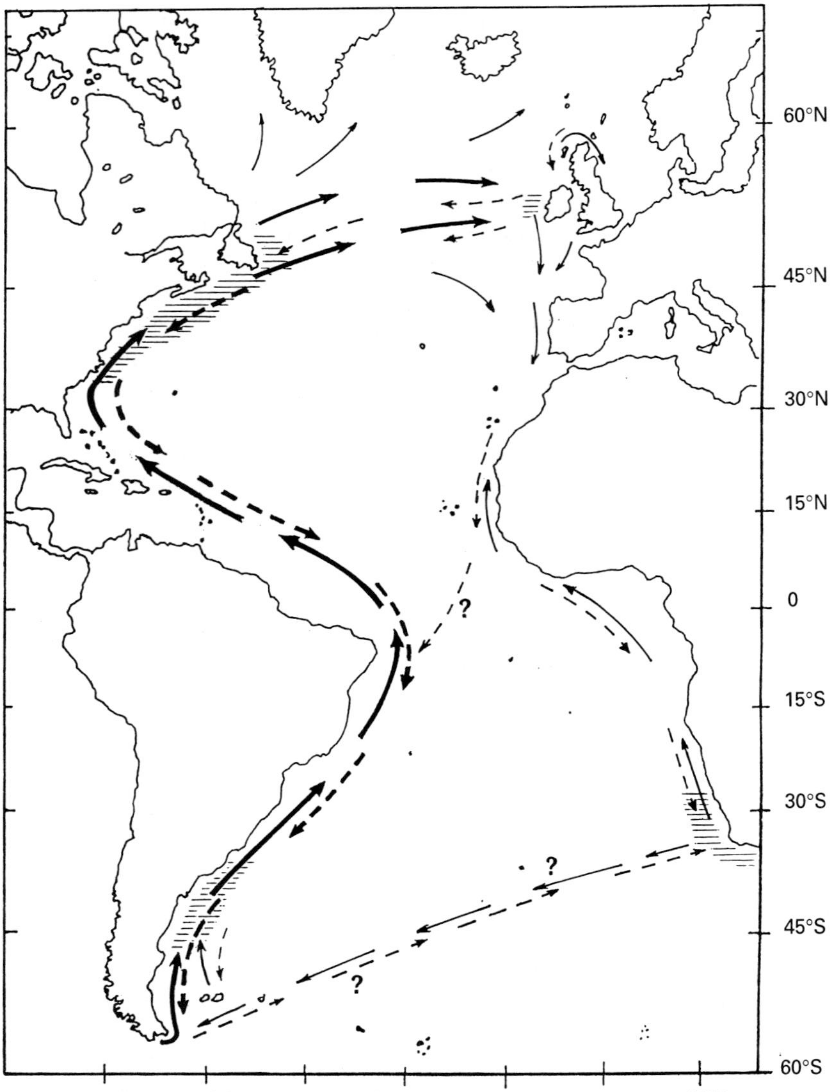

Figure 2.20 *Possible migration pathways of* Puffinus griseus *in the Atlantic Ocean. Main areas with concentrations out of the breeding season are shaded including one off the Valdez Peninsula that may be a contranuptial feeding ground.*

northward movement in the central Pacific, being most abundant in March, numbers declining in April, and the movement having almost ceased in May. The birds move through this region *en masse* (Ainley, pers. comm.).

Most birds reach the North Pacific between the end of April and the beginning of June, and Watabe *et al.* (1987) found many off Honshu 3–5 May, more than a month before the main body of *P. tenuirostris*, which tended to concentrate further north off

Hokkaido. Kuroda (1960a) also noted that *P. tenuirostris* preferred SSTs of 3.5–7°C, *P. griseus* 11–12.5°C, the coastal waters off North Honshu, where currents meet, being the main wintering quarters for *P. griseus*. Watabe *et al.* enccuntered some here in June that were emaciated and floating helplessly on the sea, but more *P. tenuirostris* were affected, with 55% of 1468 birds like this.

Some *P. griseus* enter the Sea of Japan and the Okhotsk Sea and reach both sides of southern Kamchatka and the Aleutian Islands (Ogi *et al.*, 1981). A major 'wintering' area lies east of Japan and off the Kurile Islands. By June many are in Alaskan waters and south of the Aleutians, but these could well include South American birds.

The birds from the islands near Cape Horn apparently also fly north in the Humboldt Current, pass through Californian seas and may appear off Oregon and British Columbia, then reach the Gulf of Alaska, presumably returning by the same routes. Numbers off California reach a peak in May to July, then tend to decrease only to peak again in August and September, perhaps marking the northern and return movements (Serventy, 1953), and/or the passing of different age classes.

Guzman and Myres (1983) saw the northwards movement in waves off British Columbia in May. They were feeding furiously, even taking discards from fishing vessels. As their peaks in numbers dwindled, groups of 1–100 birds flew low over the water heading northwest to north, in long, snake-like flocks.

However, the occurrence of South American birds off California has been questioned because sightings off Central America are rare (Jehl, 1974a), while between March and June many feed off Peru and Chile (Jehl, 1973b). Thus birds from colonies on the west coast of South America may not move far. Possibly the huge flocks flying off southern Chile at 44°S on 25 March were heading to join these feeding aggregations (Bourne and Dixon, 1973).

The numbers off California can be vast. An 'instantaneous' estimate of 2.7–4.7 million during the early summer was given by Briggs *et al.* (1987) who pointed out that the true figure could be 10 times more. Briggs *et al.* found rather few birds more than 100 km out, and in June to September 1982 inclusive they saw none from 200 to 400 km offshore. Jehl (1973a) over the 1000 fathoms mark in mid-October, saw virtually no *P. griseus* during cruises off southern California going out to 320 km. Therefore, unless major migrations travel much further out, or 1982 was atypical, most birds pass down the coastal upwelling zone rather than within the California Current beyond the continental slope.

The southward movement from Alaska and Japan evidently begins at the end of August. The great numbers passing down the coasts of western North America and the recoveries of Australasian birds, show that these are not all from the South American colonies. Shuntov (1974, p. 87) pictures the southward stream as dividing off California with one arm proceeding down the South American coast, the other southwest towards New Zealand, crossing the Central Pacific on a wide front. There Pitman (1986) found many along the Equator between 160 and 130°W. In October, King (1970, pp. 60,79) recorded the highest numbers flying south past Hawaii, but there were only about half the birds that had been seen during the northward passage and the return flights had almost finished in December.

Sightings of Sooty Shearwaters in the North Atlantic were reviewed by Phillips (1963) and Cramp and Simmons (1977). We cannot be sure that the North Atlantic birds all come from the South Atlantic, and the only recovery to the north to date is a

Falklands juvenile ringed on 3 May off Barbados, 9100 km away, on 1 June the same year. There is at least one behavioural difference, however, in that the birds have already moulted their flight feathers by the time they reach eastern North America; those migrating to the North Pacific moult there (Brown, 1988a).

The Atlantic birds start to move north in March (Fig. 2.21) and evidently follow the east coast of South America where flocks have been seen in April and May. The species reaches the waters off eastern North America in April and most appear to be concentrated off New England and Newfoundland in June. There is an eastwards drift across the North Atlantic in July through into September. Some birds pass round the north of Scotland to enter the North Sea (see Fig. 2.1), a movement that may have increased in recent years. The position there was reviewed by Hall *et al.* (1987) and Rasmussen (1985), among others. Most are found in the western North Sea with a peak in August. Some evidently leave via the English Channel but most cross back into the Irish Sea and are in good numbers off western Ireland in September and October, possibly feeding up in readiness for the return flight.

From eastern North America the southwards movements begins in August and by late August and early September birds are flying south off South America. There is also a southerly passage off Africa, perhaps of birds that crossed the Atlantic earlier and/or leave the North and Irish Seas. Where these birds come from is unclear.

Phillips suggested that all the Atlantic transequatorial migrants were non-breeding birds. This raises the question of the status of those that Jehl (1974b) found in June–August over the shelf waters off Argentina, mostly north of 45°S, and those off the Valdez Peninsula in early August, with flocks as far as 150 km offshore (Fig. 2.20).

Some *P. griseus* occur at sea off the southwest corner of Australia and small numbers reach the northwest reaches of the Indian Ocean, even as far as Eilat in the Gulf of Aqaba (see Fig. 4.1). They also inhabit South African seas, being rare off Natal

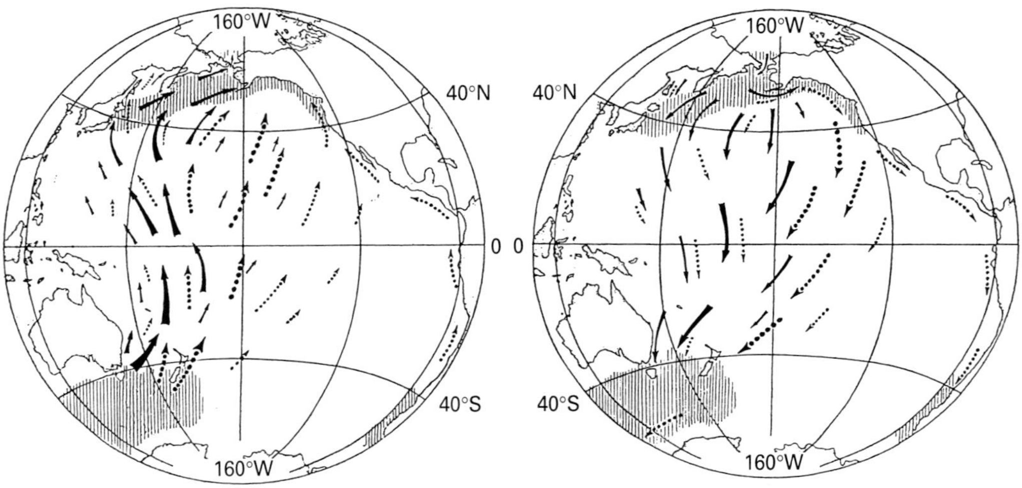

Figure 2.21 *Possible migratory routes of* Puffinus tenuirostris *(heavy arrows) and* P. griseus *(dotted arrows) in the Pacific Ocean, nuptial and contranuptial feeding areas shaded. Partly from Shuntov (1974) and Maruyama* et al. *(1986).*

but present throughout the year off the south and west coasts and abundant there in the austral winter (Ryan & Rose, 1989). The provenance of these birds is quite unclear. Some must be non-breeders or failed breeders, but whether of South American or Australasian affiliations, or from the flocks off Antarctica, has yet to be discovered. Liversidge and LeGras (1981), for example, noting the considerable variation in numbers between years, suggested that birds from the South Atlantic colonies moved first to Cape seas, then across towards South America with the prevailing winds, thence to the North Atlantic.

Sooty Shearwaters are common among South African beachwashed debris. An examination by Cooper *et al.* (1991) of 244 freshly dead revealed that 86% were in primary moult, a process which these authors calculated began, on average, on 2 January and was completed on 10 June, so that, unless they had been moulting while feeding chicks, the birds cannot have bred that year and they deduced that they were prebreeders. This is not surprising: most beachwashed petrels are immature, which does not mean that the flocks offshore are all of immatures too.

Cooper *et al.* (1991) hypothesized that the birds in the Atlantic, including those off western South Africa are all prebreeders, including those migrating to the North Atlantic, where Brown (1988) examined 603 taken at sea and deduced these as being pre- or failed breeders. Cooper *et al.* further suggested that immatures of all populations spend the austral winter in the Atlantic whereas all breeding birds do that in the North Pacific. However, at least one of the banded birds recovered there was noted as probably a non-breeder when ringed (Warham, 1964) and only one of the other nine recovered in the North Pacific has been a known breeder. Furthermore, although *P. griseus* is a common bird in summer in the huge 'aquatory' between Australasia, Africa and Antarctica, there seems to be no direct evidence of Australasian birds moving to South Atlantic seas, unless Falla's birds flying westwards between 50° and 60°S in the Australian sector were such. The true picture awaits more data, perhaps as a result of consistent banding particularly of the South American birds, and identification of the various populations perhaps by biochemical 'fingerprinting'.

The status of the flocks of Sooty Shearwaters off Antarctica (Fig. 2.22) is still unresolved, although Falla (1937, p. 205) reported them long ago. He had problems in identifying them, referring to 'the supposed *P. griseus*', but later workers who have seen dark shearwaters there have mostly described them as this species, unaware of the presence of *P. tenuirostris* in much the same areas. Given good views, the birds can be separated at sea on a range of characters. However, the pale underwing lining of *P. griseus* also occurs in some *P. tenuirostris* (Serventy *et al.*, 1971). Of 586 beachwashed *P. tenuirostris* examined in Japan, 14% had light underwings (Oka, pers. comm.).

Falla's (1937) records were at comparatively low latitudes—30–62°S. Although recent sightings have confirmed the presence of *P. griseus* in these flocks, no specimen seems to have been taken, whereas *P. tenuirostris* has been collected. Shuntov (1974, p. 77) reported *P. griseus* plentiful in the Australian sector of Antarctic seas and (p. 89) believed that some immatures of both species proceed immediately south to the Antarctic Ocean on their return to the breeding grounds. He thought that the earlier *P. griseus* in the south originated from Macquarie Island, but the population there is too small for that role. Others who have plotted this shearwater

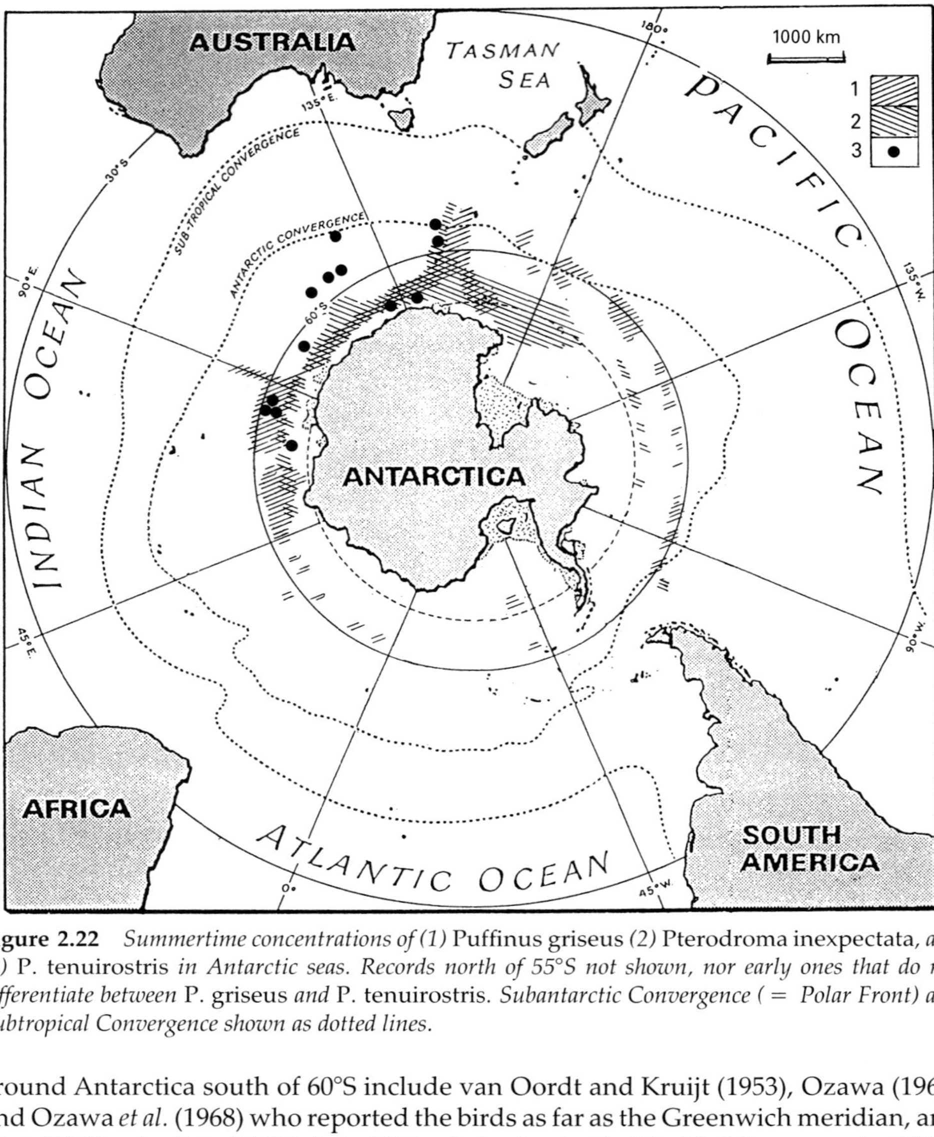

Figure 2.22 *Summertime concentrations of (1)* Puffinus griseus *(2)* Pterodroma inexpectata, *and (3)* P. tenuirostris *in Antarctic seas. Records north of 55°S not shown, nor early ones that do not differentiate between* P. griseus *and* P. tenuirostris. *Subantarctic Convergence (= Polar Front) and Subtropical Convergence shown as dotted lines.*

around Antarctica south of 60°S include van Oordt and Kruijt (1953), Ozawa (1967) and Ozawa *et al.* (1968) who reported the birds as far as the Greenwich meridian, and Zink (1978) who listed 24 birds in 100 h of viewing in the Weddell Sea, some to 41°W.

More recent reports from observers separating these two shearwaters include Ainley *et al.* (1984), Mochizuki and Kasuga (1985), Stahl (1987) and Ryan and Cooper (1989). Stahl plotted >15 birds per 10-min count at around 60°S, 100°E and Ryan and Cooper (1989) found it the most abundant bird in the Prydz Bay Region. These birds were not seen to feed but were moving mostly from east to west as were those seen by Falla (1937), by Naito *et al.* (1979) and by Ohyama *et al.* (1981). Flock size was variable, for example a mean of 9.18 ± 22.0 birds during one survey. However, not all the movements of flocks off Antarctica are in a westerly direction.

Veit and Hunt (1991) identified many *P. tenuirostris* among icebergs south of

Tasmania and some 100 000 on 28 January off Wilkes Land, 30% of which had dark underwings. They questioned previous identifications of *P. griseus* close to the continent and pointed out that in the North Pacific *P. tenuirostris* selects colder waters than *P. griseus* and would be expected to be the south polar shearwater as well, whereas, if previous identifications are correct, the reverse is true. However, as usual, no specimens were collected and both species could well be involved.

As the shearwater flocks are off Antarctica when the breeders are rearing chicks, those of both species have been assumed to consist mainly of younger birds and failed breeders, because the mean time between chick feeds of 4 days would seem too short for the parents to forage off Antarctica. However, Ainley *et al.* (1984) found *P. griseus* the most abundant bird along the Polar Front north of the Ross Sea and as far south as the first icebergs, reaching nearly 90 birds km^{-2} in December. As this front is here some 1700 km south of the southern New Zealand colonies, it may well be within range of some birds with chicks.

viii. Puffinus tenuirostris *Short-tailed Shearwater.* The ringing of Short-tailed Shearwaters has been extensive, with contranuptial recovery rates correspondingly higher than those for *P. griseus*. The classical figure-of-eight loop pattern of the movement, as deduced by Serventy (1953), was questioned by Shuntov (1974, p. 74), Maruyama *et al.* (1986) and others and, like *P. griseus*, the birds are thought to fly on broad fronts on both outward and return legs. In both directions *P. tenuirostris* appears to take more westerly courses than does *P. griseus* (Fig. 2.21). In the Okhotsk and Bering Seas it reaches higher latitudes than that species and it is unusual among migrants in moving to seas cooler than those in which it breeds.

Immature birds are believed to lead the northwards flight around the end of March, postbreeders in April and fledgelings from the end of that month into May. Usually they fly well offshore in southeastern Australian seas and at first seem to be in flocks, for example up to 1000 birds min^{-1} passed Lord Howe Island on 25 April moving at 40–50 knots and 3–15 m above the sea (Cheshire, 1980). Whether flock structure is maintained through the tropics seems to be unknown. A large number passes between New Zealand and New Caledonia and then swings north to reach the temperate North Pacific in mid-April, the South Okhotsk Sea in the third week of April and the Aleutians about late April. Ainley (in Chu, 1986) saw part of the northerly movement in March when vast numbers passed his ship in the central Pacific heading northeast.

In the Sea of Japan, most stay in the northern part, and they are also plentiful off Kamchatka and around the northern Kurile Islands. Around the Aleutians and in the Bering Sea they are more abundant than *P. griseus, P. tenuirostris* generally choosing higher latitudes. Late in the boreal summer many are in the eastern Bering Sea but then move from the shelf there to deeper waters between the Pribiloffs and Unimak Island, and some go as far north as Wrangel Island and the Chutki Sea (Shuntov, 1974, p. 86). Captain James Cook found them in August among the ice-floes. Shuntov describes a shift of the population in April and early May north and northeastwards to the Bering Sea, the Aleutians and the Gulf of Alaska but Guzman and Myres (1983) believed that the birds approached the Gulf near Kodiak Island in late April and May directly from the central Pacific. When wind-driven, they occasionally reach the coasts of British Columbia.

The southward movement begins in August. Shuntov suggested that adult birds were in the van and that those still near Cape Barrow (71°N) in mid-November and the Gulf of Alaska in December are immatures, the departure of non-breeders being extended over a long period, perhaps in waves. King (1970, p. 60) saw the movement in November east and south of Hawaii, when they comprised 32.5% of the seabirds seen. The movement ended in December and the shearwaters were not recorded passing through on the northwards leg. Morzer Bruyns (1965), when at 7°N and between 178°E and 166°W, estimated up to 400 000 birds in a day flying about 40–50 knots and heading southwest by south (190–200°) towards Fiji.

Shuntov's ideas are supported by the sightings during 12 voyages between Japan and North America in the boreal spring and autumn by Maruyama *et al.* (1986) who encountered many birds moving through the western Pacific. Suzuki *et al.* (1986) refined these findings using a computer simulation of the energy consumptions involved for birds travelling peripherally and centrally. They concluded that the loop pattern described by Serventy is probably used by birds of the year and requires less energy per kilometre, with the adults and subadults taking more direct central courses.

The return through Australian seas has been well documented and in most years part of it is visible from the mainland in late September to October, though many birds also pass further east, crossing the Tasman Sea on an oblique northeast–southwest course.

Some *P. tenuirostris* also reach the Indian Ocean as far as Sri Lanka and the Andaman Sea but whether this is part of a true migration, perhaps of birds from the small western colonies or Antarctica, or just part of the exploratory wanderings of young birds, is unclear. However, the recovery of a nestling from Ceduna that reached the Bering Sea in about 6 weeks (Serventy *et al.*, 1971, p. 130) shows that some from the westerly colonies still enter the Pacific Ocean.

Routh (1949) was the first to report this species off Antarctica. He saw 'many vast flocks' in the Indian Ocean sector between 64 and 65°S mostly in the open sea 30–80 km from the pack-ice edge. They were feeding on krill and resting on icebergs blackened by their numbers. Routh's identifications were questioned when others started reporting *P. griseus* in the same general area, a species not mentioned by him. Shuntov (1974, p. 77) noted huge quantities of both species in the Australian sector with some as far as 70°S in the Ross Sea and with some there as early as mid-November, but he did not separate the two species. First proof of the presence of *P. tenuirostris* came with the handling of 34 birds, 15 of which were collected, between 61°S and 65°30′S and 84–133°E in February and March, with thousands of similar birds in the same general area (Kerry *et al.*, 1983). On runs from Australia to east Antarctica in January and February Mochizuki and Kasuga (1985) recorded both species, finding *P. tenuirostris* very common between 53 and 57°S along 116°E and 150°E, rather north of the area where the specimens were taken (Fig. 2.23). Stahl (1987) found that along his transects *P. griseus* was much the more abundant but he also reported 11 *P. tenuirostris* at 61°S, 85°E in January and February, some close to where Kerry *et al.* collected theirs. There is no suggestion from the records of any segregation between the two species as in subarctic seas.

Thus *P. tenuirostris* does occur off Antarctica in late summer, and Routh could indeed have seen them, and perhaps both species. Naarding (1980) also suggested

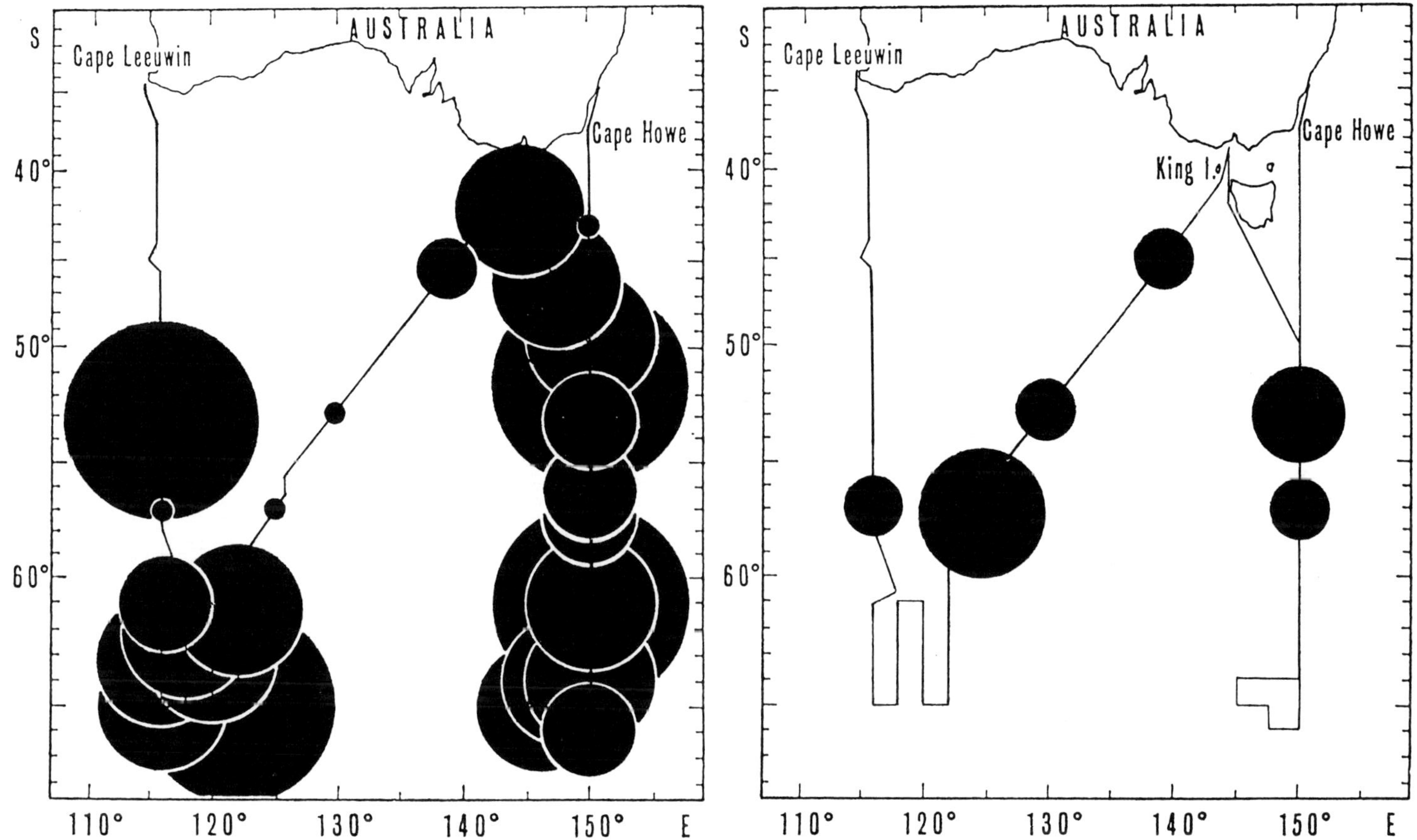

Figure 2.23 *Occurrences of the Sooty (left) and Short-tailed Shearwaters along parallels 116°E and 150°E between southern Australia and Antarctica in January and February 1984. Largest discs > 100 birds per 6-h watch, smallest discs 1–2 birds per 6-h watch. From Mochizuki and Kasuga (1985).*

that during their 3-week prelaying exodus most Tasmanian birds feed in Antarctic seas. Sightings are plotted in Fig. 2.22.

None of the observers reports any specific direction taken by the flocks—the 15 specimens had regressed gonads and could have been non-breeders, failed breeders, and perhaps even birds on 'sabbatical leave'. Such birds could possibly try to migrate north in the austral winter to form the source of those collected off Thailand, Sri Lanka and Pakistan (Frith, 1978). That some birds do not migrate has long been known and T. L. Montague (pers. comm.) has seen mid-winter flocks in Bass Strait.

d. The Gadfly Petrels. In recent years intensified studies at sea have revealed that in addition to *Pterodroma inexpectata* and *P. cookii*, a number of other gadfly petrels migrates to the northern hemisphere. All breed on Pacific Ocean islands but whether the whole or only part of their populations moves is not known. There are no records of recoveries of known breeders, so these more experienced birds, perhaps more able to feed in tropical and subtropical seas, may stay there and only the younger ones migrate. Spear and Ainley (in prep.) classified *P. externa cervicalis*, *P. e. externa*, *P. longirostris* and *P. l. pycrofti* caught in the ETP, as adult and 'subadult', these latter recognisable from their uniformly fresh plumage. Adults varied from 43 to 74% of the populations according to this criterion.

The status of some species that occur in the Pacific Ocean north of the Equator was reviewed by Bailey *et al.* (1989), Roberson and Bailey (1991), Spear *et al.* (1992b), Bartle *et al.* (1993) and Spear and Ainley (in prep.). Some specimens have also been taken at sea and around Hawaii. However, migratory pathways between the often widely separated nesting places and northern waters have yet to be determined. The distinction between migration and dispersal is blurred here: perhaps those migrating to northern waters are prebreeders during their exploratory years whereas breeders disperse in 'home' waters. Establishing the times of their appearance in northern seas is complicated by the extended breeding seasons of some populations such as *P. neglecta*. Breeding places and other data were summarized in *The Petrels*, pp. 94 and 95.

Whether any of the gadfly petrels of the South Atlantic shift north is unknown but there seems to be little evidence that southern *P. mollis* or *P. incerta* regularly cross the Equator.

i. Pterodroma inexpectata *Mottled Petrel.* The movements of *P. inexpectata* are the most extensive among gadfly petrels as the species ranges down to the Antarctic pack-ice in the austral summer and makes the long haul to the subarctic for the boreal one. Presumably the whole population undertakes this migration but we have neither recoveries of banded birds from the northern hemisphere nor evidence that a substantial number remains in southern waters.

Unlike many *'Cookilaria'* petrels, *P. inexpectata* is easily identified at sea. The birds are found widely and commonly in the northern and eastern North Pacific and around and south of the Aleutian Islands between May and October. Kuroda (1955a, 1960a, 1991) found them in cold seas (5–6°C) east of southern Kamchatka in late June and early July in company with *Oceanodroma furcata* and *Puffinus tenuirostris*. Ainley and Manolis (1979) and Bartle and Stahl (in press) reviewed the oceanic distribution and possible migratory pathways. In the Bering Sea during July to September inclusive, Ogi *et al.* (1986) found them abundant in pelagic waters at SSTs 5.0–10.9°C, but not in the Arctic Ocean; that is, in warmer seas than those favoured around Antarctica. They estimated some 1.5 million birds in the Bering Sea alone (Fig. 2.24). Furthermore, Nakamura and Tanaka (1977) also encountered high densities in July and August south of the Aleutians and Alaskan Peninsula. Birds in Gulf of Alaska waters from November to May must be non-breeders and these are joined by breeders from June onwards (Bartle *et al.*, 1993) to reach perhaps 110 000 birds in the summer (Gould *et al.*, 1982).

Ainley and Manolis (1979) deduced the routes to and from the North Pacific from sightings and specimen records. They believed that the pathways lay through the central Pacific diagonally from New Zealand, the return being by approximately the same route. Passage through the tropics is rapid. The birds have been seen flying north through Hawaiian seas abundantly in the last week of March through to April and May and going south in October and November. Pitman (1986) plots them along the Equator mainly between 160 and 150°W and small numbers have been seen flying north in March to May and south in October near Tonga, Samoa and Fiji (Jenkins, 1980).

This petrel is also common during its breeding season off Antarctica from about 135°W to about 45°E, particularly along the ice edge and among the icebergs and as far as the −0.5°C isotherm—details in Falla (1937); van Oordt and Kruijt (1953); Ozawa *et al.* (1968); Watson *et al.* (1971); Rogers (1980); Nakamura (1982); Hunt and Veit (1983); Ainley *et al.* (1984); Mochizuki and Kasuga (1985); Stahl (1987) and Bretagnolle and Thomas (1990) (Fig. 2.22). It appears to be absent from waters between Cape Horn and about 45°E.

Mottled Petrels are common south of New Zealand during the breeding season but more so in Antarctic than in subantarctic seas. Warham *et al.* (1977) suggested that these included off-duty breeders during their 12–14 day incubation spans and Ainley and Manolis (1979) agreed that the 2200–4000 km trip would be easily possible, although Harper *et al.* (1990) doubted its necessity. Three birds collected by Ainley *et al.* (1984) at 68°41′S, 171°49′E in December were judged to be non-breeders.

According to Bretagnolle and Thomas (1990) the records reveal a shift to the east of the birds off Antarctica between December and January, a concentration closer to the Continent in February, followed by a return movement as the birds withdraw from the east and the northwards shift begins, leaving Antarctic seas in March.

In the Australasian sector *P. inexpectata* is segregated on SST, being found mainly in waters at 0.7–2.0°C, whereas *P. lessonii* is most plentiful in waters at 8–10°C. *Lugensa brevirostris*, while regularly common down to the pack-ice in the Atlantic, is not common in the Australasian or Pacific sectors, and neither it nor *P. lessonii* are found in flocks off Antarctica. Bretagnolle and Thomas (1990) suggested that *P. inexpectata* monopolized a rich feeding area in the Pacific segment of Antarctic seas in summer in the absence of congeneric competitors.

ii. Pterodroma neglecta *Kermadec Petrel.* This occurs over a wide range but mainly to about 28°N in the east and central Pacific (Gould & King, 1967; King 1970, p. 21; Pitman, 1986; Spear *et al.*, 1992b). Birds have also been seen in the northwest Pacific

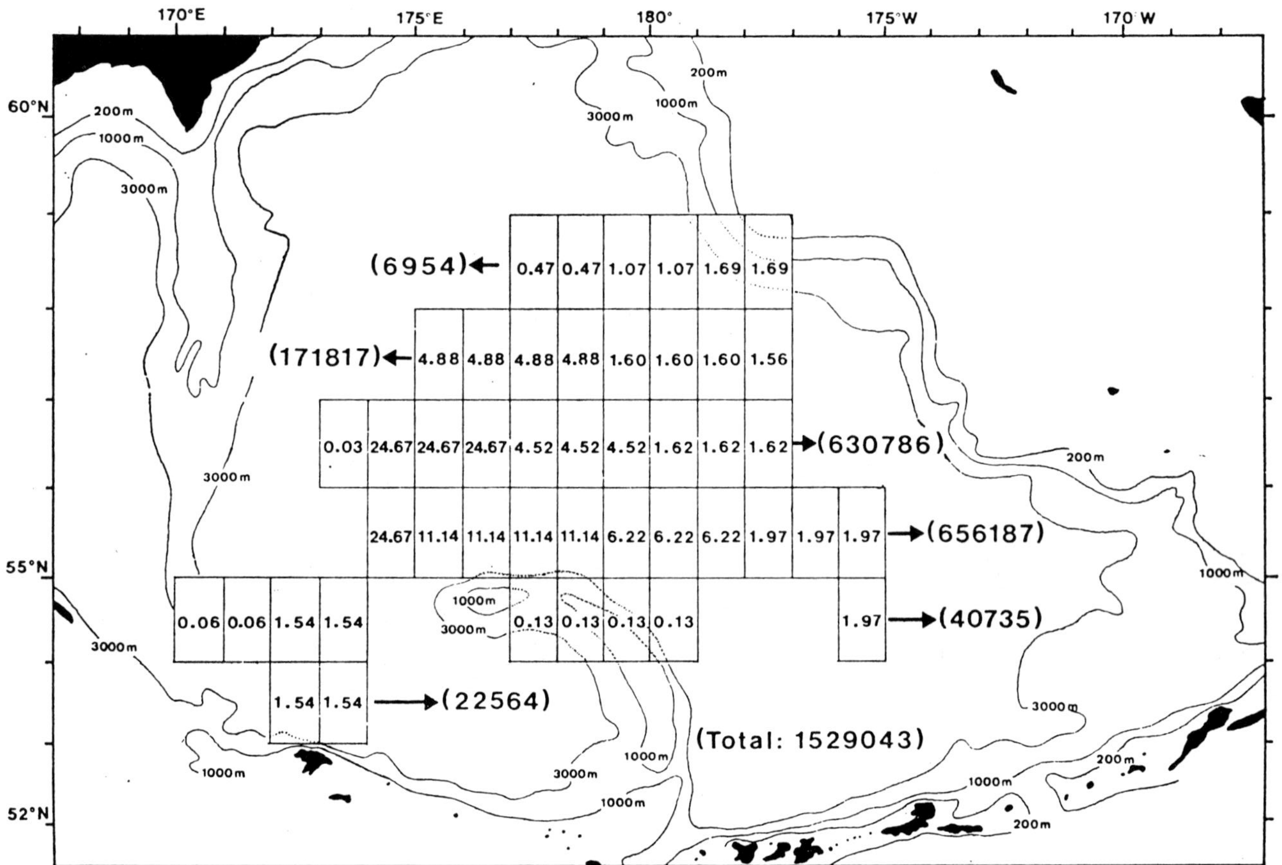

Figure 2.24 *Estimated Mottled Petrel* Pterodroma inexpectata *numbers during late (northern) summer 1985 in pelagic waters of the Bering Sea. Figures are average densities km^{-2} in 1° latitude × longitude blocks. Arrowed figures are estimates of numbers by latitude. From Ogi* et al. *(1986).*

and Japanese seas at 37°N by Nakamura and Tanaka (1977). These latter also reported an October recovery in the Philippines at 18°N of a bird banded as an adult at the Kermadecs nearly 8 years earlier. Gould (1983) saw a few along the 158°W meridian as far as 39°N with SST 15.6–24.7°C in early November and most sightings north of the Equator have been from September to January.

iii. Pterodroma externa *Juan Fernandez and White-necked Petrels* Both this species and *P. neglecta* were recorded as transequatorial migrants as early as 1921 (Loomis, 1921) and King (1970, pp. 12–14) found both subspecies and collected *P. e. externa* in his study area. Within 10–27°N and 160–145°W numbers in January to the end of April were small, increased in May and climbed to a postnuptial peak in October but fell rapidly thereafter as the birds left the area, presumably withdrawing to their southern headquarters. Further south, Pitman (1986) and Spear *et al.* (1992b) plotted 1063 and 6470 *P. e. externa*, 73 and 244 *P. e. cervicalis* respectively in the ETP, mostly between 0 and 20°N and 100 and 160°W.

Birds of the race *P. e. cervicalis* also reach northwest Pacific seas, perhaps flying directly from the Kermadecs, a route that would help explain their apparent low numbers in ETP counts. Tanaka and Inaba (1981), during voyages between 15 July and 1 December, found these birds common to the southeast of Japan as far as 10°N and 175°E but found none during May and June. The birds were most plentiful in August and where the SST was 27°–30°C, often feeding in small groups with *Puffinus pacificus*.

iv. Pterodroma solandri *Providence Petrel.* This species has been known from Japanese seas since the sighting of birds and the capture of two moulting females off northern Japan on 9 July 1954 (Kuroda, 1955a), and along the Kuriles Convergence in June 1954 (Kuroda, 1960a). Subsequent sea counts by Nakamura and Tanaka (1977), Wahl (1978), Tanaka (1986), Kuroda (1991) and Bartle *et al.* (1993) show that this bird is widely distributed in the North Pacific. During the austral winter, the breeding season, non-breeding *P. solandri* are plentiful to the east of Honshu where SSTs are 16–21°C, and along the Polar Front and associated confluence zones, although Tanaka (1986) saw 352 birds in May when the SST was only 8°C. During this northern summer the range extends across the northeast Pacific and Gulf of Alaska with sightings as far as 56°N, 145°W in July and August. In August and September, Tanaka (1986) found the species in the northwest Pacific in warmer seas (14–26°C): these may have been mainly breeding birds.

v. Pterodroma ultima *Murphy's Petrel.* Although a few specimens were collected over the past 30 years in the North Pacific, *P. ultima* was generally thought to be a vagrant there (e.g. Bourne, 1967). Then between 1981 and 1988 three more specimens turned up on Oregon beaches and in subsequent cruises off the western United States the species has been found to be common, notably in waters deeper than 1900 fathoms along the eastern edge of the North Pacific Gyre. The records were summarized by Bailey *et al.* (1989). They counted 98 birds on a single trip and caught one with the help of a cod-liver oil slick. Some were resting on the sea among Sooty Shearwaters, but most were flying north–northwest parallel to the coast, suggesting a northwards migration. Spear *et al.* (1992b) plotted 131 birds in the ETP.

Sightings have been mostly from April to June inclusive, that is, during the breeding season, and with some specimens completing wing moult, so that all may have been non-breeders.

The most northerly records are of three taken in Hawaii and its Leeward Chain with one even further north at 34°N about 630 km east of Santa Barbara (Clapp, 1975). From the timings of sightings Bartle *et al.* (1993) suggest that these birds have an anticlockwise circulation in the northeast Pacific during the breeding season.

vi. Pseudobulweria rostrata *Tahiti Petrel.* The status of this large petrel north of the Equator is conjectural. Specimens have been taken as far as 25°N off Taiwan and birds collected in the central Pacific in November, June, and April (Gould and King, 1967). At sea the species is easily confused with *P. alba*, but Spear *et al.* (1992b) separate the two and plot 916 sightings of *P. rostrata* between 170 and 100°W, as well as 192 observations of *P. alba* between the Equator and 10°N and 170° and 130°W, overlapping the range of *P. rostrata*.

Some *P. rostrata* reach eastern Australian coasts, mainly from December to April, again presumably non-breeders. Birds have also been identified off northwestern Australia in October, off the northern Australian coasts in July and August, off Papua New Guinea in November, and in the Coral Sea between May and November. The extent of the movements and what part of which population is involved awaits further investigation.

vii. The 'Cookilaria' *petrels.* At least parts of the populations of these small gadfly petrels reach the North Pacific. The status and characteristics of the six species that cross the Equator to the eastern Pacific are analysed by Roberson and Bailey (1991) and Spear *et al.* (1992b).

Pterodroma cookii Cook's Petrel

Loomis (1918) recounted the shooting in July of 19 birds at 22°25′N, 112°40′W that he considered were on their 'exodus' migration. All were passing through a complete moult and birds taken subsequently from Peru to Alaska have been in heavy moult from late April to the end of July, suggesting that their migration had ended. Pitman (1986) mapped a major concentration off Baja California, where 3000–4000 birds were estimated over a 16-km stretch on 31 July 1989, supporting the belief that these seas form a major overwintering place. Further north off central California a cruise to 145 km offshore on 29–30 April raised 98 *P. ultima* and 113 *P. cookii*. The latter were seen closely and possible *P. defilippiana* and *P. l. pycrofti* eliminated (Erickson *et al.*, 1989). Others have been seen in the central Pacific between 35 and 45°N and 160°E to 160°W and two came on board ship south of Adak Island, Aleutians, during the first week of August. Bartle *et al.* (1993) interpreted sightings in the central Pacific as marking movements between temperate and subarctic seas. Most sightings of this species have been over deep water beyond the continental shelves and around upwellings.

That large numbers of Cook's Petrels are found in northern seas is unexpected considering their depleted populations at the breeding places. With recent conservation measures they may be expected to pick up rapidly and perhaps the sightings off California reflect this.

Pterodroma l. longirostris Stejneger's Petrel

This petrel has long been known as a transequatorial migrant, Rollo Beck having collected five birds about 1000 km off California in 1906 (Loomis, 1918). The birds also appear to cross the Pacific diagonally to reach Japanese seas where several specimens have been taken.

Tanaka *et al.* (1985) analysed the distribution east of Japan between 140° and 180°E. They were first seen there in May and by June were in the area 50°N, 150°–180°E, but by late July the centre of distribution had shifted west to seas off northeastern Hondu where in mid-August 60–90 birds hr^{-1} could be seen feeding on the surface and tending to flock in the evening with *Oceanodroma castro. O. tristrami* and *Calonectris leucomelas*. In September the birds moved west between the 35 and 40° parallels with flocks of 200 seen. One of about 500 birds in November was at 32°12′N, 171°23′E, which these authors interpreted as part of the October/November return migration from Japanese seas. The numbers involved suggest that offshore Honshu is a major wintering area for the species. The birds spend the contranuptial season in warmer waters than those around its nesting colonies. On arrival in May the SST was 19–21°C, in June 18–24°C, in August off the coast 25–27°C, but in September as they spread east it was 23–26°C. The birds were densest in seas around 24°C.

Roberson and Bailey (1991) envisaged a clockwise movement through the North Pacific bringing the birds heading southeast past Hawaii by September and off California in November, and Spear *et al.* (1992b) plotted 579 birds in the ETP south and east of Hawaii. The bird breeds only on Mas Afuera Island off Chile during the austral summer.

Pterodroma nigripennis Black-winged Petrel

This easily recognized petrel nests in the austral summer and some are seen in the central Pacific from May into November. King (1970, pp. 16–19) plotted it throughout the year but it was uncommon between December and April, building up its numbers in May and June, peaking in October, then declining rapidly. King detected a tendency for the birds to arrive from the southeast and leave towards the southwest. Overall, it was the most abundant small *Pterodroma* during his study. Pitman (1986) scored 691 sightings in the ETP, particularly around 10°N; Spear *et al.* (1992b) 1930 birds in the same general area. Many occur in Japanese seas, which Tanaka *et al.* (1985) believed are reached towards the end of June and into July but probably most were to the east of their study area. They detected a spread northwards in September but in late November 40 at about 19°N, 146°E were believed to be heading back towards their breeding grounds.

Other '*Cookilaria*' species

Of the remaining small southern *Pterodroma, P. leucoptera brevipes* is said to disperse into the central Pacific and to occur regularly in October and November in the Equatorial Countercurrent southeast of Hawaii to about 10°N and 130°W (Roberson & Bailey, 1991, Spear *et al.*, 1992b). Pitman (1986) plotted no *P. l. brevipes* north of the Equator but 705 he identified as *P. leucoptera* in the eastern Central Pacific, some to 10°N. Meeth and Meeth (1983) logged birds in September and October mostly moving south or southwest between 100 and 130°W and 0 and 10°S, none north of the Equator. Spear *et al.* identified 1334 *P. l. leucoptera* in the ETP, mostly within 10° on either side of the Equator. These, and those seen by Pitman, presumably were *P. l. caledonica*, as the numbers recorded are too great to stem from the tiny Cabbage Tree Island colony. None of the 1788 birds banded there to 1986 had been recovered at a distance. Imber and Jenkins (1981) already deduced that *P. l. caledonica* undertook a broad front migration between the western and central South Pacific.

The populations of *P. l. pycrofti* are also small, its contranuptial feeding grounds undetermined, and although it has occurred in the North Pacific, its status there is also unclear but Spear *et al.* (1992b) recorded 100 sightings in the ETP and 223 birds further north, mainly off South and Baja California. These birds, *P. defilippiana* and *P. l. longirostris*, are very easily confused particularly if flying with *P. cookii.*

XII Vagrancy

A very small proportion of mostly immature birds turns up quite outside the species' usual range. Various factors have been invoked as facilitating escape to pastures new, including displacement by storms, being caught up in mass migrations of other tubenoses, or simply by disorientation due to failure of their navigation systems. Bourne (1967) reviewed the topic and points out (pers. comm.) that birds capable of undertaking long migrations may also have a greater capacity for displacement than those only shifting for shorter distances.

All else being equal, the chance of vagrancy is greater with those species with huge populations. Body size may also be a factor, fasting capacity being greater for bigger species. The high powers of endurance when crossing unsuitable habitat may help explain how southern mollymawks can reach the North Atlantic, perhaps without feeding through the tropics, whereas prions, far more abundant but small, do not.

Away from the nesting grounds, petrels sometimes fly over land either accidentally like the *Puffinus puffinus* found on Lake Geneva (Poncy, 1926), or apparently deliberately, as with the Laysan Albatrosses and Sooty Shearwaters on the Salton Sea evidently by flying north from the head of the Gulf of California, and the Sooty Shearwaters seen to be heading for the Mediterranean by flying inland from Eilat towards the Dead Sea (Shirihai, 1987). This ability might explain some of the vagrancy to the 'wrong' ocean. Did the New York State Mottled Petrel (*The Petrels*, p. 105) cross the Isthmus of Panama to get into the Atlantic Ocean or the Greater Shearwater that reached Monterey Bay (Stallcup, 1976) cross in the opposite direction, or did they simply start their northwards flights from the wrong sector of the Southern Ocean?

Vagrants can become imprinted on features of their new 'homes' and return there year after year, demonstrating that the marine habitats within range are suitable at least for their survival, and probably also for breeding. Examples include the *Diomedea melanophrys* mollymawks that have summered in the Shetlands and other northern locations—'half pairs' as Bourne (1977) put it. The pair of *O. leucorhoa* that occupied a burrow on the Chatham Islands and on St Croix Island off South Africa (Imber & Lovegrove, 1982; Randall & Randall, 1986), were thousands of kilometres from the nearest colony and even in a different hemisphere, suggest how great changes in distributions can occur.

Even in today's world of fast shipping, the possibility still arises of petrels being carried out of their usual range by seamen and ships; 'rains of petrels' attracted by ship's lights on misty nights still occur, the bodies, dead or alive being sluiced through the scuppers in the morning—if not eaten! Odd birds may be missed, hiding by day in a lifeboat perhaps, to get overcarried before escaping, possibly even into a different ocean.

Many further examples of long-distance vagrancy have been reported since Bourne's review. These include two cases of *D. immutabilis* over the Indian Ocean, *P. gravis* from the South Atlantic reaching south Australia, California and Central America, *P. puffinus* from the North Atlantic in New Zealand and California and *Bulweria fallax* from the Indian Ocean in the northwestern Hawaiian Islands.

Finally, one can anticipate some vagrancy not so far reported, for example *P. tenuirostris* in the Atlantic. This bird reaches the Indian Ocean and probably sooner rather than later one will turn up on a beach in the North Atlantic where any previous examples at sea will have been logged as *P. griseus*.

CHAPTER 3
Feeding and Foods

Contents

In earning a living from the sea, tubenosed birds employ a range of techniques to find and capture their prey. Food seems to be detected visually either directly or by noting the behaviour of other birds, occasionally perhaps by hearing their calls or the breaching of whales. The diets and feeding ecologies were reviewed by Prince and Morgan (1987) and Imber (1991) and many food analyses tabulated in Marchant and Higgins (1990) and Ridoux (1994).

I Feeding behaviour

A *How do petrels find their prey?*

It is often by no means clear how petrels find their prey. Sometimes searching is hardly necessary. Breeding birds may be able to get an appreciable part of their food close at hand where headlands and tidal fluxes over irregular bottoms cause 'boils' and upwellings, carrying plankton to the surface.

Away from predictable resources and outside the nesting season, the clues used appear to depend on a range of factors—the degree of patchiness and dispersal of the food species, its association with other animals, the conspicuousness of the prey, whether the petrels forage solitarily or in their own or mixed flocks, and whether the preferred foods are associated with detectable physical features of the marine habitat.

When foods are concentrated in time and place each year, as with the spawning Capelin *Mallotus villosus* off Newfoundland eaten by *Puffinus gravis* or *Sepia apama* off New South Wales favoured by *Diomedea exulans*, or over permanent upwellings over seamounts, then presumably the memory of past feeding is involved. And that albatrosses from as far away as South Georgia come regularly to take *Sepia* off eastern Australia shows how petrels can exploit very localized food resources.

For the many adapted to foraging in pelagic seas whose food is patchy and scarce, trial and error searching seems indicated. The characteristic sinusoidal flight paths of tubenoses may help them scan wide areas of surface water. When such birds are encountered feeding, however, several species are often involved at a restricted resource. It seems probable that a gadfly petrel towering into the sky scans the horizon to follow the movements of others: when any of them drops to the surface on sighting food, neighbours converge to partake of the feast, their movements in turn drawing in more distant birds. If the resource is large or persists long enough a melee may result. Haney *et al.* (1992) have calculated the theoretical upper (20–30 km) and lower (0.7–6.2 km) limits to the horizontal distances from which seabirds could be drawn to flocks in the open ocean.

Satellite telemetry has confirmed this picture of far-ranging petrels searching for widely separated food patches, at least for albatrosses, for example the *D. exulans* that ranged up to 3600 km from their nests to feed on spawning squid which were distributed independently of oceanographic features (Weimerskirch *et al.*, 1993).

The colour of the preferred prey may also play a part. From experiments Bretagnolle (1993) found that *Pagodroma nivea* strongly preferred red, orange, yellow and beige but ignored blue, green, white and black prey, and Harper (1979) found that all eight species studied, except *Phoebetria palpebrata*, first investigated orange or red objects, those of pink, yellow, blue, white and green being in descending order of preference. Blue items attracted *P. palpebrata*, the colour of that bird's sulcus. Harper pointed out that krill swarms appear as reddish patches when seen from above; thus the responses to orange and reddish lures are not unexpected. These colours are probably also significant to mollymawks as they occur in their bills. That grey pumice seems so attractive (Section E below) is perhaps unexpected in view of Harper's findings.

Although the movements of conspecifics often alert petrels to feeding opportunities, they may also use other birds, predatory fish, seals and cetaceans to guide them to surface swarms of potential prey, or to flying fish. Hoffman *et al.* (1981) described how a flock of *Puffinus griseus* streaming in long lines through Alaskan waters diverted to investigate a plunging Kittiwake *Rissa tridactyla*, the whole of the following stream perhaps following suit (Fig. 3.1). Similar behaviour occurs in southern seas although here common indicators, at least in inshore waters, are often Red-billed Gulls *Larus novaehollandiae*.

Some concordance between petrel prey and petrel abundance has been demonstrated, for example by Ryan and Cooper (1989) and Veit *et al.* (1993) of *Thalassoica antarctica, Fulmarus glacialoides, Daption capense, Pagodroma nivea, Diomedea melanophrys, Fregetta tropica, Macronectes* and *Pelecanoides* spp. and krill. Similar correlations have not been demonstrated in other studies. Obst (1985), for example, found that the presence of tubenoses was a good indicator of the presence of krill, but there was no correspondence between the density of the prey and of the predators. Similar results were obtained by Heinemann *et al.* (1989). They established that significant spatial associations between krill and petrels occurred only with *D. capense* and *F. glacialoides*, with only the former numerically concordant with krill density, these birds being krill specialists. These workers considered that most krill patches were unexploited at any given time—foraging petrels simply missed them as was concluded by Obst (1985) and McClatchie *et al.* (1989) and Hunt (1991a) believed that seabirds have great difficulty in finding a patch once they have left it.

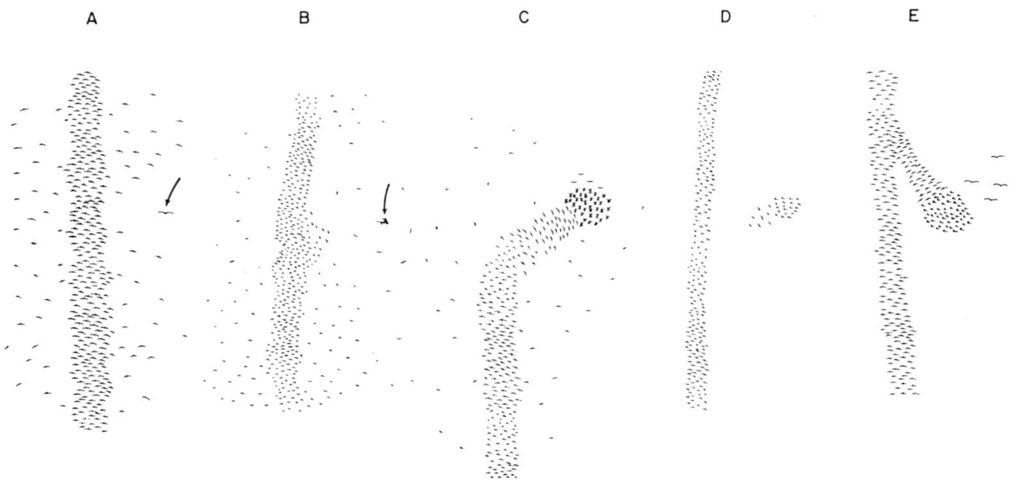

Figure 3.1 *Response of a stream of shearwaters, (A) moving towards the top of the figure to the plunging of a Kittiwake or shearwater nearby (arrowed), (B) nearby birds turn aside and (C) plunge for food while following ones rejoin the column (D) as do the emerging birds when the food is exhausted (E). From Hoffman* et al. *(1981).*

It has also been proposed that seabirds may use physical features as clues for food such as the temporary smear lines along which *Oceanites oceanicus* may congregate or birds like *T. antarctica* and *P. nivea* that fossick along the edge of the pack-ice concentrating on old, multi-year intrusions of ice where many prey animals tend to be concentrated (Fraser and Ainley, 1986).

B *Feeding equipment*

In capturing live elusive prey a petrel uses its wings, feet and particularly the bill, its only holding device. Bill shape and structure is variable and has been related to diet, from the razor-sharp nails of albatrosses which may produce stab marks on intact fish and squid, and their tomia that seem ideal for slicing through squid flesh and tentacles, to the thin bills of fish-eating shearwaters and the pincer-like mandibles with which storm petrels snip single zooplankters from the surface. Apart from such general observations there are no detailed studies of the relationship between bill structure and food and feeding methods, although Imber (1981) made some observations on the relationship of bill form with food for some prions. Many questions may be asked on bill form/food relationships. What, if any, is the significance of the difference between the bill shapes of *Diomedea irrorata* and the similarly sized southern mollymawks? Is it food related?

Food, once caught, has also to be 'handled' and the tongue is obviously important here. Its structure is more varied than that of the beak (see Fig. 10.8) but the role of the tongue in controlling caught prey has not been investigated apart from its use by shearwaters (*The Petrels*, p. 165).

Small food particles may be pumped into the oesophagus directly, but large prey may need dismembering. Pieces are torn off with the bill and the webs and feet may be spread and braced against the sea or ground to increase leverage. The bird bites and pulls backwards, shaking its head to tear and cut flesh and skin with the sharp hooked upper nail. In swallowing, the gape is expanded and the head raised allowing gravity to help the process. Diving species like shearwaters often bring their catch to the surface where it is swallowed hastily—if it has company. Presumably fish are taken head first but Rees (1961) found that the squid *Illex illecebrosus* in the stomachs of *Puffinus gravis* had evidently been swallowed tail first as all lay with head and arms towards the bird's mouth.

Attempts have been made to relate plumage colour to feeding styles, for example by Jouventin *et al.* (1981) and Simmons (1972), with dark dorsal plumage and pale underparts from the 'plimsoll line' down being regarded as an adaptation to make diving species less conspicuous to their prey. It is difficult to see this applying to petrels. The fish-eating shearwaters, for example, include countershaded forms such as *Calonectris leucomelas, P. puffinus* and *P. assimilis,* but also dark plumaged ones such as *P. griseus* and *P. tenuirostris,* both of which seem extremely competent in underwater pursuit (p. 397). Dark dorsal surfaces may have evolved as an anti-predator device to decrease conspicuousness when on land after dark.

C *Feeding styles*

The various ways in which seabirds capture food once they have found it were categorized by Ashmole (1971) and *The Petrels,* p. 12, and the feeding styles of the various petrel genera summarized in chapters 2–10 of that work. Ashmole's terminology has been largely followed, although sometimes elaborated, for example by Ainley (1977), Jouventin and Mougin (1981), Cramp and Simmons (1983, pp. 2–3) and Harper *et al.* (1985).

Ashmole was also the first to tabulate the feeding styles used in different marine habitats. Later presentations include those for 10 spp. of the North Pacific (Ainley and Sanger 1979), for 6 spp. of the Ross Sea (Ainley *et al.*, 1984), for 46 southern petrels by Harper *et al.* (1985), and for 20 and 76 spp. respectively by Harper (1987) and Prince and Morgan (1987). Harrison *et al.* (1991) detailed the styles for eight petrels in multispecies flocks around South Georgia.

About 80% of the 103 species take their food by surface seizing. A variation is to pick at floating debris—dead fish, Sargasso Weed and algal rafts. *F. glacialis, Pachyptila desolata* or *O. oceanicus* peck items, probably small crustacea, from kelp. Forster, in his *Resolution Journal* at 71°S on 30 January, 1774 wrote 'The evening before a bundle of Seaweed with some Barnacles on it passed the Ship & some Petrels were seen feeding on it' and recently grazed *Lepas* barnacles have found to be significant in the diets of *Garrodia nereis, Puffinus griseus* and *Pachyptila crassirostris.*

Other styles widely used are momentarily submerging from a floating position to grab subsurface food—'surface diving'—or, from the same position, chasing the prey under water—'pursuit diving'—and picking food at or near the surface with at most only the bill, breast or head momentarily touching the water—'dipping'. Rather fewer species use 'pattering', familiar of storm petrels that push off from the sea with their feet to maintain a more or less constant distance from it. Their wings are raised into a 'V' so that when feeding over a concentrated food patch 'they appear like so many butterflies on a wet sand bar'. (Crossin, 1974). Leach's Petrel does not normally dabble its feet but feeds while swimming with wings raised ready to shift to a new position.

'Pursuit plunging' from flight into the sea directly to follow prey under water is rather unusual but reported for *Puffinus tenuirostris* and *T. antarctica* (Morgan, 1982; Ainley *et al.*, 1992), and by the gadfly petrels *Pterodroma inexpectata* and *P. hypoleuca*, perhaps reflecting their use of fish. Shearwaters generally splash down, breast and spread feet hitting the water, and then submerge immediately. Much surface diving by shearwaters is done after the swimming bird has put its head below the surface with its wings raised, presumably to look for suitable prey within striking distance (see Fig. 9.5). The water volume scanned is much greater than that visible to a bird flying overhead. This style is also used by Bulwer's, Snow and Antarctic Petrels.

Ainley *et al.* (1993) introduced the term 'ambush feeding' for a specialized mode used by *Pagodroma*. Resting undetected at ice-floe edges, they drop down to take surfacing prey. High densities of copepods and amphipods are found under old, corrugated ice, and these become available to the birds as the ice disintegrates.

Although feeding styles for some 30 or 40 species are either not or very incompletely known, none appears to rely solely on one method, but quantitative data on the relative usage of the various styles are scarce. However, Ainley *et al.* (1984) and Harper *et al.* (1985) analysed 201 and 4926 feeding attempts respectively of a range of southern species and Prince and Morgan (1987) tabulated diets and methods for 76 of them.

Until recently, only *Diomedea irrorata, Pterodroma ultima* and *Puffinus pacificus* have been recorded as attempting piracy, but Spear and Ainley (1993) report kleptoparasitism by Kermadec Petrels *Pterodroma neglecta*. About half of their feeding attempts were to rob other petrels—*P. externa, P. rostrata* and *Puffinus pacificus* and, once, *Oceanodroma hornbyi*. Spear and Ainley, pointing out the similarities in flight style and plumage of *P. neglecta* to those of skuas, particularly the white patches at the bases of the inner primaries, hypothesized that this was a case of Batesian mimicry.

Switches of feeding modes can readily occur, for example a Cape Pigeon industriously pecking surface copepods like a pigeon picking up seeds, abruptly surface dives when some larger morsel appears. Harper (1987) gave some figures for a number of species. For example, of 1709 observations of *Pachyptila turtur* feeding, 901 (53%) were by surface seizing, 756 (44%) by dipping, 48 (3%) by pattering and 4 by surface diving.

All albatrosses and giant petrels can 'surface dive' and Harrison *et al.* (1991) measured 80 such dives by *D. melanophrys* to last from 0.5 to 5.8 s, mean 2.9 ± 1.3 s, a method used in 32% of 431 feeding attempts. Hunger may also elicit unusual feeding styles (see Fig. 4.1), but generally the human observer cannot judge how hungry a petrel is except that those resting or sleeping when food abounds around them are presumably satiated.

Dipping is the mode favoured by gadfly petrels and by fulmars such as *Daption capense* and *Pagodroma nivea* making only momentary contact with the water and swallowing their catch on the wing. In dipping, the petrel must identify surface items as food very quickly, so it is not surprising that unsuitable flotsam such as plastic are taken nowadays. Birds such as *Puffinus pacificus* that dip to take flying fish in mid-air may not even touch the water, but if the prey is too large to carry the petrel must alight to secure and disable it. Aerial capture by dipping has been recorded of some apparently unlikely species, for example by *Diomedea cauta*, *D. chlororhynchos* and *Calonectris diomedea* taking leaping mackerel, pilchards and flying fish respectively. Giant petrels, the only tubenoses regularly feeding on land, are helped to do that by the nimbleness of their footwork. Small fulmars rarely venture onshore, but Longstaff (1924) described *Fulmarus glacialis* in the Aleutians landing on ice and hobbling on their tarsi with wings extended to scavenge bear or seal carcasses. They mantled over these with extended wings evidently in the style of *Macronectes*, keeping competing gulls at bay. Both *T. antarctica* and *Pagodroma nivea* will also land on ice to manipulate food caught nearby.

Hydroplaning, where the petrel rests lightly on the surface facing the wind, wings raised and half spread to create lift, while being propelled forwards with the feet and snatching food items from side to side, was first described by Murphy (1936, p. 621) of *Pachyptila desolata*. This method is used also by *P. vittata* and *P. salvini*, occasionally by *Daption capense* and the shearwaters *Puffinus lherminieri*, *P. griseus* and *P. tenuirostris* and probably by others (see also Chapter 9.VII.E).

Filter feeding by straining water against the bill lamellae is used by prions and by fulmars such as *Macronectes* and *Fulmarus* (*The Petrels*, fig. 3.2.). Kritzler (1948) noted that by 'treading water' his captive fulmars created a swirl below them and he speculated that in the wild this might be a means for drawing water and food particles within reach. These actions were common after food had been swallowed, with the belly raised from the surface, and, as Kritzler surmised, could have helped the passage of large lumps into the proventriculus. Similar foot treading actions by *D. capense* may have been for the same purpose.

There is evidence that certain feeding styles are used more often and are more appropriate in some marine habitats than in others. Ashmole (1971) noted that in polar and subpolar seas most birds fed at or just below the surface on fairly passive prey like euphausiids and pteropods, and Ainley and Boelkelheide (1984) noted that volant pack-ice species tended to feed by dipping. At low latitudes dipping is also very widespread, perhaps because surface prey are scarce and many petrels follow tuna and cetaceans that drive small fish to the surface and flying fish out of it, making them vulnerable to *Pterodroma* petrels and *Puffinus pacificus*. Harrison and Seki (1987) suggested that the feeding methods used by tropical seabirds minimize the risk of being eaten by predatory tunas, sharks and billfish, plentiful in these waters. Storm petrels, *Procellaria parkinsoni*, *Puffinus nativitatis* and *P. pacificus* use dipping in tropical seas but hydroplaning is not commonly seen there. This feeding method seems to require a reasonable wind for giving lift to raise the body and reduce its drag against the water and, in tropical seas, water-rich plankton suitable for filter feeders is seldom available.

D *When do petrels feed?*

There are few data on the times when petrels feed, particularly when not breeding. Kuroda (1957a) thought that *P. griseus* fed close to the Japanese coast morning and evening, spending the rest of the day well offshore. Hoffman *et al.* (1981) also found *P. tenuirostris* and *P. griseus* flocking to take swarming euphausids from the surface in the Gulf of Alaska in the evening and early morning. An analysis of seabird activity in the Pacific during austral summers and autumns showed that these too tended to feed morning and evening, when light levels were low and micronekton, zooplankton and their predators were still near the surface (Ainley and Boelkelheide, 1984). In the Ross Sea, Ainley *et al.* (1984) made 16 847 sightings under overcast skies of the feeding of *F. glacialis, T. antarctica, Pagodroma nivea, Pterodroma inexpectata, Oceanites oceanicus* and *Fregetta tropica*, and found that peak feeding occurred from 0600 to 1100 h and from 1800 to 2300 h. Evidently here at *c.* 75°S the reduction in light at 'night' is enough to maintain the daily vertical migration. Only in open waters near the shelf break did the birds feed to any extent during the middle of the day.

That petrels are active at sea by night has long been known, sailors finding that they could often be captured more easily after dark than by day, and Fisher (1904) suggested that *Diomedea immutabilis* fed after dark. The topic was reviewed briefly by Brooke and Prince (1991).

Harper (1987) watched petrels feeding at night beyond the ship's illumination with the aid of a signalling light fitted with shutters that he could suddenly open up on unsuspecting birds.

Much other evidence is circumstantial, as with observations on circadian activity of *Fulmarus glacialis* by Lovenskjold (1954) in the dark nights of the Spitzbergen winter that suggested nocturnal foraging. There are many observations of petrels feeding by night at sea but as the observers could see them with the help of the ship's lights, presumably the birds were also helped in that way, particularly as squid, a major food, also tend to be attracted to lights. Harper (1987) saw feeding *D. exulans* chopping their cephalopod prey to pieces with their sharp bills and head shakings. In 11 cases it took 30 s (14.5 s–2.6 min) to handle the squid before swallowing. He noticed that on calm nights these albatrosses hunted by stealth, remaining quietly on the water and grabbing their prey with sudden snaps at the sea.

The other main line of evidence for night-time feeding as normal practice is inferred from a consideration of the prey found in stomachs (e.g. Imber, 1973); Harrison *et al.*, 1983). These include many deep-water forms believed to be unavailable at the surface except after dark. For example, Ealey (1954) found large numbers of *Euchaeta* copepods were eaten by the diving petrel *Pelecanoides georgicus* off Heard Island but he noted that as this is only in deeper water in daylight its presence implies some feeding at night, whereas the crustacean prey he found in *Pachyptila crassirostris* captured at dusk were so well preserved that they must have been taken that day. Further evidence is provided by albatrosses fitted with activity recorders (Prince and Morgan, 1987). These birds spent more time on the sea by night and, as they must alight to take their prey, this implied that they mostly fed or slept then.

Petrels that have been seen feeding by night either by moonlight, ship's lights or spotlights by Harper (1987) include *D. exulans, Macronectes, Daption capense, P. inexpectata, Lugensa brevirostris* and *Pachyptila vittata*. He also considered *P. belcheri* a predominately nocturnal feeder (Harper, 1972), but *P. desolata* entirely, and *P. turtur* (99% of 1709 observations) mostly diurnal ones. He saw *Phoebetria palpebrata, Procellaria cinerea, Fregetta tropica* and *O. oceanicus* feeding only by day. Other night-time sightings by various observers have included *Puffinus pacificus, P. carneipes* and

Procellaria westlandica, but often only doing so when ship's lights attracted plankton to the surface. In the north, *Diomedea immutabilis, Pterodroma hypoleuca, Oceanodroma tristrami, Bulweria bulwerii* and *D. irrorata* have been regarded as nocturnal feeders (Harrison *et al.*, 1983). Many of the fish taken have photophores.

Species that dive for their prey are presumably less likely to do so at night as those of their foods that migrate vertically will be within reach longer than for species using surface-seizing or dipping. Furness and Todd (1984) invoked nocturnal feeding to explain the invertebrate diets of St Kilda fulmars, which fed their chicks shortly after sunrise, whereas Foula birds fed their chicks fish throughout the daylight hours.

From an analysis of 9994 cephalopods of 61 species given by *D. exulans* to the chicks, Imber (1992) found a disproportionate frequency of bioluminescent vertical migratory species and he concluded that the birds had been feeding at night. Previously he had deduced that *Pterodroma macroptera* and *Procellaria parkinsoni* were nocturnal feeders, again from analyses of squid beaks (Imber, 1973, 1976a). He suggested that the light of their photophores could help the petrels detect the squid at night when they reached the surface.

Further circumstantial support for night-time feeding comes from the finding that the retinae of *D. immutabilis* have more rhodopsin than do those of *D. nigripes* (Silliman in Harrison *et al.*, 1983) and so are better adapted for night vision, which might explain the greater proportion of squid, many of which are bioluminescent, in the Laysan's diet. The same investigator found quite high values of rhodopsin in the retinae of *Pterodroma hypoleuca* and Harrison *et al.*, (1983) considered that this bird's reliance on mid-water fish—unusual in gadfly petrels—also suggested nocturnal feeding.

Many species deduced as feeding mainly by night also do so by day, for example *D. exulans, O. leucorhoa* and *Procellaria parkinsoni*, and thus it has been questioned whether nocturnal feeding is widespread. For example, Gibson and Sefton (1955, 1959), watched hundreds of *D. exulans* devouring *Sepia apama* by day off New South Wales. One bird, flying about 5 m above the water, suddenly baulked, dropped down and almost submerged to grapple with a writhing cuttlefish that vigorously ejected ink into the bird's face. Wood (1993) noted that *S. apama* taken by *D. exulans* over the eastern Australian shelf were dead or moribund. Other accounts also show that squid may be plentiful at the surface by day (e.g. Collins, 1884, p. 316).

A study by Lipinski and Jackson (1989) based on the cephalopods taken from the stomachs of 279 petrels of 14 species showed that many kinds of squid float when dead or dying and they hypothesized that such animals form the main source of squid in petrel diets and are usually taken by day. Croxall and Prince (1994) concluded that surface scavenging such animals may be important to *D. exulans* and *Pterodroma palpebrata*, but was unlikely to be so for *D. melanophrys* and *D. chrysostoma*, at least at South Georgia.

Further light is being thrown on this problem with the use of stomach sensors which record abrupt falls in temperature when the bird ingests food or water (Fig. 3.2) and of activity recorders (e.g. Afanasyev & Prince, 1993). Stomach sensors were used with *D. exulans* by Weimerskirch and Wilson (1992), Cooper *et al.* (1993) and Weimerskirch *et al.* (1994b). In one study over 29 days, the birds took a food item every 3.6 h, an estimated 89% of the total mass being ingested by day and from areas

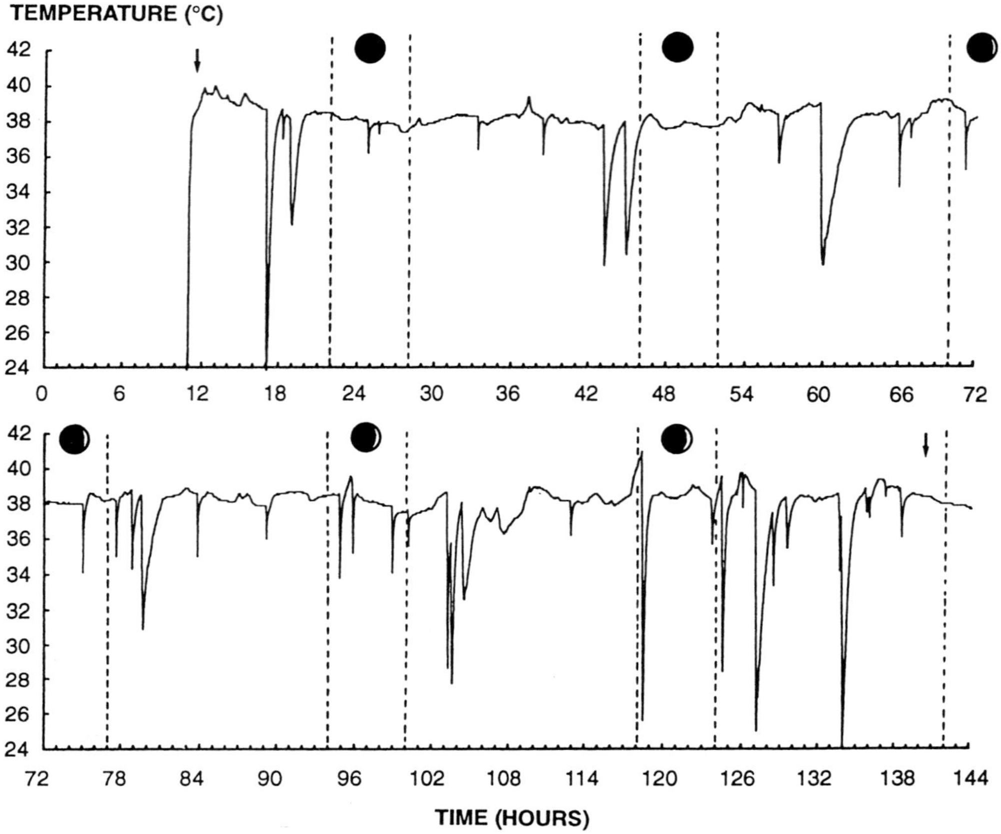

Figure 3.2 *Stomach temperatures of a male Wandering Albatross during a 6-day foraging trip, the start and end of which are arrowed, dashed lines periods of darkness. Sudden declines represent ingestion of cold food followed by slower rises as ingesta warm to body temperature. From Weimerskirch and Wilson (1992).*

not used by fishing vessels, so presumably being of natural foods. The work of Croxall and Prince (1994) also indicates that most albatrosses feed mainly by day, resting for most of the night on the surface.

E *Feeding flocks*

Petrels often forage in single species flocks, a familiar habit with shearwaters, prions and storm petrels, but also reported of *Pelecanoides urinatrix* by Weimerskirch *et al.* (1989a) who saw flocks of >1000 feeding over shallow waters off Kerguelen. Crossin (1974) used as an 'Index of Sociability' the number of birds seen divided by the number of sightings of the species in question. The figures for the 14 hydrobatids he examined ranged from 1.1 to 6.7, this latter being for *Oceanites oceanicus*. The next most likely to be seen with another of its own was *Oceanodroma furcata* (4.0). The figure of 1.6 for *O. leucorhoa* (14 355 birds scored) was more typical of the storm petrels generally.

Petrels that are usually dispersed can aggregate quickly when a food patch is revealed by roughened water, and melees may develop as birds 'belly flop' to grab their prey. In the Gulf of Alaska, Hoffman *et al.* (1981) noted that these onslaughts usually only last for 20–60 s in each location, suggesting that the birds' activities disperse or decimate the prey. The shearwaters also suppress flock formation as other bird species also disperse. Or the shoals may move and the birds move with them, shearwaters often forming long windrows that creep forwards as birds surfacing at the rear swallow their catch and leapfrog forward to flop down towards the head of the column and resubmerge. Social feeding in this way is very common with *Puffinus tenuirostris* and *P. griseus* in both hemispheres and has been described of *P. puffinus, P. lherminieri* and *P. gavia*. Feeding flocks of *Daption capense* may creep forwards in the same way, evidently tracking the movements of their prey.

Interspecific flocking is common and several species of petrel may feed with pelecaniforms, auks, gulls and terns. If tubenoses do use olfaction to find prey, they may be the initiators of some mixed feeding flocks. In northern waters the petrels tend to be on their postbreeding migration, the rest from closer breeding places; in the south, tubenoses generally predominate. Multispecies flocks are common in the tropics, perhaps because of the greater patchiness of prey there. However, even when several different petrel species are swimming or feeding within sight of one another, each may retain its integrity for much of the time. For example, in the North Pacific *P. griseus* and *P. tenuirostris* tend to form separate flocks but mingle when food is abundant (Gould *et al.*, 1982).

Aggregations often form when a localized source of food is discovered. Anthony (1895) recounts how the cries of gulls hovering over a floating cod attracted a Short-tailed Albatross which 'with a groaning note settles down by the floating fish, keeping all trespassers away by a loud clattering of the mandibles'.

Ainley and Boekelheide (1984) worked out 'Coefficients of Interspecific Association' for seabirds in the South Pacific. Most were tubenoses. Where the sea surface temperature (SST) was <4°C, that is, in the pack-ice, the significant associations were negative—the birds avoided each other, although with *Pagodroma nivea* and the skua *C. maccormicki* this may have resulted from deliberate shunning of the skua, and few flocks were seen in Antarctic pelagic seas. In the Subantarctic Zone (SST 3.0–13.9°C) most associations were positive with slightly more species in feeding flocks there than in the Antarctic. In the subtropics (SST 14.0–21.9°C), 11 species were positively associated with *Puffinus creatopus*, 13 with *P. griseus* and 14 with *Procellaria aequinoctialis*, while in tropical seas (SST >22°C) 11 species were associated with *Puffinus pacificus*. In all four regions, shearwaters, especially *P. griseus*, were important and Ainley and Boekelheide (1984) considered this species' role as an initiator of flock formation more important than its role as a suppressor. Briggs *et al.* (1987) performed similar calculations for seabirds off California, showing that *P. griseus, P. creatopus* and *P. bulleri* often associated together but were avoided by phalaropes and alcids.

That flocking over food may have drawbacks is illustrated by the feeding of a Grey Seal *Halichoerus grypus* on adult *P. puffinus*. Four birds were taken from below while the rest of the flock, engrossed in feeding, ignored the disturbance (McCanch, 1981).

F *Roles of tubenoses in seabird communities*

Just as some kinds of seabird tend to nest in the same communities or 'guilds', so species favouring particular kinds of ocean habitat tend to be found together and can be regarded as forming marine communities.

Hagerup (1926) and Murphy (1936) and many later workers identified the species characteristically exploiting various marine habitats, usually taking the large-scale perspective and often recognizing similar communities in opposite hemispheres or oceans. Bourne (1963) compared the compositions of two subtropical communities, one off West Africa, the other off southern Arabia, Hunt and Nettleship (1988) high-latitude northern and southern ones, and Duffy (1989) those of two southern upwelling regions.

The members of these groups are drawn together by a community of interests, seeking the same or similar foods in the same places and/or being physiologically adapted to similar environments. Some are more or less permanent members, others may be transients or seasonal members only. This is particularly true of the migrant shearwaters whose presence or absence because of their numbers, may have large effects on community structure and biomass (Fig. 3.3). For example, in the Bering Sea from June to the end of September *P. griseus* and *P. tenuirostris* are dominant,

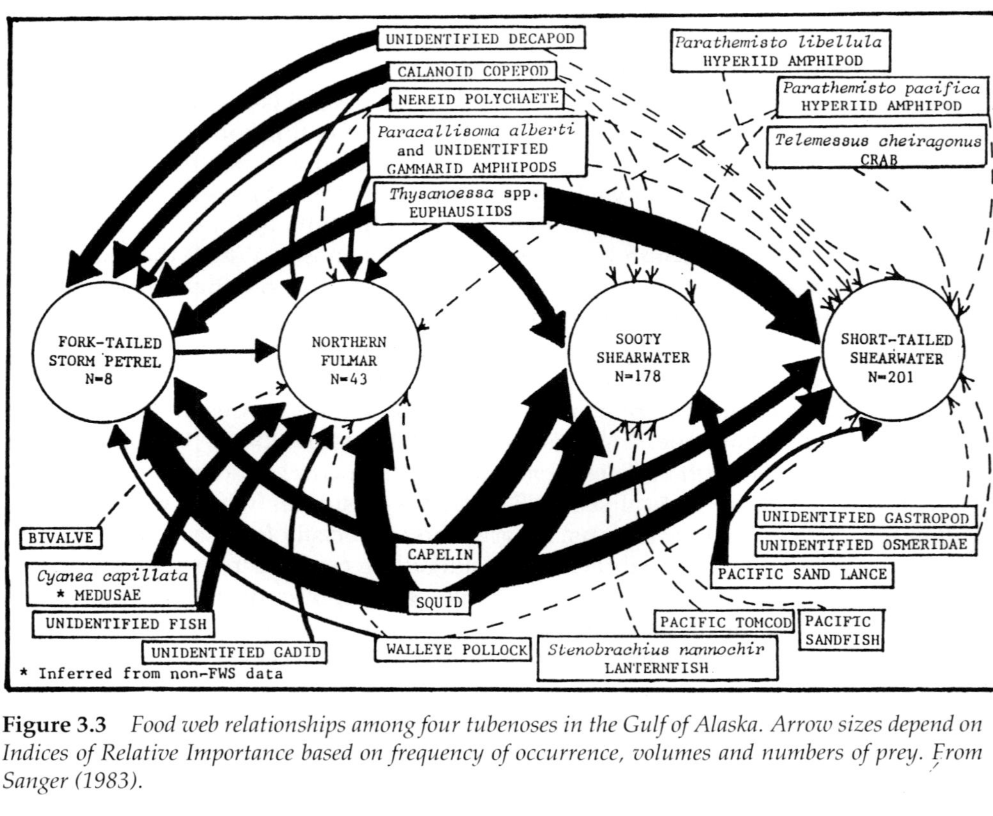

Figure 3.3 *Food web relationships among four tubenoses in the Gulf of Alaska. Arrow sizes depend on Indices of Relative Importance based on frequency of occurrence, volumes and numbers of prey. From Sanger (1983).*

totalling some 8.7–20 million birds according to Shuntov (1974) and Hunt *et al.* (1981) respectively.

Another factor affecting community composition is the presence or absence of whales and large predatory fish such as tuna or of man-made diversions such as fishing boats (Section II.C.4 below) to which many petrels are attracted. Such variables and seasonal changes tend to make marine seabird associations dynamic in space and time (e.g. Fig. 3.4).

At megascale considerations associations can also be long-lasting as shown by Ribic and Ainley (1988/89) and Ainley *et al.* (1994). The community of the Humboldt Current (Peruvian upwelling region) was spatially confined, consisting of Salvin's Mollymawk, Waved Albatross, Peruvian Booby and four local storm petrels which remained together even during the 1976 El Nino Southern Oscillation (ENSO). Also persistent was the association of six open-water species of subantarctic and Antarctic waters—Black-browed Albatross, Shoemaker, Antarctic Prion, White-headed Petrel, Sooty Shearwater and Black-bellied Storm Petrel.

Ainley and his colleagues (Ainley *et al.*, 1992, 1993, 1994) performed cluster analyses of the foods of seabirds in winter and summer in the Antarctic pack-ice (where food is plentiful year-round) and in open water to the north. They found that the diets broadly overlapped regardless of species, habitat or year, with little segregation of prey size according to predator size. When the birds of the open water shifted north in winter those of the pack did not move into the vacated habitat. It was concluded that the specialized feeding styles of *Pagodroma nivea* and *Thalassoica antarctica* were ineffective in open water and, as the open water birds ate the same foods, the petrels were selecting for the habitats to which they were best adapted rather than for the types of food found there. It would be interesting to learn how *P. nivea* breeding distant from the pack at South Georgia take their prey: there can be no 'ambush feeding' in the open sea.

In their study of the foraging zone of seabirds around The Crozets, Stahl *et al.* (1985) identified a correlation between ocean depth and the distribution and composition of petrel flocks feeding over the shelves and deep water (Fig. 3.5). Bird abundance and biomass were higher over the shelves and the composition of the flocks differed with *Diomedea melanophrys, D. chrysostoma* and *Procellaria aequinoctialis* being found mainly over the shelves, *Fregetta tropica, Pterodroma mollis* and *P. macroptera* mostly over deeper water.

Petrels are carnivores and/or scavengers, and once their diets have been ascertained the trophic levels can be used to throw light on community organization, their partitioning of food resources, competition and the like.

Ainley and Sanger (1979) made a provisional examination of the trophic levels at which seabirds of the northeastern Pacific, Bering Sea and coastal waters of western America feed. Many species are involved, but by virtue of their numbers the Procellariiformes are ecologically dominant. As with many birds, some petrels fed at more than one trophic level. The levels used were secondary carnivores—birds that fed on crustacea, polychaetes, coelenterates and fish, tertiary ones those taking fish and squid, and scavengers those eating carrion, offal and detritus.

Feeding at secondary and tertiary levels as well as scavenging were *D. nigripes, Fulmarus glacialis, O. furcata* and *O. leucorhoa*; at secondary and tertiary levels *Puffinus griseus, P. tenuirostris, P. carneipes* and *P. bulleri*; at tertiary level *D. immutabilis* which

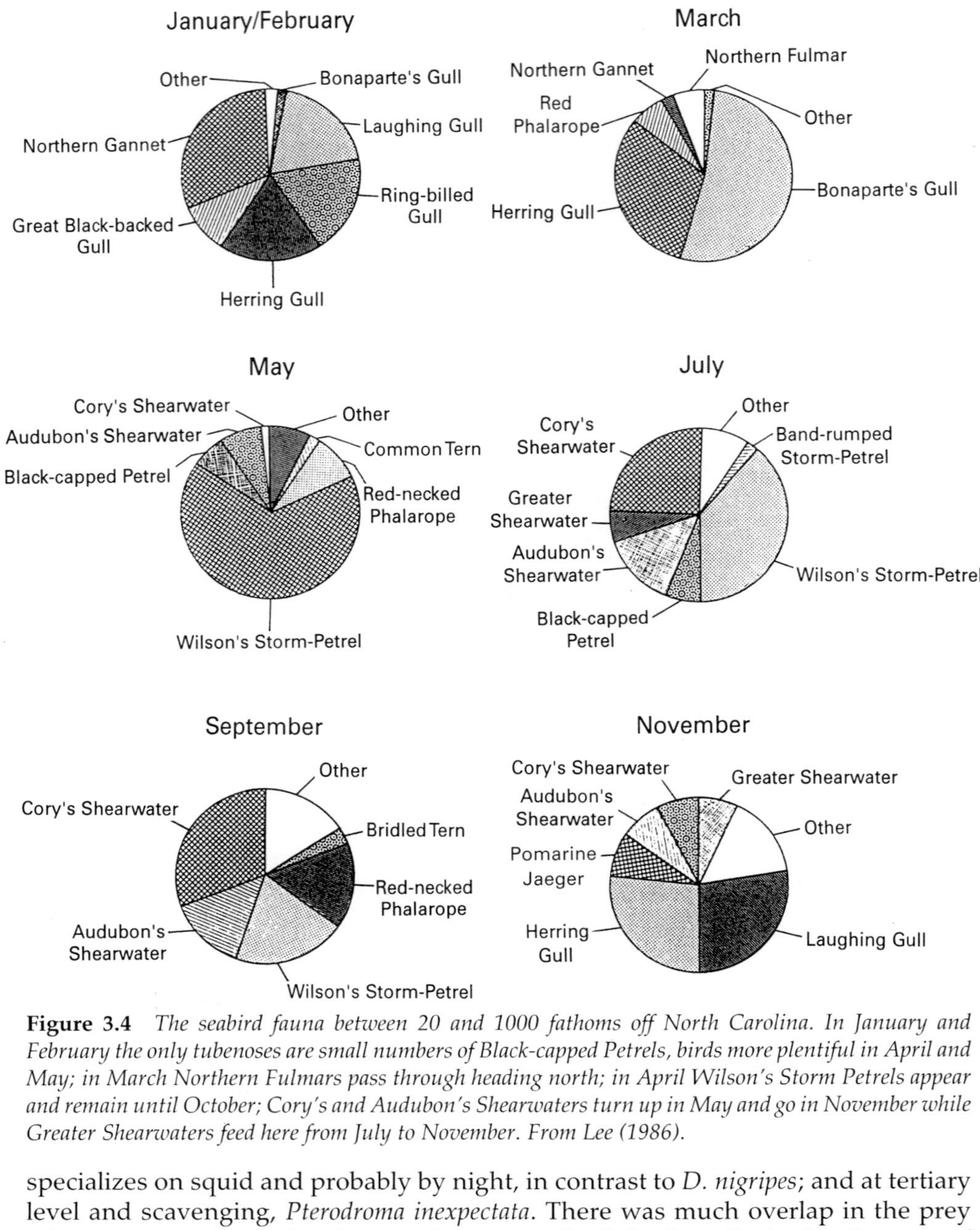

Figure 3.4 *The seabird fauna between 20 and 1000 fathoms off North Carolina. In January and February the only tubenoses are small numbers of Black-capped Petrels, birds more plentiful in April and May; in March Northern Fulmars pass through heading north; in April Wilson's Storm Petrels appear and remain until October; Cory's and Audubon's Shearwaters turn up in May and go in November while Greater Shearwaters feed here from July to November. From Lee (1986).*

specializes on squid and probably by night, in contrast to *D. nigripes*; and at tertiary level and scavenging, *Pterodroma inexpectata*. There was much overlap in the prey taken, competition being lessened among mixed feeding flocks by the use of different feeding methods, the choice of different sized prey and by feeding at different depths and at different times, for example *O. furcata* and *D. nigripes* took surface prey, shearwaters took these but also descended several metres into the water column, but at shallower depths than puffins and guillemots, and so on.

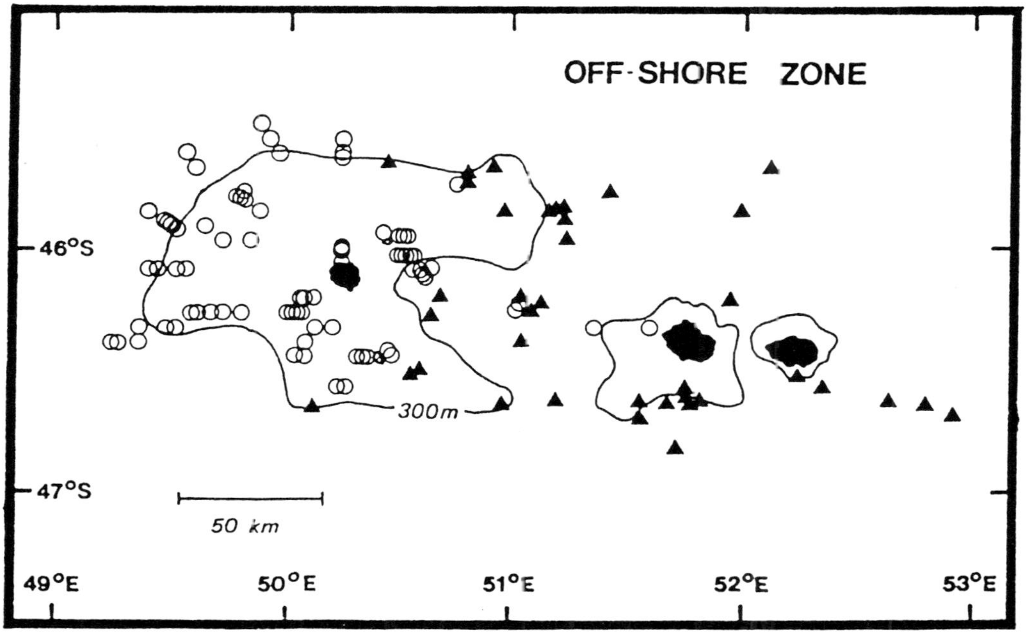

Figure 3.5 *Feeding distributions of two associations of petrels off the Crozet Islands. Open circles,* Diomedea melanophrys–D. chrysostoma–Procellaria aequinoctialis *association, mainly in shelf waters; closed triangles,* Fregetta tropica–Pterodroma mollis–P. macroptera *association over deeper water; 300 m contour is contour marked. From Stahl* et al. *(1985).*

The technique of determining trophic levels by measuring the ratios of isotopes ^{13}C and ^{15}N to ^{12}C and ^{14}C was used with Weddell Sea vertebrates by Rau *et al.* (1992), and included analyses of muscle from nine kinds of tubenose. Values for ^{15}N overlapped, indicating shared prey and trophic levels. The data for *Oceanites oceanicus* and *Lugensa brevirostris*, however, showed little overlap and the authors surmised that this reflected earlier feeding outside the Weddell Sea, perhaps during their migration. (See also Minami *et al.* (1995) for the North Pacific.)

The communities of the Benguela Current were analysed by Abrams and Griffiths (1981) and Abrams (1985a), that of the African segment of the Southern Ocean by Griffiths *et al.* (1982). The birds were classified on their main diets into planktivores (mainly prions, Blue Petrel, Cape Pigeon and storm petrels), squid eaters (albatrosses, gadfly petrels and *Procellaria* spp.), fish eaters (shearwaters, terns, cormorants and gannets) and omnivores (giant petrels, gulls and skuas).

The Southern Ocean studies extended throughout the year, and of the 42 volant species only eight were not tubenoses. The prions amounted to 34% of the 75 779 birds counted (estimated biomass 39 662 kg), *Halobaena caerulea* 13%, *Puffinus gravis* 14%, *L. brevirostris* 7%, *T. antarctica* 6.6% and *Pterodroma mollis* 6.8%.

When crossing the Subtropical and Antarctic Convergences species richness and diversity reached their peaks (Fig. 3.6A) and abundance and biomass also reached their maxima near these current boundaries (Fig. 3.6B). Numbers were at their lowest near the warm Agulhas Current at only 7.2 birds per station. Near the

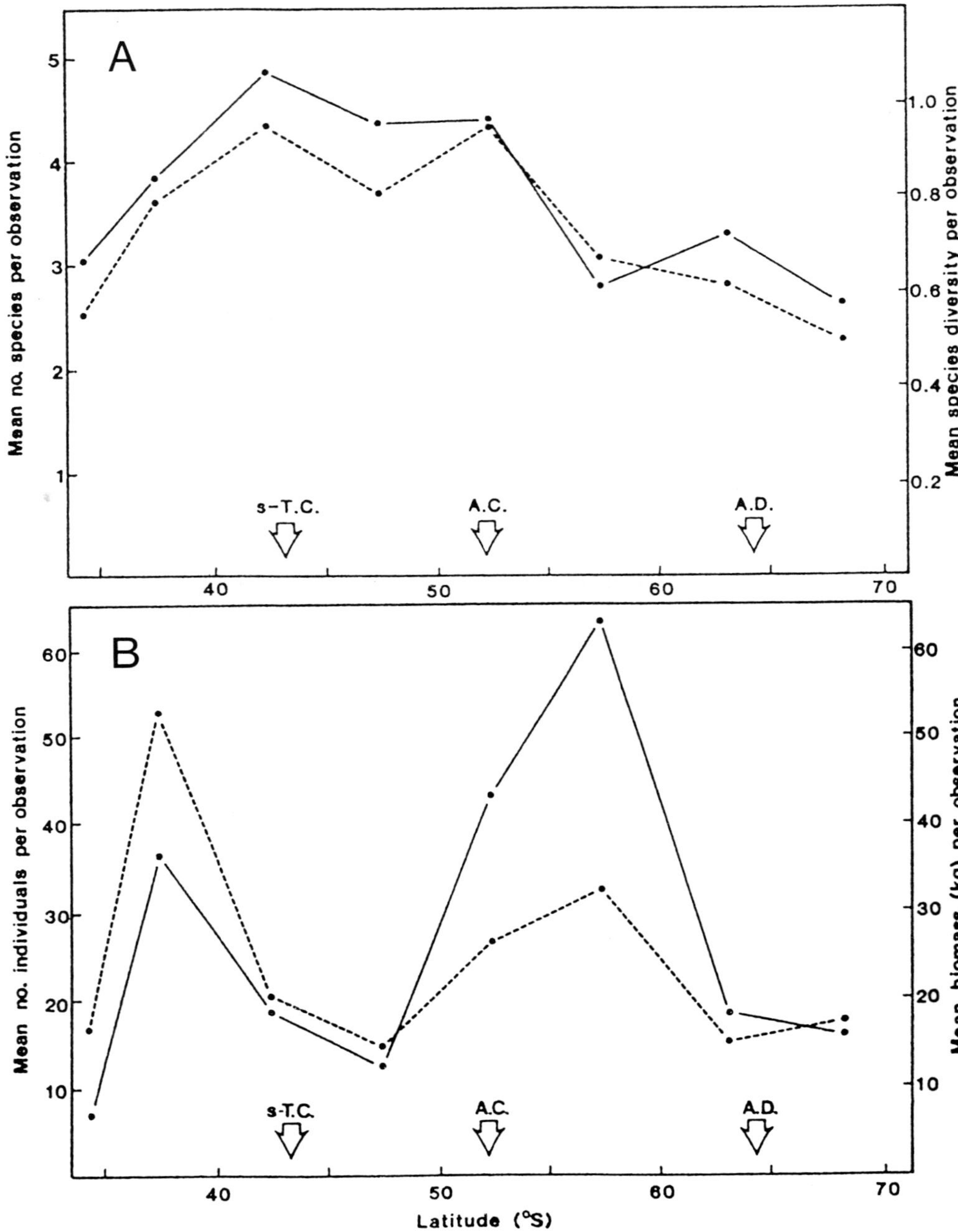

Figure 3.6 *The structure of the volant seabird community in the African Sector of the Southern Ocean, approximate annual positions of the Subtropical Convergence (s-T.C.) and Antarctic Convergence (AC) and Antarctic Divergence (AD) arrowed. (A) Latitudinal distribution of species richness (dashed line) and species diversity (continuous line). (B) Latitudinal distribution of total biomass (dashed line) and total abundance (continuous line). From Griffiths* et al. *(1982).*

Subtropical Convergence large squid-eating birds, albatrosses and gadfly petrels predominated, but near the Antarctic Convergence most were small prions and storm petrels taking planktonic animals. Griffiths *et al.* (1982) speculated that the squid eaters, being larger, could fast longer and are better equipped for surviving on scattered, irregular food resources and may avoid high latitudes in summer as the short nights restrict nocturnal feeding on surfaced squid. The planktivores, on the other hand, being small, needed the reliable and renewable resources obtaining near the convergence where birds such as *O. oceanicus* and *Fregetta tropica* appear to feed almost continuously. The omnivores were most plentiful near the Antarctic Convergence, the main player here being *L. brevirostris*.

At each station in the Benguela Current birds of one guild dominated, interspecific competition being reduced by differences in feeding behaviour and temporal separation. For example, *Puffinus griseus* chase underwater fish unavailable to gannets and surface seizers such as gadfly petrels and albatrosses, while the latter can swallow prey too big for shearwaters.

Following enhanced productivity in spring and summer due to nutrient upwelling, fish and squid numbers peaked and then mainly subantarctic species such as *Procellaria aequinoctialis* and albatrosses were most plentiful rather than planktivores and omnivores (Abrams and Griffiths, 1981).

However, a major 'unnatural' factor has operated in recent times here, and in most oceans—the vast development of commercial fisheries: in the present instance a stern-trawl industry. Abrams (1985a) attempted a comparison of the ecological structure of the avifauna of the southern Benguela Region in 1950 with that of 1980. The distribution of seabirds has always been patchy here but more food is now available, though only around the trawlers where Southern Ocean species—*D. melanophrys, D. chlororhynchos, D. exulans, P. aequinoctialis* and *Macronectes* spp. predominate, together with some Cape Gannets, cormorants, terns and gulls. At least 1000 birds escort each trawler, while away from them multi-species 'rafts' of >50 birds are scattered in the fishing zone.

Ryan and Moloney (1988) were critical of the 'guild' approach. These workers counted the seabirds and seals attending trawls by their research vessel in midwinter. They collected a range of environmental variables, introduced an index of recent trawling activity near each observation station, and calculated a 'radius of attraction' for every species. Their analyses suggested that the distributions of *D. melanophrys, D. cauta, Daption capense* and *P. aequinoctialis* are influenced by the fishery, prions, *Puffinus griseus* and *O. oceanicus* little affected. *Diomedea melanophrys, Daption capense* and *P. aequinoctialis* were most abundant round trawlers near the shelf edge, but *O. oceanicus, Diomedea cauta* and *D. chlororhynchos*, although more plentiful near there, were best correlated with distance offshore. Ryan and Moloney agreed that trawling has altered the distributions of some seabirds but they questioned whether it has increased the population sizes.

G *Feeding competition*

Interference between petrels and other seabirds is obvious when several forage at a restricted and localized food resource like a swarm of krill or shoaling fish. A hierarchy is established where the largest species dominates the rest. In the Southern Ocean this will often be an albatross, usually *D. exulans* or *D. melanophrys* with smaller birds like *Daption capense* dodging about between them (Table 3.1). The very nimble prions and storm petrels often work assiduously on the fringes without danger of attack but those swimming too close may be killed (and perhaps eaten) by a sideways sweep of an albatross's or giant petrel's bill. Shoemakers and giant petrels are particularly pushing, use their weight to good advantage, and may dispute possession with mollymawks. Prince and Morgan (1987) suggested that the more aggressive *D. melanophrys* might be able to exclude *D. chrysostoma* from krill swarms off South Georgia and that this may help explain why the former ate more krill, the other more squid.

Voisin (1991) drew attention to the relative shyness of *Macronectes giganteus* compared with *M. halli*: he considers that this is one reason why the latter can dodge among Elephant Seal harems and feed more efficiently on dead pups and placentae than the other. On land, a typical hierarchical sequence is male giant petrel, female giant petrel, skua, Weka/Dominican Gull *L. dominicanus*. Athough both sexes will feed at carcasses, the smaller billed and presumably weaker females are often absent, and Hunter (1983) showed that during their breeding seasons the males of both species ate more carrion, the females more marine food (Fig. 3.7).

Whether competition plays a major role in regulating population sizes or distributions at sea is debatable. For example, there seems no evidence that the locations of colonies of tubenoses are sited to lessen prey depletion within feeding range as Furness and Birkhead (1984) suggested for some northern seabirds. However, this idea has not been tested for tubenoses.

There are many mechanisms making for a reduction in competition for food at sea. Those operating during the breeding season were discussed by Croxall and Prince (1980). The range of bill shapes and body sizes reflects their differing diets and feeding methods, the prions forming a classic example (Harper, 1972; Harper *et al.*, 1985). Petrels, such as shearwaters, that take much of their prey underwater would appear to compete little with surface seizers or dippers. The preference of *Puffinus tenuirostris* for euphausiids and of *P. griseus* for fish may reflect the comparatively short bill of the former and the longer, more pincer-like one of *P. griseus*. Congeners

Table 3.1 Competition between large petrels for a food item—as the percentage of observations when a 'dominant' species excludes another from the food. Number of observations in parentheses. From Weimerskirch *et al.* (1986)

	Dominant				
Excluded	*Diomedea exulans*	*D. melanophrys*	*D. chrysostoma*	*Phoebetria fusca*	*Macronectes* spp.
Diomedea exulans	–	0(99)	0(17)	0(8)	9(64)
D. melanophrys	100(99)	–	0(13)	0(13)	61(38)
D. chrysostoma	100(17)	100(13)	–	(0)	100(4)
Phoebetria fusca	100(8)	100(1)	(0)	–	100(7)
Macronectes spp.	91(64)	39(38)	0(4)	0(7)	–

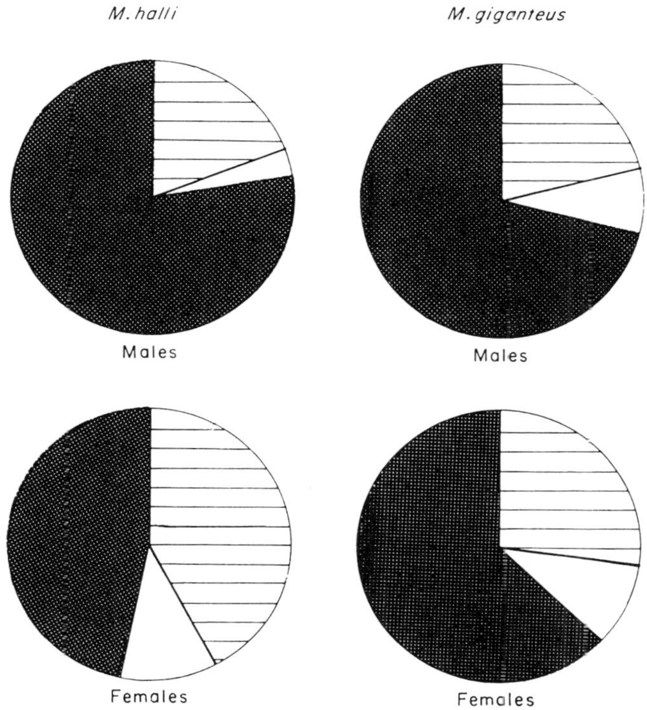

Figure 3.7 *Diets of giant petrels at Bird Island, South Georgia during the 1980–81 season. Stippled areas, penguin and seal carrion; lined areas, krill, squid and fish; blank areas, small birds. From Hunter (1983).*

of similar size may search in different places or take different prey. Others feed at different times, and pressure on resources within feeding range is obviously lessened by the different species having different breeding timetables (*The Petrels*, p. 218), so that all are not raising chicks simultaneously. The case of the two giant petrels illustrates this. Although often breeding sympatrically and, sex for sex, taking much the same foods in similar places, they lessen competition when nesting by well-marked differences in phenology (*The Petrels*, p. 55).

The mass migrations of high-latitude southern populations to rich feeding areas in the northern hemisphere lessen the pressures on the reduced surface foods consequent on the winter descent of the plankton for birds such as albatrosses, prions and fulmars that remain. Seasonal distributions may be complementary: Bourne (1963) pointed out that when *Oceanites oceanicus* off West Africa moves south during the northern winter, its place close to the shore is taken by *Hydrobates pelagicus* and probably by *Oceanodroma leucorhoa* in deeper seas, both from the North Atlantic.

Some dietary differences between petrels of similar size seem to be because the birds foraged in different places. Prince (1980a) found that, while both *Halobaena caerulea* and *Pachyptila desolata* took mainly *Euphausia superba* in the breeding season, the prion also caught many copepods (one bird contained *c.* 41 118 of them), apparently by hydroplaning. Prince deduced that the prion was feeding offshore, the Blue Petrels further out, thus reducing competition between them.

H *Associations of tubenoses with marine vertebrates*

That marine birds often associate with whales, seals and fish has long been known and fishermen still use seabirds as indicators of fish shoals. For example, the Manx Shearwater was known to Irish fishermen as 'Mackerel Cock' from their association (Watters, 1853) and the term 'whale bird' for prions and Blue Petrels seems to have arisen not only because these were among the most abundant birds flocking to feed around captured whales, but because 'it is of no use looking for whales where no birds are seen' (Bierman and Voous, 1950, p. 11.).

Although Bent (1922, p. 145) reported flocks of shearwaters and Leach's Petrels with schools of whales in Alaskan seas, Wynne-Edwards (1935) the association of pilot whales and *Puffinus gravis*, and Routh (1949) that of whales and *P. tenuirostris* in Antarctic waters, it is only in recent years that the widespread natural association of tubenoses with marine mammals and fish has become evident, although not all cetaceans attract seabirds. Petrels mainly involved are albatrosses, fulmars and shearwaters, some gadfly petrels, *Bulweria fallax*, and occasionally storm petrels. The associations at times seem accidental, with both petrels and mammals looking in the same places for the same prey, but more often tubenoses follow whales to feed on their faeces and vomit of partially digested squid and other prey, or even on air bubbles which trap food particles. Such associations arise because both partners prey on the same foods although they do not necessarily select prey of the same size. The behaviour is more common in tropical seas, but is also seen at high latitudes in both hemispheres.

These relationships are one-way, facultative commensal ones: although some get eaten by seals and Killer Whales, the benefits seem to accrue to the birds alone. There usually seems to be enough for both partners and should the prey sink beyond the petrels' reach, it will not be beyond that of the mammals or fish. At times the available food appears superabundant. Au and Pitman (1986) reported apparently satiated *P. pacificus* resting on the sea near dolphin herds while others continued to feed.

Instances of birds associating with cetacea were reviewed by Evans (1982), Enticott (1986) and Pierotti (1988) and Au and Pitman (1988).

Whales and dolphins are conspicuous animals and presumably easily recognized by seabirds and they may fly around the pods even when the mammals are not feeding. Sometimes only a single species is attracted. Martin (1986) described the feeding of *Calonectris diomedea* with dolphins, mainly *Stenella frontalis*, around the Azores. He noticed how the shearwaters would cease random flying and head directly to where the dolphins were feeding as indicated by a halo of circling birds. The aggregations were sometimes large, for example about 1000 shearwaters and 200–500 dolphins around a shoal of Horse Mackerel *Trachurus picturatus*. Underwater observations revealed a dark, spherical mass of fish. The dolphins took fish from below, the shearwaters hovering overhead pecking at surface items or landing to feed with submerged heads. Food fragments left by the mammals were eaten rather than whole fish.

Around the Salvage Islands, *C. diomedea* flocks are good indicators of Tunny and Bonito (*Katsuwonus pelamis*), and are so used by local fishermen (Zino, 1971).

In the Bering and Chutki Seas seabirds associate with Grey Whales *Eschrichtius robustus*. In summer an estimated 10 000 of these feed on the sea floor, slurping the muddy sediments through their baleen to remove benthic amphipods and when they breach the mud plumes they carry with them are conspicuous and provide ephemeral feeding opportunities for gulls, auks and petrels, particularly *Fulmarus glacialis, P. griseus* and *P. tenuirostris* (Harrison, 1979; Grebmeier and Harrison, 1992). Obst and Hunt (1990) noted that the fulmars stayed longest at the plumes, often >30 min, paddling along slowly and pecking amphipods from the surface film (mean 75 pecks min^{-1}, n = 4, SE = 11).

In tropical seas not only cetaceans but predatory fish, particularly tuna, that force smaller fry to the surface and excocoetids and some ommastrephid squid to flight, are attended by marine birds. As Ainley and Boekelheide (1984) pointed out, having such predator assistants driving deep-sea animals into range, compensates to some extent for the lack of diving seabirds and makes these seas more productive for surface feeders. In the central Pacific, tuna schools are frequently accompanied by the shearwaters *P. pacificus* and *P. nativitatis* together with Common Noddy, Sooty and White Terns (*A. stolidus, S. fuscata* and *G. alba*).

Seabird interactions with tuna and cetaceans were described by Au and Pitman (1986, 1988) and Au (1991). Flocks of birds accompanying herds of five species of dolphins are common in the eastern tropical Pacific (ETP) and are used by tuna fishermen to locate the schools and the Yellowfin Tuna *Thunnus albacares* and Skipjack *K. pelamis* that swim with them. Both birds and cetaceans apparently take food made available by the tuna and between 5 and 30°N, 43–53% of bird flocks scored by Au and Pitman (1986) were associated with dolphins.

In contrast to the flocks accompanying the fast-moving cetaceans, the very pronounced, almost obligatory, association of Parkinson's Petrel revealed by Pitman and Ballance (1992), was almost entirely with slower Melon-headed Whales *Peponocephala electra* and False Killer Whales *Pseudorca crassidens*. The mammals apparently dismembered large, subsurface prey, providing debris available to the petrels. Of 618 of these birds seen, 76% were associated with the whales, with from 1 to 300 birds to a herd. The largest flock of 300 birds represented nearly 10% of the world population of *Procellaria parkinsoni*.

Pitman and Ballance (1992) described the petrel's scavenging style as landing near the mammals and submerging its head with wings raised and partly extended, often surface diving to retrieve bits of dismembered prey. Pitman and Ballance considered that *P. parkinsoni* was particularly well adapted for this feeding role as it could stay submerged for up to 20 s and was estimated as descending to at least 10 m.

In the Southern Ocean, Enticott (1986) tabulated seven albatrosses, four fulmars, four gadfly petrels, prions, two *Procellaria*, three *Puffinus* shearwaters and *Fregetta tropica* associated with seals and cetacea. Associations with the Right Whale Dolphin *Lissodelphis peronii* were the most common, prions and Antarctic Petrels its most frequent companions. The largest group consisted of *c.* 220 cetaceans (*c.* 20 Pilot Whales *B. acutorostrata* and *c.* 200 *Lissodelphis*) and *c.* 130 petrels—*Phoebetria palpebrata, Pterodroma macroptera, L. brevirostris, H. caerulea, Procellaria cinerea, P. aequinoctialis* and prions. However, at least over the 8 years of this survey that involved 13 767 10-min observations, only in 0.18% of these were seabirds noted as accompanying whales or dolphins, so that this is not a particularly common feeding behaviour, at least for petrels in the Africa Sector of this ocean.

Around the coast of Possession Island, The Crozets, associations of seabirds with Killer Whales *Orcinus orca* were described by Ridoux (1987). Seven tubenoses and prions were involved when the whales were feeding, *Macronectes* sp., *Daption capense, Diomedea melanophrys* and *P. aequinoctialis* being the most usual.

Near Bird Island, South Georgia, the non-volant associates were seals *Arctocephalus gazella* and penguins, particularly *Eudyptes chrysolophus*. Penguins and seals were part of 97% of the multi-species flocks seen 25–26 February 1987 by Harrison *et al.* (1991). Black-browed Albatrosses, prions and giant petrels were in every flock, the albatrosses usually being the initiators of these, but subsurface seals and penguins seemed essential to flock formation. These may have been driving the euphausiid prey to the surface giving the birds access to them or to debris of damaged or dead prey. These appeared to be available only briefly so that immediately penguins or seals surfaced there was a rush of birds (Fig. 3.8) and a feeding frenzy, but prions and the storm petrels *F. tropica* and *O. oceanicus* continued to take small items long after the rest had dispersed.

Another kind of association between fur seals, penguins and giant petrels around South Georgia was described by Bonner and Hunter (1982). They watched *A. gazella* taking *E. chrysolophus*, many of which were not eaten but left to be scavenged by the petrels with Cape Pigeons often as hangers on. The seals were never seen to attack the petrels. Bonner and Hunter thought that, in the breeding season at least, the penguins were first maimed or killed by the seals.

Some associations are unexplained. Why, for example, would resting flocks of *Puffinus griseus* form around sunfish *Mola mola* as reported by Burger (1988)? In Monterey Bay, 85% of the shearwater flocks were associated in this way and particularly over the larger specimens.

The benefits to the petrels of feeding in association with other birds, fish and mammals seem to be most evident where prey are only patchily available, where there are other indicator species whose movements can be seen from afar and can lead the petrels to patches they might otherwise miss, and/or where the prey species are normally too deep to be reached.

II Foods

To be available to petrels, prey must be at or reasonably close to the surface, hence the importance in high latitudes of the diurnal migration of mesopelagic plankton and their predators to and from the surface layers. Breeding petrels need to have an assured supply within range of their nests but may need long flights to reach preferred feeding grounds (e.g. see Fig. 2.12). When feeding chicks many seem to work closer to home but parents may still be away for days. In such circumstances how can they ensure that the food is not digested before reaching the chick? Does the stomach oil expressed from captured animals delay digestion (Warham *et al.*, 1976), or, if digestion has occurred, does the chick then have to rely on the oil still retained in the parent's proventriculus? In general, one would expect the amount of oil to be higher and the protein lower the longer the parent has been away.

In recent years, data on foods from casual and opportunistic sources, such as the

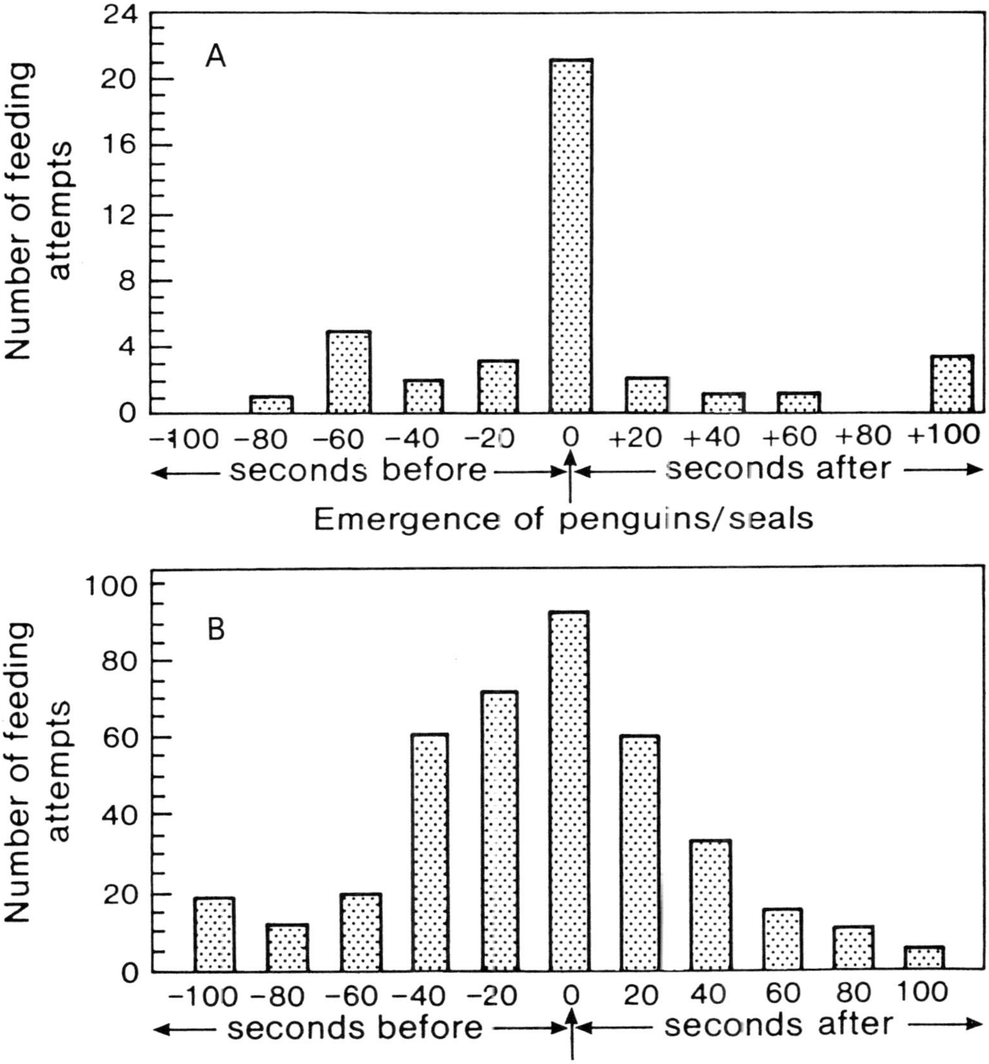

Figure 3.8 *Numbers of feeding attempts by (A) giant petrels and (B) Black-browed Mollymawks relative to the emergences of penguins and seals. From Harrison* et al. *(1991).*

examination of stomach contents of birds collected as museum specimens, have been superseded by systematic sampling and a number of detailed listings of prey animals have been published.

A *Sampling*

Many analyses derive from regurgitations from parent birds as they arrive to feed their chicks or from the chicks themselves. Fewer are from the stomach contents of

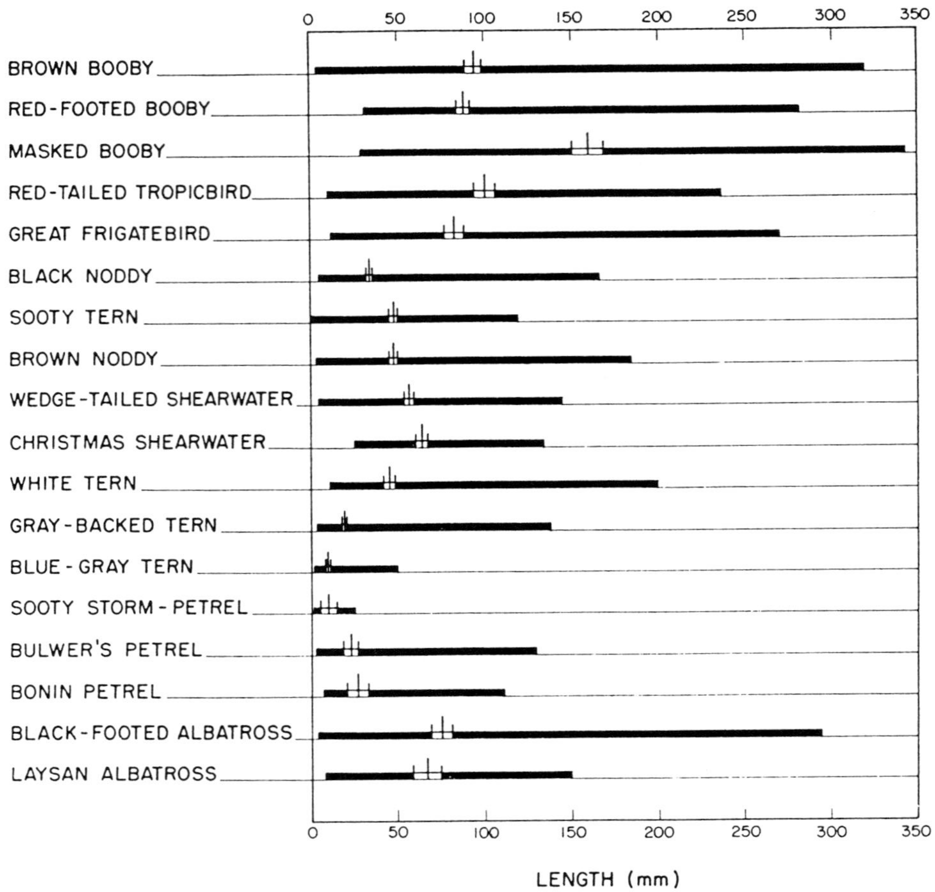

Figure 3.9 *Measurable prey lengths of Hawaiian seabirds showing means, 95% confidence limits around the means, and ranges. Larger birds, including petrels, tended to take larger prey. From Harrison* et al. *(1983).*

birds caught at sea. None provides a detailed all-year-round picture of a petrel's diet and clearly the foods taken during breeding may tell little about those used between breeding seasons, times when food may be critical for survival. Albatrosses also throw up castings of undigested matter that can be used to identify the animals from which they originate, as done by Berruti and Harcus (1978) and Brooke and Klages (1986).

Ashmole and Ashmole (1967) recommended summarizing the data under three heads: (i) by total number of prey items in each food class; (ii) by weight or volume of the different types of food in each sample; and (iii) by frequency of occurrence (f.o.o.); that is, the proportion (%) that each kind of prey bears to the whole. Their system has been widely followed and some workers have also measured identifiable and intact organisms. Figure 3.9 provides such data for Hawaiian seabirds.

Diet analysis from live birds induced to vomit, for example Schramm (1986), by

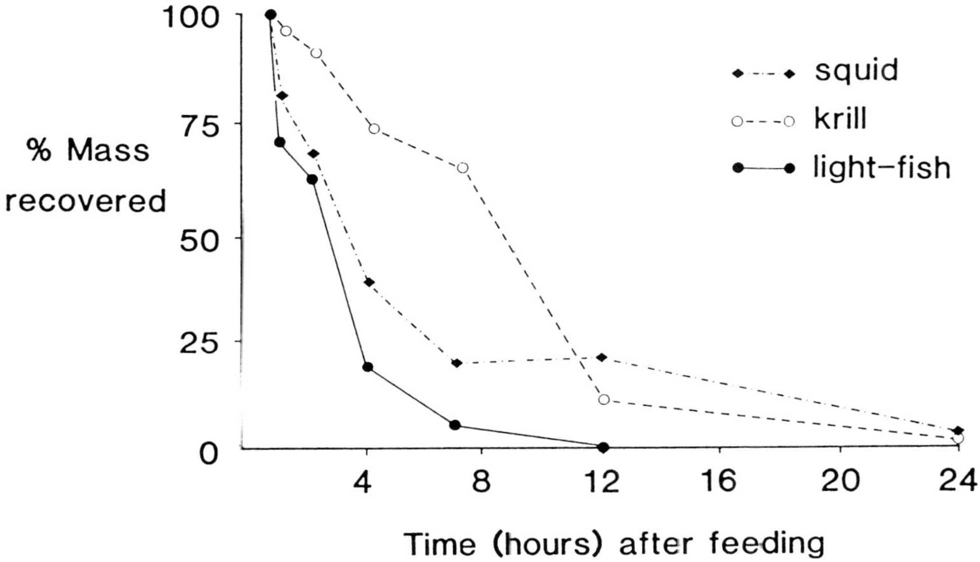

Figure 3.10 *Percentage of the mass of squid* Loligo reynaudi, *Light-fish* Maurolicus muelleri *and krill* E. superba *recovered from Shoemaker stomachs at increasing intervals after feeding. Fish could not be counted after 4 h, krill were digested slowly at first but after 12 h proportionately more squid remained and beaks recovered after 3 weeks were unworn and must have been swallowed before captivity. From Jackson and Ryan (1986).*

water offloading (Wilson, 1984, Ridoux, 1994), or by fibre-optic endoscopy (Jackson and Cooper, 1988) all have limitations. These and other problems have been discussed by Ashmole and Ashmole (1967), Harrison *et al.* (1983), Sanger (1983), Duffy and Jackson (1986), and Croxall *et al.* (1984a, 1988a). For example, most techniques do not reveal gizzard contents, so data on these are only available, if at all, from killed birds. Ainley *et al.* (1984) found that the oesophagus and proventriculus held everything from fresh prey to exoskeletons and otoliths but few squid beaks, the gizzard held the ubiquitous squid beaks and crustacean carapaces, indicating that using gizzard contents alone would overestimate squid, while using proventricular contents alone would underestimate squid.

The varied digestibilities of the different prey pose a major difficulty. Fish otoliths, for example, evidently disappear from proventriculus and gizzard within 24 h of ingestion and they are also lost in formalin-preserved material. The chances of finding recognizable coelenterates, salps and similar soft-bodied prey also seem low unless captured shortly before sampling.

On the other hand Furness *et al.* (1984) found that ommastrephid squid beaks lasted at least 50 days and *L. reynaudi* at least 38 days in a captive *Diomedea cauta*, with little signs of digestion. Jackson and Ryan (1986), after a 12-day period of conditioning to force feeding, gave captive *Procellaria aequinoctialis* chicks food of their natural prey—the fish, squid and *Euphausia superba*. The fish were the most rapidly digested with nothing identifiable after 12 h, the squid and crustacea taking longer (Fig. 3.10). No doubt the repeated stomach washes were traumatic for the birds but even if this led to digestive changes presumably it would not affect the relative rates of digestion.

B *Weights of stomach contents*

Figures for the total stomach loads carried by tubenoses—that is, food plus liquid—show no clear relationship to body weights. Mean values given by Prince (1980a, 1980b) for *Pachyptila desolata* (10.9 g), *Halobaena caerulea* (18.5 g), *Diomedea chrysostoma* (512 g) and *D. melanophrys* (433 g) represent 7.0, 9.2, 14.1 and 11.7% of their body masses. But at the extremes of the ranges some figures are very high—a 200-g maximum for *Puffinus tenuirostris* given by Montague *et al.* (1986) represents nearly 32% of the body weight of 634 g, a maximum of 850 g for *D. chrysostoma* about 23.5% of a 3625-g bird (Hunter and Klages, 1989), and a 17-g load for 59-g *Oceanodroma furcata* nearly 29% of its weight (Vermeer and Devito, 1988).

Ainley *et al.* (1992) estimated reconstituted food masses for Antarctic species: the figures rarely exceeded 25% of body mass, mean values varying between 1.5 and 7.9%. Very high maxima explained instances of gorged birds unable to fly. These workers also gauged feeding success from fullness of stomach and body mass and concluded that birds feeding in their preferred habitats did best: that is, *Daption capense, H. caerulea, L. brevirostris, Pachyptila desolata* and *Oceanites oceanicus* in the open sea; *T. antarctica* and *Pagodroma nivea* in the pack-ice.

C *Prey categories*

1 *Fish*

Few petrels ignore fish if readily available, and even in Antarctic seas, where krill have customarily been thought the mainstay of avian populations, the work of Ainley *et al.* (1992) shows that the myctophid *Electrona antarctica* may be more important, at least in the Scotia–Weddell Sea Confluence Region.

All albatrosses seem to include some fish in their diets. Among the North Pacific ones, Harrison *et al.* (1983) found that *Diomedea nigripes* ate fish from four families. Eggs of flying fish were the most prominent items, a food on which this bird appears to specialize, at least when breeding: the eggs occur in clusters, the largest weighed 314 g and contained >156 000 ova. *D. immutabilis* also took these eggs as well as a wide range of fish from at least nine families. Fish remains were also identified in 40% of the 259 chick regurgitations of *D. irrorata* obtained by Harris (1973): most were surface-living carangids and clupeids, 30–340 mm long, and they were important foods for the species.

In the southern hemisphere the fish diets of albatrosses show marked variations between species. At the Crozet Islands chicks of *D. chlororhynchos* and *D. melanophrys* were mostly fed fish, with reduced amounts evident in regurgitations from those of *D. exulans* and *D. chrysostoma* (Weimerskirch *et al.*, 1986). Analyses for South Georgian birds are given by Croxall and Prince (1980), Prince (1980b), Prince and Morgan (1987) and Croxall *et al.* (1988b) (see Fig. 3.14).

Table 3.2 Inter-island and inter-year variation in the composition of Grey-headed Albatross samples at the Prince Edward Islands. From Hunter & Klages (1989)

	Proportion of total mass (%)			Fish/ cephalopod ratio
	n	Fish	Cephalopods	
1985				
Marion Island	20	77.1	17.2	4.48
Prince Edward Island	21	44.3	39.7	1.12
1987				
Marion Island	26	51.5	39.6	1.30
Prince Edward Island	21	53.7	46.3	1.16

Based on the identification of otoliths, chicks of *D. exulans* were given 45% of fish by weight and Croxall *et al.* (1988b) reasoned that most were not obtained from fishing boat 'by-catch'. Only six species were identified, all common over the local continental shelves, but Croxall *et al.* noted that myctophids, also common offshore, were surprisingly absent from the samples.

Fish amounted to 48% by f.o.o. in 27 regurgitates from adult and chick *D. bulleri* (West and Imber, 1986) and very similar figures were gained for *D. melanophrys* and *D. chrysostoma* at South Georgia (Prince, 1980b) and for *D. cauta* at Albatross Island (Green, 1974). Hunter and Klages (1989) found that fish were important for *D. chrysostoma* at the Prince Edward Islands. Nearly 60% by weight of its diet there was fish but there were differences between Marion and Prince Edward Island, with a much higher proportion of fish from Marion in 1985 (Table 3.2). Fish were also the dominant prey of this bird at The Crozets (Weimerskirch *et al.*, 1986). Hunter and Klages suggested that the differences from the foods of the South Georgian *D. chrysostoma* arose because the Prince Edward ones fed to the north whereas the South Georgian ones fed further south where abundant *E. superba* provided an alternate food.

Fish do not feature prominently in the diets of *Phoebetria* and Thomas (1982) thought that the myctophids he found in samples from *P. palpebrata* might have been ingested only incidentally.

For some fulmars, fish constitute an important resource. Little quantitative data are available on the 'natural' food of *Fulmarus glacialis* but Furness and Todd (1984) collected adult and chick regurgitations over several years at two widely separated colonies—on St Kilda and at Foula. Despite some offal from fishing boats, most of the food identified seems to have been caught by the birds directly. The diets at these two places differed remarkably: 72% of those from Foula had eaten sand-eels *Ammodytes marinus*, whereas only 8% of the birds from St Kilda had disgorged fish: these were either Herring *Clupea harengus* or *Sprattus sprattus*. The Foula birds were reliant on the sand-eels, the St Kilda ones on invertebrates such as *Meganyctiphanes norvegica* and *Munida bamffica*. These authors judged that the dietary differences arose because of differences in the resources of the marine habitats within range during the nesting season. Basically, however, this bird is an omnivore here and in the North Pacific. There fish may feature but little in the diet or be quite important and feature Sand Lance, Capelin, Walleye Pollock and lanternfishes (Schneider *et al.*, 1985; Hatch, 1993).

Winter diets may be rather different. In the Barents Sea fish were of major importance to *F. glacialis* and other seabirds, the petrel taking cod (*Gadus morhua*) and *Boreogadus saiga*, and particularly Redfish *Sebastes marinus* and *S. mentella* (Erikstad, 1990). Those eaten tended to be from the smaller and younger size classes, median length 51.0 mm ($n = 155$), and this worker related prey sizes to the gapes (taken at the feathers) of predator species, 20.8 mm for *F. glacialis*, which presumably does bear some relationship to the opening when the gape is fully extended.

Excepting giant petrels, which in captivity may refuse fish, the southern fulmarine species have considerable dependence on adult *Pleuragramma antarcticum*, a surface-living notothenid. Arnould and Whitehead (1991) analysed theirs and earlier studies based on chick and adult food samples. While raising chicks, fish tended to dominate the diets of *F. glacialoides* and *T. antarctica*, but were less important for *Daption capense* and Whitehead (1991) found that *Pagodroma nivea* differed from *T. antarctica* and *F. glacialoides* in consuming the myctophid *Electrona antarctica* which was also taken by *D. capense*.

The Southern Fulmars ate significantly larger *P. antarcticum* (mean length 142 mm) than did the Antarctic Petrels (125 mm) and the difference seems most likely to arise because the latter fed closer to the ice and *F. glacialoides* in deeper water where the fish were larger. Antarctic Petrels can take the bigger-sized fish and do so further east where they give their chicks the same fish averaging 193-mm long (Klages *et al.*, 1990). There are other regional variations, for instance *F. glacialoides* at Adelie Land relied less on fish during a study by Ridoux and Offredo (1989), and no differences in fish sizes were apparent in a parallel study of *T. antarctica* and *F. glacialoides* by Norman and Ward (1992) (Fig. 3.11).

Snow petrels foraging off Adelie Land ate mainly fish. These were too well digested for Ridoux and Offredo to identify, but analyses of birds in the Ross Sea showed them as extensive feeders on *P. antarcticum* when over the continental shelf and slope, but not when they fed in deeper water (Ainley *et al.*, 1984).

Prions are customarily regarded as crustacean feeders but adult *P. salvini* at Marion Island in the summer contained 42% by weight of fish (Gartshore *et al.*, 1988). Fish also occurred in 83% of the samples from *Halobaena*, amounting to perhaps 50% by weight (Prince, 1980a). Most were small animals, probably myctophids, and these were also identified in Marion Island birds by Steele and Klages (1986) whereas they were rare from those at The Crozets and absent from birds beach-wrecked in Australasia.

Most gadfly petrels take fish from time to time although they seldom feature prominently in stomach analyses. Exceptions are *Pterodroma hypoleuca* and *P. inexpectata*. The former appears to be a major fish eater with at least 12 identifiable families in 144 samples, mainly from adult regurgitations (Harrison *et al.*, 1983). In the vomits of 69 *P. inexpectata* chicks, myctophids at 39% were the major items by weight (Imber, 1991).

Bulwer's Petrels in Hawaiian seas ate mainly fish, seven families being represented, amounting to 71% by volume of the stomach contents (Harrison *et al.*,

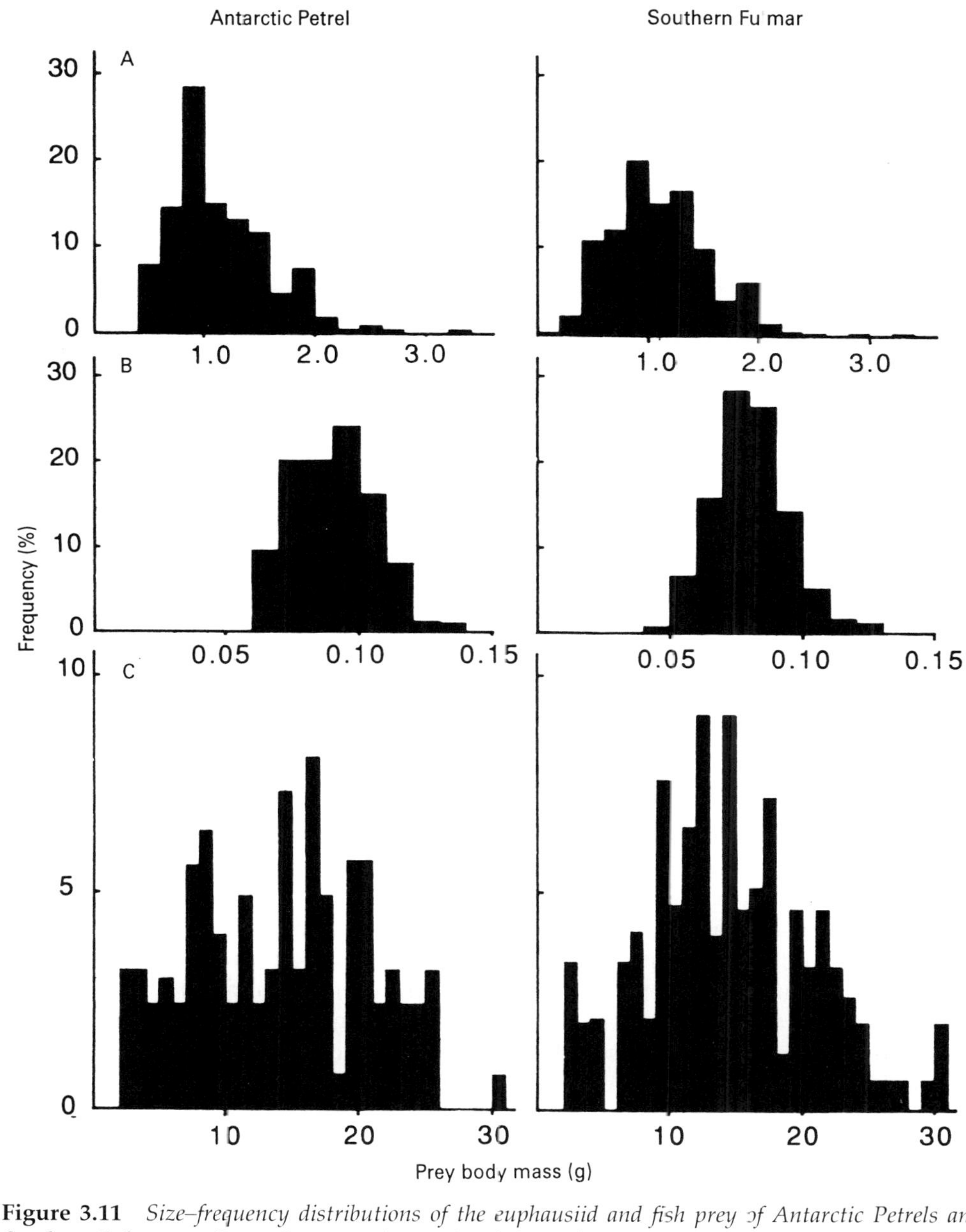

Figure 3.11 *Size–frequency distributions of the euphausiid and fish prey of Antarctic Petrels and Southern Fulmars at Rauer Bay, east Antarctica. (A)* Euphasia superba *(n = 241 & 1166), (B)* E. crystallorophias *(n = 75 & 249), and (C)* P. antarcticum *(n = 92 & 107). From Norman and Ward (1992).*

1983). There were lanternfishes *Argyropelecus* spp. and flying fish up to 130-mm long.

Imber (1976a) found a significant use of fish by three *Procellaria* petrels. The species were mostly from bioluminescent families, supporting Imber's view that *Procellaria* spp. are primarily nocturnal feeders.

In Hawaiian seas Harrison *et al.* (1983) identified 19 species from 14 fish families, as well as seven identified only to family level, eaten by *Puffinus pacificus*. By volume the diet averaged 66% fish, 28% squid and 1% crustacea. Ranked according to numbers, proportions or volumes, goatfishes Mullidae were the most important followed by Carangidae, with filefish, flying fish and gobies also prominent.

Of the two tubenoses studied by Ashmole and Ashmole (1967) at Kiribati only *P. nativitatis* was a major fish eater, taking especially flying fish, but around the Hawaiian Islands this took mostly squid although fish were still important there with 17 families represented and Mullidae, Carangidae and Exocoetidae the major components (Harrison *et al.*, 1983).

For Manx Shearwaters, clupeids, particularly *C. harengus*, *S. sprattus*, sardines *Sardina pichardus*, and anchovies *Engraulis encrasicolus* are important. Thompson (1987 in Brooke, 1990) reported that 66% of 76 stomachs held clupeids and sand-eels, 47% squid, the fish seemingly more prominent during chick rearing. Large flocks of *P. puffinus yelkouan* were also reported as feeding on big *E. encrasicolus* shoals off the eastern shore of Crimea (Dement'ev *et al.*, 1968, p. 377).

The voracious eating habits of the 'hagdon' *P. gravis* on the Grand Banks were graphically described by Collins (1884) where the birds arrive as Capelin *M. villosus* start to spawn. Brown *et al.* (1981) compared the diet of *P. gravis* with that of *P. griseus* which reaches the waters off New England first, in early- to mid-May. *P. gravis* appears a fortnight later, and then moves up to Labrador seas apparently following the spawning fish. Brown *et al.* (1981) nevertheless considered that there was unlikely to be significant competition for the shared foods because, when the birds arrive the cephalopod *Illex* and Capelin are evidently superabundant and later, when these resources decline, the shearwaters go their separate ways, *P. griseus* to British and Irish waters, *P. gravis* to Greenland, Norway or the Biscay area.

In the Gulf of Alaska the fish-eating shearwater is *P. griseus* whereas *P. tenuirostris* concentrates on euphausiids, the former feeding at the higher trophic level (Krasnow and Sanger, 1982). Further west, Ogi (1984) examined the foods of *P. griseus* north of the Subarctic Convergence (Boundary) between 150° and 190°E and Ogi *et al.* (1980) that of *P. tenuirostris* in the Okhotsk and Bering Seas and east of the Kurile Islands to 170°E. The most important food for *P. griseus* there by weight was the Pacific Saury *Cololabis saira* in all months and all regions, although squid were sometimes more important locally. The shearwaters and the fish occurred in the same areas but as the days lengthened the birds tended to move north ahead of the fish, suggesting that both were reacting to remaining within a preferred SST of 9–13°C, rather than the petrels following the shoals. In the Bering Sea, the average diet of *P. tenuirostris* based on 216 samples was 40.4% by weight of squid, 25.6% euphausids, 20.9% amphipods and only 12.6% fish, but in the North Pacific 62.5% of the average diet in 125 stomachs was fish, 18.8% squid, 8.8% euphausiid and 9.3% amphipods and copepods. Even more variable were the diets over time within different areas, emphasizing the opportunistic nature of the feeding process and the bird's capacity to maintain itself in a range of marine habitats.

In Californian waters, *P. griseus* took mainly Rockfish *Sebastes* sp. from May through July but fattened and completed their moult on anchovies *E. mordax* in August and September (Chu, 1984). By the latter month their energy-rich diet had raised their average fat content to 42% of their body weight in readiness for the southwards migration.

Off South Africa samples taken by Jackson (1988) throughout the year in the Benguela Current showed that *P. griseus* also largely depended on fish and, unlike the *Procellaria aequinoctialis* sharing these seas, *P. griseus* is not usually attracted to fishing vessels, so that the stomach contents should reflect a natural diet. The diet varied with time and Jackson suggested that many of the fish, being mesopelagic, were taken by day when forced to surface by seals, dolphins or tuna.

Fish are important prey for storm petrels, particularly for the short-legged *Oceanodroma* spp., although *Oceanites oceanicus* also feeds myctophids to its chicks (Croxall and North, 1988). Croxall *et al.* (1988a) found that this petrel also gives them relatively large fish—60–85 mm long (1.8–4.0 g), substantial items for a bird with a culmen length of only 12 mm. Fish are prominent elements in the diets of *Oceanodroma furcata* and *O. leucorhoa*. Both took a range of myctophids (Vermeer and Devito, 1988), lampfishes being of increasing importance as the season progressed (Fig. 3.12). In Japan, the fish component in the meals of *O. leucorhoa* also increased with time (Watanuki, 1985). The gonostomatid *Vinciguerria lucetia* was preyed upon by a big concentration of *O. leucorhoa* in the Central Pacific. Many birds were resting, evidently satiated, their bellies crammed with fish. Stomach contents were 15.6–24.4% of the bird's body weight (Pitman and Ballance, 1990).

Agnatha are also eaten; the lamprey *Geotria australis* by *D. chrysostoma* (Tickell, 1964; Prince 1980b), and occasionally by *D. melanophrys*. Prince estimated that the chicks of these two birds consumed over a million subadult lampreys in a season. This fish is also eaten by *Procellaria parkinsoni* (Imber 1976a) and the Arctic Lamprey *Geotria japonica* by *Puffinus tenuirostris* (Ogi *et al.*, 1980). Lampreys are a much richer source of energy than bony fish or squid. Elasmobranch material has not been reported as petrel food, but the presence of squalene in stomach oil from chicks of *D. epomophora* (Warham *et al.*, 1976) and in other petrels may have arisen from scavenging on dead sharks.

2 *Squid*

An early reference to these molluscs as foods for petrels is that of Ramirez (1603 in Beebe, 1935) who noted that *Pterodroma cahow* ate squid. Nearly all tubenoses seem to take them at one time or another and for many they are the principal food. They are seldom recorded for prions, diving petrels and the storm petrels *Garrodia nereis, Pelagodroma marina* and *F. grallaria,* and generally appear to be of minor importance to the fulmarine species and *Oceanites oceanicus* breeding around Antarctica (Whitehead, 1991).

The evidence of their use is mostly in the gizzard which is often jammed tight with their keratinous 'jaws' that are either slowly digested and eroded away or disgorged. Because the shapes and sizes of these hard parts vary according to the species and age of the molluscs it has been possible to identify many to family and even species. This has been done notably by Imber (1973, 1975a, 1978, 1992) and Imber and Berruti (1981) (Fig. 3.13).

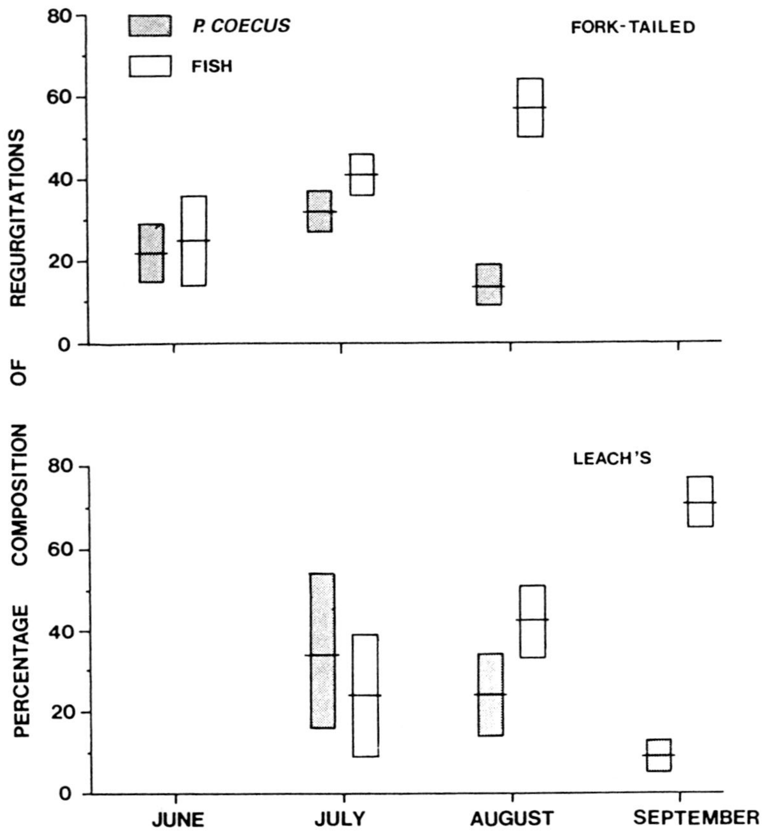

Figure 3.12 *Comparisons of the average compositions by wet weight of the amphipod* Paracallisoma coecus *and myctophid fish in the regurgitations of adult storm petrels* O. furcata *and* O. leucorhoa *breeding sympatrically in British Columbia in 1983. Horizontal lines show means and rectangles 95% confidence limits. From Vermeer and Devito (1988).*

The most detailed analyses of the role of squid in the diets of petrels are those of the Wandering Albatross by Rodhouse *et al.* (1987), Imber (1992) and Cooper *et al.* (1992) and of Black-browed Albatross by Rodhouse and Prince (1993). Rodhouse *et al.* sampled 3421 beaks representing 35 prey species and calculated the biomass contributions of each and their size–frequency distributions. Imber (1992) analysed the squid eaten at six circumpolar breeding stations based on 9994 beaks belonging to 61 species, including some undescribed. There were some marked variations between populations. Many squid species decreased in importance southwards, but some increased with increasing latitude; some were more important in mid-sub-antarctic locations, and generally their distributions agreed with those previously known from oceanographic studies. There were also seasonal variations. The ommastrephid *Illex argentinus* was 85 times more numerous in diets from 1983–84 than from 1975–78 at South Georgia and, as this species is important in the Falkland

Islands fishery that developed since the mid-1970s, Imber suggested that the later birds were scavenging around the trawler fleet some 1200 km from their nests. Regurgitation of a pair of cuttlefish (*Sepia* sp.) at Antipodes Island supports the view that breeders from here can forage as far as eastern Australia as this is where the nearest known population of cuttlefish occurs.

Kondakovia longimana is an important squid for sooty albatrosses, notably for *Phoebetria fusca*, making 61% of the squid mass in the diet (Imber, 1991). He deduced that this meant a more northerly feeding range than that of *P. palpebrata* which ate less *K. longimana* and more *Psychroteuthis* and *Alluroteuthis*, which is consistent with this bird foraging far to the south, as also deduced by Berruti and Harcus (1978) and confirmed by Weimerskirch and Robertson (1994).

All albatrosses so far studied eat squid (Fig. 3.14) and some large animals have been identified. Harris (1973) estimated ommastrephids taken by *D. irrorata* averaging 2091 ± 2779 g (100–11 000 g), the biggest three times the weight of the bird and, again, presumably scavenged when dead or moribund.

Using 5651 identified beaks enabled Rodhouse *et al.* (1990) to compare the squid intake of *D. exulans* between years when breeding success was low and high. They found that the total number and biomass eaten was significantly lower in the bad season. The ommastrephid *Martalia hyadesi* contributed 73–79% of the biomass of chick meals and the size of the squid increased as the chicks grew—they averaged

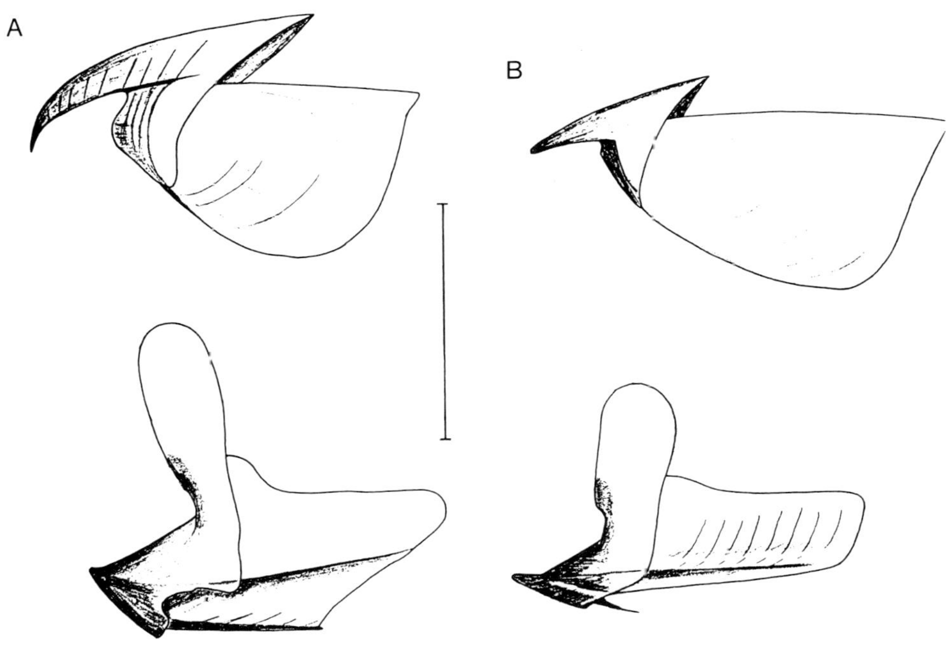

Figure 3.13 *Examples of the upper and lower beaks of squid eaten by Wandering Albatrosses. (A)* Moroteuthopsis *sp. (B)* Chiroteuthis *sp. Scale = 1 cm. From Imber (1992).*

3.2 ± 0.5 mm lower rostral length for 0–20 day chicks to 4.9 ± 0.7 mm for 120–140 day chicks, suggesting that the squid were growing too.

For *D. immutabilis* in the Hawaiian Islands squid were the highest ranking prey by volume, particularly ommastrephids, estimated as amounting to 65% of the 183 samples inspected. *D. nigripes* ate fewer, only about 32% of 172 samples (Harrison *et al.*, 1983). Both ate animals of similar size, 71–86 mm long, *D. nigripes* probably by day, the other more by night.

Squid also appear unimportant to breeding *F. glacialis* in the North Atlantic but seem to be important for fulmars off the west coast of North America and the Gulf of Alaska. Hills and Fiscus (1988) found the cranchiid *Taonius* sp. the most abundant component, a genus also important in the diets of *Pterodroma macroptera* and *D. exulans* in the south (Imber, 1973, 1992).

In Antarctic waters, squid seem of low significance for nesting fulmars but are more conspicuous in birds taken in deep water, particularly with *F. glacialoides*, *T. antarctica*, *Daption capense* and *L. brevirostris* (Ainley *et al.*, 1984; Ridoux and Offredo, 1989; Imber, 1991).

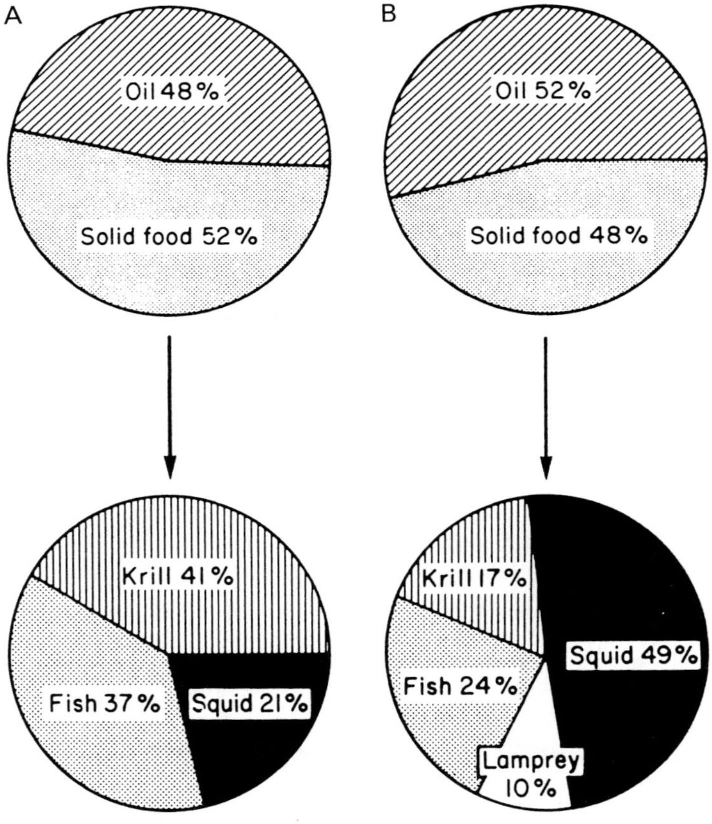

Figure 3.14 *Composition by wet weight of diets of adult mollymawks at South Georgia. (A)* Diomedea melanophrys. *(B)* D. chrysostoma. *From Prince (1980b).*

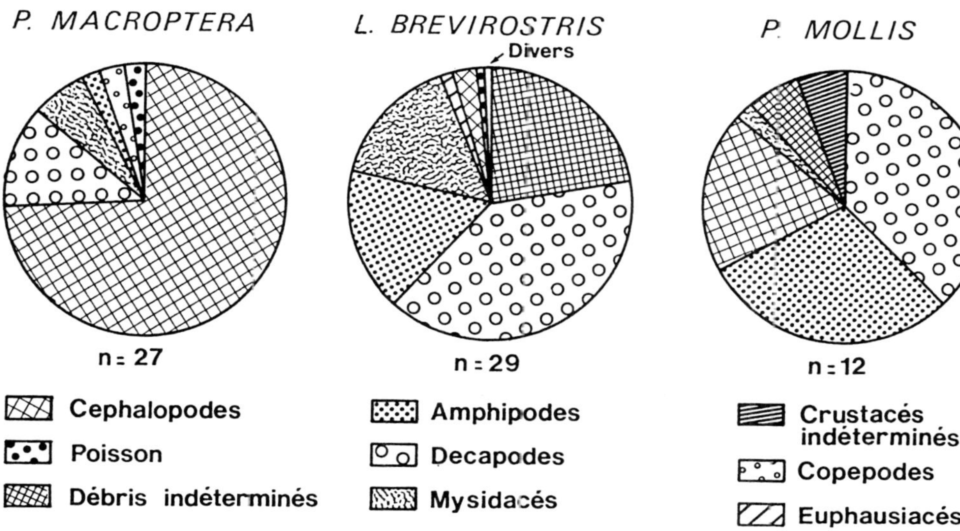

Figure 3.15 *Foods of three gadfly petrels at the Crozet Islands, by volume. From Jouventin* et al. *(1988).*

Most gadfly petrels take squid and for some they comprise their main food. Good data are available for only a few species. They are of prime importance to *P. mollis*, *P. lessonii*, *P. incerta*, *P. macroptera*, *P. phaeopygia*, *P. cookii* and *P. alba* (Ashmole and Ashmole, 1967; Imber, 1973, 1985b, 1991; Schramm, 1986; Jouventin *et al.*, 1988) (Fig. 3.15). At the Prince Edward Islands, *P. macroptera*, *P. brevirostris* and *P. mollis* ate mostly the same squid (Schramm, 1986), but as their breeding cycles differ Schramm considered that interspecific competition for them would thereby be lessened, added to which the crustacean elements differed significantly between them.

There are no quantitative data on the foods of *Procellaria cinerea*, but determinations have been made for *P. aequinoctialis* (Croxall and Prince, 1980; Prince and Morgan, 1987; Croxall *et al.*, 1995), and for *P. westlandica* and *P. parkinsoni* by Imber (1976a). In the Shoemaker data of Jackson (1988), much of the food was scavenged from fishing boats. Squid tend to dominate the diets, nine families of squid and octopods being identified for *P. westlandica* and eight for *P. parkinsoni*. In the composition of the squid taken, Imber saw a transition from the cooler subantarctic waters to tropical seas in the change from histioteuthids and crachiids to more ommastrephids, the latter being the most important foods for *P. parkinsoni*.

In the Benguela Current, Lipinski and Jackson (1989) found squid in 75% of 161 samples from *P. aequinoctialis*. Some were Antarctic endemics, but it was thought that this was the result of beak retention rather than that the birds were ranging that far.

Squid are also eaten by *Puffinus* spp. but again, are usually only of secondary importance to fish. Squid in the diet of *P. griseus*, when in the North Pacific were examined by Baltz and Morejohn (1977), Krasnow and Sanger (1982), Chu (1984), Ogi (1984) and Sanger (1987). Data for the North Atlantic birds were given by Brown *et al.* (1981). No quantitative data are available for breeding birds, the most detailed

Table 3.3 Major prey of Sooty Shearwaters from Monterey Bay (1978–79) and southern California (1977). Percent occurrence combines data from stomachs and gizzards; percentage volume and percentage number are based on full stomachs only (MBay, Monterey Bay; SoCal, southern California; n, number of birds). From Chu (1984)

	% occurrence		% volume		% number	
	MBay ($n = 154$)	SoCal ($n = 37$)	MBay ($n = 87$)	SoCal ($n = 29$)	MBay ($n = 87$)	SoCal ($n = 29$)
Fish*	97.4	94.6	91	84	33	28
Engraulis mordax	41.6	27.0	44	53	6	1
Sebastes spp.	22.7	62.2	44	17	20	25
Merluccius productus	—	51.4	—	—	—	—
Scorpaenichthys marmoratus	—	8.1	—	<1	—	<1
Cephalopods*	94.8	74.8	5	6	4	2
Loligo opalescens	68.8	35.1	5	6	3	2
Onychoteuthis/Gonatus spp.	47.4	35.1	—	—	—	—
Gonatus spp.	40.3	18.9	—	—	—	—
Onychoteuthis borealijaponicus	29.2	8.1	—	—	—	—
Octopus rubescens	19.5	2.7	—	—	—	—
Histioteuthis heteropsis	9.1	2.7	—	—	—	—
Crustaceans*	17.5	18.9	3	10	63	70
Thysanoessa spinifera	4.5	10.8	3	7	59	65

*Unidentified and identified material.

information from the south coming from birds taken off South Africa (Jackson, 1988; Lipinski and Jackson, 1989).

In the western subarctic Pacific these shearwaters take squid which comprised 7.3% by weight (73% by f.o.o.) in Ogi's study, with up to 290 beaks in a single bird. Their abundance varied with latitude, with fewer in warmer waters. Further east, in the Gulf of Alaska, there were squid in 76 and 68% of adult stomachs sampled in 1977 and 1978 and judged to be the second most important food item during those years (Krasnow and Sanger, 1982; Sanger, 1987).

Off northern California, *Loligo opalescens* were rated the second most common prey by volume and occurrence (Table 3.3) with 94% of the gizzards having squid beaks and up to 73 sets in one gizzard (Chu, 1984).

Off southwestern Africa *Puffinus griseus* is much less dependent on cephalopods although 61% of the gizzards of 48 birds contained beaks of animals smaller than those eaten by *Procellaria aequinoctialis* foraging in the same area.

The squid diets of *Puffinus tenuirostris* are better known than those of *P. griseus* with data from both hemispheres. In the northwest Pacific, Okhotsk and Bering seas, Ogi *et al.* (1980) estimated that squid (mostly unidentified) were the major items on a wet weight basis, but with much variation between zones and with fewer of them in birds from shelf regions and none where euphausids dominated, as in Bristol Bay. In the Gulf of Alaska, Krasnow and Sanger (1982) found 38% of the birds collected contained cephalopod beaks in 1977 but with none or few in other years.

Skira (1986) identified 20 kinds of squid from adult birds on the breeding grounds. Both adult and juvenile squid were used and over the 7 months' study period the proportions of squid in the diets varied little, their estimated weights being 32–75 g. On a f.o.o. basis squid were only second in importance to *Nyctiphanes australis*.

Squid are important for *P. gravis* off northeastern Canada (Brown *et al.*, 1981) and in its home waters of the South Atlantic. Lipinski and Jackson (1989) found at least four species in the stomachs of *P. gravis* from the southern Benguelan Region with every stomach containing beaks.

Christmas Shearwaters *P. nativitatis* sampled in Hawaiian seas and in the Central Pacific were using squid as a major food (Harrison *et al.*, 1983; Ashmole and Ashmole, 1967). Remains of 304 were found in 80 food samples from Kiribati and from 182 samples, Harrison *et al.* ranked ommastrephids as the most important food. Overall, the Hawaiian birds took less squid (48% vs. 71% by volume) than did the Christmas Island (Kiribati) birds, presumably reflecting differences in food availability and different sampling seasons.

Among storm petrels, squid have been reported from *Oceanodroma castro* and *Hydrobates pelagicus* and unidentified species comprised 29% of the food volume in 10 samples from *O. tristrami* (Harrison *et al.*, 1983). Ainley *et al.* (1984) estimated that 51% of 37 stomachs of *Oceanites oceanicus* collected in southern seas contained squid but they amounted to a mere 2% of 80 regurgitations from adults at South Georgia (Croxall *et al.*, 1988a). British Columbian *Oceanodroma furcata* and *O. leucorhoa* both included squid in their diets and without much change in their importance over the course of the breeding season but they were not the dominant prey (Vermeer and Devito, 1988) as was also true of *O. leucorhoa* studied in Japan by Watanuki (1985).

Pteropod molluscs are also eaten. They have been identified from *Pterodroma phaeopygia, F. glacialis, F. glacialoides, T. antarctica, Daption capense, Halobaena caerulea,* and from all *Pachyptila* except *P. belcheri,* but in no case do these appear to be important in relation to the total diets.

3 Crustacea

Euphausiids, amphipods and copepods feature in the diets of most tubenoses to some degree, some species specializing on them. Smaller forms may originate from the stomachs of larger prey, others, like the parasitic isopods in the diets of *Diomedea irrorata* and *D. bulleri*, are evidently ingested accidentally. In the Southern Ocean the swarming krill *Euphausia superba* and *E. chrystallorophias* are major foods for petrels and despite a great range of bill sizes most of those at South Georgia concentrated on mature animals (Croxall and Prince, 1980).

Crustacea seem only to be secondary foods for most albatrosses. For example, diets of the Chatham Island and Taiaroa Head *D. epomophora* included 19 and 32% by frequency of occurrence, mainly *Munida* (Imber, 1991), and *D. exulans* evidently takes even less crustacea (Prince and Morgan, 1987). In the north Harrison *et al.* (1983) found eight crustacean groups in foods from *D. nigripes* including large, deep-water mysids *Gnathophausia gigas* (average length 112 mm) and *G. ingens*, as well as a big isopod *Anuropus branchiatus* (average length 106 mm). For *D. immutabilis*, crustacea occupied 9% by volume of 183 samples. They were also in half the chick regurgitations of *D. irrorata* (Harris, 1973).

West and Imber (1986) found few crustacea in vomits from *D. bulleri* but for *D. melanophrys* and *D. chrysostoma* however, *E. superba* was important and the main food taken by breeding *D. melanophrys* (Tickell, 1964; Prince, 1980b) (Fig. 3.14).

The diets of *Phoebetria* albatrosses seem to vary in different regions (Ryan and Cooper, 1989; Imber, 1991) and only the South Georgia *P. palpebrata* have been found to take many crustacea. Thomas (1982) recorded 36% of mature *E. superba* by weight in its diet, average length 53 mm, together with some prawns and mysids. Crozet Island *P. palpebrata* also took some *E. superba* but *P. fusca* very few (Weimerskirch *et al.*, 1986).

The role of crustaceans as food for fulmars is generally greater than it is with albatrosses. Even giant petrels take some, mainly *E. superba*, and apparently directly (Johnstone, 1977; Hunter, 1983).

In Antarctic waters, the smaller fulmars also use the abundant krill extensively as demonstrated by the analyses of Ainley *et al.* (1984); Montague (1984); Ridoux and Offredo (1989); Klages *et al.* (1990); Arnould and Whitehead (1991) and Norman and Ward (1992) (Fig. 3.16).

The stomachs of 90–100% of Antarctic Petrels and Southern Fulmars contained crustacea, either *E. superba* or *E. chrystallorophias*, although on a weight basis fish were more important, at least during chick rearing. However, in birds collected over the slope and deeper water during incubation, the crustacean component was low, squid predominating (Ainley *et al.*, 1984).

Crustaceans are particularly important for *Daption capense* (Fig. 3.16). In Adelie Land these amounted to 64% by weight with more *E. chrystallorophias* than *E. superba*, whereas in Prydz Bay these petrels took none of the latter and when raising chicks took 86% by weight of *E. superba*. In Elizabeth Land, Antarctica these birds had 76% crustacea, mostly *E. superba*, by weight, the rest being fish (Green, 1986a). In deep water, *D. capense* caught fewer crustacea but more squid (Ainley *et al.*, 1984). At Signy Island, *E. superba* was in 98% of 94 samples. The deep-sea amphipod *Eurythenes* sp. was also found in good condition, probably caught alive (Beck, 1969).

In the North Pacific, *F. glacialis* seldom depends on crustacea (e.g. Fig. 3.3), though at the Semidi Islands they were significant prey in the summer (Hatch, 1993), and along the Arctic ice edges copepods at 67% were the commonest food although only 1% of the dry weight (Bradstreet and Cross, 1982). Hartley and Fisher (1936) found birds in West Spitzbergen in mid-summer were eating mainly *Thysanoessa inermis* and *Parathemisto libellula*. In the Barents Sea in the winter, crustacea did not feature in the diets analysed by Erikstad (1990) but those of Furness and Todd (1984) showed that in the North Atlantic the mainly fish diet of birds feeding chicks was supplemented by crustacea, particularly *Meganyctiphanes norvegica* and the mysid *Gnathophausia zoea*. For chicks of *L. brevirostris* at Kerguelen the figure for crustacea was 24% and all 22 of the samples examined contained some.

Although gadfly petrels tend to rely on squid, all those examined to date have been found to take some crustaceans and to feed them to their chicks (Fig. 3.15). The Bonin Petrel *Pterodroma hypoleuca* took only about 7% by volume of species from at least seven families, and some may have been secondarily derived from the fish prey (Harrison *et al.*, 1983). For the tropical *P. alba* only 2% by volume was crustacean (Ashmole and Ashmole, 1967). Among subantarctic species the highest use was from chicks of *P. inexpectata*, 30% by weight of whose identified food was crustacean

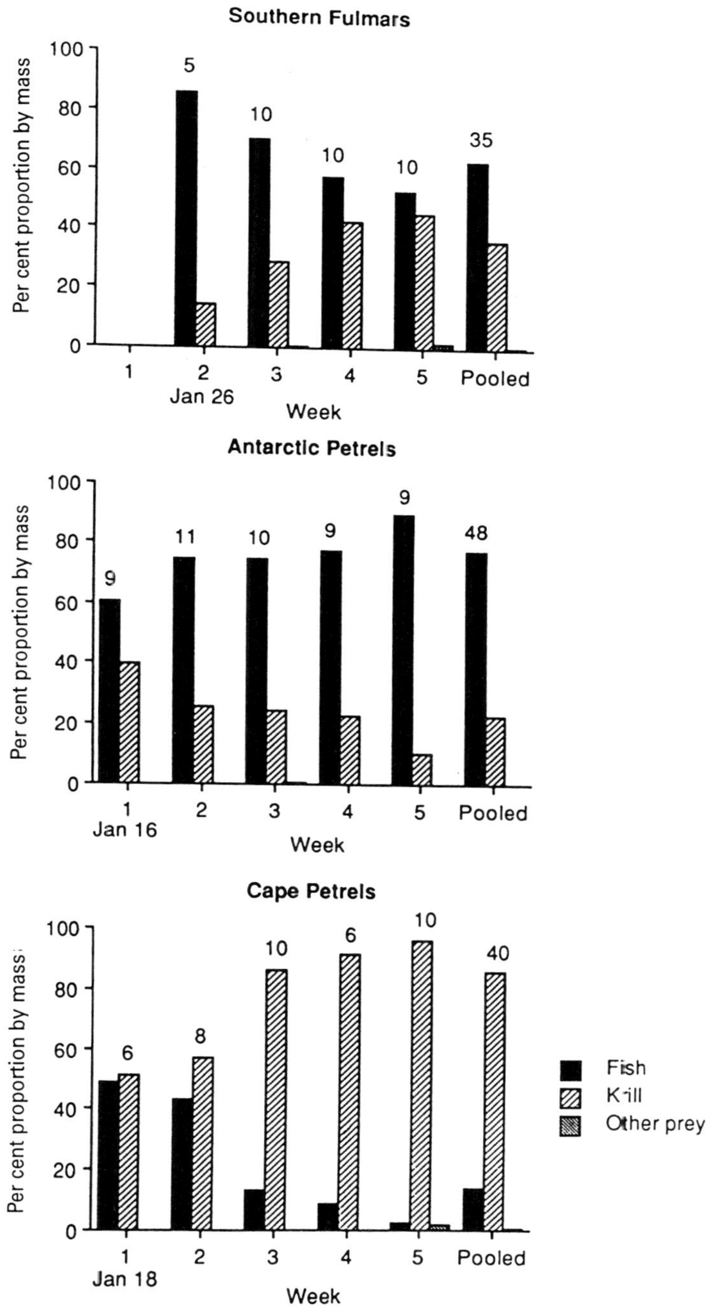

Figure 3.16 *Proportions of major food types eaten by three fulmars over 5 weeks while feeding chicks. The raw figures were weighted to allow for the apparent contributions of each prey to the whole, sample sizes at tops of bars, dates when sampling began shown below bars for first week. From Arnould and Whitehead (1991).*

(Imber, 1991). He gave lower figures of 7 and 12% for *P. lessonii* and *P. incerta*. Crustacea were less numerous in regurgitations from chicks of *P. macroptera* and *P. mollis*. In both species the animals were mostly whole amphipods, decapods and mysids, with no clear differences in diet composition or prey sizes (Schramm, 1986). In the 100 samples collected by Harrison *et al.* (1983) from Bulwer's Petrel at Laysan Island there were only 3.9% by volume of crustaceans.

Crustaceans are very important for prions, small copepods being taken by filter feeders such as *Pachyptila vittata* and euphausiids by *P. belcheri* and *P. turtur*.

In 57 food samples from Chatham Islands' *P. vittata*, the copepod *Calanus tonsus* amounted to 70% by weight, 20% of the rest being made up of at least nine kinds of amphipod, *Platyscelus ovoides* the most important, but also with about 6% of the euphausid *N. australis* (Imber, 1981).

For the next most specialized prion, a filter-feeder and hydroplaner, *P. salvini*, data are available from Marion Island (Gartshore *et al.*, 1988), The Crozets and Kerguelen (Bretagnolle *et al.*, 1990). Amphipods figured largely in all their diets, with *Themisto gaudichaudii* by far the major prey. At South Georgia the food of *P. desolata* was found to be 98% crustacean by weight and 99% by f.o.o., with 56% *E. superba* and 32% the copepods *Rhincalanus gigas* and *Calanoides acutus* plus various amphipods such as *T. gaudichaudii*.

There are quantitative data on *P. turtur* from New Zealand and South Georgia (Imber, 1981; Prince and Morgan, 1987). Euphausiids dominated the diets. At South Georgia this was *E. superba*, in New Zealand seas *N. australis*, with the next most numerous prey the amphipod *T. gaudichaudii* at both places.

T. gaudichaudii was identified from *P. belcheri* taken at sea and Bretagnolle *et al.* (1990) showed that, as expected of a thin-billed bird, fish play a major role, being present in at least half the 20 stomachs he examined.

Ealey (1954) reported crustacea from the stomachs of *P. crassirostris* at Heard Island, again mostly *T. gaudichaudii*, with up to 600 in a single petrel. The Chatham Island birds, however, ate adult *Lepas* barnacles (Cirripedia) (Imber, 1981). The stout, rather powerful bill seems well designed for pulling at and crushing barnacle 'shells'.

Crustacea form the major food of the Blue Petrel *H. caerulea* being 60% by weight of the 49 regurgitations examined by Steele and Klages (1986) at Marion Island and 92% of 156 from South Georgia (Prince, 1980a). At this latter station the euphausiid *Thysanoessa macrura* was the most abundant but the bulk by weight were *E. superba*. At Marion Island *E. vallentini* predominated, and at both locations a range of amphipods was also eaten.

Procellaria petrels seem to take few crustacea but Shoemakers in the Benguela Current did eat some stomatopods *Squilla armata* and large numbers of *E. lucens* were in birds collected at a swarm of these animals by Jackson (1988). Furthermore, South Georgian birds tending chicks took 47% by weight of krill, 33% of fish and only 19% of squid (Croxall *et al.*, 1995). These authors suggested that these differences were in line with the Shoemaker's status as a food generalist.

Wedge-tailed Shearwaters appear to take very few crustacea but *Puffinus bulleri*, at least when breeding, eats many *N. australis*, 76% of 30 regurgitations of adult birds being composed solely of this euphausiid (Harper, 1983).

Although in the North Pacific *P. griseus* is mainly a fish eater, it too relies on euphausiids at times. Around Kodiak Island, for example, while euphausiids were not taken in 1977, in June 1978 *T. inermis* amounted to 5–10% by volume and was estimated to be the second or third most important prey by Krasnow and Sanger (1982). In July and August, however, crustacea were displaced by squid or fish. Further west, in the subarctic, Ogi (1984) also found crustaceans far outweighing fish. One bird had 11% by weight of barnacles.

Off Brier Island in the North Atlantic in August and September before the southward migration the euphausid *M. norvegica* outnumbered fish remains in 29 *P. griseus* proventriculi, being 64% of the wet weight of the whole contents with up to 534 animals in one stomach, average 181 (Brown *et al.*, 1981).

Good quantitative data for the foods of *P. griseus* in the southern hemisphere are lacking, and *P. tenuirostris* is perhaps the only tubenose for which there is year-round dietary information. In general the figures show that *P. tenuirostris* favours crustacea while *P. griseus* selects fish, but both are adaptable and switch foods evidently according to what is available.

In the Okhotsk and Bering seas and the western North Pacific the diet sampled by Ogi *et al.* (1980) was very variable from one marine zone to another. For example, in Bristol Bay 99% by weight of 80 stomachs were euphausiids, mainly *Thysanoessa raschii*. In birds from other areas at similar dates there were none. Over the shelves of the Bering Sea amphipods were important, for example 60% by weight of *P. libellula* but only 16% *Thysanoessa longipes* with up to 656 of the former and 1000 of the latter in one stomach. In the western subarctic Gyre region *T. longipes* comprised 19% by weight with amphipods like *Parathemisto japonica* less important but with, on average, 65 per stomach (1–518). Copepods, *Calanus cristatus*, were the fourth in importance, averaging 190 per stomach (5–366). Over the whole vast area sampled using 296 stomachs, the fish amounted to 12.6%, euphausiids 25.6%, amphipods 20.9% and squid 40.4% by weight, so that crustacea were the dominant prey overall.

In the south during the breeding season crustacea again form the main foods with a decline by the time of chick rearing as the take of fish increases (Fig. 3.17), perhaps because *N. australis* has finished spawning by then and may have decreased in near-surface waters (Montague *et al.*, 1986).

The shift to fish during chick rearing (Fig. 3.17) was also shown in the compositions of the chick regurgitations, *N. australis* being the main food in January with postlarval fish dominant thereafter (Fitzherbert, in Marchant and Higgins, 1990).

Of the remaining shearwaters for which reasonable data are available, the case of *P. gravis* feeding off northeastern Canada again illustrates the flexibility of food choice. In 1974, euphausiids (*M. norvegica*) made up 30% by weight of the whole meal but in the next year, when *M. norvegica* did not swarm, its role fell to about 0.3%, squid taking over as the main constituent (Brown *et al.*, 1981).

Storm petrels seem more dependent on crustacea than the rest, judging from the diet analyses of *Oceanodroma leucorhoa, O. furcata, O. tristrami, Oceanites oceanicus, Garrodia nereis, Pelagodroma marina* and *Fregetta tropica*, although because of differing methods comparisons are difficult.

For Wilson's Petrel *O. oceanicus* off Adelie Land, where their food was 41% by weight crustacean, the local euphausian *E. chrystallorophias* dominated, being 37% of the diet (Ridoux and Offredo, 1989). Further west at Prydz Bay *E. chrystallorophias* was 42% by weight in 50 stomachs and *E. superba* 25% (Whitehead, 1991). In the Ross Sea crustaceans averaged 79% by f.o.o., of which 49% were euphausids (Ainley *et al.*,

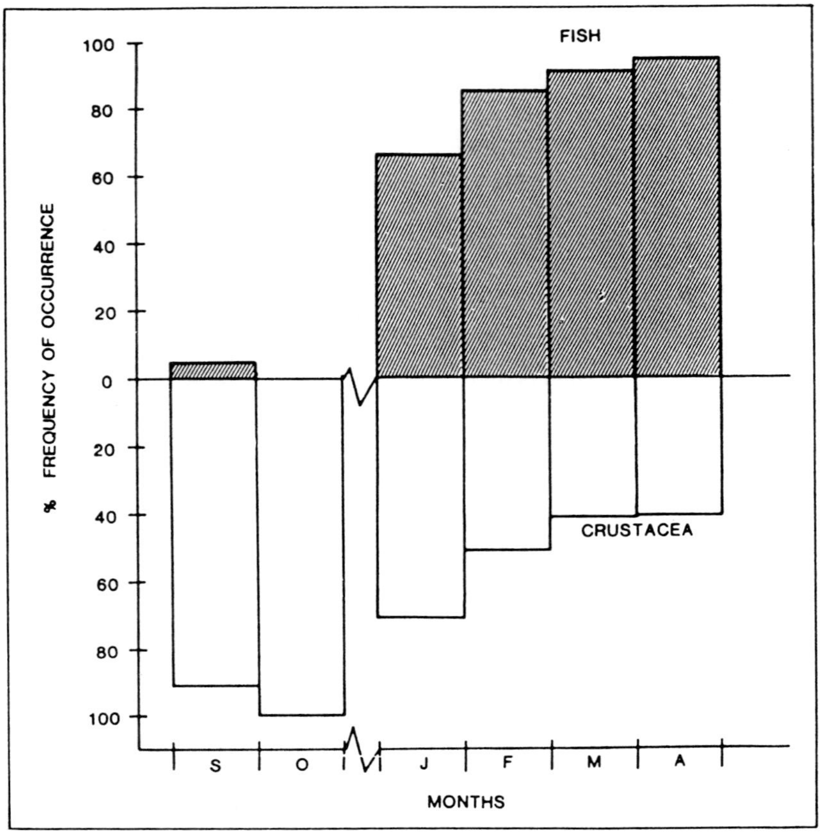

Figure 3.17 *Frequency of occurrences of crustacea and fish in 307 diets of adult Short-tailed Shearwaters sampled before egg laying (September and October) and after hatching in January during 1980 and 1981. From Montague* et al. *(1986).*

1984), mostly *E. superba*. Again, quite large animals were consumed, 35 ± 5 mm, about the same size as those taken by *T. antarctica*, a bird 20 times as heavy. Further west still, off the Antarctic Peninsula, Obst (1985) found that *E. superba* formed 85% by weight of the food, and a little further north at the South Shetlands a sample of 61 adults yielded 90% *E. superba* and 10% *E. chrystallorophias* by f.o.o. (Wasilewski, 1986). South Georgian birds ate 68% by weight of crustaceans, 52% of which were mainly juvenile euphausiids but 90% by numbers were *T. gaudichaudii*, but these accounted for only 44% of the weight of crustaceans eaten (Croxall *et al.*, 1988a). Quite different was the diet at The Crozets where, while crustaceans were still the main foods, cirripedes and copepods exceeded euphausiids by volume (Jouventin *et al.*, 1988) (Fig. 3.18).

Wilson's Petrel's close relative *Garrodia nereis* seems to specialize in snipping up cyprids of stalked barnacles *Lepus australis*, as first shown by Imber (1981) from material collected at the Chatham Islands. This animal has been found in birds from New Zealand and Marion Island and cyprids evidently dominated the diet of those at The Crozets (Fig. 3.18). Imber noted that this specialized diet should lessen competition with sympatric species such as *O. oceanicus* and *F. tropica*, but, as Fig. 3.18 shows, all three birds at The Crozets take cyprids and they breed at about the same time. Cyprids are believed to be picked from and around rafts of floating algae where metamorphosis takes place.

Apart from *F. tropica* (Fig. 3.18), the only other long-legged hydrobatid for which good data have been published is *P. marina*, a bird which, at least in New Zealand seas, also takes cyprids but, on the basis of 22 samples from the Chatham Islands, relies heavily on *N. australis* (Imber, 1981). This made up 35% of the total sample weights, another euphausid *Nematocelis megalops* about 10%. Also significant was the amphipod *Cyllopus magellanicus* (5.2%) and there were smaller amounts of seven other amphipods. The diet was quite species-rich with a mysid, an isopod, two stomatopods, for example *Squilla armata*, a larval brachyuran *Nectocarcinus antarcticus*, the balance being fish, mainly juveniles.

Of 307 regurgitations of Leach's Petrel *Oceanodroma leucorhoa* studied by Watanuki (1985) in Japan, 127 consisted of nothing but oil, the incidence of which declined as breeding progressed. Crustaceans formed important foods, mainly *E. pacifica, T. longipes* and *T. inermis*. Their frequency of occurrences increased over the season as *E. pacifica* increased at sea, reaching 55% by f.o.o. in September.

In the eastern North Pacific, Vermeer and Devito (1988) found the amphipod *Paracallisoma coecus* the most important item after fish for both *O. leucorhoa* and *O. furcata*. Both took much the same sized animals, averaging 16.9-mm long in August, for although *O. furcata* is the heavier bird, their bills are of similar sizes.

The importance of *Paracallisoma coecus* varied in the course of time (Fig. 3.12). The fish in both diets increased as the chicks were being raised, as in Watanuki's study

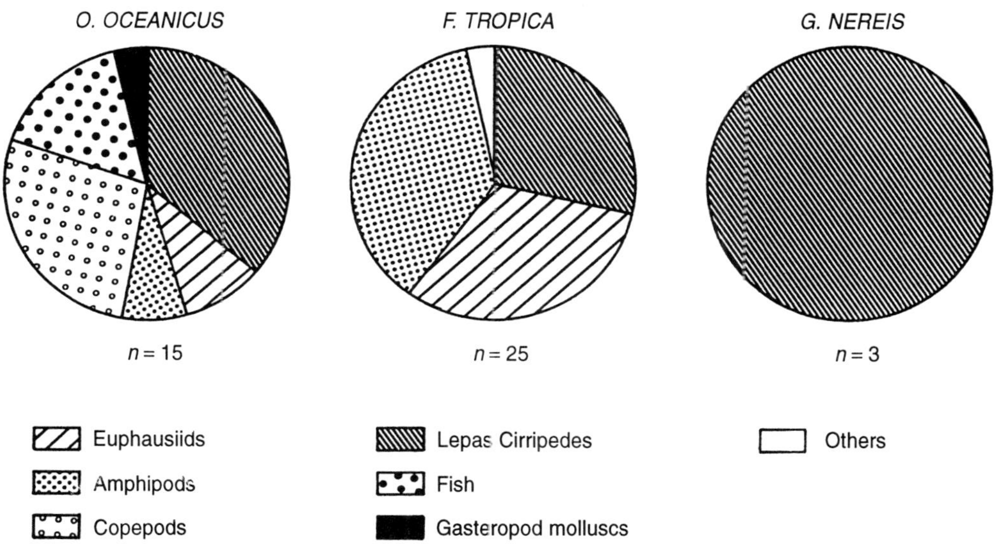

Figure 3.18 *Food regimes of Crozet Island storm petrels based on the volumes of the taxa in stomach samples. From Jouventin* et al. *(1988).*

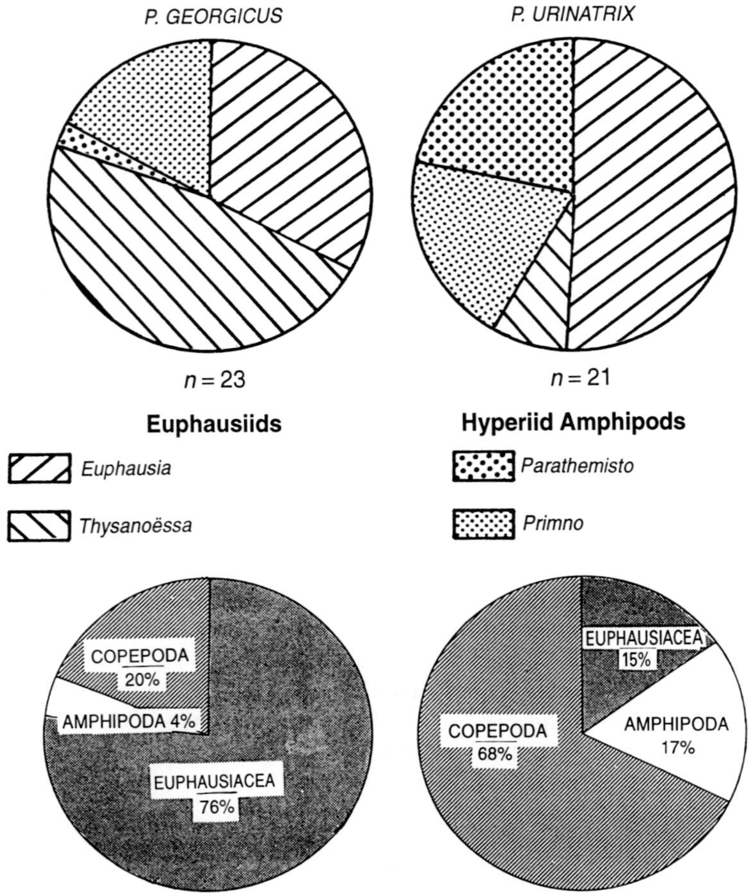

Figure 3.19 *Diets of South Georgian and Common Diving Petrels at the Crozet Islands (upper) and at South Georgia. The copepods that feature prominently in the latter birds were absent in the samples from The Crozets. From Jouventin* et al. *(1988) and Payne and Prince (1979).*

and again, with both species, the amount of oil in adult regurgitations declined with time.

Data are available for the diving petrels studied at South Georgia and The Crozets (Fig. 3.19). Payne and Prince (1979) emphasized the overriding importance of postlarval *E. superba* to *Pelecanoides georgicus*, supplemented with a range of copepods and amphipods whereas the sympatric *P. urinatrix* was more of an amphipod and copepod eater, at least as far as the chick diets went. However, these authors found that the diets of both species changed with time despite the same prey being still available, so that the change must have had some other cause. Jouventin *et al.* (1988) suggested that the dietary differences might be the result of *P. georgicus* foraging over deeper water than *P. urinatrix*, but they had no data on food availabilities at sea to check on this idea.

4 Carrion

All tubenoses appear to eat dead animal flesh when the occasion arises and the bigger species like albatrosses and giant petrels will also kill and eat other seabirds—that is, they are both predators and scavengers. A bird as small as *Fulmarus glacialoides* has been found with a prion in its stomach. Stomach contents of avian prey may also be taken but not exclusively, as the flesh, skin and feathers may be eaten, with the undigested material often being disgorged as cylindrical boluses. The catholic diets of some fulmars have long been known, *F. glacialis* being dubbed a 'feathered pig', and this and other tubenoses have long been associated with whaling fleets for the fat and stomach contents released during flensing.

On the Grand Banks fulmars and 'hagdons' selected the fatty parts of any offal and particularly the oily livers (Collins, 1884), while the activities of *Oceanites oceanicus* and prions in taking up oil globules is common knowledge. Wood (1993), however, found that *Puffinus pacificus, P. carneipes, P. tenuirostris* and *P. griseus* preferred fish offal to fat. Roberts (1940, p. 190) described how *O. oceanicus* pierces an oil globule, which, as it sucks, visibly shrinks. Dead fish and squid taken from commercial trawl, long-line, and other fisheries provide important foods, for example for *Diomedea melanophrys* in the Benguela Current (Ryan and Moloney, 1988), and around the Falkland Islands (Thompson, 1992, Thompson and Riddy, 1995). Hudson and Furness (1988) analysed the use by seabirds of fish discarded behind trawlers around the Shetland Islands. They found that *F. glacialis* got most of the offal—that is, the discarded livers and intestines—mainly because they were the most numerous birds in attendance, but when it came to swallowing whole fish they only got 2.3% of an experimental discarding: gannets *M. bassana* and the gull *L. marinus* got the bulk of those meals. In Australian seas *L. novaehollandiae* also often competes successfully with *P. pacificus*, its lower wing loading enabling it to hover close to the stern and grab food before the shearwater could alight and take it (Wood, 1993). Around the Falklands, *D. melanophrys* did much better, taking about half of the waste discarded by the squid/finfish fishery (Thompson, 1992). She estimated that waste taken from *Loligo* trawlers came to 10–15% of the total food needs from February to mid-April of the Beauchene *D. melanophrys* population. The sizes of the squid taken were very similar to those harvested by the trawlermen.

Both *Macronectes* feed penguin and seal flesh to their young, and there is an overlap in their diets then as shown by the work of Conroy (1972), Johnstone (1977), Hunter (1985, 1987) and Voisin (1991). In general, the data suggest that *M. giganteus* relies more on penguins than *M. halli* (see Fig. 3.7). This takes more seal flesh, dead pups and afterbirths.

5 Other prey

Many marine animals, not normally major prey of petrels, have been found in their stomachs. Among these are coelenterates such as medusae and ctenophores. Such soft-bodied invertebrates are not easily detected in stomach contents unless freshly taken. That *F. glacialis* eats jellyfish was reported by Faber back in 1822 and Fisher identified *Cyanea capillata* and *Aurelia aurita* in the diet (Fisher, 1952a, p. 430). The

North Pacific subspecies also takes jellyfish, especially the large, brown-rayed types such as *Chrysaora* spp. and Hatch (1993) found that some chick meals were solely of jellyfish. Anthony (1895) described these birds off southern California in the autumn gorging on the abundant jellyfish, so that when lifted from the water by their feet 'half a pint of the slimy mass will often run from their mouths'.

In northern waters both *P. griseus* and *P. tenuirostris* take some jellyfish and Ogi (1984) found the former turning to barnacles and jellyfish when saury and squid disappeared. The wind sailor *Vellella vellela* has been identified among the stomach contents of *Pterodroma phaeopygia*, *O. leucorhoa*, *D. nigripes* and *D. immutabilis*, and was the second most important prey for *O. tristrami* (Harrison *et al.*, 1983). These may have been the jellyfish on which about 1500 *P. griseus* and 500 *P. inexpectata* were gorging in the Pacific at 34°N, 174°E (Mobberley, 1974).

Harrison (1984) found scyphozoan jellyfish in all four tubenoses she collected in the Bering Sea—*F. glacialis*, *P. griseus*, *P. tenuirostris* and *O. furcata*. She noted that those that had taken jellyfish had also eaten a greater diversity of animals and suggested that some of the other foods such as hyperiids, known to live under medusae, had been ingested with them: medusae with their various associated organisms being viewed as a highly localized food patch. Pitman and Ballance (1990) likewise thought that *O. leucorhoa* were attracted to *Physalia* for the prey on their tentacles. Coelenterates would appear to be rather poor food, yielding only about 2.9 kcal g^{-1} dry weight compared with *c.* 4.8 kcal g^{-1} for squid and 5.1 kcal g^{-1} for fish, but if abundant and readily available, clearly worth using.

Annelida captured include a range of polychaetes by *F. glacialis* and *B. bulwerii*. Sometimes large numbers are present, up to 60% by volume of two samples from *F. glacialis* with 91 pairs of mandibles in one stomach (Cottam in Fisher 1952a, p. 428) and probably derived from a surface-swarming species.

Protochordates reported include tunicates. These have been identified in the meals of *D. epomophora*, *D. melanophrys*, *D. immutabilis*, *D. nigripes*, *Pterodroma hypoleuca*, *Pachyptila crassirostris*, *Procellaria parkinsoni*, *P. westlandica* and *P. aequinoctialis*. Tunicates, mostly *Pyrosoma* and up to 150-mm long, were 2% by weight of the food of 145 samples from *Pterodroma macroptera* (Imber, 1973). Salps are also encountered occasionally as also are chaetognaths.

The only insects that appear to be normal prey of petrels are the marine sea-striders (Heteroptera: Gerridae) such as *Halobates sericeus* which is quite important in the dietaries of *P. hypoleuca*, *B. bulwerii*, *O. leucorhoa* and *O. tristrami*, and they even turn up from time to time in food samples from *Puffinus pacificus*, *D. nigripes* and *D. immutabilis* (Cheng & Harrison, 1983; Harrison *et al.*, 1983).

These insect prey were first noted by Ashmole and Ashmole (1967) when they found 13% by volume of the stomach contents of *Pterodroma alba* composed of water striders, which, though small, were evidently worth collecting, no doubt by dipping. These insects often occur in patches in tropical seas so would be expected to be found by birds like gadfly petrels but, although they may usefully complement more substantial prey, their total contribution to the diets appears to be minor, on present evidence.

Other insects ingested are land species that have apparently been blown out to sea and are picked up by the birds from the surface film. The Blue Petrel seems particularly prone to take these—the gizzards of 45 'wrecked' on a beach in Victoria,

Australia held 3.3 insects per bird, mostly Coleoptera (Brown *et al.*, 1986). At Marion Island, 13 out of 25 stomachs of adult birds held assassin bugs (*Nabis* sp.) and unidentified noctuid moths, 2.1 moths and/or 6.7 bugs per stomach (Steele and Klages, 1986). The Victorian insects could have been taken in local seas but those in the Marion Island petrels must have come from far away and as chitin may be digestible by petrels (Ch. 7 VI.B), insects could provide an extra but irregular food source. There are even reports of storm petrels apparently hawking for insects over beaches and inland lakes.

6 *Stomach oil*

In many stomach analyses the liquid fraction has been decanted and only solids used in determining the proportions of food animals. The liquid contains both water and oil, the latter derived from the prey (Ch. 7 VI.A), of high calorific value (perhaps as much as 35 times that of the prey), and digestible.

At Marion Island 58% of adult *Halobaena caerulea* held oil, averaging 1.0 ± 1.2 ml, maximum 5 ml. At South Georgia the oil in these birds came to 39% by weight (7.9 g) of stomach contents or 3.7% of body weight; for *Pachyptila desolata* the mean figure of 2.1 g was only 1.3% body weight (Prince, 1980a). In consequence, Prince suggested that the prion was feeding closer to the nest than *H. caerulea* then was. Many figures for other species are within similar ranges of 1–2% body weight.

Hunter and Klages (1989) found that 55% of 10 stomach contents (average 343 g) of *Diomedea chrysostoma* was liquid and Prince (1980b) noted that about half the meals of this albatross and of *D. melanophrys* were liquid. This was mostly water except for meals of *D. chrysostoma* associated with lampreys. These have a high lipid content. Could the high levels of water in these mollymawks be the result of digestion of the oil during the birds' 3–4 days feeding forays?

Vermeer and Devito (1988) found a decline in oil in the stomachs of *Oceanodroma leucorhoa* and *O. furcata*, probably linked with a declining intake of the oil-rich *Paracallisoma coecus*. Stomachs of petrel chicks tend to contain falling amounts of oil as the season advances and they may have very little by the time they fledge (Warham, 1962; Serventy *et al.*, 1971).

D *Food consumption*

Attempts have been made to estimate the total biomass eaten by a bird, a colony or a population, or a whole region (e.g. Mougin & Prevost, 1980), often to evaluate the effect on a commercial fishery. For example, Thompson (1992) calculated that when rearing chicks *D. melanophrys* in Falklands seas consumed 1000–2000 t of offal, mainly *Loligo* and notothenids. Such exercises involve many assumptions on matters such as energy needs, assimilation efficiences, ambient temperatures and distances travelled, but may furnish useful guides within orders of magnitude.

Hunter (1985) calculated the food eaten by giant petrels at South Georgia, Signy Island, Terre Adelie, The Crozets and Macquarie Island, estimating the take of squid, penguins, small birds (that is, petrels), crustacea, fish and 'other'. Assuming 75%

assimilation efficiency, his total for the estimated 38 000 *Macronectes giganteus* and 8600 *M. halli* adults (breeding or otherwise) was 15 113 and 3037 t respectively. Hunter allowed for sexual differences in energy needs, the males being the bigger needing more—1.14 times more with *M. giganteus*, 1.18 times more with *M. halli*.

Others have estimated food consumption over the breeding season, for example Ridoux (1989) for The Crozets and Croxall *et al.* (1984a) who calculated food consumption (t day^{-1}) for the populations of *D. melanophrys*, *D. chrysostoma*, *Pelecanoides urinatrix*, *P. georgicus*, *H. caerulea* and *Pachyptila desolata* at South Georgia. Here the main consumer was the penguin *E. chrysolophus*, the next biggest demand being from *P. desolata* at 2320×10^3 t, all the rest being quite minor consumers, totalling 1548 t. There was a high demand during the prelaying stage, dropping sharply at the lay, increasing again at the hatch but declining slightly thereafter throughout chick rearing to return at fledging to the prelaying figure. Ainley *et al.* (1984) estimated the daily needs of euphausiids, squid and fish of nine Antarctic tubenoses at 0°C and Fefer *et al.* (1984) attempted an estimate of the total consumption of all the breeding seabirds of the northwestern Hawaiian Islands: this included seven tubenosed birds. The major consumer was *D. immutabilis* taking 250 760 t, the next heaviest demand coming from *Sterna fuscata* (71 840 t), then *Puffinus pacificus* at 35 440 t, *Pterodroma hypoleuca* 15 000 t and *D. nigripes* 13 000 t, the rest a mere 23 780 t from a grand total of 409 820 t.

E *Gizzard stones*

Miscellaneous inedible mineral and plant material are often present in the stomach and gizzards; some such as nuts, cinders and pumice, obviously picked up as flotsam. Pumice found in 40% of the stomachs of *H. caerulea* at South Georgia was thought to have come from the South Shetland Islands, 1350 km south–southwest, suggesting that these birds had been foraging in that direction (Prince, 1980a).

Heavier stones and gravel are also common, often of unknown origin. They are familiar from *P. griseus* and *P. tenuirostris*, known to the 'muttonbirders' as 'ballast stones', and sometimes thought to be swallowed to reduce buoyancy in species that feed by pursuit diving. However, they also turn up in storm petrels such as *Oceanites gracilis* which seldom submerge. Many of these objects must be swallowed when the birds are ashore, as some of the plant material clearly is, perhaps when adults and chicks pluck and mandibulate nearby vegetation. Whether the stones serve any useful purpose is unclear, but they can be very common—in 12.5% of the stomachs of *H. caerulea* at Marion Island, 66% being pumice, the rest laval gravel and, in 76% of beach-washed birds, again mainly pumice (Steele & Klages, 1986).

F *Variability*

That individual petrels may have individual tastes and specific search images has not been examined. Shiomi and Ogi (1992) did, however, gain evidence that adult *P. griseus* fed more on fish, immatures taking more squid and barnacles, the difference being ascribed to feeding experience.

Variation in diets between populations is, of course, widespread (e.g. Fig. 3.19), and reflects different availabilities of prey species as Hunter (1985) showed for *Macronectes* spp. and there may be greater interpopulation differences even among these birds. For example, the *M. halli* petrels on the Chatham Islands not only have no *M. giganteus* to contend with but there are no penguin colonies on which they can prey. Hence their diets, unstudied so far, can bear little resemblance to those summarized in Fig. 3.7.

1 *Variations within seasons*

Changes of diet, in the course of breeding seasons, are often detectable, as with *Pachyptila desolata* and *H. caerulea* (Prince, 1980a). Ealey (1954) deduced switches in the foods of *P. crassirostris*; the pteropod *Clio sulcata*, the main food in July and August, being replaced by the amphipod *Hyperiella antarctica* in September; yet in January and February the stomachs were distended with *Euthemisto antarctica*, evidently taken from local swarms.

Variations often correlate with the breeding stages. During the rearing of *Daption capense* and *F. glacialoides* chicks the fish *P. antarcticum* declined in importance while that of *Euphasia superba* increased, whereas with *T. antarctica* nesting nearby the reverse was found (Arnould and Whitehead, 1991).

The amounts of food carried may also change with time. Regurgitations during incubation tend to be smaller than those during chick rearing, as exemplified by *Puffinus tenuirostris*, where the mean weight of food before laying was 19 g, that during the chick stage 72 g, $P < 0.001$ (Montague *et al.*, 1986) (Fig. 3.20).

2 *Variations between seasons*

Several instances have been quoted, for example the situation with *Diomedea chrysostoma* summarized in Table 3.2. Whitehead (1991) gave good data on the foods of *Daption capense* over 3 years. The diets in the 1987/88 and 1989/90 seasons differed in their main components. In the earlier season all were *E. superba*, in the later one about the same biomass was taken but 76% of the whole was made up almost equally of *E. superba* and *E. crystallorophias*, the difference presumably reflecting the availabilities in these two seasons.

One of the main conclusions from these studies is that many petrels are opportunistic feeders, and that even those specialized for particular ecological niches can adapt to a good degree if needs be. This ability to exploit the heterogeneity of the marine habitat must be a major factor in allowing tubenoses to survive temporary food shortages, and helps account for their success. Furthermore, while shortages may affect one population, other populations may be unaffected, and even if these too are decimated there may still be a 'stategic reserve' of immatures searching seas far from the affected area.

The food spectra are complex and one of the biggest problems is to find how to correct for the different digestion rates of the various prey so that proventriculi and gizzard material can be evaluated, as well as to determine diets throughout the year.

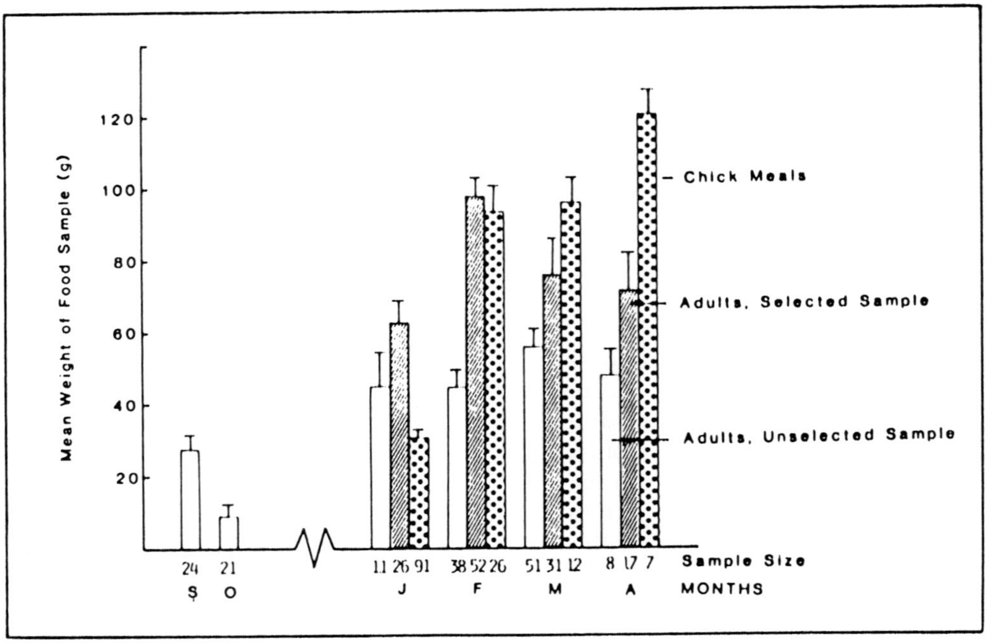

Figure 3.20 *Mean weights of food carried by 126 breeding Short-tailed Shearwaters (selected sample), and 153 adults of unknown status (unselected sample), of 45 adults during the prebreeding period, and of 136 chick meals. Lines above columns are standard errors of sample means. From Montague* et al. *(1986).*

Identification of prey from the composition of stomach oils as examined by Horgan and Barrett (1985) seems unlikely to provide a short cut, partly because the lipid composition of invertebrates changes according to breeding condition. Development of remote sensing acoustic methods may help in measuring the abundance and availability of prey, particularly for diving species.

CHAPTER 4
Behaviour and Vocalizations: A General Introduction

Contents

The communication system of tubenoses comprises vocal, visual, and, possibly, olfactory signals. A particular call is often associated with a particular behaviour although some calls may convey different messages in different contexts. Repertoires tend to be most complex in large species with large brains that display by day. As such birds are the longest lived and must make many breeding attempts to replace themselves, an extensive visual and vocal vocabulary would seem appropriate, whereas for smaller birds with fewer prebreeding years and shorter life expectancies, a more restricted repertoire is not unexpected.

Furthermore, the large species may display over several years before the pair bond is forged and even then take a long time to produce their single chick, so that the partner's co-operation is even more important for success than with smaller species. The great albatrosses must even maintain their pair bond after a year's 'sabbatical', and thereby save energy, as little activity is needed to initiate the next breeding episode. Presumably the best communicators are the best adapted, co-ordinate their activities most efficiently, and leave the most progeny, while the absolutely larger brains of the bigger birds should have a greater capacity to store and respond to complex information than those of the smaller species. A prime function of the communication system seems to be to establish compatibility, so that once a bond has been forged efficient breeding is possible. A change of partner inevitably results in loss of nesting opportunities.

Sexual, nuptial or courtship activity involves a series of displays to communicate intentions or states and they are built from visual or vocal elements which, at least in albatrosses and giant petrels, are often given in a particular sequence or 'set'. Most such activity is incomplete and copulation does not result. When it does and pair-bonds have been established, much display behaviour ceases.

Non-sexual behaviours include 'comfort' movements for the care of the body surface and agonistic, defensive and appeasement displays used in the defence of the territory or person, in hierarchical positioning, and in avoidance of predation and attack.

Apart perhaps from the yellow flush to the heads of *Diomedea albatrus* and *D. irrorata*, tubenoses lack colourful plumes and base their visual displays on exaggerated movements of body parts—expanded wings, fanned tails, waving heads and so on. However, the bills and tarsi of *D. epomophora* are said to flush seasonally (Richdale, 1939, p. 477) and eyelid colours of *D. exulans* may also alter. According to Voisin (1968, p. 97) the bill colour of *Macronectes giganteus* dulls towards the end of chick rearing and bill colours of *Fulmarus glacialis* change with the season (Dement'ev *et al*., 1968, p. 363). Even the blue legs of *Pelecanoides* intensify in nesting birds (Falla, 1937, p. 214). The extent and significance of such changes await examination.

Because of their relative tameness, the diurnally active tubenoses lend themselves to the study of their displays and vocalizations but concealment is advisable to ensure relaxed subjects. Although even nocturnally active species, typically the more timid, can be tame if hungry—wild *Puffinus griseus* have taken food from the hand (Fig. 4.1) as has *M. halli* (Horning and Horning, 1974). The most widely used behaviours are duetting and allopreening. In allopreening the active bird nibbles the head, cheeks, neck, throat or flanks of another, inducing submissiveness in that bird. The recipient turns its head to one side so as to expose the part to be preened (see Fig. 5.9C). Preening can be quite fierce and when the probing extends around the eye the nictitating membrane slides across as a protective shutter. Eye closure may perhaps help reduce aggression in the preener. The bill of allopreeners often appears oily, but whether this is from stomach oil exuding from the nostrils (as often asserted) or, more probably, excretion of the salt glands, is unknown. Courtship feeding, sometimes reported, has not been verified for any tubenose.

Allopreening may involve visual as well as tactile stimuli, but within the nest chamber of a burrowing species only tactile, aural and olfactory senses can be involved, as when parents preen their chicks in the dark. Mutual and reciprocal allopreening between partners and between parent and chick is also common. Allopreening occurs throughout the courtship cycle and precedes much more active display. It appears to signal the acceptance of the preener by the preened and helps pair-bonding.

The calming effect of head preening is readily demonstrated if a finger or small stick is used to scratch the crown or neck, when even the most aggressive of petrels becomes docile. Some will respond by gently grasping the finger with their bill and running the opened mandibles down it as they do when mutual preening: 'the bird responds to the finger as if its mate is courting it' (Warham, 1956, p. 177).

Figure 4.1 *Even species normally timid may become tame when hungry. Sooty Shearwater* Puffinus griseus *accepting food by hand. It had been neck-ringed when caught previously. Photo. Y. Gollun, from Shirihai (1987).*

Another pattern of behaviour common to all is ritualized nest building, where a bird picks up soil, grasses or other vegetation in its bill and reaches round a shoulder to drop rearwards and to one side, the 'Sideways Throwing' of Harrison (1967), the 'Construction Activity' of Jouventin and Lequette (1990) (Fig. 4.2).

No examination of the possibilities of olfactory signals, for example sex pheromones or naso-vomerine activity has been made, although the manner in which petrels reach forwards to investigate others or their nests would be expected of a bird seeking olfactory clues. And if olfactory (or even gustatory) information aids individual identification, this would be most accessible during allopreening with bills in the other's feathers and nostrils in close proximity, and with extensive inhalations during calling.

Burrowing species do not use dramatic visual signals when ashore except perhaps when fighting. What visual displays are given are effective only at close quarters and consist mainly in variations in the orientation of head and body in respect of the other bird. Communication still occurs even in total darkness and that visual

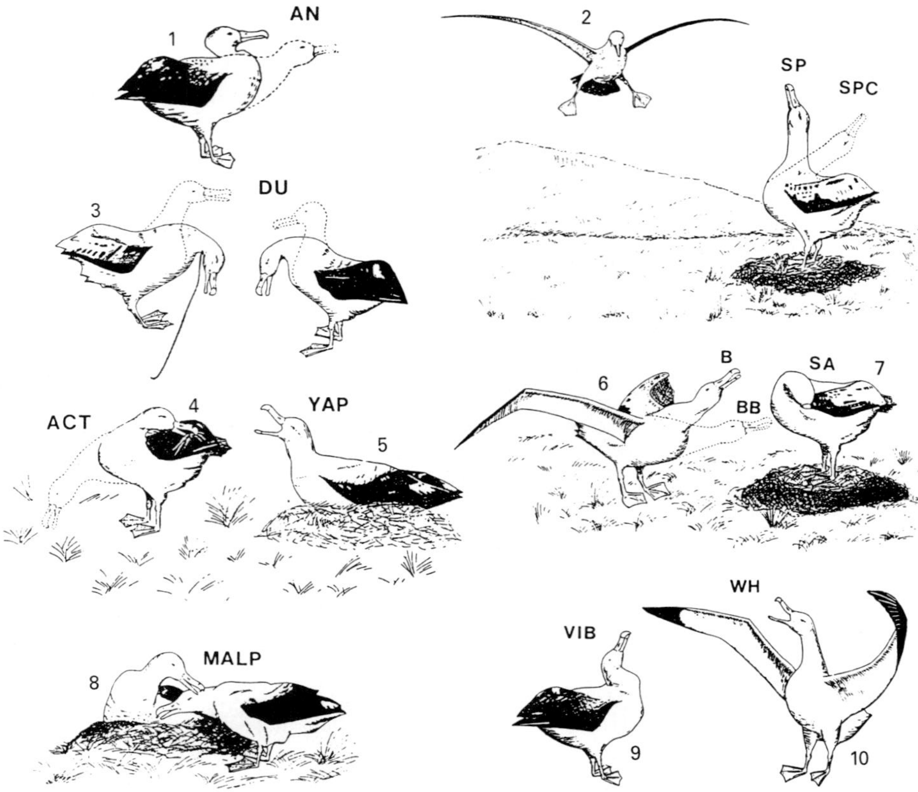

Figure 4.2 *Postures used in the 'dance' routines of Wandering Albatrosses* Diomedea exulans[1] *1, 'Arched Neck' (AN), most often used by males before moving towards or around another bird. 2, 'Sky Position' (SP) and 'Sky Position Call' (SPC) by nest holder, usually a male, to another approaching bird. 3, 'Duetting' (DU)—simultaneous 'Bowing' with 'Yapping'. 4, 'Construction Activity' (ACT), collecting vegetation and dropping to a flank or near a nest site, most often by females. 5, 'Yapping', loud, harsh calling with visual signals from the head and neck, mostly performed by males. 6, 'Billing' (B) and 'Breast Billing' (BB); a bird reaches towards another's bill or breast. When both reach forward their bills may touch. 7, 'Scapular Action' (SA), the beak is dipped to one side and hidden in or rested on the scapulars. Used in many situations including nest relief. 8, 'Mutual Allopreening' (MALP) and 'Allopreening' (ALP)—simultaneous preening of each other's heads and necks, or self preening, both used in many situations. 9, 'Vibration' (VIB) head slowly drawn back and neck folded while a strange, rapid vibratory sound issues from the vibrating but blurred mandibles, usually leading to 'Billing'. Mostly by males. 10, 'Whine' (WH) head is moved laterally and then swung upwards so that beak points to the sky, wings widely expanded and a long, piercing, far-carrying scream is given.*

[1] Terminology and figures from Jouventin and Lequette (1990).

information is unnecessary is shown by the responses to playbacks where sound alone is implicated. On the other hand, some evidence suggests that for diurnal species vocal cues alone may be ineffective, for example in experiments with *D. bulleri* (see Chapter 5.II.I).

Gadfly petrels of the genera *Pterodroma* and *Lugensa* tend to form an intermediate group between the daytime-active albatrosses and fulmars and the night-time shearwaters and storm petrels. Some such as *P. alba*, *P. ultima* and *P. arminjoniana* are active by day and visual clues may be important for them, but no special postures have been described. Other species such as *P. solandri* are crepuscular and displaying birds may be able to see one another and their species-specific plumage patterns (*The Petrels*, p. 80). Still others, such as *P. mollis* and *P. inexpectata*, are truly nocturnal and visual clues are only available at close quarters or in aerial displays under the light of the stars or moon. In such circumstances guidance by calls seems important; or do they lay olfactory trails in the sky?

During the dance routines of albatrosses intruders are often threatened by one or other dancer, males being the more aggressive. Conflicts occur over ownership of nest sites or potential partners. Rapid bill clappering is used by albatrosses, oil spitting by fulmars, hissing with head jerking by gadfly petrels and bill jabbing and wrestling by shearwaters, gadfly petrels and storm petrels. Some fighting takes place and may involve serious wounding, but its incidence is greatest in crowded colonies and probably more so where nest sites are in short supply.

Petrel vocalizations range from the quiet churrings and whistles of some storm petrels to the pigeon-like cooings of prions and the brayings of *Procellaria* spp. Shearwaters are particularly noisy and the cacophony at an active colony at the height of the season must have been terrifying to primitive peoples at first encounter, when at a low ebb—after shipwreck perhaps. Jennings (1983) suggested that the 'night-flying ravens' that frightened the Norse chieftains Ospak and Brodir when off the Calf of Man in 1014 were probably *Puffinus puffinus*.

Many vocalizations take the form of 'simple cries' as defined by Davis (1964) who noted that the resulting chevron-shaped patterns on sonagrams correspond with the opening of the beak (frequency rises) and its closure (frequency falls), an acoustic structure on which many bird calls are based. Some are built from very brief notes extending over a wide frequency range and help to locate the singer, for example the clacks of *Procellaria aequinoctialis* (see Fig. 6.23) whereas the whistle of *Fregetta tropica* is said to be ventriloqual.

Among mollymawks a range of sounds results from bill to bill contacts and by the lateral movements of the upper mandible against the lower. Warham and Fitzsimons (1987) listed four types of such sounds (see Fig. 5.10). Similar percussion sounds are heard from other petrels, for example from *P. aequinoctialis* under attack from skuas. An unusual example is when incubating Laysan Albatrosses snapping at falling raindrops, sound like the sustained slow clapping of an audience signalling a poor performer to leave the stage!

Although Brackenbury (1989) stated that all experimental studies have shown that birds produce sounds only on expiration, many petrel vocalizations include both inspiratory and expiratory phrases, as noted for *Puffinus nativitatis* by Wetmore (1923 in Amerson and Shelton, 1976), in *P. bulleri* by Falla (1924), and with *D. exulans* by Matthews (1929). This is particularly evident with albatrosses and shearwaters, the inhalatory sounds generally being low pitched and 'noisier' than those resulting from the expulsion of breath and the pulsations of the singer's throat reveals which components are expiratory and which inspiratory.

Many of the sounds used are hard to convey in words, others lend themselves to onomatopoeic representation. Examples are the Maori names for *P. gavia* of 'hakoakoa' and 'pahaka' and the sealer's 'pee-oo' for the cry of *Phoebetria* albatrosses. The Maori 'ti-ti' for the New Zealand mutton-bird *P. griseus* is not the cry of that species but that of gadfly petrels also eaten, such as *Pterodroma inexpectata*, as was pointed out by Richdale (1944a). Other folknames refer to characteristics of the calls. In the Tokelau Islands the Wedge-tailed Shearwater is 'Manutagilua'—the bird with the double cry (Huntsman, in Wodzicki and Laird, 1970).

The first comprehensive account of a vocal repertoire was that of Sparling (1977) on the albatrosses *D. immutabilis* and *D. nigripes*. Warham and Fitzsimons (1987) presented some analyses of the vocalizations of *D. bulleri* and Grant *et al.* (1983) described and figured many of the calls of *P. hypoleuca*, but their subjects were not sexed. Lequette and Jouventin (1991a) provide a detailed breakdown of the vocal repertory of *D. exulans* and link this with the sexual displays (Jouventin and Lequette, 1990; Lequette and Jouventin, 1991b).

Bretagnolle (1995) divides the calls used in sexual and agonistic contexts into two classes; those using a single call, such as *Calonectris diomedea* (Bretagnolle and Lequette, 1990), and those with two, such as *Oceanites oceanicus* (Bretagnolle, 1989a). In the first group the call is used by both sexes, in the second at least one is shared, the other call restricted to the male, or used by both.

Much of the unravelling of the petrel repertoire has been by playing back recorded calls to birds in the field. Detailed analyses of differences in temporal, frequency and amplitude characteristics and the responses to playback, allow the meanings of the vocalizations to be inferred. For methods see Bretagnolle (1995a).

Vocalizations are illustrated here by audiospectrographs ('sonagrams') all, unless otherwise stated, made with a narrow bandfilter to emphasize frequency resolution.

I Constancy and variability of vocalizations

For any particular type of call, each adult petrel tends to have its own characteristic 'voiceprint'. Birds having peculiarities of voice, perhaps being higher pitched than the majority, or with unusually fast or slow delivery, are often readily identifiable by the human ear, and this individuality holds even when the song changes according to the circumstances. For example, the pitch and volume of duetting shearwaters increases towards the climax (Fig. 4.3) but the basic repeatability of the unstressed part of the vocalization remains.

The movements associated with the calls may also be distinctive; a hurried song delivery, for example, may be accompanied by rapid movements of shallow amplitude. The extent of the within-bird acoustic variation has been measured in some species. Using both frequency and syntactical variables, Bretagnolle (1989a) found that the variations in the chatter call of individual male *O. oceanicus* was not much less than that of the population as a whole, but for the second major call, the 'grating' call, the between-bird variation for both sexes was much greater than that for the individual one. Brooke (1978e) found that spectrographs of male *Puffinus puffinus* could be correctly matched by inspection. The individual differences applied to both sexes and were additional to the sexual dimorphism in voice.

Individual variation in postures has not been quantified, published descriptions

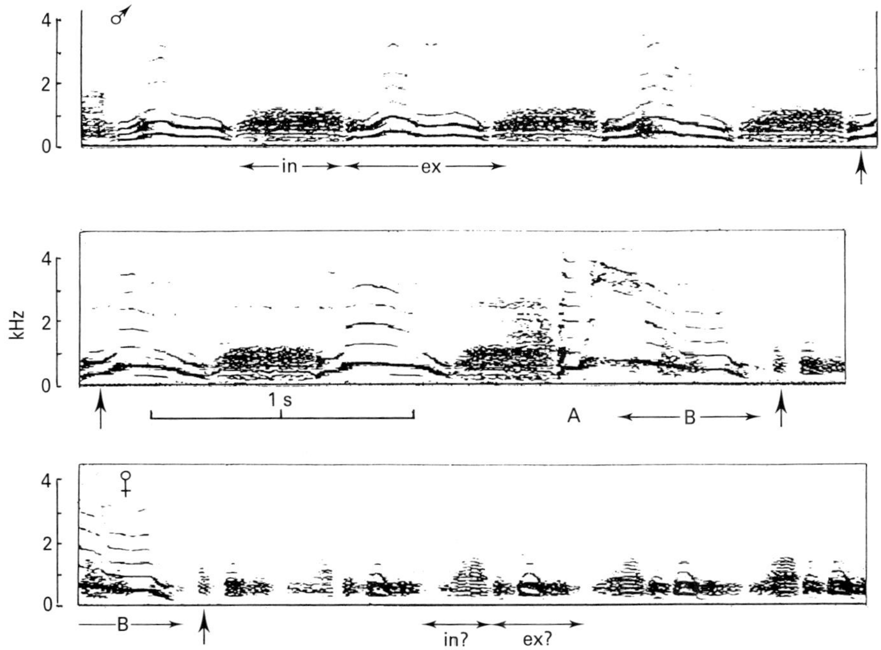

Figure 4.3 *Duetting by Sooty Shearwaters* Puffinus griseus. *The male starts in full song (top), builds to a climax with increasing volume and rising pitch to (A), when the female adds her voice at (B). She continues as her mate falls silent. Her calls, of low frequency, also tend to increase in loudness as her song develops. The male's inspiratory (in) and expiratory (ex) elements are distinct, those of the female less so. This female's song was built from six elements that resulted in a pulsed, rhythmic sound very characteristic of her sex. Points of overlap of the sonagrams are arrowed.*

covering the average extent of the movements. As with the calls, these may peak during a display and then decline.

II Intercolony variation in calls

With species such as shearwaters, whose vocalizations are very variable, elaborate analyses would be needed to detect colony-specific features, and it is seldom that the human ear detects consistent differences. For example, Guillotin and Jouventin (1980) detected no vocal differences between the Snow Petrels of Signy Island and of Terre Adelie despite substantial disparities in their body sizes.

In looking at geographical variation in the calls of *P. puffinus* and *Hydrobates pelagicus*, James (1985b) found no consistently detectable differences in the calls of birds of either sex of the shearwater from different parts of a small island (Skomer). However, ANOVA and discriminant function analyses of calls from Skomer, Skokholm, Rùm (Scotland), Puffin Island (Ireland) and Myggenaes Holm (Faroes) using 13 syntactical and frequency variables provided some evidence for 'dialect' development, but with too much overlap to assign the voice of any individual bird to any particular population.

In contrast, according to Bretagnolle (1989a), the calls of Wilson's Storm Petrel populations at Terre Adelie, the South Sandwich Islands and South Georgia, normally classified as *O. o. exasperatus*, were similar. All differed in their temporal and frequency patterns from those of the typical subspecies from the Crozet Islands and Kerguelen, thus supporting the current classification. The birds at Kerguelen used a long chattering call not heard in Antarctica.

In a detailed examination of interpopulation variation in the aerial calls of the gadfly petrel *Pterodroma phaeopygia*, Tomkins and Milne (1991) analysed large samples of aerial calls heard over four colonies less than 170 km apart in the Galapagos Islands. This revealed the existence of population 'dialects' involving frequency and temporal patterns and harmonic structure, and parallel differences were also found in a range of mensural characters and in egg volumes.

Bretagnolle and Lequette (1990) suggested that the structure of a petrel's calls are controlled genetically, the differentiation between populations being enhanced and maintained by philopatry and site tenacity and by the way in which most populations are discretely concentrated on islands.

If there is variation in non-vocal displays between populations of any species of petrel this has yet to be described, let alone quantified. Petrel colonies, of course, give off a strong odour. It is possible that each colony has its own characteristic smell used by the birds as a cue for homing in the dark.

III Ontogeny

The ontogeny of visual behaviour has not been followed closely for any petrel chick: general changes were noted in *The Petrels*, Chapter 15. Nor have changes over time in the displays of the prebreeders, if any, been evaluated. Are those displays at sea (Section X below) given by birds too young to venture on to the colonies? What is clear is that the non-breeders are those most active in visual, tactile and vocal displays on the nesting grounds. Once paired, overt sexual display is greatly reduced.

Nestlings gradually develop some forms of sexual behaviour and their voices change. Young petrel chicks produce the usual piping, cheeping and chirruping sounds (Fig. 4.4) with wing waving when begging for food. Presumably, these help to reinforce the chick's pecking at the side of the parent's beak and stimulate it to regurgitate.

Richdale (1952, pp. 79–92) followed behaviour in a *D. epomophora sanfordi* chick during a 16-day vigil. The 100-day-old chick defecated 48 times by day, 40 times by night. It left its nest 31 times, travelling a total distance of 65 m. It waved its wings 154 times and stood up 47 times. In the whole 176 h of daylight it slept for 58 h.

At age 178 days this chick 'Yapped' in response to the 'Yapping' calls of its parents on their arrival, and, if not on the nest, hurried back there to be fed. About 9 days later, on failing to induce the male to disgorge when its bill was pecked, the chick gave a 'Wing Stretched' display (Fig. 4.5) accompanied by a whine (Fig. 4.4A). The chick was then fed.

The piping calls of petrel chicks are heard from within the egg for several days before the hatch. Harper (1976) noted how the cheeping of the unhatched prion

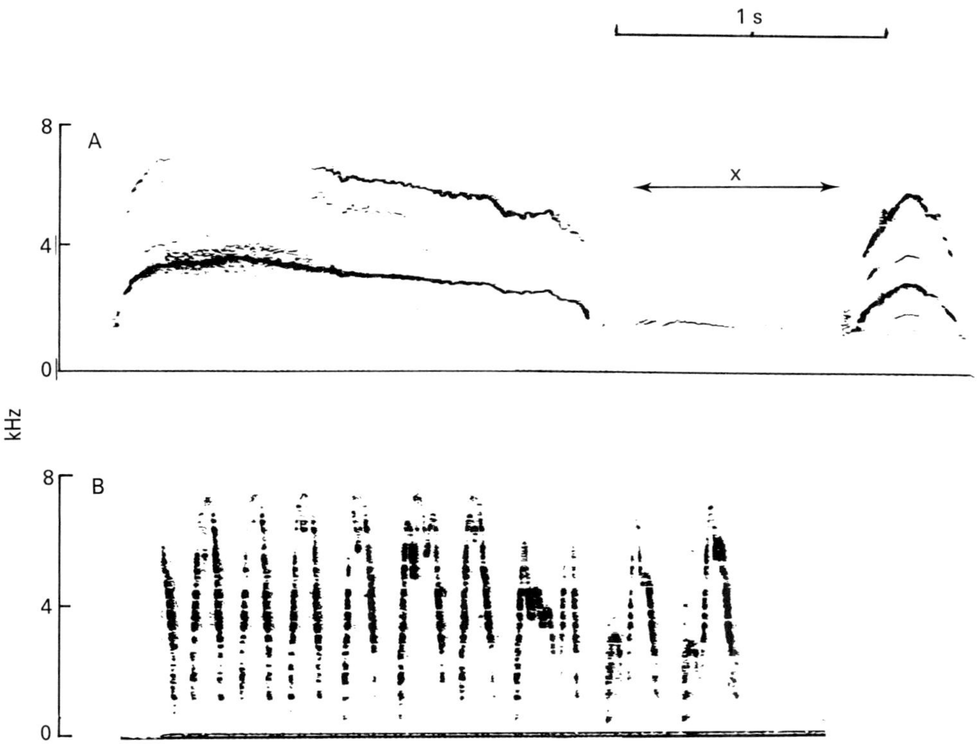

Figure 4.4 *Vocalizations of chicks: (A) Fully fledged* Diomedea epomophora *chick gives 'Wing Stretch' scream after unsuccessfully supplicating from parent (see Fig. 4.5). 'x', Inhalatory phrase. (B) Calls of young Fairy Prion (*Pachyptila turtur*) chick begging for food. Note extremely rapid frequency sweeps and changing structure of the syllables.*

Pachyptila turtur elicits calls from the parent and he believed that failure of an embryo to call resulted in egg desertion, preventing 'over incubation'. Fisher (1972) suggested that the 'eh-eh' and 'haw-haw' croaks of *D. immutabilis* and *D. nigripes*, used as the parents bend down to 'address the egg', enable the advanced embryo to learn the basic call of its species. He pointed out that, if this is correct, then it may be a powerful factor hindering interbreeding.

Calls of newly hatched petrels are short, high-pitched notes that show on sonagrams as chevrons, usually with clear harmonics. Some show extremely rapid changes of frequency, the prion chick in Fig. 4.4B providing a rather extreme case. The frequencies reached increase with heightened excitement, as when the hungry chick begs for food.

Field workers have noted changes in voice as chicks grow. The young *Fulmarus glacialis* has a soft quacking call which develops into a monosyllabic bray (Fisher, 1952a, p. 329) and later a repeated shrill note is heard with the full adult cackle being achieved before fledging. Indeed, Kharitonov and Tihkonov (1982) compared calls of adult fulmars with those of their advanced embryos and reported that these gave most of the calls used by parents, including the cackles and squeaks associated with

Figure 4.5 *Fully feathered 226-day-old Royal Albatross (*Diomedea epomophora*) chick, having failed to persuade male parent to feed it, performs 'Wing Stretch' display with accompanying scream (Fig. 4.4A), at which the parent gives the 'Gawky Look'. After Richdale (1950, p. 64).*

oil spitting. The pitch of the embryonic calls was higher than that of the adults' but the frequencies fell towards the hatch. The quiet peep of the hatchling giant petrel develops into a deep, guttural snort in the postguard stage and later a bray similar to that of the adult appears (Warham, 1962). Tennyson (in Marchant & Higgins, 1990) noted that young *Pterodroma nigripennis* gave the territorial call when but 8 days old and suggested that vocal sexual dimorphism was detectable before their fledging. Other petrels whose chicks acquire adult calls just before fledging are *Halobaena, P. salvini, P. belcheri, Calonectres diomedea, Puffinus puffinus* and *Pagodroma nivea*. Chicks of *Pachyptila turtur* have the full repertoire of the species almost on hatching (Tennyson, unpubl.).

Shallenberger (1973) followed the maturation of calls of *Puffinus pacificus* chicks. Most used three-syllable chirps with two peaks whose harmonics reached beyond 16 kHz; their fundamental frequencies ranging from 3 to 8 kHz. At 2 and 5 days old the chirps were simple cries. At 10 days, these calls had lengthened and an inhalatory section had appeared. The pattern continued to change towards that of the adult birds, with further development of the low frequency inhalant component and a flattening of the frequency peaks.

The simple cries were used when food begging throughout the chick's stay in the nest; the longer transitional and near-adult calls only when defending it from intruders or other sources of disturbance (Shallenberger, 1973; Swanson & Merritt, 1974).

Nagler and Smith (1992) recognized three types of calls used by chicks of *Oceanodroma leucorhoa*. Their 'Rhythmic' calls were typical simple cries with harmonics. Their 'Short' calls were *c.* 50 ms long, spread over a wide frequency range, and used with head thrusting and bill snapping in defensive actions and fully developed in 7-day chicks. The 'High' call was a complex simple cry easily derived from the 'Rhythmic' call, more drawn out and with harmonics. Nagler and Smith found that the 'Short' call was used when the stimulus was non-tactile, such as a disturbance at the burrow entrance, the 'High' one when touched. They did not follow the ontogeny to fledging at about 61 days.

The voices of prebreeding petrels might be expected to differ from those of the breeding birds (as do yearling penguins) and Bretagnolle (pers. comm.) believes that this is true of *Procellaria westlandica*. And if a bird such as *D. exulans* takes many years for its plumage to attain the adult state, why might not its voice also mature over a long timespan?

What, if any, of the repertoire is learnt is unknown, but when *Pagodroma* chicks were raised by *Daption* adults, the chicks only developed the calls of their own species (Bretagnolle, in press, a).

IV Sexual dimorphism

With most tubenoses examined so far the voices of the sexes have been found to differ, the females usually the harsher—see reviews in James and Robertson (1985a,c) and Bretagnolle (1995a) (Figs 4.3 & 5.7).

In *C. diomedea* sex dimorphism in size and calls is attained before the nestlings fledge (Bretagnolle and Thibault, 1995), and would be expected in birds such as *Macronectes* whose male chicks are the bigger. Indeed, vocal variation may sometimes be a linked to variation in body size, for example in *Pagodroma nivea*, where the fundamental frequencies of the songs of males (the heavier sex) are lower than those of the females (see Fig. 6.8). This could allow females to select heavier (higher quality?) males as Genevois and Bretagnolle (1995) hypothesized for *Halobaena*. In this species the heavier birds did not have deeper voices, but some temporal parameters of their calls correlated with body mass and could function similarly.

Apart from the heavier presence of filoplumes on the heads of many male tubenoses, burrowing petrels are monomorphic in plumage. Vocalizations to advertise the caller's sex would seem obligatory and in some situations—under heavy forest (e.g. *Procellaria westlandica*; Warham, 1988b), and in burrows and caves—there is no light, leaving only olfaction and sound as possible clues to the whereabouts and sexes of conspecifics.

The great albatrosses excepted, plumage differences are unlikely to provide clues as to sex, but head and bill shape probably do, as many such as *Diomedea immutabilis, D. nigripes* and *Macronectes* can be sexed by the human eye on the heavier bills and heads of the males. Breeding male *D. epomophora* also have pinker bills and tarsi than their partners (Richdale, 1950, p. 81). Postures such as 'Rapier Action' suggest the possiblity of sex identification by beak measuring (cf. Fig. 5.5B). Vocal differentiation could provide additional clues to sex and presumably save time and energy for birds seeking a partner. For further examples, see the generic accounts below.

In the visual displays of petrels, sex-specific postures are uncommon. There may perhaps be some cryptic signals undetected by the human eye. Sex may also be indicated by the way in which a posture is performed. Thus male Laysan and Black-footed Albatrosses giving the 'Sky Call' rise with their bill tips above those of the females, evidently signalling both dominance and sex. Sex differences may be shown only by the degree to which particular actions are associated with particular sexes—'Yapping', for example, was found to be a typically male figure in *D. exulans* (Jouventin & Lequette, 1990).

V Exploratory behaviour

The most obvious behaviour of this kind is that of birds probing among colonies in attempts to find unoccupied nest sites, for example the 'unemployed' shearwaters that wander in and out of burrows in the early stages of the nesting cycle. Richdale (1963, p. 90) found that some non-breeding burrows held up to 10 different *Puffinus griseus* during a season and Harper (1976) provided a picture of the rather frenetic explorations of Fairy Prions and of their reception by established pairs.

Although petrels are strongly philopatric, not all return to their birthplaces even if these still offer suitable breeding habitat. Mist-netting of marked *Hydrobates pelagicus* has shown that some visit colonies distant from their places of birth (Fowler *et al.*, 1982). They analysed 16 423 captures and concluded that all were non-breeders. What proportion of the original populations settle elsewhere is unknown but the success of Podolsky and Kress (1989) in attracting *O. leucorhoa* to nest on an uncolonized island (see Ch.12.VII.E.1) suggests that the wandering habit may be widespread. It is obviously essential for extensions of range (see Chapter 1.VII.A)

Birds such as *P. puffinus* and *P. griseus* will circle over uncolonized headlands after dark calling without alighting. Established colonies are usually not far away and the birds involved must be non-breeders and perhaps too young to obtain even air space at their natal colonies. Conceivably, such birds could found new colonies were there suitable habitat below.

The general pattern of recruitment to the breeding stock is that on their first return to the nesting grounds the non-breeding birds land late in the season and leave early while in later years they arrive earlier and stay longer (see Chapter 1.IV.D).

Displays and vocalizations of the giant albatrosses *D. exulans* and *D. epomophora* mostly take place where two or more birds associate in 'gams' (see Fig. 5.2). With *D. exulans* such groups consist mainly of younger birds identified by their darker plumage, and these provide the bulk of the calling and displaying, the breeding birds mostly ignoring this activity. Pickering (1989) found that *D. exulans* in their first year on land were mostly alone, with the males staying ashore the longer, and by their third season (at 8–10 years old) most stood by a nest site; whereas the females moved around to display with a succession of temporary partners. Observations of marked and sexed *D. immutabilis* showed that the established breeders did not join in the extensive dances and associated calling of the prebreeding birds (Meseth, 1975).

A good deal of evidence has also accumulated on the dominant role of the non-breeders in the chorus of nocturnally active species, for example by Richdale (1944a) and James (1985a). Their vocalizations reach a climax around nightfall when many birds are alighting; peaks in the chorus mark peaks in activity in the ground and air. After about 2 h the din dies away and often by midnight few calls are heard.

Playbacks tend to pull down birds with feathered brood patches rather than breeders (Furness & Baillie, 1981; Podolsky & Kress, 1992). Later in the season when the non-breeders have left, arriving and departing birds tend to be silent.

In a detailed study of the Manx Shearwater, James (1985a) followed non-breeders by glueing luminescent lights to their central tail feathers. Both sexes left their burrows to engage in calling overflights, the females spending 2.4 times more time aloft than the males. In March, when most were already on land, the level of song was extremely low (Fig. 4.6A), with very few immature birds ashore.

VI Functions of vocalizations

The visual displays of petrels are augmented by their associated vocalizations. For example, the noise of an albatross 'gam' signals its whereabouts to other non-breeders seeking partners even when the displaying birds are hidden by 'dead' ground and, at least among albatrosses, most visual elements of the courtship displays are associated with particular calls. For birds active after dark, vocal communication appears essential for bringing the sexes together.

The same display may have more than one function, serving a sexual role in one context and an agonistic one in another, for example not all bill rattlings and snappings have aggressive connotations, and female *C. diomedea* respond differently to male calls according to whether they are breeding or not (Bretagnolle & Lequette, 1990).

Bretagnolle considers that even petrels with but a single major call, such as *Halobaena*, can communicate a range of messages—individual signatures coded in temporal characters at the end of the call, geographical information in temporal variations at the start of the call, and sex in syntactical variables (Bretagnolle, 1990a). He also suggests that, while the wide frequency range and repeated syllables help the bird's location by potential partners, the unstructured 'noise' may reduce their detectability by skuas.

A *Species recognition and isolating mechanisms*

Where several congeners share a breeding place it might be expected that their voices would differ to help reduce mismatches, more particularly between nocturnally active ones. For example, four gadfly petrels nest on the Crozet Islands—*Lugensa brevirostris, Pterodroma macroptera, P. lessonii* and *P. mollis*. Two of these (*P. macroptera* and *P. lessonii*) have very similar voices, and although the first is a winter breeder, the other a summer one, there is some overlap in their times ashore. The voice of *P. mollis* is quite different from that of those two and from that of *L. brevirostris*. Bretagnolle (in press, a) has illustrated the considerable differences in the calls of the burrowing petrels of a typical subantarctic community.

Another example of several congeners sharing a nesting place concerns the San Benito Islands, home for the storm petrels *O. melania, O. leucorhoa* and *Halocyptena microsoma*. Anthony (1900) commented on the harsh notes of the first species, the rather similar calls of the second one, and the strange, high-pitched whirring of the third which resembled a rapidly revolving cog wheel: 'For about ten seconds the

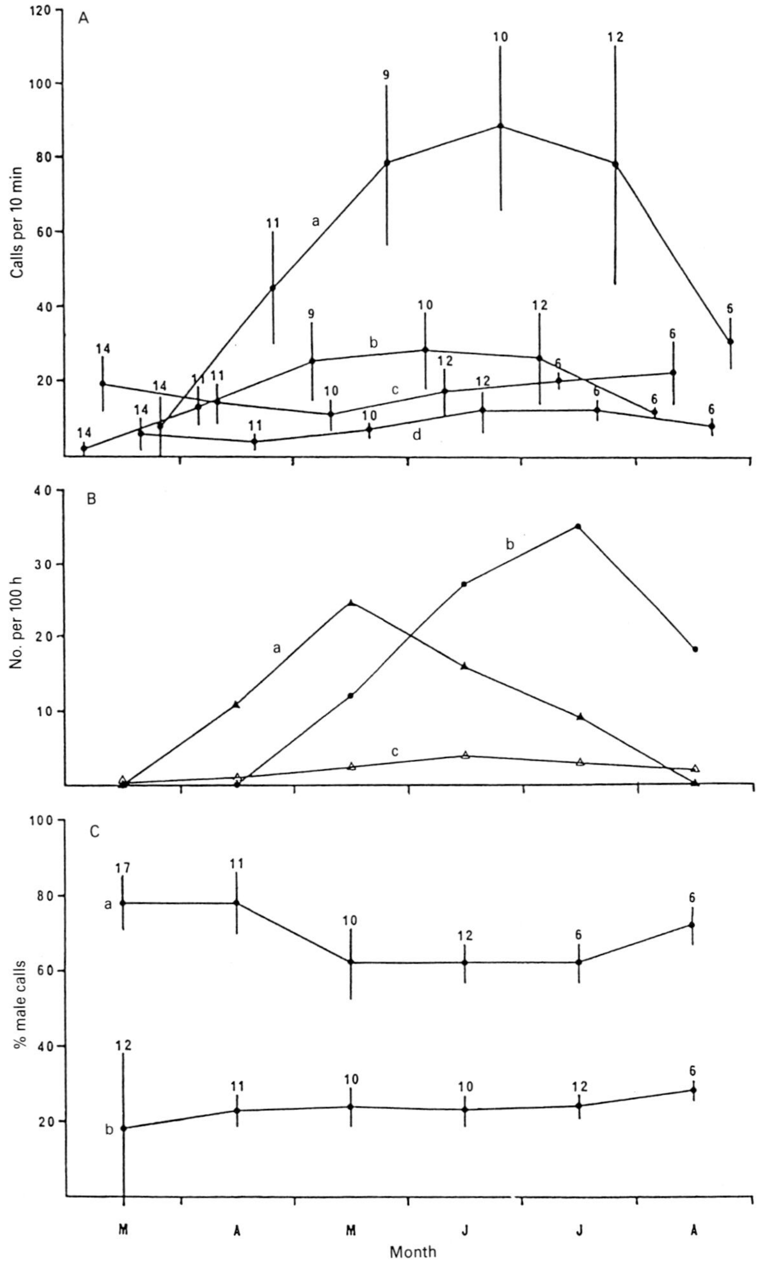
A
B
C
Calls per 10 min
No. per 100 h
% male calls
M
A
M
J
J
A
Month

whizzing continues when suddenly a note is dropped—there is a quick gasp, as for breath—and instantly the wheels begin to revolve again, giving the impression that there is a broken cog in the buried machine'. Evidently *H. microsoma* also uses 'breath notes'.

Whether voice recognition helps deter hybridization has not been examined, but with *D. immutabilis* and *D. nigripes*, albatrosses that share most of their nesting places, Sparling (1977, p. 263) suggested several potential isolating mechanisms of a vocal nature. For example, their aggressive 'noiselike ehs' differ in the number of notes in a sequence and are distinguishable to the human ear. Nevertheless, hybrids are known between these albatrosses (*The Petrels*, p. 27; Warham & Fitzsimons, 1987) and some petrels do respond to calls of non-specifics (see Chapter 6.VIII.C.2).

B *Sex advertising and mate attraction*

Courtship display concerns sex advertising, pair bonding and mate selection. Using Manx Shearwaters, James (1985a) found that more were attracted to burrows from which taped calls were coming than to control burrows without calls, suggesting that burrow calls from real birds would be highly attractive (Fig. 4.6B). Most birds caught in 'silent' traps were males—presumably the male's role is to find a burrow and, while at first these were mainly non-breeders, later in the season males entered male-singing burrows evidently to attempt evictions.

The rates of calling changed over the season, aerial ones being highest in June when prebreeders and immatures were at peak numbers. Birds of both sexes called from the wing, but females called more from the air, males more from the ground (Fig. 4.6C).

The responses to male and female calls played from the ground and from the top of a 10-m pole (simulating aerial calls) were also measured. More females than males replied and James deduced that a significant proportion of males overhead was silent. He suggested that by staying quiet, these could assess the settlement opportunities below from the amount of noise in the different areas, alighting where the prospects seemed best.

Using similar playback systems from burrows and raised positions with prebreeding *Puffinus puffinus*, Storey (1984) found that test calls of both sexes increased the frequency of calls from male birds, whereas only those of males produced increased

Figure 4.6 *(A) Call rates (mean ±1 SD) by month of Manx Shearwaters* Puffinus puffinus: *(a) Flying females; (b) Flying males; (c) Grounded males; (d) Grounded females. All show significant seasonal differences. Numbers above each point are the numbers of counts per month. Note the low rates during the prelaying stage in March and April and the increase in May around egg laying due to influx of the non-breeders. (B) Manx Shearwaters caught in artificial burrows each month per 100 h effort in (a) those from which female calls were played; (b) those playing male calls, and (c) those trapped in burrows without playbacks. Male and female calls were equally attractive, both significantly more so than 'silent traps'. (C) The mean (±1 SD) proportion of Manx Shearwater calls heard each month that were male, (a) from the ground; and (b) from the air. Numbers above each point are the numbers of counts per month. From James (1985a).*

responses from females. Storey concluded that male calls induce competing replies from other males attempting to attract females to their burrows while also stimulating unpaired females to call to gain information on the whereabouts of the advertising male by inciting it to continue singing. Playback of male calls from burrows induced females to land nearby when some prolonged 'duets' with the taped call ensued.

Male *Halobaena* singing from outside or near their burrows attracted overflying females. These called before landing and only the females gave aerial calls (Bretagnolle, 1990a). This author also found that only the male Wilson's Storm Petrel uses the chatter call which serves for self-advertisement and is highly attractive to conspecifics, particularly to females (Bretagnolle, 1989a).

The flight calls of the storm petrel *O. monorhis* elicit responses from grounded conspecifics of either sex, but mostly from their own. Tape playbacks from the ground attracted fliers of the opposite sex who called in response, indicating a sex-recognition role for the flight call and the extent of the responses showed that the birds made few errors in determining the sexes of calling birds (Taoka *et al.*, 1989b). These authors concluded that males on or below ground gave flight calls to attract mates while flying females used flight calls to elicit male responses.

C *Mate and individual recognition*

The value of being able to recognize their partners to long-lived seabirds such as petrels, which reproduce better when they retain the same nest site and mate, has been emphasized (e.g. by Brooke, 1978a; Guillotin & Jouventin, 1980). Even for such diurnally active birds as albatrosses, which can use visual clues unavailable to nocturnal species, vocal clues may help speed the partners' reunion, especially if the nest has disappeared during the non-breeding season.

When a burrowing petrel is visited by its partner, the partner often calls as it enters the access tunnel and a brief duet may ensue, for example with *P. tenuirostris* and *P. carneipes* (pers. obs.), with *O. monorhis* (Taoka *et al.*, 1989b) and others.

The vocal response of a petrel to its mate's call as it enters the burrow may be different from that given to an intruder, so that vocal and/or olfactory clues seem to be involved. With *P. puffinus*, the use of playbacks (which eliminate any olfactory clues) has shown that the females at least can recognize the voices of their mates (Brooke, 1978e; Storey, 1984; James, 1985a). Brooke concluded that female *P. puffinus* in their nests needed to know their partner's voices in order to recognize and then repel foreign males. Males in burrows, however, responded equally to the calls of their own or those of strange females, suggesting that the males did not recognize their mates by voice.

D *Neighbour recognition*

Field observations of the responses of nearby singletons or pairs to extensive calling and displays suggest that individual recognition may extend to birds other than the partners. This could be particularly appropriate for diurnally active species and for one such as *M. giganteus* which nests at high densities and in overdispersed patterns. They also visit their nests outside the breeding season, so are in contact with neighbours more often than are most other tubenoses. Neighbour recognition may help returning birds in prompt location of the nest, so reducing attacks when edging past the beaks of their fellows.

Neighbour recognition may be unimportant for burrowing species. Tests have been few, the results equivocal. Shallenberger (1973) found slight evidence for reduced responses to recorded calls of neighbours by *P. pacificus* and Bretagnolle (1995a) insignificant differences in similar tests on *Halobaena*.

E *Social stimulation*

The contagious nature of calling and displaying on the breeding grounds, and the attractiveness of playbacks suggest that mutual stimulation may affect gonad development. This could be particularly important with highly migratory species such as *P. tenuirostris* and *P. gravis* whose laying dates are closely synchronized (Marshall and Serventy, 1956b; Brown and Baird, 1965). However, such effects could not be on ovulation as this takes place at sea during the prelaying exodus, nor, at least with the albatrosses *D. melanophrys* and *D. chrysostoma*, does it directly influence yolk deposition as, by the time that these birds arrive on their colonies, rapid yolk deposition has been under way for 1–2 weeks (Astheimer *et al.*, 1985; *The Petrels*, fig. 12.9). For the noisy, active non-breeders and prebreeders however, the stimulation could help advance their sexual cycles towards those of the breeding birds. This possibility has not been studied in petrels. However, Marshall and Serventy's findings that in captive *P. tenuirostris* such development was controlled by an internal rhythm, suggests that established nesters may not be immediately affected by the noise around them.

F *Agonistic functions*

Threat and defensive postures are used by surface-nesting species but visual threats are evidently little used, at least on land, by those active at night The postures tend to make the birds look bigger to the potential aggressor (e.g. Fig. 4.7). Oil spitting may follow. This attitude was even adopted by fledgling *Daption capense* swimming in the sea with skuas hovering overhead (Burton, 1968). Tail cocking is a common feature of threat posturing, particularly well shown by *Macronectes* (see Fig. 6.1). Young albatrosses draw themselves up to their full height and rotate on their nests to face the intruder meanwhile jerking their heads and gulping in anticipation of throwing up stomach oil and stomach contents. Except for diving petrels, oil spitting is widespread, but used mainly by chicks and it is well aimed only by fulmars.

Threatening calls are used by all petrels for warning off potential attackers. They include the gurgles of fulmars priming their pumps in readiness for firing oil salvoes, the bill snaps of albatrosses that draw attention to their beaks as defensive weaponry, and a range of low pitched, 'noisy' sounds that birds underground use to deter intruders. With Laysan Albatrosses, Sparling (1977) suggested that the 'eh' calls were of two types, used in different contexts. One with harmonics in the frequency structure was used by the males towards receptive females, the other with few harmonics—Sparling's 'noiselike ehs'—given in response to an intruder of its own sex. He also found that other calls—'whines' and 'yammers' were associated with agonistic behaviour. Other examples of calls accompanying threats are the 'whinny' of *Macronectes* (Bretagnolle, 1988b) and growling calls of *P. assimilis* (Warham, 1955).

Although the effectiveness of threat calls and displays by petrels has not been evaluated, observations suggest that the chicks of surface nesters, such as albatrosses in the postguard stage, are seldom attacked by natural predators such as skuas, or even by unnatural ones such as cats. Their large size may be the major factor here, for smaller species are mostly vulnerable to such predators despite their threats, so that the main function of these calls is presumably to deter conspecifics, with surface nesters relying on oil spitting for protection, burrowing ones on darkness and underground inaccessibility.

The various cries heard during fighting seem to be a jumble of threat and alarms. Providence Petrels fighting after being lured to the ground (Section XVII below) used a cacophony of cries as they closed with each other and fighting *D. bulleri* distorted 'wails' (see Fig. 5.7), with the pitch increasing apparently in accord with the stress involved (Warham and Fitzsimons, 1987).

Alarm calls are also mostly high-pitched, used when birds are startled, caught for ringing, being eviscerated by a skua, etc. When mollymawks alight in crowded colonies they may give a loud 'Wail' apparently in stress, because at touch down

Figure 4.7 *Threat posture of Cape Pigeon (*Daption capense*)—wings fluffed out and held from body, tail cocked and fanned, making the bird appear larger than normal. From photo by F. Topliffe in Pinder (1966).*

they are particularly vulnerable to attack by conspecifics whose individual distances they may transgress.

G *Colony and nest location*

Diurnally active petrels no doubt locate their own colonies by sight but on dark nights nocturnal species may use the general hubbub to find their nesting areas. Local dialects could conceivably provide homing clues (Ch. 6.VI.B.2), particularly for immatures seeking their natal colonies for the first time. The occurrence of wandering birds by night at colonies far from where they were born and their later appearance at their home colonies, suggests that their visits may have been responses to the noise of the strange colony, for example with *P. puffinus* (Harris, 1972).

The ability of advanced petrel chicks that have wandered from their nests to return to them demonstrates a sense of position. Similarly, even when nests have been obliterated by ploughing or snow cover there are reports of their owners digging down to them.

More intriguing is the ability of adults to locate their own burrows from an apparent maze of similar ones. Lockley (1932) and Fisher and Lockley (1954) speculated that the incoming birds calling in flight were using echo-location. However, as pointed out by Warham (1955) and Serventy *et al.* (1971, p. 27), overflying shearwaters that are silent still find their nests. Nor is there evidence for the use of ultrasonics by the few species so far studied—*O. leucorhoa* (Griffin in Grubb, 1974) and *Hydrobates pelagicus* (Ranft & Slater, 1987). Wink *et al.* (1982) did propose that *C. d. diomedea* flying to nests in cliffs or caves, used the echoes of their voices to help their orientation. The calls were not ultrasonic and their hypothesis not tested experimentally. This situation is quite different from that where reflections are from a sound-absorbing medium such as vegetated and burrowed ground.

That incoming birds are guided by the calls of their mates below ground as suggested by Glauert (1946) for *P. assimilis* is also incorrect and breeders still reach their nests when the occupants are silent or absent (Warham, 1955; Brooke, 1990).

Shearwaters nesting in the open seem to depend mainly on visual clues to find their burrows as proposed by Warham (1958b, 1960) and supported experimentally by James (1986a) and Brooke (1990, p. 209) for *P. puffinus*. The evidence for an olfactory role is weak (see Chapter 7.IV.A) and the possibility of proprioception not investigated. This still leaves unexplained how petrels find their burrows in the darkness of caves or on moonless nights, heavy overcast and closed forest canopies.

VII Ground calling

All petrels give some calls from, on, or below the surface of the ground, in conjunction with bill and body movements. Most vocalizations delivered from the ground for advertising or sexual attraction seem similar to those given from the wing, for example with *Pterodroma hypoleuca* (Grant *et al.*, 1983). However, ground calls tend to be longer. Thus the 'ti-ti-ti' calls of gadfly petrels such as *P. macroptera* and *P. nigripennis* are shorter, more staccato when from the air. Songs delivered from flight also sound different because of Doppler-shifted changes of pitch (see Fig. 6.13H). For the storm petrel *H. pelagicus* an unbroken run of 983 'chikka' flight calls was recorded from a burrow by Oldham (in Witherby *et al.*, 1944, p. 26).

Storey (1984) found that the sound volume at the burrow entrance was much greater than on the surface, the burrow acting as a megaphone: surface calls reached a wider audience. She suggested that the complex calls of grounded male *Puffinus puffinus* are more easily pinpointed than the simpler ones of the flying females.

Petrels in burrows may respond to others calling outside by coming to the entrance to call themselves, as if to 'see off' the intruder, or perhaps to reply to calling overfliers of the opposite sex—usually the females.

In her playback experiments with *P. puffinus*, Storey (1984) found that the calling of the males increased when other males called, evidently competing for the available females. These approached in two stages, first landing in the general area and then listening for more male calls as guides to the burrow.

Established pairs call little from their nests except when the pair-bond is reformed at the start of the season. However, when returning to take over from their partner on egg or chick, there may be a brief 'dialogue'.

Of course, burrow calling advertises the caller's whereabouts. Predators may take advantage of this, and the risk of attack was cited as the probable reason why on clear nights male *P. puffinus* called only from below ground (Storey, 1984).

VIII Aerial displays

Most tubenoses call in flight at some time, and a few such as albatrosses and giant petrels, incorporate body movements into distinct aerial displays. Calls given from the wing carry farther than do those from grounded birds, with the callers at little risk from predators while they remain aloft.

Gadfly petrels and small- and medium-sized puffinoid petrels—*P. assimilis, P. lherminieri, P. huttoni, P. gavia, P. puffinus*—are very noisy in flight over the nesting grounds although the level of calling can vary considerably from night to night without obvious links to the weather. Larger species like *P. griseus, P. tenuirostris, P. pacificus* and *P. carneipes* call in flight infrequently. Indeed, incoming *P. griseus* and *P. tenuirostris* at dusk are eerily silent despite their vast numbers, the only sounds the sussuration of their wings. To those accustomed to the din of flying Manx Shearwaters at their smaller colonies, the silences of the inflying southern birds come as a big surprise. Similarly, two *Procellaria*—*P. westlandica* and *P. parkinsoni*—call from the air, *P. aequinoctialis* rarely calls, and *P. cinerea* not at all.

Why are the incoming southern shearwaters so quiet? Most may well be the established breeders that have no need to advertise themselves and what aerial singing occurs later in the night probably originates from non-breeders. Why there is so much more flight calling by the smaller species, except at new colonies such as that studied by Storey (1984), is unclear. Nor does there appear to be fewer non-breeders or a greater danger from predation at many southern colonies. *Puffinus tenuirostris* often shares its breeding habitat with *L. pacificus*, the biggest of gulls, which takes its toll as do other predators, corvine and raptorial. These losses have not been quantified, but are appreciable, although arguably no higher than those experienced by the smaller species. And at some southern sites the predation pressures are so low that many *P. griseus* land silently and in broad daylight. Calling before alighting would then appear to involve no added risk, at least at the present time. Larger shearwaters are less agile than the smaller ones, of course, but better able to defend themselves from attack.

James and Robertson (1985a) pointed out that male competition for nests could make aerial calling and advertising inefficient. Immature *P. puffinus* comprised the majority of those calling on the wing. The males defending and holding burrows left periodically to call in the air, presumably to attract partners. However, such behaviour involves the risk of losing the burrow to another prospecting male, so that where burrows are in short supply or non-breeders particularly numerous, aerial calling may have been abandoned. This could help explain the comparative silence of the two southern species as these typically nest at high densities. There might even be an inverse relationship between the incidence of aerial song and the proportion of eggs laid on the surface (*The Petrels*, p. 300).

The hypothesis of James and Robertson (1985a) and Brooke (1990, p. 202) that petrels with sexually dimorphic calls tend to call in the air and vice versa is not supported by the silence of inflying *P. griseus* and *P. tenuirostris*, which have sexually dimorphic calls (Fig. 4.3). Again, the storm petrel *Oceanites oceanicus*, whose sexes call differently, is rarely heard from the air (Bretagnolle, 1989a). Furthermore, while James and Robertson (1985a) list *H. pelagicus* as not singing in flight, Davis (1957, pp. 95–96) described its pursuit displays with vocal accompaniment over the nesting grounds.

That there is often more aerial calling on dark nights may be because it is safer for the non-breeders to do so then as they are at less risk to predators like gulls (Storey and Grimmer, 1986) and at very dense colonies such calling may lessen the risk of collisions: the short ejaculatory notes of *P. tenuirostris* perhaps function in this way (Warham, 1960).

IX Duetting

The 'dances' of albatrosses, well described by Meseth (1975) and Lequette and Jouventin (1991a,b), are accompanied by mutual calling, and non-breeding shearwaters, prions and other burrowing species also engage in extensive duets when courting on the surface or in their nests (Fig. 4.8). Typically the male begins and his partner joins in, so that there is some overlap when both are singing simultaneously, perhaps in unison, after which the first falls silent and the second completes the verse, as in Fig. 4.3. Such calling may be very prolonged and resumed on successive nights. Limited observation of sexed birds indicates that the males tend to be the more vocal, the females' contributions briefer. At other times there is some alternation between the voices—sequential 'singing'—but usually the two fall out of step.

Figure 4.8 *Little Shearwaters* Puffinus assimilis *duetting at the entrance to their nest cavity. Their heads are close together, throats inflated and crown and nape feathers slightly raised. Photo: Author.*

X Displays and vocalizations at sea

The sexual and agonistic posturings are not restricted to the land; a wide range of tubenoses has been reported displaying to conspecifics and, aggressively, to these and to other seabirds. Most records refer to albatrosses, partly because these are easily seen. Albatrosses are solitary birds when flying pelagically but, like other petrels, when they alight on the water they tend to form small groups and then often indulge in seemingly casual billing, bill snapping, sky pointing, allopreening and other postures from typical 'dance' sequences. Cooper (1974) reported such behaviour from *D. exulans* and mollymawks, and deduced that most were non-breeders. Harper (1987) saw pairs of *D. exulans* using the 'Sky Position' display: these are usually accompanied by calling. Wandering Albatrosses use deep, throaty barking cries when attacking baits (Willis, 1954), bill clapping when fighting off competitors such as giant petrels, while mollymawks will puff out their cheeks to expose their gape stripes. Cheshire (1990) reported calling at sea from *D. immutabilis* and *C. leucomelas*, and Black-footed Albatrosses gave high screams with wide-opened bill and extended wings to keep competitors at bay (Yocom, 1947).

D. immutabilis has used the decks of freighters as mobile display grounds, even performing full-blown 'dance' sequences (see e.g. Campbell *et al.*, 1986). Still more surprising is the case of large numbers of *P. lherminieri* landing on the deck of a research vessel to sing, display and copulate (Murphy, 1958).

Albatrosses will also display on the water near their colonies. Richdale (1939) described *D. epomophora* using nuptial displays including 'Wing Stretching' in these circumstances, and Buller's Albatross also performs 'Croaking' displays while rafting off the breeding places. Richdale (1950, p. 67) thought that *D. epomophora* could communicate the loss of an egg or a chick to its partner while at sea. There seems to be no clear evidence for such contacts (Sudbury *et al.*, 1985), but Mougin (1975, p. 81) pointed out that the existence of local dialects could provide clues for birds from the same colony to meet up at sea.

Records of smaller petrels displaying at sea are scarce, but among the shearwater rafts that form off the nesting islands and at the contranuptial feeding grounds a good deal of bill fencing and allopreening takes place. Correia (in Murphy, 1924b) noticed this of the 'cagarras', *C. diomedea edwardsi*, off the Cape Verde Islands. Such birds also call when on the water, for example *P. griseus* and *P. gravis* (pers. obs.; Anon, 1974). Among gadfly petrels, *Pterodroma externa cervicalis* and particularly *P. nigripennis*, indulge in their high-speed aerial chases with much calling far out at sea by day and by night, as reported by Jenkins (1970) and others. *P. hasitata* has been heard using a low 'growl', a 'bark', and a 'purr' at sea off Cuba (Lee & Vina, 1993).

Tubenoses can also be active and noisy in disputes over food. Aggression establishes the hierarchical status of the different species and individual birds when these congregate to scavenge at a dead animal like a seal. These actions in the fulmar *Fulmarus glacialis* have been analysed by Enquist *et al.* (1985) (Fig. 4.9). Nowadays such interactions are seen commonly among petrels crowding for offal around fishing boats when albatrosses such as *D. albatrus*, *D. melanophrys* and *D. cauta* bray and clatter their bills to harass smaller birds such as Northern Fulmars and Cape Pigeons.

Aggressive displays between *Procellaria aequinoctialis* were seen regularly by Summerhayes *et al.* (1974). Two birds faced each other, back feathers ruffled, bills ajar, heads and necks low and thrust forward and upwards while both gave loud chattering cries. Finally, one would break away to be chased briefly by the other.

Northern Fulmars are well known for the cackling groups that collect to display on the sea in the vicinity of the breeding cliffs (e.g. Pennycuick & Webbe, 1959) and similar behaviour of small groups of Manx Shearwaters chasing, displaying and calling below the cliffs at Fetlar, Shetland, was described by Tulloch (1977). Comparable behaviour occurs elsewhere, for example of *Puffinus opisthomelas* off Guadalupe Island, Mexico (Anthony, 1896).

Presumably these displays at sea are mainly between prebreeding birds, perhaps of the younger classes, as there would seem to be little value for lone breeders to display with strangers when they already have partners and nest sites.

XI Copulation

Copulation is not often seen except at high-density colonies when most participants are probably non-breeders. I have found no detailed account of this behaviour for any storm petrel. Other burrow nesters evidently mate within their nesting chambers although in some instances breeding pairs have done so on the surface nearby. Only 19 (40%) of 47 instances of copulation by *D. exulans*, noted by Tickell

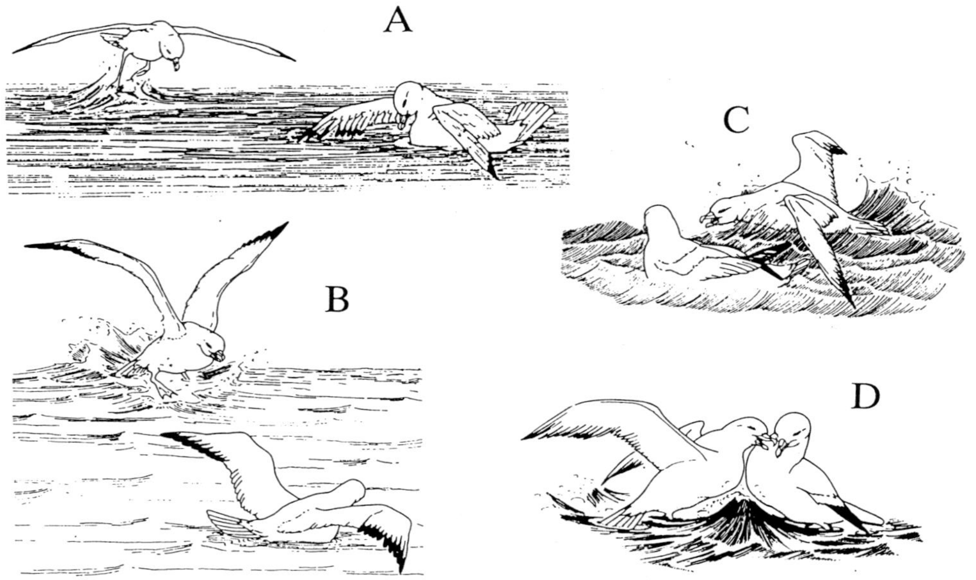

Figure 4.9 *Aggressive communication between Northern Fulmars (*Fulmarus glacialis*) competing for food. (A) Challenged from the air, a bird responds with 'Wing Raising' and 'Bend'—head and bill pointed to the water, tail fanned and cocked. (B) Threat involving 'Bill Pointing' with exposure of the pink buccal cavity and accompanied by a sharp call. (C) A challenger 'Charges' across the water to displace a bird using the 'Bend'. 'Charging' is often done from behind the challenged bird. (D) 'Breast-to-breast' action often follows on 'Charge'. Both birds lift from the water pushing against one another; fighting may follow. From Enquist* et al. *(1985).*

(1968) were by mated or prospectively mated pairs. Among albatrosses and fulmars most displaying does not lead to coition. In their study of *D. exulans* Jouventin and Lequette (1990) scored only four copulations in 12 027 observations of display postures (cf. Table 5.1).

Many copulations are not preceded by any overt actions, the male abruptly jumping on to the female's back. This is usually prefaced by vigorous allopreening which may be initiated by the female turning her head away from her partner and offering her nape and cheeks to his bill. He dibbles at her cheeks with the tips of his beak, around her (closed) eyes and nape, before reaching over to preen the far side of her head and breast. If receptive, a female may adopt a squat, submissive posture and allow the male to climb on to her back or to walk up the ramp provided by her depressed tail. His weight tends to push her wings down to her flanks while his claws curl round the humeral edges to maintain a firm grasp and balance (see Fig. 6.26C). He taps her bill vigorously with his own, producing the 'Tattoo' characteristic of coition in mollymawks, shearwaters and fulmars.

With *D. bulleri*, 'Tattooing' begins as soon as the male has a firm hold. He strops his bill across the top of the female's through an arc of 1–2 cm at *c.* 7 s^{-1}. After 10–15 s the male begins to swish his tail in a wide arc while working backwards to attain apposition of the cloacae. His wings tend to droop and the tips of the primaries are spread (see Fig. 5.9F). The male's tail motions push the female's into a near vertical position, the male meanwhile increases the amplitude and frequency of the 'Tattoo' and tail wagging. As the cloacae touch for 2–3 s bill stropping ceases but may be resumed after contact is broken off. Two or three cloacal contacts are made and the whole process is repeated several times. The male may remain on the other's back for 1–2 min after coition and after dismounting both may self- and allopreen. An unreceptive female simply raises the front of her body and unceremoniously dumps the other off.

Similar procedures hold for other mollymawks such as *D. chlororhynchos*, *D. melanophrys* and *D. chrysostoma* (Rowan, 1951; Tickell, 1984), the fulmars *Macronectes* spp, *F. glacialoides* and *Pagodroma nivea* (Prevost 1953; Warham, 1962; Brown, 1966), and for shearwaters such as *Puffinus pacificus* and *P. griseus* (Shallenberger, 1973; pers. obs.). Male great albatrosses *D. epomophora* and *D. exulans* do not strop their bills across those of the females creating a 'Tattoo' (Richdale, 1950), nor do the North Pacific albatrosses, but after cloacal contact the female *D. immutabilis* turns her head round to touch the male's bill and signal the end of the sequence. Similar bill thrusts are used by female giant petrels during coition.

Fisher (1971) noted that within a hour of Laysan Albatrosses being reunited on their return to their nesting territories, the males mated with their partners, and that this only occurred on their territorial grounds, whereas a male awaiting his partner would copulate with any bird, male or female, whose posture indicated submission. He distinguished such promiscuous and unsuccessful copulations and 'rapes' from successful ones, which are quiet and undisturbed acts. These attract no attention from other birds; rapes do (*The Petrels*, p. 256). Most successful copulations of established pairs lasted 2–3 min.

This species uses well-marked precopulatory signals. The male approaches on tiptoe, breast thrust out and with his head held well above that of his partner. If willing, she may also rise and perform 'Sideways Throwing', but she then signals her consent by squatting, flattening her back and spreading the scapular and primary feathers to either side. Fisher confirmed experimentally that a male will not try to mount a female who cannot spread and droop her wings. Furthermore, a bird of either sex that adopts this invitatory posture, through accident or other causes, will stimulate copulation from nearby males.

Reversed mounting has been seen in *P. pacificus*, particularly between immature birds with unswollen cloacae (Shallenberger, 1973).

There were 44 extra-pair copulations out of 1823 recorded between the Northern Fulmars studied by Hunter *et al.* (1992)—that is, 2.4%—but DNA fingerprinting showed that none of the strange males fathered another's chick. The established male always performed the most copulations and was the last to do so before his mate laid. These authors discussed how his sperm might achieve priority over that of the foreign male's in the storage tubules

Some vocalizations especially associated with copulation have been described. In *D. bulleri* 'Tattooing' is accompanied by a series of groans made up of many very short notes (see Fig. 5.10). During the copulation of *D. epomophora* the female uses quiet 'yapping' calls, the male *Pagodroma nivea* a series of soft clucking sounds, his partner a high-pitched churring (Richdale, 1950, p. 40; Brown, 1966). At the mating of *Puffinus pacificus* the male gives a continuous low call with a tremolo effect caused by the bill movements, the female also occasionally using a short call. This call tends to attract nearby conspecifics. If these try to interfere the mounted male will respond with defence calls (Shallenberger, 1973). In the prion *Pachyptila desolata* a special high-pitched whistle is heard during copulation, probably given by the female (Tickell, 1962, p. 18).

XII Behaviour and vocalizations of nestlings

Chicks of burrowing species spend much of the daylight hours sleeping. The young shearwater wakes up when the evening chorus begins, stretches its wings, and may respond to a noise at the entrance tunnel by calling. When it starts to venture out after dark it may crop nearby grasses and plants to take back to the nest, dropping much on the ground on the way. Young giant petrels also build up their nests and may wander off for some distance during long waits between meals.

Nestlings defend themselves by lunging with opened bill, hissing, and some discharge stomach oil. Their bills are soft and even well-grown chicks have no effective defence against a skua digging them from a shallow burrow, much less against a cat or a mongoose that can reach into the nest chamber or snap up any fledgling during the final exploratory phase of its stay ashore. According to Imber (in Marchant & Higgins, 1990), the chicks of *Garrodia nereis* are unusually quiet, perhaps to avoid detection by skuas, some nests being merely tucked into the skirts of tussocks and, in effect, above ground (Fig. 1.18).

Oil spitting forms the main defensive armoury for fulmar chicks. With *Macronectes* spitting ability declines as the chicks feather and some regularly handled never regurgitate oil or food, but one female *M. halli* did so when 112 days old despite not having been fed for about 14 days. Furthermore, at Point Geologie, Antarctica, oil spitting by *M. giganteus* is rare, whereas, for undetermined reasons, in chicks of the same species at The Crozets and Macquarie Island it is common (Warham, 1962; Voisin, 1978; Bretagnolle, 1988b, p. 117).

Oil spitting by *F. glacialis* was examined by Duffey (1951). Chicks with down still wet from the egg made the lunging actions of spitting but discharged no oil. Brooded chicks, exposed for any reason, tended, like other young petrels, to yawn expansively and were ready to spit at any disturbance although the parent might show no alarm. Lone chicks spat at any suddenly approaching bird, their own parents not excepted, and all, including gulls, tried to avoid being soiled.

Returning parents dodged these salvoes by cackling as they alighted further along the ledge and then edged forwards cackling softly until the chick calmed down and feeding could take place. Anything within about 1 m, flying or walking, was spat at and the chick only recognized its parents after 2 or 3 weeks.

XIII Parental recognition of egg and chick

A *Egg recognition*

Most petrels so far tested readily incubate eggs switched into their nests in place of their own. For example, *Pterodroma hypoleuca* accepted hen's eggs and *D. immutabilis* and *D. nigripes* plaster eggs (Grant *et al.*, 1982a,b). Indeed, albatrosses will sit on quite alien objects such as electric light bulbs, tennis balls, even beer cans (Bartholomew & Howell, 1964; Grant *et al.*, 1982b). However, Tickell (1962, p. 22) could not get the prion *Pachyptila desolata* to accept plaster eggs, attributing this to their rougher surfaces, but they did sit on eggs of Snow Petrels, Antarctic Terns and much larger hen's eggs.

Eggs outside the nest are ignored and parent birds do not try to retrieve their own eggs or chicks displaced beyond the nest's rim.

These various findings suggest that petrels do not recognize their own eggs visually, by smell or by feel.

B *Chick recognition*

From field trials, Rice and Kenyon (1962) concluded that chicks of about 10 days of age are not recognized by parent *D. immutabilis* or *D. nigripes*, whereas strange chicks aged more than 6 or 7 weeks were not accepted as replacements for those of similar ages—the parents now evidently recognized their own. When given the choice of similar-sized *D. immutabilis* and *D. nigripes* chicks, *D. immutabilis* accepted chicks of its own and of the other species about equally, but *D. nigripes* never chose *D. immutabilis* chicks, and sometimes attacked them. In the trials by Bartholomew and Howell (1964), the chicks were silent so that only visual, tactile or olfactory clues were available.

Tickell and Pinder (1972) found that adult *D. exulans* learn to recognize their chicks when these are between 8 and 20 weeks old. However, parent *D. melanophrys* and *D. chrysostoma* fed even 80-day-old aliens, so that even if they recognized these as not their own, the adults did not modify their parental care. As both burrowing and surface-nesting tubenoses seldom feed chicks except at the nest, there seems little need for parents to know their own and it also seems unlikely that philopatric offspring would be recognized if they nested near their parents.

Chicks of burrowing petrels have been switched to nests not their own and have been reared by their foster parents. Manx Shearwaters at about 35 days old were switched successfully and Harris (1969b) replaced four well-grown *Puffinus lherminieri* chicks by young ones committing the adults to feed them for a total of 103–120 days instead of the usual 75 days, so cross-fostering is evidently possible, at least with shearwaters.

Harris also transferred four young *Oceanodroma tethys* chicks to nests of *O. castro*, two pairs of which fledged them (Harris, 1969b). However, with four sets of artificial twins the strange chick was always ejected, suggesting that the adult *O. tethys* could identify their own.

Evidence for the ability of chicks to recognize the calls of their parents and for these to recognize those of their chicks is also ambiguous. Brooke (1978e) played recordings of male and female *P. puffinus* to unattended nestlings aged 30–60 days but these showed no preference for the calls of their own parents (Table 4.1). Tests with *Procellaria aequinoctialis* chicks gave similar results (Brooke, 1986).

Table 4.1 The frequency of response of nestling Manx Shearwaters (*Puffinus puffinus*) to playbacks of adult calls. From Brooke (1978e)

	Responded	Did not respond
Male parent	4	9
Female parent	4	12
Total parent	8	21
Male non-parent	2	11
Female non-parent	5	9
Total non-parent	7	20

XIV Non-sexual behaviour

A *Nest relief*

Exchanges of duty by procellariiforms during incubation or chick guarding do not involve elaborate rituals. A giant petrel will lift its wings and jump off its egg letting its relief walk on and settle down while the other leaves (Warham, 1962, p. 150). Relieved *F. glacialis* mostly depart promptly but may sit for some hours beside their partners. Simons (1985) saw the relieving *Pterodroma phaeopygia* squat beside its mate: after a few minutes of allopreening this bird stood up and allowed the other to take over.

At least in open-nesting situations, the on-duty birds may call as they recognize their partners before these reach the nest, and the relieving Snow Petrel calls persistently as it approaches, presumably, as with *Fulmarus,* to discourage oil spitting. Nest reliefs are accompanied by allopreening and both may rise and closely inspect the egg before the changeover. Manx Shearwaters allopreen vigorously during nest reliefs in the guard stage (Warham, 1952) and preening and duetting with swollen throats also feature during changeovers of *Puffinus assimilis* (Fig. 4.8).

B *Comfort movements*

There appear to be no actions special to tubenoses in their care of the body surface, nor any clear differences between the genera. Self-preening is important as is oiling of the plumage with the preen gland secretion. The back of the head is oiled by rubbing it directly on to the gland.

The propensity of giant petrels for bathing needs emphasizing. When feeding on a carcass, the whole of their heads, necks and shoulders may get plastered with blood and fat. Having fed, the birds immediately bathe vigorously to restore their plumage (Warham, 1962). How they shift the fat merely by sluicing with cold salt water is unclear. *Fulmarus* petrels and their newly fledged chicks also bathe exuberantly, and the sea beneath the big northern colonies is often covered with big rafts of fulmars, many bathing, rearing up and 'dive-dapping' so that the water pours from their backs and tails, often calling loudly meanwhile.

The most unusual aspect of feather care is that done to remove stomach oil, either that spilled by the bird itself or from strikes by others. This soiling is removed by bathing. Pryor (1968) described soiled *Pagodroma nivea* making shallow excavations in snow banks into which they rubbed their bodies, even turning on to their backs to complete the cleansing. Just how frozen rain can remove hydrophobic oils from feathers is difficult to understand.

Petrels shake their heads rapidly from side to side to throw salt gland excretions or detritus from their beaks. Sneezing may perform the same function. They also attempt the removal of debris or parasites from their heads by scratching with a foot brought up from under the wing. Incubating birds roll to one side to preen the soles of their feet.

Yawning is common and reveals the great extent of the gape. Incubating birds spend much time sleeping, and, particularly in cold weather, tuck their beaks into the scapulars so that the inhaled air is preheated. Simons (1985) found that incubating *Pterodroma phaeopygia* spent almost 95% of their time asleep, but during the hatch they slept for only 34% of the time, otherwise resting quietly. Fisher (1971, p. 47) recorded *D. immutabilis* passing into an 'incubating trance', staring ahead but with the nictitating membranes pulled across half-opened eyes. Many shearwaters sleep after feeding their chicks, resting either in the nest or tunnel, often on the ground nearby and many appear to stay there until the predawn exodus.

Grant *et al.* (1982b) monitored the behaviours of incubating albatrosses by day that resulted in the ventilation of their nests—side and foot preening, changing position, head scratching, standing up with or without wing flapping. The Laysan Albatross flushed its nest twice as frequently as did the Black-footed species.

Petrels also sleep away from the nesting colonies, for example when groups of giant petrels both during and outside the nesting season rest on open ground on windy saddles above the sea where takeoff is easy; many appear to be non-breeders. Snow and Antarctic Petrels sleep on icebergs, and Wilson's Storm Petrels may sleep on the sea during premigratory flocking (Phillips, 1954). Allison (1952) timed the activities in 24-h stints of *F. glacialis* at sea, finding that birds asleep and others resting but awake were most numerous around mid-day.

C *Responses to artificial lights*

That night-flying petrels will 'home in' to bright lights has long been known, especially in fog, which seems to disorientate them. Richdale (1952, p. 82) found that it was only on foggy days that all his Royal Albatross chicks went unfed.

Harrow (1976) reported that hunters spot-lighting for deer could draw *Puffinus huttoni* down on foggy nights, and spot-lighting has been used routinely on the nesting grounds to capture elusive species. Swales (1965, p. 26) found that at Gough Island on calm nights with low cloud or fog it was only necessary for two 750-W lights to be switched on for a few minutes before some 2000 birds were landing or circling around, creating a hazard for the researchers—a 1.2-kg Shoemaker travelling at high speed forming a formidable missile!

The effectiveness of spot-lighting varies not only with the weather conditions but also with the species. Birds that are normally active by day but that also fly overland by night, such as *D. exulans*, seem unaffected by the illuminating beams. Probably all nocturnal petrels are susceptible to some degree, but some are particularly sensitive, for example diving petrels (*The Petrels*, p. 200). And although storm petrels often alight near lights, particularly when storm-driven inland, *Fregetta tropica* and *G.nereis* sometimes disappear backwards up the beam (pers. obs.).

This phototactic behaviour has not been investigated in field or laboratory. Imber (1975b) hypothesized that fledglings are attracted because they are programmed to seek bioluminescent prey and respond as if the lights were a kind of superoptimal stimulus. He suggested that once young birds had taken such prey, artificial lights lost their attractiveness. However, adults are often highly vulnerable to ship's lights and to spot-lighting, and Bourne (1976a) gave a simpler explanation: the birds fly towards the light seeking a way out of the mist so that they can navigate in the ordinary way—from stars, moon, or landmarks.

XV Sound production

Petrels produce sounds by striking their mandibles against one another, by knocking their bills against those of other birds, or by forcing air through their syringes. Most can vocalize when air is drawn through the voice box as the lungs are inflated—the inhalant sounds or 'breath notes'—or when it is forced out producing exhalant sounds. These latter tend to be the louder. Much energy is expended in the climax of displays such as the 'Wing Spread' of the great albatrosses when the breast and belly muscles are seen to contract and expand extensively. Movements of the hyoid in altering the buccal capacity can be detected and head shaking and perhaps varying curvature of the neck all appear to influence the tone of the sounds.

The syringeal anatomy varies considerably from species to species within the order as demonstrated by Forbes (1882) and Fig. 4.10, but there are only two pairs of syringeal muscles. The mode of sound production has been examined only in *P. pacificus*. Shallenberger (1973) blew air down the throats of freshly killed or anaesthetized birds. He got no sound during lung inflation but a high-pitched whine from lung deflation which was rather like a typical exhalant call, and apparently generated from within the syrinx. Manual manipulation of the opening of the glottis affected sound quality, suggesting that glottis control may be important. It may be that much inspiratory sound production is involuntary, a consequence of the need to inhale deeply to produce high volume, far-carrying sounds on expiration.

Modification of the sound may be effected by alterations to the buccal cavity. Rapid but small movements of the mandibles may also create a vibrato effect. The tongue, of variable shape and size (see Fig. 10.8) probably also affects the transmitted sound, but no study of this has been undertaken. Indeed, the whole question of how petrels produce their calls awaits proper investigation.

XVI 'Playback'

The reactions of birds to recorded calls have been widely used in gaining insights to the functions of calls. With petrels, playback elicits responses mainly during the prelaying and incubation stages; later most breeding birds are ashore too briefly for extensive experimentation. Playbacks allow the calls of known birds to be captured, although responses are often brief so that two recorders are needed. Most studies have been of burrowing species but Warham and Fitzsimons (1987) played sounds back to *D. bulleri*.

Responses by day from burrowing species are far less than those obtained at night, with contagious effects also much reduced. Playback at high volume may provoke a greater response. Stimulated birds on the surface often attack each other briefly—cf. lured *Pterodroma solandri* (see Fig. 6.9). Responses decline with continuous playback and doubling tape speed also led to a fall off in responses from *Puffinus pacificus* (Shallenberger, 1973). Some birds adjusted their calls to synchronize with those on the tape, giving a sound an unnaturally high amplitude and pitch.

Experiments with *P. puffinus* show that paired females do not respond to the calls of strange males, but only to those of their mates and this is true also of *P. gravis* (Brooke, 1988). But paired males respond to the voices of other males (Brooke, 1978e; James 1985a), presumably to deter further investigation. In her experiments, Storey (1984) found that non-breeding female *P. puffinus* are attracted to playbacks of male calls especially if these are from burrows, calls from the surface being less attractive, as are duets from within burrows. Playbacks of male and female calls to 30–60-day-old nestlings early in the night elicited the soft cheeping accompanying feeding.

Playback has been used extensively in studies of storm petrels. Grubb (1973) provoked strong replies from *O. leucorhoa* of unknown status and sex to their 'Churr' call, with up to 15 birds hovering over the speaker, blank tape producing little response. Later studies show that within burrows these birds respond only to calls of their own sex (Taoka *et al.*, 1989a) whereas flying ones respond with 'Chatter' calls, more responses being given to calls of the opposite sex—the reverse of the situation with birds below ground (Taoka *et al.*, 1989c). Flying *O. monorhis* were also attracted to playbacks of the flight call of the opposite sex.

XVII Responses to man-made sounds

The Cahow *Pterodroma cahow* on Bermuda was eaten in great numbers by ship-wrecked sailors from the Spanish galleon of Capt. Diego Ramirez in 1603, and again in 1609 that bird supported the survivors of the *Sea Venturer*. Purchas (1738 in Jones, 1859) described how the sailors got birds 'by standing on rocks by the sea-side, whooping, hallooing, and making the strangest outcries, which attracted the birds, until they settled on the very person of the hunter'. Similar methods were used by the convicts at Norfolk Island to take Providence Petrels *P. solandri*—'When they begin to drop, we would go down into the valles, and the more we hollowed "ho, ho, ho" the birds would come running crying out "ke, ke, ke", thinking it was their mate or young . . .'. So wrote Jacob Nagle in 1790 (Dunn, 1988). Likewise the 'Ua'U of the Hawaiian Islands *P. phaeopygia sandwichensis* was also incited to reveal its nest sites by calling out in response to the hunter's imitations of its cries (Perkins, 1913; Bryan, 1908), and in foggy weather Newfoundland fishermen called up the 'hagdon' *Puffinus gravis* with imitations of their harsh cries (Collins, 1884). Bulwer's Petrel may emerge from its nest at imitations of its call (Amerson & Shelton, 1976)

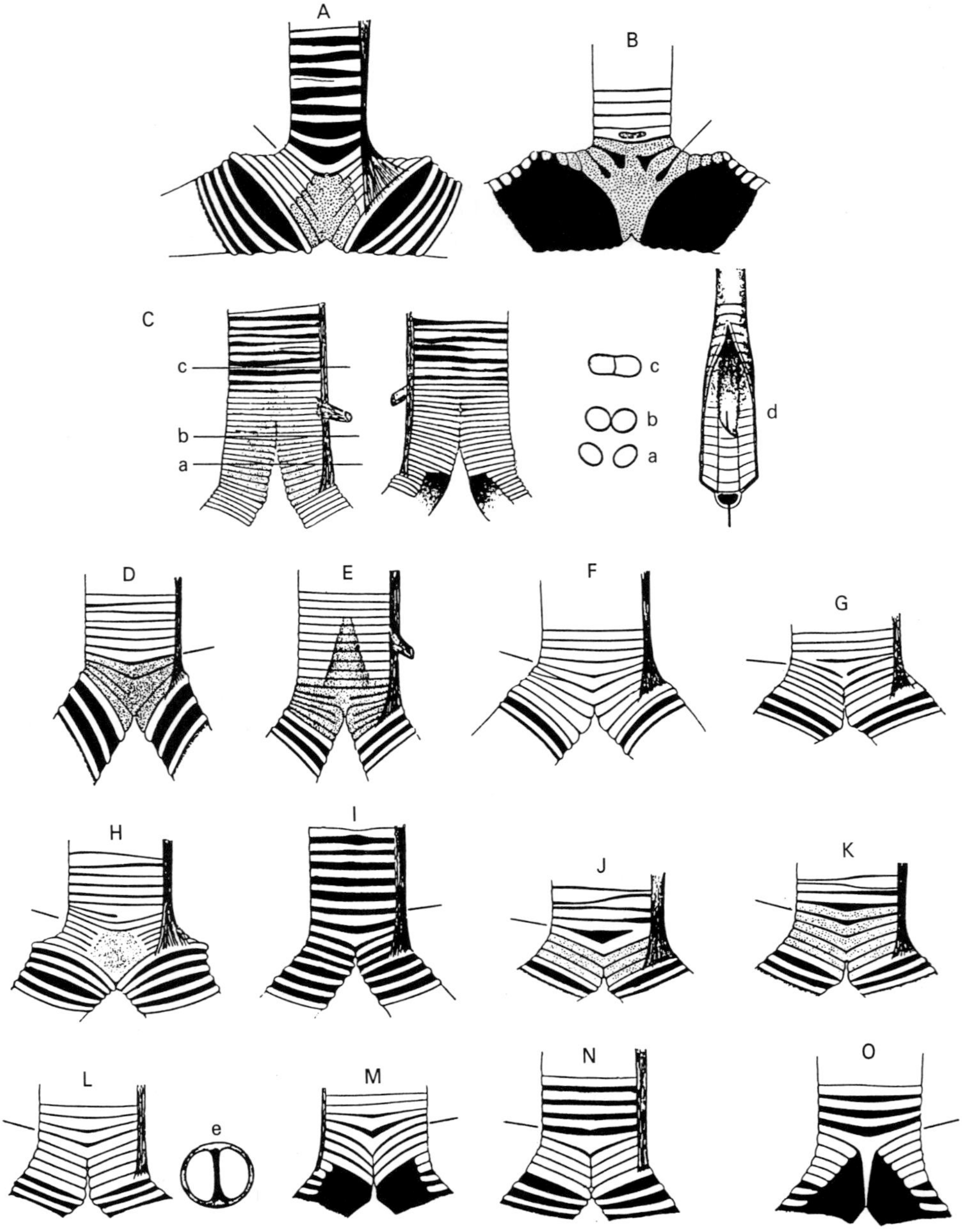
A
B
C
c
b
a
c
b
a
d
D
E
F
G
H
I
J
K
L
e
M
N
O

and *Pterodroma inexpectata* may respond similarly: in one instance an incubating female due to be relieved by its mate, even left the egg to walk towards the caller (Warham *et al.*, 1977).

Hindwood (1940) drew attention to the ease with which *P. solandri* could be brought down to earth using man-made cries. The decoy calls and the petrels' responses were analysed by Warham (1988a) (Fig. 4.11). The tests were made around the start of egg laying, the lure being a sustained hooting by flapping a hand before the mouth while sounding a steady note at around 0.5 kHz—the so-called 'war-whooping'. Many birds circled overhead in the late afternoon, with layer after layer visible from sea level up to about 800 m. They called with a harsh 'ti-ti-ti', creating a typical gadfly petrel chorus of animated chattering. Hand clapping lured some birds to the ground but 'war-whoops' provoked the most dramatic responses. Those nearest the caller replied immediately so that the general level of sound was abruptly amplified (Fig. 4.11A). These nearest birds tended to lower their feet and brake with their wings as if about to alight. Some changed their flight paths to bring them towards the source of the calls. With continued calling the petrels began to alight nearby and once down gave strident and urgent 'ti-ti' calls as they hurriedly tottered forwards over vegetation and boulders. On reaching the caller's feet they clambered on to his or her body, replying to every hoot, but now tending to peck and mandibulate the caller's clothing, using movements similar to those of gadfly petrels during mutual preening (e.g. *P. macroptera*; Warham, 1956).

The decoy calls were of lower pitch than those of the birds, but the hand movements did generate a series of shallow pulses, best shown in Fig. 4.11B, and at a rate rather similar to that of the syllables of the petrel's call. Such a pulsed pattern may be the key factor for a practical lure for these birds and perhaps help explain why hand-clapping and even playing a mouth organ can be effective.

This device is now known to work well with many other species, for example with *P. macroptera, P. externa, P. pycrofti, P. nigripennis, P. cookii, P. axillaris, P. lessonii* and *P. mollis* (Scofield, 1990; Tennyson & Taylor, 1990b; West, 1995 and Bretagnolle, pers. comm.). Tennyson and Taylor confirmed that the birds that landed appeared to be non-breeders and that, as with *P. solandri* and evidently *P. phaeopygia*, the lure calls were most effective in the prelaying and incubation stages, tailing off during the time when chicks were being reared. Some birds within burrows also replied, enabling

---

Figure 4.10 *Syringes of petrels. Lines point to junctions between tracheal and bronchial rings, as defined by Forbes. Not drawn to scale and only one syringeal muscle shown. (A)* Diomedea exulans *from before (ventral view); B.* D. exulans *from behind; (C)* Macronectes *sp. from before (left) and behind (centre), a, b and c show sections of syrinx and trachea along the lines a, b and c, to show the double nature of the tracheal tube below and its complete division by a median septum above; d, trachea opened to show median septum dividing it into two parallel tubes through the left of which a pointer is passed; (D)* Pagodroma nivea *from before; (E)* Fulmarus glacialoides *from before; (F)* Pachyptila vittata *from before; (G)* Pterodroma macgillivrayi *from before; (H)* Pterodroma lessonii *from before; (I)* Puffinus tenuirostris *from before; (J)* Hydrobates pelagicus *from before; (K)* Oceanodroma leucorhoa *from before; (L) and (M)* Garrodia nereis *from before and behind, e, the last tracheal ring from below to show the pessular bar; (N) and (O)* Pelecanoides urinatrix *from before and behind. After Forbes (1882).*

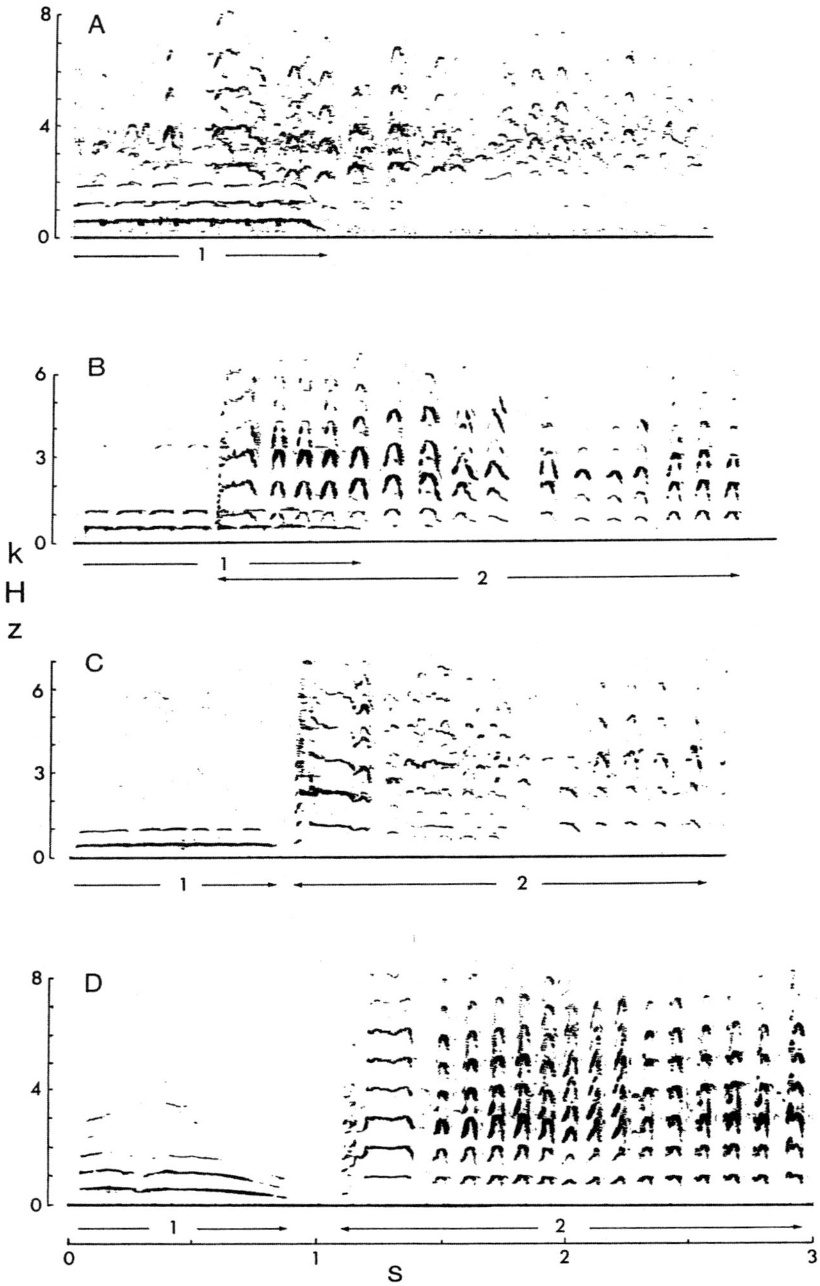

Figure 4.11 *Stimuli (1) and responses (2) of Providence Petrels* (Pterodroma solandri) *to vocal lures; (A) by a flock of flying birds; (B–D) by different grounded birds approaching the human caller. Note the rapid responses to the 'war-whoops' and their pulsed nature. The energy of these is at a lower frequency than that of the major energy bands in the birds' calls, suggesting that the timing and rhythm of the man-made sounds were more important than their frequency structure. From Warham (1988a).*

Figure 4.12 *Storm Petrels* Hydrobates pelagicus *hovering around an engine exhaust at the lighthouse at Rost, Norway. Photo: Bruno Sundin.*

occupied nests of *P. nigripennis* and *P. axillaris* to be found. Birds on eggs or chicks did not leave them but some did respond vocally. These findings suggest that the respondents are birds of either sex seeking partners, and that the calls act as superoptimal sex advertisements.

Other kinds of petrel will respond to imitations of their calls. Shearwaters will reply from their burrows, the best respondents probably being without partners and with these it may be possible to maintain extended duets. Tickell (in Beck & Brown, 1971) brought flying *Fregetta tropica* to the ground, and even to land upon him, by imitating their whistling call.

Precise imitations may not be necessary. Zonfrillo (1982a) and Clark (1985) found that *H. pelagicus* was attracted to the playback of calls of *O. leucorhoa*, perhaps responding to calls rather like their own and/or to calls associated with their natal islands where both species may have bred. In Tonga the Wedge-tailed Shearwater is said to be lured to wailing sounds and radio music, so that the people can 'whack them out of the sky to procure themselves a meal' (Jenkins, 1980).

Petrels will also respond to quite artificial sounds. Four different *Procellaria cinerea* came repeatedly to rest above or below a noisy engine exhaust at Macquarie Island, one reappearing in the same place the following year (Warham, 1969). At Great Barrier Island, *P. parkinsoni* can be found sitting near running power generators (Scofield, 1990), and on the Norwegian island of Rost, *H. pelagicus* storm petrels are attracted to the exhaust pipe of the lighthouse's diesel engine (Fig. 4.12). In all these instances the attraction appears to be the rhythmical, low-pitched nature of the sound acting as another superoptimal stimulus, probably to birds seeking mates.

Truly synthetic sounds were used by Taoka and Okumura (1988) and Bretagnolle and Robisson (1991) in investigating the characteristics of the calls of the different sexes of *Oceanodroma leucorhoa* and *Oceanites oceanicus* respectively. Taoka and Okumara used a synthetic song based on the typical rhythm of the 'Chatter' call of *O. leucorhoa*, with a tape of 'white noise' as a control. Birds in burrows responded to 'Chatter' calls of their own sex with their own sex-specific 'Chatter' call. The voice of the females was of lower frequency than than of the males. Using the same rhythm, a series of sounds at 0.5 kHz above or below 1 kHz was presented to birds of known sex in their burrows. There were no replies to tapes of 'white noise'. Otherwise all the playbacks provoked some responses, but the only significant ones were those of females to synthetic calls at 1 kHz. It was deduced that the characteristics of female calls included syllables of constant frequency at or below 1 kHz and that sex recognition is based on the call's frequency structure. Males only replied significantly to a tape closely resembling the rhythm and acoustic structure of songs of male birds. These authors suggested that the rhythm of the 'chatter' call in this species may be involved in species' recognition and that this call may carry information not only of the species involved but also of the sex and identity of the caller.

CHAPTER 5
Behaviour of Albatrosses

Contents

The dramatic displays of albatrosses attracted the attention of early explorers like M'Cormick (1842), more exact information coming from Matthews (1929) and Murphy (1936). Really detailed accounts and comparisons awaited the studies of Richdale (1949, 1950). More recent accounts include those of Fisher (1971), Meseth (1975), Sparling (1977), Jouventin *et al*. (1981), Berruti (1981), Tickell (1984), Jouventin and Weimerskirch (1984), Warham and Fitzsimons (1987), Jouventin and Lequette (1990) and Lequette and Jouventin (1991a,b).

Communication functions can be deduced from the contexts in which the postures and calls are used and what follows from them, for example with *Diomedea exulans* (Table 5.1), but no species has been subjected to field experimentation to tease out the significance of the various elements, apart from preliminary work by Warham and Fitzsimons (1987).

Copulation, 'Sideways Throwing' and allopreening are as described in Chapter 4, much self-preening appearing to be displacement activity. In 'Scapular Action' as first described by Richdale (1949, p. 12) the head is quickly turned to one side and the bill then dipped behind the folded wing on the opposite side of the body (see Fig. 5.5F). This is only used by mollymawks. It is not the same as the 'Scapular Action' described by Meseth (1975) for *D. immutabilis*, by Jouventin and Weimerskirch (1984) for *Phoebetria*, or by Jouventin and Lequette (1990) for the great albatrosses. For this action, where the bill is hidden on the same side of the body, the terms 'Beak Hiding' or 'Leg Action' are used here.

I The 'great' albatrosses

The repertoires of *D. exulans* and *D. epomophora* are basically similar. All postures are used by both sexes and most are accompanied by a special vocal signal, but some

Table 5.1 Use of postures by each sex of the Wandering Albatross. From Jouventin and Lequette (1990)

Attitude	Total number	% of total	% of each attitude between the two sexes		Exact test
			Male	Female	
Billing: 'B'	3898	32.4	52 (2013)	48 (1885)	*
Bowing: 'BW'	1501	12.2	54 (808)	46 (693)	**
Vibration: 'VIB'	1124	9.3	69 (770)	31 (354)	***
Yapping: 'YAP'	1104	9.2	72 (797)	28 (307)	***
Sing. Bill Snaps: 'SBS'	1055	8.8	60 (636)	40 (419)	***
Scapular Action: 'SA'[1]	988	8.2	58 (575)	42 (413)	***
Turn Around: 'TA'	366	3.0	70 (257)	30 (109)	***
Sky-Position: 'SP'	357	3.0	69 (248)	31 (109)	***
Walk: 'W'	296	2.5	69 (205)	31 (91)	***
Arched Neck: 'AN'	258	2.1	71 (184)	29 (74)	***
Whine: 'WH'[2]	215	1.8	83 (178)	17 (37)	***
Threat: 'THR'	187	1.6	67 (125)	33 (62)	***
Walk Away: 'WF'	155	1.3	16 (25)	84 (130)	***
Defence Billing: 'DB'	131	1.1	34 (45)	66 (86)	***
Grunt: 'GR'	105	0.9	91 (96)	9 (9)	***
Construct. Act.: 'ACT'[3]	77	0.6	34 (26)	66 (51)	***
Shaking Head: 'SH'	69	0.6	45 (31)	55 (38)	NS
Sky-Position Call: 'SPC'	67	0.6	96 (64)	4 (3)	***
Allopreening: 'ALP'	53	0.4	32 (17)	68 (36)	*
Snorting: 'SNT'	17	0.1	35 (6)	65 (11)	NS
Copulation: 'COP'	4	0.03	50 (2)	50 (2)	—
Total	12 027		7108	4919	***

Exact test: NS, non significant; *, $P < 0.05$; **, $P < 0.01$; ***, $P < 0.001$.
[1]'Leg Action'; [2]'Head Shake & Whine'; [3]'Sideways Throwing'.

elements are used more by one sex (Table 5.2) and there are sexual differences in some calls. During nuptial displays a series of postures is strung together to form a sort of 'dance' (Richdale, 1950; Jouventin & Lequette, 1990).

In the following summaries Richdale's terminology is used where appropriate.

A *Nuptial displays*

1 *'Billing'—see Fig. 4.2(6)*

The bird reaches with bill slightly ajar towards the bill or breast of another—'Bill Pointing', the male's head slightly higher than the female's. Their bills may touch or that of the active bird contact the other's breast—the 'Bill Touching' and 'Breast Billing' of Jouventin and Lequette (1990). 'Billing' is the commonest action between

Table 5.2 The use of acoustic signals by Wandering Albatrosses. From Lequette & Jouventin (1991a)

Acoustic signal	% of total (n = 6121)	Visibility of associated visual display	Main following display	% performed by males	Context	Possible function
Yapping	18.0	++	—	72	A;P;C	Sexual recognition, identification, territorial assertion
Whine	3.5	+++	Threat (89.0%)	83	D	High-intensity excitement
Threat	3.1	+++	Billing (22.2%) Whine (22.2%)	67	D;N	Threat
Grunt	1.7	++	Billing (37.0%) Arched Neck (25.9%)	91	D	Low-intensity excitement
Sky-position call	1.1	+++	Yapping (56.0%)	96	D;A;P	Territorial assertion, female attraction
Vibration	18.4	++	Billing (74.6%)	69	D;A;P	Announce a contact intention
Single bill snaps	17.2	+	Leg Action (54.8%)	60	D;A;N	Aggressiveness, announce a contact intention
Leg action	16.1	++	Billing (52.5%)	58	D;A;P	Appeasement, possibility of contact intention from the partner
Groans	20.8	+	Billing (87.4%)	52	D;P	Announce an imminent 'Billing'

By sex, main following display (when > 20.0%), context (D, dance; A, after the dance; P, pairs; N, non-displaying birds; C, between parents and chick) and visibility of the associated posture (+, rapid movement of the head or the bill; ++, movement of the head plus neck; +++, longer and ampler movement of head and neck plus wings or displacement).

courting great albatrosses (Table 5.2) and is often used in greeting or as the prelude to more intense activity.

2 'Bowing'—see Fig. 4.2(3)

The bird swings its head down towards its feet, often touching its lower breast and, if standing, may reach between its legs. This is done silently or with a low inspiratory groan, and by single birds or by pairs, when the motions of the two tend to be synchronized. Like 'Billing', 'Bowing' may have both appeasement and threatening connotations according to the situation (Jouventin & Lequette, 1990).

3 'Yapping'—see Fig. 4.2(5)

Richdale (1950, p. 16) and Lequette and Jouventin (1991a) described 'Minor Yapping' with small up and down head motions and slightly opened bill and at greater intensity, 'Normal Yapping' where the beak is opened wide and the head jerked up and down repeatedly with loud calls, rather trill-like in *D. epomophora*, but croaking in *D. exulans* (Fig. 5.1A). In *D. exulans* 'Yapping' is more a male display (Table 5.1), and in both species occurs in various contexts—by lone males, between mated pairs, between parents and chicks, concluding a courtship display, and before and after copulation. Duettists tend to synchronise their calls (Fig. 5.1B). Possible functions include individual recognition, mate attraction, and territorial advertisement (Table 5.1). With *D. exulans* extensive 'Yapping' often ends on a low inspiratory note as the bill is abruptly pulled into the breast.

4 'Bill Snaps'

The mandibles are snapped together smartly without a forwards lunge, often into the air—'Air Snaps' and usually not more than three are given at a time. 'Bill Snaps' are used during the nuptial 'dance' and often precede allopreening or mutual billing, but may also have an aggressive content—(see Section C3 below).

5 'Gawky Look'—see Figs 4.5 and 5.2A

A posture often adopted by a passive bird during a vigorous stage of the 'dance'. It stands stiffly, head pushed forward somewhat and twisted slightly while staring intensely but silently at its displaying partner. The feathers of the supraorbital ridge tend to be lifted, producing an eyebrow. The action signals its interest in the other's movements.

6 'Leg Action'—see Fig. 4.2(7)

The bird gives a sudden 'Air Snap' and then swings its head to one side to touch its flank at the top of the leg or axilla and mandibulates the feathers with 'preening

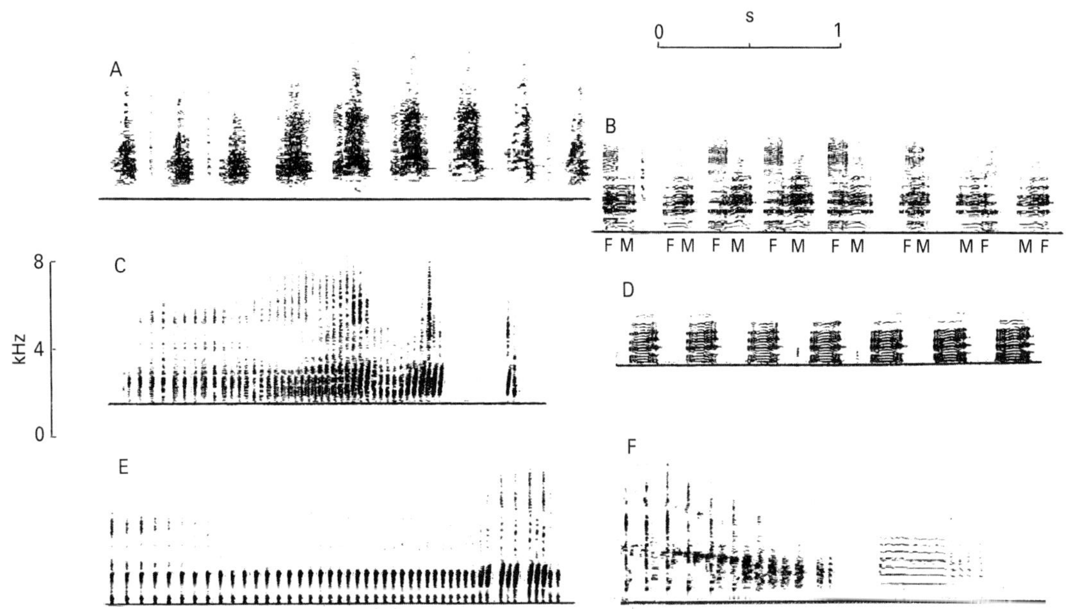

Figure 5.1 *Vocalizations of the Wandering Albatross* Diomedea exulans antipodensis. *(A) Part of a typical series of 'Croaks': note increase in pitch as call proceeds. (B) Duet of birds 'Croaking' into their nest site: the timing tends to match and the female's calls are pitched higher; F, female; M, male. (C) Vibration: note accelerating pulse rate and slurred structure of final notes and the first of the double 'clicks' ending the call. (D) Solo 'Croak' calls: note acoustic pattern distinct from that of bird in (A). (E) 'Vibratory': the note rate increases as in (C) although acoustic structure is quite different but final loud notes are also slurred over a very rapid rise in frequency. (F) A complex threat call given by a displaying bird to an intruder at a 'dance'—a thick 'Bill Clapper' followed by a 'breath' note. Bar is 1 s.*

Figure 5.2 *Displays of the Royal Albatross* D. epomophora. *(A) 'Wing Stretch' of male to female who adopts 'Gawky Look'. (B) 'Sway Walk' circumambulation of nest area by a male. (C) A small 'gam' with one bird giving 'Sky Call'. (D) Allopreening: note ruffled feathers on crowns. (E) 'Vibration' by a male: the wings are not always lifted but tails are usually cocked. Partner reaches forwards in 'Breast Billing' action.*

clops', or holds its bill along the scapulars or forearm without burying its tip in the feathers. 'Leg Action' is frequent during the 'dance' and at nest relief. It possibly functions as an appeasement signal.

7 *'Head Shake'*

The head is jerked from side to side horizontally so that the mandibles rattle loosely together. This action often punctuates nuptial displays and in *D. exulans* is used equally by both sexes (Table 5.2). Its function is unclear.

8 *'Vibration'—see Figs 4.2(9) and 5.2E*

The head is drawn back and slowly raised while the mandibles are vibrated rapidly through shallow arcs and volleys of loud, pulsed 'rubbery' sounds produced which increase in pitch and pulse speed and end with a few 'Bill Snaps' and a forwards and upwards lunge (Figs 5.1C, E & 5.3B). The first figure especially shows how the final four to six pulses seem to be overlaid by low notes on a very rapidly rising frequency, cf. the last five notes in Lequette and Jouventin (1991a, fig. 2C). These confer a peculiarly liquid timbre to the notes. During 'Vibration' the mandibles move so quickly as to appear blurred and sometimes seem not to make contact, yet the sound continues. Two different mechanisms seem to be at work and possibly the tongue is involved. 'Vibration' is used significantly more by the males and forms part of the 'dance' routine but the kind of information conveyed is unclear. Lequette and Jouventin (1991a) suggested that it announces an intention to contact the partner.

9 *'Sky Call'—see Figs 4.2(2) and 5.2C*

A loud 'Bill Snap' is given as the head is swung up with opened beak so that head and neck are at 45° or more, and a loud series of rather musical (*D. epomophora*) or ratchety braying (*D. exulans*) notes is emitted with head and bill motionless. In *D. epomophora* the wings are often spread, when the bird continues into the 'Wing Stretch' display. As the calling stops the head is abruptly lowered. Jouventin and Lequette (1990) separate the posture with loud calling and head rapidly lifted, their 'Sky Position Call', from an equally common, slower, less intense version, their 'Sky Position' movement, done in silence. 'Sky Call' is used mainly by males—when alighting, 'dancing', on nests, on the approach of another bird, and occasionally, in response to conspecifics flying overhead—evidently to call down a potential partner.

10 *'Arched Neck'—see Fig. 4.2(1)*

Described by Jouventin and Lequette (1990). Keeping its beak horizontal, the albatross withdraws its neck into its shoulders and then slowly unfolds and stretches out towards the other bird. No call is used and the function of

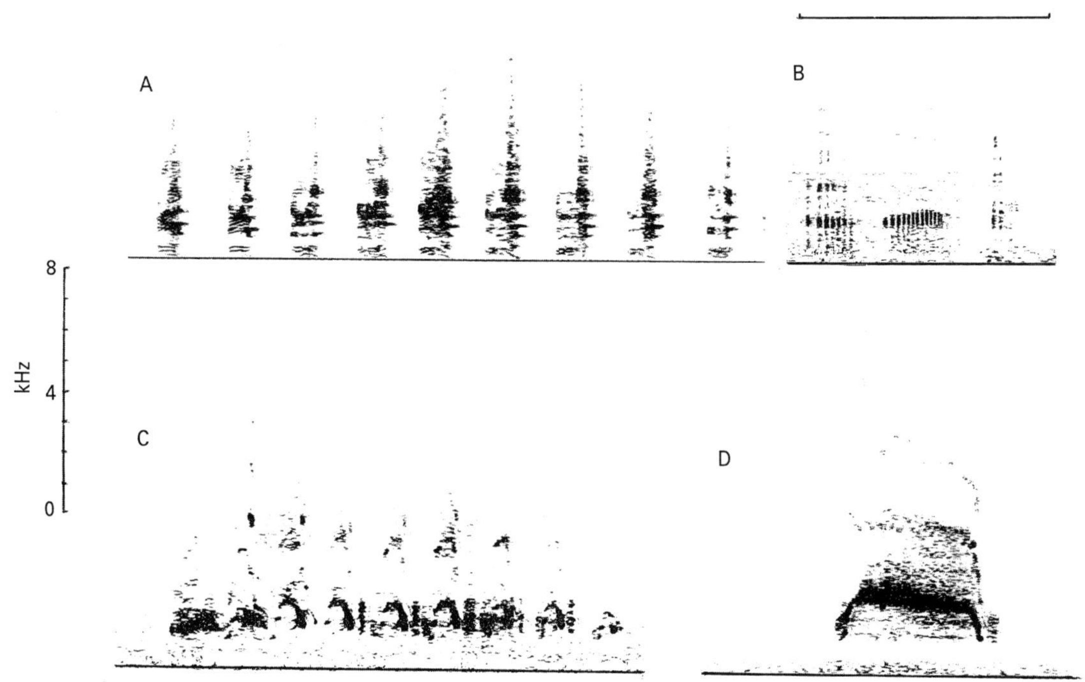

Figure 5.3 *Vocalizations of the Royal Albatross* Diomedea epomophora epomophora. *(A) Male 'Yapping'. Note typical increase and decline in amplitude and pitch during the course of the call. (B) Two bursts of 'Vibration' ending in a double 'snap'. The vibration pulses do not extend to the baseline suggesting that they are not due solely to contacts between the mandibles but that the tongue or palate may be involved. (C) Trilling call given by a well-grown chick whose 'Wing Stretch' scream was analysed in Fig. 4.4A. (D) Scream from a fully feathered chick. Bar is 1 s.*

'Arched Neck' is obscure: it may be a low intensity supplicatory action (see 14 below.

11 *'Head Shake and Whine'*

The head is raised slightly and swung from side to side about six times while a loud, far-carrying whinny is given through slightly opened bill. The wings are not extended but the breast is thrust out and the tail cocked.

12 *'Wing Stretch'—see Fig. 4.2(10)*

The 'Head Shake' leads to this spectacular visual and vocal display which is usually initiated by the male. As the 'Head Shakes' subside the bird rises to its toes, swings its head skywards, cocks its tail and fully outstretches its wings, so that their undersides face the partner with the wing tips curved towards it. That bird watches intently with the 'Gawky Look' (Fig. 5.2A). A succession of screaming whines is given in which expiratory and inspiratory phrases are prominent (Fig. 5.4), interspersed with violent downward sweeps of the head, bursts of bill clappering and gobbling sounds during which the partner's bill may be touched. The head is then swung back into the vertical and a further series of screaming whistles produced. Meanwhile, the male usually side-steps around the female, slapping down his flat feet ostentatiously.

'Head Shake and Whine' with 'Wing Stretching' form the climax of the nuptial 'dance' and comprise Richdale's 'Ecstatic Ritual'. Both vocal and visual elements are very strong, attracting non-breeders from some distance.

13 *'Sway Walk'—see Fig. 5.2B*

A walk with a strange, snake-like gait, the body horizontal, head thrust forwards and swayed from side to side at each step. Used by recently landed birds moving towards a displaying pair and sometimes by a circumambulatory male as if to demonstrate the bounds of a potential nesting territory, the female rotating to face him.

14 *Other behaviours*

Various short ejaculatory noises, groans and wheezes accompany courtship, some perhaps purely involuntary due to the effort expended. Jouventin and Lequette (1990) separate a grunting heard during 'Leg Action' but suggest no function for this. Groans are deep rhythmic noises produced in the throat according to Lequette and Jouventin (1991a), heard during 'Billing' and when the male drops nesting material beside his mate. Omitted by these workers is an apparent supplicatory display seen from both sexes of *D. exulans*, but mostly from females, in which the head is drawn back into the shoulders, bill slightly uplifted and the folded wings raised and shivered while a strange bubbling sound is given. Apart from a little 'Yapping', nest relief involves little demonstration, the relieved bird usually leaving after a few desultory 'Sideways Throwing' actions.

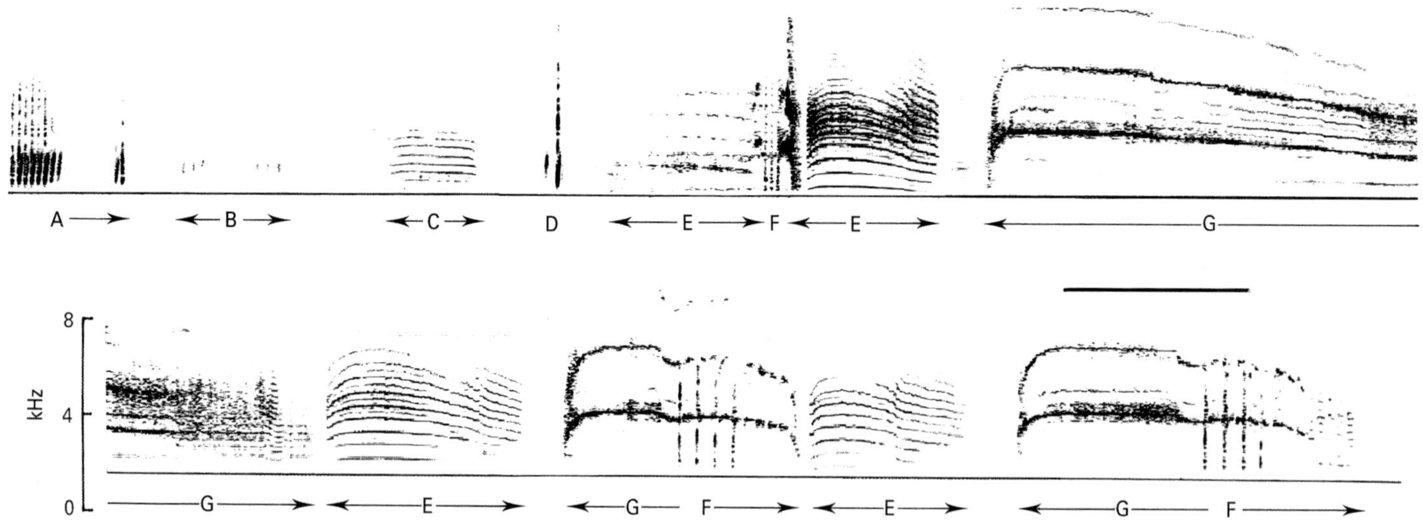

Figure 5.4 *A complete 'Wing Stretch' display by a male* Diomedea epomophora antipodensis *lasting 14.3 s. (A) End of 'Vibration' with a double 'snap'. (B) Quiet bill 'clops'. (C) Inhalatory 'wheeze'. (D) Double 'snap'. (E) Inhalatory cries. (F) Sharp 'bill snaps' from female. (G) Exhalatory 'scream'. Bar is 1 s.*

B *Aerial display*

Richdale (1950, p. 19) described aerial activity in *D. epomophora* with the birds landing to display and then taking off again. While airborne, some gave the 'Sky Call'. This also occurs when Wandering Albatrosses fly over parties displaying on the ground below (Warham, 1976). Such birds droop their wings, arch their necks, tilt their opened bills up slightly, and give the 'Sky Call'. Lequette and Jouventin (1991b) wondered whether such displays were only used by males. The Antipodes Island form *D. e. antipodensis* also forms very loose aerial parties with birds following one another in a kind of communal display as does *D. e. amsterdamensis* (Lequette & Jouventin, 1991b, fig. 6b).

C *Aggression and defence*

1 *'Defence Billing'*

The bird points its part-opened beak at another and moves back slightly at the same time.

2 *'Bill Clop'*

A response to the too close approach of another bird or of a potential enemy. The sound is softer than the 'Bill Snaps' of the nuptial display and is a graded alarm, increasing in rapidity with increasing closeness of the intruder.

3 *'Bill Clapper'—see Fig. 5.1F*

This develops from 'Bill Clop'. Volleys of violent claps of the mandibles accompanied by loud harsh calls are given and head and beak thrust towards the intruder. The threatened bird may step towards it, for example when another ventures too close during a pair's nuptial display. Most intruders move off with head downturned.

4 *'Charge'*

A clappering bird suddenly charges its adversary who retreats. 'Charge' is seldom seen and fighting is even rarer, perhaps because of the generally low nest densities.

D *Discussion*

The nuptial repertoires of the great albatrosses are as complex as those of the well-studied Laysan Albatross and are similarly built from a varied repertory. Jouventin and Lequette (1990) showed that, while the sequences of postures are not fixed, some are often linked together, for example a network comprising allopreening, 'Yapping','Leg Action' and 'Sky Call' is clearly separate from the others—(Jouventin and Lequette, 1990, Fig. 3). These authors distinguished four stages to the 'dance': (1) the male waiting at a site performs the 'Sky Call' as the female alights and walks towards him; (2) the true 'dance' begins with numerous postures and arrangements of varied intensity and duration; (3) the birds rest on or near the site interspersed with bouts of 'Yapping', 'Bowing', 'Leg Action' and allopreening; and (4) the birds separate and the female generally leaves. This is a very simplified sequence with many variants, and is often repeated with different partners several times a day.

The most obvious difference between the species is that while two or more *D. exulans* may simultaneously perform 'Wing Stretch' to one other, with *D. epomophora* that display is given by only one bird at a time (Robertson, 1985). Some detailed comparisons are given by Lequette and Jouventin (1991b). They noted that in actions involving the bill, the mandibles of *D. epomophora* were opened more widely to show the black tomia absent in *D. exulans*, and among the acoustic signals, 'Yapping' bouts lasted longer with *D. exulans*.

II The mollymawks

These medium-sized southern species have a wide repertory and, on current knowledge, many elements of their nuptial and agonistic displays are similar throughout. Details are available only for *D. bulleri, D. chrysostoma* and *D. melanophrys* studied by Richdale (1949), Warham and Fitzsimons (1987) and Tickell (1984).

Features not seen with other albatrosses include the exposure of the gape 'flashes' (*The Petrels*, Fig. 2.6), and the spreading and swivelling of the tail, seen elsewhere only with *Phoebetria*, and some mollymawk displays are sex-specific. There is no 'Sky Call'. The gape exposure is regulated by muscular control of the adjacent feathers. The broken circle of white feathers behind the eye is probably also brought into play, but is unstudied.

Here the behaviours of Buller's Albatross are taken as typical of the group, the terminology following Richdale wherever possible. His work was elaborated 34 years later by C. H. Pettigrew (unpubl.) some of whose findings are used here.

In *D. bulleri* paired birds at nests in the prelaying period are very active in display as are the non-breeders, some of whom show their immaturity by lacking the full bill colours of the adults.

A *Nuptial displays*

1 *'Pointing'—see Fig. 5.5A and C*

This takes several forms. A bird of either sex stretches forwards to point at another several metres away, holding the stance for 2–5 s (Fig. 5.5A). At closer quarters the male often starts by fanning his tail and lunging with a slight scooping and upwards movement towards the head of the other bird. Their bills may touch and his may slide along that of the other—'Rapier Action'. This often ends with a sharp 'Bill Snap'

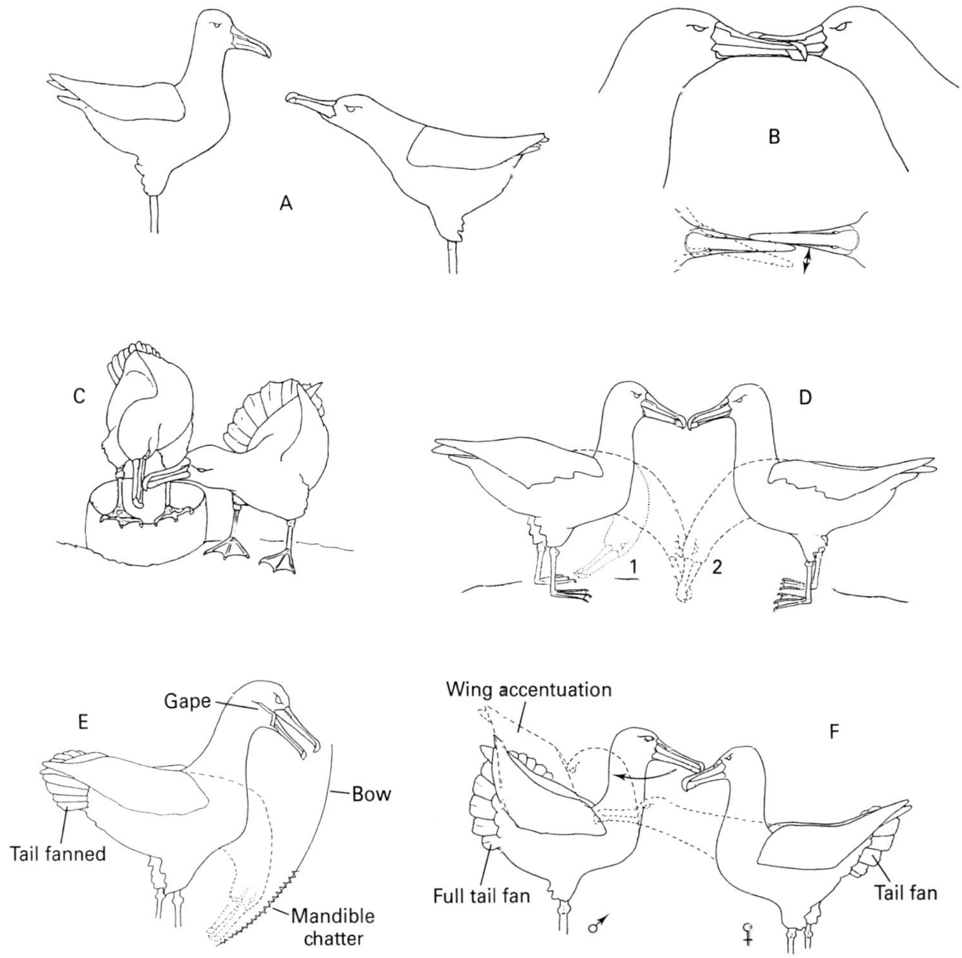

Figure 5.5 *Postures of Buller's Mollymawk* Diomedea bulleri *during sexual displays. (A) 'Pointing'. (B) 'Bill Clashing'. (C) 'Nest Pointing'. (D), 'Bowing'; 1, by a single bird; 2, by a pair. (E) 'Croaking and Nodding'. (F) 'Scapular Action'.* del *C. H. Pettigrew.*

from the initiating bird. 'Pointing' birds look straight into each other's eyes, sometimes into their nest (Fig. 5.5C). Tickell (1984) reported a barely audible throbbing accompanying it in *D. melanophrys* and *D. chrysostoma*. 'Pointing' is one of the most frequently used behaviours in the sexual repertoire (Table 5.3). Its functions are unclear, but 'Rapier Action' seems to imply the subordination of aggressive tendencies and to reflect mutual acceptance.

2 'Bowing'—see Fig. 5.5D

Performed in silence by single birds and by pairs. The head and neck are swung down until the bill points to the ground or even backwards between the legs and the

Table 5.3 Sex- and status-specific displays and behaviours of *Diomedea bulleri*

Sex and status	Display or behaviour																			Total
	Gawky Look	Bow	Croak and Nod	Scoop walk	Scapular action	Leg action	Courtship wail	Aerial display[1]	Display to aerial	Point	Bill clash	Throat expansion	Shake head	Aggression	Incubation aggression	Nest repair	Sideways throwing	Mutual preen	Copulation	
Male breeder	+	+	+	+	+	+	+			+	+	+	+	+	+	+	+	+	+	17
Female breeder	+	+	+	+						+	+		+		+	+	+	+	+	12
Male incubator	+	+	+	+	+		+			+			+		+	+	+	+	+	13
Female incubator	+	+	+	+						+			+		+	+	+	+	+	11
Male failed breeder	+	+	+	+	+	+	+		+	+	+		+	+				+	+	14
Female failed breeder	+	+	+	+				+	+	+	+		+					+	+	11
Male non-breeder	+	+	+	+	+	+	+	+	+	+	+	+	+	+				+	+[2]	16
Female non-breeder	+	+	+	+				+	+	+	+		+					+	+[2]	11
Total	8	8	8	8	4	3	4	3	4	8	6	2	8	3	4	4	4	8	8	

[1]Birds of some sex and status categories were identified in aerial display but those of other groups may also have been involved.
[2]Copulation between non-breeders was rare and thought to be mainly between older birds.

wing tips are slightly raised to *c.* 5 cm above the closed tail. The bowing of *D. bulleri* lasted 1.2 ± 0.6 s ($n = 17$), and might be repeated several times.

3 *'Bill Clashing'—see Fig. 5.5B*

This often develops from 'Rapier Action' with the active bird holding its beak parallel to and touching that of its partner. The latter then slaps its beak repeatedly (*c.* 5 s^{-1}) against that of the other creating a sharp cracking sound. Tickell (1984) suggested that 'Bill Clashing' occurs when the approached bird is rebuffed and a considerable din may result if it continues.

4 *'Bill Snap'*

Sharp closures of the mandibles produce single or serial snaps that often punctuate extended bouts of display.

5 *'Croaking and Nodding'—see Fig. 5.5E*

This is the major bisexual display. It is preceded by the 'Gawky Look' (see 8 below) as the bird turns towards the source of the stimulus. The folded wings are slightly raised, the gape stripes fully exposed, and the fanned tail twisted towards the other bird. The head is then swung down to face the feet, beak slightly ajar and bursts of croaking sounds uttered (Fig. 5.6) ending with a 'Bill Rattle' (see Fig. 5.10B). The head and body are then swung upright and the action repeated giving a rocking motion to head and body.

This display is used by all categories of birds and in varied contexts (Table 5.3) and it is contagious, a noisy pair often triggering off a chorus from neighbours. The birds are usually standing but if performed during incubation the action becomes 'Addressing the Egg', as described by Fisher (1971, p. 61) for *D. immutabilis*.

The croaks of Buller's Mollymawk vary in acoustic structure during the course of the vocalization, tending to rise in the perceived pitch. Analyses show that birds have their own distinctive note structures and groups of notes arranged in distinctive sequences (e.g. Fig. 5.6A, B). The other notes in a call do not always have a set sequence. Both the call and note lengths of males are longer than those of the females (Table 5.4 and Fig. 5.7). Warham and Fitzsimons (1987) concluded that the 'Croak' display could provide visual and aural signals about the identity of the species and individuals and, through variations in the visual movements, indicate courtship or aggressive intentions. The 'Croaks' of other mollymawks are of a similar aural nature but can often be identified to species by the human ear, for example where *D. chrysostoma* and *D. melanophrys* nest together (Fig. 5.8B, C).

6 *'Scapular Action'—see Fig. 5.5F*

This male display is directed at a female. He stands either head- or side-on to her, gives the 'Gawky Look', then skews his fully spread tail in her direction and lifts his

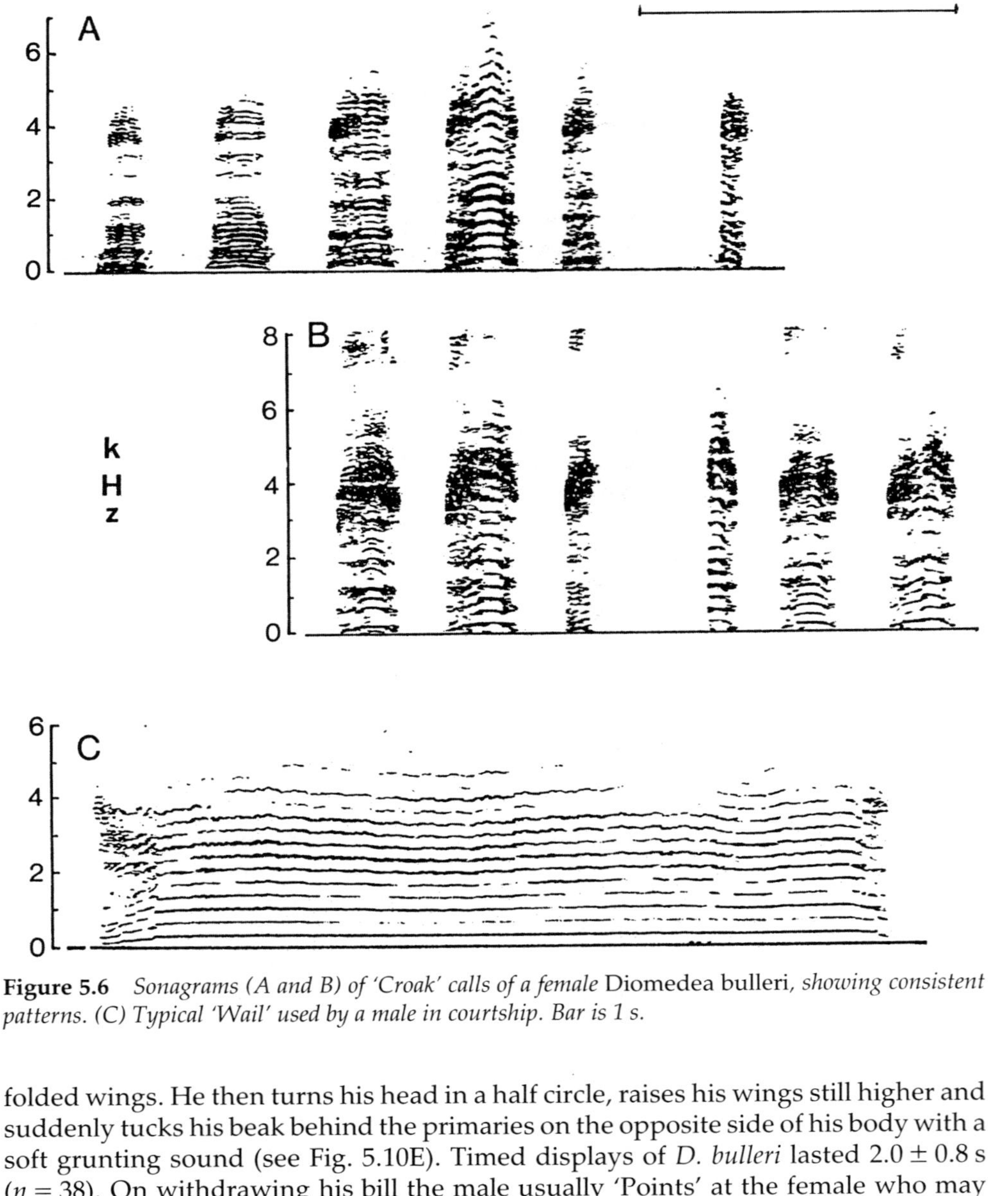

Figure 5.6 *Sonagrams (A and B) of 'Croak' calls of a female* Diomedea bulleri, *showing consistent patterns. (C) Typical 'Wail' used by a male in courtship. Bar is 1 s.*

folded wings. He then turns his head in a half circle, raises his wings still higher and suddenly tucks his beak behind the primaries on the opposite side of his body with a soft grunting sound (see Fig. 5.10E). Timed displays of *D. bulleri* lasted 2.0 ± 0.8 s ($n = 38$). On withdrawing his bill the male usually 'Points' at the female who may then also 'Point' to touch his neck or wing, usually while his beak is still hidden.

7 'Leg Action'—see Fig. 5.9D

This male posture is evidently related to 'Scapular Action'. The slightly opened beak is thrust into the contour feathers and run down the flank to the thigh, the tips of the mandibles 'mouthing' the feathers there. The tail is spread and skewed towards the

Table 5.4 **Vocalizations and noises of *Diomedea bulleri*. Durations in s, frequencies in kHz (mean ±1 SD; sample sizes in parentheses). From Warham & Fitzsimons (1987)**

	Male 'croak'	Female 'croak'	Courtship 'wail'	Scapular action 'groan'	Copulation 'groan' (♂)	'Tattoo'	Leg action
Number of notes per call	7.6 ± 3.3 (15)	6.25 ± 2.1 (5)	1	—	26.7 ± 4.3 (4)	—	—
Note duration	0.20 ± 0.29 (56)	0.16 ± 0.10 (56)	—	0.14 ± 0.08 (8)	0.54 ± 0.08 (4)	0.013	0.02 ± 0.05 (14)
Internote duration	0.20 ± 0.08 (119)	0.20 ± 0.1 (48)	—	0.33 ± 0.12 (8)	0.04 ± 0.01 (4)	0.09 ± 0.02 (12)	0.10 ± 0.05 (14)
Call duration	3.63 ± 2.18 (25)	2.53 ± 1.7 (19)	2.21 ± 0.98 (7)	2–3	0.63 ± 0.49 (2)	—	2–3
Call frequency range	4.53 ± 0.63 (111)	4.15 ± 0.66 (55)	6.65 ± 1.3 (7)	1.32 ± 0.03 (8)	4.72 ± 0.84 (4)	—	1.55 ± 0.73 (14)

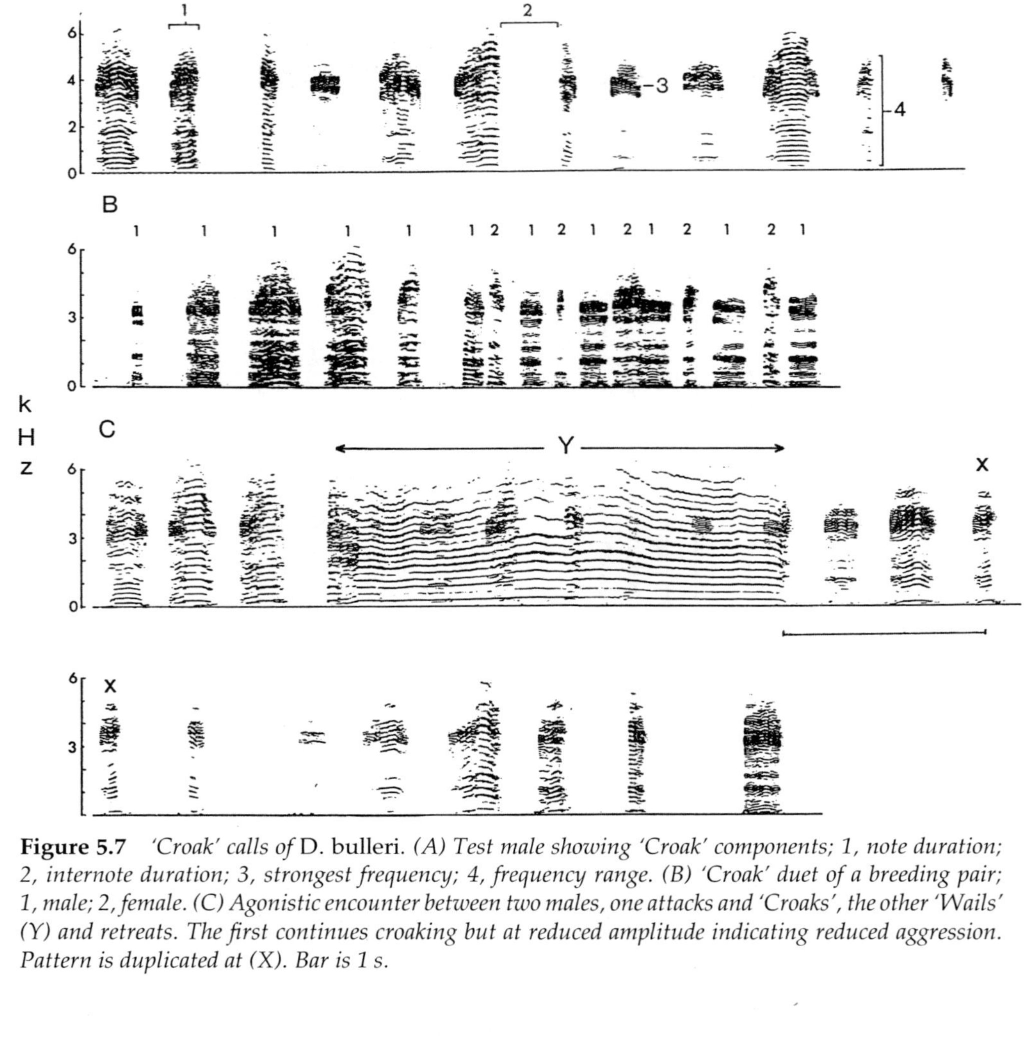

Figure 5.7 *'Croak' calls of* D. bulleri. *(A) Test male showing 'Croak' components; 1, note duration; 2, internote duration; 3, strongest frequency; 4, frequency range. (B) 'Croak' duet of a breeding pair; 1, male; 2, female. (C) Agonistic encounter between two males, one attacks and 'Croaks', the other 'Wails' (Y) and retreats. The first continues croaking but at reduced amplitude indicating reduced aggression. Pattern is duplicated at (X). Bar is 1 s.*

female, folded wings slightly lifted from the body and a grunting sound uttered (Fig. 5.10C). This display lasted 2.4 ± 1.6 s ($n = 17$).

Beak-hiding in 'Scapular' and 'Leg Action' may signal the male's non-aggressive state, inviting a closer approach.

8 'Gawky Look'—see Fig. 5.11B

One bird glares at another. Its bill is raised slightly, the head and neck feathers sleeked, the eyebrows exaggerated (especially in the male) and the gape flash exposed. The fierce look is emphasized by the white flash of feathers behind the eye (Tickell, 1984). 'Gawky Look' often precedes 'Croak', 'Scapular Action,' 'Leg Action' and 'Wail'. The posture is held for 2–3 s before another is adopted or the bird relaxes.

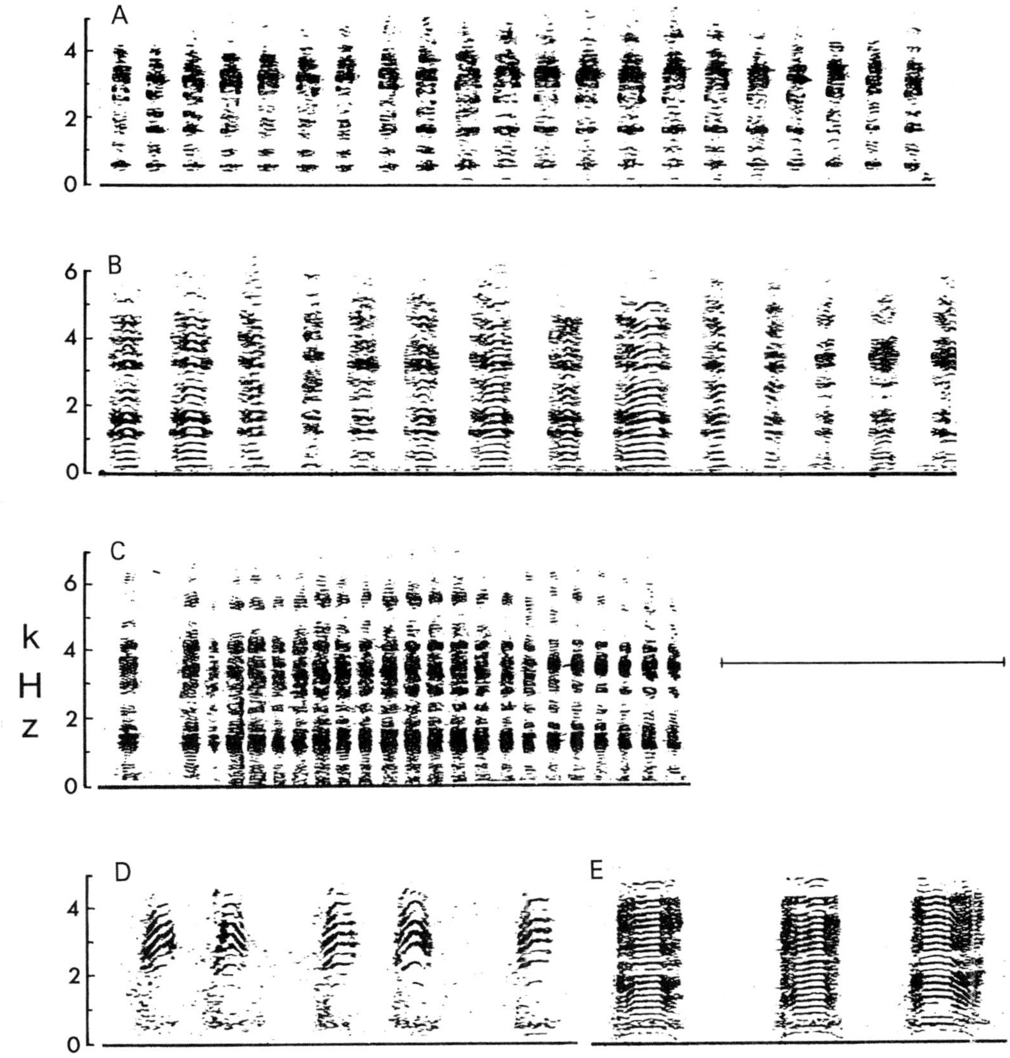

Figure 5.8 *Typical 'Croak' calls of (A)* Diomedea cauta cauta, *(B)* D. melanophrys impavida, *(C)* D. chrysostoma, *(D) and (E) two examples from* D. albatrus; *note very condensed harmonics in (E). Bar is 1 s.*

9 'Wail'—see Figs 5.7C and 5.12

This is used mainly by males in courtship when it is often associated with 'Scapular Action', but it is also given by both sexes when stressed (see Section II.D.3 below). Typically, a 'Scoop-Walking' male circling a female suddenly opens his beak wide and emits a loud, far-carrying, mournful cry as he turns to face her. Generally, while still calling, he swings his head and opened beak from side to side through *c.* 90° for 4–6 times and with fully fanned tail. The female often responds by pointing her bill

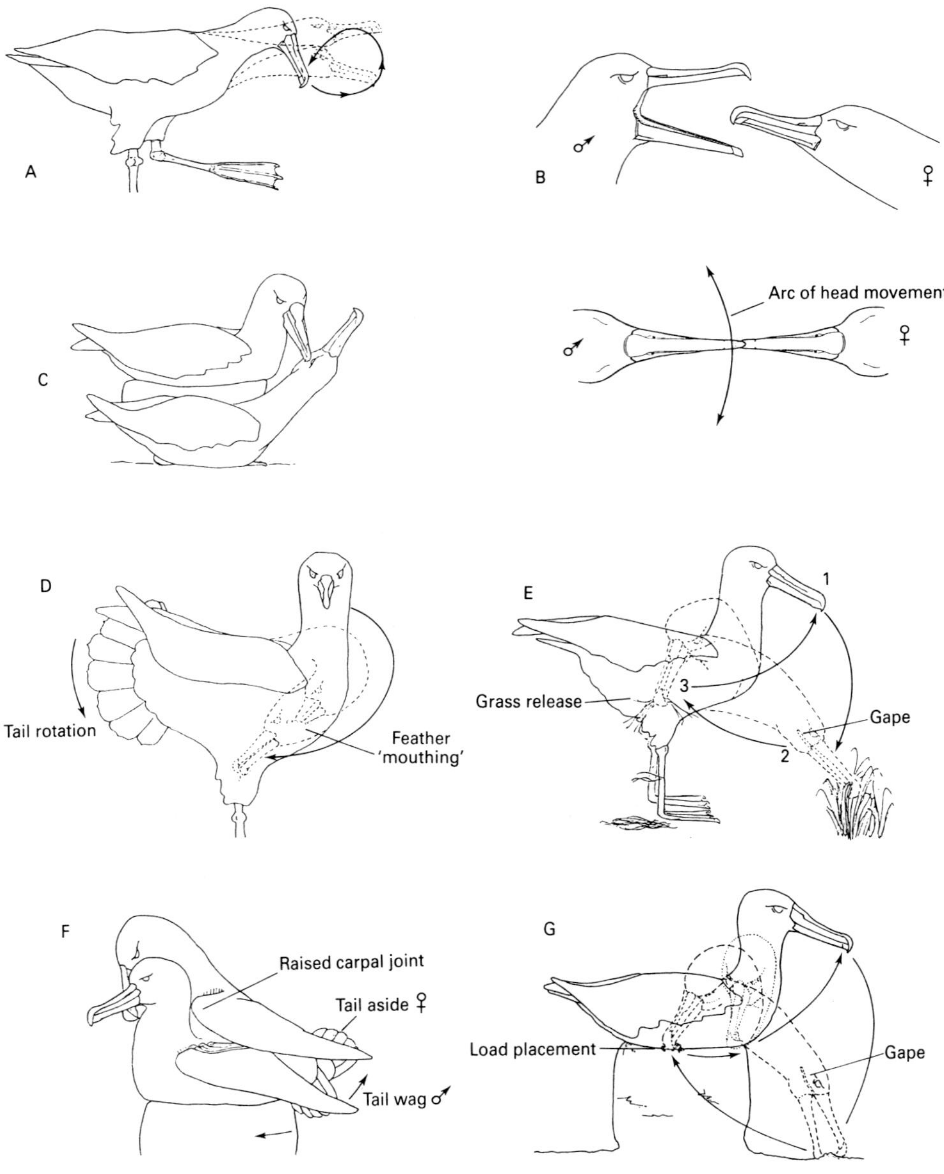

Figure 5.9 *Behaviours of* D. bulleri. *(A) 'Scoop Walk'. (B) 'Courtship Wail' in lateral and dorsal view. (C) Allopreening. (D) 'Leg Action'. (E) 'Sideways Throwing'. (F) Copulation. (G) Nest repair.* del. *C. H. Pettigrew.*

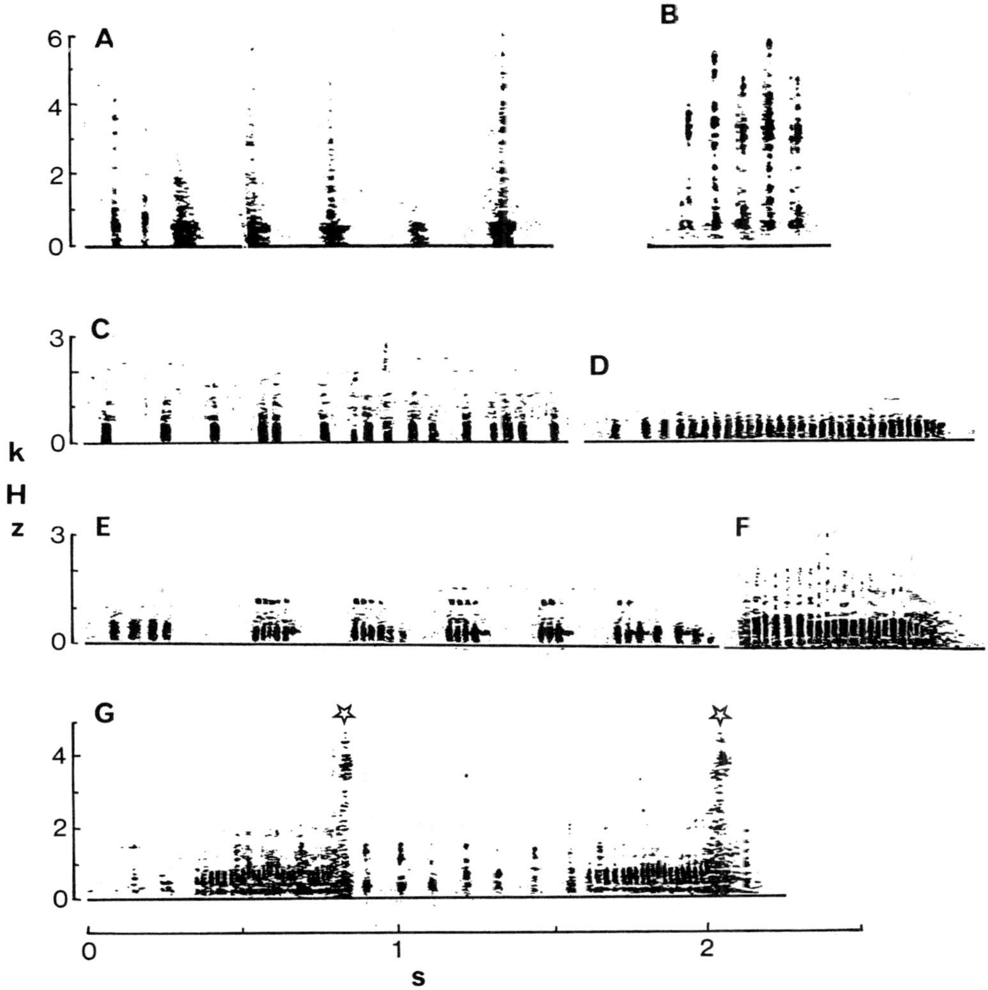

Figure 5.10 *'Groans' and 'Mechanical' noises of* Diomedea bulleri. *(A) 'Bill Clop' of mild aggression: note strong frequency component spread over several milliseconds (probably responsible for characteristic soft timbre). (B) 'Bill Rattle': the mandibles were struck together five times as the head was loosely shaken. 'Mandible Spar' sounds are similar. (C) 'Groan' of a male giving 'Leg Action'. (D) Churr-like 'Groan' of a male late in 'Scapular Action'. (E) 'Groan' of a male early in 'Scapular Action'. (F) 'Groan' by a male early in copulation: note high repetition rate of the pulses. (G) 'Groan' by a male late in copulation: pulse rate is very high and each burst ends with a squeak (*); the seven pulses between the bursts are 'Tattoo' taps caused by the male's bill striking that of the female.*

down the other's pink buccal cavity. The 'Wail' tends to climax the courtship display and such 'Courtship Wails' lasted 3.0 ± 0.6 s ($n = 24$) in *D. bulleri*.

These long calls are of a single note with many harmonics, often with frequency modulation (e.g. Fig. 5.7C). They tend to be higher pitched than other calls and, as Fig. 5.12 shows, the 'Wails' of several other albatrosses are of similar acoustic structure. The function of the 'Wail' is unclear.

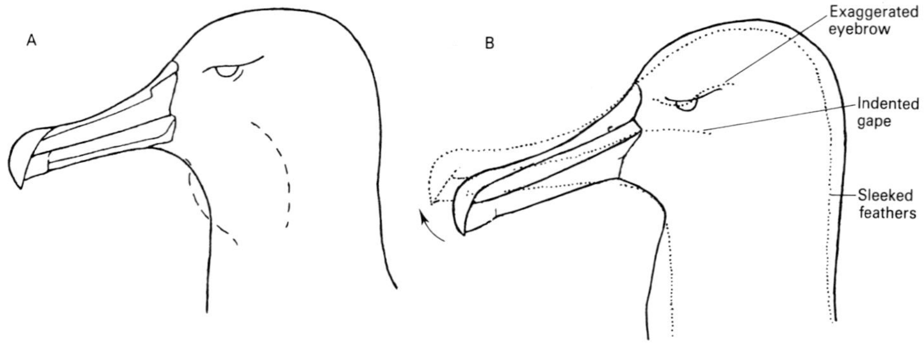

Figure 5.11 *Facial expressions of* Diomedea bulleri. *(A) 'Throat Expansion'. (B) 'Gawky Look'.*

10 'Scooping'—see Fig. 5.9A

A silent display in which the head is dropped to the chest and then swung forwards and upwards so that the bill points 20–30° above the horizontal. The head is then retracted and the process repeated resulting in a peculiar bobbing and scooping action. The tail is fanned a little and the folded wings may be raised slightly. 'Scooping' is often seen in conjunction with a stylized walk with body somewhat depressed on bent legs and feet prominently slapped down at each step—the 'Scooping and Walking' of Richdale. 'Scooping' lasted 0.55 ± 0.15 s ($n = 31$) in *D. bulleri* and was often used as a nest was circled, the partner reacting with 'Rapier Action' or 'Bill Clashing'.

11 'Throat Expansion'—see Fig. 5.11A

Not often seen and then only by males. The bird turns to face the other, raising its folded wings slightly, 'Points', and then withdraws its neck while puffing out its throat 1–2 cm and giving a guttural groan like that heard in 'Scapular Action' (Fig. 5.10D) and taking 1–2 s. Published photographs show it also in use by *D. chrysostoma*. Its function is unclear.

B *Aerial displays—see Fig. 5.13A*

Aerial displays by Buller's Mollymawk are infrequent, thought to be mostly by non-breeders, and done silently. They may be directed at conspecifics on the ground or in the air.

When directed at grounded birds the albatross approaches into the wind at a height of 5–7 m, then brakes by dropping its feet, webs spread and simultaneously rakes back its wings at the carpal joints and retracts its head into the scapulars.

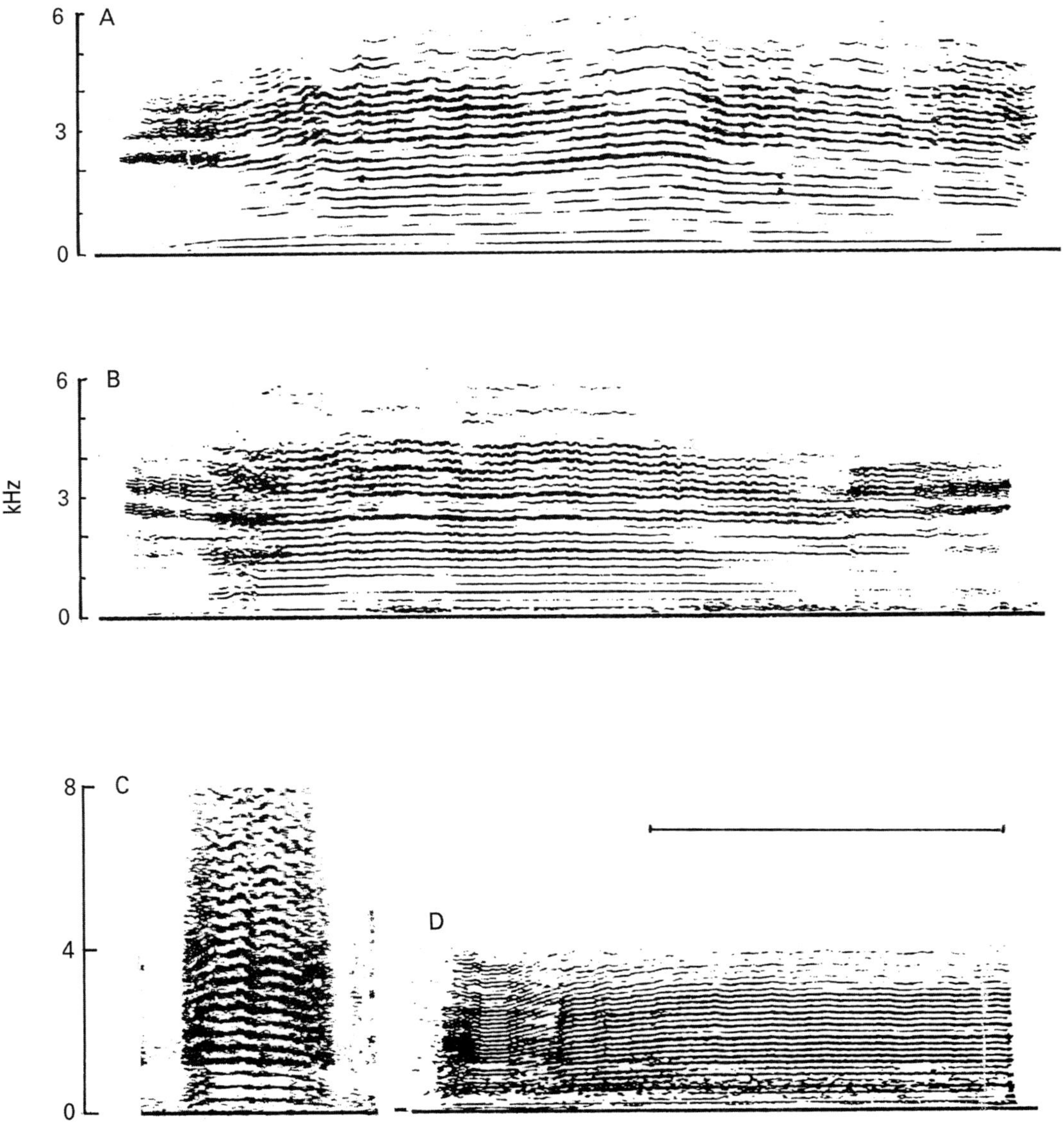

Figure 5.12 *Typical 'Wail' calls of (A)* Diomedea melanophrys melanophrys; *(B)* D. chrysostoma; *(C)* D. chrysostoma *trying to push past nesting pairs; (D)* D. albatrus. *Bar is 1 s.*

'Aerial Display' seems usually aimed at groups of non-breeders which, in *D. bulleri* at least, may turn to 'Croak and Nod' at the overflier.

Occasionally a kind of tandem flying develops with two birds keeping 2–3 m apart for up to 30 s. The leader may give the above display and the follower may copy. When synchrony is lost the display ends.

Somewhat similar behaviour may be seen with *D. chrysostoma*. Elliott (1957) noted

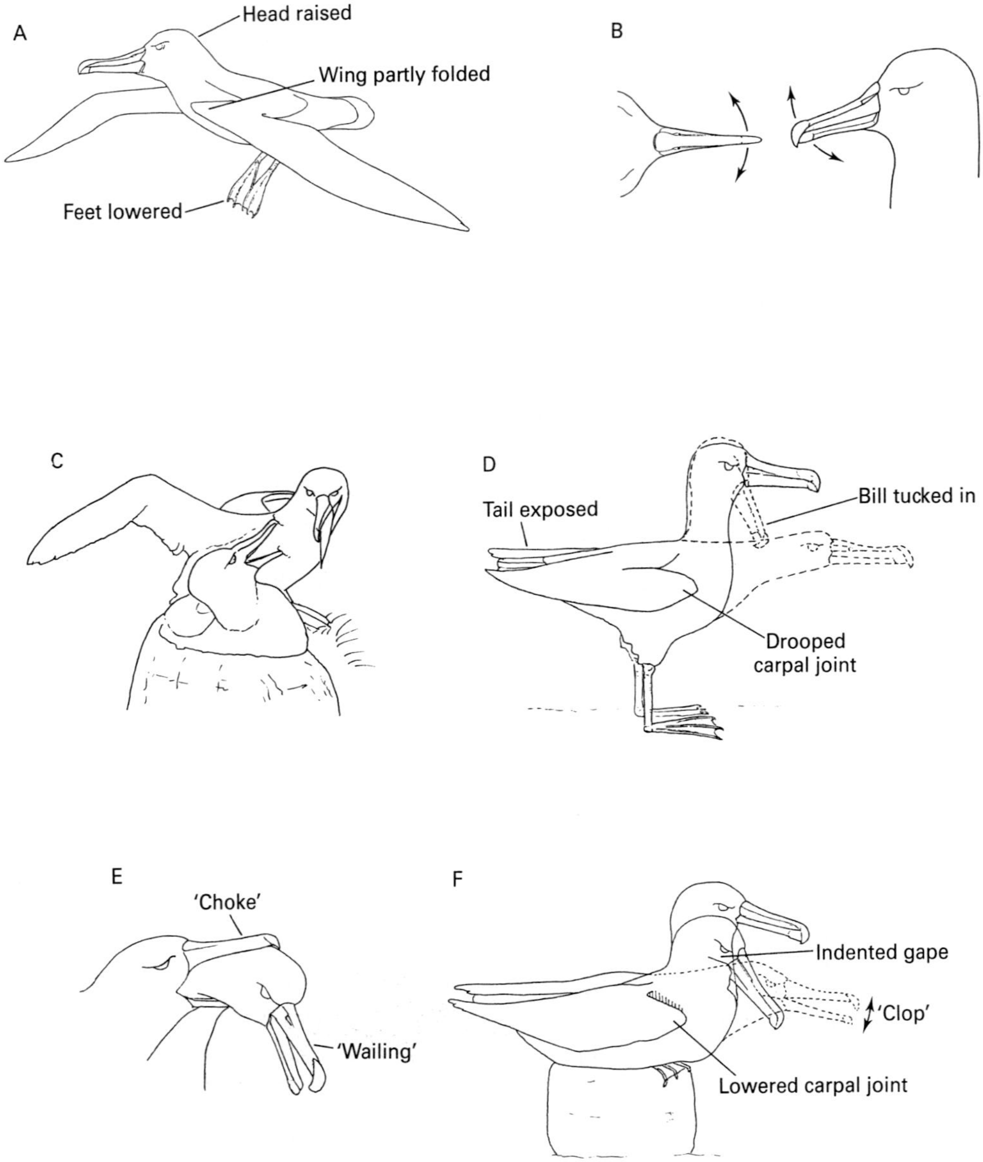

Figure 5.13 *Displays of* Diomedea bulleri. *(A) Aerial display. (B) 'Head Shake'. (C) Incubating bird attacks potential male 'rapist' who gives 'Distress Wail' before retreating. (D) 'Aggressive Posture' and 'Pointing'. (E) 'Choking'. (F) 'Defensive Posture'.* del. *C. H. Pettigrew.*

that *D. chlororhynchos* occasionally called on the wing, while Shaughnessy and Fairall (1976) saw this bird flying with arched back, fanned tail and raised head, and giving croaking cries when passing over incubating conspecifics, but this was rare.

C Other behaviours

1 Nest repair—see Fig. 5.9G

Refurbishing of old nests is mainly done by the females and much of that by sitting birds. These reach down to dredge up anything within reach, exposing the gape skin meanwhile, and tapping the load into the rim of the nest—11.3 ± 5.4 taps per load, lasting 3.2 ± 1.0 s ($n = 41$) in *D. bulleri*. This goes on throughout incubation so that nest mounds tend to heighten, for example on 27 December 1982, when few birds had arrived, average nest height was 23.5 ± 6.8 cm ($n = 142$) but by 17 February, with incubation well advanced, averaged 25.0 ± 4.8 cm ($n = 42$).

2 Nest Relief

This is preceded by 'Scooping' by the relieving bird, but 'Croaking', 'Rapier Action' and mutual allopreening are usually seen before the new bird takes over.

3 Flight Intention

'Scooping and Walking' is used in non-sexual contexts to signal an intention to take off. The bobbing tends to be speeded up just before departure. This action is used both from uncluttered ground and where departure may be thwarted by neighbours. It may help deflect interference and increase the prospects for a successful launch.

D Aggression and defence

1 Threat—see Fig. 5.13D

Most aggressive behaviour is between males. A threatening male typically adopts 'Gawky Look', fully exposes his gape flash, then drops his beak towards his chest, droops his folded wings to expose the carpal joints, the back and closed tail. The bird then either 'Points' or gives a 'Croak Display' towards its antagonist. The aggressive 'Croak' of *D. bulleri* resembles its display 'Croak' but the accompanying bow is shallower and aggressive 'Croaks' are higher pitched and last up to 15 s without any contact with the other bird (see Fig. 5.7C).

Between bursts of agonistic activity mollymawks may uproot vegetation—the 'Ground Stabbing' of Tickell—or vigorously self-preen their breast feathers.

2 *'Attack'—see Fig. 5.13*

The threatening bird grabs any part of its adversary within reach to pull and shake it. The attacked bird struggles to get away. The attacker may obtain a grip on the other's neck—the 'Choke' (see Fig. 5.13E)—which it may maintain for up to 1 min. The recipient often 'Wails' as it struggles to free itself (e.g. Fig. 5.13C). 'Charge' is a direct attack, for example when a bird gets on to a temporarily unoccupied nest whose owner then rushes in to evict the trespasser (Tickell, 1984, fig. 16).

3 *Defensive behaviour*

Incubating birds of either sex respond to mild threats by 'Head Shake' (see Fig. 5.13B) where bill and head are swung horizontally through *c.* 45°, the mandibles struck together producing a castanet-like rattle. 'Pointing' as well as 'Croaking and Nodding' are also used in defensive/threat contexts.

A bird on a nest may also attempt to deter aggressors with a 'Defensive Posture' (see Fig. 5.13F), pulling its bill into its chest and if the intruder gets closer, by 'Jabbing', jerking its head forwards and giving a rapid fusillade of 'Bill Clops' at 2–3 clops s^{-1} (see Fig. 5.10A). This may lead to 'Gulping', where the bird turns to face the threat, pulls its head and neck into its chest and makes gulping sounds as if about to vomit at the intruder. But adult mollymawks rarely eject stomach contents and if hard pressed usually resort to 'Jabbing'.

Stressed birds of either sex may use a shortened form of 'Wail' with widely opened beak and expanded gape flash, for example after being evicted from a nest by the rightful owner who may also 'Wail'. This 'Distress Wail' is also heard from mollymawks just before and immediately after touchdown among others whose personal spaces it may transgress and presumably helps deflect attacks.

'Rape' attempts are fairly common among mollymawks and males, apparently unattached, may even molest incubating females—the 'intruder males' referred to by Richdale (1949, pp. 26, 45). Females may resist by biting and 'Choking', the repelled bird usually giving a 'Wail' (see Fig. 5.13C). Something in the attitude of the incubator appears to provoke these attacks, and a pile of three on a single nest has been recorded (Warham, 1967b). Whether any effective extra-pair copulations occur seems unreported.

E *Displays of chicks*

Mollymawk chicks have a 'Submissive', beak-hiding posture, turning the head away from an intruding adult or even from a returning parent, evidently to inhibit attack. In active defence they draw up to their full height, rotate to face the threatener, clop their bills loudly in a very oily 'Gulping' and, if hard pressed, vomit stomach oils and gut contents in an ill-directed splatter. But, as Tickell (1984) put it, ensconced in their nest cylinders a half-grown chick resembles a self-righting skittle, difficult to dislodge.

Young mollymawks use the usual plaintive cries when begging but as they develop 'Croaking' becomes common.

F Non-breeders and failed breeders

Non-breeders display in small groups on clear ground away from the nesters. They use all the behaviours seen among breeders (see Table 5.3). Birds whose breeding fails during the egg stage remain at their nests for a few days before moving about the colony alone, often attracting threats from incubating birds.

G Rates of display activity—see Fig. 5.14

In *D. bulleri*, activity, as measured by 'Croaking', varies during the day and during the course of the breeding cycle. Early in the reoccupation period, with mainly males ashore 0.37 ± 0.3 'Croak' displays per half hour were scored from 11 birds in the first 4 h after their arrival, whereas late in reoccupation, with more birds, still mostly male, the rate rose to 1.61 ± 0.88 displays per half hour. Established pairs were still more active (Fig. 5.14A), but birds on eggs seldom displayed (Fig. 5.14B) and so conserved energy during their fasts, only occasionally rising from their egg to ventilate it.

H 'Posture-sets'

In Buller's Mollymawk the probability that a particular display will follow a particular previous one tends to be low, so that 'dances' such as those of *D. immutabilis* do not really develop. In an analysis of the sequential patterns in displays the correlation co-efficients between 'before' and 'after' probabilities seldom exceeded 0.5 (Fig. 5.15). However, some consistent patterns were discernible. For example, a bird performing 'Scapular Action' to one side of the body tended to repeat the action on the same side, $r = 0.50$–0.53, $P < 0.001$, $n = 323$. And, during 44 changeovers on the egg, 'Scoop Walk' was used by 42 of the outgoing birds which always performed 'Sideways Throwing'. Yet incoming birds were never seen doing this, illustrating how context can affect the frequency of a display.

I Discussion

The communication system is built from visual and vocal elements and both have the capacity to indicate the intensity of a bird's drive or response. Thus in stressed birds the 'Wail' increases in pitch and amplitude (see Fig. 5.7C) and carries further, while during displays like 'Scapular Action' the degree to which the wings are lifted, tent-like, from the body, the tail fanned, or the gape patch exposed, suggests that these allow graded responses to another's actions.

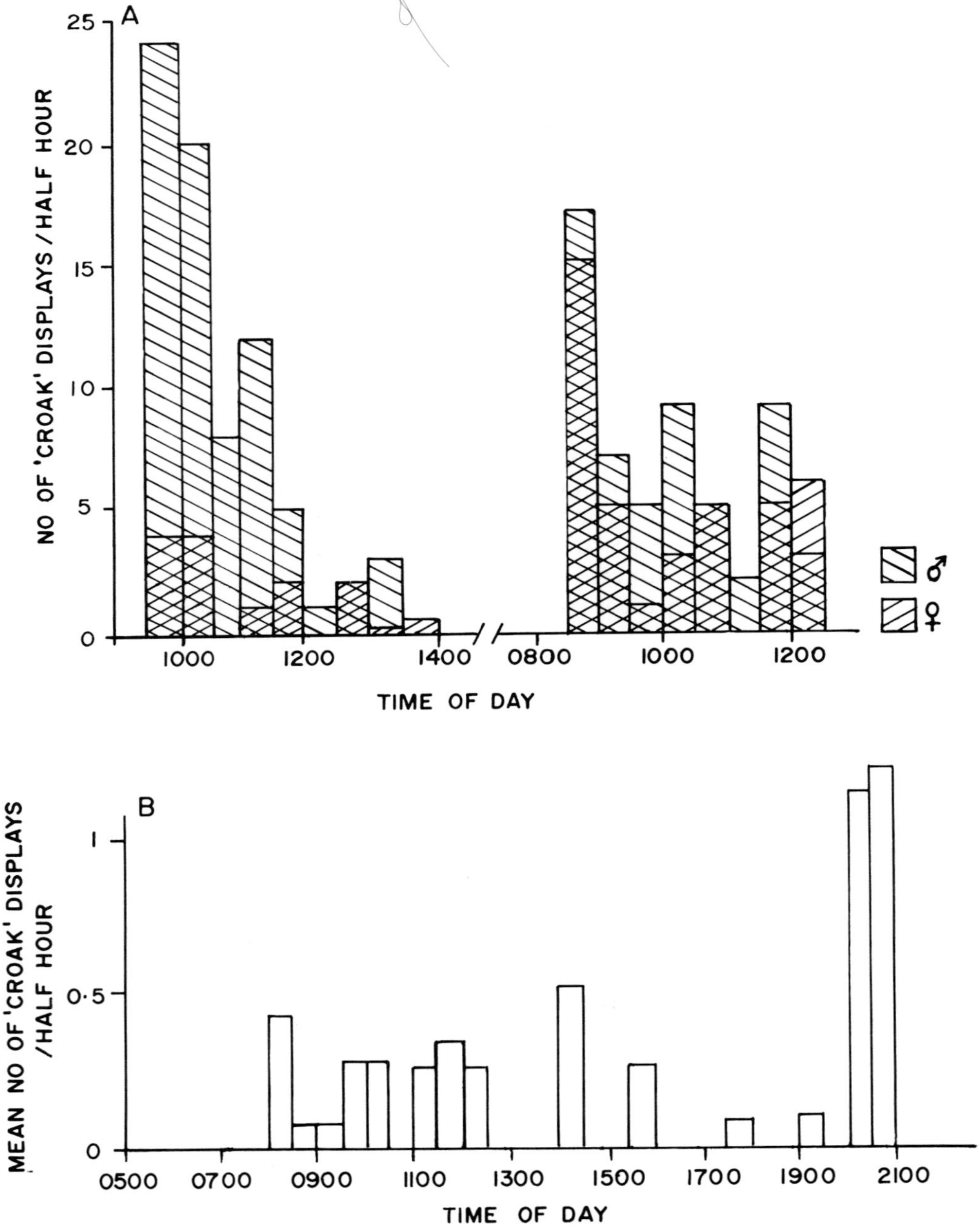

Figure 5.14 *Frequencies of 'Croak' displays from (A) 10 pairs of* Diomedea bulleri *during courtship over the first 4 h of their reunion (mean for males 5.0 ± 4.3 per half hour, for females 2.7 ± 2.1 per half hour); and (B) from 11 incubating birds (mean 0.11 ± 0.03 displays per half hour). Note increased activity in late evening with none overnight.*

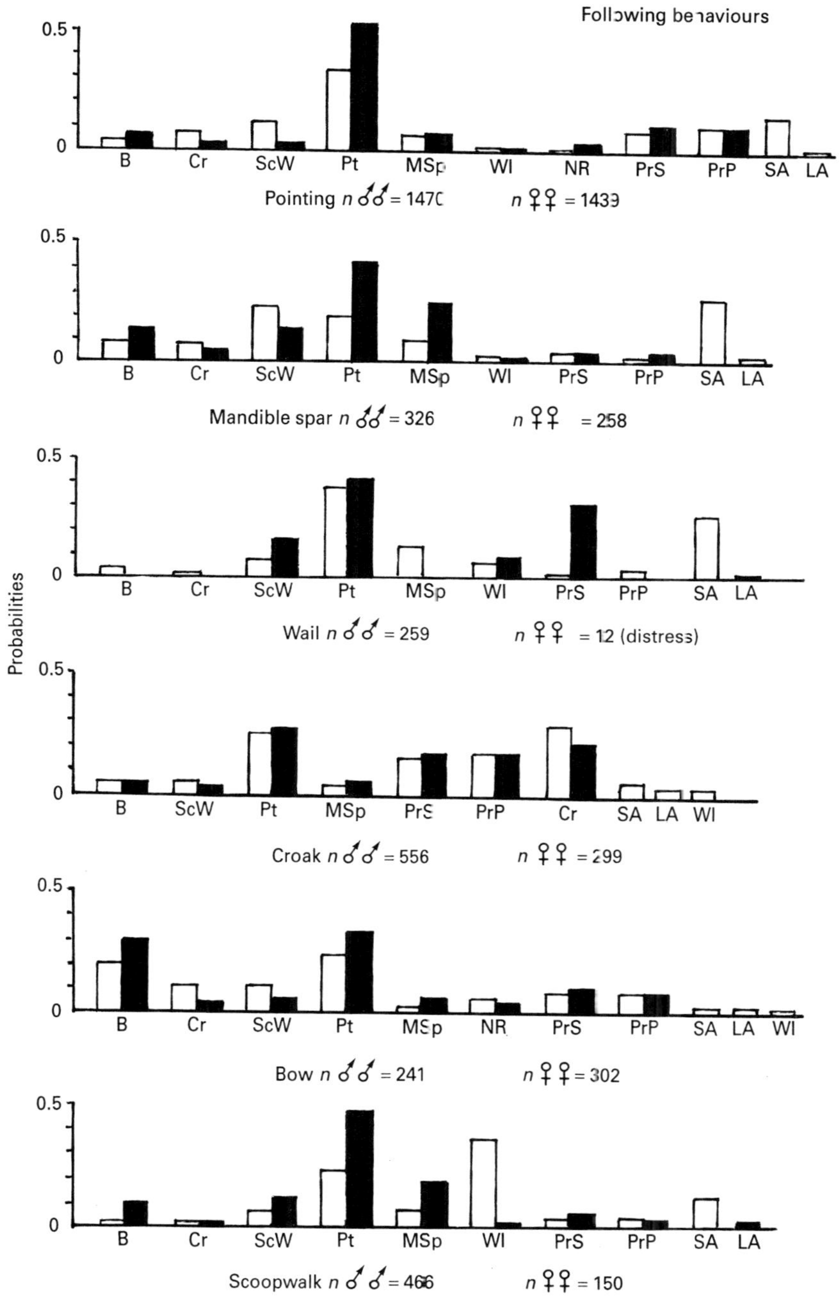

Figure 5.15 *Displays of* Diomedea bulleri. *Some probabilities that a bird will use a particular bisexual action after the performance of a previous one. Open bars males, filled bars, females. B, 'Bow'; Cr, 'Croak'; ScW, 'Scoop Walk'; Pt, 'Point'; MSp, 'Mandible Spar'; Wl, 'Wail'; NR, 'Nest Repair'; PrS, Preen Self; PrP, Preen Partner; SA, 'Scapular Action'; LA, 'Leg Action'. For example, after 'Pointing' there was about a 50% chance that a female would repeat that action. From Pettigrew (unpubl.).*

That the system uses vocal and visual components simultaneously is indicated by the results of simple tests with two-dimensional models (Warham and Fitzsimons, 1987). Few or no responses were obtained when models alone were presented to birds on nests in the prechick stage but, in the few presentations possible with a male 'Croak' call plus a model, no responses were gained using a simple model with eyes and bill whereas every presentation with an advanced one with a gape stripe plus sound resulted in a 'Croak and Nodding' response.

III The North Pacific albatrosses

The behaviours of *D. nigripes* and *D. immutabilis* (The Black and White 'Gooneys') and *D. albatrus* have many features in common, for example their use of the half-raised, half-folded wings in the 'Bill under Wing' action, an element lacking in the behaviours of the southern species. Rice and Kenyon (1962) described the actions of *D. immutabilis* and *D. nigripes*, Meseth (1975) and Sparling (1977) analysed the visual and vocal repertoires in greater detail.

The Laysan Albatross exemplifies the two 'gooneys'. Its repertoire includes 18 postures used in sexual displays and five in agonistic ones. There is no aerial display and no tail fanning, nor, except for birds nesting for the first time, do the breeders display extensively, their pair-bonding activity being restricted to allopreening and copulation. Although many of the postures used appear homologous with those of mollymawks and the great albatrosses, in the following account Meseth's (1975) terminology is used except for 'Scapular Action'.

Meseth (1968) drew attention to the use of the overhanging ridges and black feathers above the orbits which, at sea, hood the eyes against glare. During the 'dance' the ridge almost disappears as the eyebrows are raised and the head acquires a rounded appearance. In agonistic situations, however, the eyebrows are lowered distally, the dark feathered area around the orbit becomes blacker, and the bird glowers at its adversary.

As with other albatrosses, males yet to breed, or those that have lost partners, stand in a home area and advertise with a sequence of postures and calls. Wandering unpaired females may respond to these blandishments by joining in a complex, fast-moving, rather violent and noisy *pas-de-deux* in which regular 'posture sets' are discernible. Males dance with many females, and trios or quartets may form—the equivalent of the 'gams' of the great albatrosses.

The 'dance' has three stages—an invitation by the male, a 'core' section, and a short concluding one initiated by the female. The sexes are similar in plumage but displaying birds evidently recognize sexual differences: the thicker necks and bigger beaks of the males, the way in which these tend to reach above the females, and by vocal differences. No displays or calls are restricted to one sex, but some, like the 'Eh-Eh Bow', are used more often by males. The postures are copied from one partner to the other, and in extended 'dances' the female adjusts her tempo to that of the male. The choice of a mate is hers. The immediate result is not copulation but the forging of a pair-bond and the acceptance by the female of a nest site within the male's defended area.

A *Nuptial displays*

1 *'Bill Touch'—see Fig. 5.16A*

The amount of contact depends on the intensity of the situation. It may only be a delicate touching between bill tips, a vigorous nibbling by one bird at the side of the other's bill, a reciprocated side-pushing of beaks; even a brief insertion of the tip of the bill into the hollow apex of the lower mandible of the other—see Fisher (1972, fig. 4). These actions are done without calling, are seen in many circumstances and among non-breeders and breeders, and are similar to the 'Pointing' and 'Rapier Action' of mollymawks.

2 *'Eh-Eh Bow'—see Fig. 5.16B*

In advertising display this is a male posture given when stiffly erect, breast thrust out and series of strained 'Eh-Eh' calls uttered as the bird pivots forwards and down towards its feet. It raises its head in silence and repeats the performance, often bowing deeper than before. Very like the 'Croaking and Nodding' of mollymawks, the 'Eh-Eh' call is the basic vocalization of the gooneys, several other calls being elaborations of it. During incubation the display is directed towards the egg—'addressing the egg'—an action that increases towards hatching time when the chick 'peeps' from within the shell.

Sonagrams of typical series are given in Figs 5.17A,B. The croaks show as U-shaped syllables with reasonably clear harmonics, often delivered in a rhythmic series and interspersed with 'noisy' syllables that appear to be 'breath' notes. The 'Eh-Eh' calls of partners are often clearly different to the human ear.

3 *'Yapping'—see Fig. 5.16C*

Loud continuous 'Eh-Eh' calls often accompany the movements of territory holders about their defended areas. These birds usually hold their bodies low to the ground on flexed legs with their bills pulled into the neck. Sparling (1977, table 4) noted two types: 'Noiselike Eh's' with condensed harmonics used particularly in threat, and 'Harmonic Eh's' with fewer frequency bands spaced further apart, used more during the 'dance' and between members of a nesting pair. No sexual differences were revealed except that the note lengths of female *D. nigripes* were shorter than those of the males. 'Yapping' appears to be homologous with that of the great albatrosses.

5 *'Beak Hiding'—see Fig. 5.16D*

The tip of the beak is inserted in the feathers anterior to the dorsal edge of the scapula near the angle of the folded wing, the head having a slight side-to-side motion, while a short burst of soft clopping sounds is heard.

This action is not confined to males as with 'Scapular Action' in mollymawks, nor

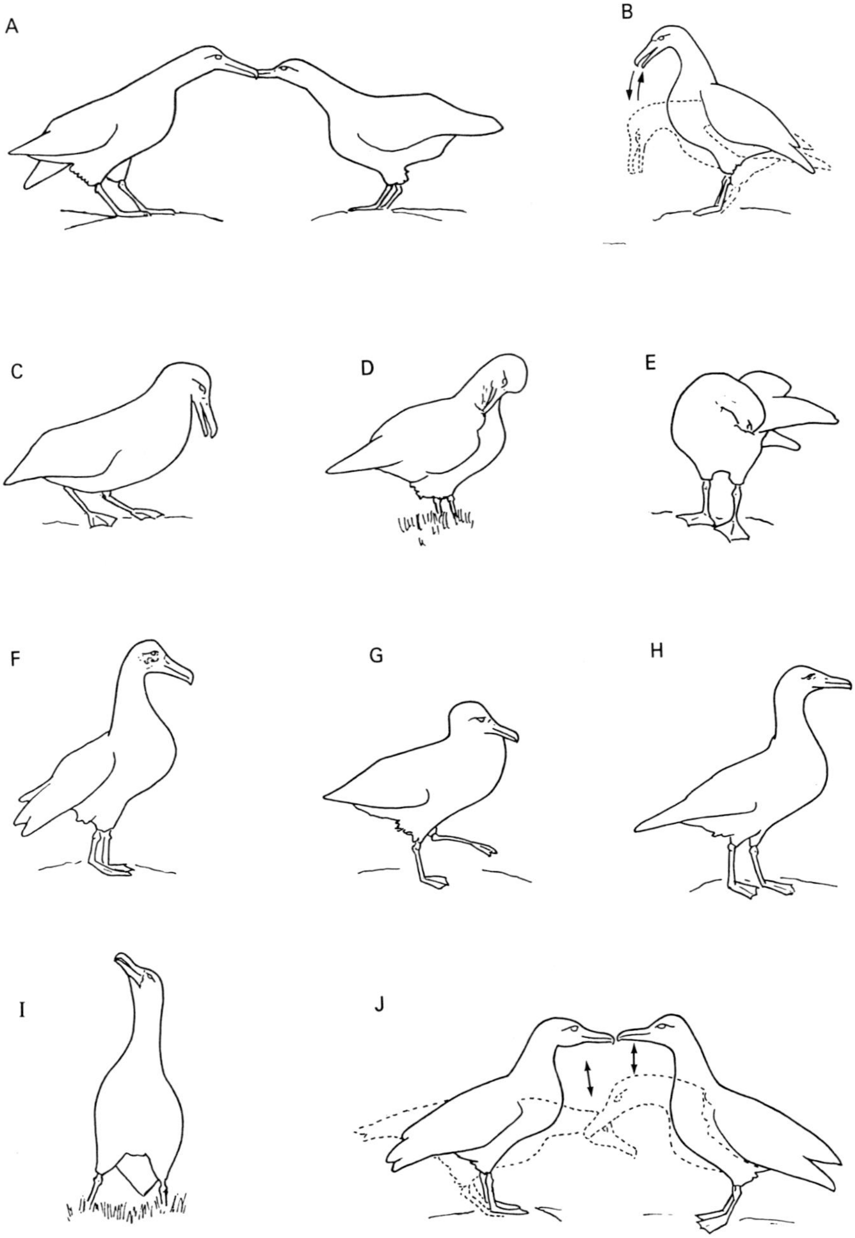

Figure 5.16 *Sexual Displays of* Diomedea immutabilis. *(A) 'Bill Touch'. (B) 'Eh-Eh Bow'. (C) 'Yapping'. (D) 'Beak Hiding'. (E) 'Bill under Wing'. (F) 'Stare'. (G) 'Bob-Strut'—start of step; and (H) end of step. (I) 'Moo'. (J) 'Bow Clapper'. From Meseth (1968).*

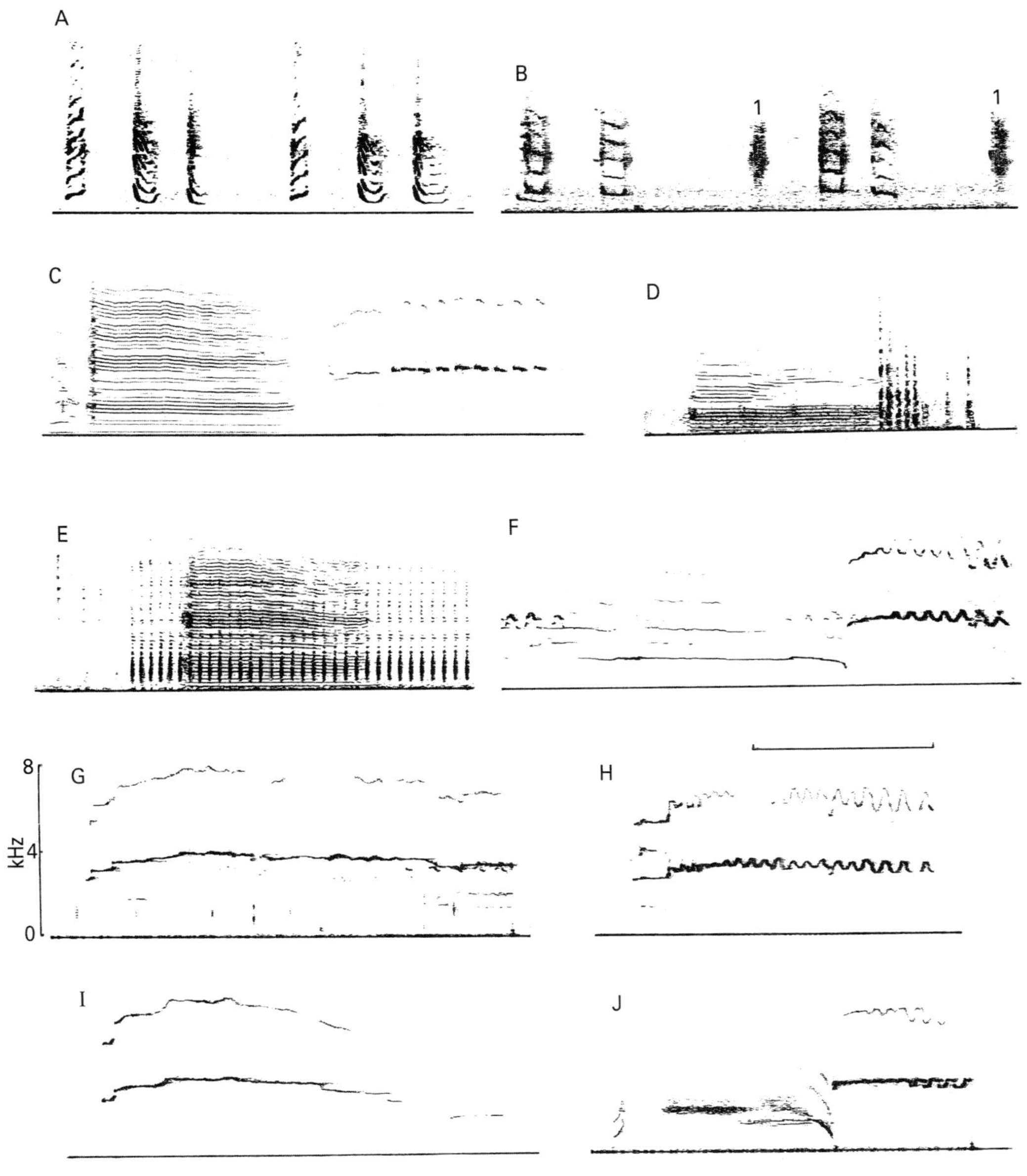

Figure 5.17 *Typical calls of* Diomedea immutabilis. *(A) and (B) 'Eh-Eh' calls, 1 'breath' notes. (C) 'Sky Call' followed by 'Whinny'. (D) 'Sky Call' with 'Bill Clapper' from partner. (E) Desultory 'Bill Clops' from female leading to 'Rapid Bill Clapper' as male gives 'Sky Call'. (F) 'Whine' followed by 'Whinny'. (G) Male 'Victory Cry' accompanying attack on intruder. (H) Complete 'Whinny'—note abrupt start. (I) 'Victory Call' following copulation. (J) Coarse 'Whine' followed by 'Whinny'. Bar is 1 s.*

is the tail fanned, but, like them, the action follows no fixed right-left pattern. The posture with bill hidden and nape exposed appears to convey a non-aggressive state.

6 *'Bill Under Wing'—see Fig. 5.16E*

In this dramatic display a standing bird quickly swings its head to one side and tucks its beak beneath the base of the uplifted wing on that side. The wing is only part unfolded and flared outward, with the carpal joint held to the body. The other wing remains unfolded. A series of muffled bill snaps is given through the concealed bill. Both sexes use this display, but only during the 'dance'. It may be the equivalent of the mollymawk 'Leg Action'.

7 *'Stare'—see Fig. 5.16F*

The bird draws up to its full height, stiff-legged and bill pointing towards the other's head, intently watching its every move. If this bird is bobbing up and down the 'Staring' one follows suit and may rise on tiptoe in the process. 'Stare' is part of the 'dance' routine, seems to be the equivalent of 'Gawky Look', and both sexes use it. No sounds are involved but the male uses it in conjunction with the 'Whinny' call when advertising (see 10 below).

8 *'Bob-Strut'—see Figs. 5.16G and H*

During the 'dance' one bird bobs its head up and down, marking time to the tempo of its partner's movements. It may also high-step around that bird—as a foot is raised the head is pulled down into the shoulders but lifted again as the step is completed, again in time with the actions of the other, cf. 'Scooping' in mollymawks and the 'Sway Walk' of the great albatrosses.

9 *'Bow-Clapper'—see Fig. 5.16J*

A 'dance' movement. The birds face one another and synchronously pivot on their feet to throw their heads down and give a loud clappering with their bills, then swing up, in silence, to repeat the action but this time sweeping their bill to the other side of that of their partner. 'Bow-Clappers' are loud, rapidly pulsed notes (Fig. 5.20D, E) and may sound very like the 'Vibration' calls of the great albatrosses.

10 *'Stare and Whinny'—see Fig. 5.17*

This is part of the male's advertising display. The 'Whinny' is also heard during the 'dance' but without the 'Stare' component. It is a forceful, staccato, ululating whistle, very high-pitched, reaching a mean of 15.1 ± 0.96 kHz ($n = 31$), according to Sparling (1977, table 3) and given with closed or nearly closed bill. Rapid vibrations of the body keep time with the ululations. The wings are closed. A complete 'Whinny' is

Figure 5.18 *Sexual displays of* Diomedea immutabilis. *(A) 'Sky Call'. (B) 'Head Flick'. (C) 'Air Snap'. (D) 'Sky Snap'. (E) 'Head Shake and Whine'. (F) 'Rapid Bill Clapper'. (G) 'Victory Call'. (H) Black-footed Albatrosses raise both wings for 'Bill under Wing'. From Meseth (1968).*

analysed in Fig. 5.17H and, following 'Head Shake and Whine' in Fig. 5.17F, J. In all these there is just one continuous note but with some birds the 'Whinny' consists of separate but simple notes as in Fig. 5.17C.

11 'Sky Moo'—see Fig. 5.16I

A standing or sitting bird slowly raises its head to about 45° and its throat swells as it gives a low 'mooo' through slightly opened beak. A few single 'Bill Snaps' follow as the head is brought down to its normal position. The 'Moo' is often used after the 'dance', following landing in a colony, or after agonistic activity. The action resembles the 'Wail' of mollymawks and the 'Sky Call' of the great albatrosses, but 'Sky Moo' has 25 or more harmonics, and does not carry far.

12 'Sky Call'—see Fig. 5.18A

A prominent movement used by both sexes in the 'dance'. The bird faces its partner, draws itself up to its full height (often standing on tiptoe), with the bill almost vertical but the neck slightly to one side. A low, drawn-out groan 'aww' is sounded through closed or scarcely opened bill (Fig. 5.17C). The other usually gives 'Sky Snap' in synchrony or another 'aww' call. Like 'Sky Moo', the 'Sky Call' is built from many harmonics, and with individual differences in energy distribution, call length, etc. 'Sky Calls' typically begin very abruptly but end more gradually as in Fig 5.17C and D.

13 'Head Flick'—see Fig. 5.18B

The bird stands stiffly upright. The head is flicked upwards and to one side, and as it is returned to its normal position a double click is heard. Meseth (1968) determined that the clicks are produced when the mandibles are pushed sideways out of alignment, when the rasping of the lower on the upper distorts the cornified tomium which then clicks back into position. 'Head Flick' is used during the 'dance' and by lone advertising males.

14 'Air Snap'—see Fig. 5.18C

Part of the 'dance' routine, precedes 'Bill Under Wing', and is done so quickly as to be easily overlooked. The bird thrusts its head forwards and upwards, neck fully extended, and snaps its beak once.

15 'Sky Snap'—see Fig. 5.18D

The bird swings its head from beneath the wing in 'Bill Under Wing' and points to the sky as in 'Sky Call' but gives a single snap, not 'aww'. When both dancers tuck their bills below their wings at the same time the 'Sky Snaps' that follow appear to help synchronize their subsequent actions.

16 *'Head Shake and Whine'—see Fig. 5.18E*

An element of the dance used more by the male. The bird bends towards its partner and waves its head vigorously from side to side while giving a loud whining cry. The vocalization tends to be prolonged and has clear frequency bands like drawn-out 'Ehs'. The posture resembles 'Head Shake and Whine' of the great albatrosses. A variant, which produces a more sustained wailing call, uses a downwards dip of the head but without any shaking, is given by both sexes when 'dancing'. This seems different from the wail of fleeing birds (see Section III.B.7 below) but may have an element of fear.

The 'Whine' is a piercing, far-carrying call with clear frequency bands. That of Fig. 5.17F is typical, the more 'noisy' example of Fig. 5.17J less so. Sparling (1977, table 5) analysed 19–22 examples which averaged 1.33 ± 0.48 s in length, had 4.04 ± 1.06 harmonics, and reached frequencies of 15.95 ± 2.2 kHz.

17 *'Rapid Bill Clapper'—see Fig. 5.18F*

A 'dance' component in which a bird, very erect, faces its partner and vigorously clatters the mandibles together so rapidly that they appear blurred, producing a buzzing sound. This appears to be the same as 'Bill Clacker' of the great albatrosses. It may contain threatening elements. Figure 5.17D and E show how even simple mechanical sounds can vary in acoustic structure.

B *Aggression and defence*

1 *'Nest Threat'*

Sitting birds discourage too-close approaches by facing the intruder, lowering the head and bill and giving simple short 'Eh' cries with small head jerks.

2 *'Bill Thrust'*

The head is thrust forward to snap violently at the head and bill of the opponent who may respond likewise. Their bills may interlock and fighting may develop with quick jerks and twists of their heads.

3 *'Glare'*

This usually follows 'Bill Thrust' and is done from a seated position. Each antagonist withdraws its neck and lifts its head to look hard at the other with their eyebrows pulled down and mandibles ajar ready for the next 'Bill Thrust'.

4 *'Charge'*

A bird suddenly rushes with extended wings and breast pushed forward to bowl over a rival. Seen of birds of either sex driving third birds from 'dances' and in sudden, apparently unprovoked, attacks on dancing partners.

5 *'Upright Threat'*

Another posture used in the 'dance' to repulse an unwanted third bird. The body is held stiffly as in 'Stare' but no sound is made.

6 *'Victory Cry'—see Figs 5.17G and I*

The winner in a dispute suddenly throws up its head and screams with opened bill. The sound is piercing and far-carrying and may last 3–4 s.

7 *'Escape Run'*

An action to avoid 'Charge'. Such birds run frantically off with head up and wings spread and often give a drawn-out 'Wail', a distress call longer than 'Victory Scream', and similar to the 'Wail' of mollymawks.

8 *'Gobble'*

This is a strangulated choking sound heard from birds trapped or alarmed and may be followed by vomiting of the stomach contents, like the 'Gulping' of mollymawks.

C *Chick displays*

In strong sun young *D. immutabilis* may travel some way from their nests to find shade. On returning to be fed they adopt a submissive attitude with the body low to the ground. The development of display postures is unstudied. Meseth (1975, p. 245) described the sole incident he saw of chicks 'dancing'. These had down only on their crowns. Both used 'Eh-Eh Bow', 'Bill Under Wing' and 'Sky Call' but the other elements were missing. 'Sky Moo' is also used by chicks still in down. Like those of the great albatrosses, the older gooney chicks often build and defend their 'sub-nests' in the parental territory. Fisher (1975b, p. 290) pointed out that hierarchies established then might be carried over into the early breeding years owing to the extreme philopatry of this species.

D *The organization of the 'dance'*

Meseth recognized three posture-sets built into the 'dance' of *D. immutabilis*: 'Stare–Sky Call', average length 2.8 s; 'Air Snap–Bill Under Wing–Sky Snap' (2.9 s); and 'Head Shake and Whine—Stare–Head Flick–Rapid Bill Clacker' (5.2 s). A fourth group, 'Eh-Eh Bow–Stare–Whinny–Head Flick–Rapid Bill Clacker' formed the basis of the male's advertising display.

The male leads in the 'dance' and may have to work up to a brisk tempo with actions such as 'Bob', 'Bob-Strut' and 'Bill Touch' to induce the other's participation. Successful 'dances' seem to be those where the tempi of movements and vocalizations of the partners are well synchronized.

Posture-sets at the core of the 'dance' are separated by the mutual actions of 'Bill Touch' and 'Bow-Clapper' which act as tempo-synchronizing devices (Table 5.5). 'Air Snap' and 'Sky Snap' may also emphasize the beat of the 'dance'. 'Eh-Eh Bow' is used at times of rising excitement and results in quickening tempo. 'Beak Hiding' by the male is immediately repeated and may be a means for marking time when the female falls behind. Otherwise it is seen only at the start and end of the dance. Some postures and posture-sets follow each other in a non-random manner (see Fig. 5.19), but there is none the less much variation in the sequencing. A bird doing 'Air Snap–Bill Under Wing–Sky Snap' tends to give the action on the opposite side of its body on the next repeat, and this posture-set constantly alternates between the partners. The end of the 'dance' comes when the female walks off, often in a crouched stance, watched by the male. The 'dance' of *D. nigripes* is more vigorous and faster than that of *D. immutabilis*, and both wings are raised in 'Bill Under Wing' (see Fig. 5.18H). Very rarely, *D. immutabilis* does that too.

Their voices are distinct, as Sparling demonstrated. The 'Eh-Eh' calls, 'Sky Call' (Fig. 5.20), 'Sky Moo' and 'Whinny' differ either in frequency structure and/or length of call. Black-foot 'Ehs' are the more harmonious, based on well-separated harmonics and each note is a 'simple cry' (see Fig. 5.20A, B) whereas those of Laysans plot as U-shaped formants starting on a falling frequency and ending on a rising one (see Fig. 5.17A, B).

E *The Short-tailed Albatross* D. albatrus

Films show that the display repertoire resembles that of *D. nigripes*, with 'Bill Under Wing' and a loud 'Sky Call' with both wings lifted and partly extended or held to the body. The common call is a double croak or 'Eh-Eh' note and a loud 'Wail' is also used (see Figs 5.8D, E; 5.12D). Perhaps suggestive is the way in which the lone bird that spent many breeding seasons on Midway Atoll persisted unsuccessfully in trying to join Black-footeds' 'dances', but made no attempt to do so with the abundant Laysans nearby (pers. obs.).

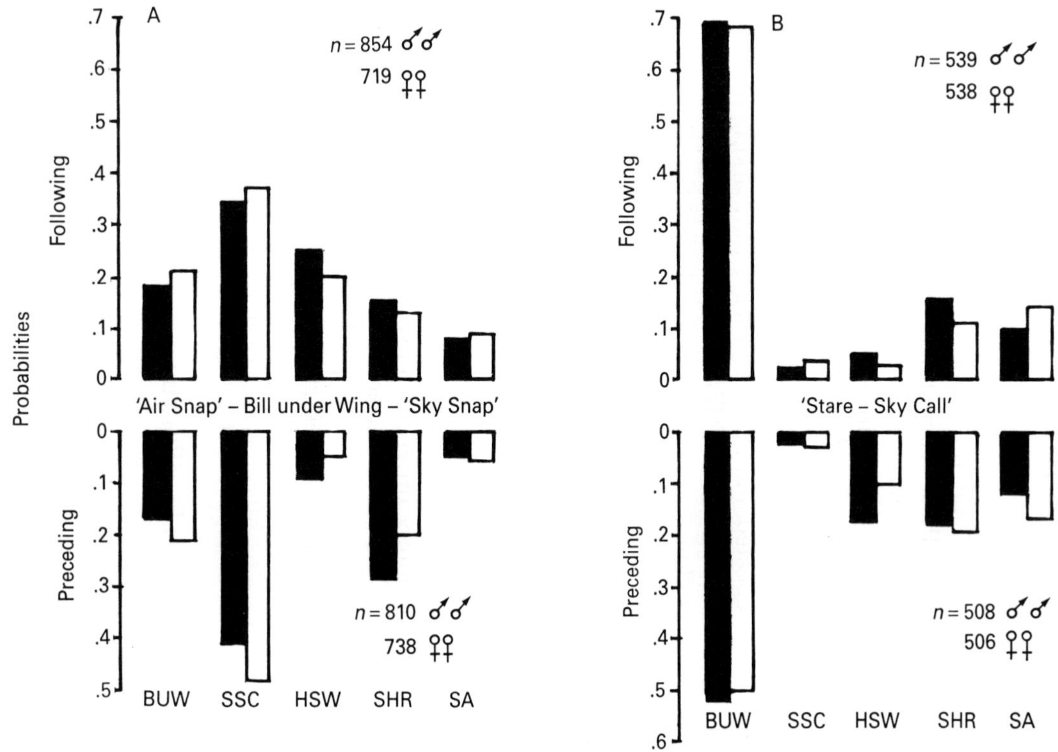

Figure 5.19 *Probabilities that certain postures and posture-sets immediately precede and follow particular postures and posture-sets in Laysan Albatross 'dances'. Filled bars, male displays; open bars, female displays. BUW, 'Air Snap–Bill under Wing–Sky Snap'; SSC, 'Stare–Sky Call'; HSW, 'Head Shake and Whine'; SHR, 'Stare–Head Flick–Rapid Bill Clapper'. SA, 'Scapular Action'. From Meseth (1975).*

Table 5.5 A typical dance sequence of *Diomedea immutabilis*. From Meseth (1975)

Male action	Mutual action	Female action
'Eh-eh Bow'		Walk to male
	'Bill Touch'	
	Throw debris	
'Beak Hiding'		'Beak Hiding'
	'Bill Touch'	
'Stare and Whinny'		
	Series of 'Bill Touches' and debris throwing	
'Eh-eh Bow'		'Stare'
	'Bill Touch' and throw debris	
'Bob-Strut'		'Beak Hiding'
Follows		Walks from male
		Stops, faces male
		'Stare-Sky Moo'
	'Bill Touch'	
Throw debris		'Stare'
'Eh-eh Bow'		
	'Bill Touch'	
'Beak Hiding'		'Stare'
	'Bill Touch'	
'Whinny'	'Bill Touch'	'Whinny'
'Beak Hiding'		'Beak Hiding'
	'Bill Touch'	
'Bow-Clapper'		'Beak Hiding'
	'Bill Touch'	
'Eh-eh Bow'		'Stare'
'Stare and Whinny'		
	'Bill Touch'	
	'Bow-Clapper'	
'Stare–Head Flick–Rapid Bill Clapper'		'Air Snap–Bill Under Wing–Sky Snap'

IV The Galapagos or Waved Albatross *D. irrorata*

The most detailed descriptions are those of Nelson (1968) which expand on earlier accounts such as those of Loomis (1918) and Beebe (1926). It is clear from these that the 'dance' of this species most resembles that of the North Pacific ones (Fig. 5.21). There is much 'Bill Touching' and 'Fencing', even between downy young and their parents (Harris, 1969c). A bowing action with croaking 'Ah-Ah' accompaniment leading to a more sustained 'Yapping', a 'Sky Call' action with a loud, prolonged 'Whoo' vocalization, 'Stare', 'Bow Clapper', 'Gawky Look', 'Bob Strut' and a flank

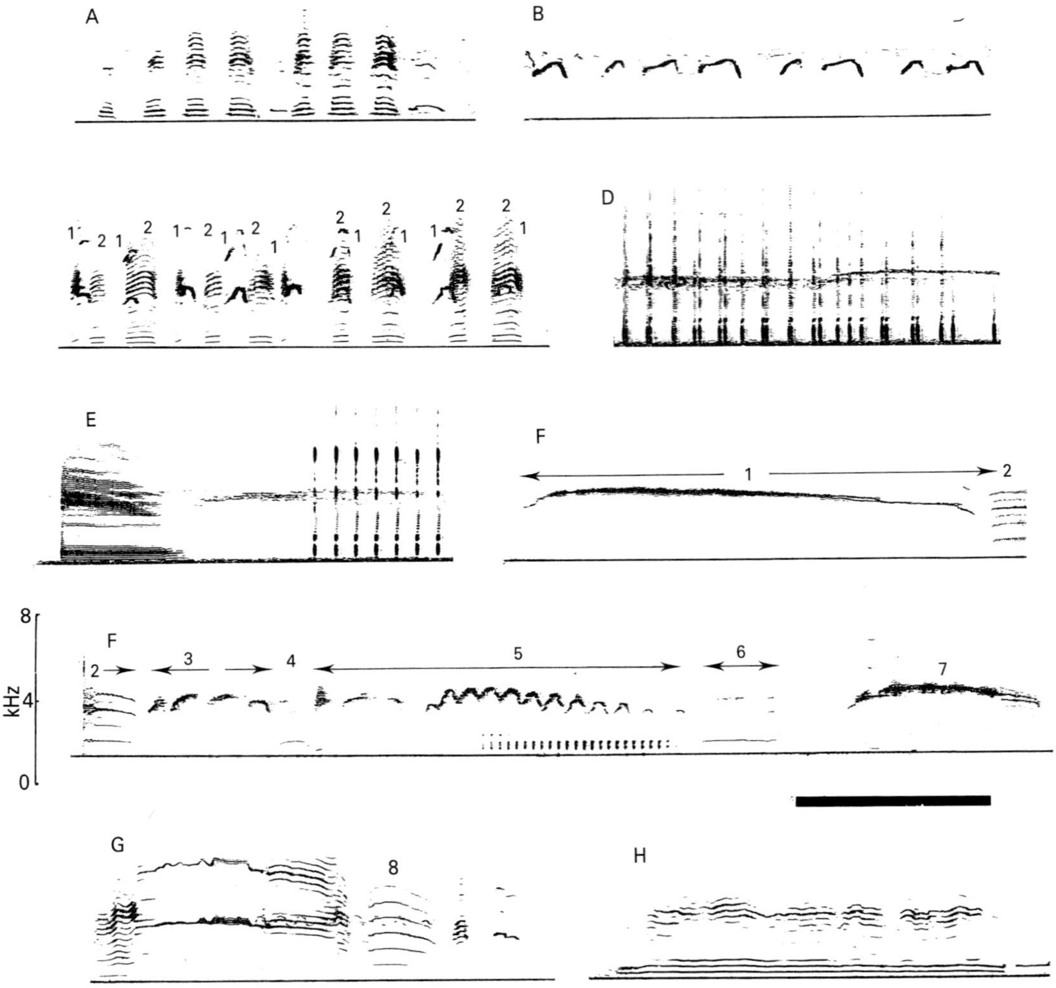

Figure 5.20 *Sonagrams from typical calls of* Diomedea nigripes. *(A) 'Eh-Eh' call with many harmonics. (B) 'Eh-Eh' call from a female based on 'simple cries' given in doublets and very like the begging calls of a chick. (C) 'Eh-Eh' duet from a pair at a nest; 1, a simple cry with few harmonics; 2, a more complex, deeper-pitched doublet call from the male: their tempi are well matched. (D) A volley of double 'clacks' while the dancing partner 'whines'. (E) 'Sky Call' followed by 'Bow Clapper'—note the different structure of these 'clacks' from those of (D) above. (F) A sequence of 'Whine' to 'Whinny'; 1, 'Whine'; 2, 'breath' note; 3, short 'Whinny'; 4, short 'breath' note; 5, 'Whinny' with 'Clapper' from partner; 6, 'breath' note; 7, final 'Whine'. (G) Postcopulatory cry followed by inhalatory call—8. (H) 'Distress Wail' from fighting bird. Bar is 1 s.*

Figure 5.21 *Display postures of the Galapagos Albatross* Diomedea irrorata. *(A) 'Bill Touching'. (B) 'Sky Moo'. (C) 'Sky Point'—partner responds with 'Bill Clapper'. (D) 'Leg Action' and 'Bill Clapper'. (E) 'Bob'. (F) 'Threat Gape' and 'Point'. (G) Aggressive encounter with 'Bill Gaping'. (H) 'Leg Action. (I) 'Sway Walk'. From Nelson (1968).*

touching action similar to the 'Leg Action' of mollymawks are all used. The 'dance' tends to finish with a 'Sway Walk', evidently the homologue of the 'Scoop Walk' of the mollymawks and great albatrosses, but differing in that the bird remains upright and proceeds with exaggerated head and foot movements (Fig. 5.21I). Aggression is signalled by 'Bill Clapper' but threat between rival males is also shown by mutual but silent gaping and the closure of their beaks with a loud 'clop' (Fig. 5.21G). Apparently missing from the repertoire is any beak hiding with wing raising, and 'Sideways Throwing' or 'Wails' seem not to have been reported, unless Nelson's 'mad laughter' is such!

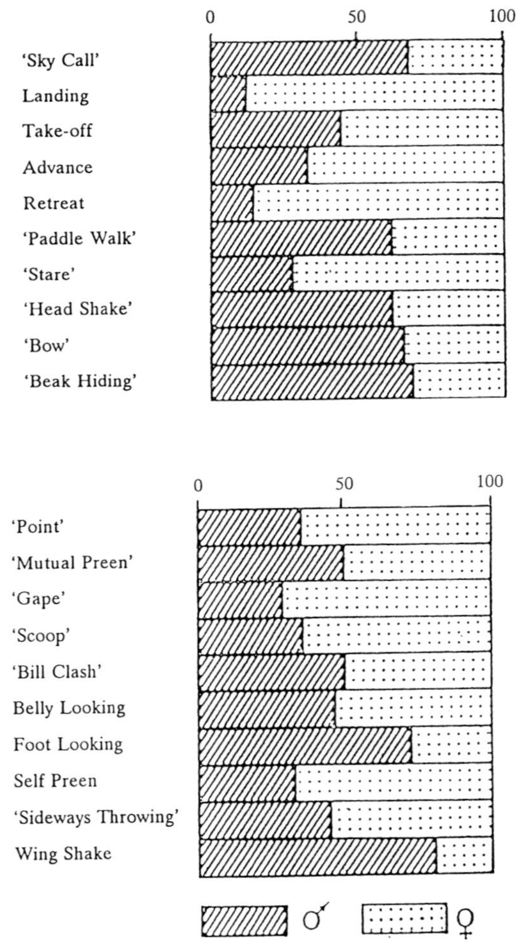

Figure 5.22 *Proportions of the different postures used by both sexes of* Phoebetria fusca*: these include behaviours giving social signals and others mainly involved in the care of the body surface. Based on 3803 observations of males, 3832 of females. From Jouventin and Weimerskirch (1984).*

Nelson's account shows that the male leads in the 'dance', that some elements are given as posture-sets, and that the actions of the partners may be linked, for example the male's arching of his neck and touching his flank signals the female to reach towards him with 'Bill Clapper' (Fig. 5.21D). The supraorbital ridges are very pronounced in this bird, and the lowered eyebrows create a beetle-browed aspect during 'Gawky Look' and 'Stare'. The crown is then also flattened producing a characteristic physiognomy (Fig. 5.21I). There are no reports of tail fanning in this species.

V The Sooty Albatrosses *Phoebetria fusca* and *P. palpebrata*

The *Phoebetria* albatrosses, nesting on cliff ledges, do not have the freedom of movement on the ground that the North Pacific species enjoy. Nevertheless, their displays still have the nature of a dance.

Many postures appear to be homologous with those of other albatrosses but in *Phoebetria* dual flights feature prominently in which the following bird matches every twist, turn, and wingbeat of the other. These albatrosses spread and twist their tails in their ground displays, and they also expose the pink buccal cavities more readily that other species do excepting, perhaps, *D. irrorata*. Foot-peering with silent bowing is common and many movements are done in a very jerky manner, crown and nape feathers often lifted to form a slight crest.

Detailed studies are few. Sorensen (1950), Richardson (1984) and others gave fragmentary information. Berruti (1981) compared the behaviours of the two species nesting sympatrically, but the accounts of *P. fusca* by Jouventin *et al.* (1981) and Jouventin and Weimerskirch (1984) provide a comprehensive analysis of the complex courtship display.

While all actions are used by both sexes, some are used more by one sex than by the other. Jouventin and Weimerskirch (1984) described 15 elements of sexual activity and five kinds of 'comfort movements' (Fig. 5.22).

Jouventin *et al.* (1981) used vector and sequential analyses to tease out the relationships of these display elements in pair formation. The advertising male attracts a female to his territory with 'Sky Call', a loud two- or three-syllable cry—the well-known 'pee-oo'—at passing females. If one alights 'Pointing', 'Bowing' and mutual preening follow. Should the female fly off, the male often pursues her and a fast synchronized flight ensues before one or both return to the ledge.

Jouventin and Weimerskirch (1984) followed the development of these nuptial displays of *P. fusca* from those of birds ashore for the first time to first breeders about 3.5 years later (Fig. 5.23). The 'dance' elements include many postures used by other albatrosses but these may not always signal comparable information. 'Beak Hiding' again seems to have an appeasing role, particularly by a male on the approach of a female. The latter reacts with 'Stare'. This seems to differ from the 'Gawky Look' of other albatrosses although perhaps having a similar function in getting eye-to-eye contact. In *Phoebetria* the 'Stare' is of birds face-to-face which turn their heads to examine their partners with each eye in turn. In 'Gaping' the bird points and then flips open its bill to display the pink buccal interior: this has aggressive and defensive

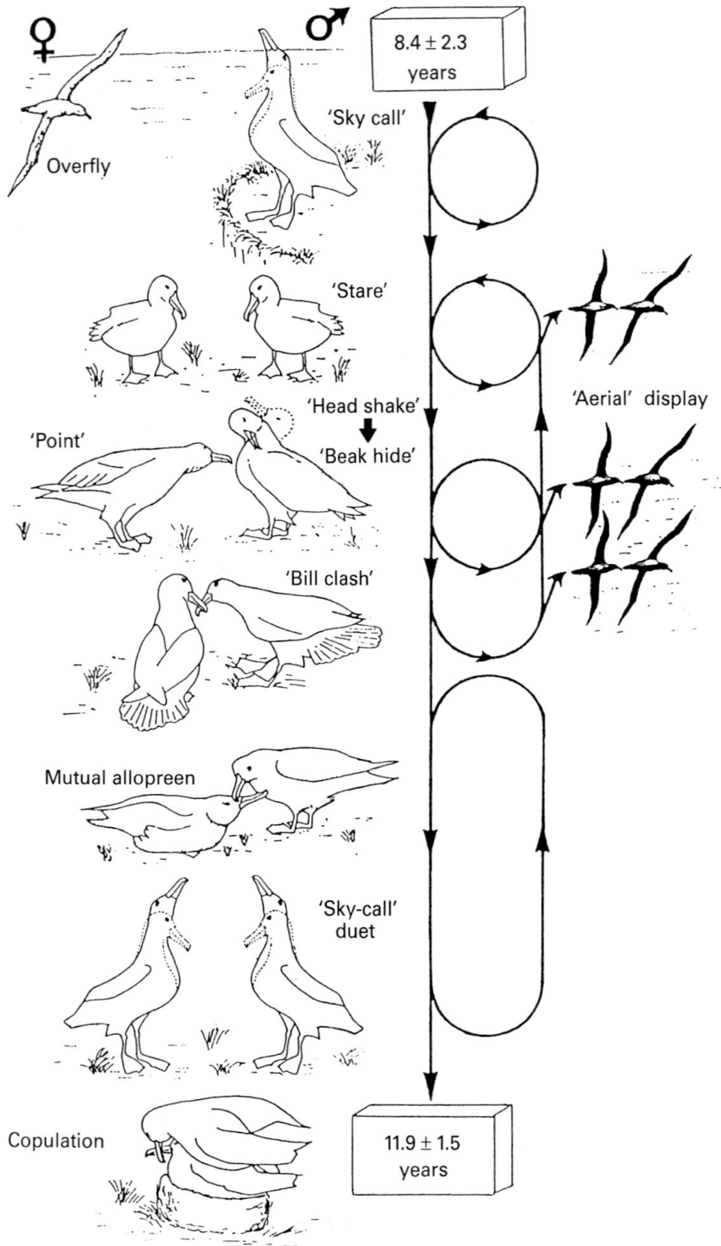

Figure 5.23 *Schematic diagram of the development of nuptial display in the Sooty Albatross* Phoebetria fusca. *Birds typically first come ashore at 8.4 years old and pass through a series of courtship stages, the female gradually being accepted by the male who finally permits contacts in ritualized combat ('Bill Clashing'). Aerial chasing has now ceased and repeated allopreening and reciprocal 'Sky Calling' tend to cement the bond. On average first breeding is attempted by 11.9-year-olds. From Jouventin and Weimerskirch (1984).*

connotations, and in 'Scooping', which involves bowing with opened beak, the aggression is evidently redirected at the ground.

Berruti (1981) identified a 'Bill Thrusting' action used by *P. palpebrata* but not by *P. fusca*. Face-to-face with their crown feathers raised, one bird waves its beak from side

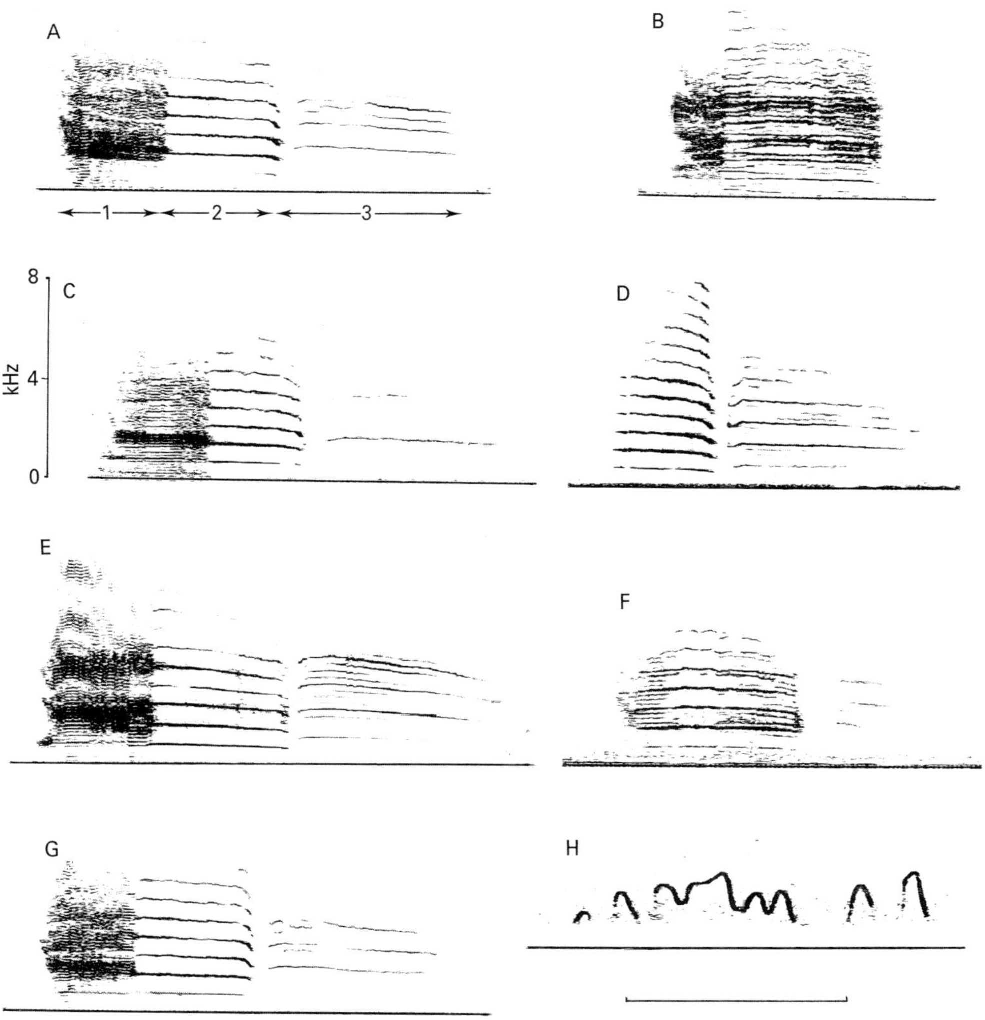

Figure 5.24 *Sonagrams from calls of* Phoebetria palpebrata. *(A) and (B) 'Sky Calls' from mated pair; A, male; B, female, Campbell Island. (1) loud section of mainly unstructured sound; (2) loud expiratory section with clear harmonics; (3) faint inspiratory section also with harmonics. (C) and (D) 'Sky Calls' from mated pair; C, probably male; D, probably female, Campbell Island. (E) Trisyllabic 'Sky Call' with quite strong final inspiratory note, Antipodes Island. (F) Female 'Sky Call', Campbell Island. The inspiratory second note is very faint. (G) Male's 'Sky Call', Campbell Island: similar to but with a different energy distribution from (E). (H) Calls of a chick begging for food. Bar is 1 s.*

to side and, at the mid-point, jerks its head upwards and makes an exaggerated biting movement. The respondent copies, alternating with the other's movements. Both have their tails fanned and skewed towards each other.

Aggression is signalled by 'Gaping' and 'Bill Snaps', with the neck hackles raised. Adults may spit oil, but it is the chicks that are especially prone to do that when alarmed.

The 'Sky Call' is the principal vocal signal, given with the bill pointing to the sky and slightly open, then jerked down and pressed to the chest when a soft inspiratory note is sounded. This call and action may be given alternately by birds in a duet (Fig. 5.23). Although so simple, the voices of individual birds vary a good deal (cf. Fig. 5.24), and those of paired ones are often easily identified (Fig. 5.24A,B). With *P. fusca* Jouventin and Weimerskirch (1984) reported the males' songs at 1.30 ± 0.12 s ($n = 9$) longer than those of the females at 1.03 ± 0.13 s (n = 9), $P < 0.001$. Birds in display flights are silent. Only Richardson (1984) has reported a muted 'aaaargh' from following birds as they touched or stretched out to reach the tail of the leader. Both this and the inspiratory coda to 'Sky Call' may be involuntary sounds due to muscular strains.

The chicks use the usual plaintive pipings but later these develop into longer warblings and thicker, adult-like squawks (Fig. 5.24H).

CHAPTER 6

Behaviour and Vocalizations of Procellariidae, Hydrobatidae and Pelecanoididae

Contents

I The fulmars

Fulmars share a number of behaviour patterns but only those of *Macronectes* have been analysed in detail. Paradoxically, no full ethogram is available for the familiar Northern Fulmar which often breeds close to towns and cities.

A *The giant petrels* Macronectes *spp.*

The visual and vocal communication patterns are similar for both species and form quite a complex repertoire but one with fewer components than those of albatrosses. The only percussion sounds commonly heard are the bill clashes of the copulation 'Tattoo'. Both species visit their colonies throughout the year and nuptial displays can be seen occasionally even in mid-winter. Some such birds appear to be paired and of breeding status, others are probably older non-breeders.

The detailed descriptions given by Warham (1962) of a composite for both species were expanded by Voisin (1978) and those for *Macronectes giganteus* by Bretagnolle (1988b, 1989b). Terminology follows Warham and Bretagnolle.

1 *Nuptial displays*

All elements are used by both sexes. As the males have much bigger bills than the females, presumably confusion over sex seldom arises.

a. 'Head Raising' (Fig. 6.1D). One bird approaches another, lifting its head to about 45° with partly opened beak, then turns it down and reaches towards the other with side-to-side sweeps of the head. The gular pouch is inflated and a soft braying given (Fig. 6.2B). The bird so addressed may respond with similar head sweeps. Their tails are usually spread but not cocked and napes mostly unruffled. Bretagnolle (1989b) separated the 'Head Raising' movement from the lateral 'Head Swinging' (his 'Sexual Low Intensity Attitude') (Fig. 6.3), both included in 'Mutual Display' by Warham (1962). Bretagnolle regarded 'Head Raising' as a signal for the start of courtship: 85% of those he scored began like this. In the prelaying period, females initiated most displays, later both sexes were equally active and yet later still the females were once more the most forward. These actions often lead to:

b. 'Bill Clashing' or 'Bill Fencing'. Facing one another, the birds rub their bills together, and may touch beaks momentarily—'Bill Touching'—before proceeding to:

c. Mutual and reciprocal allopreening, 'Sideways Throwing' and, possibly, copulation

2 *Aggressive behaviour*

Threatening, fighting and defensive activity are particularly evident among birds competing to scavenge a dead whale or seal where females of both species tend to be excluded due to their smaller size and weaker bills. The various actions were detailed by Warham (1962) and analysed by Bretagnolle (1988b).

a. 'Low intensity attitude' (Fig. 6.3). Used particularly by nesting birds. The neck is raised and head turned down towards the disturbance, the nape feathers ruffled, wings closed and a braying or whinnying sounded (Fig. 6.2B) as the head and neck are swung from side to side. This posture is very similar to the 'Head Swinging' used in courtship and Warham noted that it is often unclear where threat ends and courtship begins. In threat, however, the nape feathers are extensively raised, in courtship they are, at most, only slightly ruffled.

Figure 6.1 *Postures of giant petrels. (A) 'Upright Threat'—head swung through wide arcs, nape and back feathers ruffled. (B) 'Forward Threat' as bird runs towards competitor. (C) and (E) 'Forward Threat' before running to attack—note cocked and fanned tail. (D) Invitatory 'Head Raising' posture, nape feathers smooth. (F) Fighting between 'sealmasters'. From photographs.*

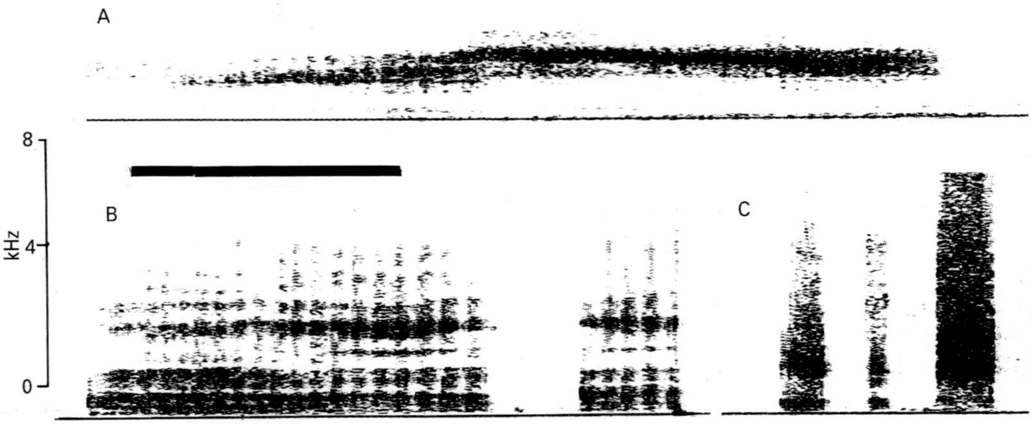

Figure 6.2 *Vocalizations of* Macronectes. *Bar is 1 s. (A) 'Whinnying' call accompanying 'Upright Threat'—*Macronectes halli. *(B) Typical growling throaty call accompanying head raising nuptial display and mild threat—*M. giganteus. *(C) Threat of* M. halli *at close approach by man: note increased amplitude and pitch as stimulus intensifies.*

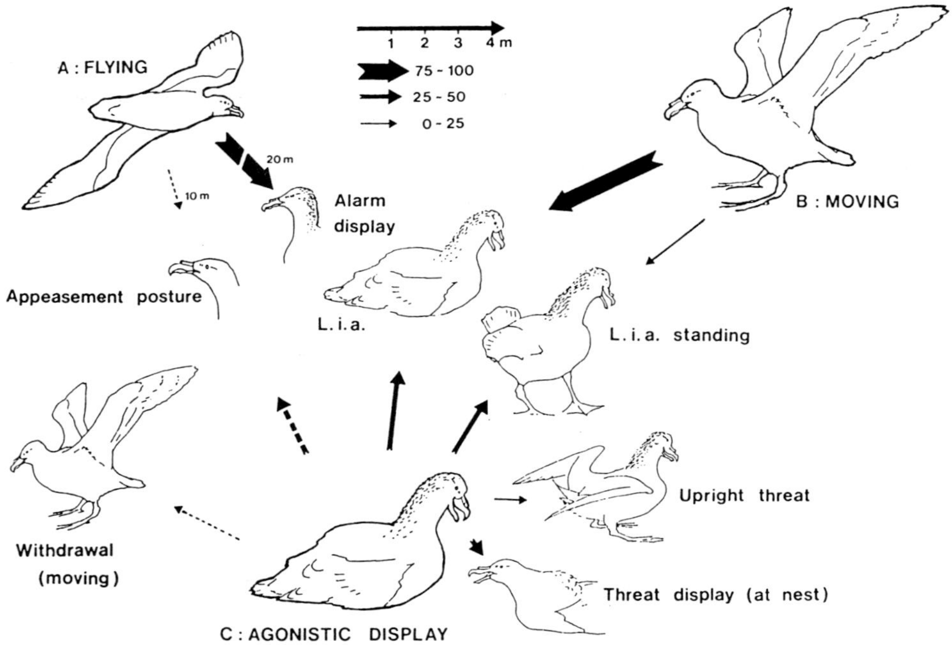

Figure 6.3 *Flying, moving and agonistic displays of* Macronectes. *Summary related to distance and type of stimulus. Arrow size indicates the distance except for (A) Flying; thickness the frequency of occurrence. Broken arrows responses limited to female birds. L.i.a. = Low intensity attitude. From Bretagnolle (1988b).*

b. 'Oil Spitting Threat'. The bill is opened, neck stretched towards the intruder, and a gurgling given that may lead to oil spitting, rather rarely by adult birds.

c. 'Forward Threat' (Fig. 6.1B,C,E). The threatening bird squats, stands, or runs at its adversary with tail fully fanned and tilted so that its tip may be directed forwards. The nape is arched, the feathers of the nape and back extensively ruffled, and the bill lowered. The wings are outstretched and arched, their tips often trailing the ground. The head is not swayed nor any sound made.

This action is common in disputes over carcasses, signals intention to attack, and is often used as the bird runs forward to do that. A dominant 'sealmaster' busily feeding, probably with neck submerged in the carcass, holds this posture with tail fully fanned and uptilted and wings extended (Warham, 1962, fig. 1). At small food sources this amounts to mantling over the meal.

'Forward Threat' often induces a dominant feeding bird to swing around and run to attack the displayer. Sometimes the dispute is only resolved after a fight (Fig. 6.1F). Birds meet breast to breast and grapple bills. Fighting may be intense and has the potential for serious injury.

d. 'Upright Threat' (Fig. 6.1A *and The Petrels*, p. 63). The wings are usually extended and may trail on the ground, the tail fanned and uptilted and crown, nape, and back feathers extensively ruffled. The neck is erect, and the downturned head is waved vigorously from side to side, rapidly or slowly according to the individual. The beak almost touches the leading edge of the wing at the end of each stroke and an expiratory neighing given (Fig. 6.2A).

'Upright Threat' is seen when the bird faces several rivals. It is as if it is trying to clear room for itself, for example immediately after alighting in a colony where others' individual distances may be infringed. By extending its wings the displayer increases its size as seen by potential rivals.

e. 'Aerial Threat' (The Petrels, p. 64). In both species birds flying over groups of conspecifics perform what appears to be 'Upright Threat' in mid-air, rising and falling like a displaying dove (Warham, 1962; Voisin, 1978). The feet are dangled, wings held out stiffly and drooped, nape ruffled and the head waved from side to side with a braying, much as with mollymawks. Two birds may perform simultaneously when flying but a few metres apart. They then appear to be threatening one another; otherwise the action seems to be directed at birds on the ground whose reaction may be to give the 'Alarm Display' or an 'Appeasement Posture' (Bretagnolle, 1988b) (Fig. 6.3). Voisin (1978) suggested that such displays were given by males.

3 *Defensive behaviour*

Bretagnolle (1988b) described reactions of *M. giganteus* to threatening situations:

a. 'Alarm'. The bird raises its head and silently watches the source of its concern, often without lifting the feathers of the head, nape or back, but, in response to a close over-flier usually does erect its hackles (Fig. 6.3). This posture is contagious, signalling alarm to neighbours (Voisin, 1978).

b. 'Appeasement' (Fig. 6.3). Like that of 'Alarm' but the beak is turned upwards, the feathers unruffled. This posture forms a silent response to other birds moving close by. Its function is unclear, but the position of the bill may signal that attack is not contemplated. A similar stance with sleeked feathers is adopted to reduce attack by birds walking through a colony, Bretagnolle likening it to the 'Slender Walk' of penguins (cf. Warham, 1972, fig. 19).

c. 'Defensive'. Nesting giant petrels may pose with wings folded but held slightly out from the body and the tail fanned, like the Cape Pigeon shown (see Fig. 4.7). It does not seem to have been recorded except in response to man. If closely approached both adults and chicks voice loud alarms (Fig. 6.2C).

d. 'Oil Spitting'. If hard pressed some adult birds will spit oil and/or regurgitate food, particularly when lightening their bodies for flight.

4 *The repertoire*

Bretagnolle (1988b; 1989b) identified a posture set 'Head Raising'–'Sexual Low Intensity Attitude'–'Bill Clashing'–'Bill Touching' to Mutual Preening. Rather unexpectedly the females, as with *Pagodroma* (Section I.E.1 below), tend to initiate courtship (Voisin, 1978). Bretagnolle found that the postures used by partners became increasingly similar in the course of time, with full displays occurring more and more often, and suggested that the males compensate for their greater aggressiveness by using more appeasement behaviours such as 'Bill Touching' and mutual allopreening. Possibly success in pair-bonding requires synchrony of movements as with *D. immutabilis* (Chapter 5.III.D).

Bretagnolle pointed out one possible reason for the heightened role of threat and appeasement displays in the repertoire of *M. giganteus*: breeding at high densities on relatively flat ground and unable to take off or alight with ease, they react frequently with conspecifics, whereas fulmars nesting on cliffs seldom infringe others' 'individual distances'. Carcass-feeders are especially vulnerable as they cannot see threats from behind when their heads are buried in flesh and blubber. As Warham (1962, p. 145) noted, such birds seem very conscious of their vulnerability and quick to swing round to attack a threatening bystander.

B *The Northern and Southern Fulmars* Fulmarus glacialis *and* F. glacialoides

Many snippets of information on *Fulmarus glacialis*, for example by Boase (1924), were drawn together by Fisher (1952a, pp. 326–336) and summarized in Cramp and Simmons (1977). Luders (1977) figured some of the postures for the southern form.

The most familiar behaviour of *F. glacialis* is the social activity in which two or more birds face one another and wave their heads and widely opened bills in sideways and upwards sweeps (Fig. 6.4A,B). Such groups appear to be of non-breeders, the fulmar equivalent of the 'gams' of albatrosses. 'Head Sweeping' is used in sexual and agonistic situations. Pairs with eggs seldom display. No action appears to be restricted to any one sex and the sexual differentiation in bill size and head shape may well be detectable by the birds themselves. Hatch (1987c) identified individual *F. glacialis* by differences in plumage patterns and marks on the culmina.

1 *Nuptial displays*

Lone birds patrolling the cliff faces may hesitate before single birds on nest sites who, if closely approached, perform cackling and head-waving displays. Presumably pairs are formed when a flier joins such a territory holder. Actions seen of paired birds on nests include:

a. 'Head Sweeping' (Fig. 6.4A,B). One bird reaches towards another with widely opened beak and swings its head from side to side and bows up and down while emitting a continuous rattling sound. The partner responds similarly. Their closed tails shake in time with the movements but the wings and feathers of crown and nape are not erected.

This is a graded action, the movements being through narrow arcs at low intensity and the rate of delivery of the syllables increases with increasing excitement, as does the degree of throat expansion. Gaping exposes the mauve or pink mouth cavity—'flashing the buccal lantern' (Selous, 1927). Afterwards the birds often swallow conspicuously and wag their closed tails before relaxing. The sweeps of *F. glacialis* are through smaller arcs than those of *F. glacialoides* (van Franeker, pers. comm.).

b. 'Bill Clashing' or 'Bill Fencing'. Two birds facing one another rub their bills together so that these touch on each side alternately. The clashing may be accelerated and accompanied by vigorous cackling. One (the female?) may place its beak momentarily within the gape of the other. This action, together with the exudation of stomach oil and/or salt gland excretion from the bill, presumably gave rise to reports of 'courtship feeding', and action unsubstantiated for any tubenose.

c. 'Head Tossing'. 'Bill Clashing' may lead to sudden backwards flicks of the head so that the nape rests momentarily on the back. This was recorded for *F. glacialis* by Bent (1922) and Boase (1924) and for *F. glacialoides* by Murphy (1936, p. 599), but not seen by Luders (1977).

These actions are accompanied by the usual mutual and reciprocal allopreening and 'Sideways Throwing' (Doran, 1963), and may lead to copulation which includes a 'Tattoo' from the male stropping his bill on that of the female. Male *F. glacialis* are recorded as making a crooning note during mounting. Hatch (1987c) found that copulation was protracted, rarely lasting less than 1 min and sometimes 8 min or more.

Figure 6.4 *Displays of fulmars. (A) 'Head Sweeping' and duetting of* F. glacialis. *In fulmars, as in other procellariids, one bird's beak may pass between the mandibles of its partner. (B) 'Head Sweeping' of* F. glacialoides. *(C) 'Head Tossing' duet of* Thalassoica antarctica. *(D) Joint nest defence by* T. antarctica. *(E) 'Head Tossing' by* T. antarctica. *From photographs and Luders (1977).*

2 *Aggression and defence*

Although there is some fighting on the breeding cliffs, for example by *F. glacialoides* (Prevost, 1953), agonistic displays seem more frequent at sea, particularly by *F. glacialis* when concentrated during scavenging (see Fig. 4.9). Duffey (1950) described a Northern Fulmar vigorously pulling up pieces of turf and moving them around with its bill after trying to perch by a nearly fledged chick. This sounds like the mollymawk 'Ground Stabbing' of Tickell (1984), and described as redirected attack, the opponent being out of reach. Martin and Myres (1969) saw 'Head Tossing' used by *F. glacialis* with a squeaky cry when defending food from gulls. Southern Fulmars may nest cheek by jowl with Antarctic Petrels *Thalassoica antarctica*. Orton (1968) suggested that *F. glacialoides* can evict the other by fierce fighting, but Pryor (1968) found the reverse was true.

a. 'Mild Threat'. The bird faces the disturbance with head thrust forward and bill widely open to display the buccal cavity.

b. 'Head Sweeping' (Fig. 4A). The same movements seem to be used in sexual or in aggressive contexts, but in threatening mode the crown and nape feathers tend to be ruffled and the voice may perhaps be different. Pryor (1968) reported that gaping threats of *F. glacialoides* were only rarely accompanied by the bird's raucous call, while Prevost (1953) regarded the repeated lifting of the head as the main defensive movement.

c. 'Defensive'. When approached by man, fulmars tend to droop their wings and spread and cock their tails before spitting oil, like *Macronectes* and *Daption* in similar circumstances (see Fig. 4.7).

d. Oil Spitting. Fulmars are infamous oil spitters, using stomach oil as artillery to repel aggressors, even from the wing (Hazelwood, 1948). The discharge is signalled by a series of head jerks as the oil is regurgitated, followed by forward lunges and twitches of the partly opened and spread wings. With *F. glacialis* a special squeak note superimposed on a low growl is given (Pennycuick & Webbe, 1959).

Most potential aggressors retreat in the face of the threatening movements and adults are rather seldom required to discharge the oil under natural conditions. However, their salvoes can displace Razorbills *Alca torda*, Kittiwakes *Rissa tridactyla*, Guillemots *Uria aalge* and even Herring Gulls *L. argentatus*, from nesting sites (Mougin, 1967; Olsthoorn & Nelson, 1990). Feather contamination with the oil can result in the deaths of birds as large as eagles (Warham, 1977b).

3 *Aerial display*

What appears to be an aerial form of the cackling display, with swollen throat and open bill, wings drooped and stiffened while gliding along an undulating course, is apparently homologous to that of *Macronectes*, and of obscure function (Warham,

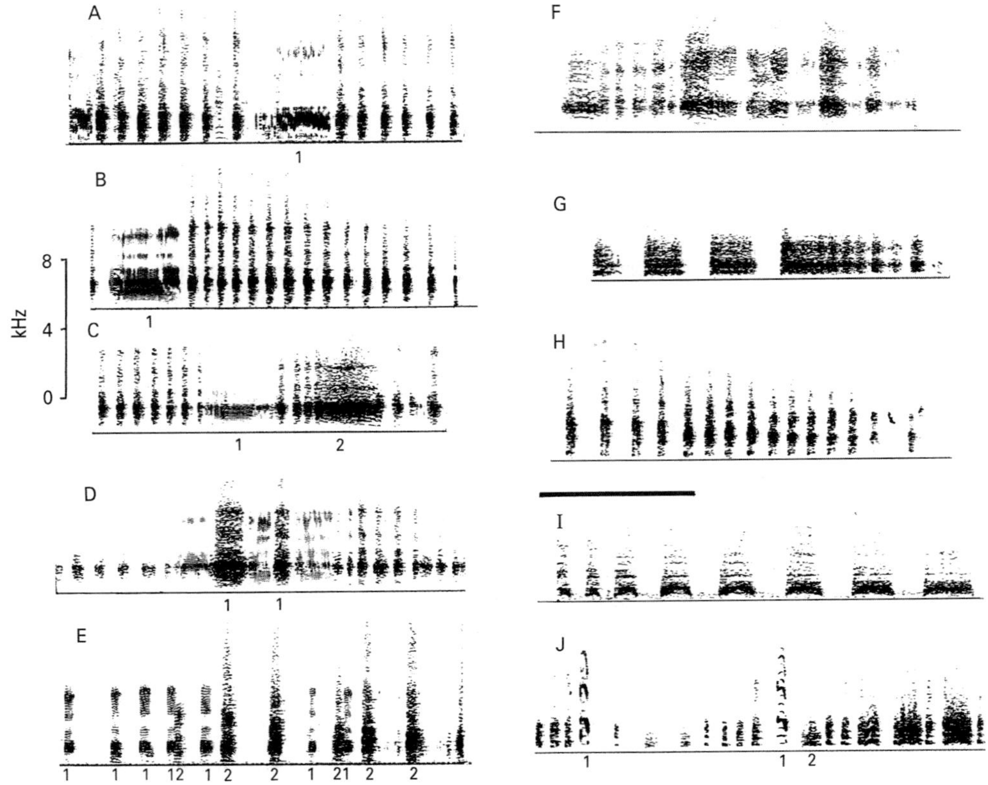

Figure 6.5 *Vocalizations of* Fulmarus, *(A)–(F)* F. glacialis, *(G)–(J)* F. glacialoides. *Bar is 1 s. (A) and (B) Part of a typical 'Cawing' sequence built from staccato notes interspersed with inhalatory rasping groans (1). (C) A second bird adds an explosive note of mainly 'white noise' (2) in response to a normal 'Cawing' with quiet inhalation (1). (D) A duet with loud input from a second bird (1). (E) A duet of simple notes but of different construction (2) going almost to the baseline. (F) Part of a duet. The notes of both sexes are rather unstructured, producing a cawing chorus. (G) A lone bird producing broad band notes that lengthen towards the climax. (H) A lone adult's call composed of typical volleys of staccato 'Caws' tailing off at the end. (I) Start of song of a single bird showing lengthening of the notes as it develops. (J) Adult calls with two high-pitched 'Pips' (1) from the chick. At (2) a second adult adds its harsh voice.*

1975). However, the cackling is also used by parents just before landing near their chick, evidently to forestall oil salvoes (Duffey, 1951) and by birds hovering over the 'gams' before alighting nearby.

4 *Vocal repertoire*

Northern Fulmars produce an amazing cacophony of hoarse cawing, quacking, rasping and jabbering sounds. Fisher (1952a, pp. 326–328) listed 26 attempts at written representations, none particularly successful. Figure 6.5 shows sonagrams of representative vocalizations of *F. glacialis*, sexes unknown. The principal note is a staccato 'Caw' given in volleys in which the speed of delivery reflects the excitement of the singer. Long trains of these are interspersed with low groaning, apparently inhalatory, syllables (Fig. 6.5A,B). In Fig. 6.5C and D second birds have joined in, with raucous bursts of 'white' noise, possibly from the males; that the sexes have different voices is suggested by analyses such as that of Fig. 6.5E, where one bird's notes are noticeably the higher pitched.

C *The Antarctic Petrel* Thalassoica antarctica

Luders (1977) described the sexual encounters *Thalassoica antarctica* which resemble those of *Fulmarus*, with all postures being used by both sexes. Nuptial displays include much 'Bill Clashing', duetting and mutual and reciprocal allopreening. A 'Head Sweeping' display with clucking or cackling calls is used with widely opened bill and Pryor (1968), and particularly Luders, emphasized the prominence of 'Head Tossing' in the repertoire. According to Pryor, one bird lowers its beak to its breast and slowly raises and extends its neck until the bill lies horizontally along the back; that is, even further than shown in Luder's figure (Fig. 6.4E). The tail may be wagged from side to side and the throat inflated. Luders found that each 'Head Toss' took about 0.7 s, with about 2 s between each and repeated three or four times with cackling from both participants. 'Head Tossing' is a graded action, tending to become less intense and the motions less extreme as the season progresses, presumably because most pair-bonds are by then established. Copulation includes a 'Tattoo'.

Intruders at nest sites are threatened with head thrust forward, neck and nape feathers ruffled and loud calls (Fig. 6.4D). Pryor (1968) noted that this species was reluctant to spit oil during the prelaying period, doing so then only when all escape routes were blocked.

The staccato notes forming their songs are very like those of the Northern Fulmar (cf. Figs 6.6A & 6.5B,H), and in duets the notes of one bird (the male?) are pitched higher than those of the other.

D *The Cape Pigeon* Daption capense

Courting pairs reach out and sweep their heads from side to side through wide arcs while maintaining a rapid chattering with their gular pouches inflated (Pinder, 1966). There is the usual mutual and self preening. 'Throw Back' has not been reported and there are no aerial displays (Bretagnolle, pers. comm.).

Figure 4.7 shows the defensive attitude and Pryor (1968) found that adults defended their nests with salvoes of stomach oil on the slightest provocation. He noted that they seemed immune to attack by skuas *C. maccormicki*, perhaps due to their readiness and efficiency in shooting oil. The birds even spit oil in disputes over food at sea (Dixon, 1933, p. 134) where they are very quarrelsome, Bennett (1927)

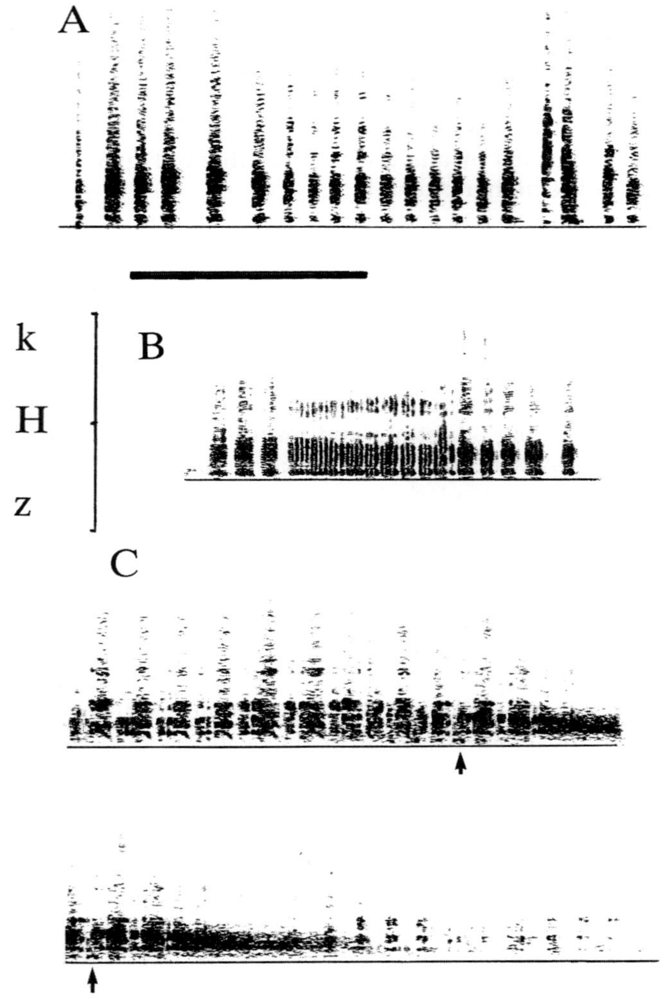

Figure 6.6 *Vocalizations of* Thalassoica antarctica. *Bar is 1 s. (A) Characteristic adult call—a series of short notes creating a sustained 'Cawing': note typical changes in timing and amplitude. (B) Normal 'Caw' calls separated by a burst of very rapidly pulsed notes creating a harsh 'Purr' of undetermined significance. (C) The end of a duet. The separate voices cannot be identified clearly but one appears to have high frequency components and one ends with a long raucous slur, leaving the other to run down quietly. Arrows mark overlap in graph.*

writing that one whale carcass was too small a morsel for two Cape Pigeons to divide in peace! When feeding at sea they keep up a harsh shrill chatter.

The voice of *D. capense* is as distinctive as its chequered plumage. It is higher pitched than that of the others, except perhaps for *Pagodroma nivea*. The most common element in the repertoire is a staccato 'Kuk'or 'Caw'. At low intensity such calls are given in slow time, but during mutual or threat displays are speeded up to produce a crackling, rippling effect (e.g. Fig. 6.7D) or a musical trill (e.g. Fig. 6.7E,F).

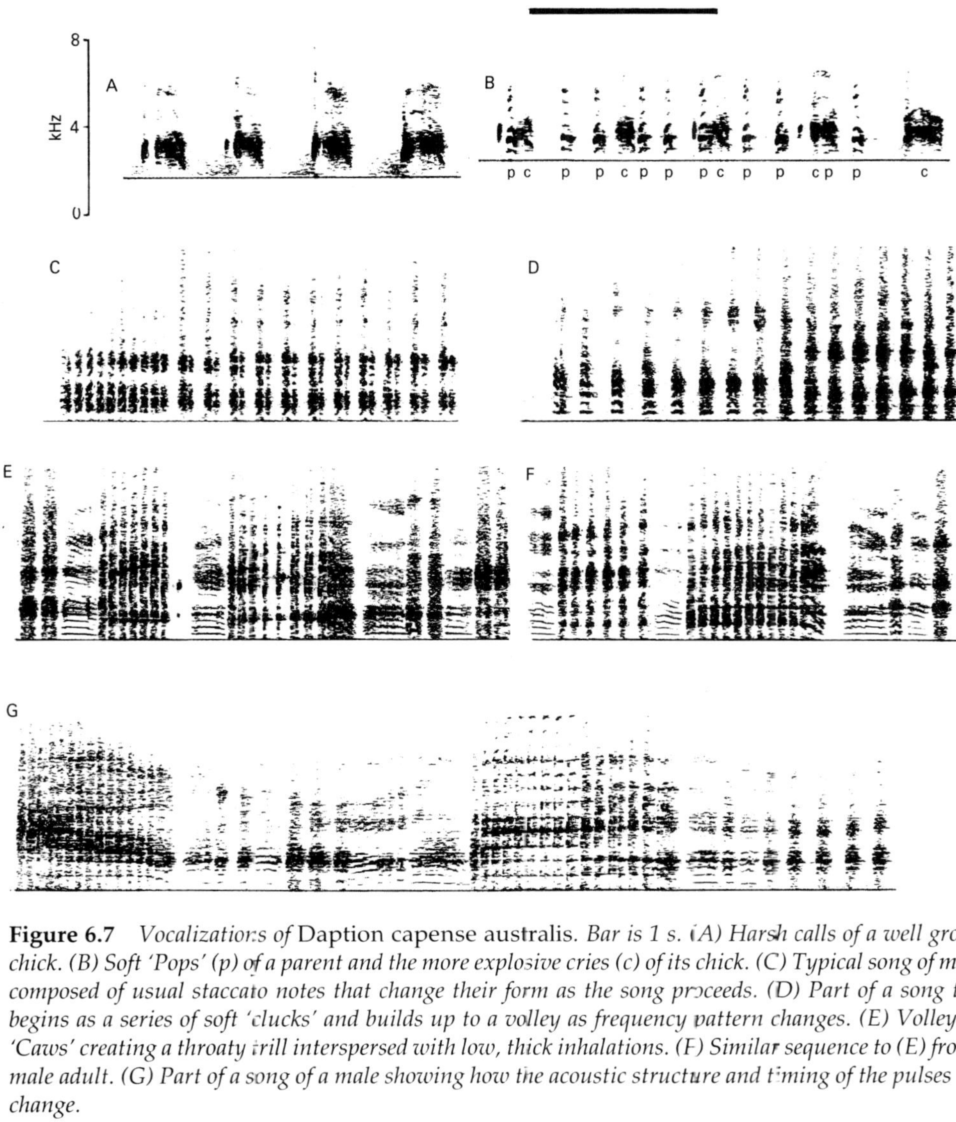

Figure 6.7 *Vocalizations of* Daption capense australis. *Bar is 1 s. (A) Harsh calls of a well grown chick. (B) Soft 'Pops' (p) of a parent and the more explosive cries (c) of its chick. (C) Typical song of male, composed of usual staccato notes that change their form as the song proceeds. (D) Part of a song that begins as a series of soft 'clucks' and builds up to a volley as frequency pattern changes. (E) Volleys of 'Caws' creating a throaty trill interspersed with low, thick inhalations. (F) Similar sequence to (E) from a male adult. (G) Part of a song of a male showing how the acoustic structure and timing of the pulses can change.*

The notes tend to change both in timing and acoustic pattern as in Fig. 6.7C, long bursts being separated by groaning inhalations (e.g. Fig. 6.7E,F). A softer, more melodious 'Kuk', due to clearer harmonics, is used by parents when with their chicks (Fig. 6.7B), which contrasts with the thick, harsh cries of the nestling (Fig. 6.7A .

Calls of males are shown in Fig. 6.7C,F and G and illustrate the acoustic variability in the one sex. In duets the calls of the female are pitched higher, but further study is needed to confirm that, as with *P. nivea*, the sexes can be separated on voice alone.

E *The Snow Petrel* Pagodroma nivea

A noisy, social species that is most active during the 1600–0500 hours of the Antarctic 'night' (Isenmann, 1970) or from 0500 to 0600 hours (Cowan, 1979), like *Oceanites oceanicus*, but differing from *D. capense* and *F. glacialoides*, which are daytime birds.

1 *Nuptial displays*

As described by Brown (1966) and Isenmann (1970), a female with a nest site is courted by males that fly past and alight to approach her cautiously. She calls and shuffles to face the male threatening with parted beak and raised head. He may take flight if heavily menaced, or may even attack. Females too may prospect for partners, being less aggressive, and drawn to males or pairs calling from nests. Once accepted, they fence with their bills and the female may place hers momentarily within his gape. Such pairs may keep up an almost constant chattering from within their nest cavity, mutual allopreening and calling at passers-by.

2 *Aggression and defence*

Snow Petrels defend their territories aggressively, often revealing their whereabouts by their calls. Perhaps more than any other adult tubenoses, they freely loose off their oily projectiles and oil splattered rocks often mark the nests. Fighting is common, with the combatants getting smeared with oil, bills interlocking and heads twisting to and fro accompanied by loud screeches. According to Brown (1966) the bird that locked on to the lower mandible of its rival invariably won. Isenmann (1970, p. 105) noted one pursuit lasting nearly 2 h. Brown described aerial pursuits with head-swayings, such as those of *Macronectes*, as nuptial activity, but according to Bretagnolle (pers. comm.), who followed 300+ such flights, they are primarily of an intrasexual and agonistic nature.

3 *Vocal repertoire*

The calls accompanying sexual and agonistic activity are built from pulsed, harsh, rather high-pitched 'Caws' or 'Clucks', often with accelerated delivery, and producing volleys of sound during peak activity. These may develop into a trill or chirrup. Such calls are also heard from birds on the wing replying to others who are often hidden among boulders.

Guillotin and Jouventin (1980) confirmed that the males' voices are pitched lower than those of their partners (Fig. 6.8)—a useful distinction for field researchers and, presumably, for the birds themselves. The fundamental frequency for six females averaged 3.43 kHz (2.7–4.0 kHz), for eight males 2.46 kHz (2.2–2.8 kHz). The lighter the bird the higher pitched its voice, presumably reflecting the smaller tracheae and syringes of smaller birds. Guillotin and Jouventin believed that the temporal patterns of the pulses furnished the most important clues to individual recognition, rather than frequency differences.

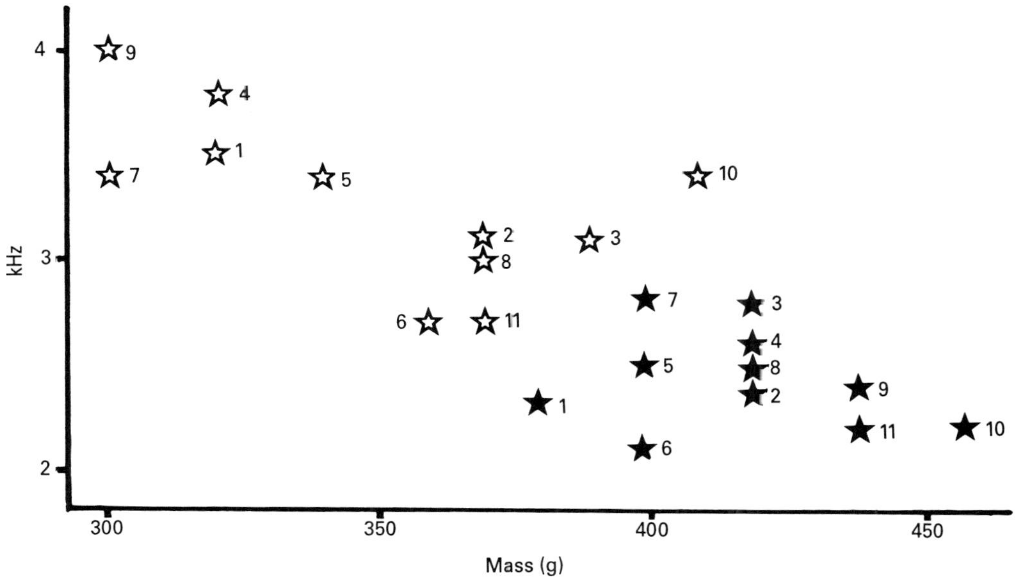

Figure 6.8 *The relationship between body mass and the fundamental frequency of the courtship 'Caw' for 11 pairs of Snow Petrels. Open stars, females; closed stars, males. From Guillotin and Jouventin (1980).*

F *Discussion*

Despite their varied appearance, the calls of the smaller fulmars are all built from very similar pulsed notes with few harmonics; all wave their heads and expose their mouth cavities in display; at least three use 'Throw Back' postures, and several, perhaps all, cock their tails in threat; and probably all are sexually dimorphic in voice.

II The Kerguelen Petrel *Lugensa brevirostris*

The behaviour of this bird is unstudied. It appears to have an aerial display accompanied by a wheezy, rather low-pitched call showing on spectrograms as chevrons with harmonics (Marchant & Higgins, 1990, p. 451; Bretagnolle, in press, a). High-pitched 'si-si' type calls have been described as well as low-pitched flute-like ones, none fulmar-like.

III The Gadfly Petrels *Pterodroma* spp.

A *Introduction*

Apart from their aerial displays, much of what is known of the behaviour of these birds resembles that of other burrowing tubenoses. Their visual display repertoire seems small whereas their vocal signals are quite complex. There are no detailed accounts of the conduct of grounded birds apart from that of Grant *et al.* (1983) on the Bonin Petrel *Pterodroma hypoleuca*.

Other gadfly petrels, such as *P. lessonii* and *P. phaeopygia*, seem to avoid displaying or calling on the surface even at night, and are often silent in their burrows, most overt activity taking place in the air. This may be related to the heavy toll exacted by predators. Such birds presumably land after their aerial displays (when they are safe from attack) and complete pair-bonding under ground. Others, such as *P. hypoleuca*, *P. nigripennis* and *P. leucoptera*, with no predators, do much of their courtship on the surface or within their burrows. *P. hypoleuca* is one with no natural enemies on land that still remains strictly nocturnal there; perhaps it evolved from a form with predators and nocturnality has been retained.

Most visible activity of a sexual nature concerns the non-breeding and prebreeding segment of the population, as with *P. phaeopygia*, where the incidence of aerial display coincides with the non-breeder's presence, with little calling once these have left half way through chick rearing (Simons, 1985). Breeding birds come and go silently except perhaps for a brief call when entering their nesting tunnels (Warham, 1956).

Gadfly petrels tend to be most active by night. Among the species of tropical regions there is usually little movement after dawn and even less around mid-day, but in the late afternoon the numbers circling overhead build up and the aerial chorus begins (Thibault, 1973; Gardner *et al.*, 1985a). Activity peaks just after dark, followed by the usual lull. With egg laying beginning, *P. solandri* start their aerial circuits of the montane nesting areas as early as 1000 h, but not until late afternoon, with the light fading, is there much ground activity, with birds landing freely. At Midway Atoll between 8 December and 28 January the first *P. hypoleuca* appeared high overhead 13.2 ± 1.3 min after sunset. At the Snares Islands, *P. inexpectata* arrives later with respect to sunset than *P. hypoleuca*. But skuas kill petrels at The Snares, whereas there are no native predators on Midway. Predators also take *P. madeira*, *P. phaeopygia* and *P. hasitata* which tend to come in 60–90 min after sunset. Yet *P. baraui* at predator-free Reunion, tends to arrive about 1 h hour before sunset, perhaps to take advantage of warm air currents overland (Bretagnolle & Attie, 1991). Clearly the factors affecting arrival times are complicated and need further examination.

There is some evidence that gadfly petrels in nests reply to the calls of overfliers, as might be expected since overfliers alight in response to man-made 'lure' calls (Chapter 4.XVII), and Vinson (1976) reported ground to air communication by *P. arminjoniana*, while Tomkins (pers. comm.) heard *P. phaeopygia* in burrows replying to others calling on the wing. Perhaps aerial displayers can also lead potential partners down to nesting sites, identifying one another by their calls learnt during the dual flighting.

1 *Aerial activity*

The most dramatic activity of gadfly petrels is their high-speed dual flight when a bird suddenly swoops close to another, raking back its wings from the carpal joints to reduce drag and increase speed, and the two jink along close together for a few seconds before parting to continue their separate circuits. As they converge their typical 'Ti-ti' call becomes distorted. Flights may be quite extended, with the pair disappearing high into the sky and as many as five birds may be involved.

Vinson (1976), watching *P. arminjoniana*, saw that although pursuer and pursued often changed positions, it was the pursuer who called, as with *P. cahow* (Wingate, in Palmer, 1962). The following bird is usually, but not always, the caller during paired flights of *P. solandri*, and through slightly opened bill. Sometimes both call. Even when watching daytime-active birds, it is usually impossible to tell the status or sexes of the participants, but Grant *et al.* (1983) deduced from their voices that the males were the chasers, the females the chased.

The petrels typically circle over a limited area—behaviour readily verified with a bird identifiable by idiosyncracy of voice, by a missing flight feather, etc. This is presumably where the flier will eventually nest or has a site. For prebreeders it could be the general area where they were born, but there is no evidence for that.

Where several gadfly petrels share nesting grounds and similar breeding timetables, different species may participate in dual flights, for example *P. macroptera* with *P. incerta* (Elliott, 1957, p. 563), despite their distinctive underwing patterns (*The Petrels*, p. 80). This could lead to hybridization and explain some of the colour polymorphism in this group, but no proven hybrids have been collected.

2 *Sexual activity on the ground*

Activity among grounded gadfly petrels is seldom as frenetic as it can be among shearwaters, perhaps because *Pterodroma* usually nests at lower densities. Grant *et al.* (1983) took advantage of the colonies of *P. hypoleuca* at Midway Atoll which could be watched under street lighting.

They found pairs and single birds scattered around after dark on the ground among the burrows during the prelaying period. One such courting bird flew up repeatedly to circle the general area several times before alighting beside its partner to rest and allopreen: sometimes both partners took off together, returning simultaneously to their starting point. A common form of sexual activity was for pairs to meander along the surface within a small area, perhaps but one or two metres across. One bird trailed hard on the tail of the other and either or both might enter a nearby burrow. Much duetting accompanied these perambulations, the pursuer the most vocal. Pursuers were on top during copulations so that these episodes were evidently of males soliciting females, perhaps showing off a nest site to a potential partner. In two instances copulation occurred within 2 m of the burrows, copulations lasting from 20 s to over 4 min and without calling.

During allopreening one bird is usually the more active. After a spate of bill fencing in which two *P. macroptera* nibbled at and along each others' bills, one preened its passive partner's head, then ran its beak quickly down that bird's neck to the shoulder and along the leading edge of the folded wing, nibbling at the feathers as it did so. The recipient remained hunched and quite submissive. This bird had previously used the 'Ti-ti' call, its partner a disyllabic braying (possibly a modified 'Kuk-u-er', see Section C.2 below).

Figure 6.9 *Fighting between* Pterodroma solandri *who met after being lured down by calling. Note the cocked and fanned tail of right hand bird and ruffled crown of the other. From a photograph.*

B *Aggression and defence*

Grant *et al.* (1983) watched Bonin Petrels emerging to patrol an area within 2–3 m of their burrow, evidently in a form of territorial activity. Birds were repelled when, with wings spread, the defender suddenly dashed forward, the attacked bird flicking up, part flying, part jumping, to land a few metres away. Many such repulsions occurred without calling, but loud 'Kuk-u-er' calls alone were often enough to cause nearby petrels to scatter. Similarly, *P. inexpectata* may leave its burrow to attack grounded birds calling near the entrance. Fights also occur, the participants pecking hard at each other's heads and napes (Fig. 6.9). Fights over nest sites also take place with *P. arminjoniana* (Gardner *et al.*, 1985a).

C *Vocal repertoire*

Much communication between gadfly petrels is by voice, which can be loud and far-carrying from the surface or air, but is often muted from the burrow.

No complete repertoire for both sexes of a gadfly petrel has been described, but audiospectrographs have been published, notably for *P. arminjoniana* (Gill *et al.*, 1970), *P. inexpectata* (Warham *et al.*, 1977), *P. mollis* and allies (Warham, 1979; Bretagnolle, 1995), *P. hypoleuca* (Grant *et al.*, 1983), *P. phaeopygia* (Simons, 1985; Tomkins & Milne, 1991), and for *P. baraui* and allies by Bretagnolle and Attie (1991).

The calls can be quite complex acoustically, with much variation even from a single bird, between birds, between the sexes and even between colonies. The following account attempts to pinpoint the main types of call with illustrations of typical examples. Because their functions have not been tested experimentally and are thus ill defined, the calls are categorized according to their aural features. The calls seem to lack marked inhalatory and exhalatory phrases.

1 *The 'Ti-ti' call*

This is the most familiar of gadfly petrel calls, being variously transcribed as 'titi', 'wik,wik,wik', 'si-si-si', 'tee-tee-tee' and so on. It is the 'high call' of Bretagnolle and Attie (1991). Such calls are known for at least 21 species, from both large and small forms and from all parts of the range. 'Titi' is the onomatopoeic Maori name for several species that are or were used for food.

The call consists of a series of short, staccato notes which, when heard *en masse* from the wing, produce a continuous chattering chorus. Wingate (1964) likened that of *P. hasitata* to the humming of bees, Bourne (1955) that of *P. feae*, as a pleasant, high pitched tittering. The same calls from the ground tend to last longer (2–4 s) and, once having begun the caller seems unable to stop until the vocalization has run its course.

'Ti-ti' calls are used by birds on the surface during mutual displays and by adults on entering their burrows, and may be heard from nesting birds by day in response to disturbances nearby. In such circumstances the call appears to function as a greeting or threat, or both. Figures 6.10–13 provide examples of analyses from a range of species.

The individual notes are basically 'simple cries', apparently developed from the cries of the chicks. In grounded birds the body pulsations, the regular swelling of the neck and of the floor of the mouth, and the opening and closing of the beak coincide

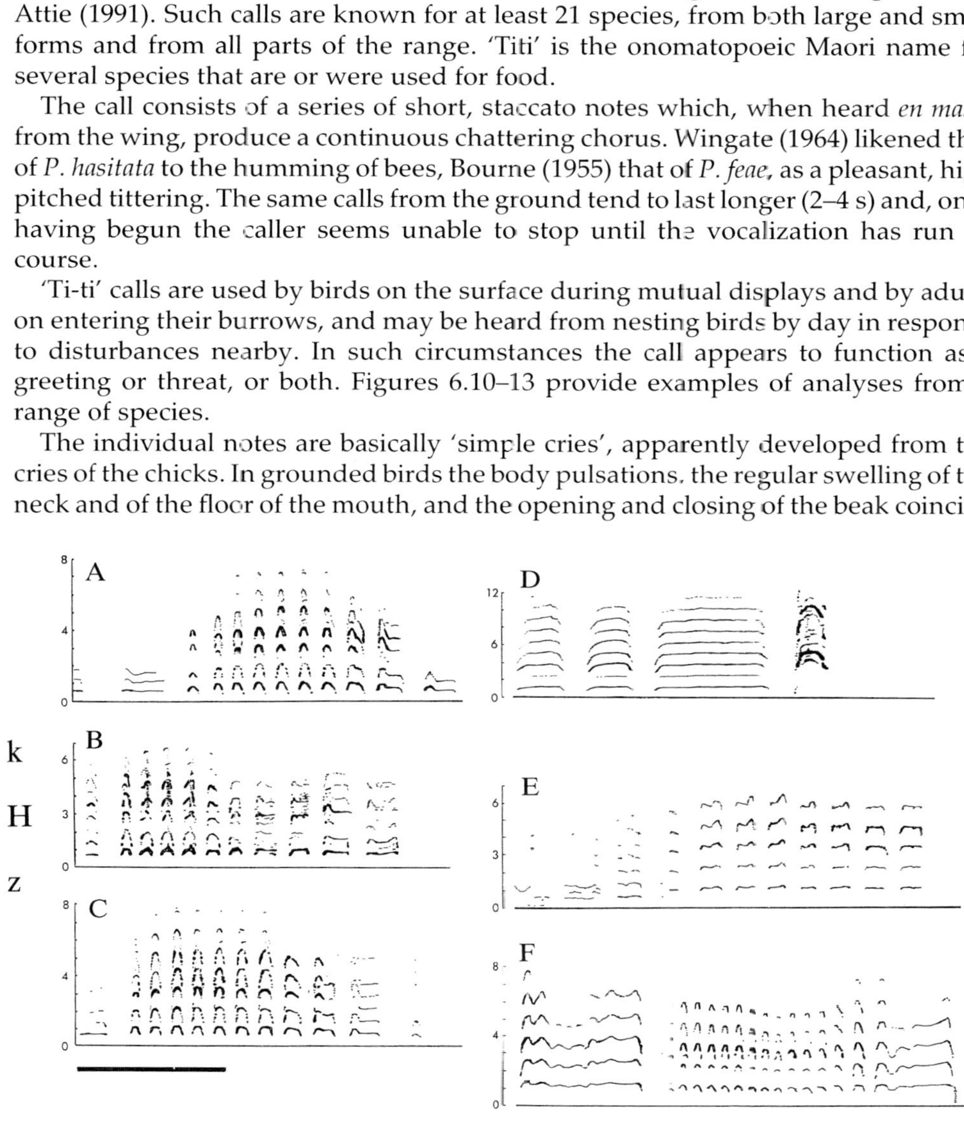

Figure 6.10 *'Ti-ti' type songs of grounded birds, (A)–(C)* Pterodroma inexpectata. *(D)–(E)* P. nigripennis. *(F)* P. solandri. *In (D) and (F) 'Ti' notes are preceded by 'Moans'.*

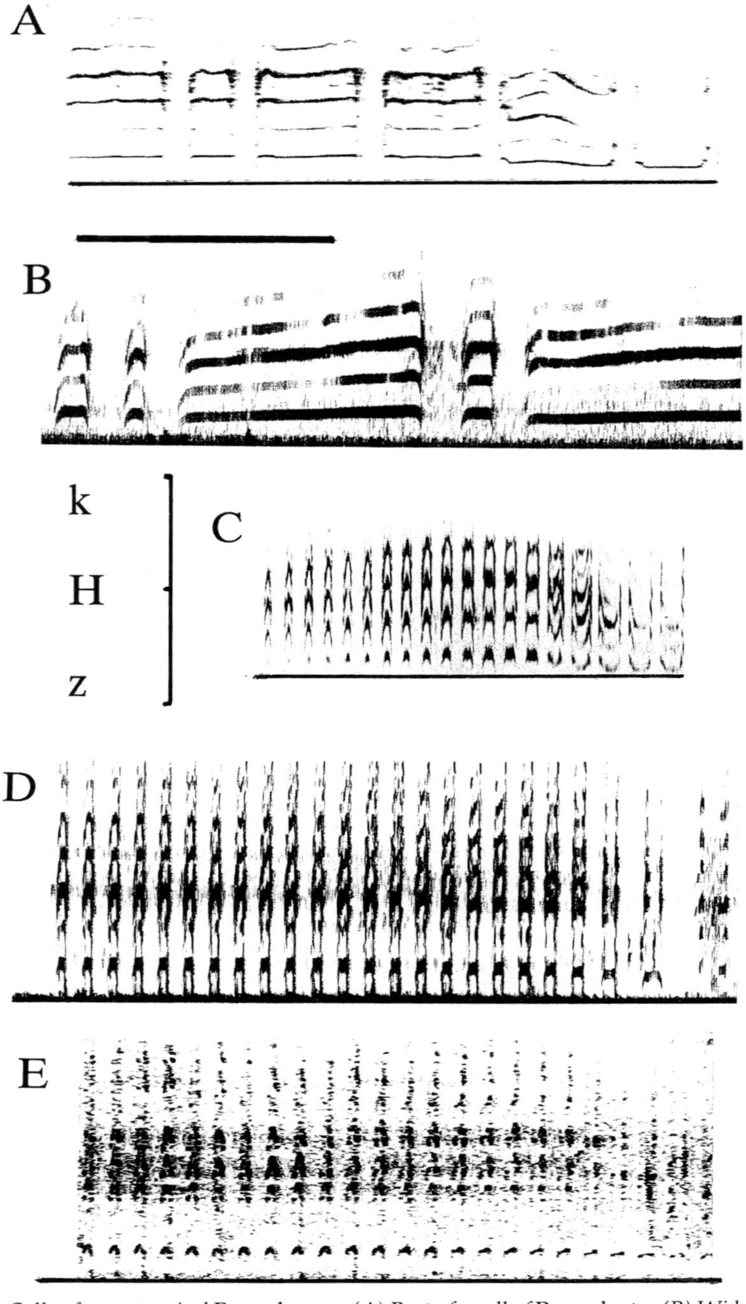

Figure 6.11 *Calls of some tropical* Pterodroma. *(A) Part of a call of* P. neglecta. *(B) Wide band-pass filter sonagram of part of a call of* P. neglecta *showing syllables plotting as inverted 'U's develop into longer ones. Note difference in energy distribution from (A). (C) 'Ti-ti' type call of* P. arminjoniana*: the chevron-shaped pulses change to 'U's in the last five notes—perhaps in intake of breath—wide-band sonagram from Gill* et al. *(1970). (D) Wide, and (E) narrow-band calls of different* P. alba.

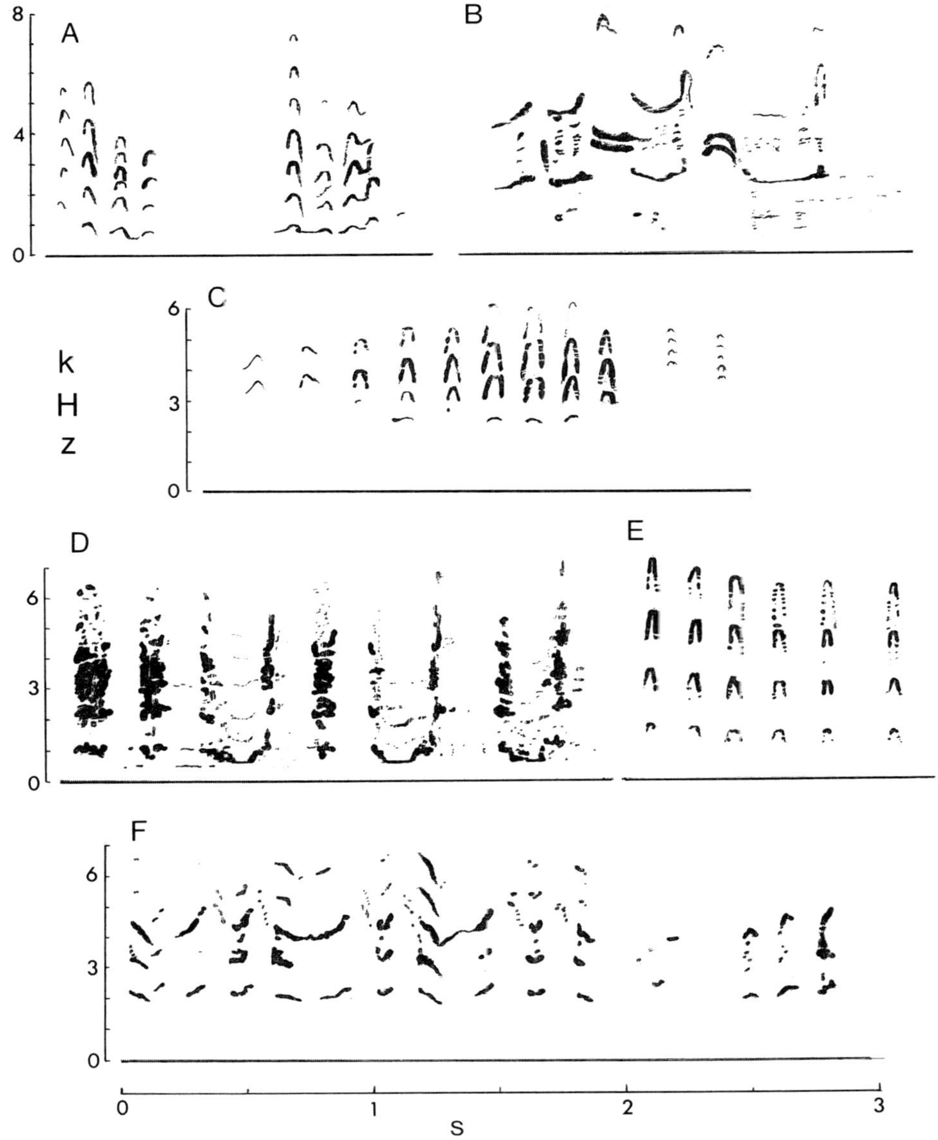

Figure 6.12 *'Ti-ti' type calls from flying birds. (A) Two bursts from* Pterodroma hypoleuca, *probably from the same bird. (C) From* P. inexpectata *flying towards the microphone—pitch increases. (E) From* P. hypoleuca *flying away, Doppler-shifted downwards. (B), (D) and (F) complex chattering cries from different* P. lessonii, *all elaborations of the 'Ti- ti' call.*

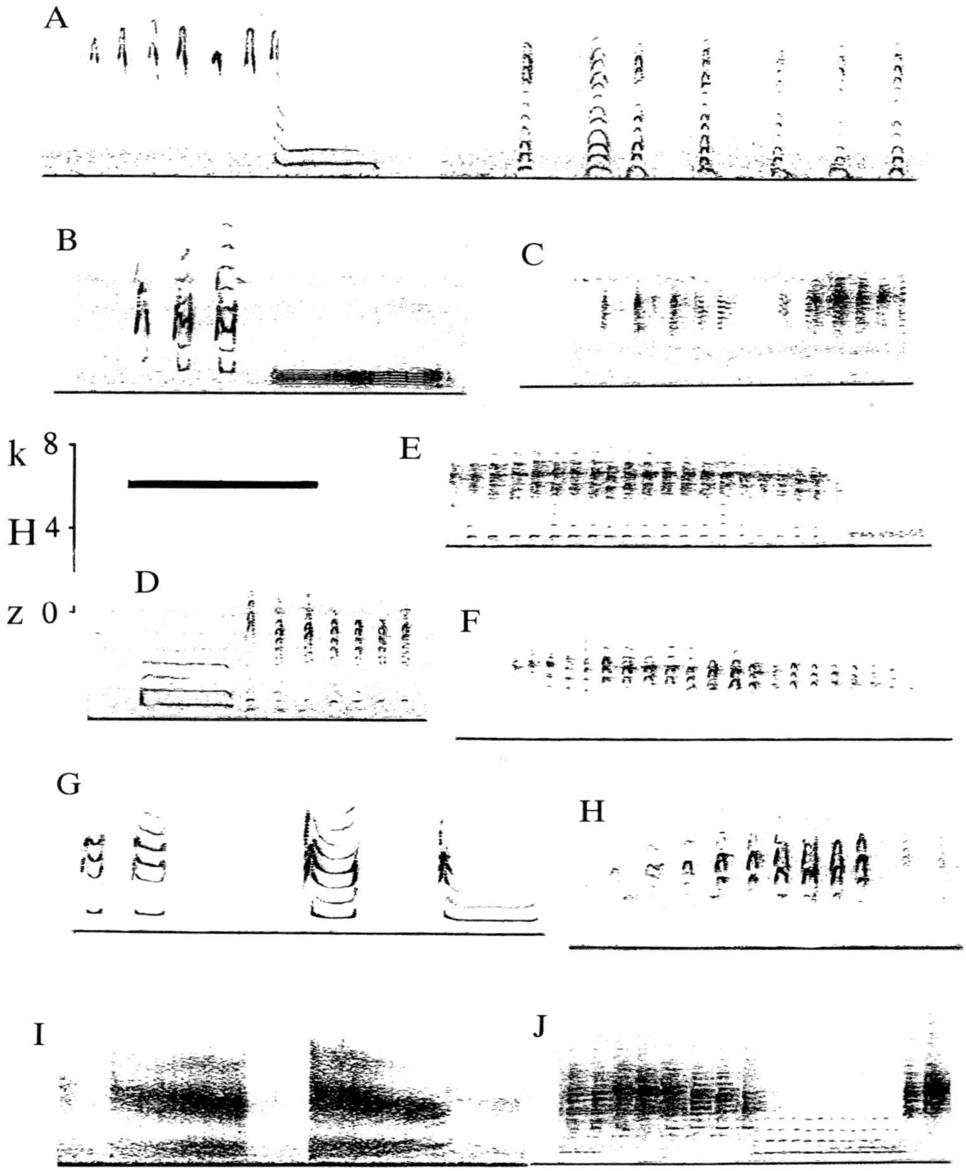

Figure 6.13 *Vocalizations of* 'Cookilaria' *petrels on the wing. Bar is 1 s. (A)* Pterodroma axillaris*: showing high-pitched 'Ti-ti' calls, a low 'Quor' type slur and lower pitched 'Ti-ti' notes with many harmonics, perhaps from another bird. (B) Complex calls of* P. leucoptera brevipes. *(C) Thick chatter calls of* P. l. leucoptera. *(D) 'Quor-ti-ti' of* P. l. brevipes. *(E)* P. cookii*—thick rasping cackle ending in low crooning note. (F)* P. cookii*—clearer 'Ti- ti' call approaching and flying away from the microphone. (G) Part of a series of calls from* P. l. brevipes*: the second pair show how short notes can be drawn out. (H) A burst of 'Ti-ti' calls from* P. inexpectata *flying towards the recorder and Doppler-shifted. (I) Two calls from* P. hypoleuca *mainly of 'white noise' and probably from the same bird. (J) Thick cackles of* P. cookii *and a low, pulsed drumming from a second bird.*

with each syllable (Warham *et al.*, 1977). With some species the harmonic structure results in a rather soft call with a liquid timbre. The quality also tends to alter during delivery, becoming louder and of higher pitch and then falling again (e.g. Fig. 6.11C). The range of frequencies may be considerable, often up to 8 kHz, as in Fig. 6.11D.

The basic arcuate formant of the notes may be considerably developed, sometimes with frequency modulation, or, as in Fig. 6.10D, the arc-shaped notes are drawn out and only the final one is a typical 'Ti'. In Fig. 6.11C–E the notes are of the usual arching type with harmonics, but given so rapidly as to sound like a twitter to the human ear. Yet again, each note may be modified as in the sonagram of Fig. 6.10F where the call began with a whoop, changed to a 'Ti-ti'-type twitter, and ended with another whooping sound.

The 'Ti-ti' call is that most commonly used from the air often as single notes or in brief volleys. Figure 6.12A shows typical bursts from *P. hypoleuca*. Because gadfly petrels travel fast, Doppler shifts may be noticeable (Figs 6.12C,E & 6.13H).

Most complex cries generally seem to be elaborations of the 'Ti-ti' call, for example in Fig. 6.12D where the notes plot as Us, as if the singer began each with open mouth, then closed it, and finally opened it again.

2 *The 'Kuk-u-er' call*

This is composed of a series of syllables of unstructured 'white noise' spread across a wide band of frequences, the last syllable possibly inhalatory. In the duets of Fig. 6.14B and D, one *P. hypoleuca* used the harsh 'Kuk-u-er', its partner the 'Ti-ti' sequence, whereas in Fig. 6.14C one gave the 'Kuk-u-er', the other a low-pitched 'Churr' (see below).

Although usually trisyllabic, single 'Kuk' notes are often given, particularly at the start of a song, as if the singer is warming up. Usually only one or two calls are given at a time over about 2.4 s, occasionally a burst will last 5 s, and the call can be varied and elaborated in length and frequency structure. Figure 6.14E records a change in this call during a sequence by one Bonin Petrel.

The 'Kuk-u-er' call has been noted only from *P. hypoleuca* and *P. inexpectata*, but may have been overlooked elsewhere.

3 *The 'Moan' call*

This is distinctive and low-pitched with notes held at constant frequency for up to several seconds—the 'drone' of Tomkins and Milne (1991) in *P. phaeopygia*. Analyses from five species are given in Fig. 6.15. Each call consists of an extended note seldom exceeding 1 kHz, often with a few harmonics. Delivered from the wing, the 'Moan' sounds particularly mournful because the frequency tends to drift down slightly. The description by du Tetre (1654) of the call of *P. hasitata* as 'lugubrious', is very apt. Otherwise the frequency is usually remarkably steady. Occasionally, as in Fig. 6.15B there may be frequency modulation. The low notes of 'Moan' are often broken up by high-pitched 'ti' or 'si' notes as in Figs 6.15B and D, and many start or end with a sudden fall or rise of frequency producing a whip-like sound.

The call of *P. ultima* (Fig. 6.15I) is extraordinary for the steadiness with which the

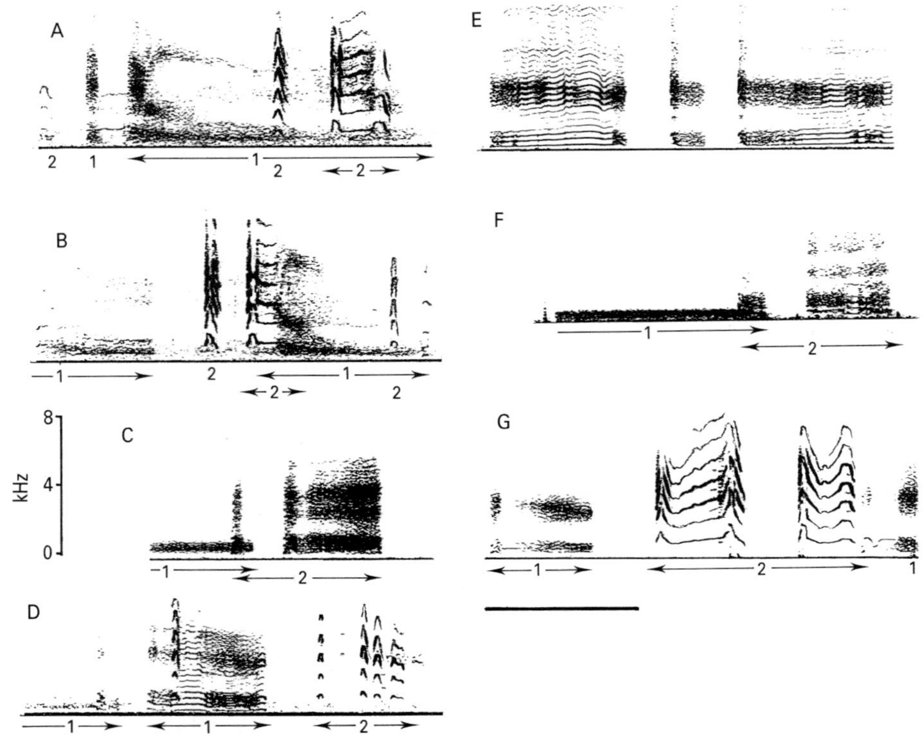

Figure 6.14 *Vocalizations of* Pterodroma hypoleuca. *Bar is 1 s. (A) A pair circling overhead, one (male?) with a harsh, unstructured call (1), the second (female?) an elaborate extended 'Ti-ti'. (B) The same pair as in (A) with bird (2) evidently more excited, its calls exceeding 8 kHz rapidly fading as they dash by. (C) The 'Purr' of bird (1) stopped as its partner voiced the 'Kuk-u-er' call. Courting on the ground. (D) Bird (1) used a harsh 'Kuk-u-er', its partner (2) a single and then a short series of 'Ti' cries. Birds on the ground. (E) Low 'Kuk-u-er' type call showing change in acoustic structure in successive notes from an excited bird. Frequency bars exceed 8 kHz. (F) One bird uses a very low pitched 'Churr' call (1); its partner (2) replies with the 'Kuk-u-er'. (G) Two grounded birds, one (1) using 'Kuk-u-er', the other complex, clear, and high-pitched notes.*

principal note is held. These notes can be short or may last for 2 or 3 s and *P. ultima* can call for about 12 s without apparently taking breath, although such songs may consist of several 'Moans' linked by shorter sections. The frequency is often slowly modulated creating a wavering like the mournful hoot of a Tawny Owl *Strix aluco* (Williams, 1960); sailors described that of *P. cahow* in similar terms back in the early 17th century (Butler, 1882).

That the lugubrious cry of *P. mollis* is used by its northern relatives is shown by the sonagrams of *P. madeira* and *P. feae* in Fig. 6.15F–H, the voice of *P. feae* on the Desertas being described as 'a long wail ending in a hiccup' (Jouanin *et al.*, 1969).

4 *The 'Quor-wik' call*

This consists of a low-pitched drawn-out note followed or preceded by one or more staccato 'ti' or 'wik' notes. Such combinations have been syllabized as 'gor-wik',

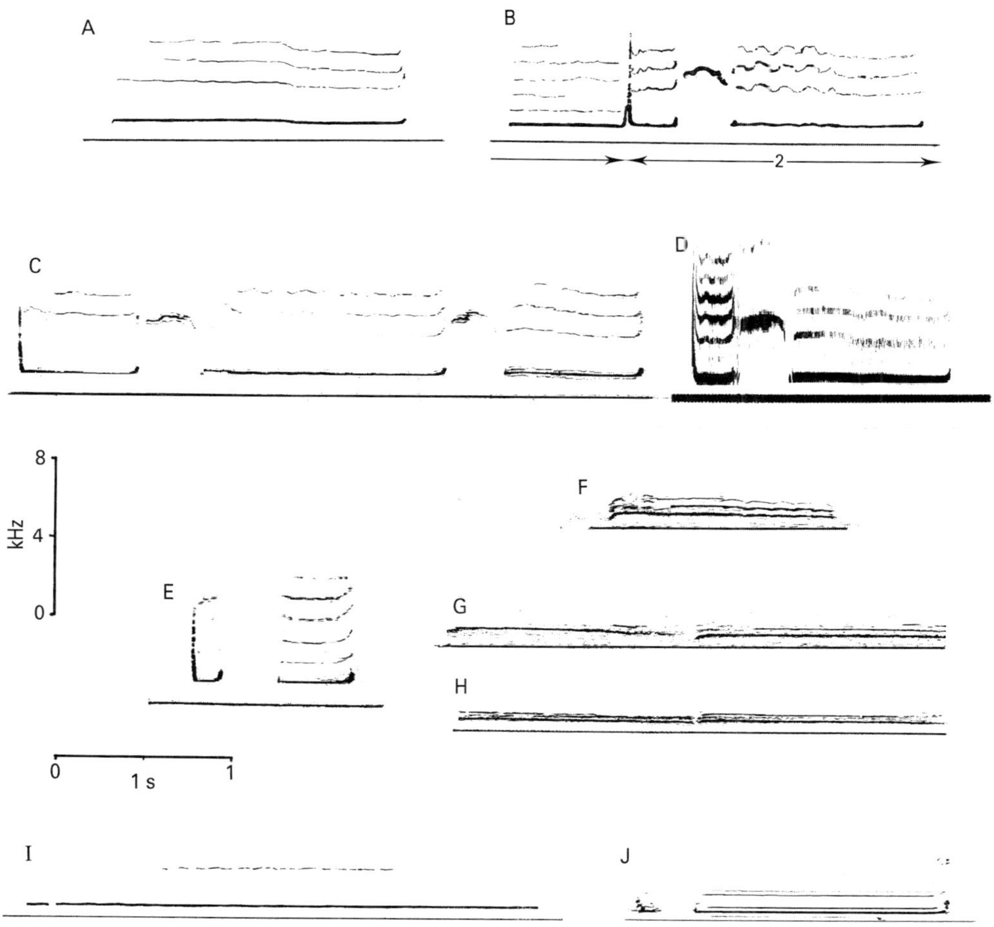

Figure 6.15 *'Moan' Calls from flying birds. Note the characteristic upturn in frequency at the end of many calls. (A) A single call from* Pterodroma mollis. *(B) Two* P. mollis *in a chase. The second cuts in with an abrupt downwards sweep of frequency immediately the first finishes, and injects a clear 'Ti' call between its two longer syllables. (C) Flight calls of a different* P. mollis. *(D) Wide-band sonagram of* P. mollis, *a clear three-syllable call with a central 'Ti' and starting with a rapid downwards glissade. (E)* P. mollis, *a two-syllable call again with an initial downwards frequency sweep. (F) Low call of* P. madeira. *(G) Two calls of* P. madeira, *the first slightly higher pitched (H) Two low calls of* P. feae, *the first slightly the lower. (I) The 'Moan' of* P. ultima *on the ground—a long clear tone at very constant frequency, although often with some slight modulation. (J) Low calls of* P. externa: *as in* P. ultima *often drawn out over 2–3 s.*

'quor-ti-ti' etc. (Grant *et al.*, 1983; Bretagnolle & Attie, 1991). The emphasis is on the first syllable which is delivered with a distended throat.

Examples in Fig. 6.16 are from the small *'Cookilaria'* species, but similar cries are heard from *P. macroptera* and *lessonii* (Warham, 1956; 1967a), *P. hasitata* (Fig. 6.18E), *P. baraui* (Bretagnolle & Attie, 1991) and *P. inexpectata*. The Hawaiian 'Ua'U for *P. phaeopygia* probably also refers to it. Grant *et al.* (1983) noted that in aerial

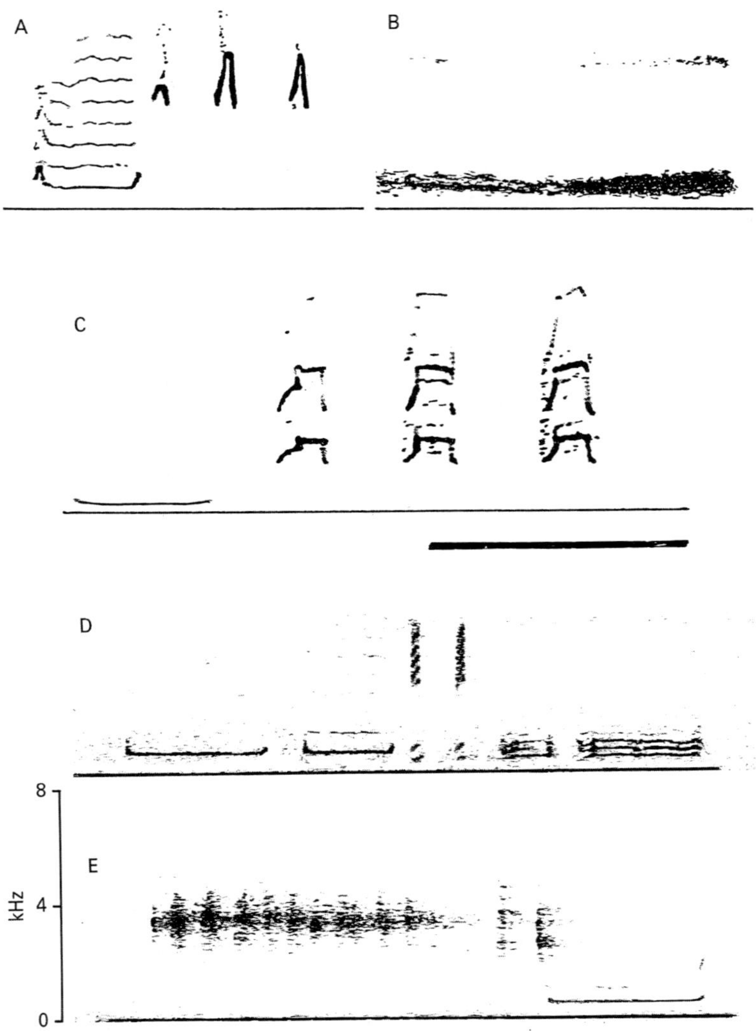

Figure 6.16 *The 'Quor-wik' call. Bar is 1 s. (A) and (C) from* Pterodroma nigripennis *on the ground—'Quor, wik,wik,wik'. (B) A low, slurred 'Moan' of* P. hypoleuca *on the wing. (D) 'Quor,quor,ti-ti' followed by two complex notes—*P. leucoptera brevipes. *(E) A flight call of* P. cookii *ending with a low 'Quor'.*

pursuits one *P. hypoleuca* used a low, slurred moan lasting 2–3 s and ending either quite abruptly or switching to a high-pitched 'Ti-ti'. The low calls were sometimes used by themselves as in Fig. 6.16B. These notes sound very like the 'Quor' calls of other species but, as the analysis shows, consist of 'white noise', so may be a different call with different functions.

The first syllable of this cry shows obvious similarities with the 'Moan', being of low and constant frequency as well as often being associated with sharp, high-pitched notes (cf. Fig. 6.16A), so that the 'Quor-wik' may prove to be variants of the 'Moan'.

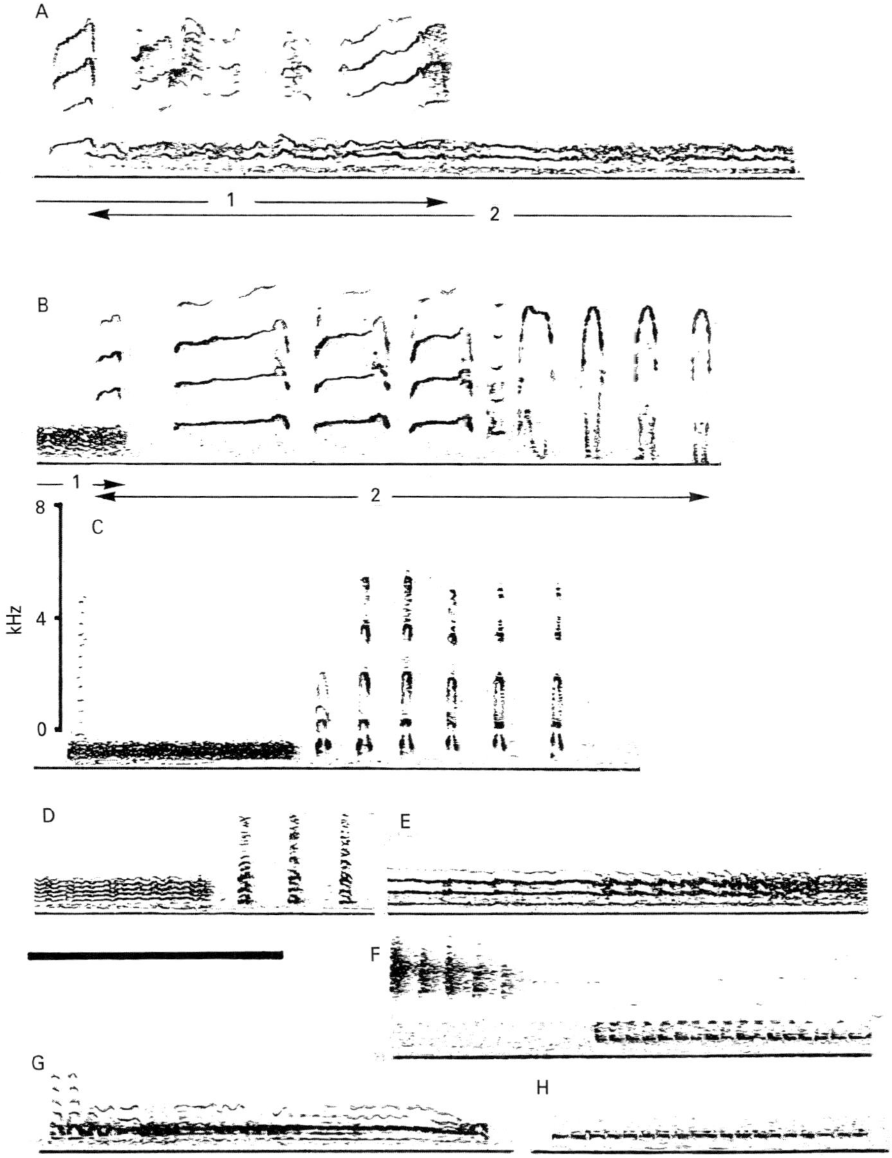

Figure 6.17 *The 'Purr' call. Bar is 1 s. (A) Two* P. nigripennis *on the ground. (2) starts to 'Purr' while (1) ends an elaborate 'Ti-ti' style sequence. (B) Two grounded* P. nigripennis*: as (1) ends its 'Purr' note the other responds with a loud, clear 'Ti-ti' vocalization. (C) A lone* P. hypoleuca *on the ground changes from 'Purr' to a 'Ti-ti' call. (D) Another example similar to and from the same species as (C) but the 'Purr' shows fine frequency bands. (E) A section of a 'Purr' call from* P. nigripennis*: note the harmonics as in (A) and (D). (F)* P. cookii*: the end of a flight call and part of a low 'Purr' call from a bird on the ground. (G) A low call from* P. l. leucoptera *with frequency modulation apparently the equivalent of the 'Purr'. (H) Another pulsed 'Purr' from* P. cookii *singing from a burrow.*

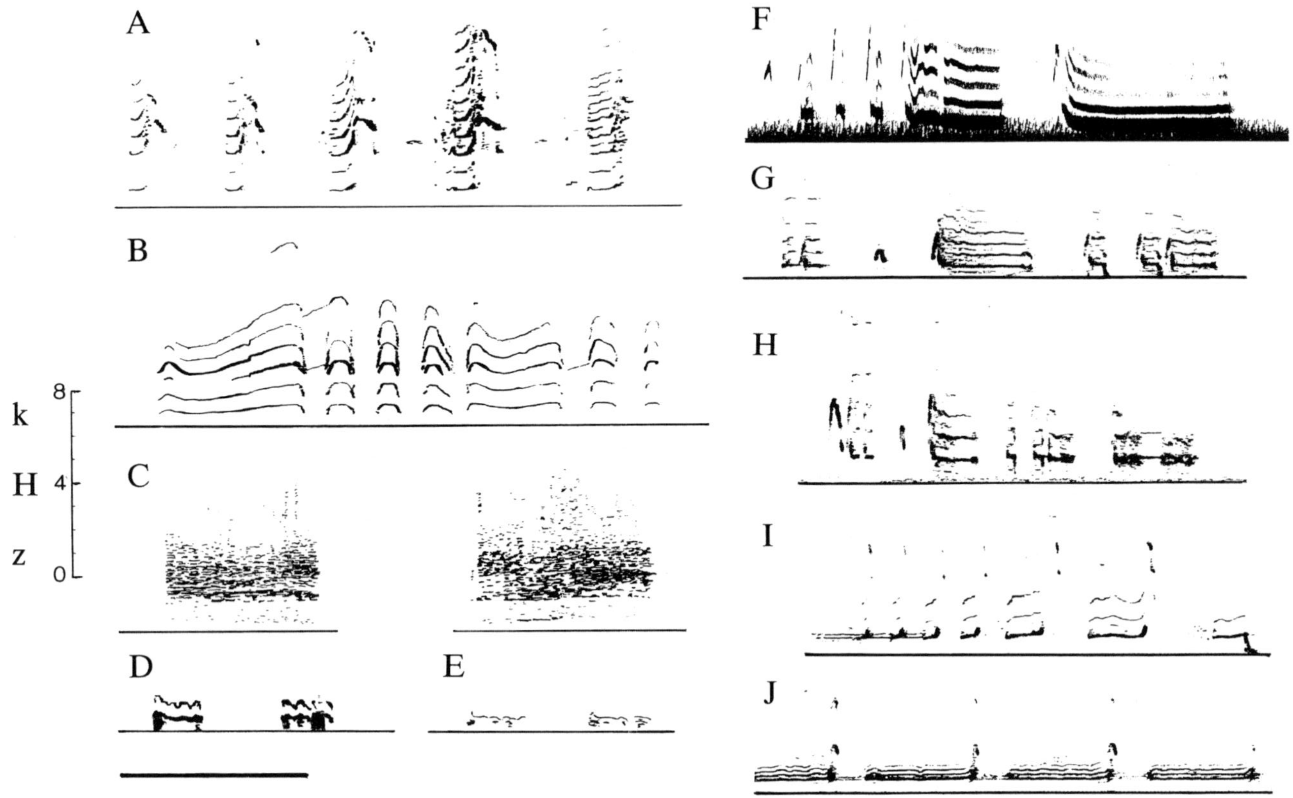

Figure 6.18 *Miscellaneous vocalizations from gadfly petrels. Bar is 1 s. (A) and (B) Elaborations of the 'Ti-ti' call as used by (A)* Pterodroma macroptera *and (B)* P. ultima, *both from the ground. (C) Two alarm calls of* P. arminjoniana. *(D) and (E) Very quiet, low-pitched calls of* P. hypoleuca *from a burrow. (F) Wide band-pass filter sonagram of a complete 'clear' call from* P. leucoptera pycrofti *in a burrow. (G) Sonagram from* P. l. pycrofti *showing varied structure of syllables. Frequency scale should be doubled for this example, time scale halved—some harmonics reach 10 kHz. (H) Another sequence of ground calls from* P. l. pycrofti. *(I)* P. hasitata: *note how low-pitched syllables lengthen as call proceeds. (J) Low, 'Quor-wik' type calls separated by possible 'breath' notes, from* P. hasitata.

5 The 'Purr' call

This is a distinctive purring not unlike the churring of storm petrels (Section VIII.C below). A single 'Purr' may run for 6 s but is often virtually continuous, broken by brief pauses, as if the bird is catching its breath, each call then running 4–5 s. On sonagrams it shows as a band of low frequency noise extending over about 1 kHz but may also have three to five harmonically related bands of low frequency. The pulse rates with *P. hypoleuca* ranged from 80 to 190 s^{-1} (Grant *et al.*, 1983, fig. 3).

Figure 6.17 presents eight analyses: one (E) part of a 2.3-s long 'Purr' of *P. nigripennis*, had four harmonics about 0.4 kHz apart and with some frequency modulation. Figure 6.17A and B are from duets between a 'Purring' bird and another using variants of the 'Ti-ti' vocalization, the two types being so distinct that different sexes seem likely to have been involved. The 'Purr' of *P. hypoleuca* is very similar, but in Fig. 6.17C the singer switches to a series of 'Ti-ti' notes, so that in this species at least, these calls are not confined to opposite sexes. Furthermore, in Fig. 6.14F, while one *P. hypoleuca* was 'Purring', its partner gave the 'Kuk-u-er' call.

The call has been heard from at least three *'Cookilaria'* species singing on the surface or from burrows. It is not loud and, if below ground, the bird is inaudible a few m away. Similar calls have been described for *P. longirostris pycrofti* (Fleming, 1941; Bartle, 1968): they may prove characteristic of *'Cookilaria'* species.

6 Threat and distress calls

Threat calls, mainly 'Kuk-u-er', are heard most often during disputes over burrows, distress calls when petrels are fighting one another, are attacked by skuas, or hauled from burrows by man. They are screams that show as unstructured sounds involving many frequencies (e.g. Fig. 6.18C).

7 Other vocalizations

Some calls are difficult to place in the onomatopoeic categories used here. This is partly because the more simple themes like the 'Ti-ti' call may be elaborately developed. In Fig. 6.18B, for example, a fairly typical call of *P. ultima* is expanded at the start and end into loud whooping cries, and in Fig. 6.18A the arcuate formants of a 'Ti-ti' call of *P. macroptera* reach high frequencies with many harmonics and with a sudden change in the last note with the harmonics now at 400-Hz intervals.

Other quiet calls from below ground are of unknown significance; Fig. 6.18D and E provide examples.

8 Sexual dimorphism in calls

Several workers have supposed that in *Pterodroma* the voices of the sexes differ (e.g. Wingate, 1964; Warham, 1967a; Jouanin & Gill, 1967; Warham *et al.*, 1977; Bretagnolle & Attie, 1991). None has established this unequivocally on sexed birds. From direct observations of birds that had copulated, Grant *et al.* (1983) deduced that *P. hypoleuca* using the 'Purr' and 'Ti-ti' calls were males, those using 'Kuk-u-er', females. In aerial duos they thought that it was the males that used the low frequency slur and 'Ti-ti' calls, the females the harsher 'Kuk-u-er' or its variants (e.g. Fig. 6.14).

Tomkins and Milne (1991) recognized two series: 'sweet' calls where the 'Moan' was a pure tone and the 'ti-ti' with clear harmonics, and 'coarse' calls of similar pattern but with little harmonic structure. They concluded that the 'sweet' calls were used by the males, the coarse ones by the females.

IV The Blue Petrel *Halobaena caerulea* and the prions *Pachyptila*

A *Nuptial displays*

All are highly social, displaying by night on the surface and within burrows. Pairs and trios are highly mobile, so that groupings change frequently, duet and allopreen. Copulation by *Pachyptila* seen on the surface is typically procellariiform except for a high-pitched whistle apparently given by female *P. desolata* and *P. belcheri* (Tickell, 1962, p. 16; Strange, 1980).

According to Bretagnolle (1990a) male *Halobaena* and prions call from burrow entrances and potential partners reply from the air before alighting, most flight calling being from females.

B *Aggression and defence*

Prions are active in defence of their burrows. Established pairs repel intruders by loud, grating, staccato calls (pers. obs.; Harper, 1976; Strange,1980). More aggressive birds are threatened by territory holders advancing with wings held from the body or even vertically and bill thrust forwards. Fighting is accompanied by loud, high-pitched screams as the birds bite and grasp each others' bills and napes. With *P. turtur* the nest owner invariably won and non-breeders were evicted from burrows by established birds (Harper, 1976, p. 367).

C *Vocal repertoire (Fig. 6.19)*

The similarity of the calls of Blue Petrels and prions was noted by Paulian (1953, p. 183) and others, the principal vocalization being described as a cooing often built from rather staccato notes of low frequency, as suggested by the 'poor popper popper popper pop' of Richdale (1944b) for *P. turtur*. Calling comes from overfliers, from those on the surface and below it and, like other burrowing petrels, both will call by day in response to a footfall, at least during the prelaying stage. Strange (1980) listed three types of cooing notes used by *P. belcheri* on the ground or air and harsh cries accompanying fighting.

In all prions the main call is built from a succession of notes to a rhythm that may characterize each species and sex, and maybe act as premating isolating mechanisms where, as often obtains, several species breed sympatrically. The acoustic structure and disposition of the harmonics, if any, vary from bird to bird and may also change in the course of a call (e.g. Fig. 6.19F).

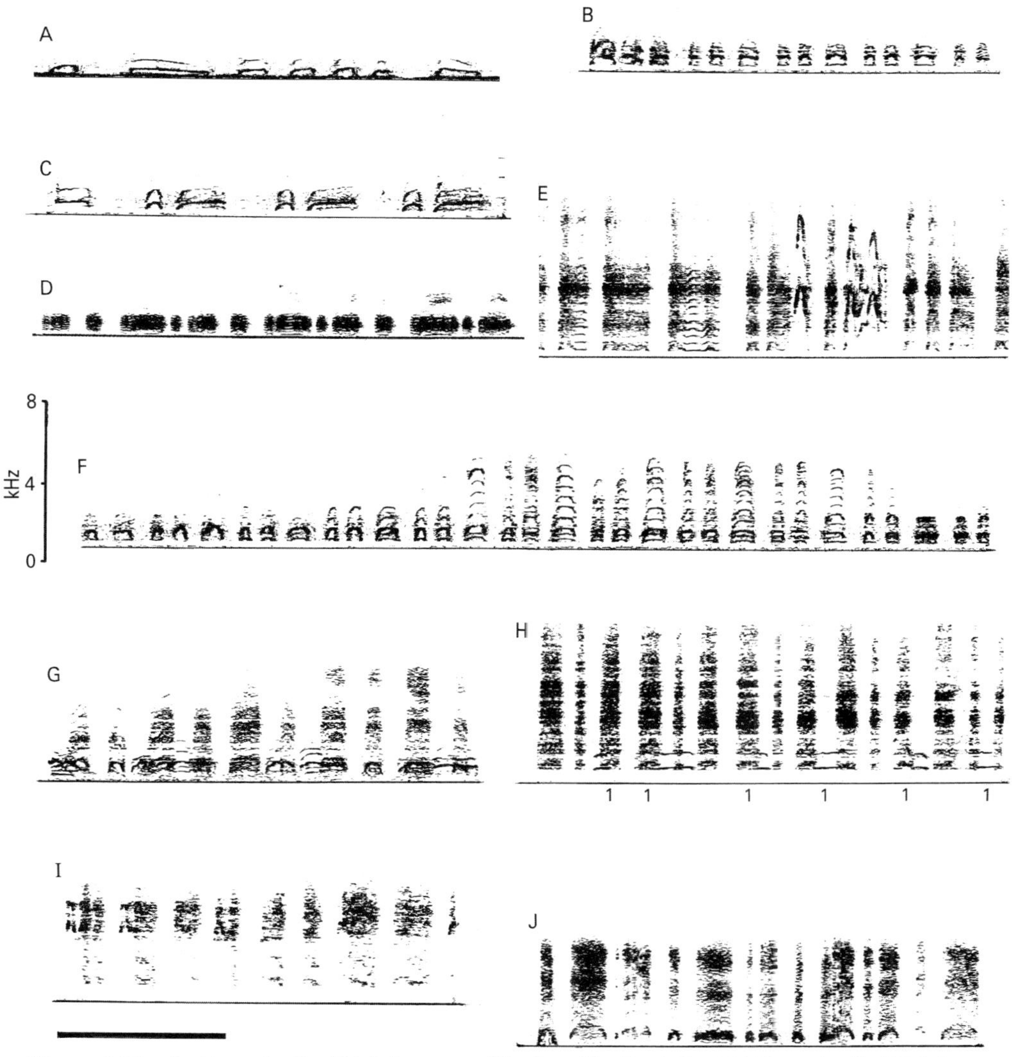

Figure 6.19 *Sections of calls of* Halobaena *and* Pachyptila. *Bar is 1 s. (A) Normal call of* Halobaena, *a low-pitched cooing—probably by a male. (B) Solo courtship song of* P. turtur *with characteristic three-syllable dot–dot–dash pattern and well-defined harmonics, probably from a male. (C) Solo courtship song of* P. turtur. *This two-note call was probably that of a female. (D) Main call of* P. vittata *with well-defined temporal pattern but mainly of 'white noise'. (E) Another song from* P. vittata *pitched much higher than in (D) and with some fine harmonics. The high-pitched acuate traces are from a chick. (F) Part of a long call of* P. turtur, *probably male, showing increase and decrease in amplitude and pitch with changing level of excitement. Dot–dot–dash pattern is retained throughout. (G) Duet of* P. turtur. *One bird uses harsh two-syllable, high-pitched notes, the other clearer, low-pitched ones, that never exceed 2 kHz. (H) Another* P. turtur *duet. As in (G) one calls very loudly and harshly, the other (1) gives a clearer call with but one harmonic and the fundamental at c. 1 kHz. (I) Hoarse alarms of* P. turtur *on being approached: high-pitched calls of a chick precede the first two notes. (J) Rhythmic calls of a well-grown* P. vittata *chick.*

Bretagnolle *et al.* (1990) figured audiospectrographs of *P. belcheri, P. desolata, P. salvini* and *P. vittata* and showed that the main calls of the sexes are distinct, both the rhythm of the notes and their sequences varying according to the sex, and Genevois and Bretagnolle (1995) found that the songs of male *P. belcheri* start with a short note, those of the female with a long one.

In the common call of *P. turtur* some notes may have a few harmonics and much 'noise', but those in many calls have purer tones and fundamental frequencies between 0.8 and 1.3 kHz, producing clearer and more piping sounds. The male version is typically composed of one long and two short notes (Fig. 6.19B) rapidly repeated, the mean call length being about 1.16 s. That of the female is a disyllabic 'Errr-err' (Fig. 6.19C) of a long and short syllable lasting 1.64 s, significantly longer than that of the male. Such calls, given from the wing or the ground, may be repeated for several minutes and are highly contagious to nearby birds. The voice of *P. vittata* is of a similar pattern to other prions but is more raucous (Fig. 6.19D,E).

Threat calls used to intruders are explosive and harsh, and Tennyson (in Marchant and Higgins, 1990) described a high-pitched 'Pihipihi' distress call of *P. vittata* and *P. turtur* when handled that was also directed by *P. turtur* at an intruding Mottled Petrel.

V Bulwer's and Jouanin's Petrel *Bulweria bulwerii* and *B. fallax*

Presumed non-breeding *Bulweria bulwerii* have been seen on Laysan Island sitting on the ground in 'clubs' of at least 1000 birds (Ely & Clapp, 1973) and Wetmore (in Amerson & Shelton, 1976) watched birds in pairs, resting, reaching towards each other and, with distended throats, giving their comical barking calls, generally rendered as 'Woof' (Fig. 6.20).

Bulwer's Petrels also call from their nesting recesses but not from the air. Wetmore's statement that the barking was used by both sexes was confirmed by James and Robertson (1985a). From their analysis of the calls of 18 males and 24 females sexed by cloacal inspection they described three versions of the 'Woof' call (Fig. 6.20) but detected no clear sexual differences. This is surprising as *B. bulwerii* remains nocturnal even on Pacific islands lacking predators, and Bretagnolle (pers. comm.) classifies all the calls as sexually dimorphic based on frequency and tempo. The call of *B. fallax* is a reiterated 'noisy grunt' (Fig. 6.20E), quite like the bark of *B. bulwerii*.

According to Thibault and Holyoak (1978), *B. bulwerii* has a strongly nauseous body odour differing from that of *Pterodroma* but only noticeable in a short period before the egg is laid: the involvement of some sexual pheromone cannot be ruled out and needs investigation.

VI The Shearwaters *Procellaria* and *Calonectris*

A *Nuptial displays*

Nuptial activity for both genera includes bill fencing and mutual preening, evidently no different from that of other tubenoses. Shoemakers *Procellaria aequinoctialis*, Grey

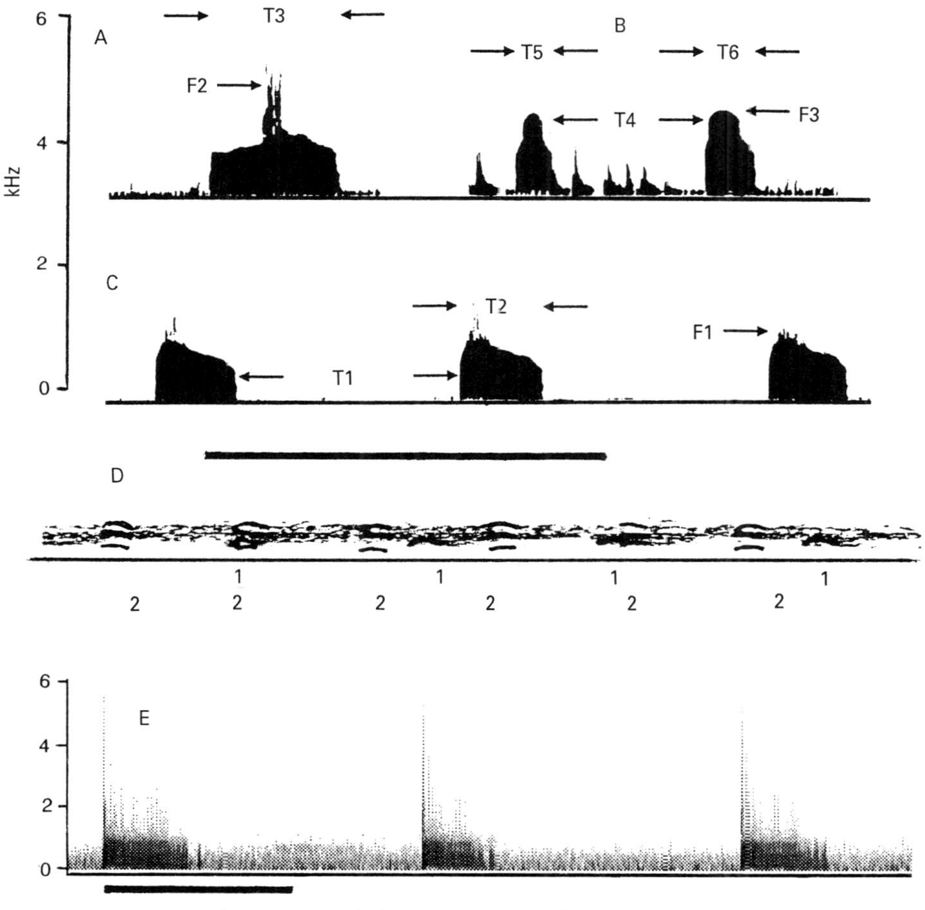

Figure 6.20 *The 'Woof' of* Bulweria bulwerii *(A)–(D) and* B. fallax *(E). Bar (A)–(D) is 1 s, in (E) 2 s. (A) 'Single Call'. (B) 'Double Call'. (C) 'Repeat Call'.* T_1–T_6, *temporal, and* F_1–F_3 *frequency variables analysed. Wide band analyses by James and Robertson (1985a). (D) A section from a duet of regularly repeated notes: one bird calls four times and its partner six times, the latter perhaps the female. Despite the slurred traces that explain the 'thick' sound of the call, there are underlying harmonics. (E) Reiterated gruff note of* B. fallax *recorded by Nanette Seto.*

Petrels *P. cinerea* and Cory's Shearwaters *Calonectris diomedea* may be active on land by day, particularly towards evening and Murphy (1936, p. 645) described pairs of *P. aequinoctialis* sitting near burrow entrances sunning themselves and 'alternately or simultaneously nibbling with their beaks and uttering their shrill warbles'. Father Schmitz in his account of the Cory's Shearwater hunt of 1892 (Schmitz, 1894), recounts that 'the couple rub their bills together in the same way as pigeons do'. Schmitz (1894) and Lockley (1952) saw the preening of females by males of this species leading to coition. All *Procellaria* and *Calonectris* species defend their nest sites against conspecifics and fighting is common at high-density colonies, sometimes resulting in death. Their chicks may spit stomach oil but this is ill directed.

In flight by day during chick rearing *P. aequinoctialis* are harried by skuas to disgorge their stomach contents, but grounded adults resist skua attack vigorously and many apparently escape, although others are killed (Moors, 1980). In contrast, during the early prelaying stage, overflying *P. cinerea* were not attacked by skuas in the daytime possibly because they had empty stomachs (Warham & Bell, 1979).

B *Vocal repertoire*

The vocalizations of *P. aequinoctialis* and *P. cinerea* have been described by Brooke (1986); of all four *Procellaria* by Warham (1988b) and of *C. diomedea* by Bretagnolle and Lequette (1990).

Birds of both genera are loud mouthed and produce inhalatory and exhalatory sounds. The accompanying muscle contractions may be obvious, the lower amplitude and more gasping 'noisy' sounds of inspiration often being deductible from spectrographs (e.g. see Fig. 6.22A).

1 Procellaria

a. The Grey Petrel P. cinerea. This appears to have a repertoire of but two components, the 'Moan' and the 'Bleat'. These create a distinctive and far-carrying chorus during the subantarctic autumn night, when newly arrived birds call from burrow entrances and, while perched on rocks or tussocks, swing their heads in wide arcs, bills fully open, 'hosing' the sound across the breeding terrain (Fig. 6.21).

The 'Moan': at Antipodes Island this comes from single birds or from pairs, but is audible only at close range. Figure 6.22A,B and E are typical analyses, illustrating the variation between birds in the disposition of energy and frequency bands and the amount of frequency modulation. The overall effect is of a shearwater-style crooning.

The 'Bleat' (Fig. 6.22E–F) is usually preceded by one or two wheezy 'Moans', the last being inhalatory, one or two short notes of intermediate length (L) and then the explosive, exhalatory 'Bleat'. Brooke (1986) called this the 'rattle cry' but it is not hard enough for a rattle, and Hutton's (1865) simile of the bleat of a lamb is nicely descriptive. It is based on notes only *c.* 0.4 s long, with strong frequency bands, separated by even shorter silences. These change to double pulses creating the impression that two birds are involved (Fig. 6.22B and G). The whole of the 'Bleat' sequence lasts 5–19 s, mean 9.4 s, $n = 37$, and it is difficult to see how a bird could sing for so long without a breath: presumably the inhalatory phase fills the lungs and air sacs and perhaps the switch to double notes marks a change to taking small breaths between each note.

Grey Petrels differ from other *Procellaria* in evidently not using bill clicks, their beaks during singing being held fully open. The tongue, however, may well play a role in sound production.

b. The Shoemaker P. aequinoctialis. The vocal repertory of this noisy bird includes a 'Clack' or 'Rattle' and sundry 'Groans' and 'Squeals'. The following data come from Warham (1988b).

Figure 6.21 *Recently arrived* Procellaria cinerea *advertising its whereabouts and availability. Photo: author.*

The 'Clack' or 'Rattle': the dominant call, given from the surface, consists of a series of staccato notes or pulses lasting about 11 s (range 2–34 s; $n = 10$), e.g. Fig. 6.23A). Both sexes use it, as solos or in duets.

As Fig. 6.23A shows, the 'Rattle' is built from two types of notes, both of very short duration, *c.* 0.03 s, spread over the frequencies and very precisely timed in respect of one another. Most of the sound energy was in note (b) and note (a) had the form of a click. The bolder trace appears similar but it is not precisely perpendicular and shows much amplitude modulation. These doublet notes were timed at rates of 5–10 s^{-1}, about the same as reported by Brooke (1986).

Figure 6.23B and C show 'Rattles' of partners at their nest. Their calls sounded different and the sonagrams show why: that of the bird in Fig. 6.23C being of lower pitch, having a slower delivery, and made from triplets of two clicks and an abrupt and remarkably rapid downwards slur (cf. Fig. 6.23D). From its more aggressive nature the bird of Fig. 6.23B was most likely the male.

Some of these pulsed clicks may have been generated by bill snaps but Imber (pers. comm) saw them calling with wide open, vibrating beaks, which were not snapped shut. On being approached an incubating bird would start 'Rattling', quietly at first, but on coming closer the pitch and amplitude of the 'Rattles' increased (Figs 6.23B,C). The 'Rattle' of the presumed male of this pair was often preceded or ended by harsh groans or squawks, as in Fig. 6.23B.

Such notes of variable structure appear to be the basis for what Brooke (1986) termed 'wheezy' calls. At Marion Island he heard these mainly from burrows and given at 3–8 notes s^{-1}; that is, at lower rates than the 'Rattle' notes. Shoemakers at Antipodes Island were not heard using such notes as 'wheezy' songs, only as preludes or codas to 'Rattles'.

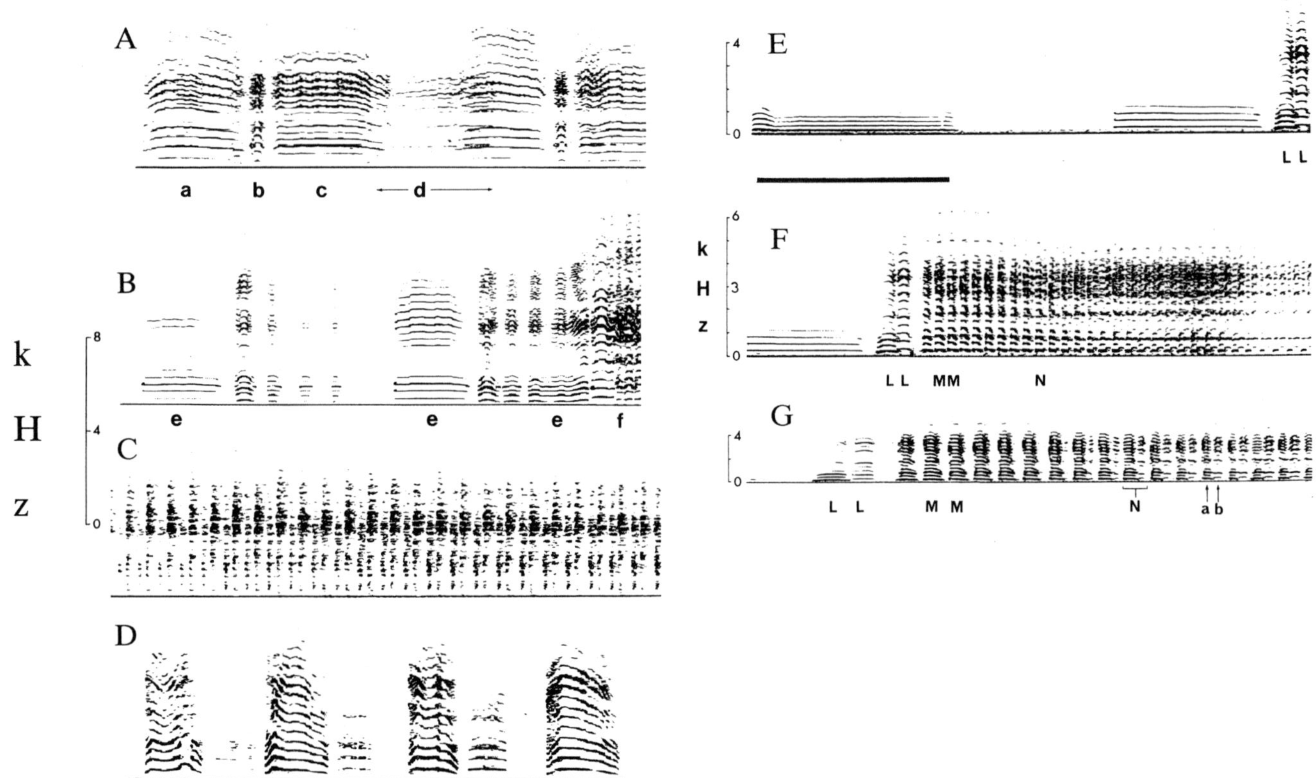

Figure 6.22 Procellaria *vocalizations. Bar is 1 s. (A) 'Moans' preceding the 'Bleat' call of* P. cinerea; *(a) inspiratory; (c) expiratory; (b) a short connecting note; (d) a call from another in the background. (B) Prelude to the 'Bleat' of* P. cinerea, *with three inhalatory 'moans' (e) and another short (expiratory?) note leading to the 'Bleat' which starts at (f). (C) Part of a 'Clack' sequence from* P. parkinsoni *singing on the ground. The notes come in doublets and the sound is spread across the frequencies. (D) Part of the threat or alarm of a grounded* P. parkinsoni. *The powerful high-pitched expiratory notes are separated by quieter, low-pitched inspiratory ones. (E) and (F) Two introductory, low-pitched 'Moans' before the 'Bleat' of* P. cinerea. *Two short notes (L) precede the fusillade of very short syllables of the 'Bleat' proper (MM). Just before (N) the notes develop into doublets as shown in the half-speed spectrogram of (G) where the switch from single to double notes (a) and (b) is clearer. From Warham (1988b).*

Alarm or Threat Calls: examples of the loud and varied squeaking cries made by a presumed male when approached on his nest are shown in Fig. 6.23F,G and H. When drawn out as in Fig. 6.23H the result was a piercing, frequency-modulated cry, the 'pig-like squeal' of Hagen (1952).

There are no records of Shoemakers calling in flight except when attacked by skuas.

c. Parkinson's Petrel P. parkinsoni. The voice of this petrel has not been detailed. Three types may be categorised:

'The Clack': the commonest call heard from birds on or below ground. It consists of strings of staccato pulsed notes which may be preceded by low, wheezy (inhalatory?) sounds (Fig. 6.22C). Here the notes came in doublets, at about 9 s^{-1} and extended over a wide frequency range, being not unlike the double pulses of the Shoemaker of Fig. 6.23A. 'Clack' calls were timed to last for 7–10 s.

'Throaty Squawks': sometimes high-pitched, these are used by grounded birds during disputes and include distress calls from birds being handled (Fig. 6.22D).

Aerial Calls: Parkinson's Petrels are not very vocal from the wing but some short 'Clack' sequences are heard of birds circling above the forests of the nesting grounds, as noted by Buller (1888, 2, p. 224).

d. Westland Black Petrel P. westlandica. Also breeding in heavy temperate rain forest, this bird has a varied repertoire of far-carrying vocalizations. Five types of call were identified by Warham (1988b).

The 'Quack': a succession of rather duck-like notes, used by both sexes, and given from the ground and air. The whole call lasted 13 s, range 5–45 s for 28 calls at 2–4 notes s^{-1}. These are simple expiratory cries, with harmonics and usually an inhalatory introduction, for example (X) in Fig. 6.24A, their acoustic structure often changing during a call and differing from bird to bird. During duets (which may develop to antiphonal singing) one bird tends to have a higher-pitched and clearer voice (Warham, 1988b, fig. 6).

The 'Jackass' Call: usually heard as the climax to a bout of 'Quacking' and sounding very like the hysterical song of the Kookaburra *Dacelo gigas* (Fig. 6.24B). The notes last only for about 0.05 s at 8–12 s^{-1}, and were timed to run for 2–56 s, mean 14 s, $n = 17$, but one bird called for 104 s before the recording tape ran out. The note rate may be changed, presumably in response to the actions of other birds.

The 'Jackass' Call is given by single birds and in duets when only one partner used it, deduced from other evidence to be the male, and Bretagnolle (pers. comm.) who recorded birds of known sex, confirmed that this call is restricted to males.

The 'Moan': low-pitched, drawn-out crooning cries are heard from birds in burrows and they sound very like *sotto voce* songs of shearwaters, but are of low amplitude and audible only at close range (Fig. 6.24C). Their significance is unclear.

Threat Calls: like Shoemakers, Westland Petrels often yell out when their burrows are invaded, their cries being of very varied acoustical structure, even in successive calls (e.g. Fig. 6.24D,E,F).

Aerial Calls: some birds call from the wing. Jackson (1958) thought that these were

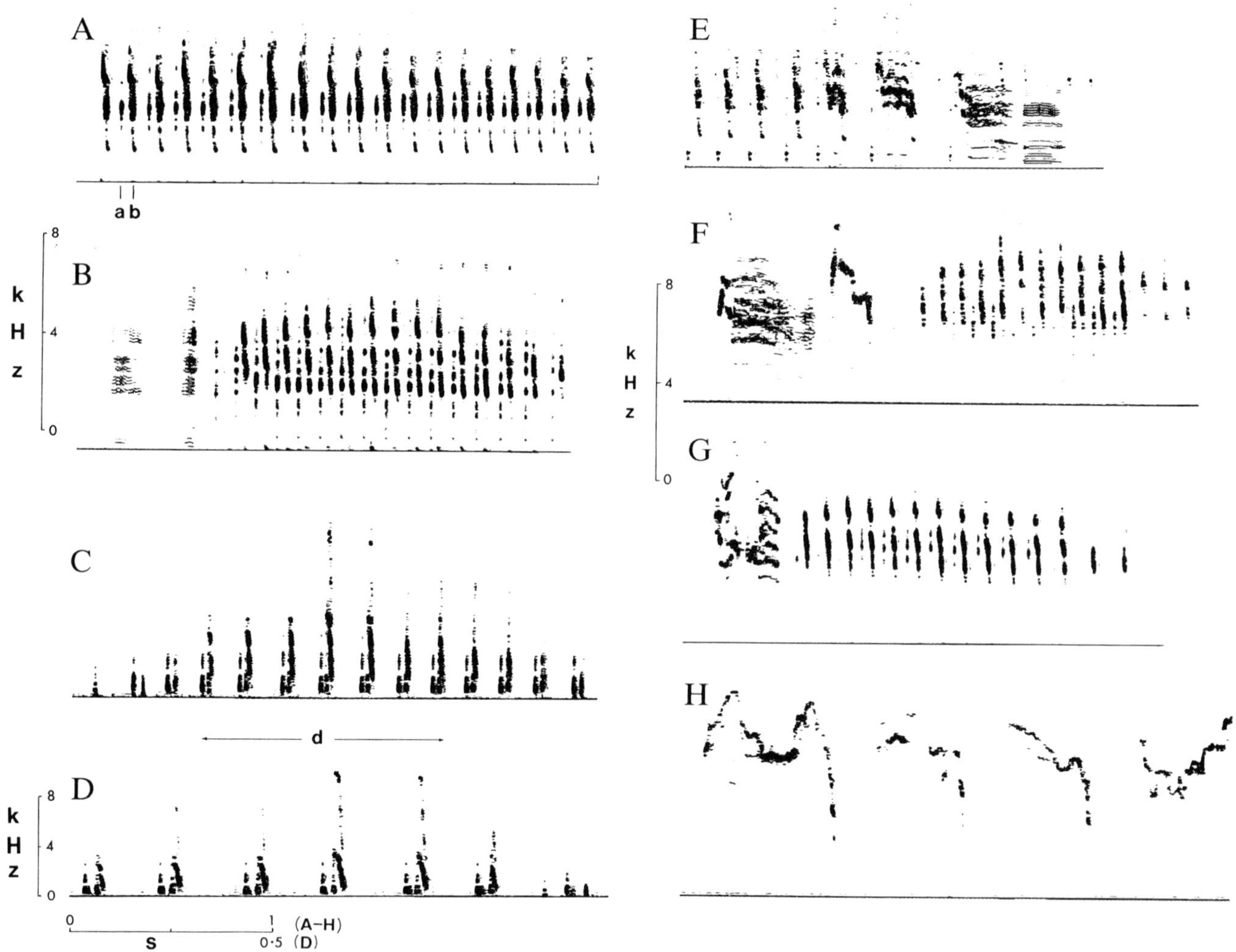
A
a b
B
kHz
8
4
0
C
d
D
0
1 (A–H)
0·5 (D)
s
E
F
G
H

answered from the ground, but this needs confirmation. The vocalization is usually a short 'Quack' with, occasionally, a deeper, repeated but short croaking sound.

e. Discussion. Even on the present incomplete data, the vocalizations of these four species can be seen to share some common features, for example having loud cries of rapidly repeated staccato notes, spread widely across the frequencies, and low-pitched inhalatory ones.

While such similarities support their classification in the same genus, some of the differences support their separation as full species. For example, calls of *P. westlandica* provide no evidence that this is but a winter-breeding race of *P. aequinoctialis*. Nor do the vocalizations suggest a very close link between *P. westlandica* and *P. parkinsoni*, but the low introductory moans and wheezes show similarities to the calls of *Puffinus*, with their marked inhalatory and exhalatory components.

Grey Petrels seem exceptional with the small repertoire, and neither I nor Brooke (1986) heard them call in flight.

Using playback to birds of known sex, Brooke concluded that the 'Rattle' of the Shoemaker, which when given from the surface was mainly used by males, served to indicate the caller's availability as a mate, and Antipodes Island birds singing in this way seemed to be advertising themselves and their burrows, although the same call was used in threat or alarm. Brooke deduced that the 'Bleat' of *P. cinerea* deterred intruders, whereas lone birds such as that in Fig. 6.21 are clearly in advertising mode.

Brooke (1986) also tested of the responses of incubating *P. aequinoctialis* and *P. cinerea* to playbacks of 'Rattle' and 'Wheezy' calls of the former and 'Bleat' calls of the latter. The results indicated that the birds recognized the voices of their partners but that there was no constant sexual response because test birds reacted as often to calls of their own as to those of the opposite sex (Table 6.1).

Brooke's conclusion that no commonly used vocalizations of these species are sex-linked is surprising in view of the sexual difference in the calls of *P. westlandica* and of the marked differences in the calls of members of pairs of *P. aequinoctialis* (Fig. 6.23). Of course, the forest-dwelling species, at times operating in almost complete darkness, need vocal clues more than *P. cinerea* and *P. aequinoctialis* breeding in open country and tending to come ashore before dark, but the matter needs re-examination.

Figure 6.23 *(A)–(D) 'Rattle' calls of* Procellaria aequinoctialis. *(E)–(H) Alarm or threat calls. (A) From a bird on the surface showing the doublet structure of the notes—a faint one (a) and a strong one (b). (B) Bird incubating in a cave. A 'Groan' prefaced the 'Rattle' which was of increasing and then decreasing pitch as the microphone was pushed closer and then withdrawn. (C) Call of the partner to the bird in (B), showing similar reaction to a close approach. (D) Half-speed analysis of section 'd' of (C) to show that the 'Rattle' notes consisted of two taps and a strong, extremely rapid, downwards glissade. (E) At the end of a 'Rattle' call, two high-pitched 'Squeals' followed by two lower pitched ones. (F) Two high-pitched 'Squeals' before a 'Rattle'. (G) The same bird as in (F) showing how the acoustic structures of alarm calls can vary even from the same bird. (H) Extended and penetrating 'Squeals' of the bird of (F) and (G) when closely approached while guarding its chick. From Warham (1988b).*

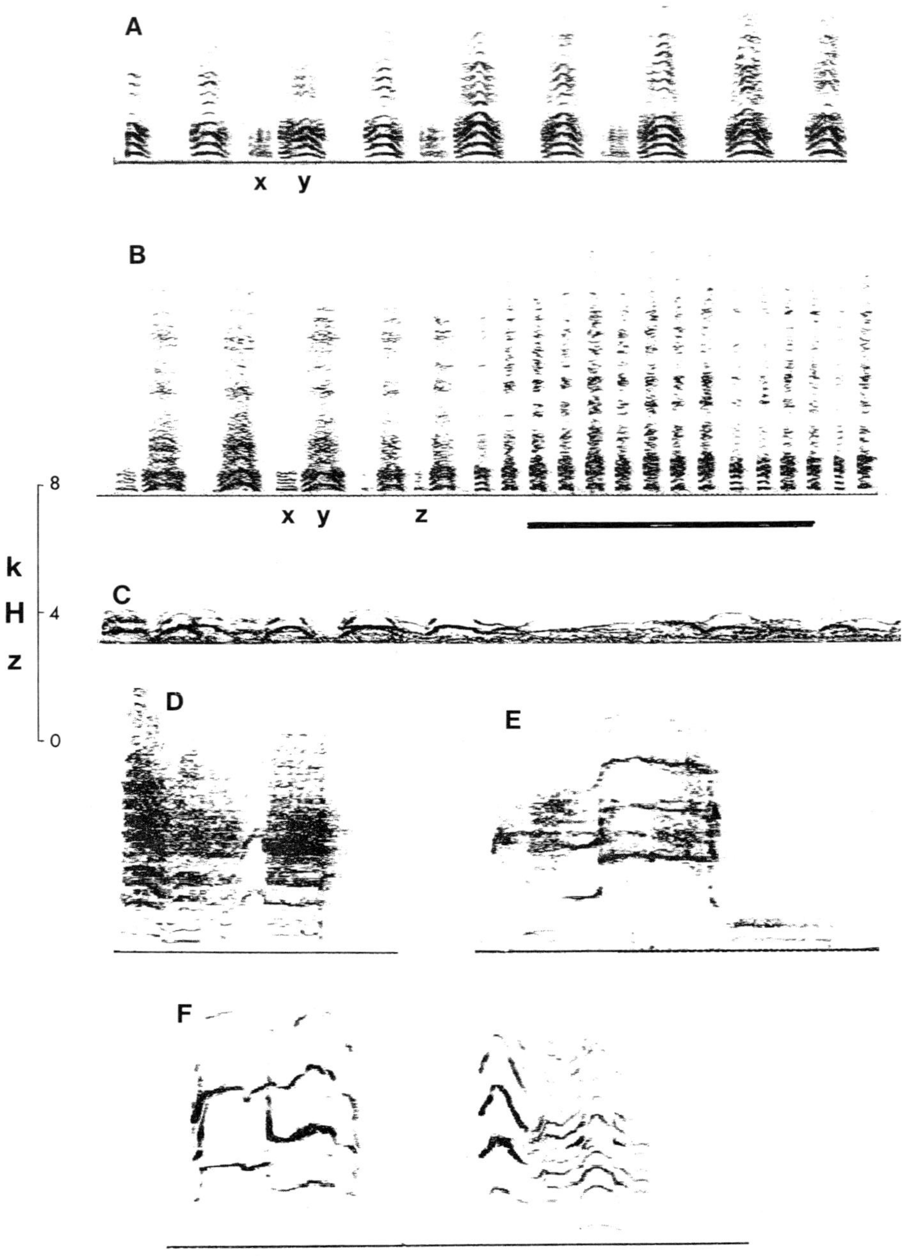
A
x y
B
8
x y z
k
C
H 4
z
D
0
E
F

2 Calonectris

The two species have quite complex calls, particularly when excited during duetting. Loud vocal displays are given on and below ground and, unlike *Procellaria*, they commonly call loudly when circling over their nesting areas. The sounds are raucous, and similar whether from ground or air.

Ristow and Wink (1980) confirmed earlier suggestions that the sexes have different flight calls. Those of female *C. diomedea diomedea* are more rasping and lower pitched with much of their energy around 1 kHz. Bretagnolle and Lequette (1990) analysed the flight calls of both *C. d. borealis* and *C. d. diomedea*, confirmed the sexual dimorphism, and found significant differences in the voices of birds of these two subspecies.

Bretagnolle and Lequette compared five temporal variables and the fundamental frequencies of the main call (their 'Duet'), of *C. d. borealis*. Only the frequencies differed; those for the males averaging 367 ± 21 Hz ($n = 12$), for the females 84 ± 7 Hz ($n = 6$), $P < 0.001$.

According to these workers the typical call of *C. d. borealis* is a repeated three-syllable sequence, two having clear harmonics, the third, with much 'white' noise, a 'breath' note (Fig. 6.25C). They considered three-syllable calls typical of the Atlantic form, two-syllable ones of the Mediterranean one. Figure 6.25A is of a *C. d. borealis* duet which begins with a female repeating a pattern of two very loud 'noisy' notes and a quieter, higher-pitched inhalatory one, until its partner breaks in with a four-syllable sequence.

Sonagrams of birds of the typical form are given in Fig. 6.25 E, F and G. Similarities of acoustic structure are apparent between D and E and G, the last being disyllabic on the criteria of Bretagnolle and Lequette (1990). The typical form also uses very deep calls at times (Fig. 6.25F).

Bretagnolle and Lequette tested the responses of *C. d. borealis* to playbacks, finding that they replied significantly more to voices of their own sex. As the order of the syllables was the most variable of the parameters measured, it was suggested that sexual recognition was based on syntactical factors.

Breeders responded more to playbacks of their own subspecies and responses, if any, to alien calls were significantly delayed. Non-breeders took little notice of alien calls, yet these are precisely those seeking partners, as Bretagnolle and Lequette pointed out. They also suggested that this sensitivity to geographical variation in

Figure 6.24 *Calls of* Procellaria westlandica. *Bar is 1 s. (A) Part of the 'Quack' of a bird singing on the ground, showing the low-pitched, low-volume, inhalatory notes (x) and the louder, higher-pitched exhalatory ones (y). (B) Part of a 'Jackass' call developing as a climax to a 'Quack', with inhalatory and exhalatory notes (x) and (y). Two shorter notes (z) link the 'Quack' to the 'Jackass' which is built on very short syllables. (C) Low moaning cries from two birds duetting from a burrow: the separate details are unclear due to overlapping. (D) A loud 'Squawk' of mainly 'white noise' given in response to a hand reached into a burrow. (E) Another alarm or threat from a different bird to that in (D). (F) Two strident screams from a lone alarmed bird: there is a great difference in the acoustic structures of the two successive notes. From Warham (1988b).*

Table 6.1 Responses of Shoemakers and Grey Petrels to playbacks of their calls. From Brooke (1986)

Test bird	Occasions (n) that test bird:			Response type		
	Responded	Did not respond		'Wheezy call'	'Wheezy', then 'Rattle call'	'Rattle call'
A *P. aequinoctialis*						
Same sex as caller	13	5	$X_1^2 = 0.86$ NS (same sex vs. opposite sex)	10	3	0
Opposite sex to caller (not mate)	14	10	$X_1^2 = 9.13$ $P < 0.01$ (opposite sex vs. mate)	13	0	1
Mate of caller	5	23		2	1	2
B *P. cinerea*						
Same sex as caller	9	2	NS (same sex vs. opposite sex)	All responses were with 'Bleat' call		
Opposite sex to caller	12	2	Fisher exact test, 1-tailed, $P < 0.05$ (opposite sex vs. mate)			
Mate of caller	4	6				

NS, not significant.

calls may enhance a female's ability to find a partner of her own song-type and thus local population.

Playbacks of male calls led to increased calling by conspecific males on the ground, suggesting that these were replying with territorial declarations to the 'new' male. The calls of flying females also increased, suggesting dialogue between them and grounded, unmated males, as with *Puffinus* (Chapter 4.VII.B).

The Streaked Shearwater *C. leucomelas* has a loud voice, its calls often very high pitched, reaching to at least 10 kHz. Bretagnolle and Lequette (1990) reported that the fundamental frequency of male calls was between 4 and 5 kHz but in many flight calls the lowest frequency is around 0.5 kHz as with the low, slurred notes in Fig. 6.25H, probably those of a female (Bretagnolle, pers. comm.).

Apart from the courtship and advertising calls, *Calonectris* have harsh, high-pitched vocalizations when fighting which, in *C. d. borealis*, are monosyllabic modifications of the 'Duet' call (Bretagnolle & Lequette, 1990).

VII The *Puffinus* shearwaters

A *Introduction*

General information on the social and sexual activities of *Puffinus*, mostly of unknown sex and status, were provided by Rowan (1952) for *P. gravis*, by James (1985a,b) and Storey (1984) for *P. puffinus*, by Warham (1955, 1958b, 1960) for *P. assimilis*, *P. carneipes*, and *P. tenuirostris*, by Harrow (1976) for *P. huttoni* and by Harper (1983) for *P. bulleri*. The detailed work of Shallenberger (1973) on *P. pacificus*, is unfortunately mostly unpublished.

As usual most ground activity is from non-breeding birds, but breeders may emerge from their burrows from time to time to join in the general vocal chorus and shearwaters may also assemble on bare, unburrowed ground to sing and court, in the style of albatrosses in their 'clubs'. There is the usual early peaking of the chorus after dark, later declining but reviving with the outgoing exodus, as shown for *P. gravis* by Hagen (1952, p. 103).

Bursts of activity are followed by rests or sleep, but the chorus is rekindled locally when particularly vociferous pairs burst into song and neighbours join in. This pool of activity quietens as the initiators fall silent, but the din kindles outbursts at the periphery so that the sound spreads outwards from its origin like the waves from a stone dropped into a pond. This stimulatory effect was noted by Rowan (1952) and Warham (1960), among others, and Harper (1983) reported that the calls of *P. bulleri* also triggered off chorusing and pair-bonding activity in nearby Fairy Prions.

B *Nuptial displays*

Shearwaters use no dramatic visual signals when soliciting partners or nest advertising, and although they may call when overflying the colonies, there appear to be no aerial pursuits as with gadfly petrels. What visual signals are used are detectable

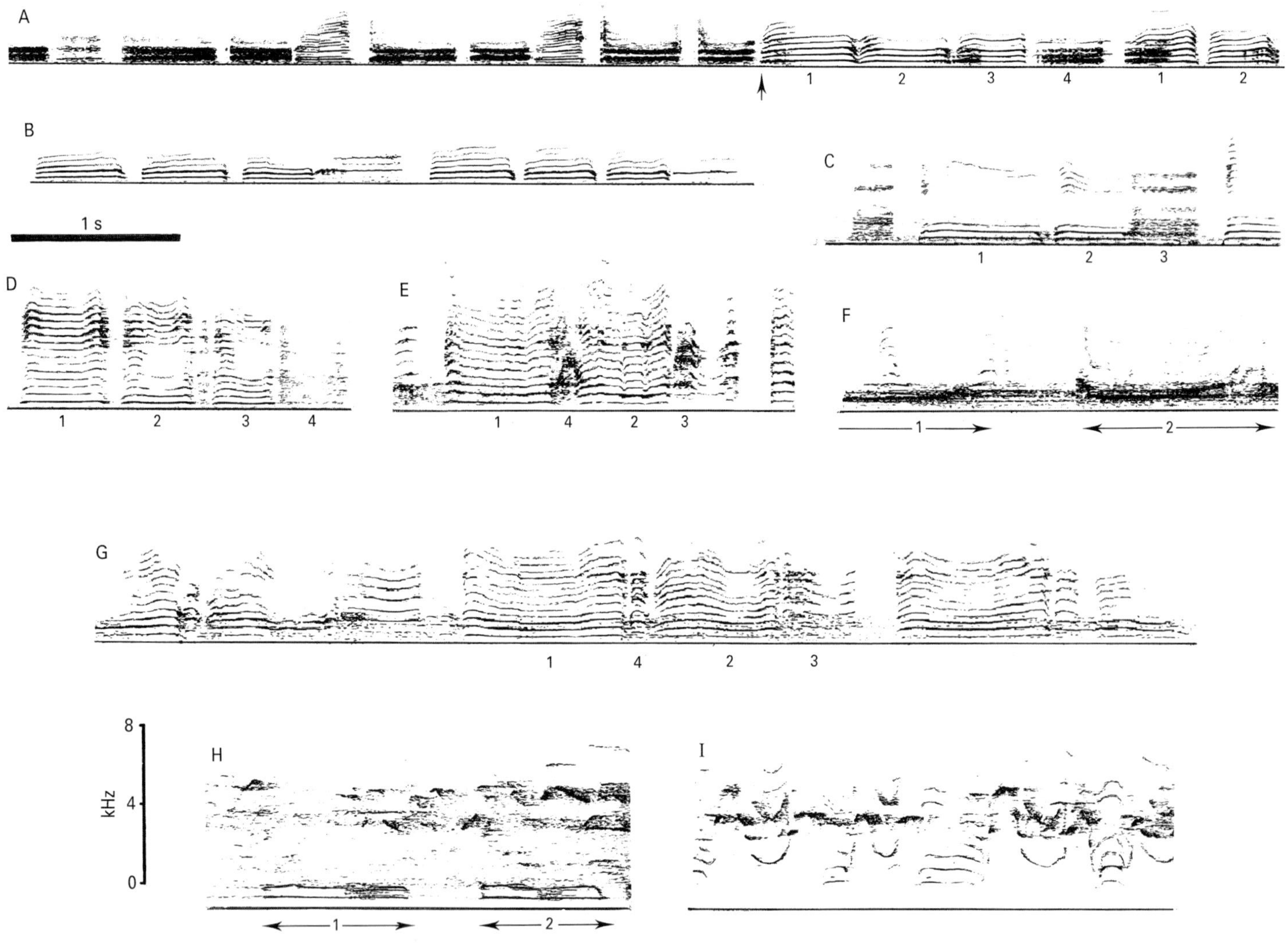

A
1 2 3 4 1 2
B
1 s
C
1 2 3
D
1 2 3 4
E
1 4 2 3
F
1
2
G
1 4 2 3
8
4
0
kHz
H
1
2
I

only at very short range, are preceded and accompanied by singing, and by tactile and possibly olfactory signals.

Typically, an active bird, often a male, approaches or trails behind a potential partner with his head held low, bill closed or only slightly ajar and calls vigorously. He reaches forwards and tries to probe the plumage of the other bird who is probably also calling. Bill-fencing may follow duetting, usually with the birds face to face (Fig. 6.26), and this may become so vigorous that both rise to their feet with one even inserting the tip of its bill inside that of the other (Fig. 6.27). During fencing the bills become shiny and bubbles may appear at the nostrils, but whether this is due to release of salt excretion or stomach oil is unclear.

Shallenberger (1973) drew attention to the concentration of preening around the eye by pale-breasted *P. pacificus*. These have white feathers in the eye socket and when raised the lower eyelid creates a white patch which Shallenberger suggested might provide a stimulatory and directive signal to the partner.

The tactile stimulation reduces aggression in the preened bird and two may sit with bodies touching, calling and allopreening for long periods. The male tends to be the most active, his calling prefaced by silent pulsations of throat and buccal cavity. The calls of both become increasingly loud, strident and higher pitched as their excitement increases, their napes are arched and the throat pulsations increase in extent with the increase in sound volume and their bodies and tails also throb in time with their calls (see Fig. 4.8). A common feature of this mutual display is for the birds to duet with their beaks pointing down a burrow entrance down which they disappear still calling (Fig. 6.26D), possibly a form of male nest presentation behaviour.

Between displays the birds often mandibulate stones, leaves and other material with 'Sideways Throwing' movements, and some material may get carried into the nest chamber.

The oscillations of the male's beak stropping that of his partner during coition tend to impart a wavering to his calls, which are now rather muted. With some species, such as *P. tenuirostris*, a special, soft, low and repeated call accompanies copulation, one distinct enough to enable the listener to locate copulating pairs (Norman, 1969). The male may grasp the other's nape in his widely opened beak (Fig. 6.26C). Calls cease as cloacal contact is made. Once hatching begins copulatory activity by the

---

Figure 6.25 *Vocalizations of* Calonectris*: A–D from the Salvages; E–G from Malta. (A) Duet of* C. diomedea borealis. *The distinctive three-syllable call of the first bird fades away as the male enters (arrow) and continues with a four-syllable response. (B)* C. d. borealis, *a four-syllable male call, the last a soft 'breath' note. (C)* C. d. borealis, *a three-syllable male call, the last inhalatory. (D)* C. d. borealis, *a four-syllable call at high intensity, cf. (B) above. (E)* C. d. diomedea, *a typical three-syllable phrase with a brief section of unstructured sound (4) possibly inhalatory and given by an excited bird. Note basic similarity to that of* C. d. borealis *in (D), but with longer syllable length. (F)* C. d. diomedea*: two low-pitched calls with much background noise. (G)* C. d. diomedea*: part of a high intensity song showing variation in the structure of the syllables in the course of a call. (H)* C. leucomelas*: an aerial chorus including two low notes (1 & 2), probably from a female with fundamental frequencies around 50 Hz. (I)* C. leucomelas*: the cacophony of the night-time flying birds: the notes are high-pitched and of varied acoustic structure.*

Figure 6.26 *Courtship postures of shearwaters. From photographs. (A) and (B) typical orientation at start of displays. (C) Copulation. (D) Duetting inspection of burrow entrance.*

Figure 6.27 *Billing by* Puffinus bulleri. *The lower bird (the female?) has the end of its upper mandible within the gape of the partner. From a photograph.*

non-breeders declines, less preening is seen, and pairs sit silently together for long periods, one or other occasionally singing quietly 'under its breath'.

C *Aggression and defence*

Established birds repel territorial intruders, mostly by loud songs but, if necessary, by fighting. Bills and claws are the offensive weapons, the wings and tail being extended to help maintain balance. The birds use loud, strident calls and tend to rise to their toes to get above their adversaries. Established nest holders generally seem to win.

Intruders may exhibit almost reptilian persistence in repeatedly trying to enter a particular burrow only to be as repeatedly repulsed, perhaps to retire bleeding. Some such birds could have formerly bred there but have somehow lost residential rights, perhaps having lost their partner.

During the predawn departure, the scene at dense shearwater colonies on calm mornings can be chaotic. Scrimmages develop near take-off points as birds struggle for a place, sometimes even leaping into the air from the back of another. As they file along the tracks leading to these favoured launching sites many call, as do birds from burrows, and the chorus may be as loud as that on the previous evening. What is the function of all this noise? Does it deter burrow owners from attacking the passers-by, and/or is it part of a territorial proclamation, as in the dawn chorus of passerines? But most of the travellers have their own nests, are often drawn from quite a wide area, and would seem to pose no threat to the nesters past which they hurry.

D *Vocal repertoire*

Special calls used during copulation, mechanical sluicing sounds during chick feeding, and howls when fighting excepted, *Puffinus* have one basic vocalization that is used during nuptial and aggressive/defensive situations.

This is composed of repeated phrases of units of 3–8 notes or syllables. Such calls have been described in many ways—for *P. puffinus* 'it-i-corka', 'it-is-your-folt'; for *P. carneipes* 'ku-kooh-ah', and so on. There is much intra- and interspecific variation but the basic structure of the song of a particular bird is consistent enough for individual recognition, for example in *P. puffinus* (Brooke, 1978e).

The exhalatory and inhalatory phrases are especially clear in shearwater calls (see Fig. 4.3), coinciding with expansions and contractions of the throat and buccal cavity. Most start quietly, build to a climax, then decline with no audible inhalations. Duets often end in a frenzied scream or a splutter, and the number of syllables may increase as in Fig. 6.28C. Duettists may sing in unison, or their syllables may alternate. Whether there is any significance in the relative timings for pair formation, as in the synchronous 'dances' of Laysan Albatrosses, or for confirming identities by matching calls as in gulls, is unstudied. Both sexes can generate a high volume of noise so that, in a burrow, with their heads close together, pressures on their ears must be very high. The energy expended seems excessive just for communication

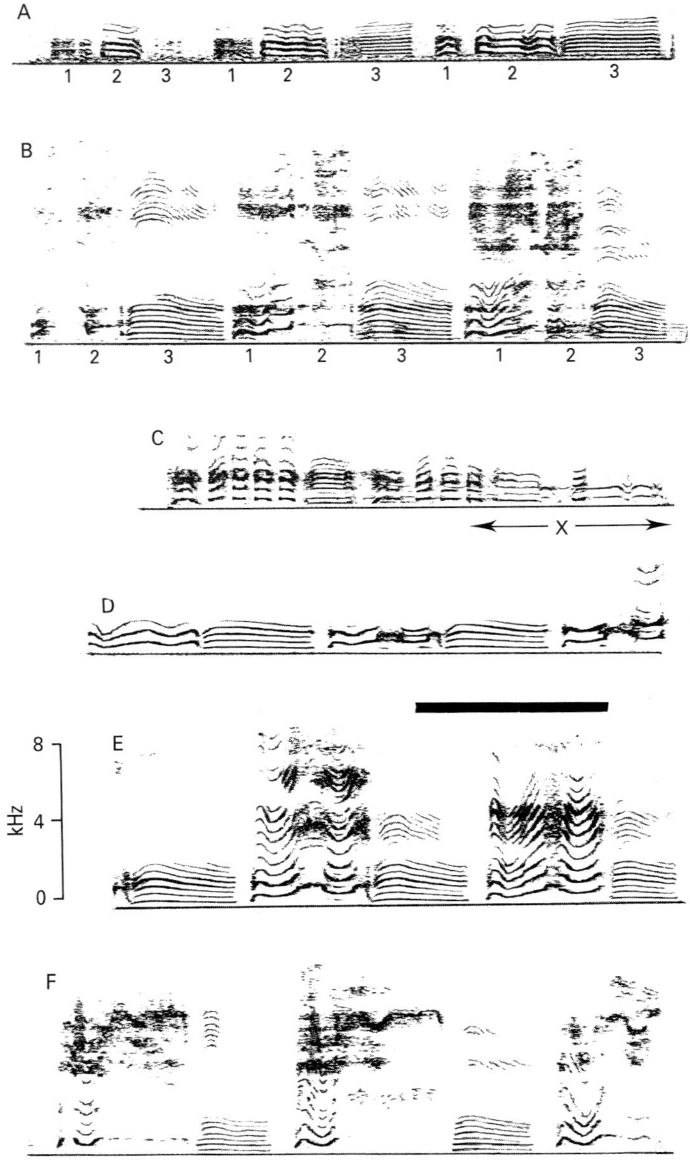

Figure 6.28 *Vocalizations of unsexed* Puffinus bulleri *calling from the ground. Bar is 1 s. (A–C) Parts of a duet that lasted 20.1 s to show call development and changes in acoustic structure. (A) Call of the initiating singer based on a three-syllable unit. (B) To show the amplitude and pitch of the same singer at the peak of its song. (C) The end of the duet showing the five to six syllable call of the second bird with the first singing quietly in the background at (X). (D) and (E) Solo birds to show changes in the sounds as the calls climax. (F) A complex two-syllable style vocalization from an excited bird with most of the acoustic energy around 5 kHz.*

between them, but perhaps such loud duets serve also to warn off intruders, mark out the occupied sites, and so reduce wasteful competition.

Calls made from the ground or from the wing are similar but aerial calls are frequently heard as mere snatches. A bout of ground calling usually lasts for 5–15 s in large species such as *P. bulleri*, *P. griseus* and *P. tenuirostris*. Aerial calling is heard mostly from the smaller species such as *P. assimilis*, *P. puffinus* and *P. gavia*; larger species tend to be silent then (Chapter 4.VIII) and the eerie silence of massed inflying *P. tenuirostris* was noted 100 years ago by Bishop Montgomery during his visits to the Bass Strait colonies (Montgomery, 1892).

The kinds of intraspecific variation in the calls is exemplified by Fig. 6.28 from unsexed *P. bulleri*, not selected to show extremes. These calls were based on two- or three-note phrases, and in all one note has many harmonics, the other notes being of widely varying acoustic structure. With excited birds amplitude, length and pitch may increase (e.g. Fig. 6.28D,E).

When threatened in their nests, shearwaters use the same calls as in nuptial contexts, but they call loudly and may come to the burrow mouth to repel the offender. If fights develop the sounds are high-pitched caterwaulings (see Fig. 6.30C) this example from *P. pacificus* being rather like a mollymawk's 'Wail'.

Female shearwaters have lower-pitched voices, often with more 'white noise', and usually shorter syllables than those of the males (e.g. see Fig. 4.3). The mean lengths of the two-syllable unit calls of *P. gravis* were 1.02 ± 0.05 s ($n = 12$) for males, 0.76 ± 0.05 s ($n = 10$) for females, $P < 0.001$ (Brooke, 1988). The general effect is that females sound harsh and asthmatical, like laboured breathing; the males' calls, with their clearer harmonics, have a more ringing, musical timbre.

As Figs 6.28–32 show, the lengths of the repeated units are usually short (e.g. in Fig. 6.28). The two-syllable, drawn-out moan of *P. pacificus* (Fig. 6.30B) is very different from that of *P. bulleri*, although these are usually thought to be closely related. Some calls of *P. nativitatis* are also quite long (Fig. 6.30A).

An analysis sex by sex of the calls of the seven subspecies of *P. assimilis* and the 12 of *P. lherminieri* (Fig. 6.31) could help unravel their systematic status, particularly if done concurrently with DNA analyses.

VIII The storm petrels, family *Hydrobatidae*

A *Aerial activity*

This is known for *Hydrobates pelagicus* (Lockley, 1932; Davis, 1957), *Oceanodroma castro* (Allan, 1962; Harris, 1969b), *O. leucorhoa* (Ainslie & Atkinson, 1937; Williamson, 1945; Waters, 1964), *O. furcata* (Simons, 1981), *O. tethys* (Harris, 1969b), *Fregetta tropica* (Beck & Brown, 1971) and *Oceanites oceanicus* (Roberts, 1940, p. 158). Details are sketchy, and it is unclear how much of the flighting is the normal circling before alighting and how much has any display content. Are the flying birds mainly non-breeders calling to stimulate burrow-holders to reply, or merely raising the sexual tone of the whole assemblage and speeding the development of their own gonads? At least with some, such as *Oceanodroma furcata*, circling birds do reply to calls from burrows below, suggesting a pair-bonding function. Chases with excited calling and

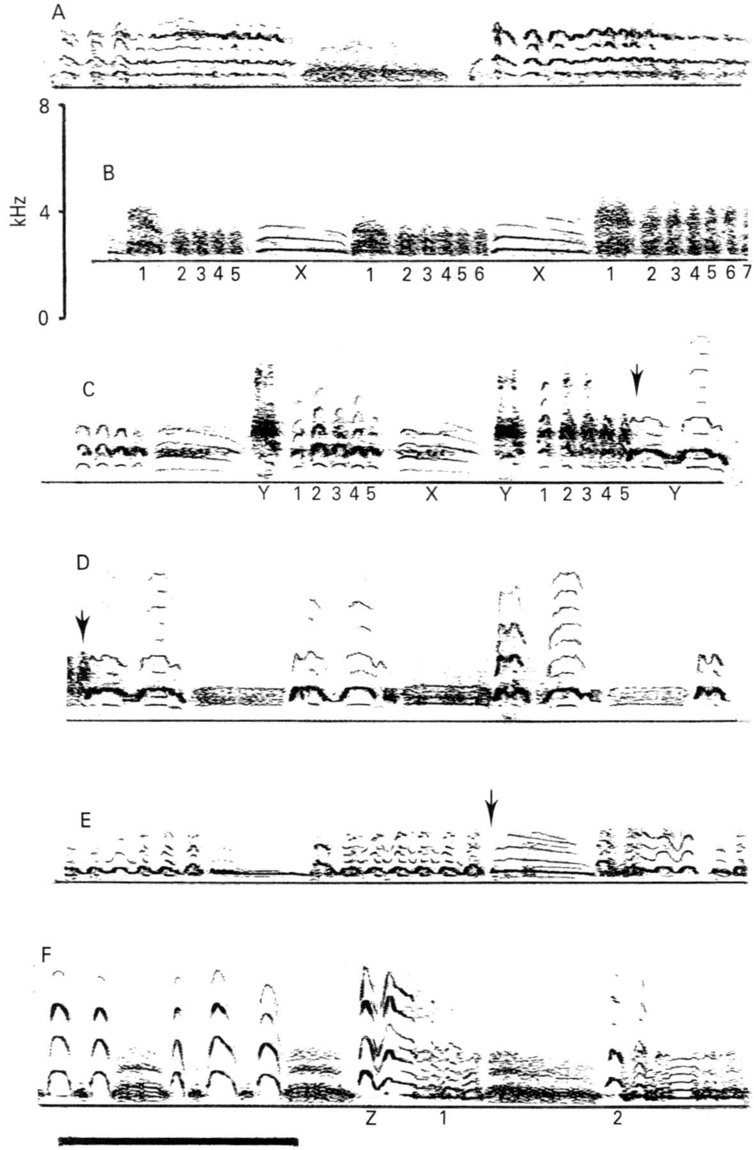

Figure 6.29 *Duetting by* Puffinus huttoni *(A–D) and* P. gavia *(E) and (F). Sexes unknown. Bar is 1 s. (A) The start of a duet: the call is based on a two-syllable unit, one high-pitched, exhalatory and frequency-modulated, the other a low 'breath' note. (B) A later stage of the same duet as (A), but of the second bird. Its song is based on harsh, low-pitched notes and one clear syllable (X) with harmonics, the whole lower-pitched than (A) and probably from a female. (C) The solo bird at the start is in full voice, its song based on a seven-syllable unit, a harsh note (Y), five short ones with arcuate formants and harmonics, and a longer clear note (X). Arrows mark the overlap with (D). (D) Shows the switch to the second bird. The pattern is now of two clear notes with harmonics, probably expiratory, and a harsh one, probably inhalatory. (E) Part of a song that became a duet at the arrow. The first bird's call was low-pitched and based on a pattern of repeated arcuate formants followed by a longer 'breath' note. (F) Part of a duet showing the climax of one bird's call at (Z) and the second bird entering at (1) as the first fades away at (2).*

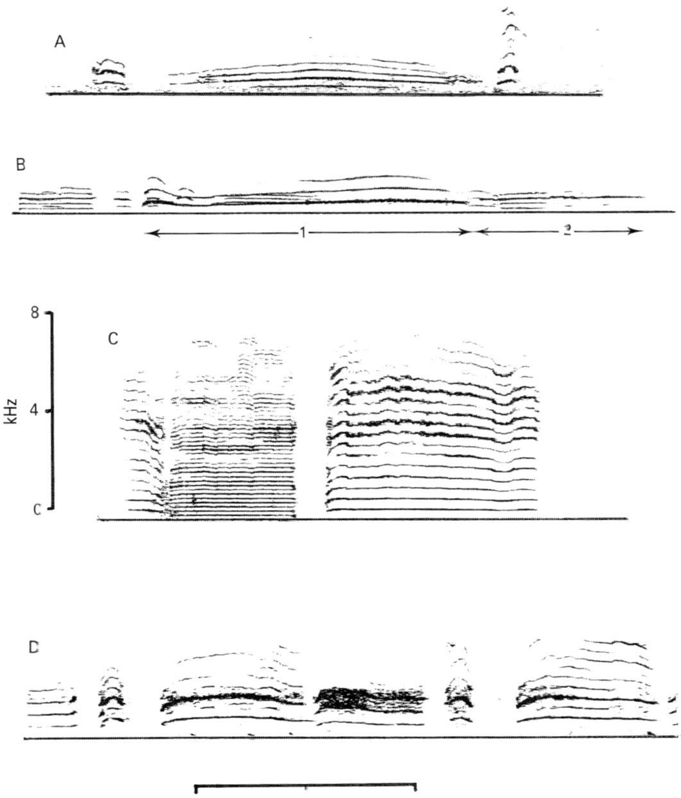

Figure 6.30 *Vocalizations of shearwaters. Bar is 1 s. (A) Two-syllable call of* P. nativitatis, *a drawn-out swelling moan, sometimes lasting 1.8 s, and a high-pitched cry. (B) A duet from a pair of* P. pacificus *in their burrow. The louder call is built from one long (c. 1.5 s) and one short (1 & 2) note with harmonics; the quieter call of the other is of shorter syllables with many harmonics delivered at a faster tempo. (C) Alarm or Threat call of* P. pacificus*: note the very sudden onset, its high pitch and many harmonics: the middle syllable is probably a 'breath' note. (D) Part of a typical call of* P. carneipes, *a measured three-syllable 'Ku-koo-ah', the long central syllable a crowing, the last a 'breath' note.*

high-speed dual zigzaggings have been described, Roberts pointing out the possible role of the white rump as a 'releaser' with *Oceanites oceanicus*. Beck and Brown (1971, p. 77) watched a pair of *F. tropica* gliding in tandem with one bird just above and behind the other while an 'oystercatcher' whistle was given. Allan (1962, p. 279) saw two *Oceanodroma castro* gliding with wings held high over their backs and engaging in a kind of leapfrogging action over one another, which attracted others to join in.

Flighting of the Galapagos Storm Petrel *Oceanodroma tethys* is particularly curious. Described by Nelson (1966) and Harris (1969b), like that of Antarctic *Oceanites oceanicus*, it is performed by day. At various times of the year thousands gathered

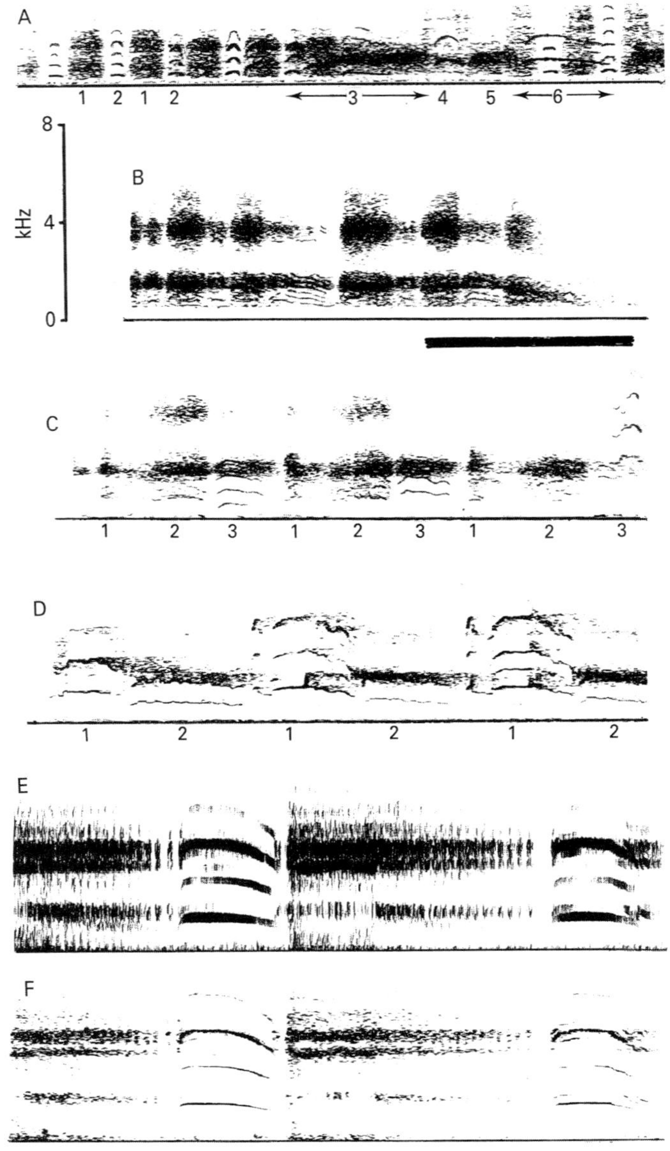

Figure 6.31 *Calls of* Puffinus assimilis *(A) and (B) and* P. lherminieri *(C–F). Bar is 1 s. (A) Part of a duet starting as a solo based on alternating short 'breath' notes (1) and even shorter, more ringing exhalatory ones (2). The partner joins in with a slurred note (3) followed by a clearer one (4), a 'breath' note (5) and a longer one (6) with clear harmonics. The first bird meanwhile has increased the volume of its call. Race* P. a. tunneyi. *(B) Harsh flight call of* P. a. assimilis. *(C) Harsh flight call of* P. l. temptator *based on a three-syllable unit. (D) Flight call of* P. a. temptator*: in contrast to (C) this call is based on a two-syllable unit and is clearer and higher pitched: syllable two appears to be a 'breath' note. (E) and (F) Ground call of* P. l. dichrous*, (E) with wide band filter to show the pulsed nature of the longer note: the shorter syllable was clear, high-pitched, almost a scream.*

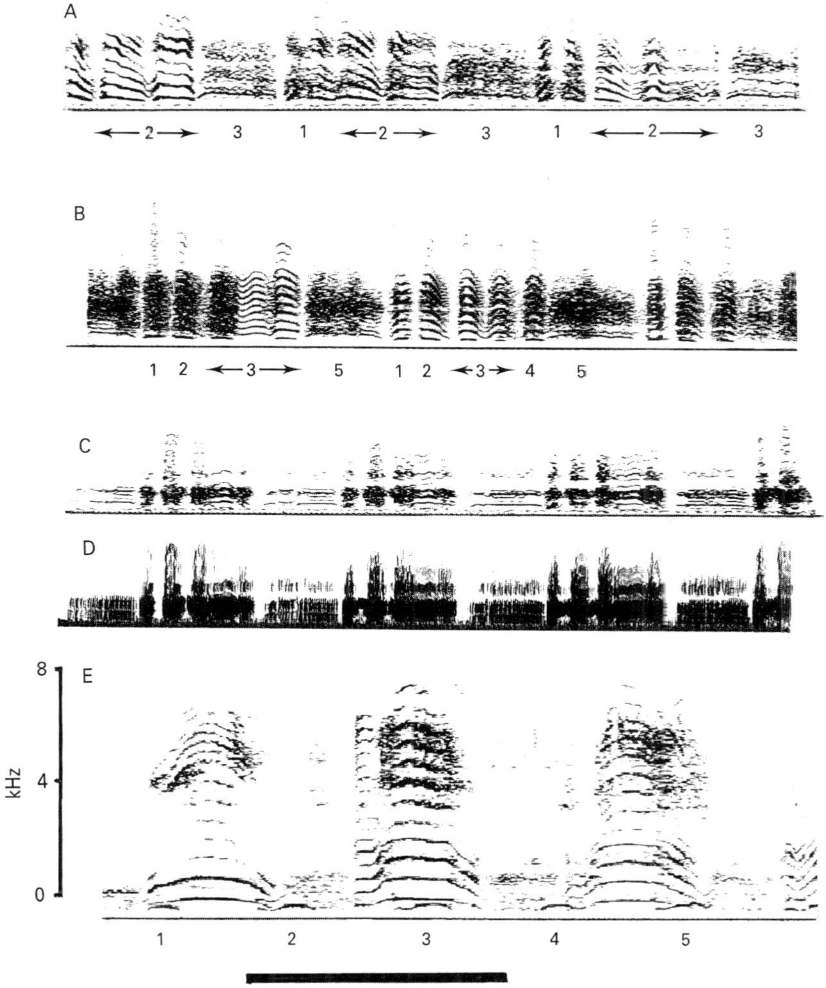

Figure 6.32 *Calls of* Puffinus tenuirostris *(A–D) and* P. griseus *(E). Bar is 1 s. (A) Part of the song of a lone bird in a burrow based on two expiratory notes (1 & 2) and a longer inhalatory one (3). (B) Part of a duet, the main call a repeated unit of four shrill exhalatory notes (1–4) and a longer 'breath' note (5) delivered rapidly and conveying the impression of breathlessness. At the start the patterns are confused by the partner's contribution. (C) and (D) A lone bird in a burrow, its song built from three short notes and a longer inhalatory one: probably a female. (D) is a broad-band analysis. (E) Part of a duet: the three loud, high-pitched syllables are from an excited male yelling at the top of his voice separated by quiet inhalations, the female's contributions (1–5) being shorter, quieter notes delivered more rapidly.*

over a restricted, non-breeding area at or near the coast of Tower Island, mostly in silence and not in pairs. Some dropped down to patter over the lava with raised wings. As the breeding season approached, these congregations shifted to be over the nesting areas, and single birds hovered above potential nesting places or repeatedly alighted there. Sometimes a bird calling from below ground was answered by an overflier. Nelson speculated on this mass flighting as being an epideictic display (Wynne-Edwards, 1962), generating stress in proportion to the density of the performers, but the birds appear to have a long nesting season so that extended flighting may reflect extended stimulatory activity for gonad development. A frenzied mass flighting of *O. oceanicus* at sea has been described (*The Petrels*, p. 181).

B *Ground activity*

Simons (1981) watched *Oceanodroma furcata* (sexed by laparotomy) in their underground nest with night-vision equipment. Newly paired birds indulged in bouts of calling interspersed with the alternate allopreening, coition taking place both in the chamber and on the surface nearby. Simons saw nest relief once. The male walked to the incubating female, gave its 3–5 note call, and preened her forehead. She rose, walked to the burrow entrance and flew off. Harris (1969b, p. 153) saw that the presumed female *O. tethys* tended to be submissive when preened, and suggested that this was necessary for her to be accepted in a colony where aggressive competition for nest sites was very high.

Storm petrels defend their burrows against other petrels mainly by calling. Harris (1969b, p. 152) described a sitting *O. tethys* fighting a persistent intruder, with the two rolling over and over on the ground. The defender eventually resumed incubation. Fighting is accompanied by harsh cries and ruffled crown feathers (cf. Taoka *et al.*, 1988, fig. 4).

When withdrawn from their burrows both adults and young birds tend to dribble stomach oil from their bills, and Graham (1890) described *H. pelagicus* 'it lies passively in the hand of its captor, gives a faint squeak, and drops a pellucid tear, in the shape of a globule of oil, from its beak'. However, oil spitting can be vigorous and possibly effective against its own kind, for example with *O. melania* (Osburn, 1911; Everett, pers. comm.).

C *Vocal repertoire*

The vocalizations are better known than other behaviours thanks to the publications of James (1984) on *H. pelagicus*, of James and Robertson (1985b) on *O. castro*, of Bretagnolle (1989a) and Bretagnolle and Robisson (1991) on *Oceanites oceanicus*, of Marks and Leasure (1992) on *Oceanodroma tristrami*, and work in Japan on *O. leucorhoa* and *O. monorhis* (Taoka *et al.*, 1987, 1988, 1989a,b,c; Taoka & Okumura, 1988, 1989, 1990).

On present information there appear to be no behavioural features separating the northern Hydrobatinae from the long-legged southern Oceanitinae, although the former seem to call more in flight than do the southern ones. The calls themselves perhaps differ more—most northern species produce whirring, churring or creaking sounds (e.g. Fig. 6.33), based on very short pulses, the southern ones use rather high pitched twitters or whistles like Fig. 6.34F or quite harsh repeated calls mainly of 'noise', as with *Oceanites oceanicus* and *Garrodia nereis* (Fig. 6.34D).

The repertories typically include a flight call, a burrow call and variable harsh cries used agonistically and high-pitched distress calls. Even the easily recognized

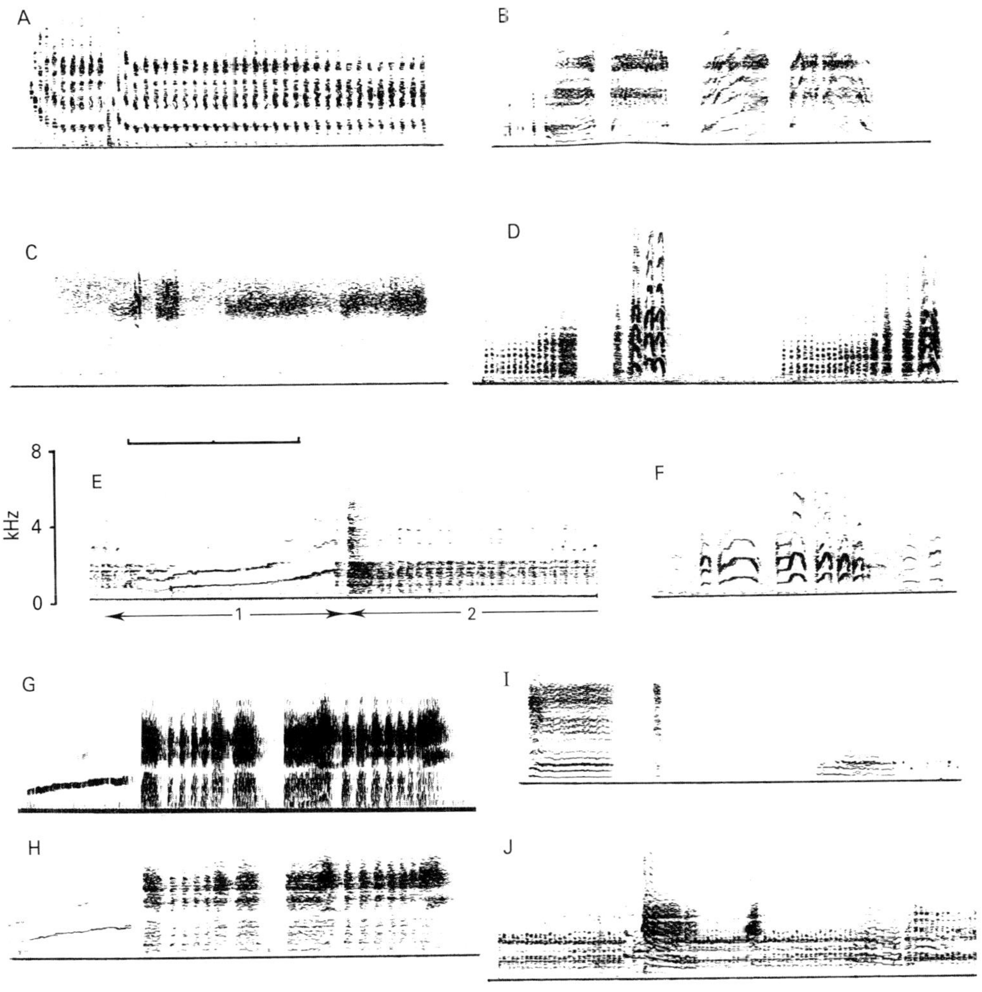

Figure 6.33 *Examples from the repertoires of northern storm petrels (Hydrobatinae). Bar is 1 s. (A)* Oceanodroma melania*: part of a high-speed 'Purr' call. The initial glissandos are unusual. (B)* O. melania*: four or five note calls of complex structure are given between 'Purrs', apparently from the same bird. (C)* O. melania*: a high-pitched, acoustically unstructured call from a flying bird. (D)* O. castro*: two calls, brief 'Purrs' followed by a high-pitched 'Chatter'. (E)* O. monorhis*: part of a series of burrow calls consisting of short pulsed 'Chatter' sections (2) separated by wheezy interludes (1) perhaps inhalatory. cf. James and Robertson (1985a). (F)* O. tristrami*—'Purring' calls typical of other* Oceanodroma *seem not to have been reported, but clear, repeated 'Crowing' cries with many harmonics like this from birds in burrows. The durations and frequency structure of the syllables vary between birds. (G) and (H) Wide- and narrow-band sonagrams of 'Chatter' flight calls from* O. monorhis *probably by a male (cf. Taoka* et al.*, 1989b, fig. 1). (I)* Hydrobates pelagicus*: calls like these are heard from burrows often as preludes to bouts of 'Churring' song. (J)* H. pelagicus*: a lone bird is interrupted by a louder one at (A).*

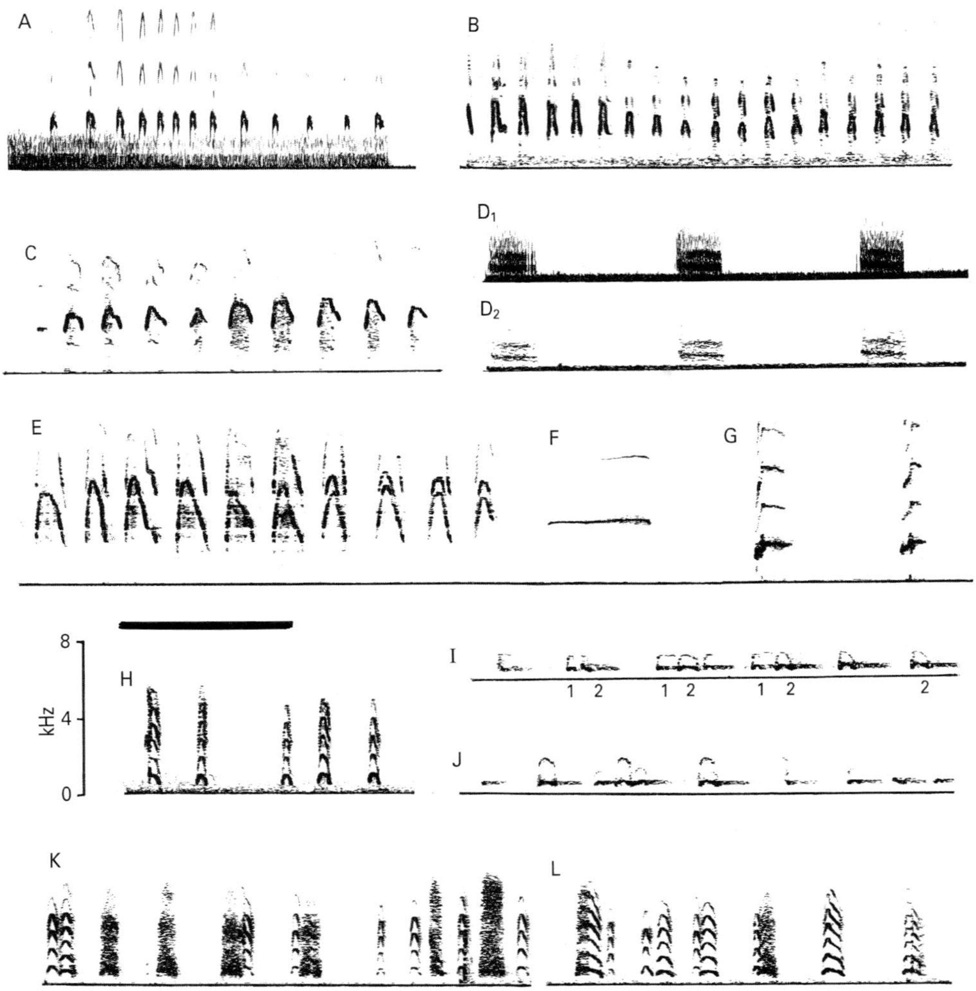

Figure 6.34 *Some calls of southern storm petrels, Oceanitinae. Bar is 1 s. (A) and (B) Wide and narrow band sonagrams of the trills of two different* Fregetta grallaria. *(C) Alarm or threat calls of* F. grallaria, *more grating than the normal trill. (D) Wide and narrow band sonagrams of the quiet, low-pitched, regularly timed, wheezy croaks of* Garrodia nereis. *This is the only call recorded but seems unlikely to comprise the complete repertoire. The acoustic structure of the notes resembles that of the 'Chatter' calls of* Oceanites oceanicus. *(E)* F. tropica*: a long, high-pitched 'Chirrup'. (F)* F. tropica*: a short example of the penetrating whistle with ventriloqual qualities. (G)* F. tropica*: another high-pitched call of unknown provenance. The frequency scale should be doubled for this graph, time scale halved; the top harmonic just reaches 16 kHz. (H)* Pelagodroma marina*: high-pitched repeated 'Chirps' from a grounded bird. (I)* P. marina*: squelchy 'Cough' notes of very low frequency from two birds (1 & 2), probably paired. (J)* P. marina*: low pitched 'Coughs' from a lone bird. (K)* P. marina*: duet from courting birds; one uses high-pitched calls with harmonics which have a 'chirpy' quality; the other's calls are of harsh unstructured sound. (L)* P. marina*: another duet: (A) has very sharp cries with clear harmonics, the other (B) a mixture of coughing sounds with 'white noise' and clearer calls.*

vocalizations include a good deal of individual variation, with the usual sundry cooing and chuckling sounds of unknown provenance heard from burrows.

1 *Sexual dimorphism*

In all five species studied closely the calls of the sexes differ, and field observations suggest that this is true of others.

The 'Chatter' calls of the male *Oceanodroma leucorhoa* analysed by Taoka *et al.* (1989a) were higher pitched, with harmonics and fewer syllables than those of the females. Surprisingly, the 'Flight' calls of the males of *O. monorhis* were harsher, lower pitched, and with more 'white noise' than those of the females (Taoka *et al.*, 1989b, fig. 1) (Figs 6.33G,H), whereas males of *O. castro* had 'Flight' calls with harmonics, so sounded clearer than those of the females (James & Robertson, 1985b).

Males of the southern species *Oceanites oceanicus* have their own 'Chatter' call used only from the ground (Bretagnolle, 1989a) and similarly male *H. pelagicus* have a 'Purr' call (James, 1984), those of *Oceanodroma furcata* a special single-syllable one persistently repeated (Simons, 1981). In all three, the call is attractive to the females and playbacks of that of *H. pelagicus* are used to lure conspecifics into mist nets.

2 *Species and individual recognition*

The range of variation in the 'Chatter' call of Leach's Storm Petrel was examined by Taoka and Okumura (1989). They measured the lengths of the syllables, of the intervals between them, as well as the fundamental frequencies, of each of the nine syllables in typical calls of both sexes. The variation with individual birds was small and when one of a pair returned to its burrow 'Chatter' calls were exchanged immediately, suggesting a role in individual recognition, since otherwise birds in nests generally responded like this only to callers of their own sex.

With Wilson's Storm Petrel, which has only a 'Grating' call used by both sexes and a 'Chatter' call restricted to males, the 'Grating' call is highly individualized and probably forms part of the system for individual recognition (Bretagnolle, 1989a). The 'Chatter' call attracts females, and from playbacks of modified computer-generated versions, Bretagnolle and Robisson (1991) concluded that species-specific information is given by the modal frequency and by the lengths of the syllables and of the silent intervals. As would be expected of a species-specific signal, the modal frequency was not part of the geographical variation between Antarctic and Kerguelen populations; so that birds of these groups would be expected to recognize one another.

However, establishing the specificity of a signal, vocal, visual or olfactory, requires subjecting the experimental animal to the signals of congeners, and this has not been done. In Antarctica, *Oceanites oceanicus* is the only storm petrel, but it shares Kerguelen with *F. tropica* and *G. nereis*. The first has a very different voice (Fig. 6.34E,F,G) and is unlikely to give rise to 'acoustic interference', whereas that of *G. nereis* is similar (Fig. 6.34D). Elsewhere others breed sympatrically, for example *H. microsoma*, *Oceanodroma leucorhoa* and *O. melania* on the islands of Baja California:

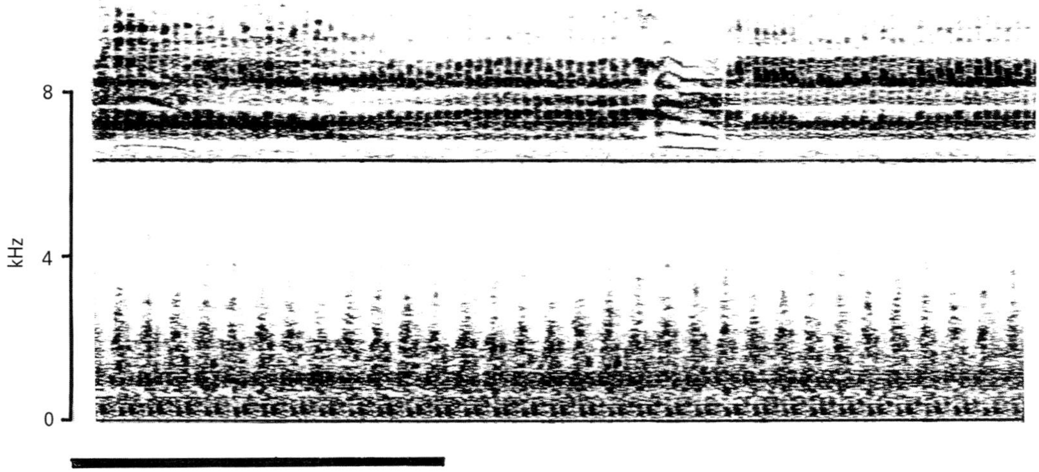

Figure 6.35 *Upper: Part of a 'Purr' call of* Hydrobates pelagicus *separated by a 'Chikka' or 'hiccough', a call restricted to males and highly attractive to conspecifics. The 'Chikka' may be a 'breath' note. Lower: Noise of engine at Rost Lighthouse, Norway, to which* H. pelagicus *are attracted (cf. Fig. 4.12). The timing of the engine pulses appears to match those of the bird's call and may form the basis for the attraction. Bar is 1 s.*

how do these maintain their specific identity? Furthermore, some storm petrels react strongly to calls of non-specifics, for example *H. pelagicus* to playbacks of *O. leucorhoa*, to those of *Oceanites oceanicus* (Zonfrillo, 1982a), even to a noise of a rusty hinge or an engine exhaust (Fig. 6.35). The attractiveness of the voices of non-specifics suggests that voice alone is inadequate as an isolating mechanism.

IX The diving petrels, family *Pelecanoididae*

Two of the four species have been studied quite intensively since the initial work of Richdale published from 1943 on, but there is little information on their behaviours or vocalizations.

Few field workers have been present during the long prelaying period, but Richdale (1943) found *Pelecanoides urinatrix* ashore then with many perched on tussocks. Thoresen (1969) saw them in mid-winter (May) sitting about in pairs 'billing and nebbing', but most were below ground, the vocal chorus being considerable. Courtship activity is quite procellariiform, with calling males attracting partners. Thoresen, watching pairs of *P. urinatrix* in nesting boxes, reported that in the prelaying period they billed and called for a few minutes and then slept side by side. Early in the morning they stirred, called to each other for half and hour or more, and spent the day at sea. Conspecifics inspecting burrow entrances were deterred by calls from within and, if necessary, combat. At a colony in Cook Strait Thoresen found *P. urinatrix* fighting for nest sites with Fairy Prions *P. turtur*, which usually emerged victorious.

Payne and Prince (1979) compared the commonest calls of grounded *P. georgicus* and *P. urinatrix exsul* nesting at South Georgia and gave sonagrams. Tennyson (in

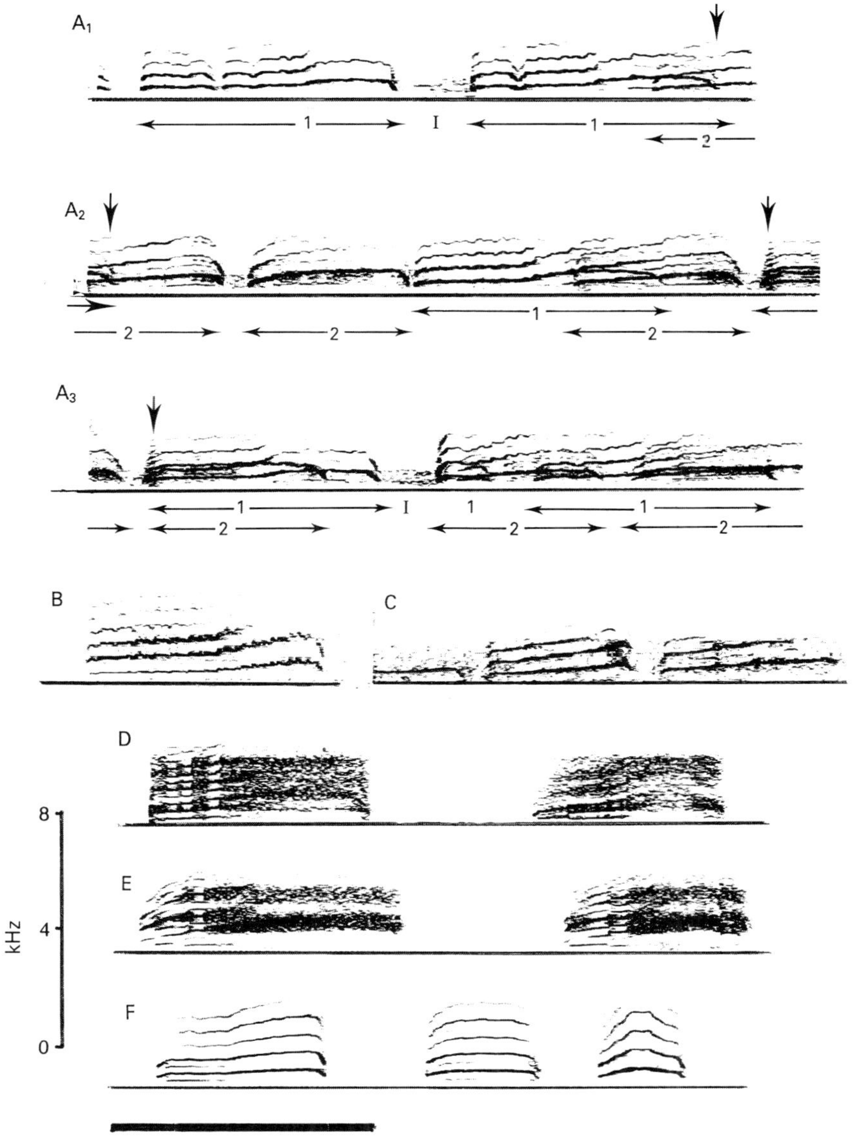

Figure 6.36 *Vocalizations of Diving Petrels. Bar is 1 s. (A_1–A_3) Duet of* Pelecanoides urinatrix chathamensis, *birds (1) and (2). (I) possible 'breath' notes. Arrows mark points of overlap. (B) A single repeated call of* P. u. exsul. *Note the frequency modulation as with the birds in (A). (C) Parts of three calls of* P. u. chathamensis*: although from the same population as those in (A) these lack frequency modulation. (D) and (E) Harsh calls of* P. georgicus, *probably males, from inside burrows. (F) Section of a call of a lone* P. georgicus *composed of a series of modified 'simple cries'.*

Marchant & Higgins, 1990, p. 728) described sexual dimorphism in the calls of *P. u. urinatrix*. The male call he syllabized as 'Kooo-ah', a note with harmonics, lasting about 1 s and rising slightly in pitch then falling abruptly at the end. This is the common call, a low moan (Fig. 6.36). The female's call 'Kua' often starts similarly but rapidly rises in pitch and then breaks into shorter syllables. The South Georgian *P. u.*

exsul calls figured by Payne and Prince are similar to those of male *P. u. urinatrix* but consist of a two-note sequence given about once per second, the first note with a rising inflection lasting about 0.7 s, immediately followed by a very short note of slightly lower pitch. As with other calls from the ground, these sequences may be repeated perhaps for half an hour or more.

The voice of *P. georgicus* differs in consisting of a series of squeaky notes showing as inverted 'U's on sonagrams (Fig. 6.36D,E,F). They are rather higher pitched than those of *P. urinatrix* and given 5–10 at a time but repeated after a pause (Payne & Prince, 1979). Persistently repeated harsher notes may perhaps be used by only one sex (Fig. 6.36D,E).

It will be interesting to compare the voices of the Indian Ocean *P. georgicus* with those of the Codfish Island population as these may have diverged perhaps 700 000 years ago (A. M. Paterson *et al.*, pers. comm.).

CHAPTER 7
Physiology and Energetics

Contents

Few of the organ systems of tubenoses have been examined critically, partly because they do not lend themselves to prolonged laboratory studies. For short-term work, however, their ability to maintain themselves on fat reserves for several days is an advantage.

I Osmoregulation

Petrels avoid desiccation by taking in water from the body fluids of their prey, by drinking seawater, and from water derived from the metabolism of fats and stomach oils. They eliminate excess salt partly through their kidneys, but more through the 'supraorbital glands', *glandula nasalis*, or salt glands, sitting in the characteristic crescentic depressions in the skull (*The Petrels*, p. 159).

A *Nasal gland excretion*

The salt gland differs from the kidney in only being active when there is an osmotic load, and its secretion is almost entirely of NaCl. Figures given by Schmidt-Nielsen

(1960) show that tubenoses produce fluids with a higher salt concentration than that of some other seabirds investigated, averaging 800–900 and 900–1000 mmol l^{-1} in *Diomedea nigripes* and *Oceanodroma leucorhoa* respectively, as opposed to 725–850 and 600–750 mmol l^{-1} in a penguin and a gull. Schmidt-Nielsen ascribed the high concentration of NaCl in the fluid from the storm petrel to the latter's diet of zooplankton which is iso-osmotic with seawater. The small *glandula nasalis* of *D. irrorata* may reflect a low-salt diet, perhaps of fish, hypotonic in respect of seawater, as Watson and Divorky (1971) suggested.

The excreted saline runs forward from the nasal tubes and along the grooves between the culminicorn and latericorns to drip from the tip of the maxillary unguis. The layout of the ducts from the gland and relationship to the air flows through the nasal tubes were described by Bang and Wenzel (1985).

With captive *D. nigripes* and *D. immutabilis*, drops of excretion usually began within 20–30 min in response to salt loads, falling to a rate of 10–20 drops min^{-1} during regular flow (Frings *et al.*, 1958). While the liquid excreted held 829 ± 7.3 mmol l^{-1} of Na^+, the blood plasma showed little fall in salt load, being 167 ± 1.9 mmol l^{-1} before and 164 ± 2 mmol l^{-1} during excretion. Stressed albatrosses also dripped saline from their bills, for example when handled or force fed, even from excitement during self-feeding—a kind of avian salivation?

Field measurements on freshly caught birds using injected tritiated water and/or $^{22}NaCl$ have provided estimates of water and sodium turnover, for example by Thomas and Wink (1985) for *Puffinus puffinus* and *Calonectris diomedea*, by Green and Brothers (1989) for *Pachyptila turtur* and *Pelecanoides urinatrix*, and for chicks of *O. leucorhoa* by Sievert *et al.* (1990).

Thomas (1988) determined that the nasal output of *C. diomedea* averaged 1797 mosmol kg^{-1}, the renal intestinal output 898 mosmol kg^{-1}—more than *P. puffinus* at 1661 mosmol kg^{-1} and renal output 596 mosmol kg^{-1}. The Na^+ excretion rates were 388 μmolNa$^+$ kg^{-1} body mass min^{-1} for *C. diomedea vs.* 280 μmolNa$^+$ kg^{-1} body mass min^{-1} for *P. puffinus*. When seawater intake was high enough to activate both salt gland and kidney excretion, 69% of the osmolytes were eliminated via the salt glands. However, Thomas deduced that the birds could only gain free water from seawater if their drinking rates were low enough to trigger only salt gland excretion; this would give them 53–60 ml free water per day, about 60% of the expected daily turnover of 110 ml day^{-1} for *Calonectris* and 80 ml day^{-1} for the *P. puffinus*. Thomas speculated that fish-feeders perhaps did not need to drink seawater anyway.

Green and Brothers (1989) found sodium influx rates for *P. turtur* and *P. urinatrix* were 174 and 170 mmol Na^+ kg^{-1} day^{-1} respectively, which they thought high, and arising from dependence on *N. australis* (160 mmol Na^+ kg^{-1}) and the needs of small birds for a higher relative intake of food and water than larger ones, with higher osmoregulatory costs. Seawater ingestion accounted for only 8.5% of total water influx for diving petrels, 17% for the prions, while only 21% of the total Na^+ influx came from seawater in *P. urinatrix*, 35% in *P. turtur*, the rest from the krill. With *D. chrysostoma*, however, Costa and Prince (1987) concluded that no significant ingestion of water occurred, which seems rather unlikely if handling awkward prey such as squid.

Sievert *et al.* (1990) found that *O. leucorhoa* chicks artificially raised on high-salt diets grew slower and survived worse than wild chicks. They also found that the latter received meals of quite low osmolality considering the high salt levels in typical crustacean prey, and they speculated that the adults selected items of low salt content so reducing the costs of osmoregulation to the chicks, and improving their individual fitness.

B *Total body water (TBW) and water turnover*

Green and Brothers (1989) tabulated data for TBW and water influx of eight procellariiforms. The TBW figures varied from 46 to 65% of body mass and in foraging birds were not related to body mass. With *Macronectes giganteus*, where the males are *c.* 1.3 × the heavier, TBW averaged 51.7% of body mass for both sexes (Obst & Nagy, 1992). It was much higher, 64.8%, in the *Fulmarus glacialis* measured by Mahoney and Jehl (1984). However, TBW decreased in a linear manner with age in chicks of *O. leucorhoa*, being about 75% of body mass at birth and falling to about 45% at 58 days old (Sievert *et al.*, 1990). It also decreased linearly in fasting adult *Pterodroma macroptera*, from a mean of 56% of body mass (Groscolas *et al.*, 1991).

The TBW of incubating and foraging Laysan Albatrosses at least 10 years old determined by Pettit *et al.* (1988) were 47.3% and 47.6% of body mass respectively, similar to that of *D. exulans* at 47.5% body mass. Pettit *et al.* suggested that their TBW values were low because their long wings give albatrosses a greater proportion of tissue of low water content, added to which the water content of avian tissue falls with age. They explained the higher TBW (57.9%) of similar-sized *D. chrysostoma* (Costa & Prince, 1987) as one for lean birds, and Roby and Ricklefs (1986) found that TBW in *Pelecanoides georgicus* was negatively correlated with total body lipid in the 12 birds they examined. However, other petrels have low values for TBW, for example *O. leucorhoa* and *Oceanites oceanicus* at 38.9–50.5% and 45.9% respectively and these have neither long wings nor albatross-like longevities. Overall, on present evidence, TBW for tubenoses seems low in comparison with other marine birds.

Water turnover rates were higher during foraging (Table 7.1). Similarly, the mean water influx for five *O. oceanicus* averaged 72 ml kg^{-1} day^{-1} while in the nest but 595 ml kg^{-1} day^{-1} for 13 foraging birds (Obst *et al.*, 1987), with water losses 600 ml kg^{-1} day^{-1} when off the nest and 128 ml kg^{-1} day^{-1} while on it. Leach's Storm Petrel chicks when unfed had very low water turnover rates (97 ± 39 ml kg^{-1} day^{-1}) but, when fed, the water flux averaged 230 ± 176 ml kg^{-1} day^{-1} (Sievert *et al.*, 1990).

In free-living adults intake varied from 878 ml kg^{-1} day^{-1} for *P. urinatrix* to 174 ml kg^{-1} day^{-1} for *D. exulans*. Green and Brothers (1989) and Thomas (1988) found allometrical relationships between water influxes and TBW or body mass. The former's allometric exponents for both relationships of 0.64 is low and unexplained. Thomas gave an exponent of 0.72 for free-living seabirds. Rates of influx for *M. giganteus* feeding chicks were higher the longer the parent had been away from its nest, and the average water input (metabolically produced) was 19 ml kg^{-1} day^{-1}, the influx during foraging trips 281 ml kg^{-1} day^{-1}, a bird maintaining constant mass needing to take in 134 ± 23 ml kg^{-1} day^{-1} (Obst & Nagy, 1992).

The calculated daily water intake of incubating Laysan Albatrosses was determined by Pettit *et al.* (1988) as 59 ml of which about 17 ml came from metabolic water and of the 42 ml some was probably due to vapour exchange with the environment

Table 7.1 Total body water (TBW) and water turnover rates in the Laysan Albatross. Mean values ±SD (*n*). From Pettit *et al.* (1988)

Activity	TBW % of body mass	Water influx		Water efflux	
		ml day^{-1}	ml kg^{-1} day^{-1}	ml day^{-1}	ml kg^{-1} day^{-1}
Incubating	47.3 ± 2.8(6)	59 ± 23(8)	19 ± 7(8)	85 ± 24(8)	28 ± 7(8)
Foraging	47.6 ± 2.3(4)	537 ± 89(4)	180 ± 47(4)	525 ± 77(4)	174 ± 37(4)

across skin and lungs. The water loss of 85 ml day^{-1} (Table 7.1.) meant a deficit of about 26 ml day^{-1} to be made good during foraging.

Groscolas *et al.* (1991) showed that TBW loss data from incubating *Pterodroma macroptera* could be used to estimate consumption of protein and lipid. This technique involves labelled water and may prove of value with other species.

II Thermoregulation

A *Body temperature*

The first published measurements of body temperatures (T_b) of petrels seem to be those of Eydoux and Souleyet (1838) who found cloacal temperatures of 11 'grand albatros' and three *D. capense* to average 39.4 and 40.7°C respectively. Data for 31 species presented by Warham (1971) had a overall mean of 38.78 ± 0.17°C, and for 41 species, including those of Warham's, Jouventin and Mougin (1981) reported a mean value of 38.3 ± 0.9°C. Data for 54 species are given in Table 7.2. The additional figures help confirm that body temperatures are not related to body size and are lower than other bird groups except for ratites and penguins. For seven of the latter Jouventin and Mougin gave a mean of 38.0 ± 0.9°C.

There is little difference in T_b when measured in the proventriculus or the cloaca, and in those birds caught on the wing or hauled from the sea the temperature increase is rather small. For example, T_b of seven procellariiforms caught flying at sea averaged 39.7 ± 0.7°C, only about 0.9°C higher than the mean for 31 species measured on land (Platania *et al.*, 1986). Likewise temperatures of telemetred *D. exulans*, for example, vary little around 38.5°C (see Fig. 3.2).

However, Farner (1956) found that 72 *Pachyptila turtur* taken on alighting had a mean cloacal temperature of 41.5 ± 1.8°C, 28 active birds at night on the ground or in burrows 39.9 ± 2.9°C, and 43 incubators by day 38.6 ± 3.0°C, all the differences being significant. Similar figures for *Puffinus griseus* ($n = 30$ in each sample) were 40.6 ± 0.45°C on landing at 1900 h, 39.9 ± 0.83°C at 2100 h, 39.0 ± 0.67°C at 2300 h, 39.1 ± 0.80°C at 0100 h and at take-off near dawn 40.6 ± 0.71°C. The differences are significant at $P < 0.001$ except for those between 1900 h and take-off and 2300 and 0100 h (Warham, unpubl.).

Variations in T_b according to activity would be expected to be greater in small species such as prions than with larger ones such as albatrosses and, as predicted (Warham, 1971), their elevation of T_b with flight seems to be low and ascribable to their energy-efficient flight style. Summarizing the data for 35 species Brown (unpubl.) found that resting tubenoses had a mean T_b of 38.6°C ± 0.8°C ($n = 60$), active ones 40.0 ± 0.9°C ($n = 44$) and incubating birds 38.1 ± 0.9°C ($n = 35$).

Proventricular temperatures of male *D. immutabilis* determined by an implanted transmitter varied little over 27 days—37.6 ± 0.33°C in one bird and 37.8 ± 0.50°C over 12 days in another (Grant & Whittow, 1983), but night-time temperatures were not recorded.

Body temperatures tend to be higher by day or night according to whether the birds are diurnally or nocturnally active. With the storm petrel *Oceanodroma furcata* Boersma (1986a) recorded a difference ($P < 0.005$) between T_bs of adults sampled by day (37.3 ± 2.3°C)—when many slept in their nests—and those at night of 39.3 ± 2.4°C. For diurnally active species, such as fulmars, T_bs should be higher by day and Howell and Bartholomew (1961a) found that in brooding *D. immutabilis* and *D. nigripes* they averaged about 1°C higher then. Part of the difference here may have been due to the higher daytime ambient temperatures (T_a). Night-time values were much lower than those of birds sitting in full sun. For *D. nigripes* the range was 5.5°C—from 36.2°C (night) to 41.7°C (sun); for *D. immutabilis* the range was 4.5°C—from 36.5°C (night) to 41.0°C (day). Howell and Bartholomew (1961b) also measured 10 each of *Pterodroma hypoleuca* by day and by night. For birds in burrows T_b averaged 38.5 ± 1.49°C by day, *vs.* 39.9 ± 1.61°C by night, again suggesting the influence of activity, since T_a was lower after dark.

Evidence for sexual differences in T_b of birds measured under the same conditions is inconclusive. Farner and Serventy (1959) found that for 15 male *Puffinus tenuirostris* T_b averaged 40.8 ± 0.97°C *vs.* 41.0 ± 0.70°C for 17 females. Wink *et al.* (1987) found that 12 incubating male *C. diomedea* averaged 36.7 ± 1.4°C *vs.* 16 females at 37.6 ± 2.36°C; and when feeding chicks 11 males averaged 38.8 ± 1.86°C *vs.* 9 females 39.5 ± 1.26°C.

That T_b may vary with latitude was suggested by Furness and Burger (1988) who used mainly my figures in proposing that high latitude tubenoses have slightly higher T_bs than boreal ones and significantly higher T_bs than tropical species. The more extensive data of Table 7.2 do not support this. For 11 truly tropical species T_b averages 38.13°C, 12 temperate zone ones 38.79°C, and 23 high-latitude ones 38.97°C. More and better data are needed to confirm such a trend. Brown (unpubl.) found no correlation with latitude ($r = 0.06$, $P > 0.50$, $n = 40$ spp.).

Most body temperatures in Table 7.2 will have been taken within the birds' zones of thermal neutrality (TNZ). The limits of these zones have been established in a few instances in laboratory conditions under changing T_as. The TNZ of *P. pacificus* was determined by Whittow *et al.* (1987) as 24–34°C (Fig. 7.1A).

B *Thermoregulation of adults*

In regulating their body temperatures to keep within their TNZs, petrels, like other homeotherms, face considerable problems at the extremes of their ranges. Species such as *Pterodroma arminjoniana*, nesting on arid stony tropical islands, meet problems in keeping cool; others in high latitudes have perhaps even more difficulty in keeping warm, particularly in open nests where chill factors are high. Antarctic

Table 7.2 Temperatures, metabolic rates and thermal conductances of petrels

Species	Temperature (°C)		RMR (kJ day^{-1})		FMR[e] (kJ day^{-1})	Thermal conductance (mW g^{-1} °C^{-1})
	Body[a] (*n*)	Egg[b]	Adult[c]	Hatchling[d]		
Diomedea exulans	39.2(10)		1755	6.62	3354 3886(M), 3680(F)	
D. epomophora	38.7(4)					
D. nigripes	38.1(10)	35.0		4.30		
D. immutabilis	37.5(10)	35.8	645	4.16	2072	
D. melanophrys	38.7(5)					
D. chrysostoma	39.7(8)		735		2396	
D. chlororhynchos	38.2(4)		481			
D. bulleri	39.5(3)					
Phoebetria fusca	38.3(?)		715			
P. palpebrata	38.1(?)					
Macronectes giganteus	39.0(25)	35.5	1154 1566(M), 1432(F)	4.28	4270	0.153
Fulmarus glacialis	38.5(41)					0.187
F. glacialoides	38.3(23)					
Thalassoica antarctica	38.5(11)					
Daption capense	39.1(29)					
Pagodroma nivea	38.7(26)					
Lugensa brevirostris	36.6(?)		195,153			
Pterodroma macroptera	37.5(5)		320,233	1.66		
P. lessonii	38.9(3)					
P. incerta	39.4(6)		213			
P. ultima	37.8(17)					
P. mollis	39.0(7)		151			
P. hasitata	39.1(9)					
P. phaeopygia	38.6(9)	34.9	469,367			
P. hypoleuca	38.2(10)	33.8	110,99,72	0.64		
Halobaena caerulea	38.4(23)		206,153			
Pachyptila vittata	38.7(16)					
P. salvini	38.9(22)		134	0.56		
P. desolata	40.3(46)			0.54		
P. turtur	38.6(43)					
Bulweria bulwerii	37.8(10)		44.0			
Procellaria aequinoctialis	39.1(9)		692,545	1.82		
P. cinerea	37.5(5)		433			
Calonectris diomedea	39.6(35)					
C. leucomelas	40.5(74)					
Puffinus pacificus	37.7(92)	36.4	119,128.5	0.78	614	0.24
P. gravis	39.8(25)		329			
P. griseus	37.8(3)		250			
P. tenuirostris	38.0(18)					
P. nativitatis	38.3(22)	35.3	127			
P. puffinus	37.8(41)		194.5 @ 23°C 298 @ 5°C	0.62		0.286 @ 5°C 0.357 @ 23°C
P. opisthomelas	37.0(4)					
P. huttoni	37.3(2)					
P. lherminieri	37.8(14)					
P. assimilis	38.6(4)		147			

Table 7.2 Continued

Species	Temperature (°C)		RMR (kJ day^{-1})		FMR[e] (kJ day^{-1})	Thermal conductance (mW g^{-1} °C^{-1})
	Body[a] (*n*)	Egg[b]	Adult[c]	Hatchling[d]		
Oceanites oceanicus	38.9(25)	35.7	37		157	0.653
Garrodia nereis	40.7(3)					
Pelagodroma marina	41.3(3)		73			
Fregetta tropica	39.8(2)					
Oceanodroma leucorhoa	39.1(42)	35.9	45.4, 61.0	0.16	142,161,124,123	0.692,0.831
O. furcata	39.7(61)	29.7	39,55	0.15		
Pelecanoides georgicus	38.7(?)		85,122	0.33	464	
P. urinatrix	39.1(6)	35.8	126	0.31	557	

[a]Mainly from Warham (1971), Jouventin and Mougin (1981), Platania *et al*. (1986) and Brown (unpubl.).
[b]Values from dummy eggs excluded.
[c]From Iversen and Krog (1972), Vleck and Kenagy (1980), Ricklefs *et al*. (1980, 1986), Bech *et al*. (1982), Grant and Whittow (1983), Ricklefs and Matthew (1983), Adams and Brown (1984), Brown and Adams (1984), Ellis (1984), Pettit *et al*. (1985), Roby and Ricklefs (1986), Whittow *et al*. (1987), Gabrielsen *et al*. (1988), Montevecchi *et al*. (1992) and Brown (unpubl.).
[d]From Klassen and Drent (1991).
[e]From Adams *et al*. (1986), Costa and Prince (1987), Pettit *et al*. (1988) and Obst and Nagy (1992).
M, male; F, female.
Several figures in one cell represent determinations from several sources.

species also face internal cooling by ectothermic prey whose temperatures may be around 0°C. The extreme case appears to be the 40-g *Oceanites oceanicus* that needs about 54 g of krill daily while incubating (Obst *et al*., 1987). They estimated that the energy to warm this krill to a T_b of 40°C would be 7.13 kJ day^{-1}, or the power needed to fuel 1.1 h of free flight. For temperature tolerance this petrel is outstanding—it raises its chick where sea surface temperatures (SSTs) average around 0°C, yet can put on weight during its migration to the north Indian Ocean where they will often exceed 20°C.

The effects of temperature extremes can be mitigated by insulative and behavioural adaptations. For example, Furness (1990) found that the weights of the insulative plumage of temperate species (*H. pelagicus, Oceanodroma leucorhoa, Puffinus puffinus*) were lower relative to that of *F glacialis* breeding at 60°N. In cold climates flying petrels lose heat through thinly insulated underwings, and reduce loss from their feet by tucking them into the belly feathers. Webbed feet are the main radiators at high T_as. Stressed chicks of *D. immutabilis* turn their backs to the sun, rest on their heels to give minimum contact with the hot ground, while their shadowed webs are spread to radiate surplus heat. In this position Howell and Bartholomew (1961a) found that with T_as of 29.4–31.0°C and the ground at 39.5–40.2°C, T_b averaged 39.3 ± 0.8°C, the feet 36.3 ± 1.1°C. Nest site selection to reduce exposure is practised by fulmars (Mougin, 1975; *The Petrels*, p. 243), while burrows provide warmer conditions in cold climates and cooler ones in hot climates. In the heat flying birds can take in seawater or bathe, while surface nesters choose to lay in shaded places.

Evaporative cooling from the moist buccal cavity by gaping (without gular flutter) is also used when under heat stress. In *M. giganteus* at T_a 10.1°C breath rate was 18.4 ± 2.7 min^{-1} rising to 54.0 ± 58 min^{-1} at T_a of 25.3°C (Morgan *et al*., 1992).

The passage of air over the nasal mucosa also conserves heat and reduces water

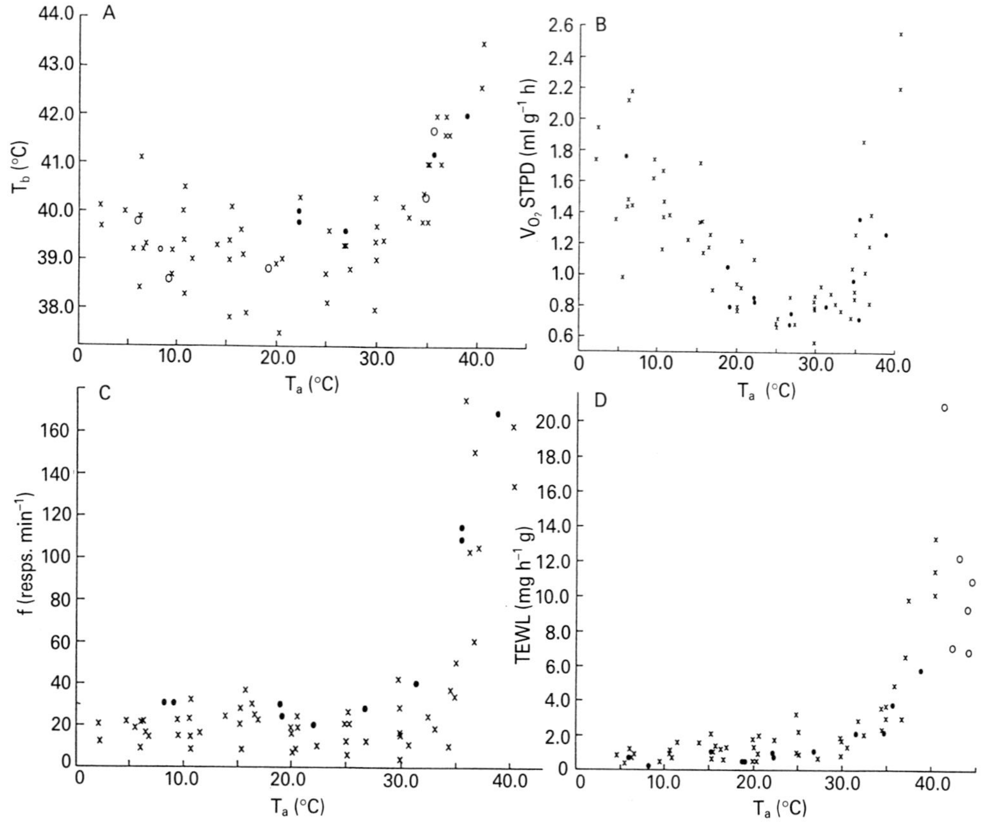

Figure 7.1 *Responses of fledgling and adult Wedge-tailed Shearwaters to different air temperatures* (T_a). *(A) Deep body temperature* (T_b). ○, *Rectal temperatures of adults;* (×), *stomach temperatures of fledglings; filled circles, stomach temperatures of adults. (B) Oxygen consumption* (V_{O_2}) *of fledglings* (×) *and adults (filled circles). (C) Respiratory frequencies* (f) *of fledglings* (×) *and adults (filled circles). (D) Total evaporative water loss (TEWL) of fledglings* (×), *adults (filled circles), and from fledglings exposed to dry air* (○). *From Whittow* et al. (1987).

loss as demonstrated by Murrish and Tirrell (1981). During inhalation the breathed air is warmed and humidified by the mucosa which is thereby cooled; during exhalation the saturated air at body temperature passes over the cool mucosa and gives up heat while some water vapour condenses there. These authors monitored the temperatures of exhaled air of *M. giganteus* and *Oceanites oceanicus* at T_a 0–30°C and at various depths within the nasal tubes, finding that something like 80% of the heat needed to warm the inhaled air was recovered by this countercurrent system.

Under cold conditions the body and back feathers are ruffled to increase insulation. If stressed, shivering generates heat in the muscles, although petrels sometimes shiver at normal temperatures when closely approached by man.

Whittow *et al.* (1987) tested temperature tolerances of *P. pacificus* by enclosing birds in a chamber for 2 h at T_as between 2 and 40°C. Oxygen consumption (V_{O_2}), total evaporative water loss (TEWL), humidity and respiration rate (f) were monitored and the thermal conductance of the bird's tissues and plumage computed (Fig. 7.1).

Values for V_{O_2} were lowest within the TNZ but increased as temperature fell until at 6.6°C it was 2.18 ml g^{-1} h^{-1}, 2.8 × the mean value within the TNZ of 0.77 ml g^{-1} h^{-1} (Fig. 7.1B). Cooled birds tended to tuck their heads into their scapulars when V_{O_2} fell. Shivering started at T_as varying between 2.3 and 13.9°C in different birds. At higher temperatures V_{O_2} again rose—to 2.57 ml g^{-1} h^{-1} at 40.4°C.

Deep body temperatures were slightly higher below the TNZ and much higher when T_a exceeded 35°C (Fig. 7.1A) and the birds became restless so that V_{O_2} increased. Within the TNZ (f) was 20.0 ± 10.9 min^{-1} ($n = 16$), but increased rapidly at higher ambient temperatures, reaching 174 min^{-1} at T_a of 35.9°C (Fig. 7.1C). Significant increases in f did not occur where T_b was less than 41.0°C. The curve for TEWL followed that for f values quite closely, being low below 35°C but reaching 13.2 mg h^{-1} g^{-1} in one bird at a T_a of 40.3°C (Fig. 7.1D). The total thermal conductance C_{total} varied little from T_as between 5 and 29°C, the mean value being 0.24 mW g^{-1} $°C^{-1}$ ± 0.05 ($n = 35$). Above 29°C conductance increased as far as 1.76 mW g^{-1} $°C^{-1}$. Whittow *et al.* (1987) concluded that the insulation of the plumage and tissues was not less than that of birds in general but less than the value of 0.29 mW g^{-1} $°C^{-1}$ for *P. puffinus* determined by Bech *et al.* (1982). In coping with T_as above their TNZ the experimental *P. pacificus* became hyperthermic before they switched to increasing (f) and evaporative water loss, but by then the metabolic heat had also increased so that this, as well as that produced in the TNZ, had to be shed to maintain thermal balance.

Morgan *et al.* (1992) collected similar data from *O. oceanicus* and *M. giganteus* to test the hypothesis that at low T_as birds reduce heat loss by increasing ventilatory oxygen extraction. This did not occur with the storm petrels which adjusted at low T_as by increasing tidal volume. For *M. giganteus* this was of secondary importance: their major route to reducing ventilatory heat loss being to increase the amount of O_2 extracted.

In birds as in mammals, it is important to keep a cool head and brain temperatures are lower than T_b in heat-stressed subjects, partly achieved by a *rete mirabile ophthalmicum* (RMO). This is a counter-current exchange system between the warm cerebral arterial supply and the cooler venous blood from evaporative respiratory membranes of the mouth and cornea. While there are no data for brain temperatures in tubenoses, Pettit *et al.* (1981) and Grant (1985) showed that RMOs are present in 16 species including albatrosses, shearwaters, fulmars and storm petrels (Fig. 7.2). This system not only allows the arterial supply to the brain to be cooled but intercarotid anastomoses permit shunting the flow into different pathways to feed the brain.

Petrels also use the highly vascularized webs of their feet as heat exchangers. The system has been investigated only in captive *M. giganteus*, by Johansen and Millard (1973, 1974) and by Murrish and Guard (1977), who followed the flow through the webs when the feet were cooled or T_b was increased. The blood enters in a single pedal artery but can return either through arteriovenous (AV) anastomoses of low resistance vessels and thence to a large medial vein, or through a high resistance web of capillaries and back in a plexus countercurrent to the leg artery (Fig. 7.3). When under heat stress, vasodilation of the AV anastomoses occurs, the vascular resistance falls, and the extremities are flooded with the warm blood so that heat is lost by conduction and radiation. But giant petrels often have their feet in wet-cold, near-freezing conditions and in dry-cold ones when on snow or ice. The experimenters

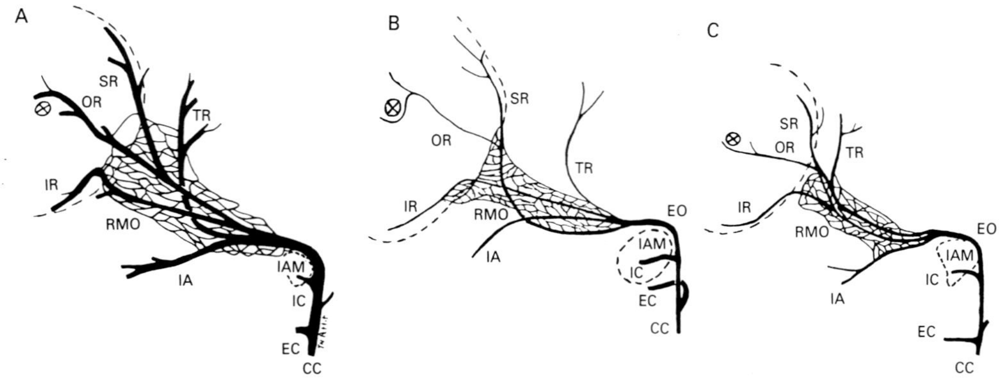

Figure 7.2 *Arterial rete mirabilium ophthalmicum in three North Pacific petrels. CC, common carotid; EC, external carotid; EO, external ophthalmic; IA, inferior alveolar; IAM, internal auditory meatus; IC, internal carotid; IR, infraorbital ramus; OR, ophthalmic ramus; RMO, rete mirabilium ophthalmicum; SR, supraorbital ramus; TR, temporal ramus; X, optic nerve. Dashed lines indicate border of orbit and IAM. From Pettit* et al. *(1981).*

investigated the vascular responses to such conditions by suddenly immersing a foot in water at −2.0–0°C. There was an immediate increase in blood flow with vasodilation and a brief increase in blood temperature—a 'cold flush' response—quickly followed by increasing vasoconstriction and blood viscosity and reduction in blood flow and tissue temperature. The reduced flow helped lessen heat loss directly, as well as shifting heat between the artery and the countercurrent vein. The reduced flow also followed the closing of the AV anastomoses forcing most blood into the high resistance web system, while the venous return was via the heat exchanger and warmed before reaching the body. Johansen and Millard (1973) found that the venous return in the ice-cooled foot rewarmed to 25–30°C in the tarso-metatarsal region. The venous temperature in the webs pulsed with each heart beat, dropping as much as 0.3°C between beats.

Murrish and Guard (1977) found that the sympathetic nervous system controlled the blood flow in the legs and feet via beta-adrenergic receptors within the AV anastomoses. Similar *venae comitantes* systems function in the legs of *F. glacialis* and *P. gravis* and were figured by Midtgard (1988).

C *Development of thermoregulation in chicks*

Since reviewed in *The Petrels* (pp. 340–342, 352–353), further work has been done by Brown and Prys-Jones (1988) on subantarctic species, by Bech *et al.* (1991) on *T. antarctica*, by Mathiu *et al.* (1992) on *P. pacificus* and by Dawson and Whittow (1994) on *D. nigripes* and *D. immutabilis*.

Some chicks of burrowing species can maintain constant T_bs once their down is dry and may only need a few days before their T_b is the same as that of their parents, for example *H. caerulea* and *P. salvini* at 3 and 5 days old, according to Brown and Prys-Jones. Yet these authors found that those of *Procellaria aequinoctialis* and

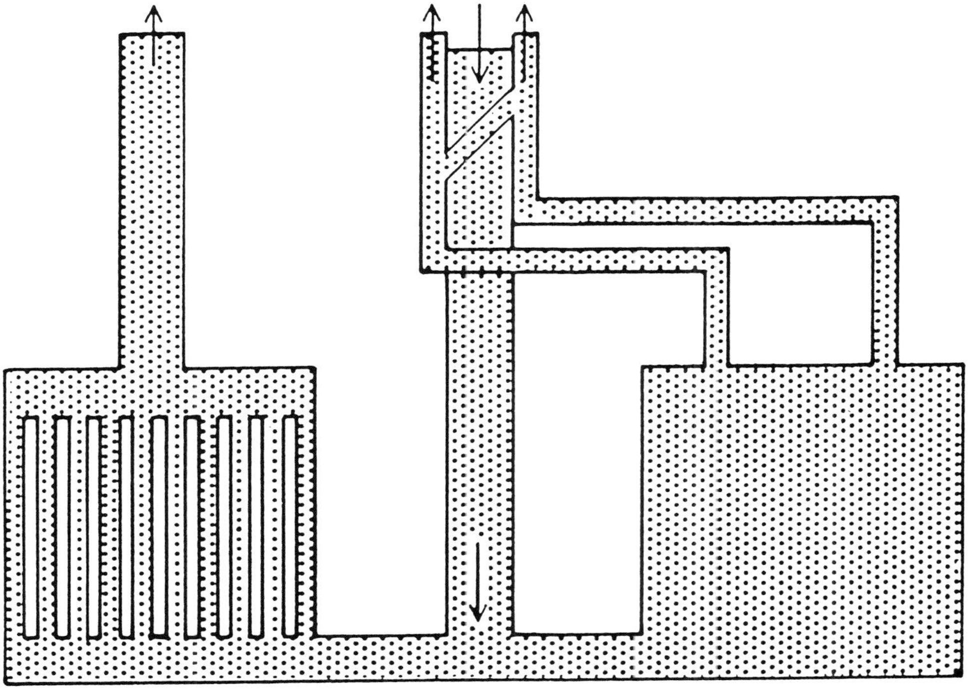

Figure 7.3 *Schematic diagram of vascular heat-exchange system in legs and feet of the Southern Giant Petrel. Central bar represents the dorsal pedal artery. Venous return from the feet can be either through a system of capillaries (right) to the countercurrent veins which exchange heat with the pedal artery, or, left, through anastomoses to a large peripheral vein. The two venous returns are controlled by the sympathetic nervous system and co-operate to regulate the temperature within the uninsulated tissues of legs and feet. From Murrish and Guard (1977).*

Pterodroma macroptera still had not reached adult values even at 20 days of age; however, they were not measured after dark, the adults were.

Mathiu *et al.* and Dawson and Whittow followed the changes in thermoregulatory capacity of late embryos and hatchlings in a mild thermal environment, Bech *et al.* from hatching to thermal independence in a severe one.

Below the 36°C at which *Puffinus pacificus* eggs are incubated, the T_b and V_{O_2} of embryos varied directly with T_a. This was true even in pip-holed eggs with improved access to oxygen. They were ectothermic (Fig. 7.4A). but once free of their shells, their down dry, 24-h-old hatchlings were endothermic. Their down averaged 3% by weight of their body mass of 38.9 g. When cooled, their V_{O_2} increased to a maximum at T_a 25°C, the heat generated maintaining a T_b of 35°C at T_as between 25 and 35°C. However, they tended to become hypothermic below 25°C (Fig. 7.4B). Above a T_a of 36°C the hatchlings increased evaporative water loss and (f) rose, from *c.* 17 breaths min^{-1} at 30°C to around 102 min^{-1} at T_a of 40°C, but this only gave incomplete control. The highest (f) for hatchling shearwaters (131 min^{-1}) at 40°C was lower than

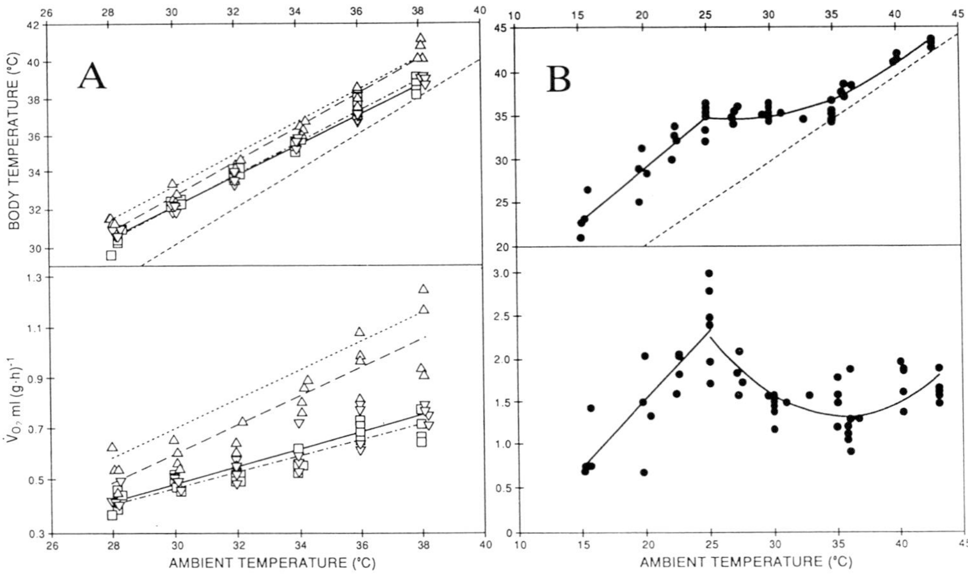

Figure 7.4 *(A) Oxygen consumption* (V_{O_2}) *and body temperatures* (T_b) *of embryo* P. pacificus *in unpipped* (▽, ——), *externally pipped* (□, .–.–.–), *internally pipped* (△ – – –), *and pip-holed eggs* (.....). *The lower dashed line in the top panel links equal values of* T_b *and* T_a. *(B) Oxygen consumption and* T_b *of hatchling* P. pacificus. *Note that 25°C represents an important boundary in the thermal dependence of both functions. Dashed line in top panel links equal values of* T_b *and* T_a. *From Mathiu* et al. (1992).

that of the adults (260 min^{-1}) at that temperature, although (f) usually tends to scale inversely with body size according to standard allometric equations. As Mathiu *et al.* (1992) point out, these limitations emphasize the benefits of the thermal protection given by the burrow.

Dawson and Whittow (1994) found no incipient endothermy in late embryos of the two albatrosses whereas hatchlings <24 h old also showed endothermic responses despite the different pipping regimes of the shearwater and the albatrosses (*The Petrels*, p. 330). The albatross's endothermic response requires a higher metabolic rate than that during the hatch, and Dawson and Whittow recalled that it is during the posthatching stage that most *D. immutabilis* chicks die (Fisher, 1975b).

The chick of *T. antarctica* is born where T_a may fall to -25°C, in open situations, and is brooded for about 11 days. The chick then weighs 150–200 g, and its T_b has climbed to and stabilized at 39.6 ± 0.9°C ($n = 35$). The TNZ of the hatchlings studied by Bech *et al.* (1991) extended from 28°C to at least 35°C. In the metabolic chamber the *T. antarctica* hatchling's oxygen consumption increased below T_a of 28°C, the thermal conductance and T_b fell, but even at T_a of 12°C it could maintain a gradient of 23–24°C between its body and that of the air. For the independent 11-day-old chick the TNZ

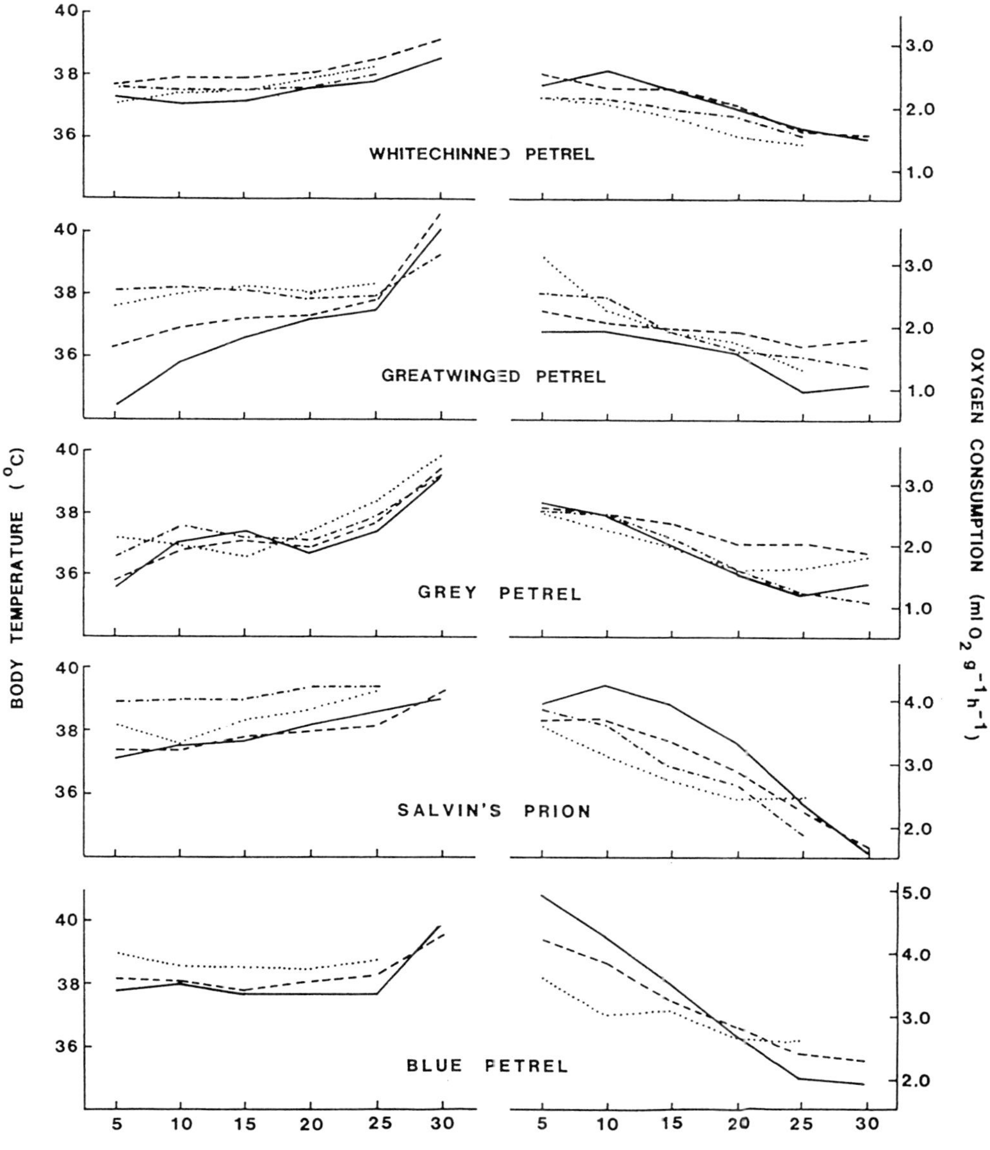

Figure 7.5 *Body temperatures and rates of oxygen consumption of burrowing petrel chicks at different ambient temperatures. (——), 1 day old; ––––, 3 days old; .–.–.–, 5 days old; ...., 7 days old. Normal burrow temperatures 2.1–12.8°C. From Brown and Prys-Jones (1983).*

was narrow, only about 24–26°C, above which V_{O_2} increased rapidly as it also did at lower values of T_a.

Oxygen consumption of 1–7-day chicks of five burrowing petrels measured in a metabolism chamber rose with decreasing temperature (Fig. 7.5) indicating some ability to control T_b (Brown & Prys-Jones, 1988). Above 25°C T_b and V_{O_2} tended to change, suggesting heat stress for chicks normally in burrows at T_as 2.1–12.8°C.

Brown and Prys-Jones ran checks on dead chicks in a wind tunnel to assess the insulating effects of their downy coats. The chicks lost heat 2.5× faster at wind speeds of 7.5 m s^{-1} than they would have in a burrow, but without their down the loss would have been 13–15× greater.

These authors also calculated thermal conductances for 1–7-day-old chicks based on a formula of Ricklefs and Roby (1983). Conductances decreased with age as in *P. puffinus* (Bech *et al.*, 1982). As Ricklefs (1989) pointed out, conductance is a two-way phenomenon, affecting not only loss of heat to the surroundings but also transfer of heat from parent to chick. Thus chicks of *Pelecanoides urinatrix* need brooding continually because they have low thermogenic capacity whereas those of similar-sized *Pachyptila turtur*, while having low conductances, have a high thermogenic capacity (Ricklefs, 1989), and need little brooding.

Figures for conductances are derived from formulae; no one seems to have measured directly heat transfer through subdermal fat, skin and feathers, or differences arising from different feather postures, of any tubenosed bird. Furness and Burger (1988) did relate plumage mass to body mass (BM) for 45 species of flying seabirds (plumage mass = $0.129\ BM^{0.129}$, both in g), and found that high latitude species had relatively heavier plumage than low latitude ones, but they did not state what petrels they included.

Bech *et al.* (1991) suggested that hatchling petrels raised in high latitudes have higher resting metabolic rates (RMRs) than those of low latitudes and Brown and Adams (1988) also deduced that high latitude embryo and neonate petrels have comparatively high metabolic rates, a high oxygen demand in cold climate species possibly preadapting the chicks to the higher thermoregulatory demands and, in burrowing forms, to the rapid development of homeothermy. Klassen and Drent (1991) with data for 16 species, determined that hatchling log RMR (ml O_2 day^{-1}) = 1.838 + 0.785 log BM (g), the rates tending to vary with latitude, petrels breeding towards the poles having higher hatchling RMRs. As the neonate petrel in these regions is as closely brooded as the egg, low T_a would seem to have little direct effect. However, high RMR in hatchlings presumably carry over into the post-brooding stage when high RMR and good thermogenic capacity may be essential for chick survival, particularly at night. Do Antarctic tubenoses cover their chicks at night even when daytime brooding has ceased?

There are some bigger differences even within regions. For the Laysan Albatross hatchling the mass-specific RMR was 45% higher than that of the adult (0.77 ml O_2 g^{-1} h^{-1} *vs.* 0.53 ml O_2 g^{-1} h^{-1}), whereas that of the Bonin Petrel nesting nearby was the same as the adult (1.1 ml O_2 g^{-1} h^{-1}) (Grant & Whittow, 1984); more data seem needed to support a general latitudinal effect.

III Incubation physiology

Petrel eggs seem to be incubated at temperatures similar to those of other birds when continually covered (Whittow, 1984), despite the parent's low T_bs. Neither brood patch nor egg temperatures are easily determined and dummy eggs monitored

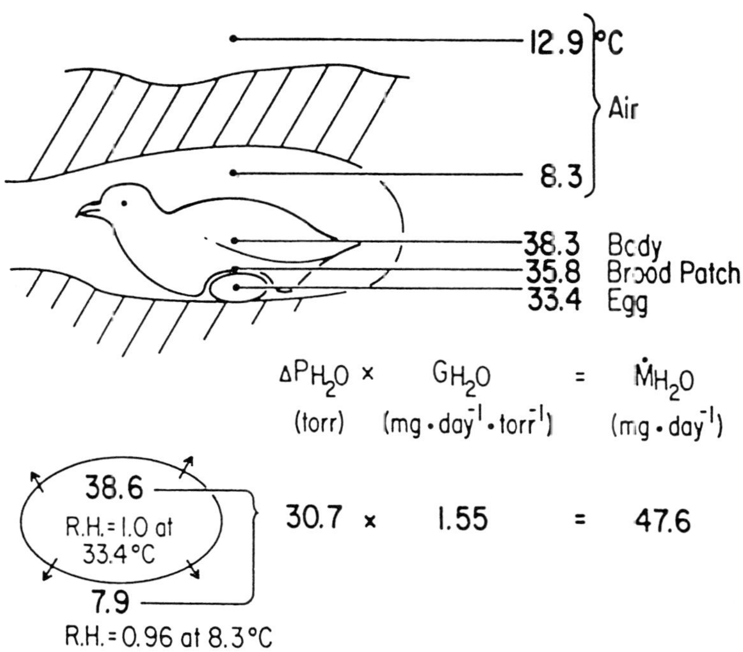

Figure 7.6 *Ambient, nest chamber, body, brood patch and egg temperatures for incubating Leach's Storm Petrels. At the lower left are shown the estimated water vapour pressures* P_{H_2O} *(38.6 torr) in the egg and in the nest chamber air (7.9 torr). The difference between these (* P_{H_2O}*) 30.7 torr × the shell conductance* G_{H_2O}*, predicts the daily rate of water loss,* M_{H_2O}*. RH, relative humidity. From Rahn and Huntington* (1988).

remotely tend to give low readings partly because of the absence of heat generated by the embryos. The difference is significant, for example the central temperatures of eggs of *D. immutabilis* and *D. nigripes* rose from 34.2 to 37.3°C and from 33.7 to 36.3°C respectively during the course of incubation (Grant *et al.*, 1982b). Petrel eggs with inserted probes have evidently never produced viable young, though this has been achieved with the gull *L. dominicanus* (Wilkinson, unpubl.).

The sparse data suggest that a gradient of 1–4°C exists between the petrel's brood patch and the mean egg temperature (Fig. 7.6 & Table 7.2) although this seems likely to decline in older embryos. The difference with *Pterodroma hypoleuca* was only 1.1°C and Grant and Whittow (1983) concluded that the incubation metabolic rates (IMRs) of this bird and of *D. immutabilis* were below or equal to their basal metabolic rates (BMRs), implying that neither needed to generate extra heat for transfer to the egg and whose thermal conductance was similar to that of the feathers shed from the brood patch.

A *Embryonic growth rates*

Partly because of their long incubation periods, the absolute embryonic growth rates (EGRs)—the hatchling's mass divided by its incubation period—of petrels are the lowest among birds, ranging from 6.5 g day^{-1} for *D. exulans* down to 0.25 g day^{-1} for *Oceanodroma leucorhoa*. A petrel embryo whose hatchling weighs 50 g has an EGR of 0.95 g day^{-1} whereas the embryo of a *Larus* gull which also hatches at 50 g grows at 1.95 g day^{-1} (Rahn *et al.*, 1984). Based on figures for 40 seabird species these workers estimated that:

$$\text{EGR (g day}^{-1}) = 0.035 \text{ Hatchling Mass (g)}^{0.82}.$$

Klassen and Drent (1991) discerned a link between hatchling RMRs and EGRs. They concluded that EGRs are higher in petrels of high latitudes than in those of the tropics, as is apparent from the more rapid development of petrels of the polar regions, most being fulmars (*The Petrels*, Chapter 14.IV). Their allometric equation for EGR based on data for 15 tubenoses was:

$$\text{EGR (g day}^{-1}) = 0.035 \text{ Hatchling Mass (g)}^{0.847}.$$

B *Water relations of eggs*

The total water loss through the pores of petrel eggs during incubation—15.7% of the fresh egg mass of 18 spp. determined by Rahn and Whittow (1988) and 15.8% for four subantarctic species measured by Brown and Adams (1988)—is similar to that of other birds. Once the eggs 'pip', however, water is lost rapidly by convection and diffusion.

Rate of water loss is governed by the water vapour conductance of the shell and its membranes and the partial pressure difference across the shell, a relationship described by Paganelli (1980).

Rahn and Whittow (1988) determined the water vapour pressure differences for the eggs of seven petrels as averaging 21.3 Torr, typical of other birds. However, daily rates of water loss for tropical and high-latitude petrels calculated by Whittow (1984) and Brown and Adams (1988) were all lower than predicted from egg mass figures (Fig. 7.7A). If allowance is made for incubation times, the high values for these bring the rate losses for tropical and cold climate species close to that of the formula for birds in general:

$$M_{H_2O} \text{ (mg day}^{-1}) = 130.4W^{0.977}/I^{0.937} \text{ (g day}^{-1})$$

(where W is fresh egg mass (g); I is incubation period (days) and M_{H_2O} is the daily rate of water loss), emphasizing the importance of the long incubation period (Ar & Rahn, 1980).

Water vapour conductances (G_{H_2O}) of bird eggs are related to egg mass:

$$G_{H_2O} \text{ (g day}^{-1} \text{ Torr)} = 0.384\text{W (g)}^{0.814}$$

and these for petrels are also lower than deduced from the above formula (Fig. 7.7B).

Resistance to the outflow of water vapour may be because of the length of the diffusion pathway, this being proportional to shell thickness, to the number of pores and the total pore area (Whittow, 1984). As, with the exception of fulmars

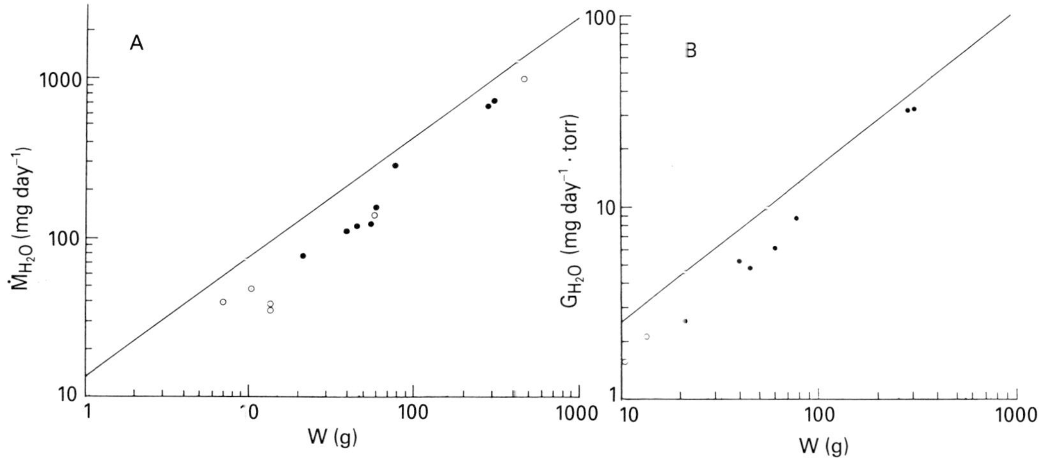

Figure 7.7 *(A) Relationship between daily rate of water loss (M_{H_2O}) and fresh egg mass (W) in tropical (●) and higher-latitude tubenoses (○). The line represents the relationship* $M_{H_2O} = 13.243W^{0.754}$ *(Ar & Rahn, 1980). (B) Relationship between water vapour conductance of egg shells (G_{H_2O}) and fresh egg mass (W) in tropical (●) and higher-latitude (○) tubenoses. Solid line represents* $G_{H_2O} = 0.384W^{0.814}$ *(Ar & Rahn, 1978). From Whittow (1984).*

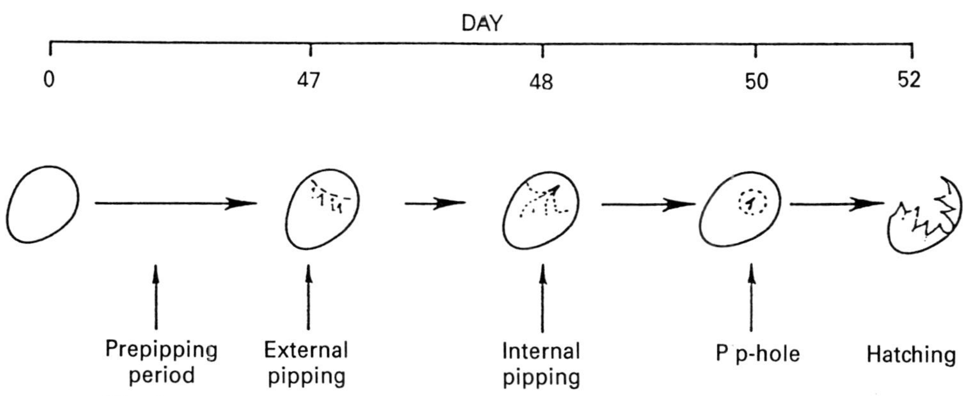

Figure 7.8 *Sequence of events in the pipping of Wedge-tailed Shearwater eggs. From Whittow (1984).*

(*The Petrels*, p. 295), procellariiform shells are thinner than those of most other birds, the low conductances apparently do not arise because of long diffusion paths. However, estimations of total functional pore areas suggest that these are smaller in petrels than in other birds, mainly by having fewer pores in their shells.

With shearwaters and gadfly petrels the shell proper is first fractured ('external pipping') then the air cell is pierced by the embryo's bill ('internal pipping') (Fig. 7.8). This start of pulmonary ventilation has little effect on water loss as the embryo is breathing the gas in the cell. When a distinct pip-hole appears a spurt in water vapour loss occurs which now includes that from the embryo's pulmonary tract. The gaseous changes associated with pipping in *Puffinus pacificus* were elaborated by Pettit and Whittow (1982a,b).

In *D. immutabilis* and *D. nigripes*, the sequence is different, internal pipping preceding star fracture, and the time between that event and the chick's freedom from the shell is correspondingly shorter than in the smaller tubenoses, the total water loss during the prepipping period relatively greater in these larger eggs.

C *Oxygen uptake of eggs*

Oxygen diffuses into the egg as it is consumed by the embryo and by the same pores through which the water vapour leaves. While the rate of water loss is constant before pipping, the oxygen demand increases as the embryo develops (Fig. 7.9). Similar curves have been given for *Oceanodroma leucorhoa* (Rahn & Huntington, 1988), for *Oceanites oceanicus* by Williams and Ricklefs (1984), for *P. pacificus* (Zhang & Whittow, 1992), and for unpipped eggs of *D. exulans, M. giganteus, Procellaria aequinoctialis, Pterodroma macroptera* and *P. salvini* by Brown and Adams (1988). The rate of O_2 uptake is boosted once internal pipping occurs, while with those species in

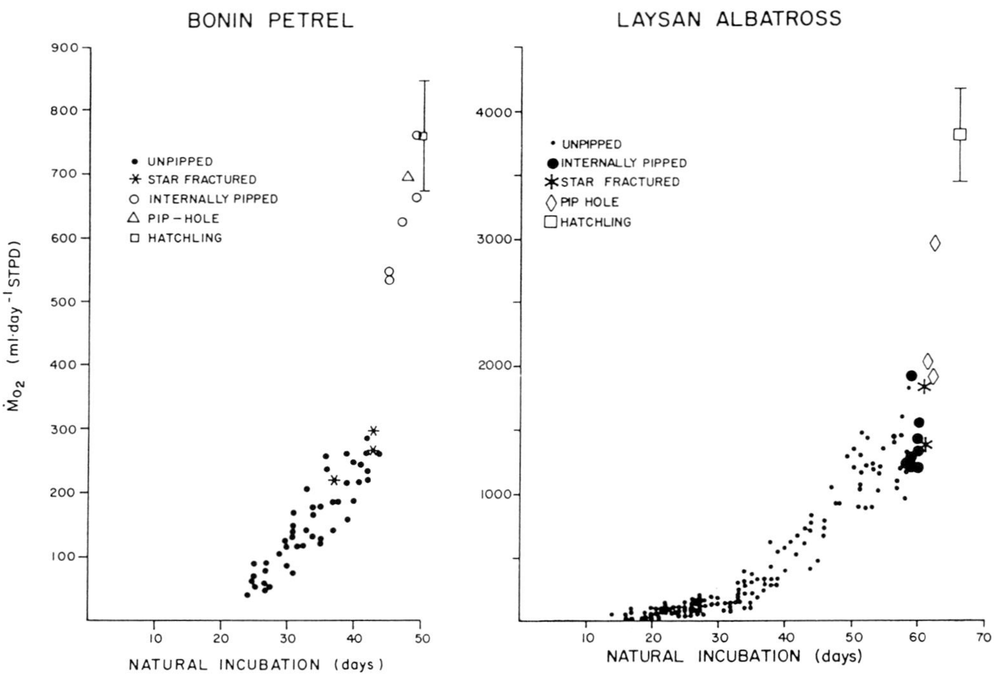

Figure 7.9 *Embryonic oxygen consumption* (M_{O_2}) *during incubation in* Pterodroma hypoleuca *and* Diomedea immutabilis. *After Pettit* et al. (1982*a*,*b*).

Table 7.3 Energetic costs of embryonic development in Laysan and Black-footed Albatrosses. From Pettit *et al.* (1982a)

	Laysan Albatross	Black-footed Albatross
Cost pre-IP incubation		
litre O_2	17.3	17.5
kJ	347.4	351.4
ml O_2 g tissue^{-1}	116.6	112.1
kJ g tissue^{-1}	2.4	2.3
Cost pipping-hatching		
litre O_2	12.4	12.8
kJ	249.0	256.0
% Total cost	41.8	41.9
Cost total incubation		
litre O_2	29.7	30.3
kJ	596.4	607.4
ml O_2 g egg^{-1}	104.2	99.2
kJ g egg^{-1}	2.1	2.0
ml O_2 g hatchling tissue^{-1}	156.3	144.7
kJ g hatchling tissue^{-1}	3.1	2.9

Note that g tissue^{-1} refers to yolk-free wet embryonic mass; g hatchling tissue^{-1} refers to total dry hatchling mass; g egg^{-1} refers to initial egg mass.

which shell fracturing is the first event, the resulting leak allows the rapid inflow of air. However, the pipping sequence is known only for four tropical tubenoses, not for any high-latitude one. In these the percentage of total O_2 consumed between pipping and hatching decreased with egg size being 50, 42.4, 41.8 and 41.9% for *Pterodroma hypoleuca, Puffinus pacificus, D. immutabilis* and *D. nigripes*, respectively, the total O_2 consumed being 5610 ml, 7776 ml, 29 671 ml and 30 246 ml respectively (Pettit *et al.*, 1982a,b; Whittow, 1984). The energetic costs of embryonic development in the two albatrosses are given in Table 7.3. These data suggest that procellariiform seabird embryos consume more oxygen (99–208 ml g^{-1} fresh egg mass) than most other birds (102 ml g^{-1} fresh egg mass), a figure said to be independent of incubation time. As in other aspects of their physiology, differences between petrels and non-procellariiforms appear to be linked to the former's abnormally long incubation periods.

Brown and Adams (1988) found that total embryonic oxygen consumptions of five subantarctic species considerably exceeded that expected from standard allometric equations. They were about 30% above those of tropical petrels with similar sized eggs but, as prepipping O_2 intake was broadly similar for both groups, the difference must be mainly due to consumption between pipping and hatching. This is not surprising. This stage sees the start of pulmonary breathing and the chicks must struggle to break free of the shell and to unfold their limbs, and as the parents tend to lift slightly, the cold ambient air will impinge on the hatchling.

D *CO_2 production of eggs*

The partial pressures of carbon dioxide in the air cells immediately before pipping have been measured in the four tropical species discussed above, together with that of the air around the egg. The air cell partial pressure of CO_2 was highest in *Pterodroma hypoleuca* (45.5 Torr) perhaps because the CO_2 partial pressure of the burrow air (5.0 Torr) was higher than for the other petrels (Whittow, 1984).

IV Chemoreception

Taste has not been investigated in petrels. That taste receptors are functional is suggested by the rejection of items such as orange peel, paint chips or mineral oil when touched by the tip of the beak, but the locations and functions of any receptors on the bill or tongue, in the oral mucosa or around salivary glands are unknown. Taste may be important for scavengers but would seem less so for species feeding by dipping, or why would plastic pellets be so readily ingested?

Olfaction, however, has long been suspected as a major sense (e.g. Lachmund, 1674), not only in detecting food, but also in short-range location of colonies, nests and even partners. Many fishermen have believed that tubenoses can smell. O'Reilly (1818), for example, thought that the 'mallemucks' (*F. glacialis*) smelled the blubber he tossed overboard when no birds were in sight. There are many reports of petrels being attracted to slicks of animal oils, including the famous case of Rollo Beck rowing into an empty sea off Peru, tossing out fat and meat scraps as he went, then doubling back to collect the petrels which 'skipped and danced like butterflies along a blossoming hedge-row' (Murphy, 1925, p. 271). Vision could obviously have played a part there, but Collins's (1884) experiment in throwing out pieces of liver in dense fog with no birds visible, suggests that the initial response was to a scent trail. As the pieces drifted astern storm petrels or a shearwater would appear, weaving from side to side 'like a dog working up a scent', until they reached the food.

Spreading 'chum' (e.g. codliver oil, mixed with cereal) is still used to toll up tubenoses for observation and collection, particularly storm and gadfly petrels.

In addition, the great development of the olfactory bulbs of the brain and the extensive mucosa of the nasal cavity (Chapter 10.VIII.A) argue for Procellariiformes as macrosmatic animals.

An early review is that of Gurney (1922) but for petrels the series of papers by Wenzel (1967, 1980, 1987, 1991) and Bang and Wenzel (1985), should be consulted.

A *Field investigations*

Trials with controls to eliminate visual and other clues were conducted by Grubb (1972, 1973, 1974, 1979) on land and at sea.

Grubb (1972) tested the responses of *Puffinus gravis, P. griseus, Oceanodroma leucorhoa* and *Oceanites oceanicus* to randomly presented olfactory stimuli produced by pairs of sponges, one soaked with codliver oil, the control in seawater, both held 1 m above the ocean surface. The oil-soaked sponge was more attractive to

O. oceanicus (115 passes *vs.* 36 to the control), $P < 0.01$); to *P. gravis* (56 to 3; $P < 0.025$); to *O. leucorhoa* (9 to 0; n.s.); but less attractive to *P. griseus* (6 to 12). Non-procellariiforms did not react to either sponge.

Leach's Petrels flying by night over a clearing some 100 m from the nearest burrows were presented sequentially with (i) the species's 'purr' call + nest material; (ii) the same call + a tray of sham nest material; (iii) nest material and 'sounds' from a blank tape; and (iv) blank tape + forest floor litter (Grubb, 1974). The 'purr' call, of course, was highly attractive, many birds spiralling down to pass and repass the speaker. Grubb scored 418 passes over nesting material as against 249 over the sham material during equal time periods ($P < 0.025$). There were 5737 flights over the playback 'purr' + nest material, more than over the nest material alone ($P < 0.005$) but not significantly more than the 4515 passes for the 'purr' call alone; but the birds flew closer to the combined stimuli. These results indicate that sense of smell played at least a secondary role in attracting the birds.

Petrel colonies generate a musty odour which may enable a boatman to locate a breeding island at night or in fog by 'following his nose' (pers. obs.). Nests are also marked by their smell, whether by the bird's musty body odour transferred to nesting materials or, with fulmars, augmented by splashed stomach oil, which tends to be long lasting and odorous, especially on warm days.

Grubb (1974) used a 'night-scope' to score the behaviour of *Oceanodroma leucorhoa* nest-finding after dark. His tests did not eliminate proprioception and ambient light values were not given, but vocal guidance, unless ultrasonic, was absent. Typically, incomers hovered over the tree canopy before alighting downwind from their nests, then shuffling upwind towards it, often swinging their heads from side to side, with beaks close to the ground. In calm air they landed closer and followed more circuitous routes and tended to bump into obstacles. Some changed course before reaching their burrow, behaviour consistent with their having passed through the odour plume and turning back to find it. Such behaviour is seen in other species, such as shearwaters, which keep their bills low and in the best position to pick up a scent, tend to travel upwind, often swing their heads laterally and frequently stick them down burrows only to withdraw them abruptly (not in response to any call from within) before turning to enter their own nests nearby.

James (1986a) stated that Manx Shearwaters do not mark the ground or rocks near their nest with body odour. This is untrue of many petrels. For example, the trunks and limbs climbed by *Pterodroma hypoleuca* and *P. cookii* for take-off are readily identified by the human nose and presumably by conspecifics. However, James found that *Puffinus puffinus* landed into the wind but approached its nest from any direction. Placing household odour wicks in burrows to mask the petrel odour (to the human nose) did not confuse returning birds, and when established and odour-marked nesting boxes were shifted, the owners returned to the original nest location, ignoring the scent of their own nests but a metre away. Brooke (1990, p. 215) concluded that *P. puffinus* does not use smell to reach its nest.

Grubb (1974) plugged the nostrils of 12 *O. leucorhoa* to see if they could get home without the help of olfactory clues. None returned to its nest but 11 out of 12 with unobstructed nostrils did so. In parallel trials with 23 whose olfactory nerves were cut none came back, whereas 17 of 23 sham-operated controls did so.

Shallenberger (1975) failed to get such clear-cut results with *P. pacificus*: in small-

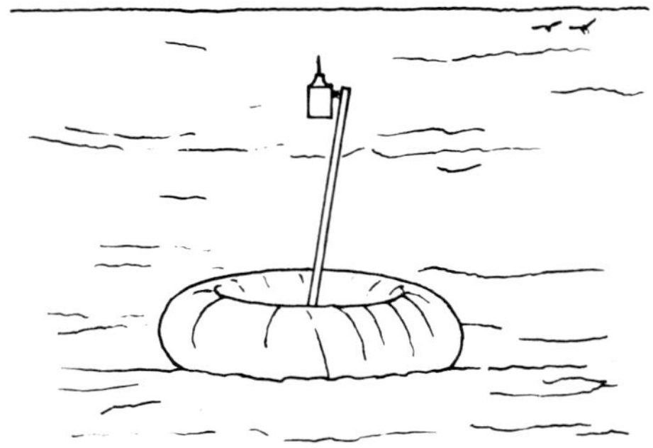

Figure 7.10 *Raft supported by inflated tyre for presenting odorous substances to birds at sea. Bottle with wick is held on gimbals. After Hutchison* et al. *(1984).*

scale trials 25% of those with transected olfactory nerves successfully returned. Evidently, visual means were mainly relied on by James's and Shallenberger's shearwaters nesting in the open, a different situation from that of Grubb's storm petrels.

Jouventin (1977) found that captive *Pagodroma nivea* could locate buried fish and identify where others were hidden in simple choice tests with visual, tactile and auditory clues absent. These petrels also persistently pecked at the one hand with which the experimenter had manipulated the fish while ignoring the other one, much as Grant *et al.* (1983) noted *Pterodroma hypoleuca* pecking the hand that had held its partner, ignoring the 'clean' hand simultaneously offered.

Hutchison and Wenzel (1980) and Hutchison *et al.* (1984) in at-sea trials, systematically presented seabirds with a range of olfactory stimuli of which tuna oil, squid and krill homogenates were food related, vegetable oil and bacon fat less natural, and petroleum, hexane and mineral oils unrelated to the natural foods. The test species were *D. nigripes, Puffinus griseus, P. creatopus, P. puffinus, P. bulleri, P. tenuirostris, F. glacialis* and a range of *Oceanodroma* spp. The materials were presented by day as surface slicks, and, at 1 m above the surface, from bottles containing the control material from which a saturated cotton wick protruded 15 cm (Fig. 7.10).

The petrels approached the food-related materials mainly from downwind, with the albatrosses, shearwaters and notably the fulmars, zig-zagging across the breeze as if sensing an odour plume. The effect was greatest during trials with the wick when many birds circled the bottle and repeated the hound-like coursing approach. None of the gulls, terns, cormorants, etc. showed any interest in the odorous substances nor were the petrels attracted to the various control and non-food materials (Fig. 7.11).

In other tests conducted during 50 cruises, Hutchison *et al.* (1984) again found that 77–100% of *F. glacialis* and *P. griseus* approached from downwind of the wick, but were sighted only once or not at all with heptane and seawater controls. The most attractive was the volatile fraction of codliver oil, vacuum-extracted from whole oil and dissolved in heptane. While fulmars seldom came near the squid or krill homogenate, *P. griseus* was strongly attracted to both, but more fulmars approached the food-related baits in foggy weather than did the shearwaters.

Codliver oil lures rested in the Southern Ocean by Jouventin and Robin (1984) and

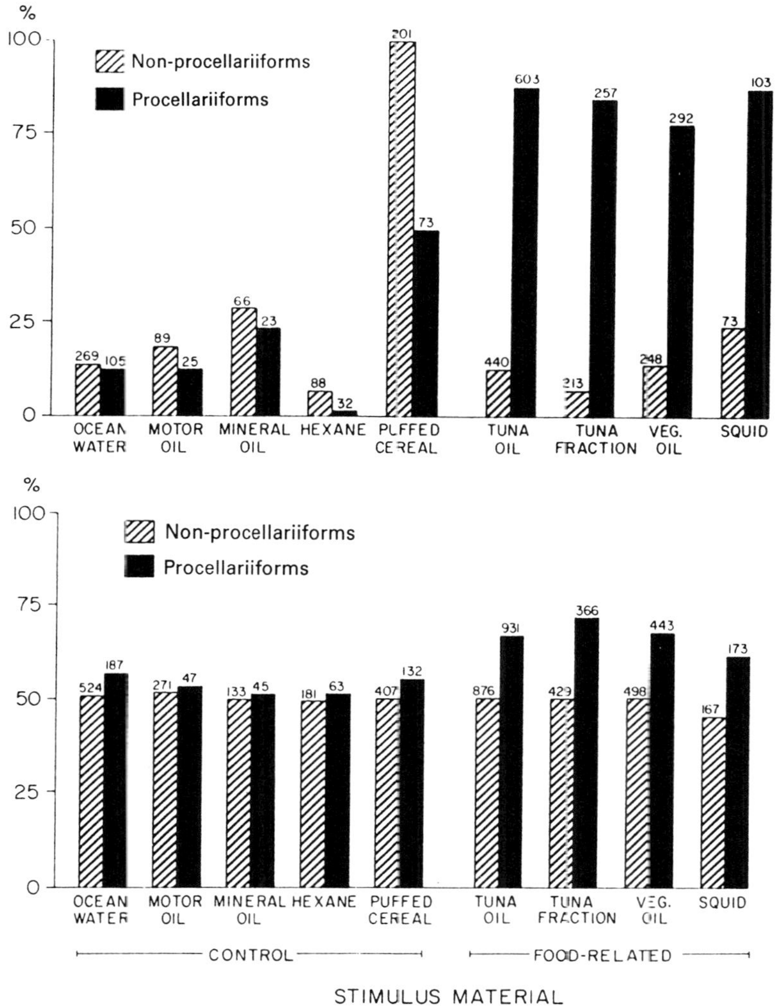

Figure 7.11 *(Upper) Percentage of procellariiforms and non-procellariiforms coming within 10 m of stimulus, landed on water and/or tried to feed near stimulus. Total number flying upwind is shown above each bar. (Lower) Percentage of procellariiforms and non-procellariiforms approaching each stimulus from downwind. Total number sighted, regardless of direction, is shown above each bar. From Hutchison and Wenzel (1980).*

Lequette *et al.* (1989) proved attractive to *Oceanites oceanicus, Fregetta tropica, Pagodroma nivea* and *D. capense*, but not to prions or diving petrels (Fig. 7.12). Those attracted approached on a serpentine course, as noted by earlier experimenters. Verheyden and Jouventin (1994) scored the reactions of seabirds to isolated codliver oil slicks in pelagic seas, and counted birds before and after releasing the odorant from an airtight box. The results suggest that storm petrels, fulmars and some shearwaters (notably *Puffinus gravis*) use their sense of smell when finding food, but that prions (five spp. tested), *Halobaena, Calonectris diomedea, Procellaria cinerea* and most albatrosses do not. Attracted birds came from upwind, 98% flying <1 m from the surface, apparently within a layer where the odorant was strongest.

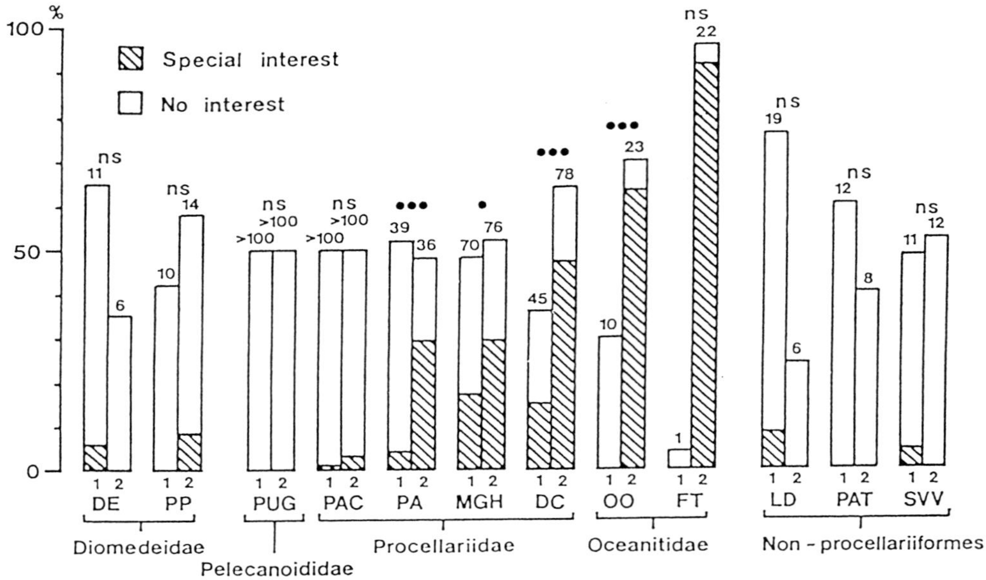

Figure 7.12 *Reactions of seabirds off Possession Island to sponges soaked with seawater (control) and codliver oil. 'special interest' = >1 flight over sponge, circling it, etc. Numbers observed over columns. DE,* Diomedea exulans; *PP,* Phoebetria palpebrata; *PUG,* Pelecanoides *sp.; PAC,* Pachyptila *sp.; PA,* Procellaria aequinoctialis; *MGH,* Macronectes *sp.; DC,* Daption capense; *OO,* Oceanites oceanicus; *FT,* Fregetta tropica; *LD,* Larus dominicanus; *PAT,* Phalacrocorax atriceps; *SVV,* Sterna vittata *and* S. virgata. *ns, not significant,* (●), $P < 0.05$; (●●●), $P < 0.001$. *From Lequette* et al. *(1989).*

Diving petrels, with smaller olfactory bulbs and feeding underwater, where the nasal valves will close automatically, do not need olfaction in foraging but the lack of response by prions, which tend to take surface or near-surface crustacea, is puzzling as these tend to take surface and near-surface crustacea which, at heavy densities, can be detected by the human nose. The hovering, hesitant flight style of storm petrels would seem to help them in picking up and following an odour trail, but with their faster flight, larger species such as *Procellaria aequinoctialis* could easily miss a small source entirely. Odours from zooplankton churned up by ships' propellers may perhaps be detected by petrels in the wake.

B *Laboratory studies*

Grubb (1974) ran a series of Y-maze tests ingeniously devised to eliminate residual odours from previous trials. The subjects, Leach's Petrels, were given choices between air that had passed over the bird's nesting material and litter from areas with no petrels. In 106 trials 36 birds moved down the arm towards their own nest material, 13 down the control arm ($P > 0.01$) and 57 did not respond. When stomach oil or preen gland secretion was presented, neither was significantly favoured.

Wenzel and her associates also ran olfactometer tests on captive *Puffinus opisthomelas* and *D. nigripes* while monitoring changes in pulse and respiration rates and

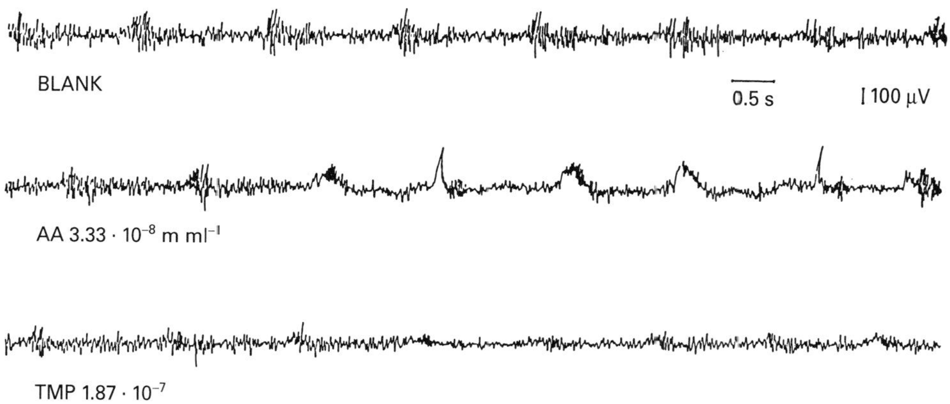

Figure 7.13 *Bulbar activity from* Puffinus opisthomelas. *Top line, intrinsic activity; centre, response to amyl acetate (AA); bottom, response to trimethylpentane (TMP). From Wenzel and Sieck (1972).*

electrical activity in the olfactory bulb (Wenzel & Sieck, 1972). The test odorants were amyl acetate (AA), pyridine and trimethylpentane (TMP)—none likely to mimic natural odours. Both species responded by changing heart rates and electrical activity, but the changes were not significantly greater than those of a range of non-procellariiform birds with smaller olfactory bulbs. Shibuya and Tonosaki (1972) also got significant responses to AA in extracellular recordings in the olfactory mucosa of *C. leucomelas*.

Figure 7.13 shows the EEGs of a *P. opisthomelas* shearwater breathing pure air, AA, or TMP. EEGs of resting birds had relatively high frequency bands (40–45 Hz) and spindle bursts coinciding with inspiration, as well as slow oscillations linked to the respiratory cycle. In response to AA, all showed pronounced increases in spike activity followed by short bursts at lower frequencies. Bilateral sectioning of the olfactory nerve abolished spindling and most high frequency activity. Respiration rates also changed (Fig. 7.14), but responses to mashed fish were not greater than to the 'unnatural' odorants.

Stimulation of the olfactory nerve of *F. glacialis* evoked responses in the bulb and forebrain projection areas—ventral hyperstriatum, parolfactory lobe and prepiriform cortex (Hutchison & Wenzel, 1980). Tested with codliver oil and krill extracts, spontaneous firing in the bulb was consistently suppressed, whereas some odours, for example cedrol, were excitatory. These authors suggested that the petrel olfactory bulb handles elaborate chemosensory input and communicates widely with the anterior areas of the brain, and in a more complex manner than in the better-studied pigeon.

Evidence for some olfactory capability has now been gained for about 27 petrels including members of all the families except the Pelecanoididae, but no gadfly petrel has been tested.

Evaluating olfaction poses problems in eliminating clues from other senses, artefacts due to stress in captive birds, etc. Despite such limitations, the evidence is strong that some petrels can respond to odours, but that they do so to help food-finding under natural conditions or use smells to find and identify nests or partners, particularly at night, needs further demonstration. The value of being able to smell

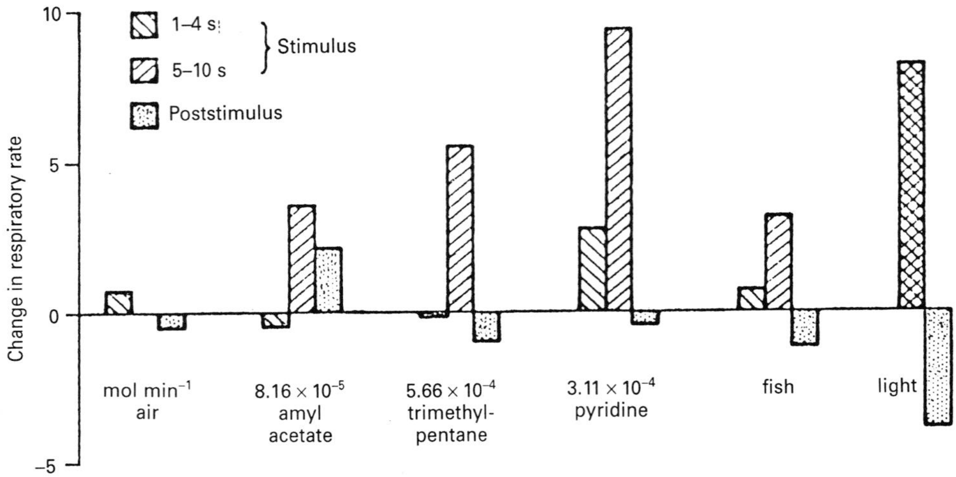

Figure 7.14 *Changes in respiratory rates of* Puffinus opisthomelas *to stimuli. The response during the 5–10 s stimulus period was significant ($P < 0.05$) to the visual stimulus and to all odours except pyridine. From Wenzel (1967).*

food after dark is obvious and that *Pagodroma nivea* has the largest olfactory bulbs sits well with its lifestyle among the pack-ice, where low temperatures must reduce chemical volatility and long winter nights visibility.

Many of the potential uses of olfaction were outlined by Bang (1966). That the general colony smell could guide birds on the final stages of their incoming flights was suggested by Grubb's work (1979) and, just as there may be colony-specific vocalizations (Chapter 4.II), similar olfactory markers for colonies would be advantageous to birds at night or in fog.

The petrel's musty body odours may also be significant if these vary sufficiently to provide olfactory 'fingerprints' for species, bird, or burrow. Olfactometer tests with a few *Puffinus pacificus* were inconclusive: some responded with increased breathing rates to odours of their mates, while others breathed more slowly (Shallenberger, 1975). As in other experiments with restrained birds, the effects of stress may have overridden other responses.

Giant petrels provide an apparent paradox. Why are these predominately diurnally active birds and their nests the strongest smelling of all tubenoses, their eggs holding the odour for 100 years or more? Is this a consequence of their carrion diet?

The possibility that pheromones are used, triggered hormonally and tied to the sexual cycle was raised by Bang (1966). That there are seasonal differences in the body odours of *Macronectes* and *Bulweria* was suggested by Shaughnessy and Voisin (1981), and Thibault and Holyoak (1978) respectively; such possibilities need following up. The petrels would also seem a suitable group in which to investigate the possibility of a vomeronasal system for signalling reproductive condition.

Although there is no doubt that many birds that feed after dark still use visual clues in home-finding (Warham, 1955; Serventy *et al.*, 1971 p. 27; James, 1986a), the additional cue from a scent trail could be of great value, lessen the risk of predation while blundering around, and help explain how birds still seem to find their homes in caves and dense forests where there may be no light; quantitative data on orientation in such situations are lacking and needed.

Wenzel (1986) and Matochik *et al.* (1991) listed a range of other aspects needing further study. For example, *P. pacificus* is the only tropical petrel so far investigated. Do other tropical species, for example *Pterodroma*, also have large olfactory bulbs and how do these compare with congeners of higher latitudes? Identification of the precise chemistry of the biogenic odours, for example of plankton-rich water or shoaling fish, to which the birds may be sensitive, could allow further field testing and perhaps even the development of a supraoptimal odour stimulus like the audio-one for gadfly petrels (see Chapter 4.XVIII).

V Photoreception

Although the gross anatomy of the eyes of several tubenosed birds has been examined over the years (see Chapter 10.VIII.B) the study of the functioning of the visual system in relation to their lifestyles has only just begun. For instance, no proper evaluation of their colour vision has been made although they appear to detect a wide range of colours.

An interesting recent discovery is of the existence of a special area in the retina on to which the narrow binocular field projects (Hayes *et al.*, 1991) (see Fig. 10.23). These authors postulate that this helps the detection of rapidly moving stimuli when the bird scans the sea's surface from a few centimetres above it. As the petrel moves forward, it is only in this limited part of the visual field that the image moves relatively slowly across the retina, increasing the chance of detecting prey near a shifting surface. However, ther are no apparent modifications to lessen interfering surface reflections, apart from the 'lens-hood' of dark feathers around the eye sockets of species such as mollymawks.

The *area centralis horizontalis* or visual streak (see Chapter 10.VIII.B) has been supposed to increase sensitivity to vertical movements of objects in relation to the horizon where most items of interest to petrels are likely to lie, and to help fixation of the head in relation to the horizon. Martin (1990) has suggested that it might also facilitate measuring the elevation of the sun or stars by the migratory species.

That most of the refraction takes place at the lens (at least in *P. puffinus*) and much less at the cornea, may be an adaptation to improve underwater performance where the corneal contribution is lost. No second fovea has been described for any petrel such as apparently allows a diving kingfisher to maintain a clear image of a target fish during the switch from air to water.

To test the visual actuity of *Pelecanoides urinatrix*, Brooke (1989) ran Y-maze tests giving them choice between a completely dark tunnel and one lit at 11.8 cd m^2 which could be dimmed with neutral density filters. When the light was reduced so that the birds no longer moved down the lighted tunnel to escape, it was presumed that they could no longer discriminate between the lit and unlit openings, and that the visible threshold had been reached. It was tentatively concluded that this lay between 1×10^{-5} and 1×10^{-4} cd m^2, that is, about equal to that of a domestic pigeon. As diving petrels seem to find their burrows on dark rainy nights and under forest canopies, they may need more than vision to find them. However, Brooke apparently made no attempt to 'decontaminate' the maze after each run, so that one or other arm could have been inadvertently tainted with olfactory markers.

Despite the prevalence of nocturnal activity, petrel eyes show no obvious external adaptations for life after dark. For instance they lack a *tapetum lucidum*, although *D. nigripes* and *D. immutabilis* are said to show a silvery-white eyeshine in a torch beam. The 'red-eye' effect in flash photographs is not due to reflection from a vascular bed, but probably from red or yellow oil droplets in the retinal cones (G. Martin, pers. comm.). Lockie (1952) did note that in the shearwater *Puffinus puffinus* the region of pure cones was small which he deduced allowed more rods for night vision, whereas in the fulmar *Fulmarus glacialis* there were more cones, as expected in a more diurnally active bird. However, examinations of the retinal cells have not indicated any great sensitivity to low illumination, the retinae not being rod-rich, while on the data presented by Martin and Brooke (1991), the brightness of the image thrown on the retina of *P. puffinus* is more like that of pigeons than of owls.

Hayes and Brooke (1990) speculate that the differences in the shapes of the high density regions in the retinae reflect feeding styles. For example, *P. puffinus* and *Pelecanoides urinatrix*, both of which have well-defined central areas, feed under water where the prey will be on roughly the same plane as the bird, so that a horizontal sensitive retinal area may be an advantage. A similar explanation is advanced in the case of *Phoebetria fusca*, also with a clear central band and likewise judged to spot its prey on the same horizontal plane. This explanation seems less convincing; sooty albatrosses tend to cruise quite high up and also tilt their heads when scrutinizing objects from above or to one side.

These authors noted a marked difference in the distributions of ganglion cells in *Pterodroma mollis* and *P. incerta* compared with that of *Lugensa brevirostris*. In the latter, the cells have a concentric distribution which Hayes and Brooke (1990) suggested is appropriate for a bird feeding by dipping which might benefit from a good sensitivity throughout its visual field. This petrel is thought to feed by night and on small prey so that its high number of ganglion cells—over twice as many as in the much larger eyes of *Phoebetria fusca*, for example—is not surprising. That *Lugensa* (body weight 330 g), has relatively large eyes is also borne out by the large retinal area (477 and 479 mm^2) in the two examples measured, compared with a bird such as *Puffinus puffinus* (470 g), two of which had retinal areas of 138 and 150 mm^2.

That interpretation of the functions of the patterns of the retinal sensitive areas is complicated is shown by Hayes and Brooke's finding that in both *L. brevirostris* and *P. assimilis*, birds with quite dissimilar feeding styles, the cell density contours tend to be concentric around the *ora terminalis*—in the first species the densities were high, in *P. assimilis* they were low. Also puzzling is the finding that the retinae of *P. puffinus* and *P. assimilis* are so different, the first having a well-localized central streak, the latter a more diffuse one, yet both having rather similar feeding styles.

VI Digestive system

Tubinares differ from most other birds in the role of the highly extensible proventriculus, the small gizzard, and the arrangement of the duodenal loop (*The Petrels*, p. 6). The histology of the proventriculus was examined by Kuroda (1957b, 1960b, 1986). In *M. giganteus* the 292-cm^2 surface area is 50% of the 584-cm^2 surface of the total gastrointestinal tract, as measured by Roby *et al.* (1989). Kuroda (1986) described the wall of the proventriculus as being paper thin when expanded with food, in great contrast to its thick walls when empty. He envisaged the helically-twisted intestine of *Pterodroma* (*The Petrels*, p. 85) as slowing the passage of digesta to allow more efficient absorption, the coils relaxing to permit the onwards flow of the remaining material. In a storm-wrecked *P. hypoleuca* the intestines were empty and the twists completely lost.

A *Differential digestion*

Many procellariiform prey, such as myctophilids, contain high lipid levels, often rich in wax esters. For example, female *Euchaeta antarctica* contain 30% by dry weight of lipid and up to 45% when carrying egg sacs, while *E. chrystallorophias* accumulates 25–35% lipid during the southern winter, consuming it steadily during the succeeding spring and summer (Littlepage, 1964). Copepod egg sacs are fragile and when swallowed by a petrel probably burst before reaching the proventriculus; other less vulnerable lipid stores are released during digestion.

Wax esters are usually regarded as indigestible by birds. How these are handled in the petrel digestive system was unravelled by Place and his collaborators (Roby *et al.*, 1986;, Duke *et al.*, 1989; Place, 1991, 1992a,b). They showed that the proventriculus acts like a separating funnel, the lighter lipid layer floating above the aqueous part of the food. The aqueous layer is the first part to pass through, and it is by this differential digestion and the slow passage of the lipid layer, that 'stomach oil' is separated out, either to be metabolized or spat out in defence—the oil first and, when that is exhausted, the aqueous layer below.

The lipid is light, specific gravity *c.* 0.83, and of low viscosity, difficult to retain in a collecting tube unless wax sealed. However, the layout of the backwards-facing pylorus and ascending duodenum ensures that the lipid layer is always uppermost except when disturbed by the refluxing described below.

Food may remain in the proventriculus for many hours, the lipid for days, and Place *et al.* (1989) confirmed earlier deductions that there is very little or no lipolysis here. Presumably this is where much proteolysis occurs and large prey items are broken down—the opening to the gizzard being too small to accept large pieces. The lipid and fatty alcohols are converted in the duodenum to fatty acids and then incorporated into triglycerides in storage depots (Place & Roby, 1986). The precise role of the tubular glands in the proventriculus still awaits attention.

The passage of the aqueous layer from the proventriculus is much more rapid than the lipid one: in 36-day-old *Oceanodroma leucorhoa* chicks the aqueous layer was held on average for 11 h in the proventriculus, the lipid one for 123 h (Place, 1992a). Once in the duodenum and upper intestine, the digesta are refluxed back to the gizzard and proventriculus. In the gizzard the digesta are mixed with powerful bile salts, the lipid emulsified into micelles and broken down in the upper intestine.

This pattern has also been demonstrated in *M. giganteus* and *Procellaria aequinoctialis* (Roby *et al.*, 1989; Jackson & Place, 1990). The system is a little different in the albatross *Phoebetria palpebrata*, the outlet to the pylorus being on the side of the gizzard and the evacuation of lipids being relatively rapid (Jackson & Place, 1990).

The arrangement in *Pelecanoides* helps explain why this is the only tubenose so far known not to store proventricular lipid (Warham *et al.*, 1976; Roby *et al.*, 1986). Here the pylorus and gizzard are rather dorsal so that the lighter lipid will tend to emerge first, the food passing through the intestinal tract much faster than in the other tubenoses, giving little time for the stomach oil to separate out. The average gastric emptying rate for *Pelecanoides georgicus* was 30% of lipid h^{-1}, mean retention time in the proventriculus 2.3 h, compared with 5.5% h^{-1} and mean retention time of 12.5 h in chicks of *M. giganteus* (Roby *et al.*, 1989).

B *Assimilation*

The refluxing system helps account for the highly efficient assimilation of non-polar wax esters. These energy-rich resources are particularly plentiful in zooplankters of high latitudes and their processing by tubenoses has been mainly studied in cold-climate species.

The ability of seabirds to use the wax ester moiety of their prey was confirmed by Obst (1986), who fed adult *Oceanites oceanicus* pure wax (hexadecyl oleate) or triacylglycerol (olive oil). The faeces of the wax-fed birds contained fatty alcohol and fatty acids, the products of wax hydrolysis, the birds themselves maintaining weight on a pure wax diet. The triacylglycerol was also digested, and more quickly than the wax.

Roby *et al.* (1986) fed ^{14}C-labelled cetyl palmitate to chicks of *Pelecanoides urinatrix*, *P. georgicus* and *Pachyptila desolata*. The proportion of the total energy given by the wax esters was 11% in *P. urinatrix*, whose main prey was *Euphausia superba*, 27% for *P. georgicus* on a diet of *E. superba* and *Thysanoessa*, but for the prion, eating krill and fish, only 8%. Recoveries of lipid from the faeces revealed that, in nearly all cases, <1% of the labelled ester was excreted.

Similar high metabolizable energy of wax esters by *Phoebetria fusca* and *Procellaria aequinoctialis* was found by Jackson and Place (1990), who also demonstrated a simultaneous capacity of these birds to take up a triglyceride—tripalmitin.

Place and Butler (1986) established the importance of bile salts in the hydrolysis of wax esters by reducing the intestinal bile salt concentrations in 35–50-day-old *Oceanodroma leucorhoa* chicks that had been fed ^{14}C-labelled triolein and cetyl oleate with the resin cholestyramine. Assimilation efficencies, based on analyses of the excreted material, were very high (94–96%). Experiments using their pancreatic extracts have shown that *Pelecanoides urinatrix, Pachyptila desolata* and *O. leucorhoa* broke down triglycerides at about the same rate as they did wax esters, unlike non-tubinare birds and mammals.

In *O. leucorhoa*, the bile salts were 95% taurine conjugates of chenodeoxycholate and cholate, with some phospholipid, cholesterol and relatively high titres of triglyceride (Place, 1992b). Place gained preliminary evidence that a bile-dependent carboxyl ester lipase in the pancreatic secretions is important in hydrolysing wax esters, allowing petrels to live off the 'fat of the sea'.

Subcutaneous and visceral deposits lack wax esters, and in *Puffinus tenuirostris*, for example, the composition of their fatty acids can be explained by the conversion of fatty alcohols derived from the wax esters of their prey, *N. australis*, into fatty acids of equivalent chain length and saturation (Bishop *et al.*, 1983; see Table 8.4).

An overall figure of 75% assimilation efficiency is often assumed in energetics calculations. Jackson (1986) fed fledgling *Procellaria aequinoctialis* known weights of their normal prey—light fish, squid and krill—to compare energy input with that lost in faeces and urine. She found 69.1, 68.0 and 67.5% energy assimilations for fish, squid and krill respectively. Jackson (1992) also looked at how long squid, pilchard and prawns remained in the guts of adult *Phoebetria fusca* which had been force fed 48 h previously. Retention times averaged 13.0 h for pilchard, 11.2 h for squid and 15.8 h for prawns.

The bony skeletons of fish, the mainly keratinous 'beaks' of squid and the chitinous exoskeletons of crustacea all appear to be ingested and their debris excreted. Although the vomited boluses of bones and feathers are found on albatross colonies, similar ejecta are seldom recorded for smaller tubenoses.

The chitin of prey such as krill (2.1–2.9% of their dry mass) contains about 17.9 kJ g^{-1} dry mass. However, their insoluble jacket hinders digestion of the arthropod tissues, so that some means of penetrating this barrier would be very useful to crustacean-feeders. Jackson *et al.* (1992) showed that chitinase from the gastric mucosa could hydrolyse chitin (a mucopolysaccaride polymer) to N-acetyl-D-glucosamine, which *P. fusca*, *Procellaria aequinoctialis* and *O. leucorhoa* could absorb.

VII Excretory system

Apart from work on heavy metals (see Chapter 8.II.F), the processes of nitrogen excretion and kidney function have yet to be studied in petrels. White faeces imply that the bird has eaten recently, pink faeces that the meal included crustacea, and green faeces that it is in a postdigestive state, its excreta tinted with bile pigments. Jackson (1992) found that *Phoebetria fusca*, starved until their faeces were green, had completed excretion by 50 h after the meal, whether this was of pilchards, squid or prawns.

VIII Cardiovascular system

Information on the working of the tubenose circulatory system is fragmentary. The circulation of the limb and of the ophthalmic rete have been referred to in Section IIB above.

Limited examinations of the properties of the blood include HbO_2 dissociation curves for *F. glacialis* (Clausen *et al.*, 1971) and for *M. giganteus* by Milsom *et al.* (1973). Both species had similar levels of haemoglobin (14.1 g% and 14.6 g%), comparable to values for other non-diving birds except penguins. Haematocrit values were 43% for *M. giganteus* and 48% for *F. glacialis*. The dissociation curves lie further to the right compared with those of mammals but the reduced affinity for oxygen cannot be explained by lower body temperatures as T_bs for the two groups are similar. The

Bohr effect in both fulmars was lower than in diving species such as penguins or grebes. Grant and Whittow (1983) gave a haematocrit figure of 39.3% for *D. immutabilis*, Block and Murrish (1974) and Kirkwood *et al.* (1995) give comprehensive blood data for *M. giganteus* and *Puffinus puffinus* respectively.

Haematological studies of *Oceanites oceanicus* were reported by Myrcha and Kostelecka-Myrcha (1979) and Kostelecka-Myrcha and Myrcha (1989). Haemoglobin content of adult blood was 16.5 ± 0.6 g% in 19 males but 17.9 ± 0.4 g% in 11 females. The storm petrel had higher haemoglobin and more erythrocytes ($4.02 \times 10^6\ ml^{-3}$) than other seabirds, which these authors related to this small active bird's high oxygen demand. Other red blood cell counts were $2.70 \times 10^6\ mm^{-3}$ for *F. glacialis* and $2.6 \times 10^6\ mm^{-3}$ for *M. giganteus*.

In nestlings of *O. oceanicus* from 1 to 61 days old a three-stage development in the respiratory role of the blood was followed (Fig. 7.15). The initial haemoglobin decrease was unexplained, being the reverse of the authors' findings for Arctic and northern birds. They noted the correspondence of the three-stage changes in blood variables with the rates of growth of the chicks, this being greatest from 10 to 30 days and thereafter falling off to fledging.

Few figures for pulse rates are available. Williams and Ricklefs (1984), studying embryos of *Oceanodroma leucorhoa* and *Oceanites oceanicus*, found that the greatest change was at low temperatures: embryos at 15°C had about five heartbeats min^{-1}, those at 35°C about 120 min^{-1}, Q_{10} values ranging from 15.0 between 10 and 15°C to 1.7 between 35 and 41°C. The heartbeats of embryo *D. immutabilis* and *P. pacificus* at 36°C became increasingly variable with age. In the albatross the rate, 195 ± 14 ($n = 172$) beats min^{-1} during prepipping, rose to 221 ± 12 ($n = 199$) beats min^{-1} after pip-holing (Tazawa & Whittow, 1994).

Five adult *F. glacialis* had pulse rates averaging 295 min^{-1} (230–330) when on the surface, falling to 200 min^{-1} (180–210) when submerged in water but rising to 400 min^{-1} in compensatory tachycardia on emergence (Galanstev *et al.*, 1979). Great variation was also found with *D. melanophrys*. In resting birds the range was 66–167 beats min^{-1} *vs.* 247–409 min^{-1} when walking a treadmill (Bevan *et al.*, 1994).

Dramatic changes in the pulse rates of breeding *D. irrorata* when approached by man were reported by Jungius and Hirsch (1979). When the observer was >18 m away the rates were *c.* 84 and 64 min^{-1}, but at 7–10 m one bird had a pulse rate of *c.* 180 min^{-1} which oscillated rapidly and approached 200 min^{-1} with the observer only 2 m away. The other bird showed a steadier increase to about 140 min^{-1} at distances of 2–3 m. Except at these close ranges the birds appeared undisturbed.

Tirrell and Murrish (1979) studied how incubating *M. giganteus* control the blood flow to the brood patch in response to different body and egg temperatures. They introduced cold, cool and warm eggs to the brood patches. The responses of hyperthermic birds ($T_b = 40.5$°C) to a cold egg were a sharp and sustained vasodilation, those of hypothermic ones to a cold egg a decreased flow to the brood patch—up to a 75% reduction. That control of blood flow to the patch was via the sympathetic nervous system was tested by injection of the alpha-antagonist phenoxybenzamine: 12 min later the flow increased by nearly 300% over normal levels. Tirrell and Murrish envisaged a hierarchy of CNS centres, one monitoring thermal stimuli from the patch regulating the blood flow to it, another adjusting cutaneous flow to help maintain optimal T_b.

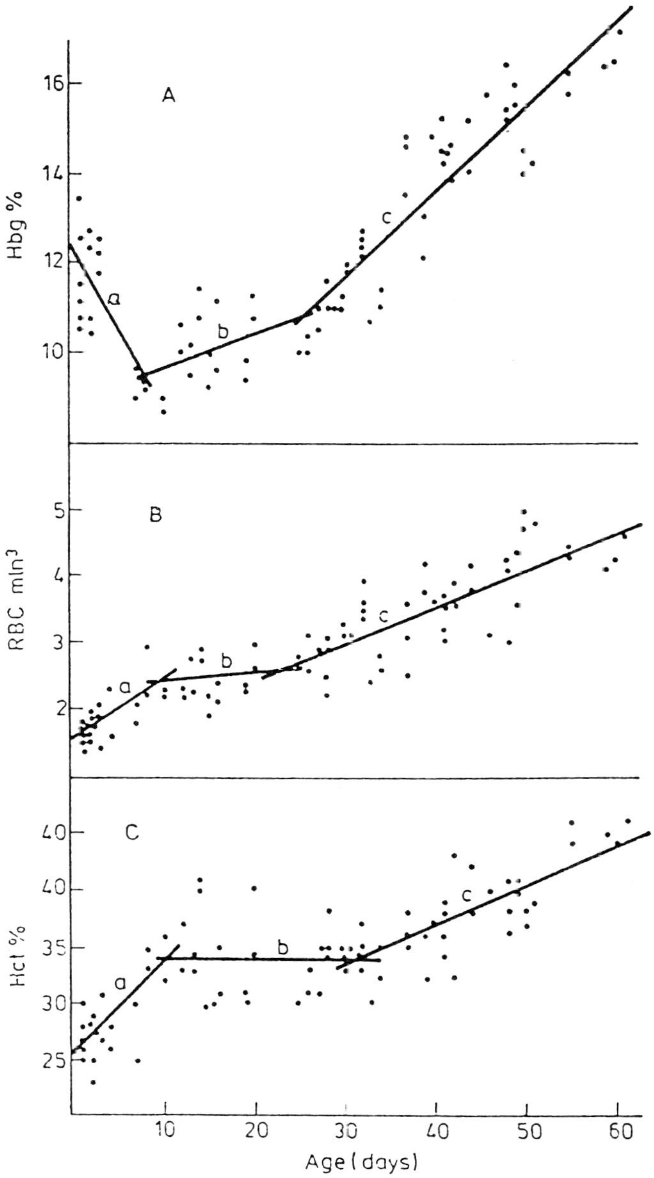

Figure 7.15 *(A) Changes of haemoglobin concentration (Hbg). (B) Number of erythrocytes* $\times 10^6$ *in 1 mm*^{3} *of blood (RBC). (C) haematocrit values (Hct), during the development of chicks of* Oceanites oceanicus: *(a), (b), (c). three phases in development. From Kostelecka-Myrcha and Myrcha (1989).*

IX Endocrine system

The hormone control system of tubenoses has been investigated during the reproduction of albatrosses, using a cannula in the intertarsal vein to withdraw blood samples in the field (Hector, 1984). Hector and his colleagues measured changes in hormonal levels over time in *D. exulans*, *D. chrysostoma* and *D. melanophrys* sexed by laparoscopy. Smith *et al.* (1994) examined the adrenocortical response to stress in *P. urinatrix*.

A *Corticosteroids*

Hector and Harvey (1986) found corticosterone levels during incubation spells similar for both sexes, and in all three species the initial titres were low but increased before showing species-specific patterns (Fig. 7.16). Despite its prolonged fasting, levels were much lower and fluctuated more in *D. exulans* than in the mollymawks, for unresolved reasons. Increased titres of corticosterone in the latter birds were ascribed to its feedback role in stimulating the mobilization of glucose and fat reserves during incubation when they neither fed nor drank.

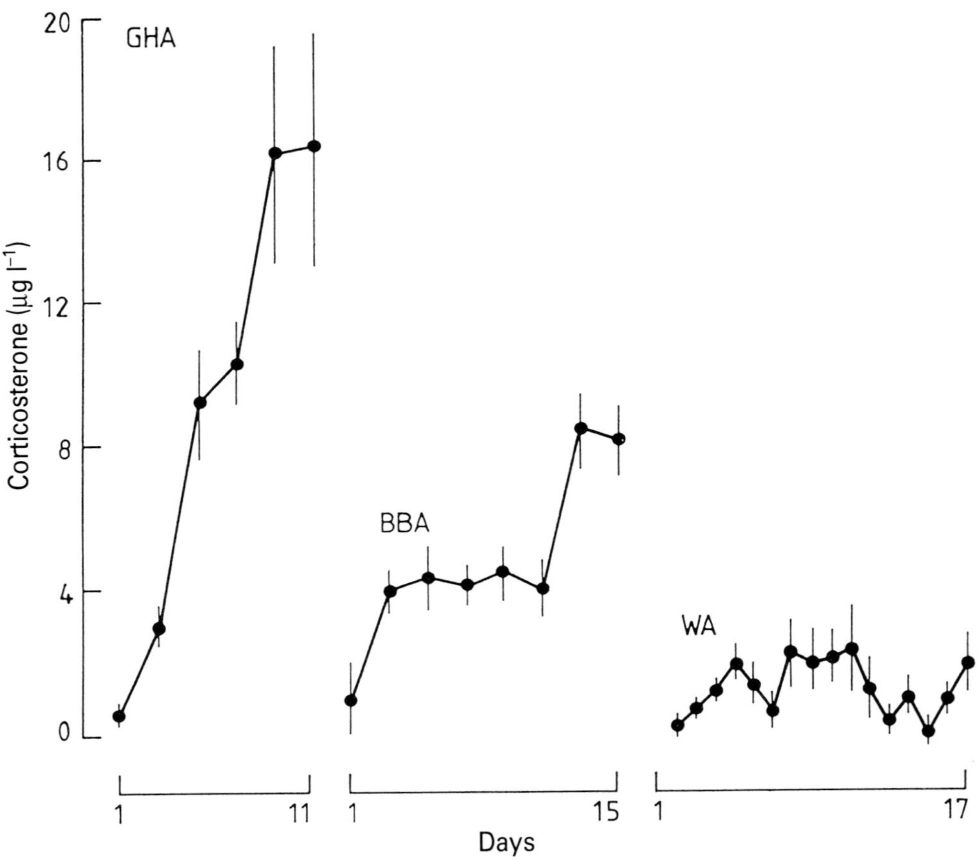

Figure 7.16 *Corticosterone levels in the plasma of Grey-headed (GHA) (n = 6), Black-browed (BBA) (n = 6) and Wandering Albatrosses (WA) (n = 9) throughout single incubation shifts. Values are means ±SE. From Hector and Harvey (1986).*

B Prolactin

In mollymawks, Hector and Goldsmith (1985) found that prolactin levels in both sexes rose at the copulatory stage of the prelaying period, remained high throughout incubation, fell around hatching, and stayed low to the end of the guard stage. In *D. exulans*, levels were higher in the females and the rise was delayed until mid-incubation. There were no significant changes during incubation nor any diel variations, and as adults coming in to incubate that had not seen their eggs still had high levels, there was no suggestion that tactile stimuli from the egg to the brood patch were involved. By experimentally altering the incubation period, it was found that the fall in prolactin was not tied in to the hatching time, so that, although in all three species the falling levels usually coincided with the end of the guard stage, the change was not initiated simply by the hatch.

C Pituitary and gonadal hormones

The males had annual gonadal and hormonal cycles whether they bred annually (*D. melanophrys*) or biennially (*D. exulans, D. chrysostoma*) (Hector *et al.*, 1986a,b). In breeding mollymawks, Hector *et al.* (1986b) found similar high levels of testosterone in both sexes during the prelaying period. The level fell rapidly about egg laying and remained low until an increase at fledging time partly coinciding with postnuptial display activity.

Progesterone levels were similar in males of both species but more variable in the females (Fig. 7.17), differing between the species. Oestradiol levels were high until midway through incubation in *D. chrysostoma* but fell earlier in *D. melanophrys*.

Some *D. chrysostoma* lay in successive years but the state of the testes and hormonal levels in the males indicate that these could breed annually. Hector *et al.* found that the ovary of a bird in the year after successful nesting was undeveloped, levels of oestradiol low, but progesterone values high. They suggested that the progesterone inhibits oestrogen production and maturation of a follicle, presumably in response to some environmental variable.

The endocrine relationships of delayed sexual maturity in *D. exulans* were also examined. The testes during the breeding season in immatures of 5 years or above were as large as in many breeding birds (Fig. 7.18), with testosterone levels within the range of those of the breeders, indicating that 5-year-old males could breed were they not bonded to immature females. Females less than 7 years old had higher and more variable progesterone and low oestradiol and luteinizing hormone (LH) levels and were physiologically immature; the breeders had low levels of progesterone and high levels of oestradiol and LH (Hector *et al.*, 1986a, 1990).

Hector *et al.* (1990) established from field experiments with *D. exulans* that high levels of progesterone were not triggered by high levels of sexual displaying, and they also confirmed that this hormone was liberated from the ovary. Only progesterone was produced in immatures, but oestradiol was also produced in mature birds. The basis of delayed maturity in females was postulated as the ovarian release of progesterone inhibiting vitellogenesis and egg formation.

The endocrine basis of biennial breeding by *D. exulans* is unclear. Whereas

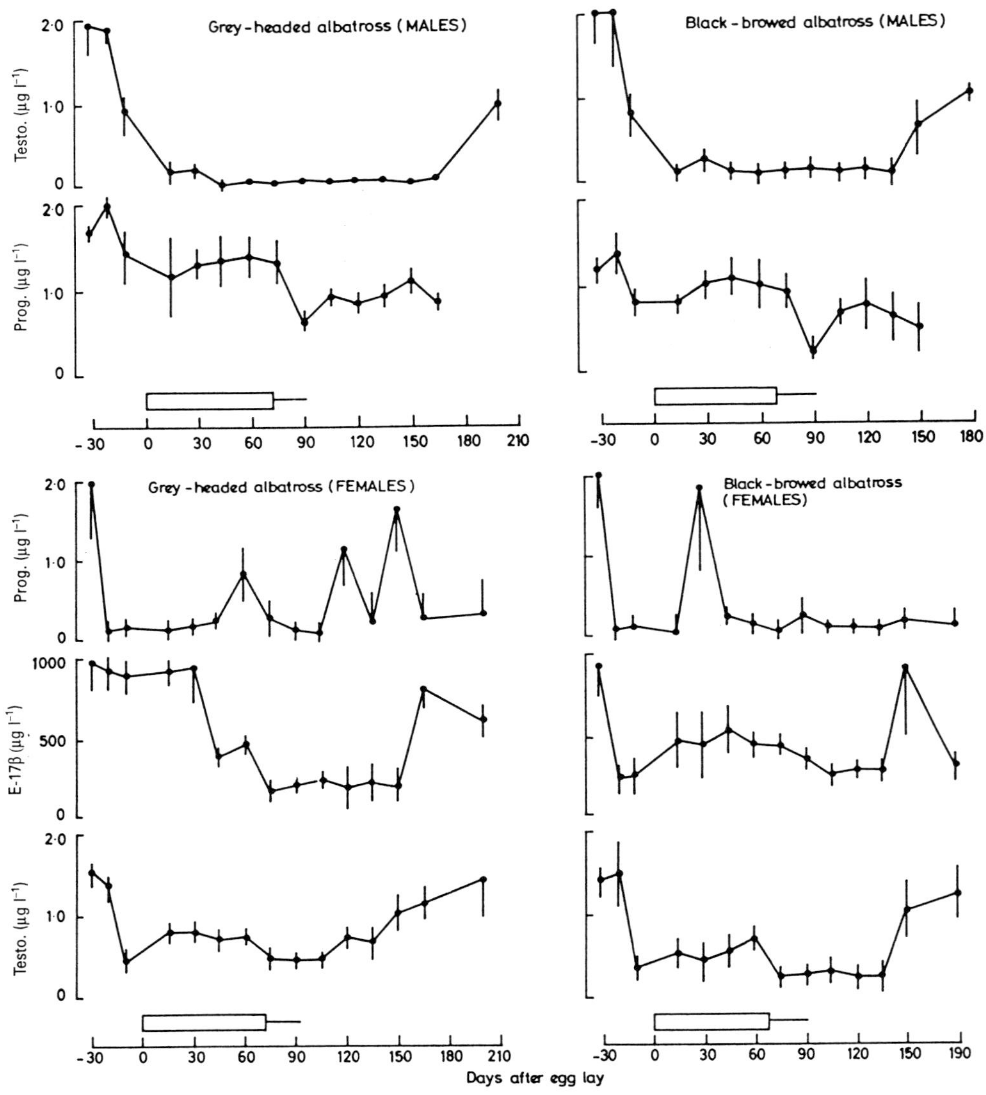

Figure 7.17 *Mean (±1 SE) testosterone and progesterone levels in male and female Grey-headed and Black-browed Albatrosses during the breeding season. Incubation and guard stages shown as a box and a bar. From Hector* et al. *(1986b).*

D. chrysostoma was regarded as basically an annual breeder with some environmental factor operating after successful nesting to inhibit secretion of oestrogens the next year, *D. exulans* may be inherently a biennial breeder with some factor associated with chick rearing inhibiting pituitary response to the external conditions normally stimulating breeding. Hector *et al.* (1986a) found that parent females around fledging time still had regressed gonads only 6 weeks before the current year's birds turned up with mature ones. For more detailed discussion of the interactions of endocrine,

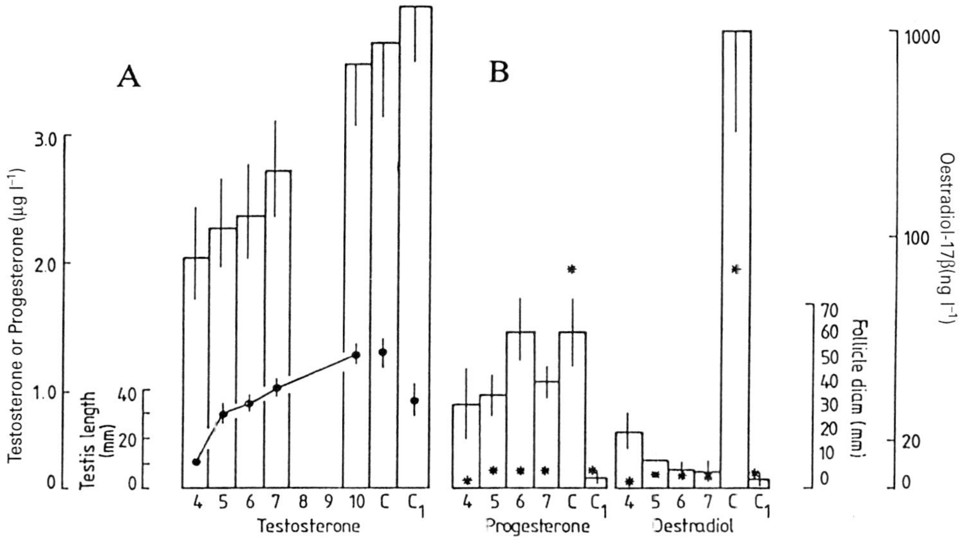

Figure 7.18 *(A) Maximum testis lengths and testosterone concentrations in different age classes of immature male Wandering Albatrosses. (B) Maximum diameter of largest follicle and concentrations of oestradiol and progesterone in different age classes of immature female Wandering Albatrosses. Data for breeders caught at copulation (C) and at the same time as the non-breeders sampled (C_1) are shown for comparison. Means are shown ±SE. From Hector (1988).*

ecological and behavioural factors that may affect reproduction in albatrosses, see Croxall (1991).

X Metabolic rates

Determinations of the energy expended per unit time under given conditions have been made for petrels. Those from birds at rest within their TNZs and in a postabsorptive state are usually called basal metabolic rates (BMRs). Many values come from birds at rest but not necessarily fasting, giving the so-called resting metabolic rates (RMRs), and a few are from incubating birds (IMRs). With the use of double-labelled water (DLW), energy expenditures of birds actively foraging or displaying can be made—the field metabolic rates (FMRs).

Figures computed from the rates of weight losses of fasting birds using standard equations may be unreliable—see Croxall (1982b), Obst *et al.* (1987) and Pettit *et al.* (1988). More direct figures based on O_2 consumption or by DLW techniques are preferred, but all methods involve a range of assumptions about the metabolites being burnt and the environmental conditions and stresses experienced by the subjects. Allowances for diel activity variations have seldom been made but Adams and Brown (1984) found that even in the complete darkness of a metabolism chamber, burrowing species were more active at night, their lowest metabolic rates reached during daylight (Fig. 7.19). And, while a 'fasting' petrel may have digested the protein and carbohydrate moiety of its meal, slowly absorbed stomach oil may

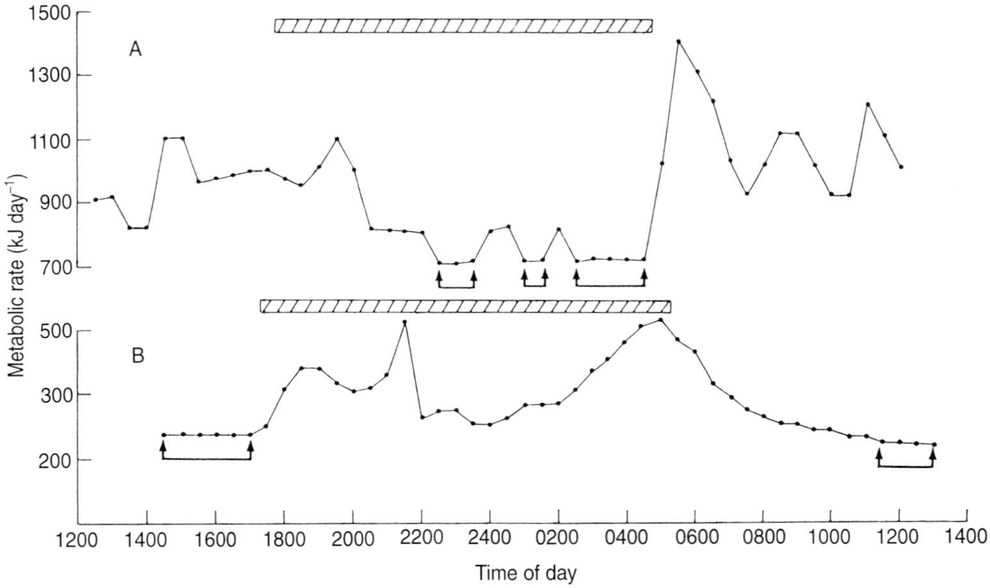

Figure 7.19 *Metabolic rates of (A) a surface nesting tubenose* Phoebetria fusca *and (B) a burrowing one* Lugensa brevirostris *over 24 h at normal photoperiod. Hatched areas show hours of darkness, stable periods (arrowed) those used to calculate metabolic rates. Lows in (A) could be while asleep. From Adams and Brown (1984).*

still remain, which, at about 40 kJ g^{-1}, is a major energy source. Values for *Pelecanoides* spp. , which do not store oil, should be free from this source of error. A major problem is the dearth of data on TNZs, some high-latitude species, especially *Oceanites oceanicus*, probably incubating below their lower critical temperatures.

Respiration chambers are usually in field laboratories where it is difficult to ensure that the subject's metabolic rate is not raised through stress. Grant and Whittow (1983) converted burrows of *Pterodroma hypoleuca* into respiration chambers, while with *D. immutabilis*, the nest, built on a plywood platform, was slid, complete with incubating bird on a plaster egg, into an airtight box to form a chamber. At times the study bird 'slept' with its bill beneath the scapulars and in both series the birds tested appeared close to their normal relaxed state. Heart beat rates were not monitored and the dramatic increases in these when *D. irrorata* were approached (Section VIII above) suggest that appearances can be deceptive with stressed petrels. However, Whittow *et al.* (1987) found that respiration rates of *Puffinus pacificus* remained relatively steady in the TNZ at $20.1 \pm 10.9\ \text{min}^{-1}$ ($n = 16$), suggesting that despite restraint, they were not unduly stressed.

The DLW method also involves assumptions, for example that body water is not diluted by the birds drinking seawater. Clearly some seawater must be ingested with the food: this may give errors in food estimates though not for metabolic rates based on CO_2 production (Nagy, 1989). That food is effectively the only source of exogenous water during foraging flights was suggested by Costa and Prince (1987). However, captive tubenoses drink seawater quite copiously, for example the fulmars studied by Kritzler (1948).

Bevan *et al.* (1994), using *D. melanophrys*, tested the use of heart rate for estimating oxygen consumption and metabolic rates, finding that:–

$$V_{O_2} = 0.0157 fH^{1.60} \ (r^2 = 0.80,\ P < 0.001),$$

where V_{O_2} is in ml min^{-1} and fH is beats min^{-1}. Thus heart rates may prove useful for determining FMRs and daily budgets in free-living birds, the results comparable with other methods.

A *Resting metabolic rates (RMRs)*

The available data are listed in Table 7.2. No BMR figures are shown because of the reservations given above: values from captive birds are considered to be RMRs. The data are somewhat heterogeneous, collected in different ways, sample sizes often small and TNZs usually unknown. RMR values from different methods should be comparable but there are very few comparative studies for petrels.

As with other animals, metabolic rates scale with body mass (Fig. 7.20). The exponent of 0.70 for the RMR data approaches the figure of 0.73 given by the formulae of Aschoff and Pohl (1970) and Kendeigh *et al.* (1977) for the BMRs of non-passerines. The equation for 'BMR' for 22 petrels calculated on the Kendeigh *et al.* formula given by Rahn and Whittow (1988)—BMR = $2.378^{0.73}$—is similar. Understandably, the scatter around the regression line is considerable, the points being mean values of small samples which themselves have high variances, so that detailed comparison with the results from 'standard' equations is probably unrealistic until more and better data are available.

The major outlying point is that of 454 kJ day^{-1} for *Pterodroma phaeopygia* from Simons and Whittow (1984), a high value for a tropical species and perhaps affected by altitude, but also not measured within the TNZ.

Ellis (1984) and Pettit *et al.* (1985) tentatively suggested that tropical seabirds have low BMRs, so reducing costs of thermoregulation and food demand in a region of low marine productivity. Brown (pers. comm.) concluded from a re-examination of the data that metabolic rates of petrels do vary with climate, those in high latitudes (particularly the smaller species) having higher metabolic rates. However, there are few figures for tropical species and in their analysis of DLW results from foraging *D. immutabilis*, Pettit *et al.* (1988) concluded that this bird's energy expenditure was not less than that of albatrosses seeking food in subantarctic waters. Ellis also suggested that in tropical seabirds low metabolic rates go with dark dorsal surfaces. But most tubenoses have dark upperparts, and those in which they are pale, such as '*Cookilaria*' and *Pachyptila* spp., may be warm- or cold-water birds.

The hypothesis that procellariiform birds might have low BMRs, partly on account of their low T_bs (Warham, 1971) has been supported by some results such as those of Obst *et al.* (1987), Gabrielsen *et al.* (1988) and Whittow *et al.* (1987), but not by others such as those of Adams and Brown (1984). The consolidated data of Fig. 7.20 do not resolve this question although, by definition, RMR values should be higher than BMR ones.

B *Field metabolic rates (FMRs)*

The FMRs in Fig. 7.20 for eight petrels are mainly from DLW estimations. In the middle of the size range FMR is about three times RMR, emphasizing the cost of activity. The figures given by Obst *et al.* (1987) for *O. oceanicus* of 37 kJ day^{-1} RMR *vs.* 157 kJ day^{-1} when flying (i.e. 4.2 × RMR), are not surprising for despite its ability to glide and slope-soar, this small bird still beats its wings and hovers a lot. Likewise, the high values of FMR for the two diving petrels must reflect the high costs of 'whirring' flight both above and below water.

Birds with chicks have higher FMRs than those active during the egg phase, for example in *O. oceanicus* (Obst *et al.*, 1987). For *Oceanodroma leucorhoa*, FMR integrated over full cycles of activity averaged 127 kJ day^{-1} during incubation and brooding

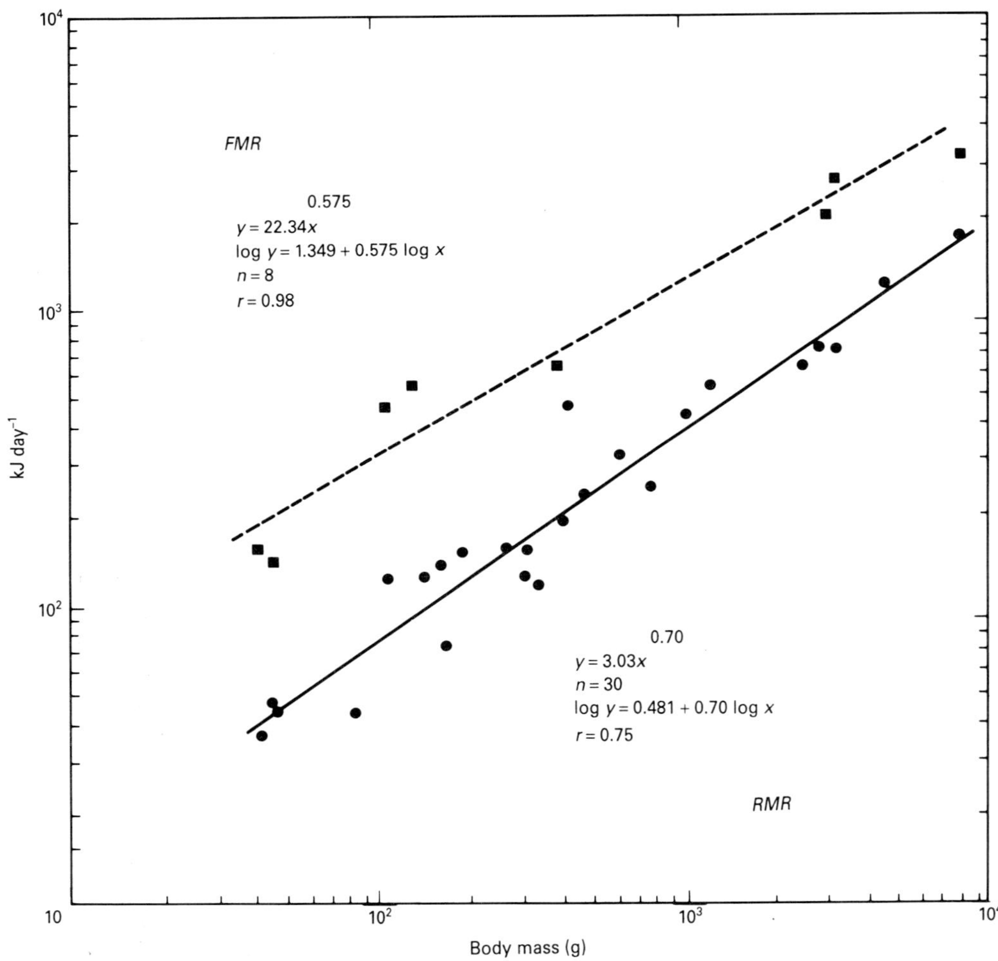

Figure 7.20 *Field metabolic rates (upper) and resting metabolic rates for tubenosed birds in relation to body mass. Data from Table 7.2.*

against 142 kJ day^{-1} when rearing chicks (Montevecchi *et al.*, 1992), reflecting the longer spells at sea by petrels raising large chicks. In this study FMR increased significantly with time at sea, and while birds in burrows lost weight, those at sea did not, remaining in energy balance. FMR also increased with time at sea with *M. giganteus* (Obst & Nagy, 1992).

The plot of FMR in Fig. 7.20 includes data for *D. chrysostoma* from DLW determinations where one bird carried an activity recorder so that Costa and Prince (1987) could compute the time on the water and in the air over a 3-day forage. The energetic cost of flying was 3.0–3.4 W or 3.0–3.4 × the BMR predicted from formulae. As these authors pointed out, this is much less than that of birds such as *Larus delawarensis* and *L. atricilla* with flapping flight (7.5–11.1 × BMR) and may be ascribed to the albatross's economical flight dynamics. The FMR of *D. exulans* on the wing was estimated by Adams *et al.* (1986) as being a mere 2.35 × BMR.

DLW figures for FMR of 4330 kJ day^{-1} for *M. giganteus* brooding and feeding chicks by Obst and Nagy (1992) are very high, estimated at 2.8 × BMR during brooding to 6.3 × BMR while at sea. Obst and Nagy attributed the high foraging costs partly to the much higher disc-loading than that of similar-sized albatrosses and partly to the fulmar's more frequent visits to the young. What they were eating was unknown but, apart from environmental variables such as wind speed and direction (*Macronectes* is a very fast and evidently efficient glider in strong winds), birds feeding on carrion may be prodigal in their use of energy as they often have to fight for their food.

C *Incubation metabolic rates* (*IMR*s)

Very few direct estimates of the energy requirements of incubating tubenoses are available. The egg itself makes little contribution to the heat generated—only 3 and 2% respectively of the demand in *Procellaria aequinoctialis* and *Pterodroma macroptera*, for example (C. R. Brown, 1988).

Figures determined include 689 kJ day^{-1} for *D. immutabilis* (RMR 645 kJ day^{-1}); 1466 kJ day^{-1} for *M. giganteus* (RMRs for males and females 1566 kJ day^{-1} and 1432 kJ day^{-1} respectively (Ricklefs & Matthew, 1983); 90 kJ day^{-1} for *P. hypoleuca vs.* RMR 110 or 90 kJ day^{-1} (Grant & Whittow, 1983); 81 kJ day^{-1} for *Oceanites oceanicus vs.* RMR of 37 kJ day^{-1} (Obst *et al.*, 1987); and for *Oceanodroma leucorhoa* 83 kJ day^{-1} *vs.* RMR 45 kJ day^{-1} (Montevecchi *et al.*, 1992).

Obst *et al.* (1987) deduced from their figures for the thermal conductance of the storm petrel *O. oceanicus*, that little heat was lost from its nest cavity so that the extra cost of incubation above maintenance and thermoregulation was negligible. However, these birds lay in arid rock crevices, whereas for many petrels burrowing in soils the chamber will be far from energetically neutral.

It would be expected that petrels would conserve energy by maintaining their metabolic rates close to basic values during their long incubation stints. However, Grant (1984) compared IMR for 14 species (calculated by the weight-loss method) with BMR predicted from the Lasiewski–Dawson equation, and assumed that the loss was of 55.5% fat, 9.2% protein, the remaining 35.3% being water. The ratios of IMR/BMR ranged from 0.85 in *P. macroptera* to 2.31 in *O. oceanicus*. The data of Grant and Whittow (1983) for *D. immutabilis* and *P. hypoleuca* were quite different, suggesting that there were no incubation costs, the calculated IMR/BMR being 1.11 and 1.54 *vs.* measured values of 0.96 and 0.82.

During incubation, energy consumption will usually vary with the sex as male petrels tend to have slightly longer shifts than the females. From DLW data, Pettit *et al.* (1988) found that although male *D. immutabilis* are bigger than their mates, they spend less time foraging so that the total energy needs of the female parents then are the higher—94.7 MJ *vs.* 76 MJ for the males. For *D. exulans* Adams *et al.* (1986) calculated that males expended 125.6 MJ, the females 75.7 MJ, partly because of the greater size disparity in the sexes of this species.

RMRs for incubating, double-labelled *O. oceanicus* were obtained by Obst *et al.* (1987) despite much nest desertion. Ambient temperatures in the nest (mean 2°C, range −6° to +11°C) were beyond the 16°C lower limit of the bird's TNZ, so that the metabolic rate probably never fell to basal levels while in Antarctica. The calculated energy expenditure of 81 kJ day^{-1} or 2.19 × BMR represents a very high cost for incubation by any bird. With a mean T_a of 13.7°C, Montevecchi *et al.* (1992) got a mean of 83.3 ± 19.4 kJ day^{-1} for eight Leach's Petrels, seven on eggs, one with a chick, IMRs being 1.89 × BMRs.

Unexamined are the effects of the ammonia and CO_2 build-up within the nesting chambers of burrowing species. Most other hole nesters move in and out many times daily, their bodies acting like pistons to eject contaminated and inject fresh air, whereas petrels remain for many hours on their egg. How do the levels of waste gases affect the birds' oxygen demands and RMRs? Furthermore, no such environment is provided when a burrow is turned into a metabolic chamber, fresh air pumped in and monitored air removed.

XI Energy budgets

Costs can be integrated to produce energy budgets for specific activities such as breeding, during a particular stage of the annual cycle, or over a whole year and, by summation, to obtain estimates of the energy used by a colony or ecosystem.

An example is that constructed by Berruti *et al.* (1985) for the Shoemaker chick (Table 7.4). The data were based on (i) regular weighings of the chicks to get the food intake; (ii) analyses of food samples from the chicks and adults; (iii) estimates of the energy contents of meals; and (iv) chick energy expenditures (RMR) from O_2 measurements in metabolism chambers. The assimilation efficiency of the food was taken to be 80% and in calculating the accumulated chick biomass the mean value of 25.4 kJ g^{-1} dry weight was based on carcass analyses.

Table 7.4 shows both measured energy expended (i.e. maintenance costs + cost of biomass synthesis), and energy in this accumulated biomass. Total energy for a chick growth over 96 days was 66 938 kJ and at 80% efficiency assimilation was 54 070 kJ. Total food consumption was 8637 g wet weight representing 62 909 kJ.

Berruti *et al.* (1985) pointed out that the semiprecocial development allows both parents to forage within 3–7 days of the hatch, the energy delivered being at first high in relation to the chick's need—the mass delivered being 46% of a 10-day

Table 7.4 Energy budget of Shoemaker, *Procellaria aequinoctialis*, chicks by 5-day age class. From Berruti *et al.* (1985)

Chick age class	Energy expenditure (kJ day^{-1})	Energy expenditure per age class (kJ)	Energy accumulated in biomass (kJ)	Total energy used
1–7	175.8	1231	478.3	1709
8–14	262.1	1835	923.3	2758
15–21	356.9	2498	1202.0	3700
22–28	451.0	3157	1360.0	4517
29–34	516.5	3099	1033.1	4132
35–40	569.3	3416	980.7	4297
41–48	630.3	5042	1068.7	6111
49–55	676.0	4732	836.3	5568
56–62	686.2	4803	190.5	4994
63–69	731.8	5123	866.7	5989
70–76	727.7	5094		5094
77–83	727.2	5090		5090
84–91	712.7	5702		5702
92–96	649.6	3248		3248
Totals		54 070	8840	62 909

chick's mass, only 10% in 92-day one. Much energy went into biomass, 30% initially but again falling with age, and being only about 4% between 56 and 62 days, most then going into maintenance. From about 70 days on no further biomass was added during the normal weight recession.

Roby (1991) used similar methods in assessing the costs of raising the chicks of plankton-feeding diving petrels *Pelecanoides urinatrix* and *P. georgicus*. The body composition of chicks of known age gave figures for lipid accumulation and lipid-free dry matter (LPDM). Energy expenditure was determined in metabolic chambers. Meal sizes came from regular weighings and their composition from analysis of adult regurgitations. Roby calculated the energy equivalents of growth at each stage from the energy equivalents of the lipid-free dry matter + the cost of its biosynthesis—0.33 × the energy equivalent of accumulated tissue (Ricklefs, 1974). Maintenance costs were the difference between the energy equivalent of oxygen consumed and that of the biosynthesis of LFDM and lipid. Total energy consumptions were the sum of growth, lipid deposition and maintenance needs (Fig. 7.21).

Rates of growth—that is, accumulation of LFDM—were the same in both diving petrels and the cost of growth relatively minor compared with that for maintenance and fat deposition. Much of the latter was wax ester, derived in *P. georgicus* from calanoids and in *P. urinatrix* from *E. superba*. Wax ester was equivalent to 59 kJ day^{-1} (29% of total energy consumption) for *P. georgicus* and 21 kJ day^{-1} (12% of the total) for *P. urinatrix*. As Fig. 7.21 shows, the latter accumulated lipid at a constant and lower rate, and Roby concluded that because of its low dietary lipid, much dietary protein must be used by *P. urinatrix*. For *P. georgicus*, however, although

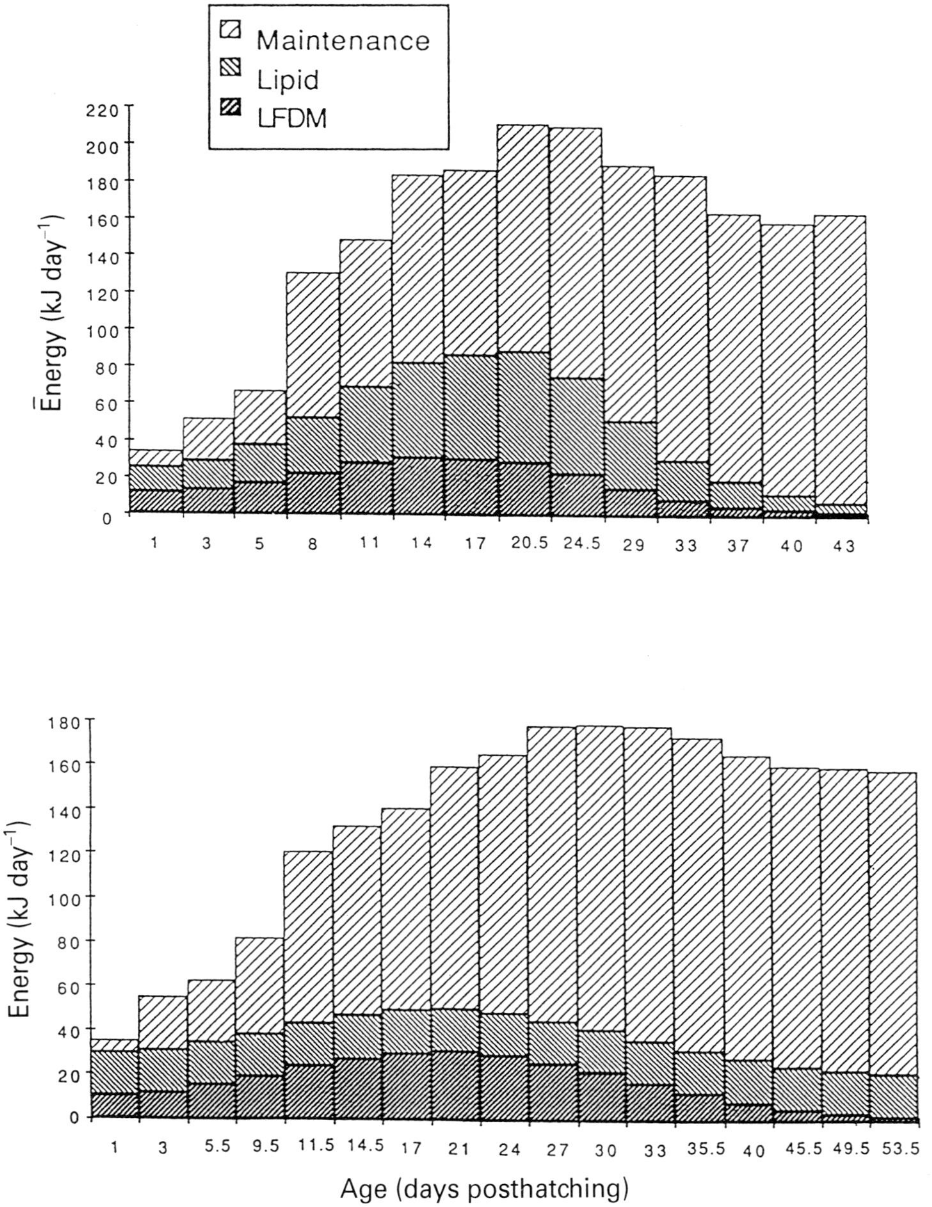

Figure 7.21 *Age-specific energy budgets of chicks of (A)* Pelecanoides georgicus *and (B)* P. urinatrix *calculated from rates of accumulation of lipid and lipid-free dry matter (LFDM) and daily respired energy. From Roby (1991).*

Table 7.5 Time and energy budgets of one pair of Laysan Albatrosses during incubation of the egg. From Pettit *et al.* (1988)

		Male (3000 g)	Female (2700 g)
On nest	Time (days)	44	21
	Energy expenditure rate (kJ/day^{-1})	760	684
	Total energy expenditure (MJ)	33.4	14.4
At sea	Time (days)	21	44
	Energy expenditure rate (kJ/day^{-1})	2029	1825
	Total energy expenditure (MJ)	42.6	80.3
Total	Time (days)	65	65
	Energy expenditure (MJ)	76	94.7

lipid deposition was faster, growth, as measured by accumulation of LFDM, was similar.

The energy budgets Pettit *et al.* (1988) developed for *D. immutabilis* during the incubation stage were estimated from CO_2 production by the DLW method with field observations of the injected birds to establish their time budgets. Calculated energy expenditure was three times greater in the foraging birds (Table 7.5), the 689 kJ day^{-1} for those on eggs, being very similar to RMR values calculated earlier of 645 kJ day^{-1}. Pettit *et al.* pointed out that these birds differ from *D. exulans, D. chrysostoma* and *D. melanophrys* whose IMRs are higher than their RMRs, perhaps because of the greater thermoregulatory costs of the southern species. Total energy consumed by the *D. immutabilis* embryo (from its oxygen consumption), 0.6 MJ, was but 0.4% of the estimated use of energy by the two parents during incubation.

Other energy budgets include that for nestling *Pterodroma phaeopygia* given by Simons and Whittow (1984) who estimated that the total needs over the chick's 110-day rearing period were about 54 000 kJ. Using satellite data Salamolard and Weimerskirch (1993), calculated that the power needed during the chick guard stage of *D. exulans* was 6.2–6.6 × 10^3 kJ day^{-1}, falling to 3.8–4.2 × 10^3 kJ day^{-1} after guarding.

Integrating individual energy budgets to assess the energy flux through a whole colony or ecosystem (e.g. Abrams, 1985b), represents a large step. Croxall *et al.* (1984a) highlighted some of the problems, while Wiens (1984) described how energetics information can be integrated into models as tools for predicting energy flows, food consumption and the like.

Table 7.6 gives estimates of the energetic needs of tubenoses of a subtropical seabird community at French Frigate Shoals (Pettit *et al.*, 1984) where, during the breeding season, some 35% of the energy passed through six species of petrel, of which *D. nigripes*, taking 25%, was the biggest consumer after the 48% used by *Sterna fuscata*.

The total energy consumed was the sum of (i) the calorific content of the average egg × cost of its synthesis × the number of eggs; (ii) the gross energy for the

Table 7.6 Total energy requirements of seabirds at French Frigate Shoals. From Pettit *et al.* (1984)

Species	Breeders (kcal × 10^6)	Non-breeders (kcal × 10^6)	Eggs (kcal × 10^6)	Chicks (kcal × 10^6)	Total (kcal × 10^6)
Black-footed Albatross	786.3	706.1	3.03	421.2	1916.6
Laysan Albatross	157.6	330.1	0.64	101.4	589.7
Wedge-tailed Shearwater	63.4	115.9	0.26	22.0	201.6
Christmas Shearwater	0.7	0.8	0.003	0.2	1.7
Bonin Petrel	1.1	1.0	0.005	0.2	2.3
Bulwer's Petrel	7.3	6.6	0.03	1.1	15.0
Red-tailed Tropicbird	27.9	27.9	0.09	7.0	62.9
Masked Booby	112.6	42.3	0.11	25.4	180.4
Brown Booby	8.2	3.5	0.009	1.5	13.2
Red-footed Booby	72.2	60.3	0.05	15.7	148.2
Great Frigatebird	53.9	122.7	0.06	10.9	187.6
Sooty Tern	1320.9	2201.2	7.20	171.6	3700.9
Grey-backed Tern	13.9	13.5	0.06	2.9	30.4
Brown Noddy	300.2	250.4	0.64	14.2	565.4
Black Noddy	22.4	38.8	0.05	1.3	62.5
White Tern	20.1	34.9	0.05	0.6	55.6

development of the average chick × the number of chicks; (iii) the total expenditures of adults while on the island; and (iv) the total expenditure of adults when at sea. Corrected for an assumed assimilation efficiency of 80%, the total energy consumption was 40 478 × 10^6 kJ.

The six procellariiforms at this breeding station were estimated to require 1660 t of food, some 26% of the 6312 t for all the resident seabirds.

A different approach is exemplified by a preliminary study of the energetics of the seabird community of the Bering Sea, of which *F. glacialis* and *Puffinus* spp. are prominent components, by Schneider *et al.* (1986). Knowing the numbers and distributions of each species month by month they calculated their energy needs from known consumptions of foods of known calorific contents. From this and standard metabolic formulae, the daily energy flux through each species was estimated. The annual flux was derived from the product of the occupancy rate (days km^{-2} $year^{-1}$) and the daily energy fluxes. For example, in the Inner Domain Region in 1981 these authors estimated that *Puffinus* spp. (mainly *P. tenuirostris*) accounted for 92% of the total flux of 30.6 kJ m^{-2} $year^{-1}$.

XII Energy contents of petrel eggs

Petrels other than albatrosses lay eggs with unusually high yolk contents (*The Petrels*, pp. 298–300), and their energy is proportional to the relative yolk content. For 10 species, Rahn and Whittow (1988) determined an average value of 7.3 ± 0.8 kJ g^{-1} egg content.

XIII Stomach oil stores in chicks

The amount of stomach oil in chicks of some, if not most, petrels declines as they approach fledging. This has long been known to harvesters of the 'mutton birds' *P. griseus* and *P. tenuirostris*, and of *H. pelagicus* (Serventy *et al.*, 1971; Graham, 1890). However, with the storm petrel *Oceanodroma leucorhoa* the reverse applies. Place *et al.* (1991) using a radioactive marker, found that the amount of oil increased during the final days in the nest when the body weight was falling as depot fat stores were consumed. For example, after 60 days of age the mean volume of oil was 6.3 ± 2.9 ml ($n = 25$), a threefold increase over that during the previous 54 days of 2.3 ± 1.8 ml ($n = 52$). The authors considered that this accumulation resulted from a slowing of gastric emptying in older chicks, and hypothesized that the oil acts as an energy-dense reserve to tide birds over fasting periods and as a source of water for the chick at fledging (cf. Warham, 1977b). They pointed out that the oil is a lighter source of energy than adipose tissue, incurs no energetic cost of remobilization, and can be jettisoned for weight reduction in emergencies. Bech *et al.* (1988) calculated that only 5–10% of the food of the *T. antarctica* chick needs to be stomach oil to meet the needs of maintenance metabolism and is a more efficient resource than having to mobilize part of the diet into fat. Obst and Nagy (1993) doubted that in the absence of stomach oil *Oceanites oceanicus* could deliver energy to its chick fast enough for successful breeding in Antarctica.

Why this system is apparently not universal among tubenoses is unclear, but no doubt storage varies from season to season, just as the chicks' weights, condition, and the numbers of surface eggs, vary between 'good' and 'bad' years.

XIV Premigratory fattening

The role of premigratory fattening to power migratory flights was evaluated by Chu (1984) who followed the changes in depot fat in *P. griseus* during their stay off California. Moult was completed between May and August and heavy moult in July coincided with a 12% loss of body weight and stored fat. Depot fat has about 39.4 kJ g^{-1} (Groscolas *et al.*, 1991). After the moult weight gain was rapid, changing from a minimum of 0.03 g g^{-1} lean dry weight to maxima of 2.0–3.5 g g^{-1} lean dry weight before departure, while non-fat body components and water content remained constant. Chu calculated that each bird must consume 324 kcal day^{-1} (1357 kJ day^{-1}) from July to August and 367 kcal day^{-1} (1537 kJ day^{-1}) from August to September to gain the additional fat. The fat increase began as their diet changed from rockfish and market squid to oily anchovies. The final stores might well enable the birds to get back to New Zealand seas without feeding *en route*, but a poor anchovy season could jeopardize that.

CHAPTER 8
Biochemistry

Contents

I Procellariiform lipids

A *Stomach oils*

Proventricular oils are found both in adults and chicks, in and out of the breeding season, being absent only in *Pelecanoides urinatrix* (Warham *et al.*, 1976; Roby *et al.*, 1986; Obst & Nagy, 1993). Eighty-three per cent of the oils from 24 species of tubenoses contained waxes, 96% triglycerides, these two being the most prominent components.

Reviews of research on the compositions of the oils were given by Lewis (1969a), Serventy *et al.* (1971, pp. 30–32), Imber (1976b), Warham (1977b) and Jacob (1982).

At first the oil was suspected of being a secretion of the highly glandular proventriculus, a view put forward by Scouler (1826), after examining the tract of *Daption capense*, and later advocated by Matthews (1949) following histological study. However, Selby (1833, II, p. 525) and Audubon (1835, III, p. 449), stated that the young *Fulmarus glacialis* was fed on oil from the adult's prey, and others such as Howell (1917) and Dawson (1923) believed it to have a dietary origin, being held in the proventriculus before being digested or passed on to the chick. Hagerup (1926), who also held this view, pointed out that when small crustaceans are crushed in water the oil that collects on the surface looks and smells like fulmar oil. Later research has confirmed that the oils are food derived.

Early workers such as Gray (1871, p. 502) on *F. glacialis* oil, Lavauden and Mourgue (1918) on that of *Hydrobates pelagicus*, Evers and Foster (1920) on *Puffinus tenuirostris* and Rosenheim and Webster (1927) on *F. glacialis*, gave figures for physical characteristics such as viscosity (0.372 for *P. griseus*), melting points and specific gravities (varying from 0.80 to 0.92 for *F. glacialis, P. tenuirostris* and *P. griseus*). Figures for *F. glacialoides* of 0.839 ± 0.025, $n = 14$, and for *Thalassoica antarctica* of 0.838 ± 0.029, $n = 18$, are given by Norman and Ward (1992).

The oils are thin, straw-coloured to reddish liquids, with a sweet, slightly fishy or musty smell, and often set to a wax at low temperatures. From fasting birds they may

Table 8.1 Energy contents (kJ g^{-1}) of stomach oils. From Warham *et al.* (1976)

Species	No. of determinations	Range	Mean
Diomedeidae			
Diomedea epomophora	3	41.71–42.26	41.99
D. melanophrys	4	39.87–40.54	40.32
D. chrysostoma	2	39.62–39.79	39.71
D. bulleri	6	39.20–39.83	39.60
Phoebetria palpebrata	2	39.33–39.46	39.41
Procellariidae			
Macronectes halli	2	40.04–40.25	40.17
Fulmarus glacialis	3	38.87–39.50	39.20
Pterodroma inexpectata	2	40.63–40.71	40.67
P. neglecta	2	39.12–39.33	39.23
Puffinus griseus	2	41.92–42.01	41.97
P. tenuirostris	3	41.76–42.47	42.17

have a green tinge, perhaps from bile pigments refluxed from the duodenum. Other pigments are the natural lipid-soluble ones of marine invertebrates, for example the carotenoid astaxanthins in oil from albatrosses, giant petrels and prions resemble those from decapods such as euphausids (Lewis, 1969a; Clarke & Prince, 1976). However, those of *P. tenuirostris* oil differ from the pigments of its favoured prey *Nyctiphanes australis*, according to Cheah and Hansen (1970a).

The proventriculi may store considerable quantities of oil: the chicks of St Kilda fulmars yielded about 250 ml each, those of *P. griseus* and *P. tenuirostris* as much as 200 ml from a single bird.

The energy contents of the oils are consistent with their lipid constituents (Table 8.1). The value for *Pterodroma phaeopygia* of 41.74 kJ g^{-1} determined by Simons and Whittow (1984) is similar. Such figures represent a considerable concentration of the energy of the prey at the time of capture.

The main constituents of oils from 25 species are given in Table 8.2.

Warham *et al.* (1976) analysed mostly pooled samples from 16 species to establish whether these were of similar make up, as would be expected from a secretory source, or different, as expected if food derived. They gave the fatty acid/alcohol compositions of the wax esters, the triglyceride compositions, as well as the fatty acid and glyceryl ethers of cholesterol esters of *Procellaria aequinoctialis* and of the diacylglyceryl ethers of that bird and of *Pterodroma mollis*. Clarke and Prince (1976) analysed 35 samples from *Halobaena caerulea* to establish the degree of intraspecific variability, as did Bishop *et al.* (1983) for *Puffinus tenuirostris*. There was much intra- and interspecific variation, and Horgan and Barrett (1985) used oils from three species to test the feasibility of deducing petrel diets from the lipid classes and fatty acid ratios of the oils and stomach contents. They found that many unknown variables prevented precise identification of lipid origins. They and Warham (1977b) discussed variations in prey lipids.

Table 8.2 The main constituents of petrel stomach oils. From Warham (1977b)

Species	WE	TG	Ch	DAG	S	Authority
Diomedeidae						
Diomedea epomophora	X	X			X	Warham *et al.*, 1976
D. exulans	X	x			x	Lewis, 1969a
D. melanophrys	x	X				Warham *et al.*, 1976; Clarke & Prince, 1976
D. chrysostoma	x	X				Clarke & Prince, 1976
D. bulleri		X				Warham *et al.*, 1976
Phoebetria palpebrata		X				Warham *et al.*, 1976
Procellariidae						
Macronectes giganteus	X	X				Clarke & Prince, 1976
M. halli	X	X				Warham *et al.*, 1976; Clarke & Prince, 1976
Fulmarus glacialis		X	x			Cheah & Hansen, 1970a; Warham *et al.*, 1976
Daption capense	X				x	Warham *et al.*, 1976
Pagodroma nivea	X					Warham *et al.*, 1976
Pterodroma macroptera	x	X				Cheah & Hansen, 1970b
P. lessonii		X				Warham *et al.*, 1976
P. inexpectata	X	X				Warham *et al.*, 1976
P. mollis	X	x	x	X		Warham *et al.*, 1976
Halobaena caerulea	X	x				Clarke & Prince, 1976
Pachyptila desolata	X	x				Clarke & Prince, 1976
Procellaria aequinoctialis	x	x	x	x		Warham *et al.*, 1976
P. westlandica	x	X			x	Lewis, 1969a
Puffinus griseus	X	x				Warham *et al.*, 1976
P. tenuirostris	X	x				Cheah & Hansen, 1970a; Warham *et al.*, 1976
P. carneipes		X		X	x	Lewis, 1969a
P. pacificus		X	X			Cheah & Hansen, 1970b
Hydrobatidae						
Hydrobates pelagicus		X				Warham *et al.*, 1976
Oceanodroma leucorhoa			x	X		Lewis, 1966

X, principal constituent; x, less abundant constituent; WE, wax ester; TG, triglyceride; Ch, cholesterol; DAG, diacylglycerol; S, squalene.

Watts and Warham (1976) looked for any intermolecular specificity by calculating carbon number distributions for the wax esters, triglycerides, and diacylglyceryl ethers assuming random esterifications. Most of the lipids did have the random structures typical of marine lipids and the intermolecular distributions of the intact lipids were unlike the specific triglycerides synthesized by the mammary gland, the mammalian gut or liver.

Table 8.3 Average chain length of the monoester wax constituents of stomach oils from tubenoses. From Warham *et al.* (1976)

	FA	Alc
Diomedea epomophora	18.41	18.48
D. melanophrys	18.61	16.92
Macronectes halli	18.38	17.43
Daption capense		
Specimen 1	17.59	20.43
Specimen 2	18.02	20.44
Pagodroma nivea		
Specimen 1	18.22	17.21
Specimen 2	18.11	17.14
Pterodroma mollis	18.80	17.59
Puffinus inexpectata	18.03	17.98
P. tenuirostris	18.48	17.96
P. griseus	18.60	17.35

FA, fatty acids; Alc, alcohols

1 *Chemical compositions*

*a. Hydrocarbons.*The proportions of squalene $C_{30}H_{50}$ and pristane $C_{19}H_{40}$ in some oils was high (Table 8.2). Pristane is prominent in copepods; squalene, an intermediate in the biosynthesis of cholesterol, could have come from scavenging a shark. Dogfish livers are highly attractive to shearwaters such as *Calonectris diomedea*. Squalene was found in *Diomedea exulans* in a wide range of body organs except the brain, although the blood held 10 ppm (Lewis, 1969b).

b. Monoester waxes. Chain lengths of the fatty acids and alcohols were between C_{14} and C_{22} and average chain lengths, with the exception of oil from *Daption capense*, were smaller in the alcohols than in the acids (Table 8.3). Monoenic components predominated with some saturated dienic and polyenic constituents, similar to other wax esters of marine origin. All the wax esters had total carbon numbers 30–44, as in those for marine invertebrates and fish (Watts & Warham, 1976).

c. Cholesterol esters. These were not common in the oils. Warham *et al.* (1976) gave the compositions of those from *Procellaria aequinoctialis*. They differed from those determined by Cheah and Hansen (1970b) for *Puffinus pacificus* and *Pterodroma macroptera* and Lewis (1966) for *Oceanodroma leucorhoa*, but could still have been derived from squid.

d. Diacylglycerol ethers. These were found in the oils of *Procellaria aequinoctialis* and *P. westlandica*, in one of the latter being 51% by weight of total lipid, with none in another. In *Pterodroma mollis* there were considerable proportions of both wax esters and diacylglycerol ethers, perhaps from a common source (Warham *et al.*, 1976)—these lipids are widely distributed in the marine environment.

e. Triglycerides. These have been found in the oils of all species studied with the exception of that of *Pagodroma nivea*. The fatty acid compositions vary intra- and interspecifically as Clarke and Prince (1976) showed. Hexadecanoic, octadecanoic and eiconsenoic acids predominate, but higher unsaturated fatty acids, for example 20:5 and 22:6, are common, averaging about 10% of the whole *vs.* 50–60% for 16:0, 18:1 and 20:1 combined. Average chain lengths were 17.91–18.71, similar to those of the wax ester acids. These are the common triglyceride fatty acids of marine lipids.

f. Di- and monoglycerides. Diglycerides reached 10% in the oils of *F. glacialis* and *Pterodroma inexpectata* and were present in all the species sampled by Warham *et al.* (1976). Smaller titres of monoglycerides were also detected except in oil from *Pagodroma nivea* where none was found.

g. Fatty acids. Present in all oils analysed. Palmitoleic (16:1) usually predominated with low values for palmitic acid (16:0). Oil from *D. capense* had low amounts of oleic acid (18:1) but high myristic (14:1), palmitic and 22:1 acids, origins unknown.

h. Alcohols. Most oils contained alcohols and with similar compositions to the fatty acids but in varying proportions in the different species, 16:0, 18:1 and 20:1 predominating as in other marine forms. Again, *D. capense* had no C_{14} or C_{16} alcohols but 72 moles% of 20:1 alcohol.

i. Phospholipids. Southern Ocean petrels were found to have only 0.2–0.9% of phospholipids in their oils despite potentially high levels in their prey (Warham *et al.*, 1976). Horgan and Barrett (1985) identified 2.0–29.7% of these lipids in oils from *Pterodroma macroptera*, 7% in one *Diomedea exulans* but none in any samples from *Procellaria aequinoctialis*. Warham *et al.* suggested that in the southern species either a highly active phospholipase is present or that phospholipids are rapidly absorbed pynocytically, perhaps to preserve them, as they are important components of plasma membranes.

j. Free steroids. Only cholesterol has been identified (Table 8.2).

Bishop *et al.* (1983) established a high correlation between the fatty acid and alcohol compositions of the triacylglycerol and wax esters of *Puffinus tenuirostris* oils and those of their prey, *N. australis* (Table 8.4). The considerable inter- and intra-specific variation support the conclusion that the oils come from the prey, the different compositions reflecting differences in meals and their varied digestion. The prey lipids also vary with their age, breeding status, and environmental conditions.

Contamination with stomach oil can be lethal to other birds whose plumage becomes matted (see review in Warham, 1977b). According to Jacob (1982), the highly unsaturated lipids are oxidized in air yielding a sticky resinous product. This effect seems supported by the experiments of Clarke (1977) on oiled and dead *Falco peregrinus*. The composition of the fatty acids of their feather lipids was consistent with soiling by fulmar oil. After 1–10 days' weathering on falcon feathers, this oil produced a mixture similar to that of the Peregrine contamination, the 20:5, 22:5 and 22:6 acids being rapidly degraded with a corresponding fall in iodine values. However, fulmars which spray each other with oil seem unharmed, apparently decontaminating their plumage by intensive preening and bathing. That non-tubinares cannot do that because their preen glands are much smaller, those of the petrels producing large amounts of wax for cleansing the feathers (Jacob, pers. comm.).

Table 8.4 Fatty acid and alcohol composition (%) of the triacylglycerol and wax ester fractions of (a) *Nyctiphanes australis* and (b) stomach oil and depot fat of *Puffinus tenuirostris* chicks. From Bishop *et al.* (1983)

(a)	Triacylglycerol fatty acids		Wax ester fatty acids		Wax ester fatty alcohols	
	Mean $n=5$	Range	Mean $n=5$	Range	Mean $n=5$	Range
14:0	6.2	4.3–11.4	2.3	1.6–2.8	12.7	11.9–13.6
16:0	24.6	23.1–26.8	2.3	1.2–2.8	59.1	54.5–66.1
16:1	10.5	8.8–11.2	14.7	11.0–18.8	8.3	6.8–9.3
16:2	3.0	0.3–5.8	14.3	4.5–31.4	—	—
18:0	4.2	2.9–5.0	0.2	0.1–0.7	2.4	1.0–3.1
18:1	40.3	27.9–46.0	41.7	31.6–53.4	7.2	5.4–8.3
18:2	1.6	1.4–2.0	1.5	0.1–2.6	—	—
20:1	5.6	3.2–6.7	1.7	0.7–2.9	5.0	2.4–6.7
20:5	+	—	6.2	3.4–9.4	—	—
22:1	0.8	0.1–2.5	0.7	0.1–1.7	2.1	0.2–4.7
22:6	+	—	0.8	0.1–2.0	—	—

(b)	Proventricular oil						Depot fat	
	Triacylglycerol fatty acids		Wax ester fatty acids		Wax ester fatty alcohols		Triacylglycerol fatty acids	
	Mean $n=3$	Range	Mean $n=3$	Range	Mean $n=3$	Range	Mean $n=3$	Range
14:0	5.1	4.8–5.4	1.5	1.1–2.1	13.0	12.3–13.8	3.6	3.5–3.9
16:0	11.5	8.9–16.1	1.2	1.0–1.5	50.7	40.6–57.9	23.4	19.5–26.3
16:1	16.8	16.1–17.2	18.4	16.5–20.5	10.1	8.9–11.8	9.4	9.2–9.6
16:2	2.2	1.5–3.5	3.8	1.3–6.1	—	—	0.4	0.2–0.5
18:0	1.2	0.9–1.7	0.2	0.1–0.6	3.5	2.9–4.3	4.7	4.0–5.2
18:1	41.8	32.8–47.5	49.6	43.2–53.1	8.8	7.4–9.8	42.1	35.4–45.8
18:2	2.3	2.0–2.5	3.0	2.8–3.4	—	—	1.6	1.2–1.9
20:1	3.1	2.4–4.2	4.0	2.3–5.0	7.0	4.0–8.9	7.8	6.8–9.3
20:5	8.9	7.3–10.9	11.2	10.8–11.5	—	—	1.4	0.1–3.9
22:1	2.1	0.5–4.0	2.7	0.2–5.7	0.5	0.1–1.6	2.7	0.5–6.8
22:6	2.0	0.1–4.2	1.7	0.1–3.6	—	—	2.0	0.1–4.4

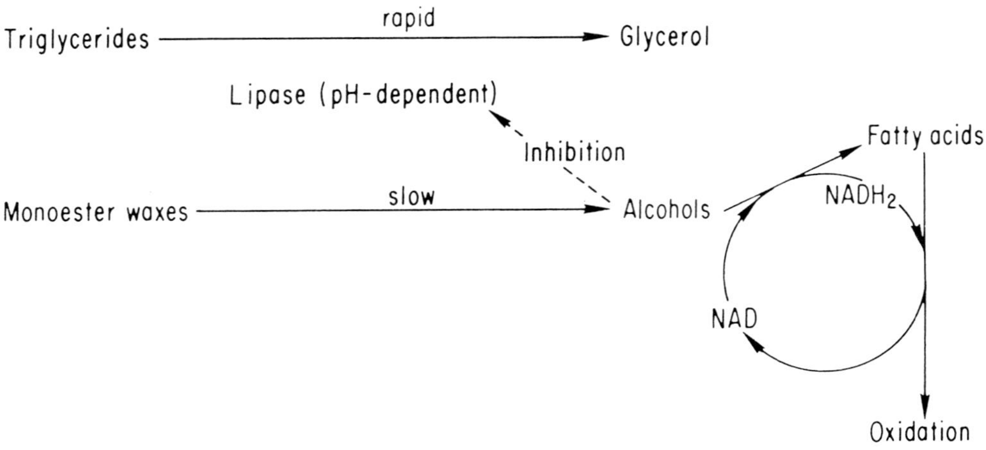

Figure 8.1 *Tentative patchways and regulatory steps in the lipolysis of stomach oils. From Jacob (1982).*

Jacob (1982) emphasized that triglyceride hydrolysis proceeds much faster in some fish than does monoester wax, while the fatty alcohols produced by their breakdown inhibit lipases so that the products of wax breakdown may retard further lipid digestion, an effect that would be more evident after the combustion of the triglyceride acids (Fig. 8.1).

B *Depot fats*

Subcutaneous and visceral deposits lack ester waxes, and in *P. tenuirostris*, for instance, the make-up of their fatty acids can be explained by the conversion of fatty alcohols derived from ester waxes or prey into fatty acids of equivalent chain length and saturation (Bishop *et al.*, 1983; Table 8.4). These authors also noted that the depot fat constitutions of the adults were more variable than those of the chicks and suggested that the foraging parent experiences a constant turnover of depot reserves whereas the chick tends to accumulate fat until needed at or after fledging.

When lipid extracts of the skin of chicks of *Pelecanoides georgicus*, *P. urinatrix* and *Pachyptila desolata* that had been fed ^{14}C-labelled wax ester by Roby *et al.*(1986) were analysed, up to 50% of label was recovered as triacylglycerols, fatty acids or partial glycerols, indicating the relationship between the stomach oil and fat reserves. Likewise, Bishop *et al.* (1983) found that the depot-fat fatty acids of *Puffinus tenuirostris* agreed well with that of their stomach oil if allowance was made for the oxidation of the fatty alcohols derived from the wax esters to fatty acids (Table 8.4).

Lill and Baldwin (1983) also monitored the accumulation of depot fat in adults and chicks of *P. tenuirostris* by assaying the levels of three hepatic enzymes (NADP malate dehydrogenase (NADP-MDH), NADP isocitrate dehydrogenase (NADP-IDH), and glucose-6-phosphate dehydrogenase + 6-phosphogluconate dehydrogenase) as indicators of fatty acid synthesis. The protein contents of cytoplasmic liver extracts did not change significantly.

The depot fat of the chicks reached at least 35% of their body weight by the time that the feeding ended. There was little increase in hepatic NADP-MDH and pentose-shunt dehydrogenases, but that of NADP-IDH fell substantially. Lill and Baldwin (1983) concluded that there was little *de novo* fatty acid synthesis, rather that the dietary acids were incorporated directly into the fat reserves.

Hosie and Ritz (1980) reported that the subcutaneous fat of both adults and chicks of *P. tenuirostris* contained pure triglycerides with high 18:1 acids while their stomach contents held lower amounts of 18:1 and more 22:6 fatty acids than in the fat or stomach oil, indicating preferential hydrolysis. The presence of large amounts of dietary fatty acids in the stomach would be expected to depress hepatic lipogenesis.

C *Preen gland lipids*

These have been described from *Diomedea melanophrys, Pachyptila belcheri, Procellaria aequinoctialis, Puffinus griseus, Garrodia nereis* and *Pelecanoides urinatrix* in a series of papers by Jacob (1976, 1978) and Jacob and Hoerschelmann (1982) tabulated the compositions of the glands' monoester waxes of 37 tubenoses in 15 genera. Their compositions were constant within a species, bore no relationships to those of the depot fats, and were quantitatively different from those of the proventricular oils (Jacob, 1982).

These tubenoses waxes were unusual in consisting of mainly complex homologous and isomeric series of multi-branched fatty acids and fatty alcohols, particularly those with a first branch in the 3- or 4- position. The co-occurrence of 3- and 4- branched acids was characteristic of these birds. However, *Diomedea* represented one end of the range of variation, having mostly 2,4,*x*-trimethyl- substituted acids but only traces of 3-methyl-branched ones, *Garrodia*, with a very high number of unbranched acids (quite different from that of *Oceanites oceanicus*), the other. A similar situation was found to hold for the wax alcohols. A possible pathway for the biosynthesis of procellariiform preen waxes was given by Jacob and Hoerschelmann (1982) (Fig. 8.2).

D *Feather waxes*

There are lipid-secreting cells in the skin of birds according to Lucas and Stettenheim, (1972, p. 630) who included two *Diomedea* spp. in their samples. Jacob (1978) extracted lipids from the plumage of *Puffinus griseus*. Their compositions resembled those of the uropygial gland lipids. However, the role of the sebaceous glands in tubenoses needs investigation; Lucas and Stettenheim suggested that lipids from the epidermis nurtured the skin itself, the preen gland secretions the feathers.

E *Bone marrow lipids*

Oka and Maruyama (1985) investigated the usefulness of changes in the weights of the marrows in the tibiotarsi and femora of *P. tenuirostris* as indices of their physical

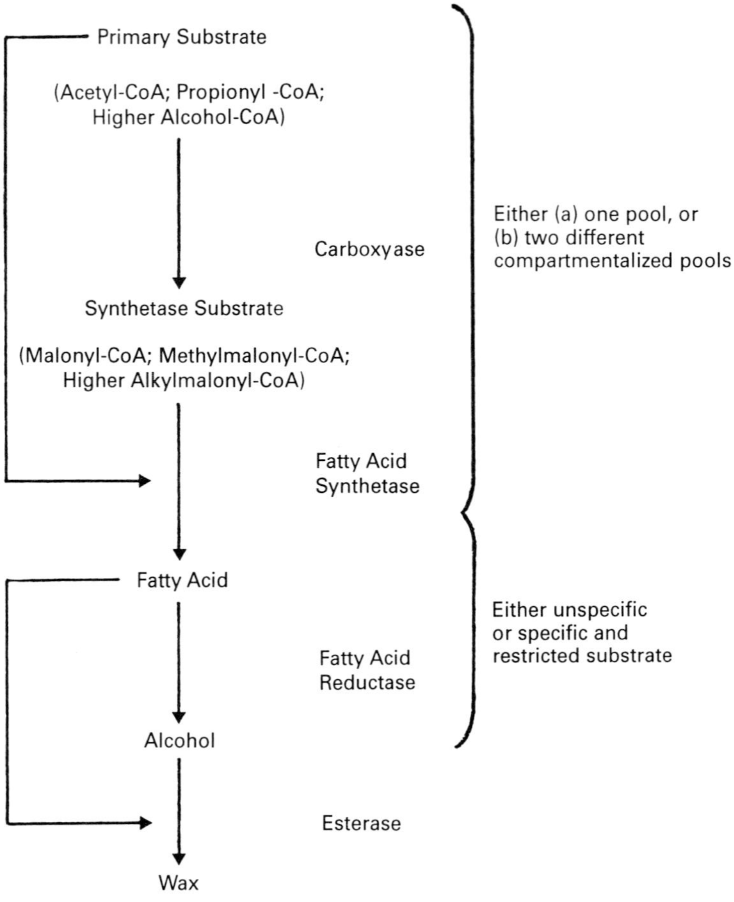

Figure 8.2 *Tentative biosynthetic pathways for formation of the fatty acid and alcohol components of preen gland waxes. From Jacob and Hoerschelmann (1982).*

condition. Lipids in these bones are drawn upon in times of need. Oka and Maruyama showed that the per cent dry weight/wet weight of the marrow was a good indicator of body condition, and they devised a method of visual evaluation from the marrow texture.

II Biochemistry of pollutants in petrels

A *Mineral oil ingestion*

Peakall *et al.* (1981) looked at the disruption of hormonal control mechanisms in a small sample of adult *Oceanodroma leucorhoa* after injection of fossil fuel hydrocarbons. When given oil orally, these birds had raised levels of thyroxine—71.1 ± 6.8 ng ml^{-1} as against sham-dosed controls with 44.7 ± 4.3 ng ml^{-1} ($P < 0.01$), but corticosterone levels were not significantly altered. Growth of *O. leucorhoa* chicks was also significantly depressed after parents had been intubated with Prudhoe Bay crude, whereas direct dosing of the chicks did not hinder their growth (Trivelpiece *et al.*, 1984). These authors deduced that the reduced growth rates of the chicks were most probably because of the parents' lowered ability to find food. In another study long- and short-term effects of oiling on *P. pacificus* were examined by Fry *et al.* (1986).

Elevated thyroxine levels imply increased metabolic activity. Butler *et al.* (1986) measured RMRs of adult *O. leucorhoa* that had been oiled externally or internally. Results were equivocal. Values from metabolic chamber tests were only slightly higher than those of mock-treated controls whereas using DLW, the oil-dosed birds had higher RMRs (75.9 kJ day^{-1} *vs.* 60.8 kJ day^{-1}), $P < 0.001$.

Trivelpiece *et al.* (1984) also compared liver, adrenal and salt gland weights of treated and untreated *O. leucorhoa* adults. There were no differences in liver weights but both glands were significantly larger in the oiled birds with $P < 0.01$. However, *O. furcata* chicks dosed with weathered Prudhoe Bay crude by Boersma *et al.* (1988), although fledging 2 days later than controls, were otherwise not perceptibly affected. These researchers speculated that storm petrels can digest long-chain hydrocarbons such as pristane and long-chain *n*-alkalenes such as C_{15} and C_{17}, because these are relatively non-toxic and in their natural diet. They also occur in Prudhoe Bay oil and chicks may produce detoxifying enzymes to render the oil harmless. Whether the tubenose kidney can detoxify contaminants from crude oil or chlorinated hydrocarbons awaits closer study. Peakall *et al.* (1987) investigated levels of six mixed-function oxidase enzymes in *O. leucorhoa* as a baseline for comparison with levels in auks living nearby. Levels in petrel livers of one of these, aldrin epoxidase, were higher than in *P. puffinus* assayed by Knight and Walker (1982), but as the petrels were not dosed, the role of these enzymes in breaking down oil contaminants, if any, is unknown.

Boersma (1986b) sampled regurgitate from adult *O. furcata* breeding off Alaska and the Aleutians, and identified components from petroleum ingestion by gas chromatography. She ranked the samples as dirty when 95–100% of chromatogram peaks matched a Prudhoe Bay standard (Fig. 8.3). The incidence of dirty samples also corresponded with the known oil spillages within range of the colonies. In only one instance out of eight did the 'quality' of the stomach oil not decline after a spillage.

B *Chlorinated hydrocarbons*

Like other seabirds, petrels have received their loads of chlorinated hydrocarbons (CHCs). Bourne (1976b) provided a detailed review of work on seabird pollution during 1971 to 1975, including data on loads of lipid, DDE and polychlorinated biphenyls (PCBs) for many species including those for 18 tubenoses. Since then, detection techniques have improved and more recent reviews such as those of Noble and Elliott (1986) for Canadian seabirds and of Elliott and Noble (1993) for the temperate North Pacific, cover a greater range of contaminants.

Comparing levels of contaminants in petrels that have been determined by different workers is difficult as methods have varied and the data presented in different ways. Also despite a common affinity for lipids, the amount of a xenobiotic in one tissue is often unrelated to that in another (Osborn *et al.*, 1987), so that levels in

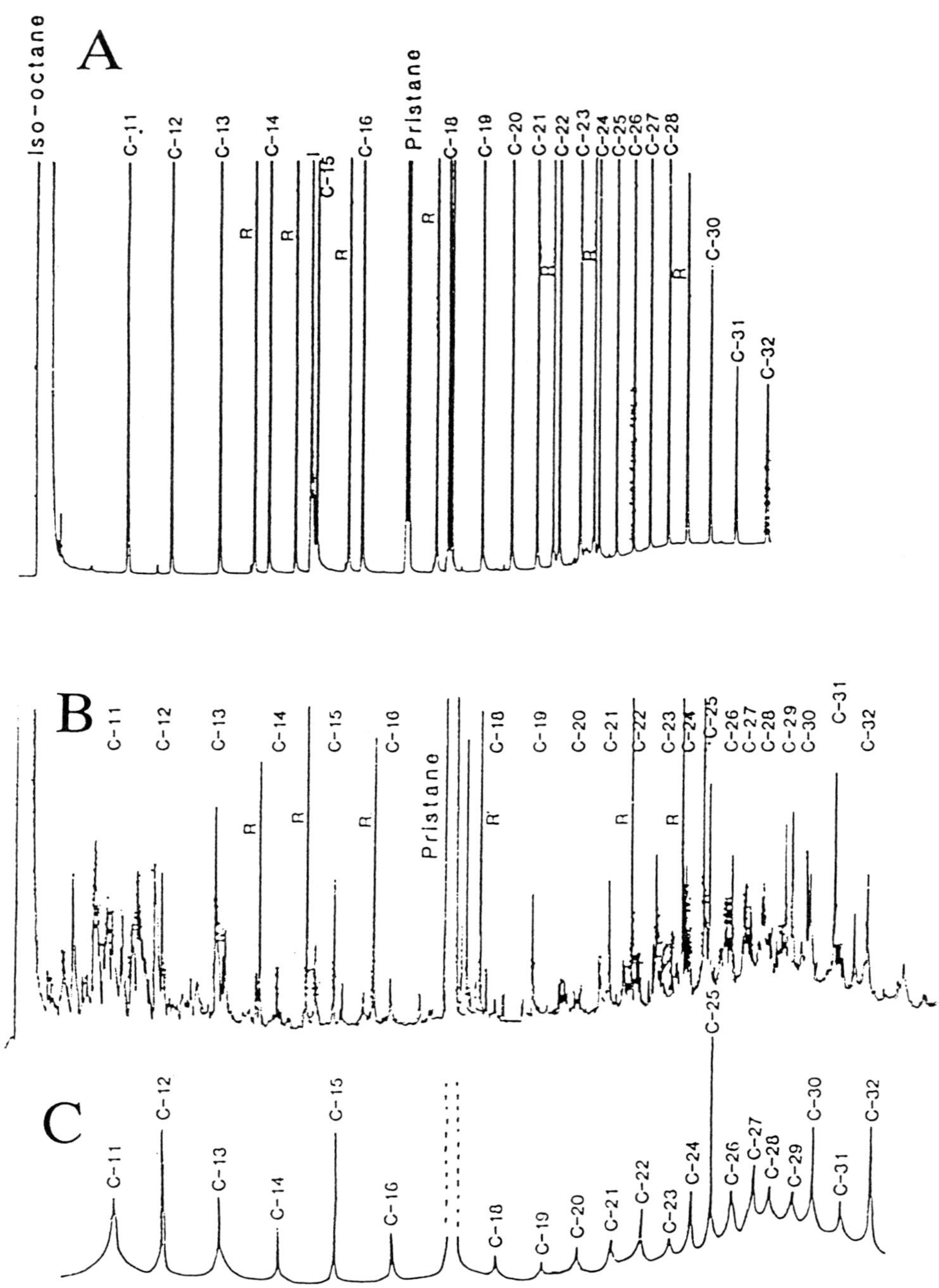

Figure 8.3 *Spectra of samples from Prudhoe Bay crude oil and the gut of the storm petrel* Oceanodroma furcata. *(A) Prudhoe Bay standard showing regular* n*-alkane peaks of carbon chains. (B) Stomach sample from petrel containing fossil fuel hydrocarbons. (C) Tracing of (B) showing 21* n*-alkane peaks matching those of the crude oil. From Boersma (1986b).*

one organ may be inadequate indicators of total contamination. In *Calonectris diomedea*, for instance, DDE levels increased from brain to kidney to liver to muscle, to preen gland to depot fat (Renzoni *et al.*, 1986).

Despite early reports of egg shell thinning due to the disruption of calcium balance, for example in *Oceanodroma homochroa* (Coulter & Risebrough, 1973), there seems to be no evidence of mortality, behavioural abnormalities or reduced breeding success following contamination with these chemicals, nor any recent demonstration of eggshell thinning. However, there are few long-term data, the best perhaps being the PCB levels monitored in Canada of *O. leucorhoa* eggs between 1970 and 1990 that showed no change (Elliott & Noble, 1993).

1 *Organochlorine pesticides*

The most abundant derivative is DDE, more prevalent than the other metabolite DDD. Other compounds of this class include dieldrin, heptachlorepoxide (HE), toxaphene, oxychlodane, heptachlorocyclohexane (HCH) and hexachlorobenzene (HCB) isomers.

2 *Polychlorinated biphenyls (PCBs)*

These consist of a mixture of isomers and have been found in the eggs and bodies of virtually all petrels examined in recent years and are widely distributed in marine ecosystems.

3 *Polychlorinated dibenzodioxins (PCDDs) and dibenzofurans (PCDFs)*

These enter the environment accidentally during chemical manufacture or waste incineration. Dioxins have been detected in some seabirds but as yet not in petrels. For example, none was found in *O. leucorhoa* eggs from British Columbia in 1987 (Elliott & Noble, 1993).

C *Levels in relation to feeding areas*

Occasionally high levels of contaminants can be traced to local sources, for example the shell thinning of the inshore feeding *O. homochroa* off California in the 1960s which was associated with a massive discharge of effluent from a DDT manufactory. In 1967, seven *P. griseus* from burrows in the South Island, New Zealand were carrying considerable loads of DDT and its derivatives as well as PCBs in their fat—17.3 ppm of DDT and derivatives (78% of which was DDE), nearly twice as much as in the same species taken off California at about the same date. However, the high levels were ascribed to a local source—aerial top dressing with superphosphate to which DDT was added for control of pasture insects (Dacre, 1974).

The effect of local conditions on CHC levels is nicely exemplified by the situation in *C. d. borealis* breeding on Grand Salvage Island and the typical race breeding in the

Mediterranean, as detailed by Renzoni *et al.* (1986). The basic information in Fig. 8.4 shows that CHC levels were much higher in the Mediterranean birds and presumably most xenobiotic chemicals were picked up in the Mediterranean, not in the Atlantic. Some of the highest values were found in birds from Majorca (e.g. 80 ppm DDE and 95 ppm PCB in fat) where there is little industry, and Renzoni *et al.* hypothesized that the contaminants had originated from industrial discharges into the Rhone and Ebro Rivers and around Marseilles. A suggestion that the thinner shells of *C. d. diomedea* eggs might be due to pesticide action made no allowance for *C. d. borealis* being a bigger bird that lays a bigger, thicker-shelled egg (*The Petrels*, p. 295). Renzoni *et al.* gave values of 0.385 mm and 0.321 mm for these two subspecies whereas larger samples of pre-DDT-era ones were 0.35 mm and 0.30 mm thick respectively.

Local sources may also explain why subantarctic petrels that stay in southern seas in the austral winter have lower levels of CHCs than those wintering north of the Convergence. Thus the eggs of *Pachyptila turtur, P. vittata, Halobaena caerulea, Lugensa brevirostris* and *Pterodroma mollis* from Marion and Gough Islands held less DDE than those of *P. macroptera, Procellaria aequinoctialis, Diomedea chlororhynchos, Puffinus gravis* and *D. exulans* when measured by Gardner *et al.* (1985b). Ohlendorf and Harrison (1986) suggested that the higher DDT residues in *P. pacificus* eggs from Hawaii than in those of other local seabirds arose because the shearwaters migrated to seas off Central America where large amounts of DDT were still being used.

Pollution levels in species measured at the same breeding colony in the same season may differ, suggesting different sources of contamination. For example, four *P. griseus* at the Snares Islands in 1971 had 1.23 ppm DDE and 2.1 ppm PCBs in their pectoral muscles whereas *Pterodroma inexpectata* sampled there had 2.1 and 5.3 ppm DDE and PCBs respectively in these muscles (Bennington *et al.*, 1975). Both probably acquired their loads when on migration in the North Pacific, *P. inexpectata* possibly getting most from prey tainted by effluent from Japan. Indeed, work by Ludwig *et al.* (in press) shows serious contamination of *D. immutabilis* and *D. nigripes* occurs at Midway Atoll with some shell thinning, apparently due to wastes discharged from Asia.

The storm petrel *Oceanites oceanicus* seems particularly prone to acquiring loads of pesticides and PCBs. The adipose tissue and liver of birds taken at the South Shetland Islands contained 15.7 and 1.8 ppm of PCBs respectively, an order of magnitude above the levels in other birds there (Lukowski *et al.*, 1987). Earlier, Risebrough (1974) had found 46 ppm in the depot fat of this species. The levels of DDT and its derivatives (16.4 ppm) were also high, with 8.4 ppm DDE, 4.9 ppm DDD and 3.1 ppm DDT in adipose tissue and lesser levels in breast muscle and liver (Lukowski, 1983). Lukowski considered that the pollutants had been acquired during the contranuptial migration, exacerbated by the bird's propensity for picking up fat globules in which organochlorines are likely to have been concentrated. Previously, Risebrough and Carmignani (1972) had found that *O. oceanicus* far apart at Cape Hallett and Palmer Station had quite different CHC loads and suggested that these populations wintered in different seas, the Hallett birds perhaps going to the Indian Ocean where the pollution levels off Arabia could be lower than off North America.

Norheim and Kjos-Hanssen (1984) found highly significant linear correlations

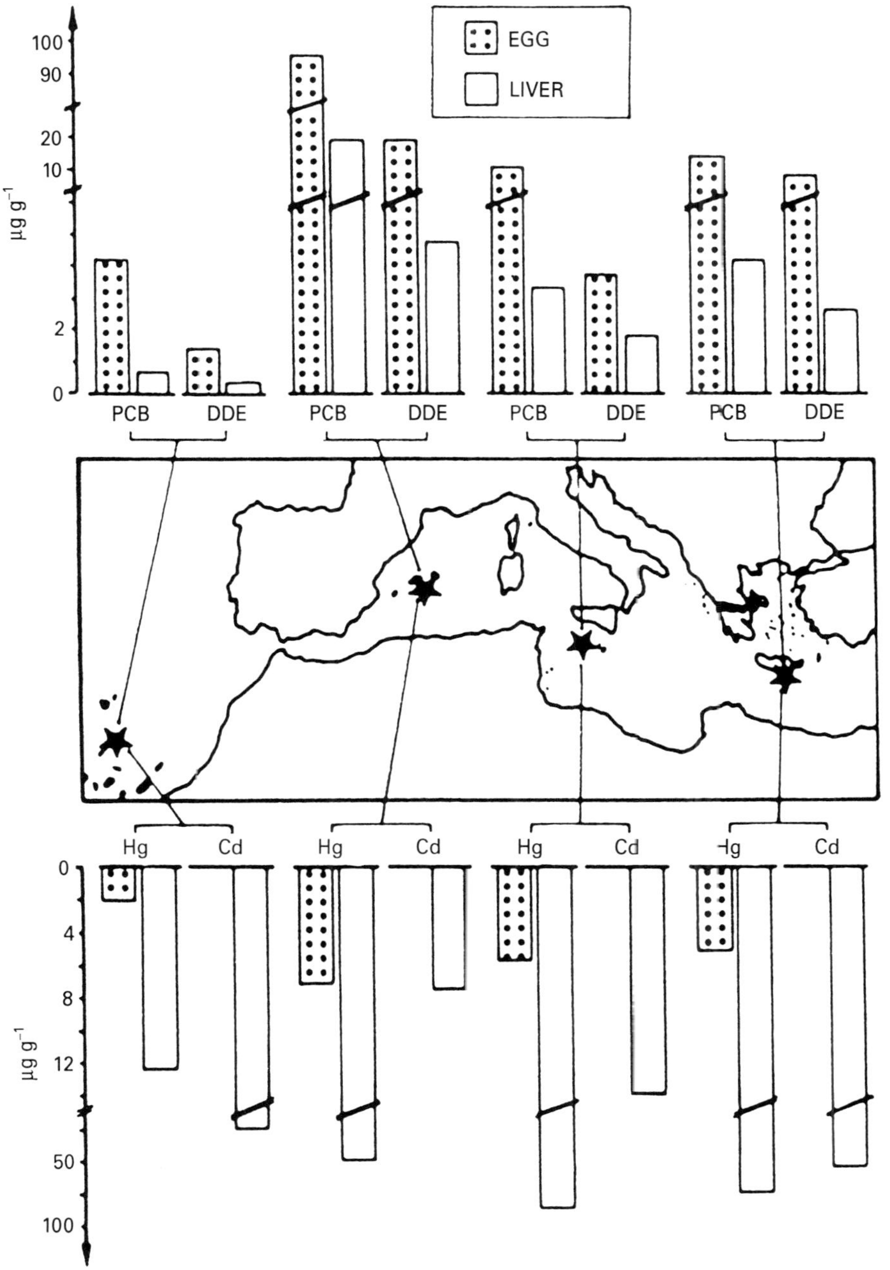

Figure 8.4 *Contaminant CHCs and heavy metals in the eggs and livers of Cory's Shearwaters from Grand Salvage Island, Majorca, Linosa and Crete. From Renzoni* et al. *(1986).*

Table 8.5 Mean concentrations ($\mu g\ g^{-1}$ wet weight) of mercury, hexachlorobenzene, DDE and PCBs in liver (L) and fat (F) of fulmars from Spitzbergen and Bouvet Island. From Norheim *et al.* (1982) and Norheim and Kjos-Hanssen (1984)

Species		*n*	Hg	HCB	DDE	PCB
F. glacialis	L	10	2.1 ± 1.2	0.049 ± 0.014	0.63 ± 0.23	1.6 ± 0.4
	F	7	na	1.7 ± 0.4	22 ± 9	59 ± 22
F. glacialoides	L	6	2.9 ± 1.9	0.005 ± 0.003	0.011 ± 0.007	nd
	F	3	na	0.16 ± 0.04	0.36 ± 0.22	0.1 ± 0.0

na, not analysed; nd, not detected.

between the concentrations of DDE, PCB and HCB in the livers of five kinds of seabird, including *F. glacialis*, caught off Spitzbergen. When the data for all 48 analyses were considered together, $r = 0.99$, $P < 0.001$, suggesting a common source in the Gulf Stream bringing a continuous supply of pollutants from the eastern Atlantic.

Contaminant levels for *F. glacialis* at Spitzbergen were significantly higher than those for *F. glacialoides* at Bouvet Island (Table 8.5). The differences presumably reflect the lower pollutant levels in Antarctic seas.

D *Effects of age and sex*

As CHCs persist in tissues, it would be expected that older birds would accumulate higher levels of contaminants. There are no good data, but this factor could be important for these long-lived birds. In a sample of 11 adult and one chick of *Macronectes giganteus* aged from 80 days to 15+ years, Conroy and French (1974) found a tendency for liver DDE levels to increase with age, but the correlation was not significant. Variation according to sex is unstudied, but Gaskin *et al.* (1978) found levels of most CHCs in female *P. griseus* from the Bay of Fundy region higher than in the males there.

E *Changes with time*

With the phasing out of CHCs such as DDT it might be expected that levels in affected seabirds would be decreasing. Again, there are few data, but DDE levels in eggs of giant petrels at Macquarie Island appear to have increased from 0.36 ppm in 1978, 0.43 ppm in 1979 to 0.62 ppm in 1983 for *M. giganteus* and from 0.81 in 1979 to 0.91 ppm in 1983 for *M. halli* (Luke *et al.*, 1989). Higher values for *M. halli* may be because its more northerly range brings it closer to Australasia and Africa.

For *O. leucorhoa* from eastern Canada, however, the picture is different. Pearce *et al.* (1989) found a general decline in DDE and PCB levels from 1968 to 1984 in eggs from the Bay of Fundy and Great Island, Newfoundland (Fig. 8.5), particularly between 1972 and 1976. Levels of other contaminants did not change significantly except that those of HCH. DDE, PCBs and oxychlorane were higher, and dieldrin, heptachlorepoxide and HCB levels lower, in eggs from the Bay of Fundy than for those from Newfoundland.

F *Heavy metals*

A range of heavy metals occurs in petrels and/or their eggs, as with other seabirds. Most appear to be natural components of their ecosystems and taken up in their food and when drinking. For example, the livers of *Pagodroma nivea* from east and west Antarctica carried 0.41 ppm and 0.65 ppm wet weight of Hg respectively, although this bird seldom leaves the pack-ice (Anderlini *et al.*, 1972; Norheim *et al.*, 1982). Elements such as copper, iron, zinc and manganese are essential to life, incorporated in enzymes, their concentration in tissues evidently regulated metabolically. Other elements are at very low levels and were not identified in earlier analyses while some, notably Hg and Cd, may reach quite high titres.

Elements like Hg and Cd accumulate up the food chain and possibly with age, as petrel chicks usually have lower levels than adults (Lock *et al.*, 1992; Monteiro *et al.*, 1995). However, at least with albatrosses, Hg burdens do not seem to increase with age (Thompson *et al.*, 1993). Although kidney damage has been described, particularly at high levels of Cd, even heavily loaded petrels have been apparently healthy (e.g. *Oceanodroma leucorhoa*; in Elliott *et al.*, 1992), and no behavioural abnormalities have been described even where levels would be dangerous in man. Loads determined in *Pachyptila turtur* by Brothers and Brown (1987) and in six petrels from the North Pacific by Honda *et al.* (1990), were all independent of the sexes.

Data for four tubenoses were tabulated by Bourne (1976b); for more recent reviews see Ohlendorf (1993) for the North Pacific, Lock *et al.* (1992) for New Zealand seas, and Thompson (1990) for marine vertebrates in general. Table 8.6 shows analyses for five metals in Northern and Southern Fulmars, both remote from industrial activity.

1 *Mercury*

This has been found in tissues or eggs of at least 48 species in all four families, in some at high concentrations, in others only as traces. Mercury is usually most concentrated in liver and kidney tissue but has also been found in the brain, muscle and blood.

In *C. diomedea*, the Hg titres increased in the order fat, preen gland, brain, muscle, kidney and liver (Table 8.7) with values in the latter two organs being highly correlated (Renzoni *et al.*, 1986). Birds from the Atlantic had but 13 ppm (dry weight) in their livers, Mediterranean birds 86 ppm at Linosa, 75 ppm at Crete and 50 ppm at the Balearics. Levels in their eggs, though lower, were in the same order. The very heavy values from Majorca could be influenced by industrial effluent into a shallow, land-locked sea which itself may have high natural Hg levels.

It has been suggested that low levels in the livers of prions and diving petrels derive from their diets of crustacea which themselves contain little Hg, whereas

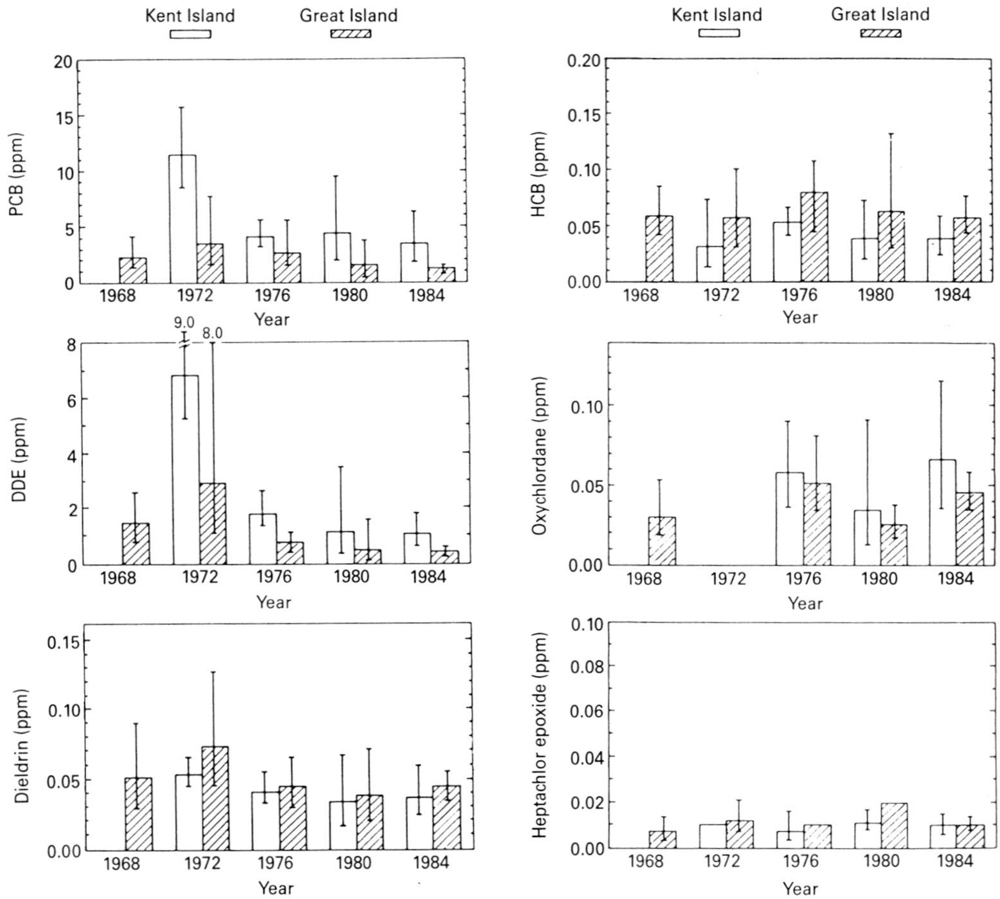

Figure 8.5 *CHC contaminants in eggs of Leach's Storm Petrels in Atlantic Canada, 1968–84. Bars show ranges. From Pearce* et al. *(1989).*

those species feeding on squid, fish and carrion ingest more (Lock *et al.*, 1992). In Antarctic fulmarines, the livers of *Pagodroma nivea* held 0.65 ppm (wet weight) of Hg, those of *Daption capense* 1.3 ppm and of *F. glacialoides* 2.9 ppm (Norheim *et al.*, 1982), this last higher than the loading of *F. glacialis* at Spitzbergen (Table 8.5). These figures contrast with the high mean value of 140 ppm wet weight in the livers of three *M. halli* from New Zealand seas (Lock *et al.*, 1992). The eggs of both *Macronectes* spp. from Macquarie Island also had quite large titres of Hg, 32 of *M. giganteus* averaging 1.26 ppm wet weight, 17 *M. halli* 1.30 ppm, but those of *P. nivea, D. capense, F. glacialoides* and *T. antarctica* from Antarctica held much less—0.22–0.34 ppm wet weight (Luke *et al.*, 1989).

Levels of 10.2 ppm and 29.4 ppm dry weight in livers of *Puffinus puffinus* and *F. glacialis* from St Kilda, regarded as high by Osborn *et al.* (1979), are low compared with those of *Pterodroma nigripennis* and *P. macroptera* with liver levels of 66 and 112 ppm wet weight (Lock *et al.*, 1992), who also recorded 135 ppm wet weight in the liver of *Procellaria cinerea.*

Highest levels are in albatrosses. Muirhead and Furness (1988) reported 268 and 141 ppm wet weight in the livers of *Diomedea exulans* and *Phoebetria fusca* respectively at Gough Island, Honda *et al.* (1990) 95 ppm in those of *D. nigripes*, Lock *et al.* (1992) 295 and 282 ppm wet weight in *D. exulans* and *D. epomophora* respectively, yet Gough Island *D. chlororhynchos* had but 7.7 ppm wet weight. The high levels in the albatrosses would be toxic to man. The petrels appear to have adapted to these metals, showing none of the ill effects seen in some land birds with lower levels. No consistent links with diet were apparent but differences in the way Hg is stored in liver and limitations in excretion via the feathers were suggested.

Mercury links to disulphide bonds in keratin and accumulates in hair and feathers, and the latter have been proposed as biological indicators of environmental levels. Furness *et al.* (1986) examined concentrations in feathers, including those of *F. glacialis, Puffinus puffinus, P. gravis, Pterodroma incerta, P. mollis* and *Lugensa brevirostris*. In all these birds levels of Hg in the primaries fell according to the moult sequence. The innermost (No. 1), being the first to be shed, had the highest levels, No. 10, the outermost and last to be shed, the lowest, the r values ranging from -0.85 to -0.98 with $P < 0.001$ in all cases. Furness *et al.* interpreted this general finding as the result of storage in and elimination of Hg via the feathers, the two innermost primaries taking the metal up from the body tissues as they grew, but as the Hg pool became depleted the later feathers only acquired smaller and smaller amounts. Between moults the metal evidently accumulated in the body tissues. Furness *et al.* recommended that body feathers be used for monitoring purposes, as their levels were more consistent than those of flight feathers.

Feather Hg in *Bulweria bulwerii, C. diomedea, Puffinus assimilis* and *Oceanodroma castro* from the Azores were given by Monteiro *et al.* (1995). Levels in *B. bulwerii* and *O. castro* were high, those from the storm petrel taken in spring *c.* 50% lower than from autumn birds, suggesting that two populations may have been sampled.

Mercury in tissues exists both in the inorganic form and as organic, lipid-soluble methyl mercury, this being the more toxic, and accumulating up the food chain. Thompson and Furness (1989a) gave the levels of both forms in the feathers of a range of marine birds that included *D. exulans, Phoebetria fusca* and *F. glacialis*. The feather Hg was of the organic type in them all.

Table 8.6 Mean concentrations and ranges (ppm wet weight) of selenium, mercury, cadmium, zinc and copper in liver and kidney from Northern and Southern Fulmars. From Norheim (1987)

		Liver					Kidney		
	n	Se	Hg	Cd	Zn	Cu	Cd	Zn	Cu
Fulmarus glacialis	10	3.0 (1.4–6.4)	2.1 (0.6–4.2)	17 (6.1–32)	73 (50–95)	6.2 (5.6–6.7)	55 (22–114)	50 (32–96)	4.1 (3.3–4.7)
F. glacialoides	6	13 (11–16)	2.9 (0.8–6.2)	5.1 (2.2–10)	42 (36–54)	4.1 (3.4–4.8)	38 (16–75)	47 (39–69)	4.5 (3.3–6.3)

Table 8.7 Trace elements in organs and tissues of five *Calonectris diomedea diomedea* from Majorca (ppm g^{-1} dry weight). From Renzoni *et al.* (1986)

	Hg		Cd		Pb		Zn		Se	
Tissue	$\bar{x}$	SD	$\bar{x}$	SD	$\bar{x}$	SD	$\bar{x}$	SD	$\bar{x}$	SD
Brain	5.45	1.79	3.00	0.73	1.56	1.15	53.11	17.66	26.64	10.22
Muscle	4.76	1.48	3.93	1.09	0.61	0.19	55.67	9.99	16.96	3.03
Kidney	14.96	4.51	42.82	5.88	0.48	0.39	100.66	17.14	89.94	11.19
Liver	49.69	23.08	7.52	2.38	0.60	0.38	133.33	40.02	81.97	16.76
Preen gland	2.21	0.25	0.54	0.20	0.67	0.20	54.09	6.82	25.38	5.48
Fat	0.63	0.34	0.33	0.04	0.42	0.11	23.79	4.00	9.14	1.17

Table 8.8 Total and organic mercury levels (ppm dry weight) in liver tissues of petrels from Gough Island. From Thompson and Furness (1989b)

Species	*n*	Total Hg Mean ± SD	Organic Hg Mean ± SD
Diomedea exulans	2	1343.0	31.3
D. chlororhynchos	9	21.9 ± 16.4	3.8 ± 1.6
Phoebetria fusca	8	472.6 ± 188.8	13.2 ± 4.9
Pterodroma incerta	11	77.8 ± 33.7	13.6 ± 2.9
P. mollis	8	50.8 ± 37.5	9.4 ± 2.9
Lugensa brevirostris	7	11.8 ± 4.0	7.9 ± 2.4
Pachyptila vittata	10	0.8 ± 0.2	0.8 ± 0.2
Puffinus gravis	11	4.5 ± 2.6	2.1 ± 0.8
P. assimilis	9	3.1 ± 0.4	2.8 ± 0.5
Pelecanoides urinatrix	12	1.1 ± 0.3	1.0 ± 0.3

Both kinds occur in the liver, but Thompson and Furness (1989b) showed that their proportions there vary greatly by species (Table 8.8). In general the higher the total Hg the lower the proportion of methyl mercury. In six species examined by Thompson *et al.* (1990) the mercury in both muscle and feathers was of the methyl form.

Thompson and Furness (1989b) hypothesized that the birds store accumulated Hg in the organic form for incorporation in growing feathers and elimination with the old. With long-lived petrels that have slow moulting cycles and whose feather coat takes several years to replace, excretion via the feathers might be limiting and the build-up of inorganic Hg could be greater than in the smaller species with shorter moult cycles. One possibility would be for a long-lived, slow-moulting bird to demethylate the organic Hg and store it in an inorganic form. This was also suggested for the two North Pacific albatrosses in which Honda *et al.* (1990) found high Hg levels, but with <10% of the whole load in the feathers. As <5% of the liver level was of the organic form, Honda *et al.* believed that demethylation and storage in the inorganic form took place there, the remaining methylmercury being shifted elsewhere. How much can be excreted directly is not known, but Thompson and Furness (1989b) envisaged the birds as having evolved a system where the Hg they ingest and accumulate is balanced by that they lose during the moult and excretion. They suggested feeding trials to find out whether and how demethylation can occur.

The usefulness of feather Hg as an indicator of the level in the whole bird or of the sea where it lives is reduced because Thompson *et al.* (1990) established that the intertissue ratios vary with the species, the state of moult, and the distributions and proportions of the inorganic and organic forms present.

There are few analyses of Hg levels of birds in particular colonies over time. Noble and Elliott (1986) reported that while concentrations in the eggs of *O. leucorhoa* at Great Island, Newfoundland varied little between 1972 and 1980, those at Kent Island rose from 0.296 ppm wet weight in 1972 to 0.378 in 1976 to 0.545 ppm in 1980, $P < 0.001$. This last figure is the highest mercury load recorded in Canadian seabird eggs, the reason unknown. Since 1980, feathers of *Puffinus puffinus* have held significantly higher Hg loads than they did before 1930; the reverse was true for *F. glacialis* (Thompson *et al.*, 1992).

2 *Cadmium*

This is widespread in seawater, generally at lower levels in invertebrates than in seabirds, as it is bioaccumulative. It tends to be high in petrels, particularly in the kidney and liver.

Bull *et al.* (1977), who detected high kidney levels in *F. glacialis, P. puffinus, H. pelagicus* and *O. leucorhoa* from St Kilda, pointed out that the birds were healthy and apparently unexposed to industrial pollutants, while kidney levels of Arctic *F. glacialis* and Antarctic *F. glacialoides* were similar (Table 8.6). In both these fulmars, there were strong correlations between the levels of Cd in the kidney and in the liver (Norheim, 1987). Other data include those for 12 tubenoses from Gough Island by Muirhead and Furness (1988), for 31 species from the southwest Pacific by Lock *et al.* (1992) and for *P. puffinus* and *F. glacialis* by Osborn *et al.* (1979)

As with Hg, Cd levels in *C. diomedea* were quite high—47 ppm wet weight in kidneys of Grande Salvage Island birds (Renzoni *et al.*, 1986; Fig. 8.4), but such levels are not unusual in tubenoses. For example, Lock *et al.* (1992) gave mean kidney Cd for *Pterodroma nigripennis, D. epomophora* and *T. antarctica* of 76, 48 and 43 ppm wet weight with *Procellaria cinerea* higher at 94 ppm, and in most species Cd levels were higher in adults than in birds <1 year old, $P < 0.05$.

Some links with diet have been indicated. For example, *Pterodroma alba, P. hypoleuca* and *B. bulwerii* eat sea-skaters *Halobates* sp., picking them up from the surface film. These insects carry considerable loads of Cd (Cheng & Harrison, 1983), its concentrations in regurgitated *Halobates* from *B. bulwerii* and *P. hypoleuca* assayed by Cheng *et al.* (1984). It seems rather unlikely that these insects form a major energy source for petrels, the Cd load more probably to be from other prey such as squid, or from seawater, and Hamanaka (1984) suggested that high kidney Cd in *Puffinus griseus, P. creatopus* and *D. immutabilis* came from the many squid they ate, with their longevity leading to accumulation.

Another probable source of Cd in tubenoses are amphipods like *Themisto gaudichaudii*, *T. compressa* and *Parathemisto libellula*, which contain high levels of the metal (Rainbow, 1989).

Notwithstanding the apparent good health of seabirds with substantial loads of Cd and the existence of an apparent detoxification system (see G below), examination of the kidney tissue of *P. puffinus* and *F. glacialis* from St Kilda by Nicholson and Osborn (1983) revealed substantial cellular damage, namely lesions with necrosis, degeneration of the proximal tubular epithelium, and the blocking of some nephrons with cellular debris. This damage was attributed to high tissue Cd and Hg, particularly the former, but there was no overt evidence of toxic effects. Nicholson and Osborn stated that the avian kidney has much spare capacity and can regenerate damaged tissue. However, Elliott *et al.* (1992) found no evidence of renal damage in 60 *O. leucorhoa* with even higher Cd levels.

3 *Lead*

This element has been identified in some petrel bones, but the only reported mortality was that of fledgling *D. immutabilis* which had eaten weathered paint chips (Sileo & Fefer, 1987). There was lead only in 25% of the 32 tubenoses examined by Lock *et al.* (1992) and the metal never exceeded 1.56 ppm dry weight in organs of the four populations of *C. diomedea* examined by Renzoni *et al.* (1986) (see e.g. Table 8.7). Bones of adult *Daption capense* and *M. halli* from New Zealand waters, containing 11 and 9 ppm dry weight and 12.3 ppm in two *M. halli* chicks, suggest uptake from garbage by these familiar ship followers.

4 *Copper, zinc and selenium*

These essential elements have been found in all petrels checked for them. The levels were similar from species to species, similar in liver and kidney, and showed little geographical variation (Anderlini *et al.*, 1972; Muirhead & Furness, 1988; cf. Table 8.7). Liver Cu in birds <1 year old was higher than in the adults, as was the tendency with Zn in kidney tissue, possibly due to take up of Cu in growing feathers (Lock *et al.*, 1992). Selenium levels for *F. glacialis* and *F. glacialoides* were determined by Norheim (1987) (Table 8.6) and this element may be transferred to the egg, at least in *P. pacificus* (Ohlendorf & Harrison, 1986).

5 *Iron*

In analyses of *D. immutabilis*, *D. nigripes*, *F. glacialis*, *P. griseus*, *P. tenuirostris* and *O. monorhis* by Honda *et al.* (1990), the highest values were in the liver, and higher in adults than in immatures, possibly due to an increase in iron-bound proteins with age. There were low Fe levels in albatross muscle and liver tissue, but as these were in moult, Honda *et al.* suggested that much Fe had gone into feather formation, moulting being said to reduce hepatic Fe concentrations.

G *Detoxification processes*

Apart from the elimination of Hg via the feathers, there appear to be protective systems lessening damage from heavy metal loads that might be fatal in other birds. For discussions see Nicholson and Osborn (1983) and Thompson and Furness (1989b).

In the fulmars, Norheim (1987) found that concentrations of Zn and Cd in both the liver and the kidney were highly correlated, as were the levels of Se and Cd in their livers. Norheim also established a general correlation in Spitzbergen and Antarctic seabirds between titres of Se and Hg, and as Se and Zn are known to reduce the toxicity of Hg and Cd in man, Norheim suggested that these act similarly in seabirds.

A role for Se in the detoxification of Hg, based on the demethylation of methylmercury and the formation in the liver of a Hg–Se non-toxic compound, has been suggested (Thibaud, in Renzoni *et al.*, 1986). The latter also pointed out that the low values of Cd in eggs suggests an efficient binding mechanism for that metal. This may involve a metalloprotein whose sythesis is induced by Zn, and which binds it and Cd to a soluble cytoplasmic fraction or metallothionen. This is believed to act as a temporary store for heavy metals, preventing them from interacting with other enzymes or proteins.

Detoxification of xenobiotic and lipid-soluble CHCs and ingested mineral oil may involve a group of hepatic enzymes, the mixed-function oxygenases. The activities of some of these were examined in *P. puffinus* by Knight *et al.* (1981) and Walker *et al.* (1984) and their activity in the presence of CHC measured in that bird by Knight and Walker (1982). Peakall *et al.* (1987) also looked at the levels of these enzymes in the liver of *O. leucorhoa*, Elliott *et al.* (1992) those in its kidneys.

In both species the overall pattern was of a reduced level of activity compared with those of other seabirds tested. This suggests that these tubenoses cannot detoxify CHCs with hepatic enzymes as some other seabirds evidently do, although perhaps they do so through direct excretion via the kidneys.

III Miscellaneous biochemical matters

The 'mutton bird' (presumably *P. tenuirostris*) was used by Laurent (1981) in experiments to develop a radioassay method for hyaluronate levels in aqueous humour. Injections of sodium hyaluronate have been used to replace vitreous humour in man and also in the anterior chamber to protect the corneal epithelium during and after surgery. The petrel contained a mean of 0.9 μg ml^{-1} of Na hyaluronate in its aqueous humour and 7–9 μg ml^{-1} in its vitreous humour as against 1.1 and 140–338 μg ml^{-1} in man.

Lawrie (1952) determined that the pectoralis of *P. puffinus* held 0.69% of myoglobin by wet weight. In this muscle the enzymic activities (as μl.O_2 mg^{-1} dry fat-free weight h^{-1}) were, succinic dehydrogenase 50, succinic oxidase 230 and cytochrome oxidase 1800. The domestic pigeon's pectoralis exhibited similar activity but with only 0.22% of myoglobin. Lawrie suggested that the pigeon's rapid wing action ensured quick replenishment of blood and lessened the need for an energy store, whereas the shearwater's slower wing strokes allowed a higher concentration of myoglobin in its wing musculature.

In their study of the development of thermoregulation in neonate *P. pacificus*, Mathiu *et al.* (1992) looked at calcium–magnesium-activated myofibrillar ATPase activity in their skeletal muscles, because in some birds there is a sharp increase in its activity in the pectoralis with the onset of endothermy. Mathiu *et al.* found that the activity of this ATPase in the shearwater hatchling averaged 0.26 ± 0.053 ($n = 6$) and 0.36 ± 0.068 ($n = 8$) μmol PO_4 (min mg myofibrillar protein)$^{-1}$ for the pectoralis and gastrocnemius muscles respectively, a significant difference at $P = 0.02$. However, there was no difference between late embryos and neonates, suggesting that in the shift to endothermy, escape from the physical bonds of the egg was more important than biochemical maturation, the leg muscles being more useful than those of the wings.

The pectoral muscles of *Oceanites oceanicus* were used by Wilson *et al.* (1963) in investigating the two forms of lactic dehydrogenase, one with the so-called M enzyme, the other with the H enzyme. The latter is thought to predominate in the heart and is inhibited by low concentrations of pyruvate; the other, found in skeletal muscle and associated with anaerobic activity arising from sudden exertion, is inhibited only at high pyruvate levels. They found that the amount of substrate inhibition in the pectoral muscles of the storm petrel was the strongest of all the 40 birds measured. The authors concluded that there was a good qualitative link between the degree to which the breast muscles of birds can keep up continuous activity and the degree to which the lactic dehydrogenase is inhibited by high pyruvate levels, the storm petrel providing an excellent example of a bird in continous activity.

CHAPTER 9
Locomotion

Contents

Some locomotory patterns were described in *The Petrels* and there are many references scattered in the literature, little on movements on land or under water, and much remains unclear. Klemm (1969) analysed the hind-limb musculature and its function in standing, walking, running and burrowing, Pennycuick (1982, 1987b) looked at petrel flight dynamics.

I Standing

The petrels best able to stand erect for quite long periods are the albatrosses, the giant petrels and *Pagodroma nivea*. Their centres of gravity then lie over their large webbed feet with the lower parts of the legs forming straight lines below them. The North Pacific albatrosses also rise on tiptoe during their 'dances' when, according to Klemm (1969, p. 121), the adductors of the femora keep the legs straight so that all the muscle power and weight is on a strong, direct support. Power to lift on to their toes is given by the strong *pars interna* of *M. gastrocnemius* that extend the tarsometatarsi, together with the flexors of the digits. The long claws gripping the substrate help stability. These seem to be the only petrels that normally lift on to their toes; others only do so momentarily at take-off.

Displaying *Diomedea immutabilis* and *D. nigripes* bob up and down while on tip-toe, providing a severe test of the skeletal and muscular system. A rigid knee joint is important here and Klemm (1969, p. 128) reported that outwards rotation of the

Figure 9.1 *Wilson's Storm Petrel* Oceanites oceanicus *normally rests with its tarsi and feet on the ground but can lift briefly on to its toes if necessary. From video film by C. M. Miskelly.*

shank may be controlled by *Mm. exterior iliotibialis, extensor iliofibularis* and *flexors cruri* which reduce wobbling during bobbing.

Klemm (1969, p. 126) found the flexors of the digits and *M. gastrocnemius* better developed in the males, particularly in *D. nigripes*, partly explaining the more vigorous dancing of this bird.

Neither albatrosses nor giant petrels stand for long unless active; otherwise they subside with the tarsometatarsi along the ground—the normal pattern with resting petrels. It is often thought that most petrels cannot rise on their fully extended legs (e.g. Klemm, 1969, p. 142). However, many can rise to their toes if necessary, some, such as prions, with ease, others only for a few seconds. Even long-legged *Oceanites oceanicus* can do so (Fig. 9.1). More often the Oceanitinae only do so with their extended wings lifting to take the weight off their legs. Others that stand readily are *Procellaria aequinoctialis* and *P. cinerea* (Fig. 9.2).

II Walking and running

Albatrosses and giant petrels have a pigeon-toed digitigrade gait, shifting their weight at each step like a goose, to keep a leg below the body. The adductors and *M. externa iliotibialis anterior* help draw the leg towards the body and eliminate a straddled gait. The leg is flexed with the help of strong *Mm. cruris lateralis* and *extensor iliofibularis*, the long 4th digit whose sharp claws can be strongly flexed, tend to anchor the leg and prevent yawing, while the action of *M. caudofemoralis* helps shift the centre of gravity and the balance on the one leg while the other is moved. Propulsion comes mainly from the *M. pars interna* and *M. gastrocnemius*, as well as the arrangements of the flexor tendons on digit III which provide a longer power arm. Abduction of the shank and the extension of the tibiotarsus is mostly from contraction of *M. extensor iliotibialis lateralis* in conjunction with *Mm. femoralis tibialis externus* and *femora tibialis medius* (Klemm, 1969, p. 123)

The walking postures of *D. immutabilis* and *D. nigripes* differ significantly. The latter withdraws its head close to its body bringing the centre of gravity more caudad and ventrad whereas *D. immutabilis* moves with head erect and slightly forward. Klemm (1969, p. 125) explained the stance of *D. nigripes* as due to a weaker *M. flexor cruris lateralis* giving less power to raise the front of the body; in *D. immutabilis* this

Figure 9.2 *Grey Petrels* Procellaria cinerea, *like other* Procellaria *spp. , appear better able to stand erect than* Puffinus *shearwaters. Photo: author.*

muscle is larger and the postacetabular ilium slightly longer relative to trunk length giving this muscle more power to lift the forebody.

Klemm (1969, p. 132) emphasized that all three digits are important for *Diomedea* when walking. The pivoting gait is helped by the long 4th toe and a modification to the flexor tendons permitting the phalages to be drawn in close contact with the ground, firming the supporting base. The flexors for digit IV tend to be bigger in the males, hence their more erect stance.

For most other tubenoses ground movement usually consists of a waddle–shuffle on the slightly extended tarsometatarsi and the tibiotarsi. The more caudal position of the legs arises because of the shorter postacetabular ilium giving a smaller area for muscle attachments such as *M. flexor cruris lateralis* which is weak in these birds. When contracted, this raises the forebody and the centre of gravity shifts forwards tending to make the bird top heavy.

When walking, smaller species such as *Pterodroma hypoleuca* have their legs splayed out more than do albatrosses. With *P. hypoleuca* the long flexors of the digits and *M. gastrocnemius*, which extend the tarsometatarsus and provide most of the power stroke, are weak, which helps explain the reduced walking ability. However, despite such limitations, probably all tubenoses can move in a true digitigrade style with closed wings, at least over short distances. Prions are very active and move freely on their toes, and even *Puffinus* shearwaters can step out with wings folded, swaying from side to side while hobbling along in a strange shoulders-hunched posture and a short, stilted stride, as when moving down their tracks to launching sites (cf. Lockley, 1931; Frith, 1976, p. 53). *Fulmarus glacialis* and *Daption capense* can adopt a digitigrade gait over short distances, *Procellaria* do so regularly, and even storm petrels such as *Oceanites* and *Pelagodroma* can do this (*The Petrels*, p. 181). Such long-legged Oceanitinae can bounce along, half running, half flying, padding the vegetation with their feet at every wing stroke, as if still at sea. Normally, storm petrels and shearwaters shuffle forward on their tarsi, extending their wings as balancers when they are in a hurry.

On unimpeded flat ground giant petrels can sprint, wings flailing, about as fast as a man can run. Male *D. immutabilis* are seen at their most sprightly when dashing to converge on a female that has adopted a precopulatory posture, or when running in a 'charge' (Chapter 5, III,B 4).

III Climbing

On the ground most obstacles are simply avoided, others surmounted. Much climbing is involved during take-off in rocky or wooded terrain where height is essential for successful launch. Climbing birds use all available means (Fig. 9.3). A captive *Hydrobates pelagicus* climbed with the aid of its hooked bill, using the carpal joints as hands, the feet having only a minor role (Mathew, 1881).

To reach launching sites in trees the wings are extended and the birds flutter up the trunks helped by beak and claws. Tree climbing is common with '*Cookilaria*' spp., with prions, and with birds as big as *Pterodroma solandri*, *Procellaria westlandica* and *Calonectris leucomelas*.

IV Burrowing

Tunnelling methods have been described (e.g. *The Petrels*, pp. 245–247). At least with *Pterodroma hypoleuca*, both sexes burrow.

Klemm (1969, p. 123) found that this bird's legs are directed rather laterally by the configuration of the antitrochanter and the femur and the strong action of the posterior rotator muscles, so that little power was wasted in raising the body. Furthermore, *M. flexor cruris lateralis* is relatively large, is attached ventrally to the synsacrum and inserted proximally on the tibia, so can pull directly on the extended shank to rake the toes through the soil. Strong *Mm. pars interna* and *externa iliofemoralis* add power to toss the spoil clear.

At the end of the power stroke the top of the foot is probably resting on the ground. The well-developed main extensors of the tibiotarsus (*Mm. femora tibialis externa* and *femora tibialis medius*) appear to power the recovery stroke with *M. fibialis anterior* helping to overcome the drag of the foot on the ground.

Excavating birds can work fast in soft soil. In hard ground it is obviously arduous, for example with *Pelecanoides georgicus* and *Pachyptila desolata* raking through stony and/or frozen substrates. The claws become abraded; so do the bills, for example in *Pelagodroma marina* (Murphy & Irving, 1951).

Figure 9.3 *Shearwaters scaling boulders or trees to reach launching sites use the hook of the bill to pull upwards, sharp claws and webs of feet to latch on to cracks and crevices, so that the digital extensors can push the body upwards while the tail and wing tips form a tripod of support:* Puffinus tenuirostris. *Photo: author.*

V Swimming

A *On the surface*

The webs are expanded on the power stroke and folded on the recovery one, manoeuvring being achieved with asymmetric strokes of the feet with expanded or closed webs. Little movement of the tibiotarsus or femur is normally detectable in surface swimmers. The main propulsive force is generated by the *pars interna* of *M. gastrocnemius*, while the strong *pars externa* together with *Mm. extensor iliofibularis* and *flexor cruris lateralis* keep the tibiotarsi flexed. The recovery stroke is mainly powered by *M. tibiotarsus anterior*. The feet are usually used alternately.

Klemm (1969) was unable to detect common muscle arrangements for swimming in genera such as *Diomedea, Procellaria, Puffinus* and *Pelecanoides*. Nor were the arrangements constant in species that swim and burrow. He noted, for instance, that the *pars interna* in *Pachyptila desolata* and *Daption capense* was the better developed and the muscle bulk placed similarly on the leg, yet the one burrows with its feet, the other not. But he also pointed out that the shapes of the synsacra differ, as does the placement of their attached muscles, factors that also affect limb movement.

Although storm petrels do settle on the sea, their surface locomotion is unstudied. A captive *H. pelagicus* swam poorly with alternate leg strokes. Diving petrels also paddle in this way but if alarmed will also use their half-expanded wings as oars.

Surface swimmers tend to splay their legs to either side, but when moving the feet are drawn towards the mid-line and the strokes of considerable amplitude drive the bird forward and generate a bow wave. Storm petrels ride high but, according to Rijke (1970), their feathers have poor resistance to water penetration, perhaps explaining their infrequent swimming. Shearwaters tend to swim with wings partly raised and outstretched and tail cocked (Fig. 9.4)—this may be a stance perhaps mostly adopted when prey is anticipated. Seen from below the shearwater contour resembles that of the hull of a boat (Fig. 9.5A).

A shearwater's plumage is very waterproof and rafting birds face wind and waves and either tuck their heads down and allow seas to break over them, emerging unruffled to shake their heads ready for the next wave, or they may ride over the surf. When feeding inshore, birds such as giant petrels may also have to cope with large waves and avoid being pounded on to rocks. They may swim towards a curling breaker and float, cork-like, over the top, may crash through the crest, or dive into an advancing wave to emerge on the far side with wings outspread. They may lift off before a breaker and flick over the crest, to alight in calmer water beyond.

Figure 9.4 *Shearwaters often swim with wings extended and tail cocked, particularly when food is sighted, evidently in readiness to grasp it.*

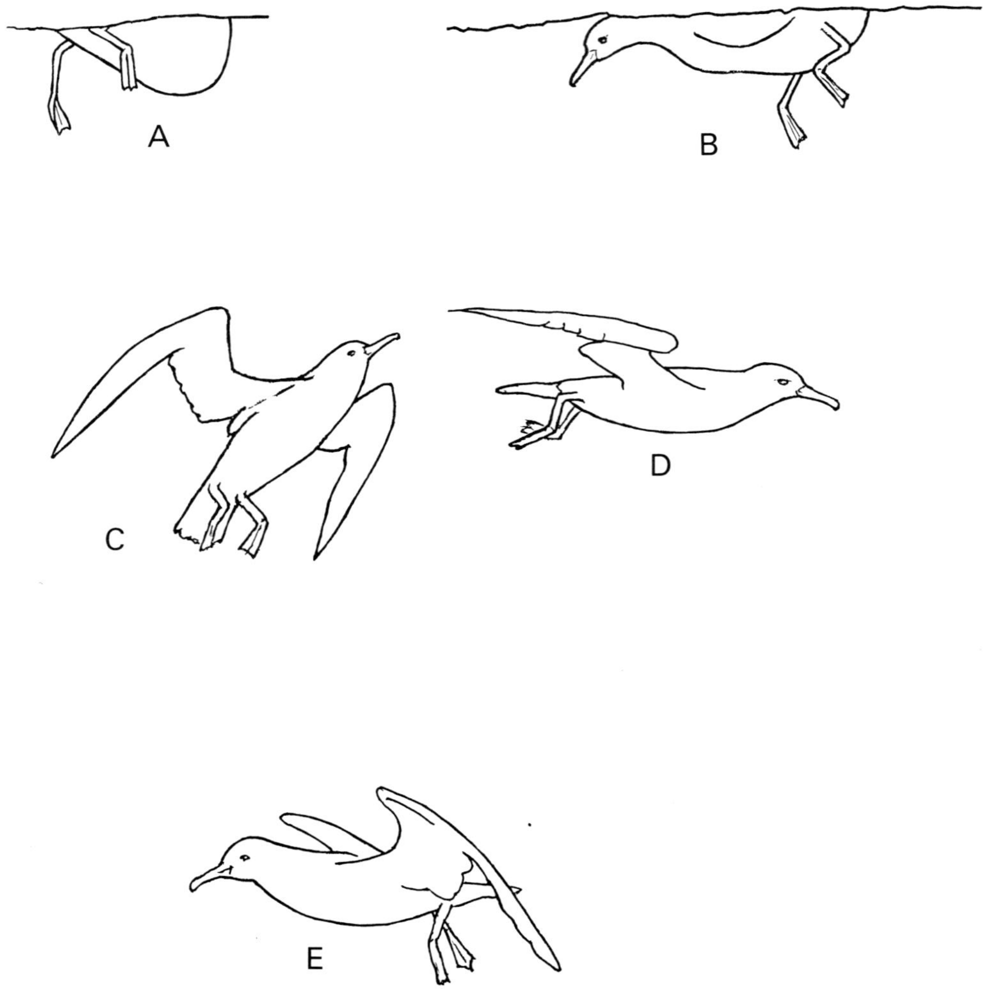

Figure 9.5 *Underwater swimming of Sooty Shearwaters attracted by fatty baits. (A) Idling on the surface, the body shape is that of the hull of a boat, the feet splayed out. (B) Ducking the head to look for prey: the feet paddle alternately. (C) Passive floating up towards the surface using natural buoyancy. (D) Level swimming, wings starting the downbeat with leading edge dipped. (E) Level swimming, wings at end of recovery stroke.*

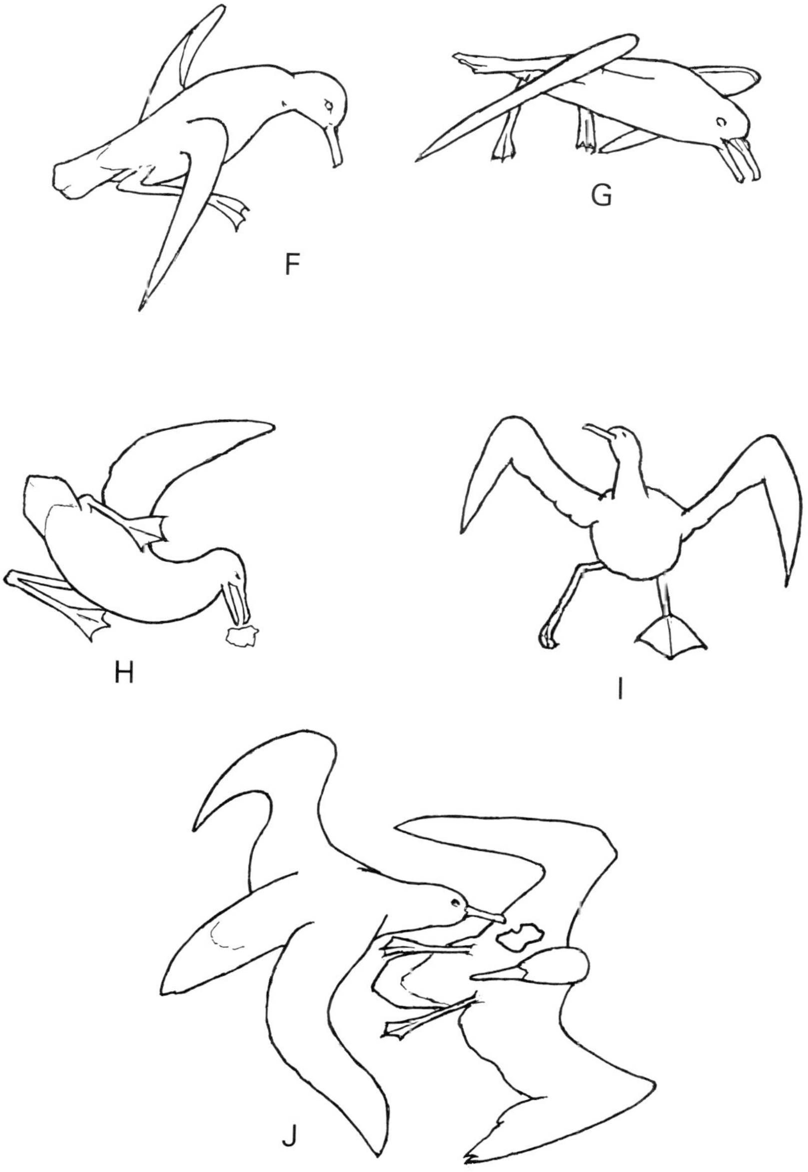

Figure 9.5 continued *(F) Peering down at possible food while drifting upwards. (G) Heading for food. (H) Food snatched before others reach it. (I) Powered ascent with deep wing strokes and twisting to the right. (J) A struggle for possession. From film by Andrew Penniket of the New Zealand Television Natural History Unit.*

B *Beneath the surface*

There are occasional records of submergence for many tubenoses, apart from the familiar examples of diving petrels and shearwaters, including some species apparently poorly adapted for diving (*The Petrels*, pp. 31,62). Even *Oceanites oceanicus* has been reported going down to take recreational fishermen's baits.

Most dives, such as those of prions, are shallow, and albatrosses and Northern Fulmars often submerge with their wing tips protruding from the surface. But both can go right under. For example, Harrison *et al.* (1991) timed *Diomedea melanophrys* under water for 2.9 ± 1.3 s ($n = 80$). *F. glacialis* has descended to 90 cm (Allison, 1952) and submerged for up to 6 s (Wahl, 1984), but shearwaters may remain under longer—*Puffinus gravis* for >12 s at times, *Pelecanoides urinatrix* for 40 s and *Procellaria aequinoctialis* for 10–45 s (Paulian, 1959). Duffy (1983) timed *Puffinus griseus* underwater for an average 7 s, maximum about 15 s; Morgan and Ritz (1982) reported *P. tenuirostris* staying under for an average of 8 s, while *P. gravis* and *P. griseus* remained submerged for a maximum of 12 s (Brown *et al.*, 1978). The latter reported that *P. griseus* submerged more readily than *P. gravis*, with *Calonectris* less well adapted to underwater activity, in line with Kuroda's (1954) evaluations on anatomical grounds.

Oatley (1979) saw *D. melanophrys* persistently diving 5 m for fishermen's baits and Skira (1979) reported *P. tenuirostris* feeding among weed on the seabed about 12 m down. Besson (1973) recorded dead *P. puffinus yelkouan* in nets set on the sea floor at 30–50 m. Wood (1993) saw *P. gavia*/*P. huttoni* swimming 2–3 m below after natural prey and remaining immersed for about 10 s, Oka (1994a) instanced *P. carneipes* 13 m down and *P. griseus* taking seaweed at 7–8 m.

More precise data come from birds carrying depth gauges. Thus maximum depths reached by *Pelecanoides georgicus*, *D. exulans*, *D. melanophrys*, *D. chrysostoma* and *Phoebetria palpebrata* were 25.7 ± 11.4 m, 0.3 ± 0.2 m, 2.5 ± 1.3 m, 3.0 ± 1.8 m and 4.7 ± 3.4 m respectively (Prince & Jones, 1992; Prince *et al.*, 1994). Seemingly, *P. palpebrata* regularly feed under water. For *Procellaria aequinoctialis*, Huin (1995) found the mean maximum depth to be 6.0 m. Weimerskirch & Saga (in press) give a mean for *P. griseus* of 38.7 m.

Methods of submergence vary with the species and the prevailing conditions. Diving petrels glide momentarily before diving, feet dropped and widely separated. Shearwaters may submerge from a floating position, often peering below before doing so (Fig. 9.5B). Others stall and belly-flop from a short height to disappear immediately.

A submerged shearwater or diving petrel is very smooth hydrodynamically with tight and very waterproof plumage. Wood (1993) related the diving abilities of five *Puffinus* to the degree of lateral compression of their tarsi; he considered that the sequence *P. pacificus*, *P. carneipes*, *P. tenuirostris*, *P. griseus* and *P. gavia* reflected their increasing skill under water related to the reduced resistance on the recovery stroke.

Tubenoses swim under water with elbow and wrist joints sharply flexed forming sculls—'wing-rowing'. Common Diving Petrels seem not to use their feet, these being stretched out from the closed tail in the style of a penguin (Miskelly, pers. comm.). For other petrels the feet seem to be very important in propulsion and steering (see Fig. 9.5I and photographs in Oka, 1994a).

Diving shearwaters trail air bubbles behind them (forced from the plumage to reduce buoyancy?) and often have a silvery appearance from air on their feathers and even caught on the bill by the nasal tubes.

Under water a shearwater is loose-limbed. Most of the wing movement is from the carpal joint outwards. The wing strokes are through arcs of 25–30° when the bird is unhurried, and on the downbeat the leading edge is dipped; on the recovery stroke the reverse applies. Typically, the power stroke is fast and decisive, the recovery one slower, so that progress tends to be jerky due to the density of the watery medium. Much movement appears to result from the propeller action of the outer wing segment.

Submerged Sooty Shearwaters can use all four limbs independently and change direction abruptly using asymmetric wing and foot strokes and twists of their tails. They can accelerate very rapidly using deep wing beats and foot thrusts (e.g. Fig. 9.5I). When unhurried, their natural buoyancy carries them up with the wings raked back and motionless and the feet dangling (Fig. 9.5C). They surface like corks and take to the air immediately unless manipulating a catch.

Birds competing for baits will struggle vigorously to grab the food and dart to the surface to swallow it, and presumably such competition obtains when feeding on natural prey. Most approaching a bait only opened their beaks to grasp the food; it seems rather doubtful that the buccal cavity is at a low pressure at the point of capture, unless the prey is moving fast (*The Petrels*, p. 165).

Pennycuick (1987b) noted a bird's problems when using the same wing in air and in water, wrist and elbows being flexed to reduce the span in the denser medium and beaten at a lower frequency then. None the less, shearwaters fit his pattern of a 'standard seabird' nicely, and are very effective under water.

Feather loss during moult must affect a petrel's diving ability, particularly with transequatorial migrants such as *P. tenuirostris* and diving petrels with nearly synchronous moult (Watson, 1968), implying considerable difficulty in getting airborne. However, it is clear that both groups can feed under water despite feather loss, food probably being abundant then, and the reduction in wing area perhaps compensated for by extending the elbow more. Studies of captive birds with clipped remiges could throw light on this.

What effects gizzard stones, which decrease buoyancy, and stomach oil which increases it, have on diving and swimming ability are conjectural. 'Ballast stones' might be of greater value in the Southern Ocean where frequent winds facilitate take-off, but less effective in the tropics where in light airs a lot of energy is used then.

VI Flying

A *Introduction*

The varied flight styles of the different groups of tubenoses were described briefly in *The Petrels*. At sea most of the Diomedeidae and Procellariidae use slope and dynamic soaring to a greater or lesser extent. Extracting energy from the wind makes gliding very efficient, permitting a bird to search vast areas of ocean without having to alight, except when food is sighted. The hydrobatids also gain lift from the updraughts on the windward sides of waves, while the Pelecanoididae have evolved more for underwater flight than for aerial progression. They may be the most dexterous petrels under water but in air seem capable of only wide-radius turns. Shearwaters are still agile when airborne, although less capable of tight manouevring then than when submerged.

Von Mangold (1948) suggested that their nasal tubes might provide gliding petrels with information about air currents and Cone (1964, p. 84), noting their similarity to Pitot-static tubes, that they might help determine air speed; these speculations are untested.

There are many descriptions of flight modes, for example of *F. glacialis*, that 'glide close to the surface and then, when they have lost their momentum, they suddenly beat the slope of an oncoming wave, and, leaving a little track of flying spray, they shoot forward and rise once more into the air' (Kennedy, 1917). Other accounts highlight subtle or obvious differences between species as aids to field identification.

The dramatic evolutions of albatrosses have challenged many researchers to attempt to work out the forces being tapped to produce apparently effortless locomotion and these have included theoretical models such as those of Magnan (1922), Idrac (1926) and analyses based on better data and often involving computer simulations. Among these are the publications of Cone (1964), Wood (1973), Wilson (1975), Alerstam *et al.* (1993), Sachs (1993), and particularly of Pennycuick (1982, 1987a,b). Pennycuick *et al.* (1984) computed a range of flight parameters and examined their relation to the power delivered to the chicks of nine species of petrel.

B *Flight around the breeding colonies*

There is usually a daily pattern to the numbers of inflying birds; those of diurnally active species tend to build up in the late afternoon, while nocturnally active ones land at, or after, dusk and into the night. Little is known of landing times in relation to status. Prebreeders and non-breeders perhaps come in later than those with nests and delay even further on moonlit nights, for example *P. puffinus* (Storey & Grimmer, 1986).

The patterns of the inward flights of petrels vary with the weather, ambient light levels, and the stage of the breeding cycle. On very dull days, under sudden overcast or eclipses, shearwaters may suddenly converge on to the land and alight, perhaps to leave again when the obscurity clears.

The situation at the Snares Islands, whose ground is densely burrowed by *P. griseus*, is typical. The birds assemble offshore in 'rafts', mostly after mid-day. These form long streams that grow as newcomers add themselves to their heads or tails as with *P. tenuirostris* (Warham, 1960). The birds ride facing the wind, the flocks moving closer in as the day progresses. Flocks disperse if the seas became confused or extensively white-capped, and in gales rafting appears to cease.

Rafting birds tend to be restless, often flying up to change their position in the flock, a restlessness that increases as the day draws on until, in the late afternoon, groups start to fly up and converge on the land. Their coherence tends to disappear as they join others already overhead, but some initial stratification can often be discerned. The lower birds tend to follow individual circulatory paths bringing them

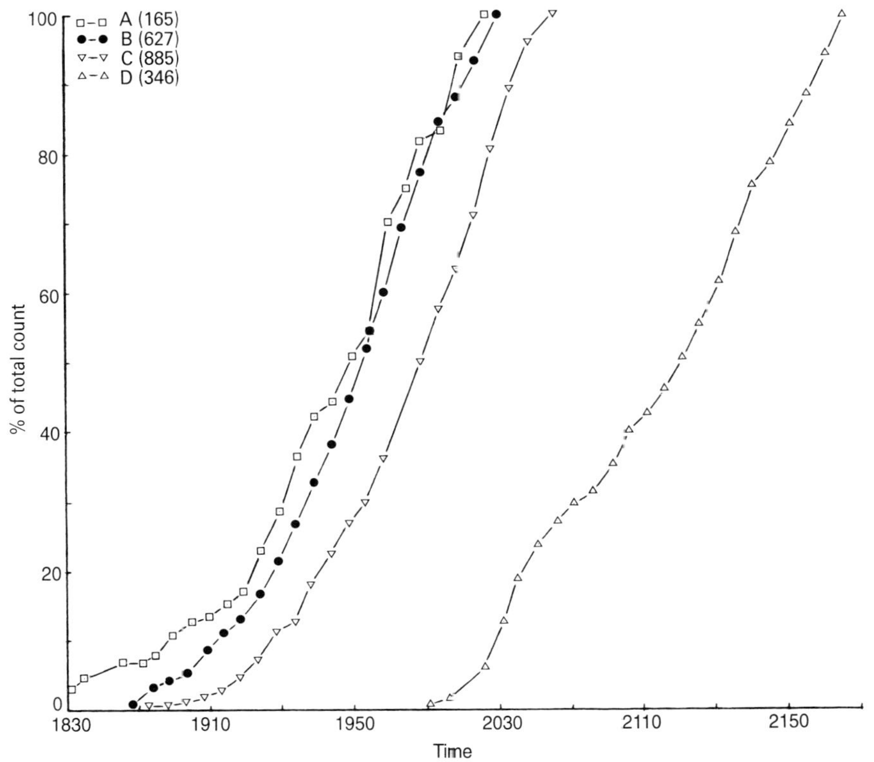

Figure 9.6 *Counts of* Puffinus griseus *landing within 10-m² quadrats on Main Island, The Snares, 18 January 1983; that is, towards the end of incubation. (A), (B) and (C) on the western side of the island, (D) at Punui Bay on the northeast coast. Total numbers of birds counted until fading light stopped observations, in parentheses.*

round to particular places where they hesitate and eventually land, lowering their feet briefly before accelerating for a new circuit. Birds with nests in open ground seem to circle less than those beneath a forest canopy.

Because a very substantial proportion of the total population at The Snares lands before visibility is lost, the course of the landings can be followed. At least with westerly winds, the birds come down in a roughly clockwise sequence, with successive waves landing on the southwest coast and spreading out from there.

Figure 9.6 shows simultaneous counts of the birds alighting at four points in comparable vegetation. The birds on the east coast at Punui Bay delayed landing for over an hour after they had started to come down in other quadrats not more than 200 m away. This pattern was typical, for example on 13 February 1983, during chick rearing, half the counted landings at Punui Bay occurred 55 min after half had landed in the other quadrats. This sequence of arrivals is unexplained as the Punui Bay colonies are shadowed early, so that landings would be expected to begin there: other potential influences like the numbers of skuas or differences in nest densities were judged unimportant.

Gadfly petrels tend not to form rafts but flocks may fly offshore to await

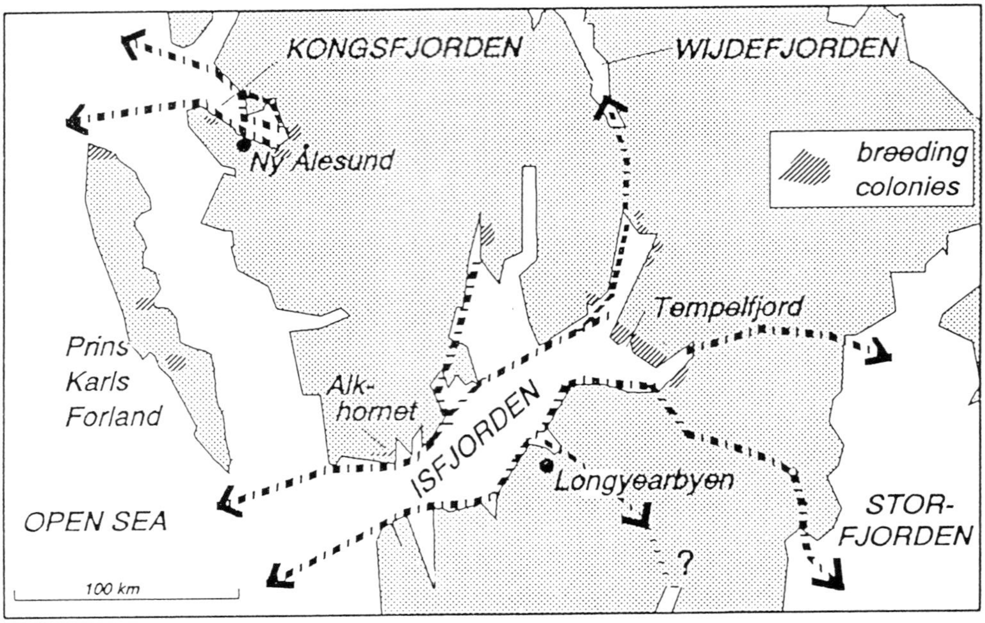

Figure 9.7 *Northern Fulmars* Fulmarus glacialis *breeding inland will fly over land to reach feeding grounds. Routes shown are of birds moving from nesting grounds deep in Spitzbergen fiords. From Camphuysen (1993).*

nightfall. Gifford (in Loomis, 1918) described *Pterodroma phaeopygia* off Indefatigable Island in the Galapagos circling in loose flocks with individual birds spiralling up several hundred feet before heading inland. However, a more dramatic pattern was described by Okamoto (1972) of *C. leucomelas* around Kanmurijima Island, Japan. Early in the evening rafts encircled it, but as dusk approached spiralling columns developed of shearwaters climbing to slightly above the height of the island (170 m), and from which birds peeled off to descend to their colonies. Possibly this was energetically more effective for gaining height in slight winds than to fly up directly, but the spirals formed over the sea with no indication of assistance by thermals or standing waves. In the predawn departures, however, the birds dispersed in all directions, as seems the general rule with tubenoses.

Petrels will fly over land, the extreme example being those *Pagodroma nivea* that breed 300 km inland in Antarctica, and some *F. glacialis* nesting deep in Spitzbergen fiords travel 100 km or more dowm valleys and over glaciers to feed in other fiord systems (Fig. 9.7). Others may fly overland on migration and there are suggestons that *Puffinus griseus* sometimes flies up the Nile Valley to reach the Mediterranean from the northwest Indian Ocean. Okamoto (1972) mapped the routes of *C. leucomelas* starting their southerly migration from Kanmurijima Island in the Sea of Japan by flying inland along river valleys across 100 km of western Honshu to reach the Pacific Ocean. If Lockley's Manx Shearwater released in Venice and recovered back at its

Table 9.1 Body masses, wing areas, loadings, spans and aspect ratios of petrels. Areas include interwing spaces. Samples sizes in parentheses. Mainly from Warham (1977a & unpubl.). Some from Pennycuick (1982) are starred

Species	Body mass (g)	Wing area (cm^2)	Loading ($N\ cm^{-2}$)	Span (cm)	Aspect ratio
Diomedea e. epomophora	8697(19)	5404(8)	158	282	14.7
*D. e. chionoptera**	8730(?)	6110(3)	140	303	15.0
D. e. antipodensis	6636(14)	4709(6)	138	271	15.6
D. nigripes	3143(306)	3382(11)	91.2	216	13.8
D. immutabilis	3092(367)	3096(19)	98.0	204	13.5
*D. m. melanophrys**	3790(?)	3560(3)	104	216	13.1
D. m. impavida	2911(16)	2932(6)	97.4	209	14.9
D. c. cauta	4030(36)	4430(1)	89.2	247	13.8
D. chrysostoma	3300(21)	2614(6)	124	200	15.3
D. b. bulleri	3012(31)	2774(17)	107	205	15.2
Phoebetria fusca	2500(176)	3036(3)	80.8	203	13.5
P. palpebrata	2918(28)	3412(3)	83.8	215	13.5
*Macronectes giganteus**	5190(?)	3310(4)	154	199	12.0
M. halli	4255(30)	3130(15)	133	193	11.9
Fulmarus glacialoides	802(21)	1210(3)	65.0	117	11.3
F. glacialis	815(350)	1240(3)	64.5	113	10.3
Thalassoica antarctica	615(20)	952(20)	63.4	105	11.6
Daption c. australe	437(95)	731(11)	58.6	88	10.6
Pagodroma nivea	225(5)	586(5)	37.7	78	10.4
Lugensa brevirostris	357(126)	611(6)	57.3	85.5	12.0
Pterodroma m. macroptera	587(61)	862(3)	66.8	104	12.5
P. lessonii	586(14)	894(6)	64.3	101	11.4
P. incerta	539(5)	1044(3)	50.7	108	11.2
P. solandri	446(9)	964(9)	45.4	103	11.0
P. magentae	465(17)	906(3)	50.3	104	11.9
P. m. mollis	276(8)	656(7)	41.3	83	10.5
P. e. externa	430(45)	869(3)	48.5	100	11.5
P. e. cervicalis	445(30)	855(4)	51.0	98	11.2
P. p. phaeopygia	408(66)	832(2)	48.1	97	11.2
P. a. heraldica	296(4)	638(1)	45.4	87.5	12.0
P. inexpectata	329(42)	632(20)	51.1	84.5	11.3
P. hypoleuca	181(144)	605(12)	29.3	76.4	9.6
P. nigripennis	178(12)	540(10)	32.3	73.3	10.0
P. cookii	190(12)	557(3)	33.5	75	10.1
P. l. leucoptera	190(21)	549(11)	33.9	73.7	9.9
Halobaena caerulea	202(215)	464(5)	42.7	68.7	10.2
P. v. vittata	196(70)	537(16)	35.7	67.6	8.5
P. salvini	159(85)	456(5)	34.2	61.1	8.2

Table 9.1 Continued

Species	Body mass (g)	Wing area (cm^2)	Loading ($N\ cm^{-2}$)	Span (cm)	Aspect ratio
*P. desolata banksi**	168(?)	460(5)	35.8	62.6	8.5
P. belcheri	145(66)	331(8)	43.0	57.6	10.0
P. turtur	132(100)	366(12)	35.4	56.0	8.6
B. bulwerii	99(191)	434(7)	22.3	66.2	10.1
Procellaria a. aequinoctialis	1134(16)	1572(5)	70.8	139	12.2
P. westlandica	1199(67)	1732(8)	67.9	140	11.0
P. cinerea	1018(23)	1170(3)	85.3	119	12.1
Calonectris d. diomedea	553(211)	1307(5)	41.5	114	9.9
C. d. borealis	946(52)	1230(3)	75.4	121	11.9
C. leucomelas	545(86)	1215(18)	44.0	110	9.9
Puffinus pacificus	342(47)	1147(13)	29.3	101	8.9
P. bulleri	432(120)	1115(13)	38.0	99	8.8
P. c. hullianus	595(17)	1127(3)	51.8	106	9.9
P. creatopus	685(3)	1125(3)	59.7	105	9.8
P. gravis	813(27)	1098(8)	72.7	115	12.0
P. griseus	819(299)	882(13)	91.1	103	12.0
P. tenuirostris	588(184)	775(6)	74.4	93	11.2
P. p. puffinus	426(15)	615(13)	67.9	79	10.0
P. a. newelli	388(32)	630(5)	60.4	75.8	9.1
P. gavia	272(19)	390(5)	68.4	67	11.5
P. huttoni	398(41)	521(10)	74.9	73.8	10.5
P. a. elegans	238(5)	429(5)	54.4	60	8.4
Oceanites o. exasperatus	38(226)	192(3)	19.4	39	7.9
Garrodia nereis	31(14)	157(4)	19.6	32	6.5
Pelagodroma marina maoriana	47(100)	222(5)	20.8	40.7	7.5
Fregetta t. tropica	54(10)	259(8)	20.5	43	7.1
Hydrobates pelagicus	28(50)	170(23)	16.2	34	6.8
Halocyptena microsoma	20(30)	152(20)	12.9		
Oceanodroma castro	45(102)	256(4)	17.4	47.3	8.7
O. l. leucorhoa	43(9)	297(10)	14.3	46.9	7.4
O. tristrami	94(32)	402(1)	22.9	56.5	7.9
O. melania	59(46)	300(5)	19.3	48	7.7
O. homochroa	38(834)	213(32)	17.5		
O. f. furcata	63(25)	303(23)	20.4	48	7.6
Pelecanoides u. exsul	119(22)	200(8)	58.3	38	7.2
P. georgicus	117(12)	195(5)	58.9	38.8	7.7

Skokholm nest 1490 km away, flew across Europe directly, it certainly took its time—14.2 days—but perhaps it stopped in the Bay of Biscay to refuel. An overland route does seem more likely than a long traverse of the presumably unfamiliar Mediterranean, as Lockley (1942, p. 177) also surmised.

C *Alighting and take-off*

The ease with which petrels can alight depends largely on the loads their wings must support (see Section VIII.A below). Tubenoses usually land into the wind, reducing speed so that they stall at touch-down. This comes easily to broad-winged species such as storm petrels, but is much more tricky for albatrosses and giant petrels with high wing loadings (Table 9.1) and consequent high flight speeds. Such birds land easiest in exposed habitats unencumbered by obstacles and where strong winds help reduce ground speeds. The crash landings and somersaults seen of *D. immutabilis* may be a consequence of the less windy conditions at the low-lying atolls where they breed.

The smaller gadfly petrels have quite low wing loadings and in open country land easily and take off without needing high launching sites, whereas diving petrels and most shearwaters find difficulty in getting airborne even in fresh winds. Such restraints may help explain why shearwaters do not make repeated landings and departures on a single night, whereas courting '*Cookilaria*', prions and storm petrels do.

The circling of breeding birds around the general area of their nests before finally alighting may form part of the bird's orientation to its nesting site from associated landmarks and neighbouring birds. Even diurnally active petrels breeding on flat open ground behave like this, particularly noticeable with cliff nesters.

At colonies under a forest canopy this circling may be to find a gap through which to descend. The opening may be small and landings hard for heavily loaded birds such as shearwaters, involving intensive wing-whirring (see below). Some do alight heavily but, unless snagged in the fork of a branch, even females heavily laden with eggs rarely sustain injury. Others crash into the branches and flutter-fall to the ground. Such haphazard approaches are common among smaller species such as *P. assimilis* and *P. auricularis*, less so with larger ones such as *P. griseus*, *P. carneipes* and *Procellaria westlandica*: even *Pelecanoides urinatrix* diving petrels are occasionally found perched on branches on their way down.

Pennycuick and Webbe (1959) described how flight stability is achieved by *F. glacialis*. Pitching was controlled by moving the wings forwards or backwards, thus changing the centre of gravity. This action also automatically altered the amount of wing extension which helped regulate yawing, but this was mainly corrected by differential flexion of the wings, producing a counter-yawing moment. Rolling was stabilized by differential twisting of the wings, fine adjustments by increased deflection of the primaries, coarser ones by rotating the whole wing at the shoulder.

The fulmar's tail can be moved up and down, twisted, closed or spread, helping control pitch and yaw, although the birds can fly and steer without their tails. Pennycuick and Webbe (1959) believed that the spread tail acts not as a brake, this being the role of the feet, but as an extra lifting surface which, when near-stalling, deflects the air stream downwards and increases laminar flow. Forked tails, as in *Oceanodroma leucorhoa*, may act as slotted flaps and, in *O. furcata* at least, they appear to serve as double rudders (Fig. 9.8).

When flying, the feet are tucked up under the flank feathers but can be variously lowered to help regulate the rate of descent and may be moved forwards and backwards in effecting these adjustments. The alula in petrels is small (Fig. 9.9), and

Figure 9.8 *Storm petrel* Oceanodroma furcata *showing forked tail evidently being used as a rudder. From photograph by Rich Rowlett.*

is not prominent as an antistalling slot, but even large albatrosses may separate the primaries slightly for this purpose.

Sooty albatrosses tend to use cliffs that face the prevailing westerlies (Mougin, 1970a) and while this facilitates take-offs it must also complicate landings. None the less, these most aerial of albatrosses seem to cope with such difficulties with ease, although the influence of eddies and upcurrents induced by the cliffs has not been studied. Furthermore, in a breeze blowing across the top of a cliff there may still be upwards draught where the wind curls round at the edge: this could facilitate landings.

Pennycuick and Webbe (1959) and particularly Rueppell (1977) described a wing action called 'fanning', seen of *F. glacialis* approaching a cliff site and used in light airs to help braking before alighting. This is a regular movement of the wings along their lengths with the axis of rotation at the leading edge. These oscillations were at 7–9 s^{-1} producing a rhythmic raising and lowering of the angle of attack. Rueppell investigated the effect using a model wing oscillated in a wind tunnel, and deduced that fanning acts as a stabilizing brake and generates some lift. Furthermore, the fanning can be of greater amplitude on one wing than on the other, allowing a slow-moving bird to turn helped by a sideways-deflected tail.

This rotatory action was also recorded by Pennycuick (1982) of albatrosses. He related it to their shoulder lock, perhaps being powered by a separate 'sprint' muscle in their deep pectoralis. However, 'fanning' appears to be used widely by petrels, with or without wing locks, landing in light winds or calms (Fig. 9.9). Further study appears justified.

Alighting at sea involves similar actions to those when touching down on land but is less hazardous: if dunked by a breaker the bird bounces free on the far side, shakes the water off its head and carries on. Petrels often launch themselves from a wave

Figure 9.9 *Touch-down in (A)* Diomedea albatrus *and (B)* Puffinus carneipes. *Both are about to stall. The upper bird is turning to the left—note asymmetry of tail and wings and small alula raised; lower bird has right wing moving back in advance of the left and both carpi supinated. From photos: (A) by F. Hasegawa and (B) by author.*

but that shearwaters can get away even in calms is shown by the *Puffinus puffinus* that took off from the water in an aviary without flapping, and in little wind. It could even hover briefly against the breeze (Swennen & Duiven, 1982).

Flight from land normally starts with a leap as the fully spread wings begin the downbeat. In strong winds even birds with high wing loadings can launch themselves from any space allowing their wings adequate clearance; in light airs and calms many need raised perches on rocks, cliff edges or tree branches for successful take-offs. The 'warming-up' exercises of shearwaters were described in *The Petrels*, p. 162. Storm petrels, prions and small gadfly petrels can usually spring into the air even in calms.

The status of the birds leaving in the early hours has not been studied but, at least with *P. griseus* checked around the end of hatching, birds with feathered brood patches, and therefore probably immature, left earlier than the bulk of the probable breeders—those whose patches were still bare (Warham, unpubl.).

In light airs albatrosses and giant petrels trying to leave from level or near-level ground run into any wind they can find, beating their wings through small amplitude until lift-off is achieved. Their long legs help keep the wings clear of obstructions, but a good runway may still be needed. This is often downhill. A similar procedure is used in calms at sea, when getting aloft involves considerable energy expenditure and is perhaps seldom attempted except when disturbed by man.

VII Flight mechanics

A *Dynamic- and slope-soaring*

Petrels have very cambered wings, best seen in the great albatrosses which have thickened and streamlined leading edges to their wings which are drooping and hollowed out behind. Much of the supporting surface consists of patagium. To withstand strong forces this may be supported by tendonous ossifications and wing locks (see Fig. 10.4). The lift created is readily demonstrated when a hand-held extended and disarticulated albatross wing is swung around by the holder spinning on their heels: even at the slow wind speeds generated, the arm is pulled up by the induced suction. Extended from the window of a slowly moving vehicle the effect is quite dramatic.

Pennycuick (1982) considered tubenose flight under three main categories, gliding, flap-gliding and flapping. Thus, while albatrosses travel mainly by gliding, most procellariids use a flap-and-glide style, planing along wave slopes or climbing through wind gradients, interspersed with wing beats to maintain momentum or to change direction. Small shearwaters and diving petrels mainly flap; the dancing flight of storm petrels involves all three methods. Most attention has been directed to gliding and to the way in which albatrosses in particular can gain energy from a sheared wind—'dynamic soaring'—a device largely restricted to tubenoses.

Dynamic soaring by petrels seems first to have been described by Baines (1889) and Hutton (1903). Early authors appreciated how the birds took advantage of the reduced wind speed near the surface and, by climbing against the wind, stored potential energy could be spent on a downwind descent accelerating into the lower stratum of weaker wind, and how they can sail longer and rise higher in a strong wind than in a weak one.

It was formerly thought that the main source of gliding flight was energy drawn from within the wind shear zone. However, soaring along the flanks of waves, gaining lift from the wind deflected upwards, was deemed by Magnan (1925) to be the major flight mode for gliding albatrosses and further evidence for this was gained by Wilson (1975) and Pennycuick (1982) (*The Petrels*, pp. 32, 33).

Slope soaring along the windward faces or crests of waves is usual with large petrels where the waves are large enough to accommodate them. Smaller species also seem to gain energy in flying along lee slopes, when, rising to the crest, they use the sudden increase in air speed generated by the wind forces in the wake to gain height before diving to continue their glide in quieter air (cf. Scorer, 1958, p. 121). Jansen (1983) described a rather similar flight style of *F. glacialis* sailing just below the wave crests and gaining height as it crossed them. Jansen postulated that on passing over the crest, part of the airstream curled back producing an updraft in the lee which the birds also exploited.

Dynamic and slope soaring are usually combined, the precise pattern no doubt depending on wind strengths, wave heights and directions. The planar path is an expanded spiral with the direction of movement at an angle to the prevailing wind. Most progress is made when close to the sea.

The pattern is often repetitive. A bird can be followed horizon to horizon, repeating the movements for as long as it remains in sight, and the pull-ups of flocks are often synchronized—all responding to the same wind forces in the same way at the same time. This is a familiar pattern with shearwaters and prions. Banking in unison, the sun lights up their underwings, throwing their many reflections back to the observer simultaneously (Warham, 1960). Idrac (1926) timed the cyclic movements: with *D. exulans* these were repeated on average every 10.7 s, for *D. melanophrys* every 7.4 s and for *Macronectes* every 9.6 s.

Pennycuick (1982) showed that energy gained from a vertical wind gradient when a bird glides downwind is associated with an increased glide ratio, the faster the glide the stronger the effect as the bird passes into regions of reduced wind strength, the larger birds with their higher wing loadings benefiting the most.

In a computer simulation of the effects of drag on albatross flight, Wood (1973) allowed for the bird's closeness to the sea and considered that when it sails along a wave the induced drag is reduced and the lift increased—the so-called 'ground effect'.

Under the same conditions pull-up heights vary with the sizes of the birds, albatrosses rising more than shearwaters which pull-up higher than prions. However, gadfly petrels have high-towering, erratic, roller-coaster flights, and although they do skim the surface it seems unlikely that their great arcing climbs are powered solely by updrafts; a good deal of wing beating seems essential. More information is needed on their flight style, particularly from tropical species that experience few high winds.

Pennycuick (1982) followed a slope-soaring *D. exulans* in almost no wind but with a swell of about 3 m amplitude and a wavelength of some 100 m. As the waves moved along they displaced the air up and down and the bird sailing along the leading edges achieved about 8.0 m s^{-1} without flapping its wings.

While gliding is aided by high winds, in the not infrequent hurricanes that sweep the Southern Ocean, the tops of the waves are sliced off and the air above obscured by spray. Visual and radar observations are impossible but presumably the birds simply drift downwind. Confused seas must also pose problems, for example when a long swell is crossed by local wind-generated waves.

In estimating straight-line distances covered by the zigzag mode of flight, corrections for the amount of deviation must be made. Both Pennycuick (1982) and Alerstam *et al.* (1993) have estimated zigzag to straight line distance ratios (Table 9.2). Pennycuick reconstructed the tracks of birds within range of his ship and Fig. 9.10 shows two examples taken over about 1 min.

The birds followed by Alerstam *et al.* (1993) were monitored with tracking radar in

Table 9.2 Flight speeds of petrels. Upper section: tracked by radar under near-calm conditions; Lower: tracked optically under variable wind conditions. From Alerstam *et al.* (1993)

Species	Groundspeed ($m\ s^{-1}$)			Airspeed ($m\ s^{-1}$)			Speed increment from wind ($m\ s^{-1}$)		Straightness	
	Mean	SD	Range	Mean	SD	Range	Mean	SD	Mean	SD
Macronectes giganteus	13.3	—	12.4–14.4	14.5	—	13.5–15.2	−1.2	—	0.84	—
Diomedea melanophrys	12.9	1.7	11.2–15.6	12.7	1.2	11.2–14.3	0.1	0.7	0.82	0.09
D. chrysostoma	13.2	1.5	11.0–15.5	13.0	1.2	11.3–15.1	0.2	1.1	0.84	0.11
D. chlororhynchos	11.1	1.0	9.9–12.4	12.0	1.0	11.2–13.8	−0.9	0.7	0.78	0.09
Fulmarus glacialoides	9.4	—	8.0–10.9	10.6	—	9.2–11.5	−1.2	—	0.94	—
Puffinus gravis	10.8	1.4	8.9–13.1	10.8	1.3	9.4–13.2	0.0	0.5	0.94	0.05
P. assimilis	14.4	—	14.0–14.9	14.4	–	14.2–14.5	0.0	—	0.94	—
Oceanites oceanicus	7.1	1.9	4.4–10.8	7.4	1.9	4.7–10.0	−0.3	1.0	0.90	0.08
M. giganteus	11.8	3.2	4.9–17.4	16.3	3.3	8.6–22.0	−4.5	5.0	0.91	0.09
D. melanophrys	12.9	5.0	7.7–25.0	13.4	2.9	9.0–19.6	−0.5	3.9	0.92	0.07
D. chlororhynchos	12.5	5.9	4.3–21.2	10.8	1.8	9.0–14.0	1.7	5.2	0.96	0.04
Calonectris diomedea	8.6	1.3	6.0–11.0	13.2	0.7	11.9–14.3	−4.6	1.3	0.96	0.03
F. glacialoides	10.5	3.0	6.2–15.6	9.8	2.4	5.8–12.8	0.6	1.4	0.96	0.04

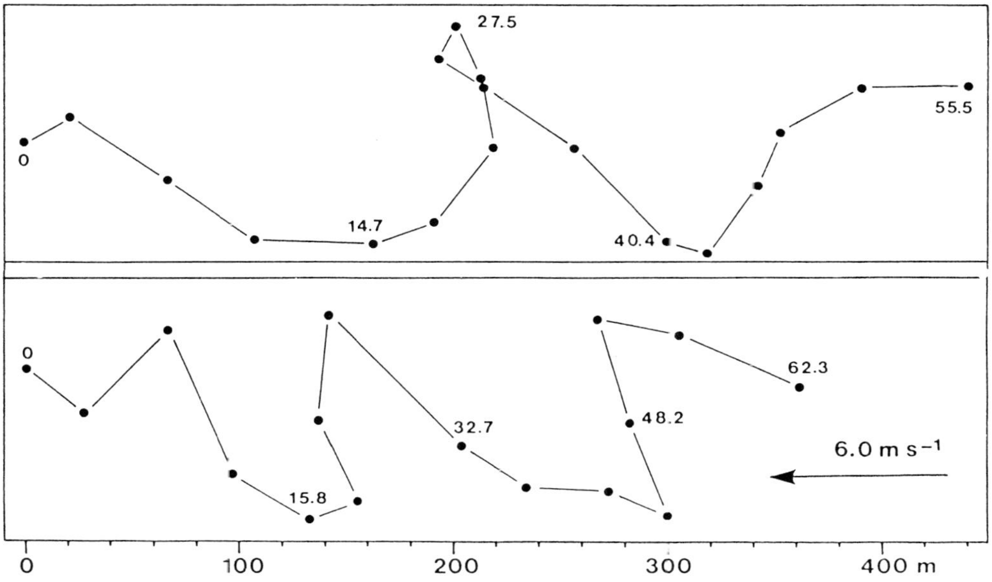

Figure 9.10 *Flight tracks from above of* Diomedea exulans *from shipboard observations. Upper: slope-soaring in near-calm but a swell of about 3 m. Birds travelled at straight-line speed of about 8.0 m s^{-1} without flapping wings. Lower: progress in a 6.0 m s^{-1} head wind showing cross-wind sections along windward slopes. Pull-ups were only 1–2 m and there was no wing flapping. Straight-line speed was about 5.8 m s^{-1}. Times are seconds from start. From Pennycuick (1982).*

calm seas or with an optical rangefinder in windy conditions. They are on a larger scale than those of Fig. 9.10, and seldom show the pull-ups or wave-riding sections.

B *Flap-gliding*

This term covers the flight styles of many small- to medium-sized petrels which progress by glides interspersed with bursts of flapping. All petrels must beat their wings in flying over flat seas with no wind. The big albatrosses probably lack the muscle volume to beat them continuously, so settle until the wind returns. None the less, some can proceed by flap-gliding. For example, Miller (1936) writes of *D. nigripes* in almost negligible wind coming 'bowling across the sea, close to the surface, flapping and soaring like gigantic shearwaters, but with, of course, a slower wingbeat.' *D. irrorata* of the Galapagos Islands is also said to beat its wings more and soar less than southern species; its flight in these quieter equatorial seas is unstudied.

In good conditions of wind and sea one or two flaps to correct a flight path may be all that is needed (Fig. 9.11). In light winds many shearwaters and prions flap and glide extensively, banking steeply at the start of the downwards dive, now to one side, now to the other. For migrating birds crossing the doldrums, extensive flapping close to the surface seems to be the norm. It is no doubt energy-expensive. Whether

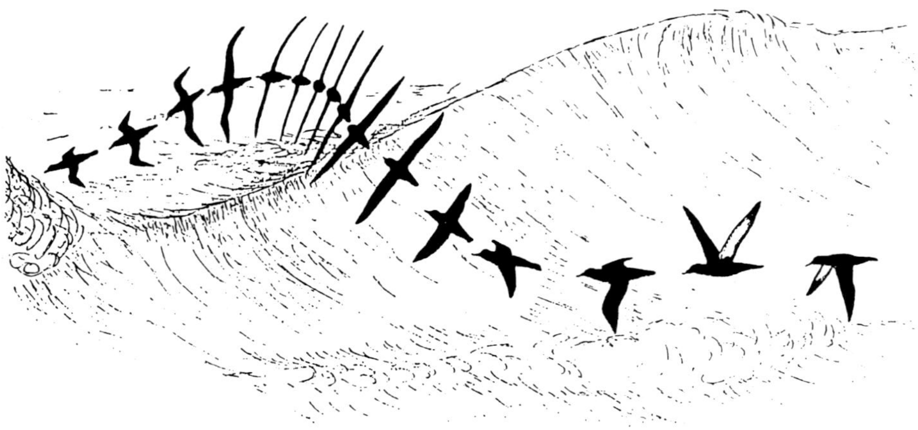

Figure 9.11 *Sooty Shearwaters flap-glide from one wave to another: note wing flexing as birds dive along the windward flank. From Jansen (1981).*

flap-gliding along waves results in long tacking movements on either side of the migration path is not clear, but a very direct line of flight is indicated when birds pass an observer for hours, all on the same bearing.

The flight styles of congeners may be quite different. For example, while *P. puffinus* used a flap-glide style close to the sea with fully extended wings beaten quickly through shallow arcs and glides with only small pull-ups, *P. griseus* nearby beat its wings less often but with deeper strokes and its glides sent it wheeling high above the waves, now tilting to the right, now to the left with the plane of the wings reaching the vertical (Devillers, 1965).

C *Flapping*

The whirr-flight of the diving petrels provide the prime example of this style among the tubenoses, and the flight of *P. assimilis* and *P. lherminieri* also approaches that mode, as does that of *P. gavia*, particularly in calm seas and enclosed sounds. All tend to stay close to the water and presumably benefit from the ground effect and gain lift from wave-deflected air when flying into the wind.

With such small species the wing stroke frequency tends to be relatively high and no doubt both this and the amplitude are varied according to the needs of the time but we lack data on wing-stroke frequencies for petrels, or on how these change with wind or sea conditions. In albatrosses and giant petrels much of the movement occurs at the carpal joint (Fig. 9.12).

D *Hovering*

Long-legged storm petrels such as *Oceanites oceanicus, Pelagodroma marina* and *Garrodia nereis* search for surface plankton along the leeward sides of the waves while facing the wind. Their feet are dangled and their wings capable of high amplitude

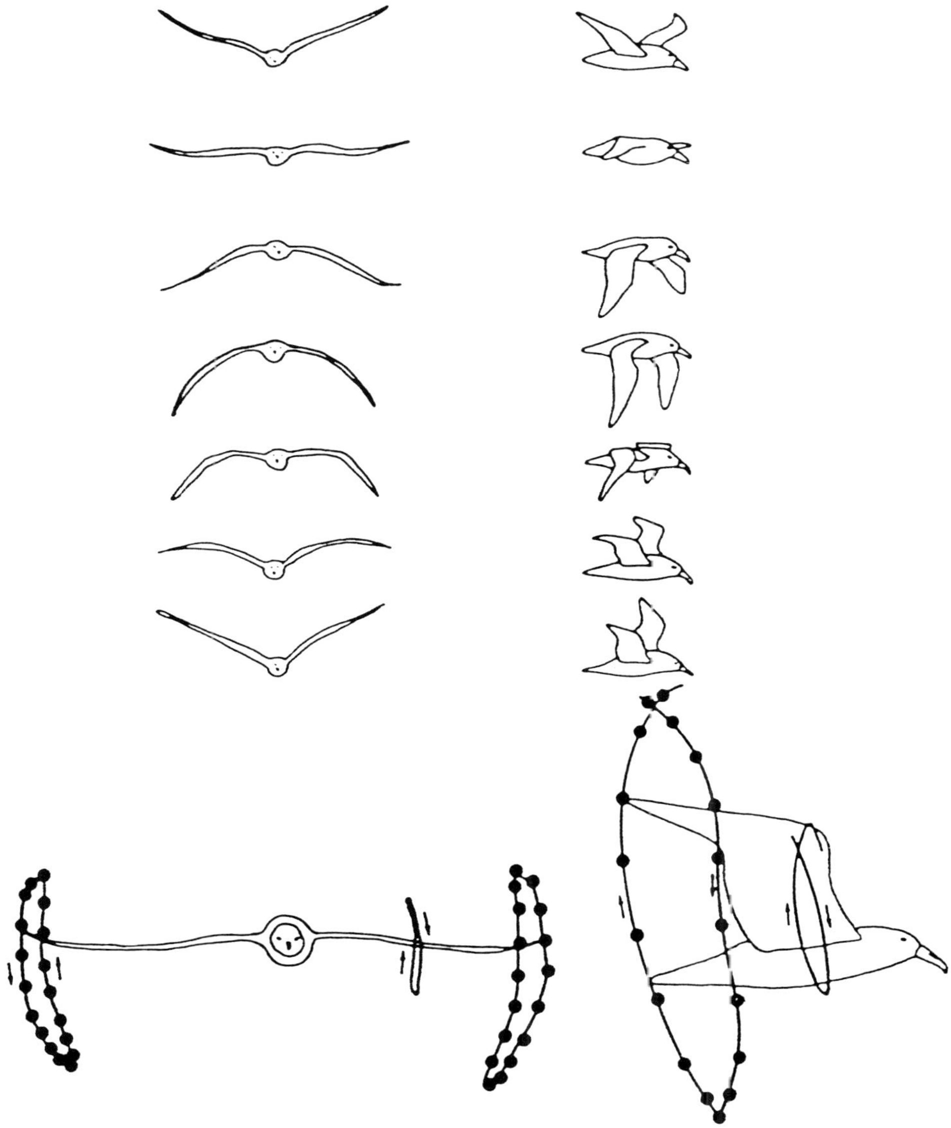

Figure 9.12 *Flapping flight in* Diomedea melanophrys *from cine film. The wings are flexed and slightly curved on the upstroke, their tips swept back to reduce the effective wing span but fully stretched on the downstroke. Bottom figures show successive positions of wing tips as if the bird was stationary. From Scholey in Rayner (1983).*

strokes, and, by side-slipping and continually adjusting their trim, they can keep their distance from the surface and, being of low inertia, are automatically pushed up with every heave of the sea. *O. oceanicus* also feeds on the windward slopes of waves where there is a comparatively undisturbed surface even in strong gales, as Roberts (1940, p. 170) demonstrated with smoke experiments. The petrels get lift from the

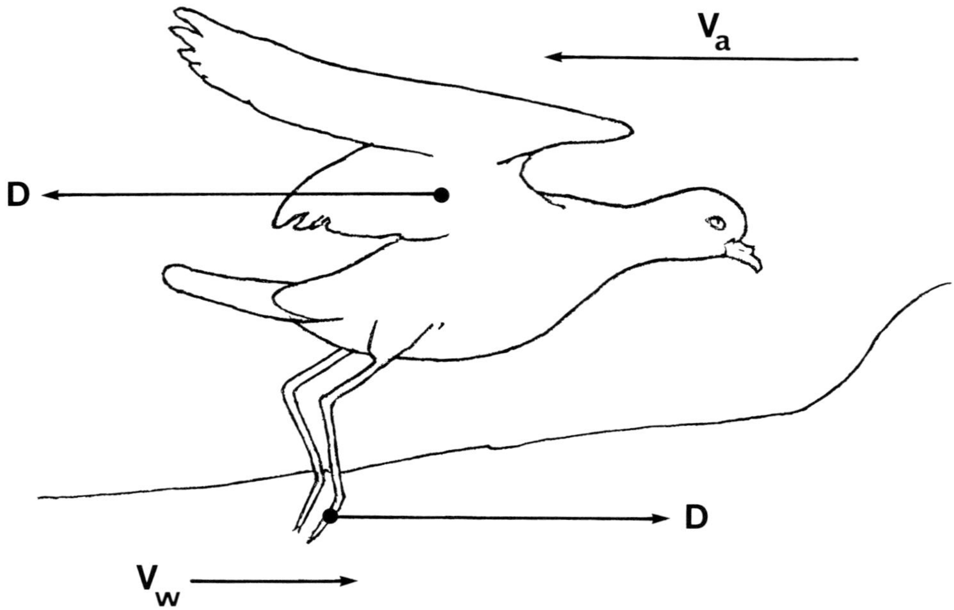

Figure 9.13 *'Sea-anchor' soaring in* Oceanites oceanicus *facing the wind. The bird's feet move backwards through the water at a speed* V_w *represented by a forward flow of water relative to the feet and produce a drag* D *directed forwards. This balances the backwards aerodynamic drag on the wings produced by the relative airspeed* V_a. *After Withers (1979) and Pennycuick (1987b).*

deflected air, but if they rise more than a few centimetres from the water in gales they are instantly blown away downwind.

Storm petrels can raise their wings so that their uppersides are angled only about 20° apart and do this when dropping to pick up food. Presumably this helps control the rate of descent, and keeps the wings above the spray, ready for a propulsive downbeat as the prey is seized.

Withers (1979) from cine film of hovering flight by *O. oceanicus* in still air, recognized two devices for remaining 'stationary' while feeding. As the bird scanned the surface its wings were rapidly pronated and supinated, evidently in a form of 'wing fanning' as already described. In the other technique, reminiscent of the 'hydroplaning' of prions (*The Petrels*, p. 124), the bird soared facing the wind with the expanded webbed feet just below the surface forming a sea-anchor and providing hydrodynamic drag to offset the aerodynamic drag on the wings and body (Fig. 9.13). The ground effect also reduces the aerodynamic drag and therefore the speed at which the feet must move backwards through the water to balance that drag.

Other petrels can hover using updraughts at cliff faces, trimming their wings so that the lift nicely balances their weight. The *Phoebetria* albatrosses are particularly agile and can examine a human observer from close range while hanging in the air curtain rising along a slope. Their wings and tails are never still, small adjustments to their trim being continually made to maintain position. Some such adjustments may be induced by the airflow itself from 'back eddies', as shown when wing feathers of fulmars lift off the trailing edge of a wing about to stall.

E *Glide-running*

Storm petrels fend off the sea with their feet while staying close to the surface. For *O. oceanicus* and *G. nereis* this involves patting with both feet simultaneously, and *Pelagodroma marina* has been described as using a hop-glide action with its long legs, webs spread, hitting the wave crests to bounce off in a series of kangaroo-style leaps. Others, such as *Fregetta grallaria*, often drop a single foot to steady them and *Nesofregetta fuliginosa* has an even more specialized foot action (*The Petrels*, p. 180) deserving closer examination.

O. leucorhoa feeding when 'walking on the water' was figured by Atkin (1979), and Erickson (1955) described a similar action of about 50 *Fulmarus glacialis* in a harbour in Labrador. The birds suddenly started to sail into the wind on stiff, outstretched wings, kicking the surface with alternate feet so that they gained speed and glided up for a short distance before descending to repeat the movement. Erickson could see no purpose in this sudden change in flight action which was neither in response to a wind shift nor to the appearance of food. It appears to have been a kind of 'hydroplaning' as used by prions, but without their bodies touching the surface.

F *Flight direction and wind conditions*

Alerstam *et al.* (1993) considered the effect of wind direction on flight paths of the mollymawks *D. melanophrys*, *D. chrysostoma* and *D. chlororhynchos* (Fig. 9.14). They found most birds travelling with the wind on their left and pointed out that this would lead them away from low pressure cells and towards the high pressure ones in which Jouventin and Weimerskirch (1990a) considered albatrosses could be trapped for days and were to be avoided. Satellite fixes show, however, that *D. exulans* drifted on the surface during calms (Weimerskirch *et al.*, 1993, p. 335), while during short foraging trips when feeding chicks they commuted to specific feeding areas irrespective of wind direction.

Blomqvist and Peterz (1984) examined occurrences over 10 years of *F. glacialis*, *Puffinus griseus*, *Sula bassana* and *Rissa tridactyla* (all flap-gliders) in the Kattegat in relation to prevailing southwest to northwest moving cyclones. In strong cyclones with the eastwards-moving depressions creating southerly winds on their eastern flanks, the birds tended to fly east to head away from the centres of the cyclones where feeding conditions were thought to be poor. As the winds veered southwest with the passage of the depression, birds in the Skagerrak tended to fly south along the Swedish coast and as the wind shift continued westsouthwest to west, many entered the Kattegat. From a point on the Swedish side 1251 *F. glacialis* and 244 *P. griseus* were counted in southwest winds, only 32 and 16 respectively in northerly ones. Neither was common under settled conditions.

Blomqvist and Peterz (1984) also drew attention to another complicating factor—the delayed orientation of the wave-train alignment consequent on the inertia of the water mass as the wind gradually swings round, so that for a while the waves travel at an angle to the wind. These authors calculated that the birds could benefit from a tail wind component by flying at an obtuse angle to it.

In pelagic seas, foraging *D. exulans* appear to keep the wind on the quarter. Two

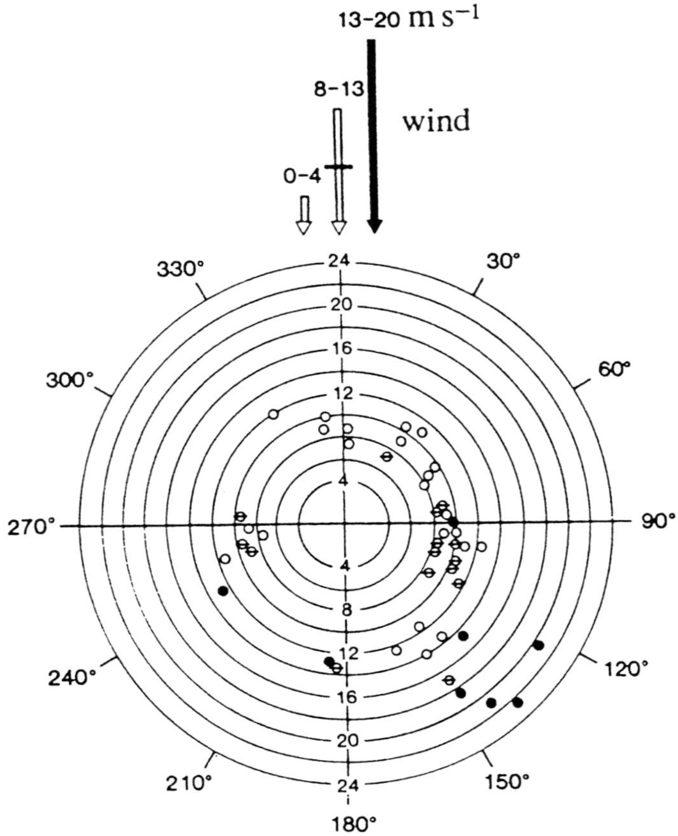

Figure 9.14 *Straight-line speeds and directions of flight of mollymawks in relation to wind direction in weak, moderate, and strong winds. Most travelled with wind on their left flank and flew faster in strong winds. From Alerstam* et al. *(1993).*

that were tracked flying on parallel courses by Weimerskirch *et al.* (1993), altered direction when the wind changed to keep this behind and to one side of them (see Fig. 2.13). Preliminary studies by Weimerskirch and Robertson (1994) also suggest that *Phoebetria palpebrata* use tail or crosswinds on their fast commutings to Antarctic seas.

VIII Petrel aerodynamics

The data needed to calculate variables such as power requirements, flight speeds, wing stroke frequencies, glide angle and the forces acting on a flying petrel, are body mass, the total area of the wings when fully spread in a natural manner, the body area between the bases of the wings, the wing span and the frontal area of the body. Table 9.1 gives some basic figures for 74 species. Pennycuick (1989) has facilitated calculations from such variables by his manual which comes with a disk containing BASIC programs that will run on PCs using MS-Disk Operating Systems.

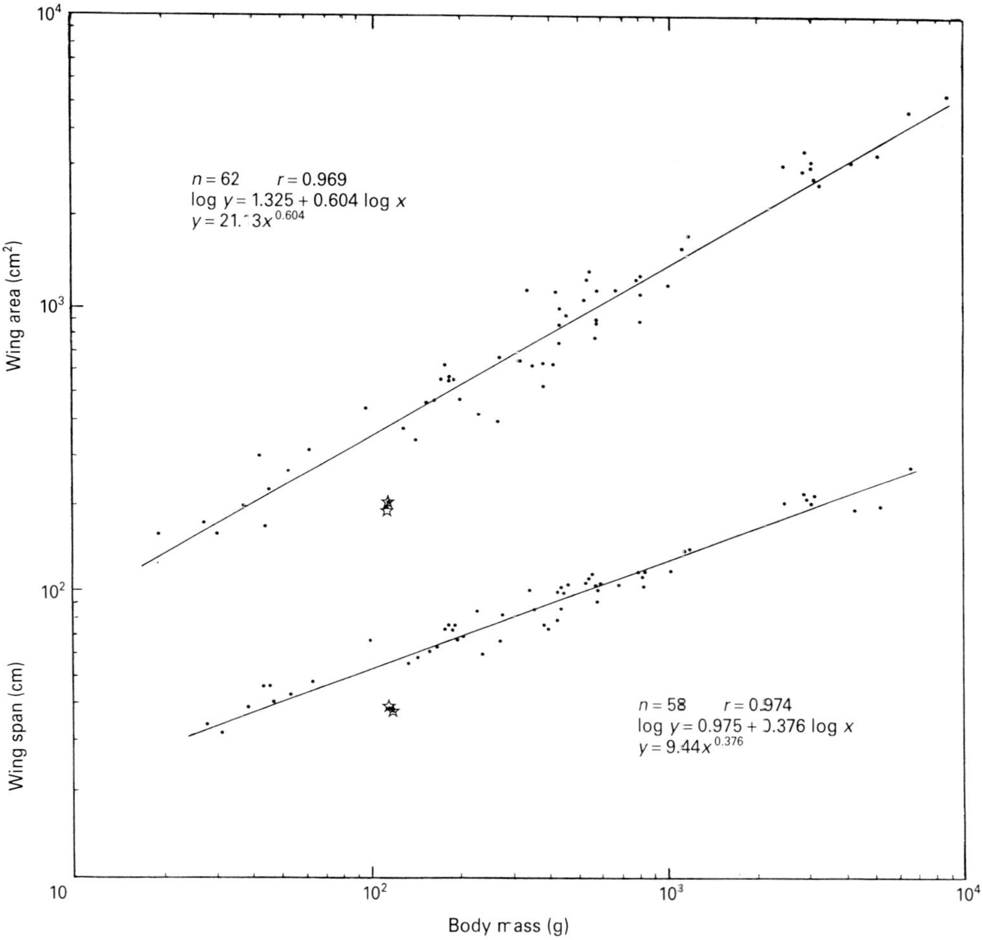

Figure 9.15 *Log:log plot of body mass vs. wing area (upper) and vs. wing span (lower). Stars are points for* Pelecanoides *spp.*

The area of a transection of the body at its widest part is needed for estimating drag and other variables. This cross-section is often assumed to be circular but in petrels it tends to be a blunt ellipsoid and figures for cross-sectional areas have not been published.

A *Wing areas and loadings*

Figure 9.15 shows the allometric relationship between body mass, wing areas and wing spans. For each point there are at least three determinations for areas or spans. As neither mass nor area is an independent variable, the fitted line represents the major axis of an envelope enclosing the plotted points. Only full-winged birds were measured, but nevertheless the data are rather heterogeneous, though less so than in an earlier plot (Warham, 1977a).

The fitted line for wing areas has a power exponent of 0.60, whereas on the basis of the law of similitude for birds of like proportions whatever their size, this should be 0.666. The exponent for wing loadings therefore is 0.30 not the theoretical 0.33. However, a big petrel is not precisely a big replica of a smaller one: with increasing size the wing areas do not increase sufficiently to maintain the theoretical relationship. Pennycuick (1987b), from a smaller data base, gave a slope of 0.63 excluding the abnormally short, small-winged *Pelecanoides* spp. If these birds are excluded, that for Fig. 9.15 reduces to 0.59. Pennycuick (1987b) also confirmed that diving petrels conform to a line for auks (see also Warham, 1977a), emphasizing their convergence for a similar lifestyle.

Gadfly petrels and shearwaters plot in the central area of the graph between 170 and 900 g. Four of the shearwaters fall above the trend line, eight below it, while the 12 gadfly petrels lie close to that line. The main outliers, *Puffinus pacificus* and *P. bulleri*, have low wing loadings for their masses and, together with their relatively long tails, may explain their aerial dexterity in taking flying fish on the wing, etc. Separate measurements of 10 *P. pacificus* also gave low figures of 39.8 N m^{-2} (J. Hatch, pers. comm.).

Most of the other shearwaters have smaller wing areas and higher loadings than similar-sized gadfly petrels. These include *P. tenuirostris, P. puffinus, P. huttoni, P. gavia, P. assimilis* and especially *P. griseus*. This probably reflects their need for reduced buoyancy with so much of their food being taken under water. Of the gadfly petrels plotted, *Pterodroma hypoleuca* deviates the most from the trend line and its loading of 29 N m^{-2} helps explain its ability to take off in the lightest of winds.

Calonectris d. diomedea and *C. leucomelas* fall above the trend line: these are familiar gliding species. The very different figure of 75 N m^{-2} for *C. d. borealis* (Table 9.1) may be an artefact of the small number of wings measured; this and the high aspect-ratio require confirmation.

Among the *Procellaria* shearwaters *P. cinerea* has the highest wing loading, again from only three measurements, but this bird too dives frequently.

Little is known of the flight and feeding of the Bulwer's Petrel, but at 22 N m^{-2} it has a wing loading approaching that of a storm petrel. Of the latter *Oceanodroma leucorhoa* has the lightest loading, less even than the very buoyant long-legged Oceanitinae. Ainley *et al.* (1975) established that the difference in wing loadings of this bird and *O. homochroa* breeding sympatrically was significant and hypothesized that the low loadings benefited the pelagic-feeding *O. leucorhoa* because it would use less energy per unit distance over the long hauls between nesting and feeding places. *O. homochroa*, however, is a coastal feeder, and could therefore cope with heavier loads of food. The lowest wing loading is probably that of the smallest tubenose, *Halocyptena microsoma* at about 13 N m^{-2}, but this is based on estimates of interwing areas—better data are needed.

At the other end of the size range the wings of *Macronectes* spp. carry weights approaching those of the great albatrosses, although Pennycuick's mass (Pennycuick, 1982) for *M. giganteus* is high for this bird—unless all were males. The figure for 170 of mixed sexes given by Hunter (1984b) of 4477 g would give a loading of 133 N m^{-2}, as for *M. halli*. Such high loadings mean fast flight and difficulty in getting aloft in calms. *Macronectes* vomit their stomach contents when stressed at such times and then run helter-skelter upwind, beating their wings frantically until these grip the air and they rise in shallow arcs. Six male and six female *M. halli* had wing loadings averaging 155 N m^{-2} and 132 N m^{-2} respectively, giving the females some advantage in take-offs, particularly if loaded with an egg.

Most of the albatrosses plot above the trend lines of Fig. 9.15, perhaps as a result of mass reduction through bone pneumatization, with the two *Phoebetria* deviating the most—not surprising with these most aerial of albatrosses.

Body masses and the wing areas of a petrel vary over time; the masses because of varying loads of food, stomach oil, egg, and especially of depot fat; the wing areas because of feather loss during moult. Fisher (1961) weighed *Pterodroma hypoleuca* at different times of the day. He found that the wing loadings of both sexes were lowest between 1400 and 1800 h, highest between 0200 and 0600 h; possibly the early morning samples included more new arrivals with full stomachs. Ainley *et al.* (1975) found that the loadings of *O. leucorhoa* fell from 18 N m^{-2} (interwing areas excluded) in the prenesting period to 16 N m^{-2} by hatching, whereas for *O. homochroa* breeding nearby the loading averaged 20 N m^{-2} throughout its time ashore. The differences were ascribed to the long, energy-consuming foraging of *O. leucorhoa* and the greater mass declines this bird experienced than with *O. homochroa*.

The effects of moult on petrel wing loadings and spans must be considerable with diving petrels and migratory shearwaters that shed their remiges over a very short period, but there appear to be no figures to relate the loss of feathers to change in flight capacity. Ainley *et al.* (1975) did report that, while *O. leucorhoa* remained full winged during breeding, *O. homochroa* shed the two outer primaries while still feeding chicks, losing about 3% of wing area, but this was compensated for by a seasonal decline in mass. With species such as *Puffinus tenuirostris* and *P. griseus* such compensation may not occur. Ogi *et al.* (1981) reported that in the North Pacific the mean lengths of *P. griseus* wings (as a measure of wing moult) was at a minimum of 277 mm in August with the birds weighing 864 g; that is, much the same as their premoult weight of 861 g in April (cf. 819 g when on the breeding grounds).

B *Wing shapes and aspect-ratios (A/R)*

Wing spans *vs.* body mass for 58 species are also shown in Fig. 9.15. Theoretically the slope should be 0.33 but it is steeper at 0.38 (cf. the plot for 22 species by Pennycuick (1987b) with a slope of 0.39). In both, larger birds have relatively longer wings than smaller ones. In consequence, aspect ratios (measures of the narrowness of the wings) increase with increasing body size (*The Petrels*, fig. 1.8). From Table 9.1 it can be seen that storm petrels not only have low wing loadings but also low aspect ratios (A/Rs); that is, they are broad-winged, A/Rs mostly 7–8. The next group, the prions, are heavier and with longer wings, A/Rs mostly 8–9. There is no consistent trend among shearwaters, even *P. griseus* which dives a lot, still having narrow wings, A/R 12, similar to those of the larger *Procellaria* spp. *Calonectris, Puffinus pacificus, P. bulleri* and *P. carneipes*, however, are rather broader-winged, A/Rs 8–10, whereas fulmars and gadfly petrels tend to be long-winged, A/Rs 10–12.5. Giant petrels and the albatrosses have the most attenuated wings, A/Rs 13.5–15, and for both sexes of *M. halli* A/Rs were 11.9, despite their different masses.

This general attenuation of the wing with increasing size arises because the relative lengths of its distal segments—the manus with its attached feathers—decrease with increasing body size (see Table 10.1). In summary, among Procellariiformes, small body size is correlated with low aspect-ratio wings, low wing loadings, low flight speeds and high maneouvrability, large body size with high aspect-ratio wings, high wing loadings, high flight speeds, and poor manoeuvrability. Intermediate body size is associated with intermediate values for these characteristics.

C *Flight speeds*

The higher the wing loading the faster the minimum speed at which a petrel must glide before stalling, the speed for minimum rate of sink or the best glide ratio and minimum estimated gliding speeds ranging from about 4 m s^{-1} for small storm petrels to some 12 m s^{-1} for the great albatrosses.

Pennycuick (1982, 1987a) measured the speeds of petrels at sea in known wind conditions during straight sections between pull-ups. These included the groundspeed and airspeed. In most of his determinations the airspeeds were greater than the groundspeeds as the birds usually flew against the wind (Fig. 9.16). From these and other data, Pennycuick (1982, 1987a) calculated the speeds that would give the maximum range for flapping flight without refuelling, the speed for the minimum power consumption and other variables.

In general, as Fig. 9.16 shows, the bigger petrels travelled faster, but when these absolute speeds were converted to dimensionless relative speeds by dividing by the minimum power speed (Fig. 9.16), the means fell with increasing body size. Only for *Oceanites oceanicus* did the mean airspeed approach that of the speed for maximum range; otherwise the larger the bird the greater the difference—they flew progressively more slowly in relation to their minimum power speeds. Pennycuick interpreted this as showing that the smaller birds fly as would be normal for powered flight, while the largest move at low relative velocities consistent with slope-soaring along waves. Pennycuick (1987a) calculated that the largest tubenoses lack the musculature to exceed their minimum power speeds, therefore they fly slower than would be most efficient and soar to avoid powered flight. Mean airspeeds when slope-soaring were lower than when flying higher above the sea, for example 14.1 ± 4.1 m s^{-1} *vs.* 15.0 ± 3.1 m s^{-1} for *D. exulans*.

The speed of a petrel on a straight leg of its zigzag flight is obviously different from the overall groundspeed, so that zigzag speeds have to be divided by zigzag:straight-line ratios to give straight-line speeds. For example, Pennycuick (1982) calculated that the zigzag ground speeds in a 10 m s^{-1} following wind, in calms, and in a 10 m s^{-1} head wind were: (i) for *D. exulans* 12, 14.5 and 23 m s^{-1}; (ii) for *Daption capense* 7, 9.5 and 18 m s^{-1}; and (iii) for *O. oceanicus* 5, 8.5 and 16 m s^{-1}, direct line speeds being reduced by about 0.66 for the large species and 0.75 for the small ones. Airspeeds generally are higher when flying into a wind, lower when flying with it, an energy-saving system (Pennycuick, 1978, 1987a). However, Alerstam *et al.*

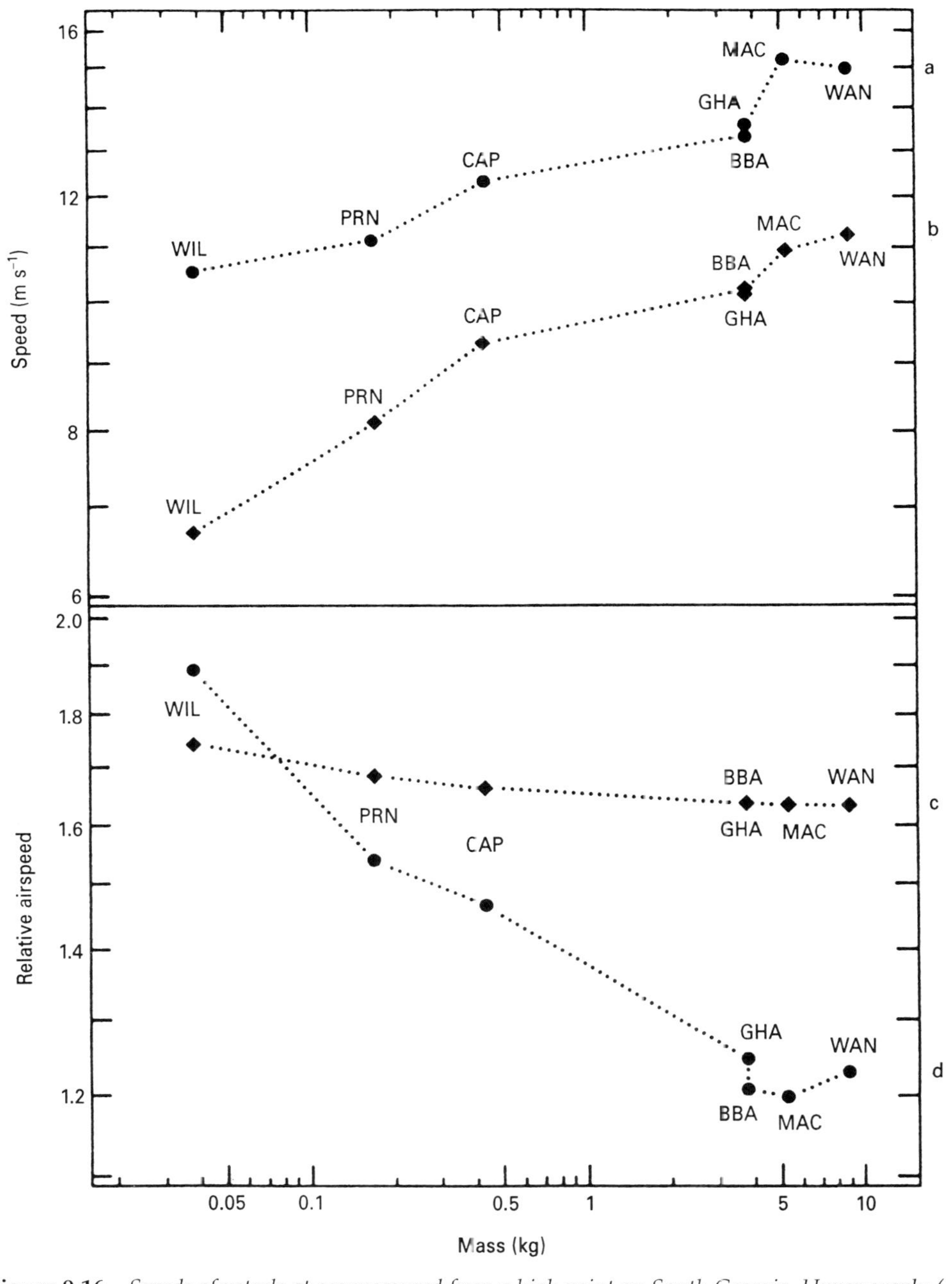

Figure 9.16 *Speeds of petrels at sea measured from a high point on South Georgia. Upper graph: (A) mean airspeeds (B) mean groundspeeds. In lower graphs speeds are expressed in non-dimensional form: (C) calculated still-air maximum range speeds and (D) observed mean airspeeds as in (A). WAN,* Diomedea exulans; *BBA,* D. melanophrys; *GHA,* D. chrysostoma; *MAC,* Macronectes giganteus; *CAP,* Daption capense; *PRN,* Pachyptila desolata; *WIL,* Oceanites oceanicus. *From Pennycuick (1982).*

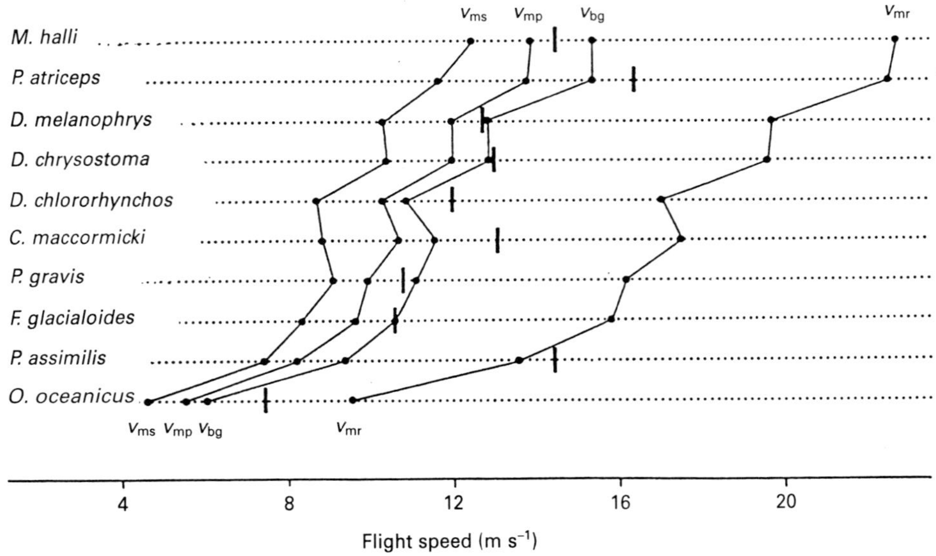

Figure 9.17 *Airspeeds of seabirds (bars) tracked by radar in near-calm conditions in relation to* V_{ms}, *gliding speed for minimum rate of sink;* V_{bg}, *gliding speed for best glide ratio;* V_{mp}, *speed for minimum power in flapping flight;* V_{mr}, *maximum range speed in flapping flight. From Alerstam* et al. *(1993).*

(1993) found no effect of wind direction on the airspeeds of the three mollymawks they tracked by radar, but there were marked effects with *Macronectes, F. glacialis, P. gravis, P. assimilis* and *O. oceanicus*.

Most tubenose flight speeds have been clocked over short distances, but, using radar and optical means, Alerstam *et al.* (1993) collected figures over rather longer runs (Table 9.2). They also compared the observed speeds with those calculated from the flight formulae of Pennycuick (1989), finding that most airspeeds were close to those giving the best glide ratios (forward speed:sinking speed). Only *P. assimilis* flew faster, close to the velocity for maximum range under flapping flight (Fig. 9.17). These authors pointed out that a factor perhaps affecting the low flight speed of *O. oceanicus* may be the costs incurred in accelerating to suitable speeds after feeding pauses. Occurring every 40 s, these extra costs would reduce the optimal speeds.

Alerstam *et al.* (1993) also separated flights of this storm petrel into those where they were feeding by picking at the surface and those where they flew directly with few pauses, apparently travelling to and from preferred feeding places. When foraging they averaged 5.5 m s^{-1} but when commuting 8.8 m s^{-1}, the latter birds flying on straighter paths.

Also examined was the relationship between flight speed and direction in relation to wind speed and direction, summarized in Fig. 9.14 for the mollymawks. Their ground speeds were corrected for straightness. In weak winds the birds flew at 10.3 ± 2.0 m s^{-1} ($n = 23$), but in various directions in relation to the wind. In 8–13 m s^{-1} winds their velocity was the same—10.2 ± 2.6 m s^{-1} ($n = 15$)—but most travelled at about 104° to left or right of the wind. In strong winds the albatrosses moved at 16.6 ± 4.7 m s^{-1} ($n = 8$), mostly in oblique following winds, seeming to avoid heading into moderate or strong winds perhaps because they could not then cover ground efficiently in this direction by soaring.

Even broader-scale data come from satellite-tracked birds. As with *O. oceanicus*, Weimerskirch and Robertson (1994) found that *Phoebetria palpebrata* flew quicker when commuting to and from the feeding areas at a mean of 10.2 m s^{-1}, but flew at only about 3.6 m s^{-1} when collecting food. The maximum speed recorded was 23 m s^{-1} (84 km h^{-1}). The night-time groundspeeds when commuting were not significantly different from those attained by day. Over 17 days of travelling between breeding and feeding places these birds spent only 2 days flying into the wind, 5 days in cross-winds and 10 days downwind. With *D. exulans*, Salamolard and Weimerskirch (1993) found groundspeeds of males were lower than those of females, both increasing with increasing wind strength.

Straight-line speeds of birds returning after long-distance displacement in homing experiments tend to be lower than those of regular commuters, perhaps because homers need to feed in what are unfamiliar seas. Billings (1968) timed a 4796-km flight of *Oceanodroma leucorhoa* that averaged 4.0 m s^{-1}. A Laysan Albatross covered 5150 km at 5.9 m s^{-1}, a Manx Shearwater 4907 km at 4.5 m s^{-1} and a Short-tailed Shearwater 15 200 km at 4.2 m s^{-1}, these being minimal figures as there is no evidence that great circle routes were followed.

D *Disc loadings*

Birds with high disc loadings require higher power output from the flight muscles and cruise most efficiently at high speeds. For the birds in Table 9.1 the highest disc loadings are at 16.4 N m^{-2} for *M. giganteus*, 12.7 and 11.3 N m^{-2} for *D. epomophora* and *D. e. antipodensis* respectively, about 8.7 and 9.6 N m^{-2} for medium-sized birds such as *Puffinus puffinus* and *P. griseus*, but significantly low for *P. pacificus* at 4.2 N m^{-2}, reflecting this bird's aerial buoyancy. The lowest value is that for *O. leucorhoa* at 2.5 N m^{-2}. The diving petrels at 9.7 N m^{-2} need high power outputs for continous flapping because of their short wings.

E *Flight ranges*

Given basic data on flight variables and estimates of the metabolic power required, the energy consumption in flight can be calculated. Pennycuick (1982) estimated the straight-line distances and times to consume fat equal to 1% of the body mass for *D. exulans* and *Oceanites oceanicus*. The former was calculated to fly 987 km and take 29.0 h to do this, the storm petrel 38.8 km in 1.64 h. From similar data, Pennycuick *et al.* (1984) estimated foraging radii for nine tubenoses from South Georgia, correcting for zigzag progression, wind speed and direction but ignoring feeding *en route*. The radii varied from 1110 km for *D. exulans* down to 219 and 212 km for *Pachyptila desolata* and *O. oceanicus*. Some of these estimates can be compared with data from satellite-tracked birds. For example, the range computed for *Phoebetria palpebrata* was 491 km, while Weimerskirch and Robertson (1994) found their birds moving rapidly to their

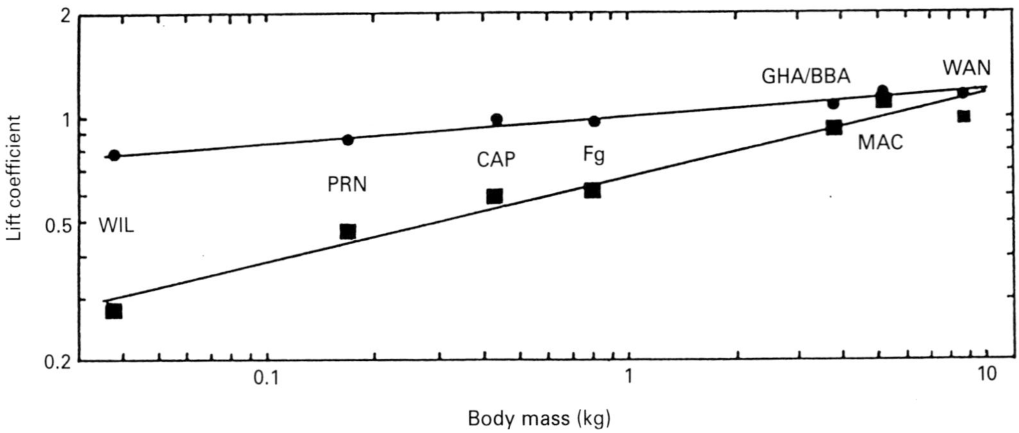

Figure 9.18 *Lower line: lift coefficients at mean airspeeds in flapping and flap-gliding for eight tubenoses. Upper line coefficients for gliding at their minimum power speeds. WAN,* Diomedea exulans; *MAC,* Macronectes giganteus; *GHA,* D. chrysostoma; *BBA,* D. melanophrys; *Fg,* Fulmarus glacialis; *CAP,* Daption capense; *PRN,* Pachyptila desolata; *WIL,* Oceanites oceanicus. *From Pennycuick (1987a).*

feeding places about 1500 km away in 2–4 days and, although some stopped for a few hours at night, they may not have fed significantly during these flights.

F *Glide ratios*

The lift generated by a gliding wing can be expressed in a non-dimensional lift coefficient (C_L) calculated from the weight, velocity, wing area and air density. The value falls with increasing airspeed and increasing lift. Pennycuick (1960) calculated the maximum C_L for *F. glacialis* at 1.8 when the fully spread wing would stall at *c.* 7 m s^{-1}. Figure 9.18 (upper) shows the C_L values at their mean airspeeds for eight tubenoses as determined by Pennycuick (1987a): they varied with the 0.24 power of body mass. Pennycuick (1960) also plotted airspeeds *vs.* sinking speeds for gliding *F. glacialis*—the 'glide polar'. He calculated their minimum rate of sink at about 1.2 m s^{-1} as against 0.8 m s^{-1} for a glider, but pointed out that a petrel is not a glider but a powered aircraft with its engine idling, and in this light its performance is by no means contemptible!

G *Power requirements for flight*

Pennycuick (1982) presented figures for estimating the rates of doing work for nine petrels. Alerstam *et al.* (1993) using Pennycuick's formulae, plotted flight power/BMR values for five species in flapping flight at different flight speeds (Fig. 9.19).

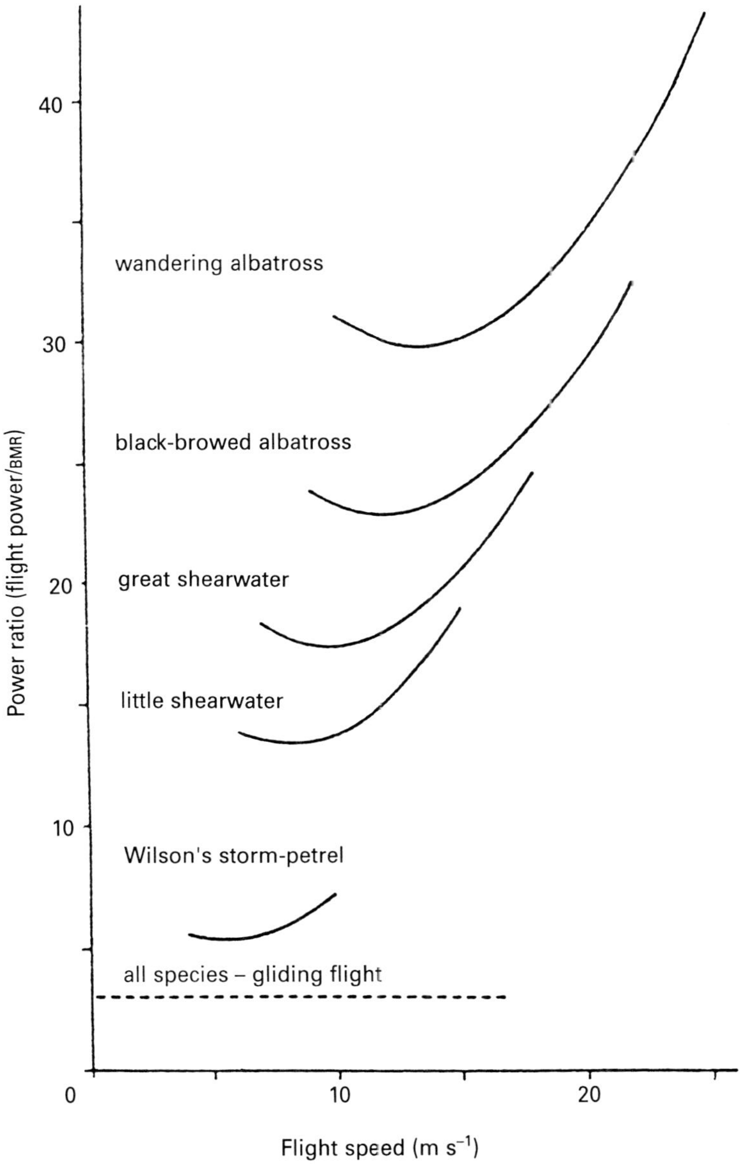

Figure 9.19 *Flight power relative to calculated basal metabolic rate (BMR) for petrels of various sizes as estimated for an intermediary range of flapping flight speeds (including the minimum power and maximum range speeds) from the formulae of Pennycuick (1989). Power expenditure for gliding is about 3 × BMR for all, whatever their size and speed. From Alerstam* et al. *(1993).*

These workers estimated that gliding flight called for power expenditure of about 3 × BMR for all species independent of size, and if albatrosses flew by flapping their power bill would be 22–35 × BMR. Only when progression by soaring was as low as 2 m s^{-1} was the travel cost by gliding and flapping about equal.

CHAPTER 10
Anatomical Matters

Contents

Detailed technical diagnoses of the anatomy of the petrels are in Milne-Edwards (1867–68), Forbes (1882) and Salvin (1896), as well as in Beddard (1898), Pycraft (1899), Godman (1907–10), Bianchi (1913), Mathews and Iredale (1921), Lowe (1925), Witherby *et al.* (1944) and Dement'ev *et al.* (1968). A few have concentrated on a single species, for example Cazin (1886) on *Macronectes*.

Tubenoses are readily defined by having external nostrils produced in tubes. Other special features include the huge glandular proventriculus and ascending duodenum, the form of the vomer, and the compound rhamphotheca built from 11 to 12 separate elements (Fig. 10.1).

I Skeleton

Skeletons have been described by many authors, for example Wagner (1840) described those of shearwaters and storm petrels, Eyton (1867) illustrated those of *Macronectes, Puffinus, Diomedea exulans* and *Hydrobates pelagicus* and Milne-Edwards (1867–68) and Milne-Edwards and Grandidier (1882), complete and disarticulated skeletons of *Pachyptila vittata, Puffinus griseus, P. pacificus* and *Oceanites oceanicus*. Forbes (1882) figured the bones of *Pterodroma lessonii, Procellaria aequinoctialis, Phoebetria* sp. and *Macronectes* and 16 sterna from all four families. Pycraft (1899) gave keys to the genera based on their skeletons, vertebrae, sterna, pectoral and pelvic

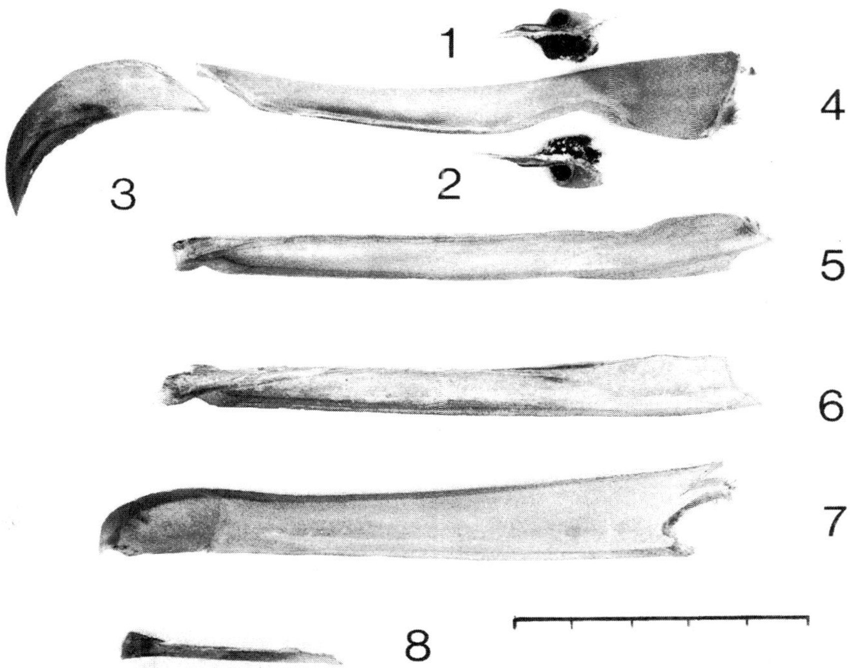

Figure 10.1 *The bill plates of* Diomedea immutabilis*: 1 & 2, naricorns: 3, superior unguicorn; 4, culminicorn; 5 & 6, latericorns; 7, ramicorn with inferior unguicorn attached; 8, inter-ramicorn. Bar = 5 mm.*

girdles, and fore and hind limbs. Other useful data are in Shufeldt (1888, 1907) and Oliver (1945). In 1922, Shufeldt described and figured the skeletons of *Daption capense, Calonectris diomedea, Pterodroma cahow, Pelecanoides urinatrix* and *Pterodroma cookii* and many bones of *Puffinus* spp. Although Kuroda (1954) was particularly concerned with *Puffinus* and *Calonectris* and figured their bones, he also described those of albatrosses, fulmars, prions, gadfly, diving and storm petrels, and tabulated their dimensions. Later (Kuroda, 1983) he added data on species untreated earlier. In both papers he was concerned to relate structures to modes of life, especially to flight style and aquatic performance. Gilbert *et al.* (1981) depicted the major bones of *Diomedea immutabilis* and *Puffinus gravis* and provided keys to separate individual bones from those of other bird orders.

Others have concentrated on the skeletons of particular species or genera, for example Kuroda (1953, 1955a,b, 1959) on *Puffinus nativitatis, P. bulleri, Pagodroma nivea, Pterodroma* spp., *P. hypoleuca* and *D. immutabilis*, Mayaud (1932a, 1934) on *Puffinus* and Kuroda (1967a) on *Pelecanoides*.

Shufeldt (1888, 1907) listed the osteological characters separating tubenoses from other avian groups. Their skulls have large and deep supraorbital depressions, the upper mandible decurved apically and very sharp, symphysis of the lower mandible also more or less decurved with the articular truncated posteriorly. They are holorhinal and schizognathous and have large and broad vomers pointed anteriorly and often depressed and arched posteriorly. The cervico-dorsal vertebrae number 21–23. The coracoid is unusually wide, the sterna broad with posterior borders entire, regularly four-notched or asymmetrically excavated. The hind toe is absent or reduced to a single joint, the patella free, small, and articulated high up in the posterior aspect of the much enlarged procnemial crest of the tibio-tarsus.

The sterna, coracoids, some of the skull bones, the humeri, vertebrae, pelvis and some ribs of the larger gliding forms are increasingly pneumatized, but not the furcula, scapulae or leg bones. Pneumatization is reduced in some, such as *Fulmarus glacialis*, where much of the skull and mandibles are pneumatic, but the vertebrae, sternum and pectoral girdles are not. There is also little pneumatization in the storm petrel skeleton.

A *Skull*

The bone architecture was described by Forbes (1882), Shufeldt (1887, 1888, 1907), Lowe (1925), Condon (1939), Kuroda (1954, 1983), Imber (1985a), and in great detail by Pycraft (1899), to whom reference should be made. Pycraft in particular examined the fine details of the bones in all the genera except *Halobaena*, *Halocyptena*, *Pagodroma* and *Garrodia*.

Crania of the small species tend to be smooth, the cerebral eminences comparatively large reflecting the relative sizes of those parts of the brain, whereas in the bigger species the surfaces are rougher, ridges and fossae being better defined, the ridges often corresponding to the underlying sulci. However, the skull bones ossify rapidly, the sutures being obliterated early and ageing thereby difficult, although the frontal bones of young birds are often appreciably thinner. Representative skulls were figured in *The Petrels*, pp. 5, 21, 83 and 159.

A bone not peculiar to tubenoses, but of unusual shape and taxonomic significance, is the prefrontal (lachrymal) forming part of the anterior wall of the eye socket. In most petrels, including *Diomedea*, this bone is not fused to the frontal, but in the fulmarine genera and *Pterodroma*, *Pachyptila* and *Halobaena* it is ankylosed to the frontals and nasals anteriorly and to the mesethmoid (ethmoid) posteriorly, and is often attached by ligaments to the jugal arch. However, in *Bulweria* (both species), this bone is free, a feature separating this genus from *Pterodroma* (Bourne, 1975; Kuroda, 1983). Its forms based on many specimens were detailed by Cracraft (1968).

Fusion of the prefrontals was used to re-establish the genus *Pseudobulweria* by Imber (1985a, p. 214), who considered that its full fusion in these birds was derived from an early ancestor of *Procellaria*: in today's species the fusion is still incomplete.

Another feature of the petrel skull is a bonelet, the *os uncinatum* or 'ossiculum lacrymo-palatinum' (Brandt, 1839–40). This paired structure articulates with the prefrontals above and is attached by ligaments to the palatines below. It is best developed in albatrosses as a tiny, rather spatulate process about 12 mm long. In the Procellariidae and Pelecanoididae it is represented by a rod or nodule in the ligament but in *Oceanodroma* and *Hydrobates* is replaced by a narrow ligament. The bone is figured by Brandt and by Forbes (1882). Reinhardt (1871) reported it in some gulls,

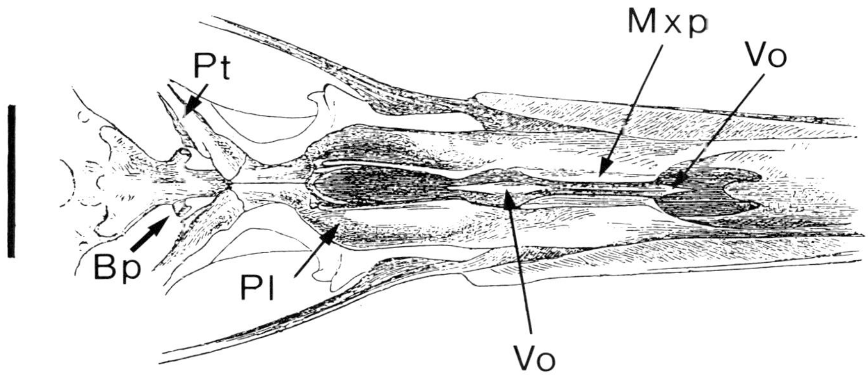

Figure 10.2 *Skull of* Macronectes *in ventral view, to show basipterygoids (Bp). Pt, pterygoid; Vo, vomer; Mxp, crista ventralis of palatine (Pl). From Huxley (1868).*

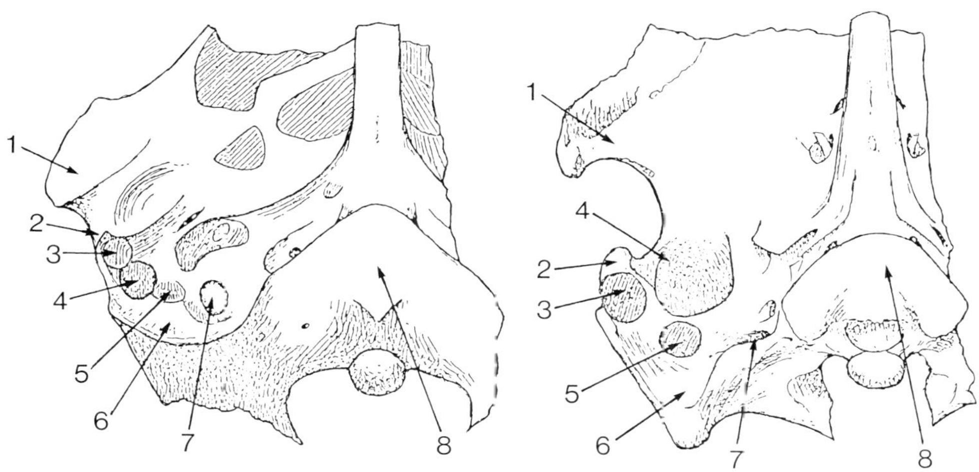

Figure 10.3 *Basal views of skulls of (left)* Oceanites oceanicus *and (right)* Diomedea exulans, *showing quadrato-tympanic relations, the quadrate itself being removed. 1, Postorbital process; 2, squamosal process; 3, squamosal facet; 4, entrance to upper tympanic recess: 5, opisthotic facet; 6, ala tympanica of the exoccipital; 7, fenestrum ovalis; 8, basitemporal. Not to scale. From Lowe (1925).*

frigate birds, auks and even cuckoos, so it is evidently of little taxonomic value, although so used by Lowe (1925).

Small, but of some significance, for example in defining Hydrobatidae (Forbes, 1882, p. 55) and Procellariidae (Lowe, 1925), are the basipterygoid facets, small nubs of bone arising from the pterygoids and articulating with the palatines (Fig. 10.2). These are well developed in fulmarine, diving and gadfly petrels, variously so in shearwaters, but missing in *Pelagodroma* and *Oceanites*, and appear as minute prickles in *Hydrobates* and *Oceanodroma* (Pycraft, 1899, p. 384).

Lowe (1925) focused on differences in the relations of the quadrate bone and the surrounding tympanic structures (Fig. 10.3). In the Hydrobatidae (*Nesofregetta* not examined), the opening into the upper tympanic recess lies directly between the facets for the squamosal and opisthotic heads of the quadrate whereas in the other families the opening to the large upper tympanic recess lies above the line joining the two facets. On these differences, the type of sternal notching, and the presence or absence of basipterygoids and uncinate bones, Lowe separated the tubinares into the Oceanitidae and Procellariidae (subfamilies Procellarinae and Diomedeinae).

Saiff (1974), who examined the osteology, soft anatomy and associated nerves and blood vessels of the middle ear of a very representative range of petrels, confirmed Lowe's findings in general. The species differed only in small details and supported the monophyly of the order.

B *Forearm*

The proportions of the wing bones change in a regular manner with body size (Table 10.1). Humerus and radio-ulna tend to be of equal length but the manus becomes relatively smaller with increasing body size. Storm petrels have the manus longer than humerus or ulna, and with its added primaries, this segment is 4.5–6.0 × the humeral one—these birds need broad and flexible distal segments to support the

Table 10.1 Lengths of wing bones and body masses of petrels: (a, humerus; b, ulna; c, manus; d, manus + primaries)

	Mass (g)	Length (mm) a	b	c	d	a:b:d
Diomedea exulans	8677	428	417	290	601	1.0:1.0:1.4
Macronectes giganteus	4549	243	236	212	498	1.0:1.0:2.0
D. melanophrys	3515	259	262	202	521	1.0:1.0:2.0
D. chrysostoma	3300	254	255	195	507	1.0:1.0:2.0
Procellaria aequinoctialis	1134	152	152	147	372	1.0:1.0:2.4
P. cinerea	1018	134	132	127	334	1.0:1.0:2.5
Puffinus griseus	849	103	98	106	304	1.0:1.0:3.0
Fulmarus glacialis	763	118	116	117	332	1.0:1.0:2.8
Thalassoica antarctica	641	98	93	100	310	1.0:0.9:3.2
Pterodroma lessonii	586	108	109	113	306	1.0:1.0:2.8
Puffinus puffinus	426	79	72	86	237	1.0:0.9:3.0
Pterodroma inexpectata	329	80	80	87	253	1.0:1.0:3.2
P. mollis	276	81	82	86	260	1.0:1.0:3.2
Puffinus assimilis	238	62	56	65	192	1.0:0.9:3.1
Pachyptila vittata	196	64	65	67	206	1.0:1.0:3.2
P. desolata	159	57	56	56	193	1.0:1.0:3.4
Pelecanoides urinatrix	124	43	33	44	123	1.0:0.8:2.9
Pelagodroma marina	47	27	24	37	161	1.0:0.9:6.0
Oceanodroma leucorhoa	39	35	35	42	156	1.0:1.0:4.6
Garrodia nereis	32	21	19	34	126	1.0:1.0:6.0
Hydrobates pelagicus	28	26	24	33	117	1.0:0.9:4.5

free, fluttering wing action. At the other end of the size scale, although humeral and radio-ulna segments are very long, the manus, even with the primaries, is but 1.4–2.0 × the length of the inner segments. This is tied in with the gliding flight, in which the main aerofoil is supported by the forearm.

In many petrels, the ulnae and radii are bowed for the placement of the forearm muscles but in albatrosses these bones are long, thin, straight and close together, reflecting the reduced musculature and their owner's reduced manoeuvrability. In some diving forms, such as *Puffinus griseus*, the humeri are flat, whereas in *Pelecanoides* they are rounded, but flattened ulnae and radii help to produce a hydrodynamically efficient wing. The humeri of *Calonectris* are unflattened, as accords with their non-diving feeding mode (Kuroda, 1954).

Some petrels have small tendinous ossifications at the elbow attached to the ectepicondylar process. These were first recorded (in *Puffinus*) in 1829, and further investigated by Reinhardt (1874). He found two of these ossicles in some albatrosses and in *Puffinus, Pterodroma* and *Procellaria*, but only one in some *Pterodroma, Bulweria bulwerii, Diomedea chlororhynchos* and *Phoebetria*. They were missing in *Pachyptila, Halobaena, Lugensa*, fulmars (including *Macronectes* and *Pagodroma*), *Pelecanoides* and storm petrels, although Kuroda (1954, p. 163) found one in *Oceanodroma castro*.

Mathews (1935, 1936) and Brooks (1937) concluded that, small as they are, these sesamoids are significant in strengthening the elbow joint when the wing is extended and act as 'spreaders' to support the patagial fan (Fig. 10.4). The bonelet reaches from the forward projecting process at the distal end of the humerus to *M. patagialis longus*, giving a strut-like support.

Yudin (1957) described how the elbow of the fully extended tubinare wing is locked in position, so saving muscular energy (Fig. 10.5).

C *Vertebral column*

Petrels have 38–42 vertebrae, and variations in the structure of these down the column, which are marked in the ventral spinous processes, were discussed by Pycraft (1899) and by Kuroda (1954, 1983). In the albatrosses and larger petrels, but not in the storm petrels, the vertebrae tend to be pneumatic, particularly the anterior ones.

Kuroda pointed out that the neck muscles are anchored to the cervical hypapophyses and that these are best developed in diving species, being long in *Puffinus griseus, P. tenuirostris* and *P. puffinus* but short in *Calonectris* and absent in the albatrosses. He associated their better development with the need for mobility of the neck when under water, but noted (1954, p. 11) that they are also well developed in *Fulmarus* and storm petrels which seldom dive. The hypapophyses of *Procellaria cinerea*, a diving shearwater, are also longer than those of *P. aequinoctialis*, extending to the 6th thoracic vertebrae in *P. cinerea* but only to the 5th in *P. aequinoctialis*.

D *Rib cage*

The ribs, together with the sternum, form a protective box enclosing the viscera; this box is rather square in storm petrels, albatrosses and fulmars but in the more aquatic

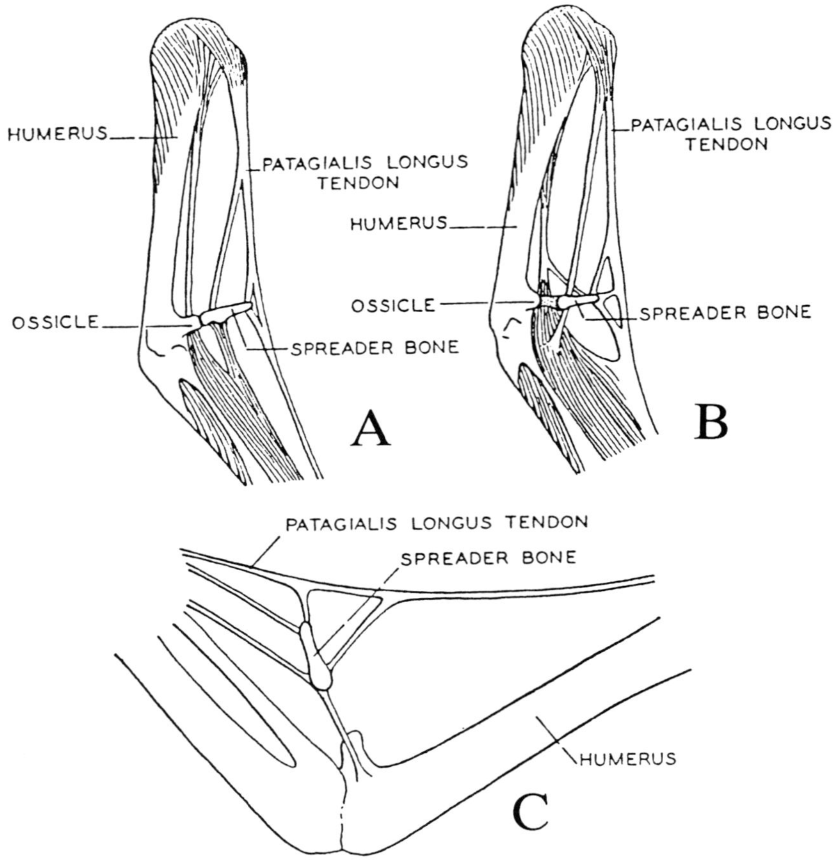

Figure 10.4 *Sesamoid ossicles in (A) the right wing of* Puffinus bulleri; *(B) the right wing of* Pterodroma inexpectata; *and (C) the right wing of* Diomedea nigripes. *From Brooks (1937).*

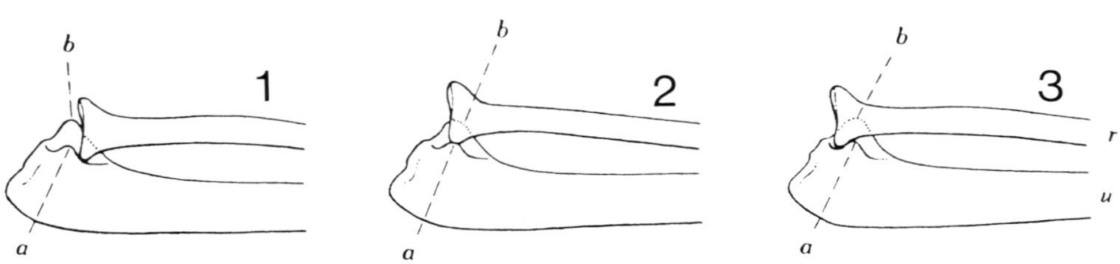

Figure 10.5 *Scheme of movement of the radius along the ulna as the elbow of* Puffinus griseus *is progressively extended (1–3): r, radius; u, ulna; a, articular surface forming a seat for the head of the radius; b, processus lateralis ulnae. From Yudin (1957).*

forms the ribs are long, oblique in position and extend backwards further. This is true of *Procellaria cinerea* and *Puffinus griseus* more than of *Procellaria aequinoctialis*, an arrangement Forbes (1882, p. 46) believed helped resist water pressure when submerged. However, Kuroda (1967a, p. 123) noted that despite its diving habit,

Pelecanoides has a rather storm petrel-type rib cage with only five ribs attached to the sternum, whose rib bearing edge is very short.

E *Sternum and pectoral girdle*

The shapes of the sterna vary considerably between species, most having the posterior border excavated by one or two holes on either side, sometimes shallow, sometimes deep, while, in diving and storm petrels, the border is entire (*The Petrels*, pp. 176, 196). The state in the latter is like that in passerines and may reflect the considerable musculature needing accommodation for their energetic flight styles. Conversely, the sterna of albatrosses are relatively short, gliding flight needing reduced pectoral muscles and reduced body mass. In gliding shearwaters the sterna and girdles approach the albatross type, whereas storm and diving petrels have long sterna and short coracoids and clavicles, the latter curved forwards to increase attachment for the large pectoral muscles.

F *Pelvic girdle*

Mayaud (1932a, 1934) and Kuroda (1954, 1955b, 1983) described the pelvis. It tends to be broad in albatrosses and fulmars that need a wide support for walking, and narrow with long ischia for attachment of the muscles driving the feet in diving shearwaters. The diving petrel pelvis is not narrow but more resembles that of '*Cookilaria*' and *Pagodroma*, an unspecialized condition according to Kuroda (1967a, p. 131).

G *Pelvic limb*

The leg bones of tubenoses are non-pneumatic presumably for added strength—to allow albatrosses to alight heavily or nocturnal species tumbling through trees to hit the ground with a thud unharmed. The most marked feature is the enlargement on the tibio-tarsus of the *crista cnemialis cranialis* (processus rotularis). This is relatively small in albatrosses, fulmars and storm petrels but large in diving shearwaters (*The Petrels*, p. 84), rising above the articulating surface for the femur, less expanded in *Puffinus carneipes* and reduced again in *Calonectris*. There is a tiny patella embedded in tendon behind the root of the crista, almost vestigial in albatrosses and storm petrels. The tarsometatarsus is variably flattened laterally in the aquatic shearwaters and diving petrels, less so in *Calonectris*: albatrosses, fulmars, gadfly and storm petrels have rounded legs.

II Musculature

Garrod (1874) used four hind limb muscles and *M. ambiens* as taxonomic indicators. These were the femoro-caudal (A) and accessory femoro-caudal (B)—*Mm. caudo-*

femoralis, and *iliofemoralis* respectively in modern terminology—and the semitendinosus (X) and its accessory (Y); that is, *Mm. flexor cruris lateralis pars pelvica* and *pars accessoria*.

Garrod (1881) found *M. ambiens* (Am) missing in *Pelecanoides* and *G. nereis* although present in *Diomedea, Pterodroma, Daption, Pagodroma, Macronectes, Bulweria, Puffinus* and *Pachyptila*.

He separated the tubinares into the 'true' petrels (ABX+Am), and the storm petrels (ABXY, Am variable). He found (X) in all petrels examined, but placed *Pelecanoides* with the 'true' petrels as it lacked (Y). It was exceptional too in that as well as (Am), (B) was missing. He also found (B) absent in *Bulweria* and Klemm (1969, p. 194) showed that some *Pterodroma* also lacked this muscle, having the same formula, namely (AX). Klemm extended Garrod's analysis, adding five more muscles to support his own taxonomic analysis.

Forbes (1882) described and figured the principal muscles of the pectoral region, shoulder, elbows and thighs. He drew attention to the two-layered *M. pectoralis* with its different insertions on sternum, clavicle and humerus. *M. supracoracoideus* is also well developed but short, in some species only one-third of the length of the sternum, but in diving petrels reaching almost to its posterior margin, and being nearly as extensive in storm petrels. A third muscle, his 'pectoralis tertius' is a broad thin band parallel with the coracoid and between it and the furcula and attached to the membrane between them. Kuroda (1960c) also recognized three parts to the tubenose pectoralis and believed that the middle layer was important in producing the propelling action of the wing's distal segment. He found that the proportional mass of the pectorals varied from 4.0% in *D. immutabilis* to 9.2% in *Pterodroma solandri*. *M. supracoracoideus* was smallest in *D. immutabilis* and *D. nigripes* (0.27–0.25% of body mass), but between 0.6 and 1.0% in the smaller species he sampled. He noted that the third layer consists of white fibres suited to frequent brief contractions. Later, he argued that its possession of this middle layer, although very thin (0.9 g), by *Pelecanoides*, was evidence for an aerial origin for this genus (Kuroda, 1967a).

In probably all adult petrels *Mm. pectoralis* and *supracoracoideus* are deep red, as Pennycuick (1982) found in *Diomedea* and *Macronectes*. Pennycuick also reported a wing lock at the elbow of these birds formed by the fan-like tendon of the most superficial section of *M. pectoralis*. The tendon originates along the carina, converging to link up with the rest of the pectoralis's deltoid insertion. Pennycuick found this lock absent in *Procellaria aequinoctialis* and *Pachyptila desolata* where the deep pectoralis amounted to 9–14% of the total pectoral muscles as against 23 and 31% in *D. melanophrys* and *Phoebetria palpebrata* respectively. He suggested that, while this may be a tonic muscle in smaller petrels, it may have another function in larger ones with wing locks. However, at least two other wing locks have been described; by Hector (1895) and by Yudin (1957; see Section I.B) and the whole question of these devices in tubenoses requires further attention.

Gadow (1891) included some petrels in his study of avian myology, Mitchell (1913) examined the peroneal muscles of Forbes's specimens and Hofer (1950) the jaw muscles of *Pachyptila vittata*. More recently Yudin (1961) detailed the jaw musculature of tubenoses, and showed that the lower jaws in tubenoses and other fish-eating birds, with their ankylosed mandibular symphyses, have two pairs of flexion zones within the rami which can be bent outwards in the caudal zones and inward in the rostral ones. Such mandibular spreading enables the swallowing of bulky food. Dzerzhinsky and Yudin (1979) also included *F. glacialis* and *P. vittata* in their comparative study of the jaw muscles of the *Sphenodon* and birds.

The complicated cervical musculature was dissected out by Kuroda (1962) using *Pagodroma nivea*, and Popova (1972), in a comparative study of the neck in diving birds as a 'beam construction', included those of *D. exulans, D. melanophrys, D. chrysostoma, D. chlororhynchos, Phoebetria palpebrata, M. giganteus, Procellaria aequinoctialis* and *Pelecanoides urinatrix*.

Klemm (1969) compared the anatomy and functioning of the hind limb of *D. immutabilis* to a wide range of species looking especially at intraspecific variation in muscle size and arrangements (see Fig. 10.6). He tabulated data for 33 tubenoses noting much variation from bird to bird and from species to species. He used interspecific differences to construct a classification based on the limb musculature. However, similar muscle arrangements were not always evidence for close relationships, as Fig. 10.6 demonstrates. McKitrick (1991a) provided detailed descriptions of the forelimb muscles of *Pelecanoides garnotii*, finding several small ones hitherto unrecorded and also (McKitrick, 1991b) scored the presence or absence of 66 hind limb muscles in a range of species (see Chapter 11.IV.A & Fig. 11.3).

After analysing the muscles of *Puffinus griseus* and *P. tenuirostris*, Yoshida and Ogi (1987) concluded that the former was adapted more to efficient underwater swimming, *P. tenuirostris* more for long distance migration (Fig. 10.7).

III Digestive system

The main characteristics were set out in *The Petrels*, p. 6 and in Chapter 7.VI. Many of these were noted by the early anatomists, such as Lesson and Garnot (1828) and Audubon (1835–39), the latter figuring the alimentary tracts of *Oceanites oceanicus* and *Hydrobates pelagicus*. Forbes (1882, pl. 2) showed the liver, stomach and duodenal loop of *Procellaria aequinoctialis* and the internal structure of the gizzard of *F. glacialis*. Matthews (1949) figured the alimentary tracts of *F. glacialis, Puffinus puffinus* and *H. pelagicus*, McLelland (1979) that of *Procellaria parkinsoni*, Roby *et al.* (1989) that of *M. giganteus*. Jackson and Place (1990) described and measured those of *Procellaria aequinoctialis* and *Phoebetria fusca*.

The accessory organs—pancreas, spleen, gall bladder and bile ducts—seem not to show any special features. Broughton (1994) found that in *D. immutabilis* and *D. nigripes* the size of bursa fabricii correlated with that of the gonads and the bird's age.

A *Mouth*

A very variable feature of the buccal cavity is the tongue (Fig. 10.8). The form of the palate above it tends to reciprocate that of the tongue and invests it when the mouth is closed. For example, the palate of *Pelecanoides* is smooth, lacking ridges and spines (except around the interior nares which are ringed by sharp spines) as is the tongue (Fig. 10.8.19). Audubon (1835–39, pp. 313, 645) described the tongues and palates of *H. pelagicus* and *O. oceanicus* and the variability in lingual anatomy was used by

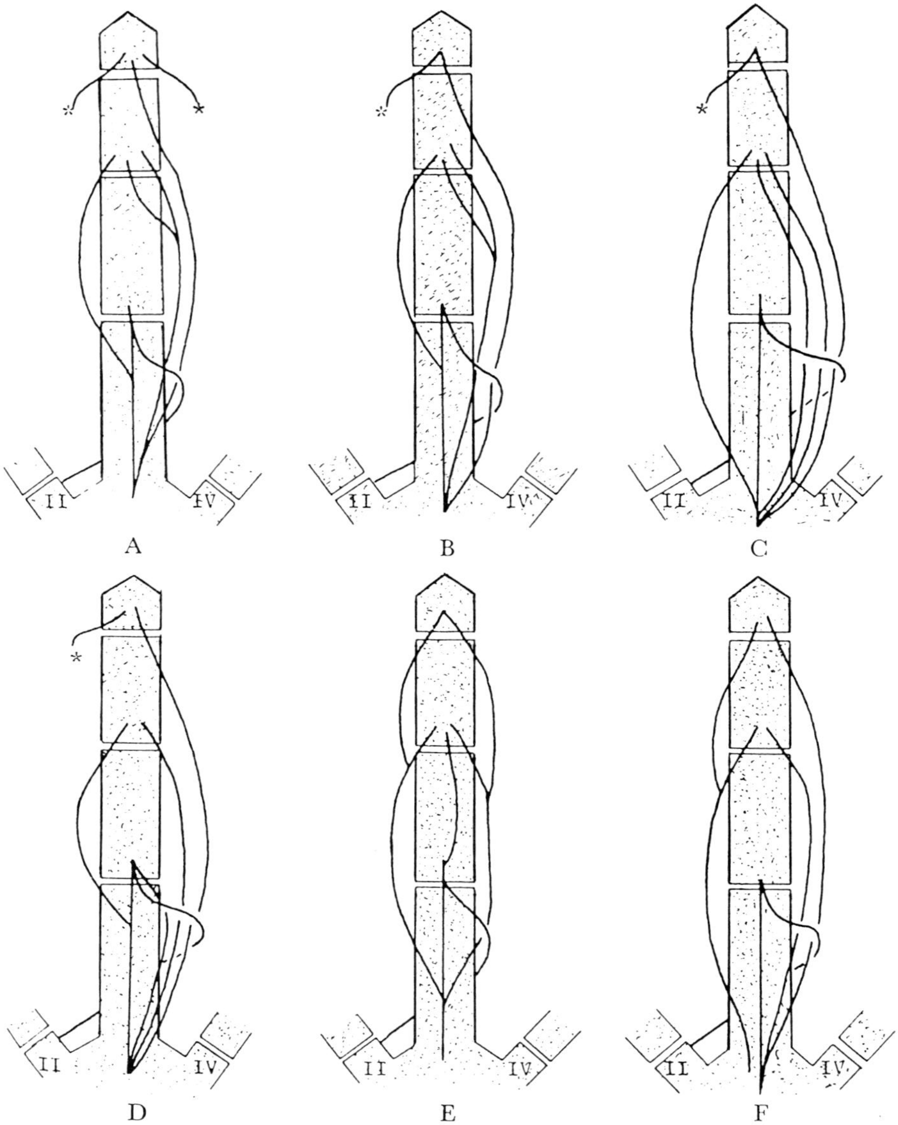

Figure 10.6 *Variations in the insertion of* M. extensor digitorum longus *on digit III. Diagrammatic, right foot. (A)* Procellaria cinerea *and* P. aequinoctialis. *(B)* Puffinus carneipes, Bulweria *and* Pelecanoides urinatrix. *(C)* Puffinus nativitatis *and* P. gavia. *(D)* P. pacificus. *(E)* P. assimilis *and* Fulmarus glacialis. *(F)* Pachyptila vittata *and* P. desolata. *From Klemm (1969).*

Jacquinot and Pucheran (1853) to define their tubinare taxa. They pointed out that in *Diomedea* the tongue is short, as also in *Puffinus*, whose tongue they dubbed '*en fer de lance*' from its many retrose papillae, whereas that of hydrobatids was relatively large: in fulmars thick and as long as the bill.

The tongues of all species studied are naked at their tips but, except for those of

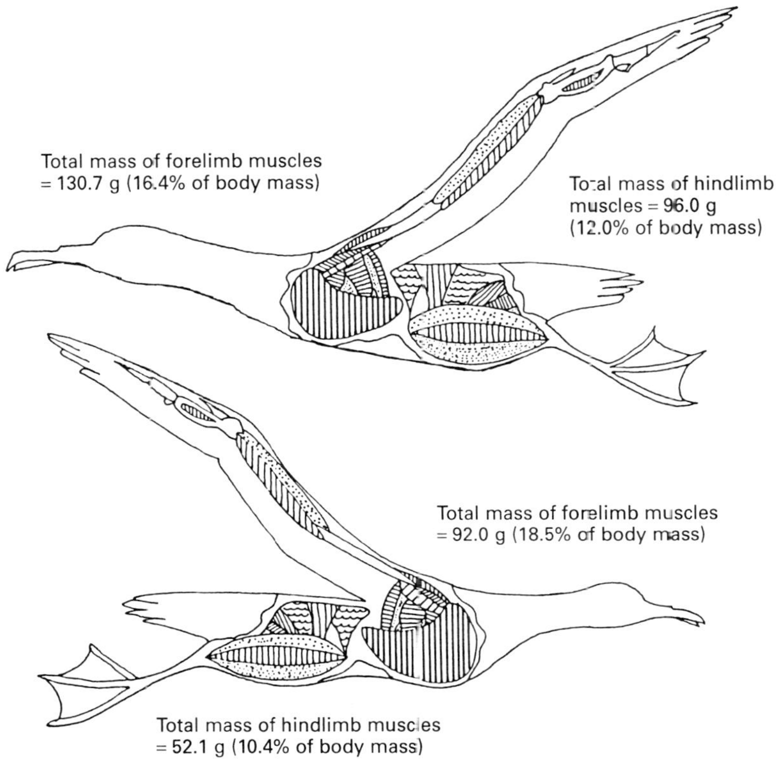

Figure 10.7 *Wing and leg muscles in* Puffinus griseus *(upper) and* P. tenuirostris. *From Yoshida and Ogi (1987).*

storm and diving petrels and the fulmars, which are free of papillae except at their bases, the others are mostly invested with retrose papillae along their margins and in some *Puffinus* over the surface generally. These stiff, spine-like processes are clearly used to help hold slippery prey and shift them towards the oesophagus. The extent of the attachment of the tongue to the floor of the mouth varies: in *Pelecanoides* and *Pachyptila* it is free only towards the tip.

As shown in Fig. 10.8.1–5, while the tongues of some *Diomedea* spp. are broad, heart-shaped and spiny, those of mollymawks are longer with few or no papillae, and that of *Phoebetria* is long and pointed, the surface glabrous with spiny processes only at the base.

The tongue of *Macronectes* (Fig. 10.8.12) is long, narrow and deeply grooved, its base fringed with papillae which are continued along its edges for some distance. The tongues of other fulmars (Fig. 10.8.6–9), particularly *Fulmarus* and *Daption*, are also grooved dorsally and can perhaps be folded at their apices and even appear bifid. Tongues of prions are fleshy, the spines small and mostly at the base. That of *Pachyptila vittata* fills the whole space between the mandibular rami.

In *Pterodroma, Bulweria* and notably in *Puffinus*, tongues are rather fleshy, slightly grooved and more or less fringed with triangular backwards-directed papillae.

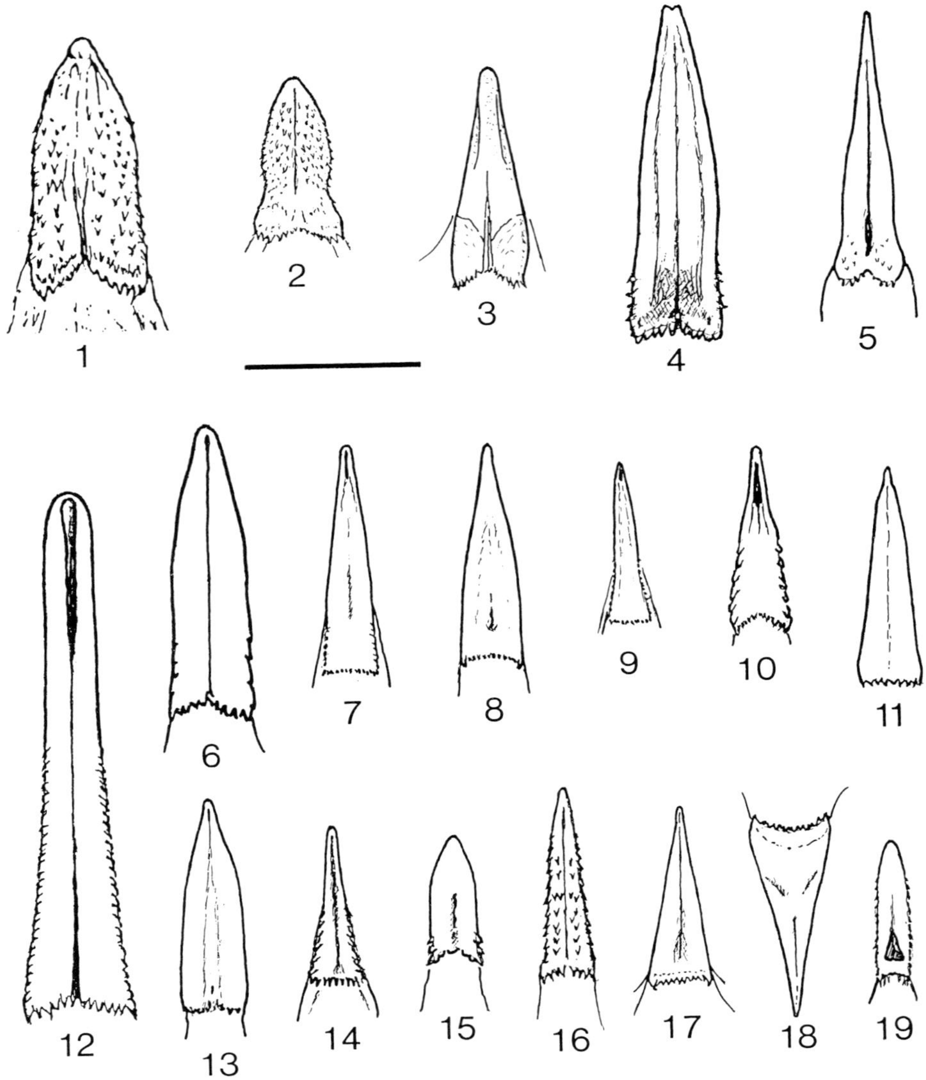

Figure 10.8 *Tongues of: 1,* Diomedea immutabilis; *2,* D. albatrus; *3,* D. cauta; *4.* D. chlororhynchos; *5,* Phoebetria *sp.; 6,* Fulmarus glacialis; *7,* Thalassoica antarctica; *8,* Daption capense; *9,* Pagodroma nivea; *10,* Pterodroma lessonii; *11,* Halobaena caerulea; *12,* Macronectes *sp.; 13,* Pachyptila desolata; *14,* Procellaria aequinoctialis; *15,* Calonectris leucomelas; *16,* Puffinus tenuirostris; *17,* Garrodia nereis; *18,* Fregetta tropica; *19,* Pelecanoides urinatrix. *Bar is 2 cm but 17 and 18 are twice the scale of the rest. Partly from Forbes (1882).*

These reach their peak in *Puffinus tenuirostris* with papillae on the dorsal surface more or less arranged in rows, while in *P. griseus* and *P. carneipes* these are mainly along the margins and *P. assimilis* and *P. huttoni* have about nine along each edge. Those of *Calonectris* are quite distinct (Kuroda, 1954, p. 174; Fig. 10.8.15).

Tongues of storm petrels, including *Oceanodroma* and *Halocyptena*, are triangular, tapering and thinning anteriorly with pointed and rather membranous tips and with only a basal fringe of retroverted papillae (Fig. 10.8.17,18).

Although these considerable variations have been known for 100 years, no studies of the relationship of forms to functions have been made apart from that of Morgan and Ritz (1982) (*The Petrels*, p. 165). The variability suggests equally varied functions in collecting, manipulating and swallowing food, and possibly their naked tips bear organs sensitive to taste and touch.

The entrance to the large, distensible oesophagus, that passes without restriction into the proventriculus, is marked by a ring of papillae often continuous with the quite variably shaped laryngeal pad.

B *Digestive tract*

Forbes (1882) figured tracts of *Procellaria aequinoctialis* and *Fregetta grallaria* and Cazin (1886) described that of *Macronectes* and the types of proventricular glands. Forbes found a zonary concentration of glands in the narrower ascending part of the proventriculus and Pernkopf (1937) described their zonation in *F. parkinsoni*. According to Kuroda (1960b) the proventriculus of *Diomedea* is lined with a thin layer of small, dense, uniformly distributed and unspecialized glands.

The gizzard is small, globular with muscular walls and with a dark, corrugated lining often of a corneous nature but softer in albatrosses and storm petrels.

The small intestine passes upwards as the duodenum whose loop encloses the pancreas in the normal way. The large intestine is much shorter than the small one, for example 3% of the whole in *Diomedea exulans*, 1% in *Daption capense*, 2% in *Macronectes* but 19% in *Pelecanoides urinatrix* (Forbes, 1882, p. 23).

In Diomedeidae and Procellariidae there are two small caeca fairly close to the rectum. Kuroda (1954, p. 174) figures those of *Calonectris* and *Puffinus*. The caeca of *Puffinus griseus* differ in pointing caudally. Among storm petrels, *Halocyptena* and *Oceanites* lack a caecum, while *Oceanodroma leucorhoa*, *O. monorhis* and *O. matsudairae* have but one.

As with other birds, the intestines lie looped into a compact mass whose arrangements in birds generally were teased out by Gadow (1890). He put the tubenoses in his cyclocoelous (mesogyrous) category, but Kuroda (1986) found several arrangements present, even within a genus. Mitchell (1901) noted that the chief difference between the intestines of *Diomedea exulans* and *Fulmarus glacialis* was that the three minor loops in those of the fulmar were represented in the albatross by two long loops with shorter ones between them. Mitchell found a *diverticulum vitellinum* in the long loop of the upper intestine in *D. exulans*—this must have been a young bird.

Imber (1985a) examined the gross anatomy of the intestinal tract in a wide range of species from all families. In all albatrosses, fulmars, *Procellaria* spp., shearwaters, prions, storm and diving petrels the intestines were basically simple tubes. In many gadfly petrels, however, the upper portions were twisted, a condition first noted by J. R. Forster during Cook's voyages, probably in *Pterodroma lessonii* and *P. inexpectata* (Olson, 1990). Imber used the presence or absence of this helicoidal condition in reclassifying these birds (*The Petrels*, pp. 85, 88). He categorized a range of species on the number of spirals and the proportion of these that were anticlockwise and whether continuously or only partly twisted. Smaller species (<100 g) had 40–65 spirals, those >400 g, 70–106 spirals. A hatchling *P. l. pycrofti* already had a twisted upper intestine.

Kuroda (1986) examined the guts of *P. hypoleuca*, *P. solandri* and *P. externa*. Four of five *P. hypoleuca* were strongly twisted. In the fifth, a storm-killed bird, the untwisted, thin-walled gut was evidence of the bird's exhaustion. Both *P. externa*, of each subspecies, had untwisted alimentary tracts although Imber separated them as full species, *P. cervicalis* having simple, *P. externa* having helicoidal intestines.

Kuroda's histological examinations revealed that gut twisting in *P. hypoleuca* was associated with depressions on opposite sides of the intestinal wall. Between these inner and outer depressions ran transverse circular muscles which evidently caused spiralling when contracted. He envisaged that their tightening and slackening creates and relaxes the coiling, in the latter event allowing the food to continue flowing onwards. Kuroda hypothesized that helecoidal intestines are a feature of warm-water species where the density of food tends to be low, so that if the food is retained as long as possible the maximum energy can be absorbed.

Kuroda used alimentary tract lengths divided by wing lengths as relative indexes of tract development. He found that the index increased with increasing body size but suggested that within a group, for example of shearwaters, the more northern species, including southern transequatorial migrants, have longer intestines, the resident tropical ones, shorter ones. For example, *Puffinus griseus* and *P. tenuirostris*

Table 10.2 Intestine lengths *vs.* body mass of petrels. Mainly from Forbes (1882) and Kuroda (1986)

Species	Body mass (g)	Intestine length (cm)	Species	Body mass (g)	Intestine length (cm)
Diomedea exulans	8677	319	*P. carneipes*	649	83
D. nigripes	3212	208	*P. griseus*	787	90
D. immutabilis	3092	205	*P. tenuirostris*	559	64
Phoebetria fusca	2500	152	*P. nativitatis*	324	51
Macronectes giganteus	4467	240	*P. puffinus*	420	61
Fulmarus glacialoides	802	126	*P. lherminieri*	168	44
F. glacialis	813	140	*Oceanites oceanicus*	38	30
Thalassoica antarctica	707	131	*Garrodia nereis*	32	22
Daption capense	452	86	*Pelagodroma marina*	47	31
Pterodroma solandri	439	89	*Fregetta tropica*	52	20
P. inexpectata	329	74	*F. grallaria*	53	33
P. hypoleuca	182	43	*Hydrobates pelagicus*	28	23
Pachyptila desolata	159	44	*Oceanodroma castro*	44	23
Bulweria bulwerii	92	29	*O. leucorhoa*	48	27
Procellaria aequinoctialis	1394	143	*O. tristrami*	89	31
Calonectris leucomelas	545	77	*O. matsudairae*	62	32
Puffinus pacificus	390	50	*O. furcata*	59	38
P. bulleri	437	55	*Pelecanoides urinatrix*	124	42

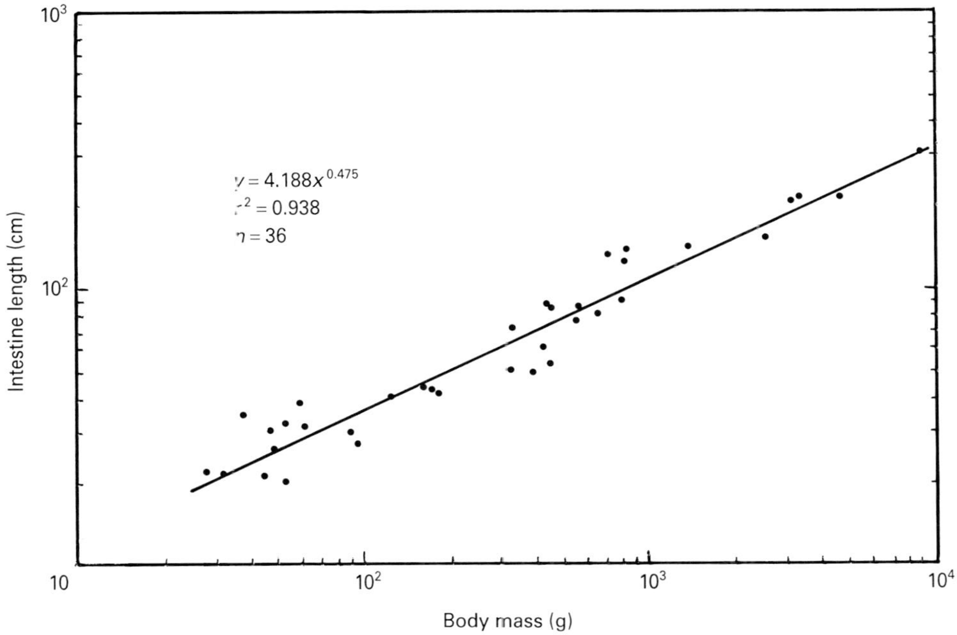

Figure 10.9 *Relationship of intestine length to body mass in 36 petrels. Data mainly from Table 10.2.*

had relatively the longest intestines, next was *P. carneipes* which does not migrate so far north, while in *Calonectris leucomelas*, *P. pacificus* and *P. nativitatis* as well as *P. bulleri* (which keeps to warmer seas) the intestines were relatively short.

Intestine lengths also scale allometrically with body mass (Fig. 10.9), a somewhat surprising relationship between a single- and a three-dimensional variable, and suggests that length is a reasonable measure of intestine volume. Jackson (1992) found that intestine length scaled significantly with body mass$^{0.37}$ for nine flying seabirds including three tubenoses.

The right hepatic lobe is usually the larger in birds and petrels but *P. puffinus*, with the right one 6 × larger, is exceptional (Gadow, 1891) and with little asymmetry in other shearwaters studied. In albatrosses such as *D. immutabilis*, the right lobe is about twice that of the left; in *O. leucorhoa* the lobes are of similar size.

Liver weights of *Pterodroma hypoleuca* returning to their nesting grounds were found by Fisher (1961) to follow a daily cycle. They fluctuated by 37% over 24 h in males and 47% in females, the major increase being between 1600 and 2000 h, decreasing most rapidly between 0400 and 0800 h. Liver weight (less gall bladder) varied from 2.3 to 5.2% of body mass in males, from 2.0 to 4.2% in females, and these differences were suggested to relate to different diets and different times of feeding.

IV Excretory system

This is little studied in tubenoses apart from the salt gland system (see Chapter 7.I.A). The kidneys are flattened and deeply recessed in depressions on the ventral

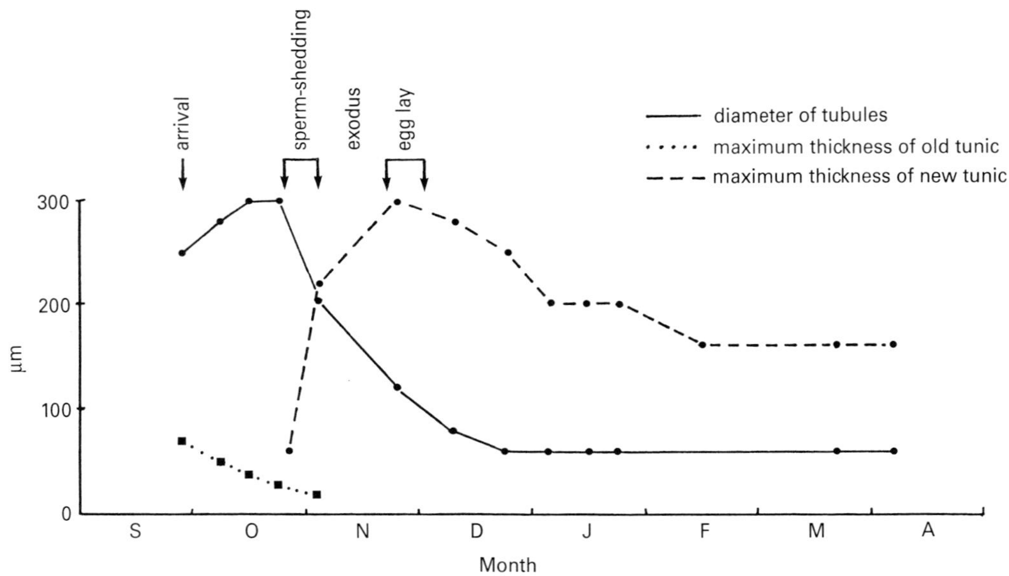

Figure 10.10 *Seasonal variations (µm) of testis tunic in* Puffinus tenuirostris. *From Marshall and Serventy (1957).*

surface of the synsacrum and traversed by nerves and blood vessels. They are usually clearly separated into three divisions, for example in shearwaters, but Johnson (1968) found that while this was true for *P. hypoleuca* and *O. leucorhoa*, in *D. immutabilis*, *D. nigripes* and *F. glacialis* the middle division was ill defined, apparently fused to the others, particularly to the caudal one, as in many passerines. Kuroda (1963) figured the arrangements in *C. leucomelas* and *Pagodroma nivea*.

Nicholson and Kendall (1983) found that the ultrastructure of the dark or intercalated cells from the distal and collecting tubules of *Puffinus puffinus* and *F. glacialis* was similar to that of other birds. They suggested that the cells of the distal avian nephron are involved with proton and potassium pumping and control of urinary buffer concentrations.

V Reproductive system

The procellariiform system is typical of other birds, including the use of cloacal sperm-host or storage glands (*fossulae spermaticae*) in the females. The most unusual feature is the rapid regression of the gonads following copulation and ovulation.

The microanatomical changes in the gonads have not been examined in recent years. Wynne-Edwards (1939) described changes in the ovary of *F. glacialis* and figured atretic and ovulated follicles. Marshall (1949) looked at the testis of this species, and especially the role of the lipoid Leydig cells in the interstitium and their seasonal changes. He found heavily lipoidal and new interstitial cells with spermatogonia in the regenerated tubules as the petrels shift to sea from the colonies in the autumn, testis regeneration taking place while the days were shortening. Similarly,

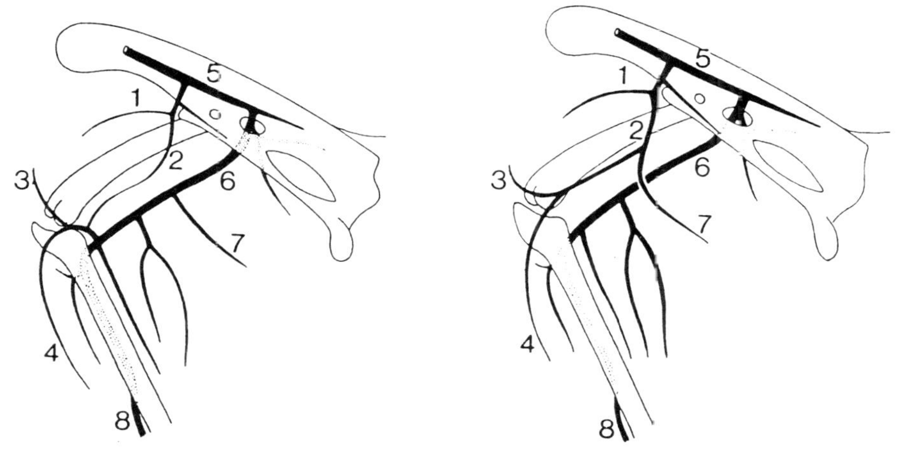

Figure 10.11 *Semischematic drawings of the arterial patterns in the hind limb as seen in medial view. Left of* Puffinus *(also Anseriformes, Galliformes etc.); right* Fulmarus. *1,* A. femoralis; *2,* A. femoralis med.; *3,* A. genic. med.; *4,* A. cruralis med.; *5, aorta, 6,* A. ischiadica; *7,* A. fem. caudalis; *8,* A.tibialis cran. *From Midtgard (1988).*

in *P. tenuirostris* the spring sexual cycle is initiated during decreasing day lengths while the birds are north of the Equator and nearly 2 months before the lay, so that when they reach their nesting places gametogenesis is well advanced. The ovaries, however, are then much less advanced (Marshall & Serventy, 1956b).

The male gonad cycle is notable mainly for the speed with which testis metamorphosis follows shedding of spermatozoa. In *P. tenuirostris*, the postnuptial rehabilitation of the testis tunic can occur while the seminiferous tubules are still unchanged and the Leydig cells inexhausted (Marshall & Serventy, 1956b, 1957; Fig. 10.10).

In the ovary the changes are less dramatic. On arrival in late September the largest *P. tenuirostris* oocyte averaged 3.4 mm across, by 2 November before the exodus it was 7.1 mm but on return by 25 November the largest was only 3.0 mm in diameter. Immature birds at the colonies in mid-January had testes with primary spermatocytes but no spermatozoa or lipoidal metamorphosis and no new tunic had been formed. Tickell's (1962) study shows a rather similar cycle of changes in *Pachyptila desolata*. In *Puffinus tenuirostris* the cycle appears to be under the control of an internal rhythm (*The Petrels*, p. 225).

VI Circulatory system

This is little studied in tubenoses. The four-chambered heart weighs 9.5 g in *D. immutabilis* (0.31% of body mass (BM)), 3.32 g in *Puffinus huttoni* (0.79% BM), and 2.1 g in *Pterodroma hypoleuca* (1.2% of BM).

The general arrangement of the vascular system shows only minor variations from normal avian patterns but few species have been examined critically. Garrod (1873/74) found both carotid arteries present in all he inspected (*H. pelagicus, G. nereis, B. bulwerii, Pterodroma lessonii* and *Pachyptila vittata*, as did Glenny (1955) for a

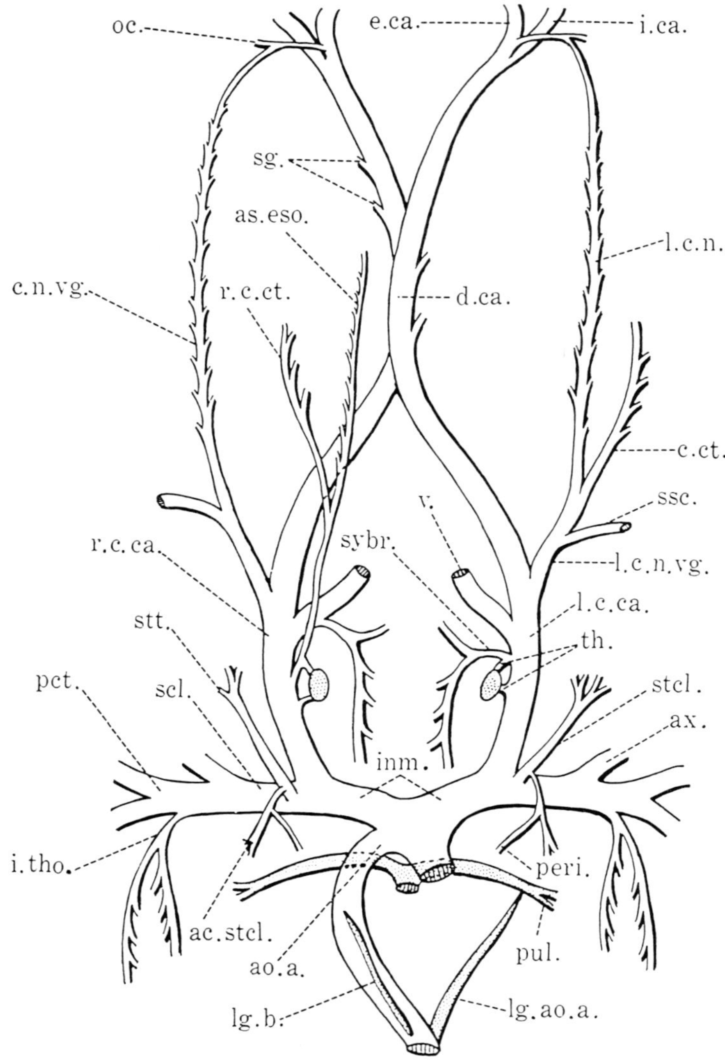

Figure 10.12 *Main arteries in neck and thorax of* Puffinus gravis *in ventral view: ac.stcl., accessory sternoclavicular; ao.a., aortic arch; as.eso., ascending oesophageal; ax., axillary; c.ct., cervical cutaneous; c.n.vg. comes nervi vagi; d.ca., dorsal carotid; e.ca., external carotid; i.ca., internal carotid; i.tho., internal thoracic; inm., innominate; l.c.ca, left common carotid; l.c.n.vg., left comes nervi; oc., occipital; pct., pectoral; peri., pericardial; pul., pulmonary; r.c.ca., right common carotid; r.c.ct., right cervical cutaneous; scl., subclavian; sg., segmental; ssc., subscapular; stcl., sternoclavicular; stt., sterno-tracheal; sybr., syringeobronchial; th., thyroid; v., vertebral. From Gobeil (1969).*

range of species from all four families. Garrod (1881) also found that in *Pelecanoides*, alone among petrels, instead of running deep of the femoro-caudal muscle, the femoral vein was superficial to it (see Forbes, 1882, pl. V4). All species also appear to have two jugular veins, the right generally the larger.

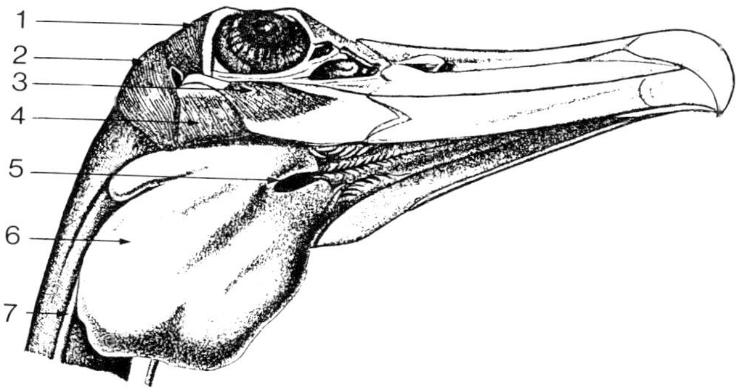

Figure 10.13 Saccus cervicocephalis *of* Diomedea exulans. *1*, M. add. mand. ext. temp.; *2*, M. add. mand. ext. vent.; *3*, M. dep. mand.; *4*, M. pterygoideus?; *5*, *choana*; *6*, saccus cervicocephalis; *7*, V. jugularis. *From Ulrich (1904).*

Vessels of the *rete mirabile ophthalicum* in several species and the rete of the leg of *M. giganteus* have been described (see Chapter 7.II.B). Saiff (1974) found an ophthalmic rete in all 10 tubenoses he examined except for *Pelecanoides garnotii*. He also described the circulation of the arteries of the middle ear and the entry of the circulation into the head. Arteries of the hind limbs of *Puffinus* and *Fulmarus* were detailed by Midtgard (1988) (Fig. 10.11), arrangements differing in the two species. How extensive such differences are among tubenoses is unrecorded.

Particulars of the arteries in the neck and thorax of *Puffinus gravis* were presented by Gobeil (1969) (Fig. 10.12), arrangements that appear similar in *Puffinus* and *Procellaria aequinoctialis*. Vestiges of the *ligamentum aortae* still persisted in these and the species examined by Glenny (1955).

VII Pulmonary system

In contrast to that of penguins, the petrel lung has not been examined, but some details are available of the air sacs that take up a considerable space in the body and help cushion the viscera from shocks, Hamlet and Fisher (1967) finding only minor inter- and intraspecific variations in *D. immutabilis*, *Puffinus pacificus* and *P. nativitatis*.

Earlier, Forbes (1882, p. 32) had noted that in the Procellariidae the two interclavicular air sacs are separated for most of their length by a double-walled septum which is imperfect behind, giving free communication between them, but in the Hydrobatidae and Pelecanoididae they are completely fused together. Cazin (1886) reported the great development of the sacs in *Macronectes*, Bignon (1889) described those in the head and neck of '*D. fuliginosa*' (*Phoebetria fusca?*), while Ulrich (1904) gave a detailed account with life-size figures, of the air sacs of '*D. fuliginosa*' and *D. exulans*. He also described the cervicocephalic air sacs and the pneumatic foramina of sternum, coracoid and humerus (Fig. 10.13).

Hamlet and Fisher (1967) identified the five main air sacs as a single clavicular one with medial and lateral chambers each with three diverticulae, paired cervical sacs with two intervertebral diverticulae, paired anterior thoracic sacs with postcardial diverticulae, paired posterior thoracics with no diverticulae and paired abdominal sacs with three diverticulae. These authors also described the extensive and complex ramifications of the diverticulae and their relationship to lungs and pneumatic bones, especially in *D. immutabilis*.

The syringeal anatomy was shown in Fig. 4.10 and the structure of the nasal apparatus is dealt with in Section V.III.A below.

VIII Nervous system

There is only incidental information on the central and peripheral nervous systems. For instance, Murrish and Guard (1977) found that vasodilation in the legs of *M. giganteus* could be stopped by blocking the tibial nerve and beta-adrenergic receptors with procaine and they believed that the AV anastomoses in the feet (see Fig. 7.3) are centrally controlled through the sympathetic system.

A *Brain*

As with other birds, petrel brains scale allometrically with body size (Fig. 10.14), ranging from 0.2% of body weight in *D. epomophora* to 0.4% in *Macronectes*, 0.5% in

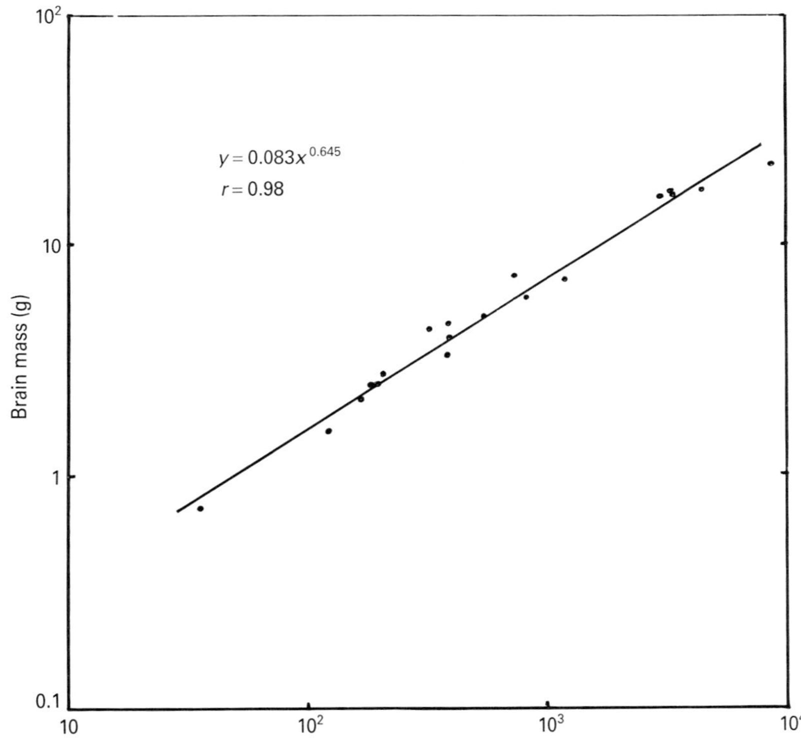

Figure 10.14 *The relationship between wet brain mass and body mass in 19 species of tubenosed birds. Data from Table 10.3.*

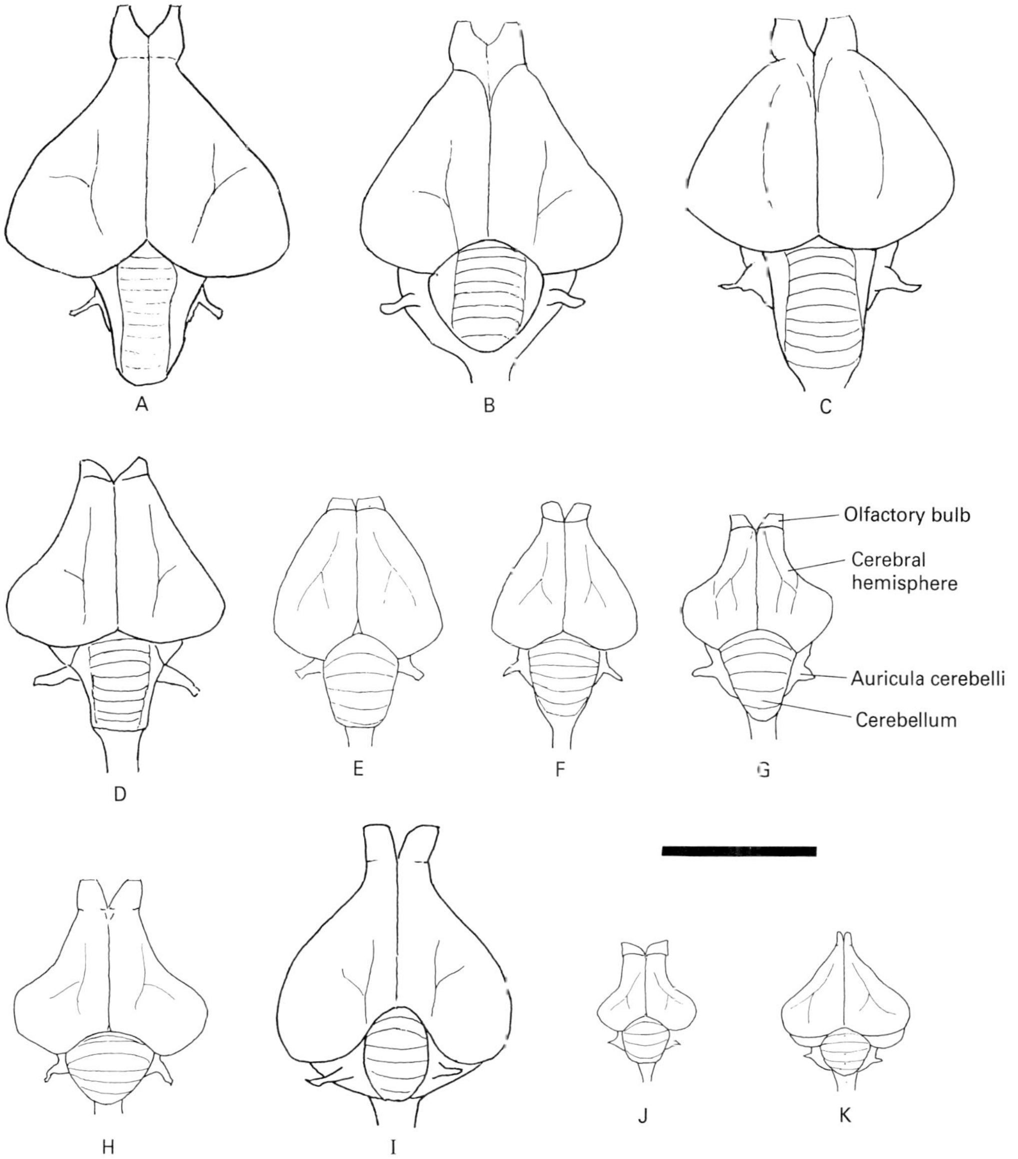

Figure 10.15 *Petrel brains in dorsal view. (A)* Diomedea immutabilis. *(B)* Phoebetria palpebrata. *(C)* Macronectes halli. *(D)* Fulmarus glacialoides. *(E)* Daption capense. *(F)* Halobaena caerulea. *(G)* P. salvini. *(H)* Pterodroma inexpectata. *(I)* Procellaria westlandica. *(J)* Oceanites oceanicus. *(K)* Pelecanoides urinatrix. *Bar is 2 cm.*

D. immutabilis, 0.7% in *Puffinus griseus*, 0.9% in *P. tenuirostris*, 1.2% in *P. inexpectata*, 1.4% in *Pagodroma nivea* and *P. salvini* to 1.8% in *Oceanites oceanicus*. The equation for Fig. 10.14 may be written:

$$\log_{10} \text{brain mass} = 0.645 \log_{10} \text{body mass} - 1.081$$

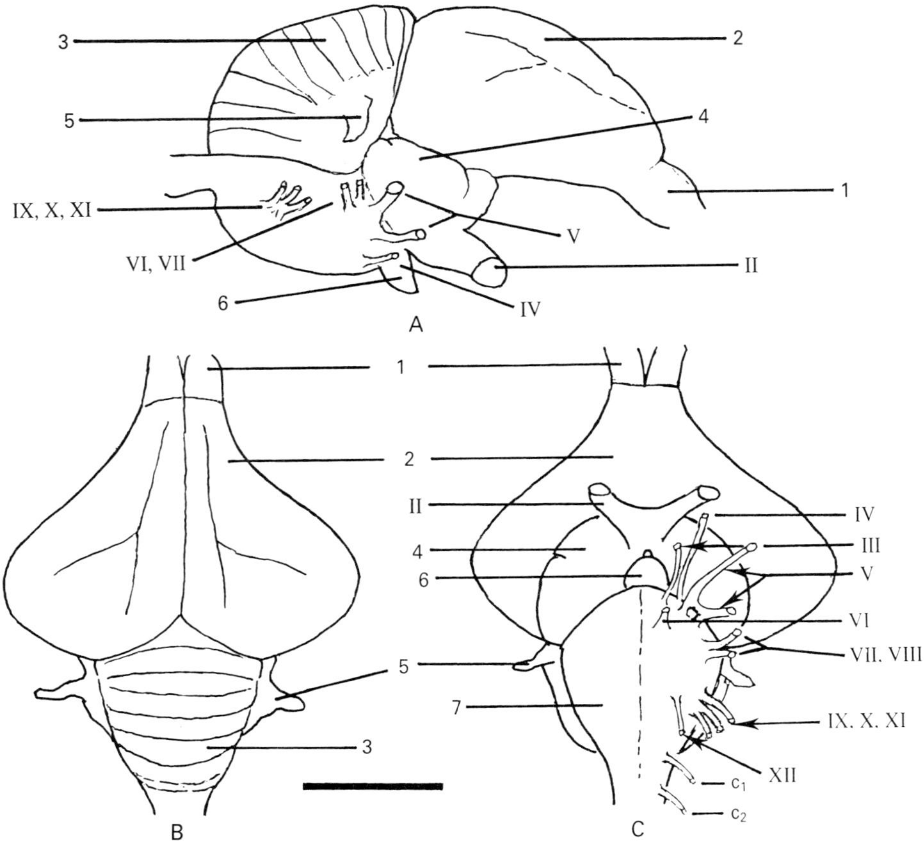

Figure 10.16 *Brain of* Puffinus griseus *in (A) lateral; (B) dorsal; and (C) ventral view. 1, olfactory bulb; 2, cerebrum; 3, cerebellum; 4, tectum mesencephali including optic lobe; 5, auricula cerebelli; 6, hypophysis; 7, pons. II-XII, Cranial nerves;* c_1, c_2, *cervical nerves. Bar = 1 cm.*

which has the same slope as that calculated by Bennet and Harvey (1985) for 30 species of semiprecocial, non-procellariiform, birds. Early embryos also have relatively larger brains than late ones—about 3% of body mass in a 62-day-old *D. bulleri*, down to 2.6% in a 66-day-old one and only 1% in a chick about to hatch.

Figures 10.15 and 10.16 show typical petrel brains. Little work has been done on their anatomy. Brandis (1896), figured the cerebellums of *Diomedea exulans, F. glacialis, Daption capense* and a storm petrel, noting the increasing complexity with increasing body size, and a detailed stereotaxic atlas of the brain of *F. glacialis* has been published by Matochik *et al.* (1991).

Quay (1972) examined the pineal glands of *Diomedea immutabilis, Puffinus pacificus, Pterodroma leucoptera* and *Oceanodroma homochroa*, and found them atrophied in the last three species. The pineal was large and obvious in the albatross as it is in median sagittal sections of brains of *D. bulleri*, but no stalk is visible in gross dissections of *Puffinus tenuirostris, Oceanites oceanicus* and *Pelecanoides* (Warham, pers. obs.). Quay suggested that pineal atrophy was linked to nocturnality in Procellariiformes, but

Table 10.3 Dimensions of petrel brains: a, Bulb diameter (mm); b, hemisphere diameter (mm); c, ratio (%); *n*, number measured; d, mean body mass (g); e, brain volume (ml); f, wet brain mass (g)

	a	b	c	*n*	d	e	f	*n*
Diomedea epomophora	8.5	33.0	26	2	8697	20.8	21.9	2
*D. nigripes**	8.0	28.0	29	2	3212	16.5	16.9	2
D. immutabilis	8.4	30.9	27	4	3258	15.7	16.8	4
Phoebetria palpebrata	8.7	28.5	31	1	2969	15.5	16.1	1
Macronectes halli	8.9	29.6	30	1	4390	16.0	17.2	1
Fulmarus glacialoides	6.8	22.7	30	6	735	6.8	7.2	5
*F. glacialis**	5.7	21.0	27	2	—	—	—	—
*Daption capense**	5.5	20.0	27.5	1	—	—	—	—
Pagodroma nivea	6.8	18.9	36	4	388	4.38	4.43	4
Pterodroma inexpectata	7.4	19.4	38	3	320	4.03	4.32	1
P. hypoleuca	—	—	—	—	181	2.22	2.47	1
Halobaena caerulea	5.8	18.1	32	1	202	2.50	2.73	2
Pachyptila vittata	4.8	16.9	28	1	196	2.27	2.49	1
P. salvini	4.2	16.1	26	2	164	2.05	2.17	3
P. desolata	4.1	14.0	29.5	1	—	—	—	—
P. turtur	4.9	15.0	33	2	—	—	—	—
Procellaria westlandica	6.8	25.5	27	1	1200	6.93	6.98	1
*Puffinus pacificus**	5.5	17.8	30	2	390	3.60	3.97	1
P. griseus	5.9	20.6	29	8	819	5.01	5.85	8
P. tenuirostris	4.9	19.8	25	4	560	4.68	4.93	7
*P. opisthomelas**	5.0	17.0	29	1	—	—	—	—
*Oceanites oceanicus**	3.6	10.8	34	4	35	0.93	0.72	3
*Oceanodroma leucorhoa**	3.3	10.0	33	1	—	—	—	—
*Pelecanoides georgicus**	2.0	11.3	18	1	121	1.32	1.53	1
P. urinatrix	2.7	11.8	23	4	—	—	—	—

*From Bang and Cobb (1968).

D. immutabilis is supposed to be a nocturnal feeder, the rest being nocturnal or crepuscular only when on their breeding grounds.

The main interest has been on the unusual development of the olfactory tissue: in a few species the olfactory bulbs exceed in size relative to the whole brain those of all other birds with the possible exception of the Kiwi *Apteryx australis*. Klinckowstrom (1890) depicted the brain of *F. glacialis* and estimated the mass of the bulbs to the whole brain as 1:22. By contrast, that of a goose was 1:67, that of a buzzard 1:513. Strong (1911) also commented on the large bulbs of this fulmar and Wood Jones (1937) described and illustrated the brains and olfactory apparatus of *D. cauta* and *P. tenuirostris*, noting the probability of their having a fully functional sense of smell.

Cobb (1959/60, 1960) used the ratio between the longest diameter of the bulb and that of the ipsilateral hemisphere as an index of olfactory bulb development, finding that those of *O. oceanicus*, *D. nigripes*, *Puffinus opisthomelas* and *P. gravis* ranged from 29 to 30% of the hemisphere diameters, just below that of the Kiwi at 33%.

Following Cobb's work the complex anatomy of the olfactory systems of a wide range of birds was described by Bang (1960, 1965, 1966, 1971), Bang and Cobb (1968) and Bang and Wenzel (1985).

Bang and Cobb (1968) found that the large development of their olfactory bulbs was characteristic of tubenoses, with those of *Pagodroma nivea* (index 37) even exceeding those of the Kiwi (34). Theirs and later data are given in Table 10.2. The bulbs of *Pelecanoides georgicus*, index 18, are the lowest for any petrel examined, but those of *P. urinatrix* seem more in line with the rest. However, measuring of formalin-fixed brains whose parts differ in shape so that their maximum diameters vary in orientation, and whose wet weights, volumes and shapes vary with conditions of fixation, lack of uniformity in drying, etc., is difficult, and the data from the small samples in Table 10.2 need using with care. More figures for great albatrosses, storm and diving petrels are needed.

Bang (1966) unravelled the complex structure of the nasal organs of eight petrels including the histology of the respiratory and olfactory tissues. The air inhaled through the nasal tubes passes over highly vascularized epithelia into a series of elaborately scrolled chambers or turbinals covered with sheets of mucus-producing and ciliated cells. Here the incoming air is filtered, warmed, humidified and sampled chemically. There are three chambers or conchae: the small anterior one helps to warm the air, the middle concha is lined with respiratory-type tissue, the posterior one with olfactory tissue. Petrels, however, have another concha rising from the internasal septum. This scroll interdigitates with the coils of the olfactory concha to provide an extra sensory surface. Bang worked out that the incoming air is directed in the nasal tubes past the nasal valve to produce a 'nozzle' effect as it impinges on the olfactory epithelium further inside. The salt gland discharges its secretion at the root of the anterior concha which flows forwards and out along the beak groove (Fig. 10.17). This anterior concha in *Pelecanoides* reaches almost to the nasal openings and forms the 'paraseptal process' of Murphy and Harper (1921). Diving petrels are unusual in that the septum is perforated, although it is unclear how this affects air flow to the chambers.

The nasal valve (*valvula nasalis*) cuts access from the anterior to the middle concha. It is a flap formed from a double fold of mucous membrane fastened to the nasal roof at one edge, the rostral edge free. When open, the free edge sticks to the nasal septum but when submerged the pressure activates it to seal the opening to the olfactory concha, thereby protecting it from inrushing water (Fig. 10.18).

The mucous membrane of the caudal, olfactory concha houses glands that secrete continually replaced fluid in which odorous molecules must be dissolved to stimulate receptor cells (Bang & Wenzel, 1985, p. 206). The receptor cell axons are unmyelinated and, bundled together, form the left and right olfactory nerves which pass into the olfactory bulbs. In *Pagodroma nivea* the olfactory concha is very large, no bundles form, and the receptor-cell axons pass directly into the bulbs. The olfactory cavity is also innervated with branches of the trigeminal nerve which run back into the brain stem. Wenzel (1980) suggested that odorous molecules may stimulate these nerve endings directly and comprise a second pathway for olfactory information to higher centres.

The bulb itself is composed of distinct layers of cells (Fig. 10.19) six of them in

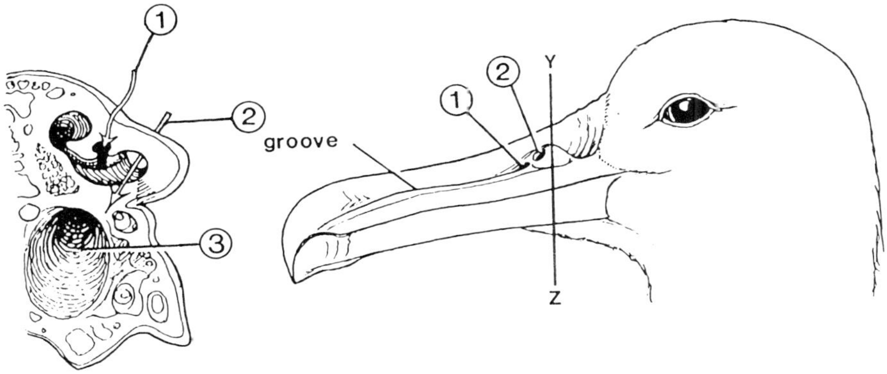

Figure 10.17 *Air and fluid flow in the nostril of* Diomedea nigripes. *Incurrent air is pulled into the short tube (2), excurrent salt gland fluid discharged through the channelled opening down the beak groove to drip from the tips of the ungues, (3) is air space in the upper mandible. Left-hand figure is a transverse section at Y–Z. Scars at base of culmen may be residues of previous moults. From Bang and Wenzel (1985).*

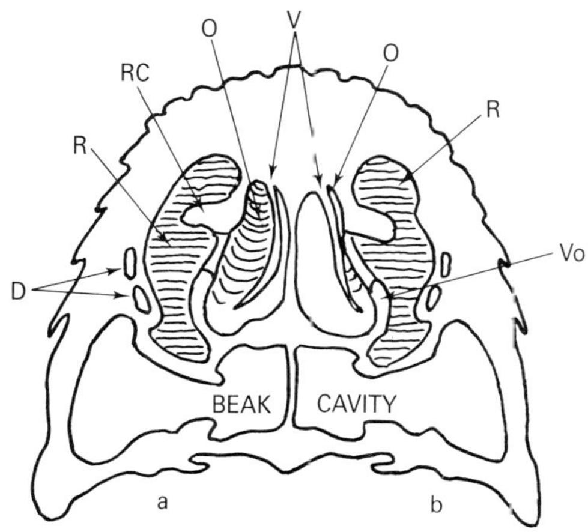

Figure 10.18 *Section through bill of* Pagodroma *at valve level to show (a) valve open for inhaled odour reception; (b) closed to seal olfactory mucosa from water during submergence. Note role of vomer in separating olfactory and respiratory channels and supporting valve action. D, lateral salt-gland ducts; R, respiratory air channel; RC, respiratory concha; O, olfactory air channel; V, valve; VO, vomer. Beak cavities connect with air sacs. From Bang (1965).*

F. glacialis including an outer glomerular layer where incoming olfactory nerve axons synapse with the mitral cells whose axons in turn project to higher centres. Bang (1966) found that the cells in the mucosa of the main concha were larger in larger species. Wenzel and Meisami (1987) estimated that *F. glacialis* has about 120 000 mitral cells compared with 20 000 in the pigeon. In general, the arrangement of the

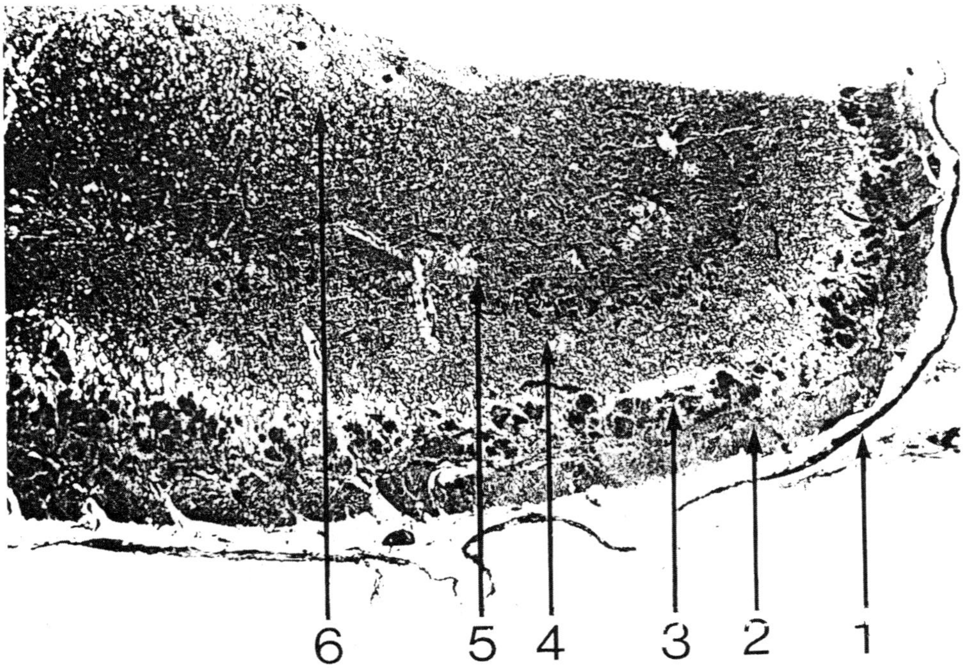

Figure 10.19 *Parasaggital section through olfactory bulb of* Puffinus tenuirostris. *1, Dura mater; 2, olfactory nerve layer; 3, glomerular layer; 4, external plexiform layer; 5, mitral cell layer; 6, granular cell layer.* ×*90.*

tubenose olfactory nervous system appears very like the macrosmatic mammalian one.

Wenzel and her colleagues (e.g. Wenzel & Meisami, 1990), have described the various layers in the bulbs of adult and nestling Northern Fulmars. They estimated a total of 25×10^3 glomeruli per bulb in the external plexiform layer, 20×10^6 cells in the granular layer, and 120×10^3 mitral cells, more than in macrosmatic rat and rabbit. However, the granule cells are loosely scattered, not arranged in highly organized manner as in *Rattus*, and periglomerular and external tufted cells are under-represented, so that the glomerular layer is relatively undifferentiated as in lower vertebrates. The fulmar chick's olfactory mucosa was estimated as covering *c.* 250 mm^2. In comparing cellular patterns of this bird with that of mammals, Wenzel and Meisami speculated that the petrel system should provide high sensitivity in odour detection but less sensitive odour discrimination.

B *Eye*

Until recently, what little was reported on the structure of the petrel eye was of the retina of *Puffinus griseus* (Wood, 1917), *D. cauta* and *Macronectes* (O'Day, 1940), and *P. puffinus* and *F. glacialis* (Lockie, 1952). The fundus of *P. griseus* (Fig. 10.20), has the typical central streak *(area centralis horizontalis)* where cell densities are high, as do *P. puffinus* and *F. glacialis*. However, *D. cauta* has a *fovea centralis*—figured by O'Day. Lockie found that, compared with the diurnal fulmar, *P. puffinus* had larger rod paraboloids and more rods in the central streak. Lockie examined rods and cones of light-adapted retinae of *F. glacialis* and of *P. puffinus* in light- and dark-adapted states. He detected little movement in the cones but the shapes of the rods changed greatly in the light-adapted condition, the paraboloids becoming much smaller and the outer segments of the rods shifting deep into the pigment layer past the oil droplets of adjacent cones.

Our knowledge of the tubenose eye and vision has been enhanced greatly by detailed examinations of a range of species by Hayes and Brooke (1990), Hayes *et al.* (1991) and Martin and Brooke (1991).

Martin and Brooke (1991) showed that the eye of *P. puffinus* is of the 'flat' type with the equatorial diameter greater than the axial length. It is unusual in that the corneal contribution of 68 diopters (D) to refraction is much less than that of the lens (109 D), which Martin and Brooke pointed out might aid vision under water where the lost corneal contribution would be compensated for by the high lens refraction.

Martin and Brooke (1991) also found that the total visual field in the plane of the optic axis of two sedated 70-day-old *P. puffinus* subtended 285°, with only 11° of binocular vision. The blind area behind the head extended over 75° (Fig. 10.21)—one catches these birds at night more easily when approaching from behind.

Hayes and Brooke (1990) examined the retinae in *Phoebetria fusca, Lugensa brevirostris, Pterodroma mollis, P. incerta, Pachyptila vittata, Puffinus gravis, P. assimilis, P. puffinus,* and *Pelecanoides urinatrix*. They mapped the distribution of the Nissl-staining ganglion cells that form the first cell layer to receive light transmitted through the lens (Fig. 10.22). All species had a pecten and an *area centralis horizontalis*, but no definite *fovea centralis* as in *D. cauta*, while the extent and shape of these central bands varied between the species as did the cell densities there; for example, in *P. fusca* the maximum density of the ganglion cells was 8900 mm^{-2}, in *P. puffinus* 21 500

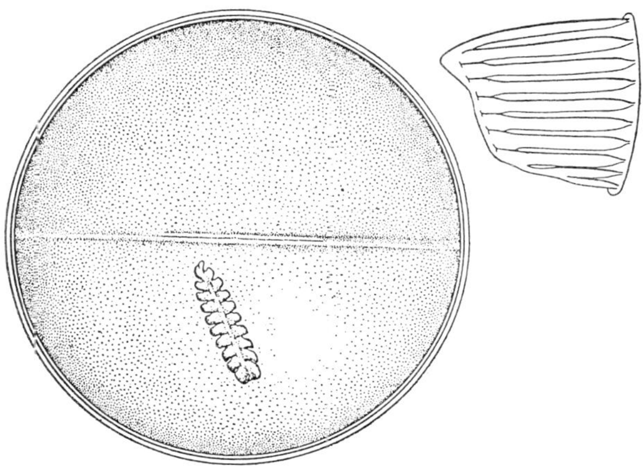

Figure 10.20 *The fundus of* Puffinus griseus *as seen through the ophthalmoscope. A lateral view of the pleated pecten is given on the right. From Wood (1917).*

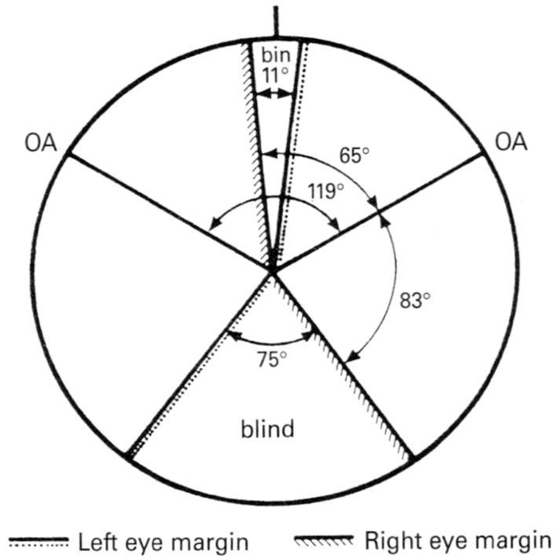

Figure 10.21 *Representation of the retinal visual field of* Puffinus puffinus *in an approximately horizontal plane containing the optic axes (OA). Bar at top indicates the direction of the beak. Note very narrow binocular field, and the rather wide blind area behind the head. From Martin and Brooke (1991).*

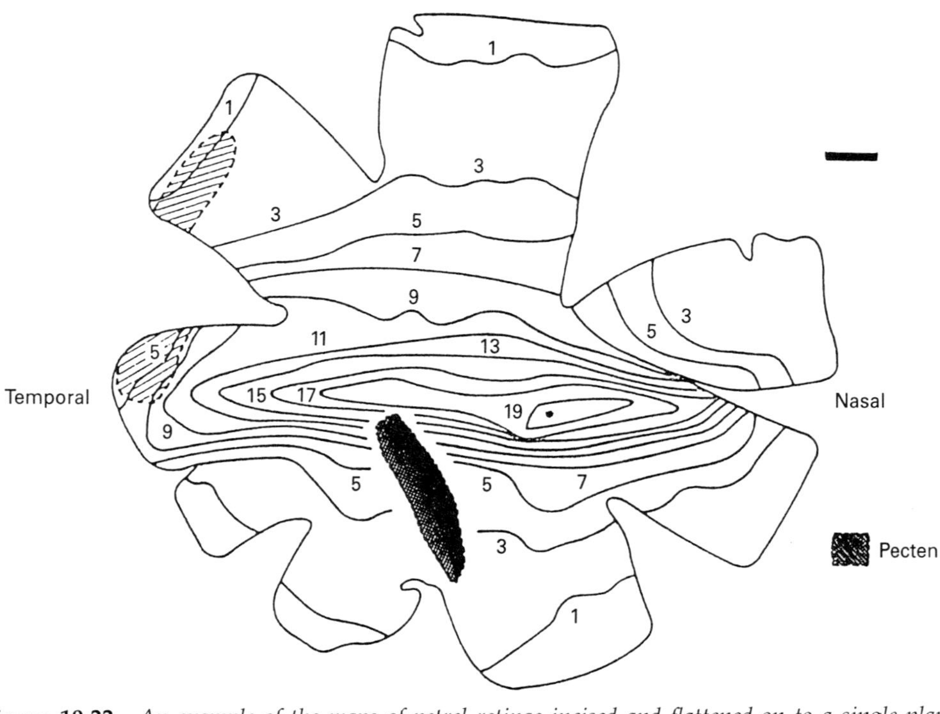

Figure 10.22 *An example of the maps of petrel retinae incised and flattened on to a single plane. Contour lines join points of equal ganglion cell densities, the numbers the thousands of nuclei cm^{-2}. The dot marks the area of peak cell density; the special region with the large ganglion cells is shaded. Bar = 1 mm. From Hayes* et al. *(1991).*

mm^{-2}. Estimates of the total number of ganglion cells in each retina ranged from 3.8 $\times 10^6$ for *L. brevirostris* to 568 000 for *P. urinatrix*, and, over all, the species averaged 1.3 $\times 10^6$ of these cells. Total retinal areas ranged from 97 mm^2 for *P. urinatrix* to 734 mm^2 for *P. fusca*.

The retinae of at least five (*P. puffinus, L. brevirostris, P. gravis, P. vittata* and *P. urinatrix*) examined by Hayes *et al.* (1991) had an area of large, regularly arranged ganglion cells forming a narrow, low-density band close to the edge of the retina from the temporal horizontal meridian to the dorso-temporal region. Each cell had six primary dendrites which extended almost beneath the cell body of its neighbours. Hayes *et al.* named this specialized region *area giganto cellularis* to distinguish it from the *area centralis*. Hayes *et al.* noted that these ganglion cells appeared similar to the alpha-ganglion cells widely distributed in the cat's eye which are thought to be sensitive to movements in the visual field. In the tubenoses examined they are localized so that the visual projections of the narrow region of binocularity below the bill fall there (Fig. 10.23).

Although hatchlings may have smoky-blue irides, fledglings leave with the brown eyes which most species retain throughout life. With many *M. giganteus* and most *M. halli*, the brown pigment becomes increasingly speckled with whitish as the birds age, the patterns in the two eyes seldom being identical (Warham, 1962, p. 156). The pale hazel irides of *D. melanophrys impavida* are acquired before the bill has attained full adult colour (*The Petrels*, cover photo).

IX Integumentary system

The integument lacks ornamental excrescences such as wattles or bare flesh (except for the oral flanges of mollymawks and their variously coloured bills) and lacks decorative plumes, the golden feathers on the heads of *D. albatrus* and *D. irrorata* perhaps excepted. The most unusual feature of this system in the petrels is their compound rhamphotheca.

A *Rhamphotheca*

The proteinaceous keratin covering of the bill in tubenoses is the most complex of compound rhamphothecae. It is hard and cornified and is presumably generated from cells of the dermis that separate the jaw bones from their horny coating, as in other birds. Evidently, dilation of the blood vessels in the dermal layer causes the flushing of the beaks of the great albatrosses.

The shape of the rhamphotheca closely follows the underlying skeleton, with variations in plates such as the naricorns helping the final shapes that characterize the various taxa. The plates themselves total 11, but 12 in albatrosses, because of their single median inter-ramicorn (see Fig. 10.1). These bill plates have been little studied. Lonnberg's (1904) derivation of them from reptilian shields is outlined in Chapter 11, Section IV.A, otherwise the only detailed work is in the descriptive studies of von Boetticher (1930, 1934, 1955). In the last publication he devised a nomenclature differing from that of Coues used here.

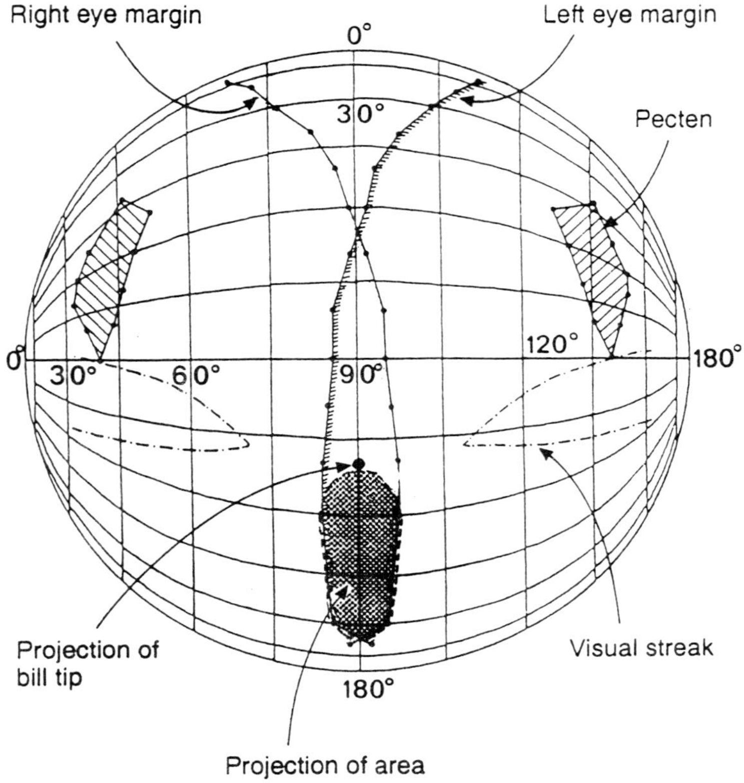

Figure 10.23 *Retinal visual field map of both eyes of a Manx Shearwater showing the visual projection of the novel area. The map shows the visual field as if projected on to the surface of a transparent sphere centred on the bird's head. The bird is placed at the centre of the projection facing the viewer. The co-ordinates are conventional latitude and longitude but with the Equator aligned vertically in the median sagittal plane of the bird. From Hayes* et al. *(1991).*

The nasal tubes are formed by rolling of the skin during the chick's development and early days in the nest (Fig. 10.24) and the whole rhamphotheca remains soft and often darker than in the adult, colour changes being marked in mollymawks such as *D. chlororhynchos*, which fledge with glossy black bills, the bright yellow of the adult's culminicorn taking some years to develop. The origin of the yellow in the bills of this and other mollymawks is unknown, nor is that of the red suffusion seen in the ungues of some species and in *M. halli*: this does not appear to alter seasonally and may not be due to suffusion by blood.

The tomia and the cutting edges of the ungues are kept razor-sharp by the thegotic action of the upper against the lower. Presumably the keratinous material is produced continuously towards the tip and edges replacing the wear. The process and timing of moulting of the plates is unstudied. Captive *D. immutabilis* and *D. nigripes* shed the outer layer of the beak during feather moult and Iredale (1955) had a prion in which 'the upper covering of the mandible' slipped off disclosing a new

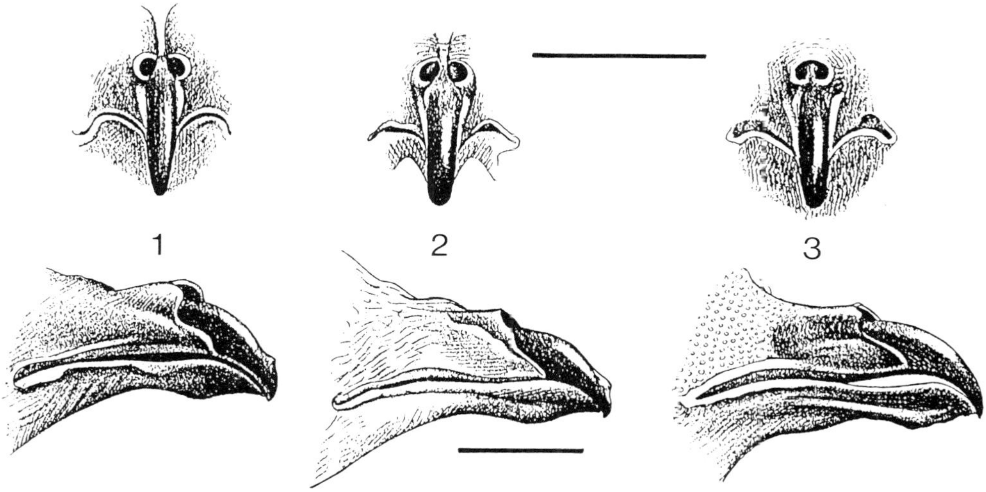

Figure 10.24 *Development of the nasal tubes in* Lugensa brevirostris, *seen from the front and side: 1, in a late embryo; 2, in a 4-day-old chick; 3, in a chick* c. *10 days old. Note the rolling of the soft rhamphothecal material in forming the tubes that are initially separated by the culminicorn (1 & 2), later to be fused together on top of it (3). The older chick has shed its egg-tooth. Bars are 1 cm. From Studer (1889).*

covering below. Biermann and Voous (1950) found *Macronectes, F. glacialoides* and *T. antarctica*, taken at sea had bill plates crumbling or throwing off horny flakes. In *T. antarctica*, the shedding began with the ramicorns and apparently ended with the nasal tubes, this evidently coinciding with moult of the flight feathers as was desquamation of the leg scales. Scars at the bases of the bills of giant petrels and albatrosses (Fig. 10.17) may be remnants of earlier layers, but the whole process of regeneration of the petrel rhamphotheca requires investigation.

Another unexplained feature is the function of the small foramen at the mandibular symphysis in albatrosses and some fulmars such as *Macronectes* (Fig. 10.25). This seems to have been figured previously only in *F. glacialis* (Kay, 1953). The opening appears to lead into tubular perforations of the dentary bone suggesting a link with taste sensilla or even a pathway for the passage of rhamphothecal keratin.

B *Scales and claws*

The modified integument of petrel feet, the podotheca, is horny and of varying flexibility. Although the scalation has been used as a taxonomic character, particularly in storm petrels (Mathews, 1928; *The Petrels* p. 178), the scales themselves have not been closely examined except by von Boetticher (1932) who considered that the small hexagonal ones on the proximal parts of the toes of albatrosses were primitive (see Chapter 11, IV.A). In some tubenoses the tarsi are unscaled. How often ecdysis of this part of the integument occurs is unclear. Verwey (1927), who described a

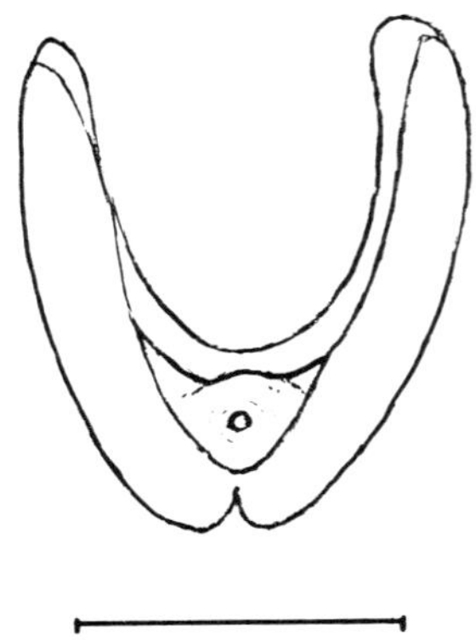

Figure 10.25 *Mandibular ungues of* Macronectes *seen from in front showing the small circular foramen of unknown function. Bar is 1 cm.*

specimen of *F. glacialis* in moult with feathers growing from its tarsometatarsi and third and fourth toes, found that not only were the wings and body in moult, but the tarsal skin too was desquamating.

Dark webs should wear better than pale ones and holes and lacerations in webs are very common, but if resistance to wear is important, why are the feet of some fulmars and albatrosses pale? The last cannot take to the wing instantly if attacked, so perhaps pale feet are less conspicuous when seen from below by predatory fish.

The blue legs of *Pelecanoides* and *Puffinus assimilis* are presumably the result of Tyndall scattering of reflected light, blue pigments being unknown in birds. In polychromatic species, pale plumaged birds tend to have pale feet and webs (e.g. in *M. giganteus*, *Pterodroma neglecta* and *P. arminjoniana*). Variations in foot colour are not rare, for example the blue legs of some *P. nigripennis* (Serventy *et al.*, 1971, p. 103), now known to be the extreme of variation, at least in birds from the Kermadec Islands.

C *Glands*

The preen or uropygial gland is the main cutaneous gland. It is bilobed, each lobe drained by a duct that releases the liquid, waxy sebum for waterproofing the plumage, the papilla between the ducts bearing a ring of down feathers forming a wick. Excluding Paris (1913), who included *Puffinus gravis* and *H. pelagicus* among his study birds, the histology of this gland in tubenoses has not been worked on, but Jacob (1978) showed that the secretion has antimicrobial properties (see Chapter 8.I.C).

Petrels evidently also have anal glands although only reported (in *O. homochroa*) by Quay (1987). They occur in the integument around the vent but are of unknown function.

There are probably also sebaceous glands scattered over the skin, possibly linked with its musty smell, but these have not been demonstrated.

X Plumage

In their gross morphology, petrel feathers are typical of birds: the flight ones strong and stiff, the pennaceous and plumaceous ones on the body with after-shafts which, with the plumaceous proximal vanes of the contour feathers, form an insulative under-shirt. After-shafts are minute in *Diomedea*, but very well developed in *Pelecanoides*. Wing formulae for many species are given in Cramp and Simmons (1977) and Marchant and Higgins (1990). Typically there are 12 rectrices but 13 and 14 have been reported in albatrosses, gadfly petrels and shearwaters (e.g. by Loomis, 1918, p. 36).

A *Feather structure*

Feather structures were closely examined by Chandler (1916) using *D. exulans, Puffinus griseus, Oceanodroma melania* and *Pelecanoides urinatrix*. She found several peculiar features. The barbules of the remiges and body feathers have all the various types of barbicels seen elsewhere but the bifurcate ventral 'teeth', dorsal prongs at the base of the distal barbules of the remiges, and the elongated, wavy ventral 'teeth' of the proximal barbules are found solely in Procellariiformes, or are best developed in them (Fig. 10.26). The flexules of the body feathers are also very highly developed. Only the loons out of the many species examined showed great similarities in feather structure (Fig. 10.26.4).

On the structure of its barbules, Chandler (1916) considered *Diomedea* the most specialized tubenose, those of the remiges having more 'frills' than any other feather known (e.g. Fig. 10.26.1C). The barbules of the remiges of *P. griseus* and *O. melania* are similar, with the same twist at the junction of the base and pennulum, but in *O. melania* some other processes are minute or lost. In *P. urinatrix* the breast feathers have rather weak barbules and lack the flexules and the wavy, slender ventral 'teeth' on the proximals, which Chandler considered the more specialized, and which are present in the other species. The barbules of the down feathers of the adults are notable for their forward-curving prongs, which may be forked (Fig. 10.26.5).

B *Pterylography*

The body feathers are arranged in tracts (pterylae) separated by apteria, arrangements unstudied in tubenoses since the work of Nitzsch (1867). The pterylosis of *F. glacialis, Daption capense, Macronectes, H. pelagicus, Puffinus lherminieri, Diomedea exulans* and *D. chlororhynchos* was rather similar, but the albatrosses were peculiar in having the anterior dorsal pterylum strong and coarse and completely separated from a weak posterior part. The wing feathers are arranged in the usual manner: they are diastataxic but tubenoses lack the subalar apterium of many other birds.

Small plumules are also evenly distributed over the body, in the apteria and between the contour feathers. These plumules are shed locally to form the brood patch during incubation. In the dark eye patches of *F. glacialis*, Zonfrillo (1982b) described dark feathers in which the distal third of the rachis is bare producing a bristle-like structure, perhaps helping in the reduction of glare or controlling the

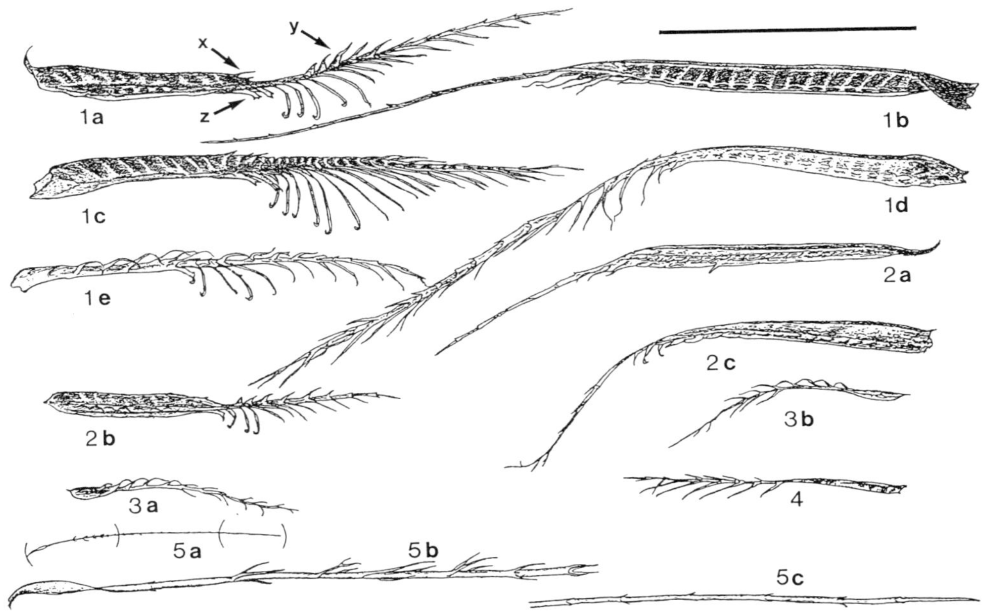

Figure 10.26 *Feather barbules from tubenoses. 1,* D. exulans. *1a, Distal barbule from inner vane of remex showing characteristic features of forward projecting prongs (x), short, dorsal cilia, triangular basally (y), and unusual bi- or trilobed ventral 'teeth' (z). There is also a pronounced twist between the base on the left and the pennulum at about (x); 1b, proximal barbule as above with very long base and ventral 'teeth' differing from other birds in being slender, flexible and wavy; 1c, distal barbule from outer vane of remex; 1d, proximal barbule from outer vane of remex; 1e, distal barbule from outer third of barb of breast feather. 2,* Puffinus griseus. *2a, Distal barbule from inner vane of remex; 2b, proximal barbule from inner vane of remex; 2c, proximal barbule from outer vane of remex. 3.* Oceanodroma melania. *3a, distal barbule from near tip of barb of breast feather; 3b, proximal barbule from near tip of barb of breast feather. 4,* Gavia immer—*proximal barbule from breast feather for comparison with 3b. Flexules, like 3b of the body feathers of tubenoses, are more developed than in other birds. In both petrel and* Gavia *the curved dorsal barbicells or flexules are developed while the base is reduced. 5,* Puffinus griseus. *5a, Barbule of down feather; 5b, basal portion; 5c, terminal portion. (Bar is 0.5 mm and refers to 1–4. 5a × 27, 5b,c × 190.) From Chandler (1916).*

flow of air over the eyeball. Similarly modified contour feathers are known in other birds (Chandler, 1916, p. 274).

C *Filoplumes*

Filoplumes are scattered among the contour feathers and may be particularly prominent on the head and nape (*The Petrels*, p. 2), a feature recorded at least 100 years ago (*contra* Imber, 1971; James, 1986b). Loomis (1918), for example, found filoplumes prominent on the crown, nape, cervix and sometimes the back of *Pterodroma externa*, *P. phaeopygia*, *Puffinus griseus*, *P. pacificus*, *P. bulleri* and *P. creatopus*, and on chicks and adults of *Diomedea irrorata*.

Imber (1971) found filoplumes present on the heads of many procellariids and hydrobatids, except in *F. glacialis, F. glacialoides, Pterodroma lessonii* and *Oceanites oceanicus*. In breeding *P. macroptera*, the number was significantly greater on the males than on the females, and better developed in breeders than in immatures. James (1986b) found that among breeding *Puffinus puffinus* more males than females had filoplumes on their heads and necks but many of both sexes had none, sexing on this feature being unreliable. James also counted significantly more filoplumes sex for sex on the heads of breeders from Skomer than of those from Rùm with $P = 0.008$. Plant (1989) reported that the number of filoplumes on storm petrels was very variable, often lacking, but conspicuous in living *Garrodia nereis* and *Fregetta tropica*. However, Imber's data suggest that, at least for some gadfly petrels, filoplume abundance could be a useful adjunct to other sexing techniques.

Imber (1971) hypothesized that the protruding silvery-white filoplumes act as secondary sexual characters, clearly visible at least on moonlit nights, which would explain their absence on the white crowns of *Pterodroma lessonii* and the fulmars. He found them on *P. neglecta* but did not state the situation in the white morph of that species. That much courtship in gadfly petrels occurs on dark nights is a problem, although not negating their role with tropical diurnally active birds such as *P. arminjoniana*, or for possible ancestral ones.

However, filoplumes occur in feather tracts alongside other feathers, not by themselves, nor on the apteria, and it has been proposed that they act as proprioceptors providing sensory input for the control of the adjacent feathers, facilitating posture adjustment. In some non-procellariiforms, free nerve endings have been found in the follicle wall with sensory corpuscles embedded beside them (Lucas & Stettenheim, 1972). Such a sensory network could be activated when female petrels preen the heads of potential partners. The more aggressive males then become quite submissive (p. 174). Likewise, the males often nibble the napes of the females during copulation. Such a role, could, of course, be complementary to sexual signalling.

D *Feather coat*

As far as is known, all tubenose chicks, except those of the Oceanitinae, grow two coats of down (*The Petrels*, p. 335 & Fig. 10.27). These natal downs are usually quite copious although the throats and facial areas of storm petrels are bare at first and become thinner ventrally. The mesoptile coat tends to be shaggier and longer, more so while the protoptiles are still attached. Why the long-legged storm petrels have but one generation of down is a mystery to be investigated.

The protoptiles may be paler or darker than the mesoptiles; sometimes both are of similar shades. Paler mesoptiles have been attributed to fading, but without experimental proof. In *D. immutabilis, D. nigripes* and some *D. irrorata*, the protoptiles are white tipped giving a pepper-and-salt effect. Down colour may be linked with adult plumage colour in polychromatic species, as with pale-phase *M. giganteus* which have white down as chicks.

There is usually some variation in down colours among chicks, but with *Fulmarus glacialis, M. giganteus, D. bulleri, Pagodroma nivea, Pterodroma neglecta*, and especially in *D. irrorata*, this is quite marked. Harris (1973), classified *D. irrorata* chicks as dark,

Figure 10.27 *Dorsal contour feather of* Pachyptila turtur *showing relationship between pseudo-protoptile, pseudo-mesoptile and teleoptile. Arrows mark the junctions of the two downs. Bar is 1 cm, terminology that of Percival. From Percival (1942).*

medium and pale, and found that their proportions varied between colonies and, within colonies, from year to year. Possibly some extra-pair copulations had occurred.

Gadfly petrels such as *Pterodroma mollis* have been alleged to have unusually soft plumage (hence the name), and the juvenile plumage of shearwaters softer than that of the adults, but these claims have not been verified quantitatively. Nor are there any data on plumage weights or feather numbers, although Furness and Burger (1988) reported that for a variety of flying seabirds (including four unspecified tubenoses) total plumage mass was allometrically related to body mass$^{0.927}$, with high latitude species having heavier feather coats than low latitude ones.

Plumages are mainly in shades of brown, grey and whitish, often paler below. Melanistic, leucistic and albinistic examples occur, including completely white feathered specimens, all commoner in museum cabinets than in the wild. The juvenile plumage and those of immatures are usually similar to those of the older birds but the juvenile coat lacks the old feathers usually scattered among the new in the adult plumage. Minor differences in plumage patterns occur, for example in its juvenile plumage *P. inexpectata* has more white on rectrices 2–6 than the adult birds. Sexual dimorphism in plumage is also rare, being most noticeable in the great albatrosses where females are darker than their partners.

Species in which the younger birds have the darker plumage are mostly albatrosses—*D. exulans, D. albatrus* and *D. irrorata*—the dark feathers tending to be on their crowns and dorsum, cheeks and underparts being lighter. The differences are less pronounced in mollymawks but these too tend to be darker on the head and underwings when young.

Changes to full adult plumage take years in *D. exulans*. Adult males of *D. e. chionoptera* and *D. e. gibsoni* may be very white on wing coverts, bodies and tails, lightening of the females ending at an earlier stage. The plumage development of *D. e. antipodensis* and *D. e. amsterdamensis* is the most retarded for they breed in a plumage similar to that of a young *D. e. chionoptera*: as far as their plumage goes they are neotenic. Adult plumage is also acquired slowly by *M. giganteus*, apart from white-phase birds. The all-dark examples that are occasionally found breeding among normal plumaged birds can be regarded as also in neotenous plumage and this condition is widespread among Chatham Islands *M. halli*, where many breeders are dark-eyed and retain the dark brown juvenile plumage.

In *D. exulans* progression to the whitest plumage takes at least 20 years (Robertson *et al.*, unpubl.). Weimerskirch *et al.* (1989b) analysed the plumage changes of Crozet Island *D. exulans* of known age and sex using Gibson Code scores (*The Petrels*, p. 24). On first return at about four year the males are already whiter than the females and this dichotomy continues over the next 15–18 years. These authors had good samples of old birds which indicated the females reaching their final stage when 20–25 years old, the males taking at least 35 years to do that. Sex for sex, 7–10-year-old non-breeders were darker than the breeders. There are no data for the rare very white 'snowy' birds such as the *D. e. gibsoni* figured in Bailey and Sorensen (1962, p. 143f) and, contrary to Weimerskirch *et al.* (1989b) *D. exulans* does not follow Gloger's Rule precisely because at Campbell Island dark *D. e. antipodensis* breed further south than lighter *D. e. gibsoni* in warmer seas at the Auckland Islands.

Weimerskirch *et al.* (1989b) also looked at the plumage scores of Wandering Albatrosses sighted in the Indian Ocean where dark birds would probably be immatures or *D. e. amsterdamensis*, as *D. e. antipodensis* seems not to reach there. They found pale plumaged birds absent north of 42°S, dark ones missing below 59°S and concluded that, while immatures tend to forage in subtropical waters, the most southerly feeders must be mainly mature males.

Polychromatism is occasionally common in Procellariidae and Hydrobatidae and appears not to be sex-linked. Colour morphs of *Fregetta* caused problems to systematists (*The Petrels*, p. 175), and are still not fully understood. The great variation seen in the disposition of black and white feathers on the abdomen of *F. tropica* (see Marchant & Higgins, 1990, p. 705) is exceptional for a tubenose. As with *D. irrorata* chicks, the proportions of colour types, for example in *Pterodroma neglecta*, seem to vary with time, but in others such as *P. arminjoniana* (Gardner *et al.*, 1985a), they apparently do not.

Recently, Bretagnolle (1993) examined possible factors affecting the coloration of petrels with the help of correspondence analysis. His variables included colour patterns coded into nine categories: body size, diet, feeding techniques, feeding flock sizes, sea-surface temperature (SST), and latitude of the breeding places (as an index of solar radiation), and taxonomic status. Plumage weight or thickness or diurnal *vs.* nocturnal activity ashore were not included. No one factor was highly correlated with colour pattern, but feeding style and flock size were important. Most species with dark upper- and underwings inhabitated low latitude, warm seas; light-winged ones inhabited high latitude, cold seas. Bretagnolle discussed possible links between predation and coloration, citing the cryptic plumage of prions and '*Cookilaria*' as possibly lessening predation on these small flocking species, which either feed without alighting or are very mobile and can more easily escape if attacked. In contrast, diving petrels, counter-shaded to deceive their prey, can escape by submerging.

However, the main link was with the feeding ecology; categorizing the species on this factor or on plumage colour gave similar results, suggesting that these two variables have a common basis. Bretagnolle (1993) did not relate colour patterns to species recognition as Murphy and Pennoyer (1952) did for gadfly petrels. For earlier speculation on links between tubenose plumage and ecogeographical rules see von Boetticher (1932).

Moult and plumage wear can change the appearance of a petrel. For example, storm petrels that have shed P9 and P10 have shorter, more rounded wings as P7 and P8 are then the longest. In *Puffinus gravis* a white wing bar may appear when the shed coverts expose the white bases of the primaries, and the glossy, pale-edged contour feathers that create a scaled effect on the upperparts of some species (notably *Pterodroma*) are lost with wear and their general appearance darkens. Old feathers, dotted among the new, are duller or bronzed and brown, a condition usually ascribed to fading, but has a true chemical change occurred or is a structural physical one responsible? Primaries take the most wear, their tips often broken and edges abraded. Alleged bleaching of the head and neck feathers of giant petrels from immersion in blood from scavenged carcasses could not be demonstrated (Warham, 1962, p. 156).

XI Moult

A *Introduction*

Most feather replacement in petrels occurs at sea, either before the attainment of breeding condition or between breeding episodes. The timing and extent of moult varies with age and/or status but only studies on albatrosses have outlined the moults of known-age birds. Others have drawn their conclusions from museum or beach-washed birds, mostly of unknown provenance, birds caught at sea, and of those on the breeding grounds.

Despite the definite sequence and timing of feather replacement, there is often much intraspecific variation even with birds of the same sex and status. Moult may also be asymmetrical, particularly in the tail, but also in the primaries, for example only 41.4% of 152 *D. chlororhynchos* had identical patterns of primary moult on both wings (Furness, 1988b), 52% of *D. exulans* (Weimerskirch, 1991).

The sequences of plumages and moults are simple, apart from non-annual breeders, or those such as albatrosses and giant petrels that pass through distinctive intermediate plumages before maturity. The definitive plumage does not indicate sexual maturity, for most petrel chicks moult from mesoptyles into the first juvenile plumage, this being similar to that of the adult. Thereafter, this basic plumage is renewed at each moult for the rest of the bird's life except that after the juvenile coat, made up of 100% new feathers, a few, perhaps many, older ones may be retained, sometimes leaving a melange of old and new, particularly on the body.

Moult regimes enable established breeders to be in full feather by their landfall to breed, although sometimes wing and tail feathers may still be regrowing, for example in *Hydrobates pelagicus* and *Oceanodroma castro*.

Factors influencing the sequence and timing of moult include whether the species is sedentary or migratory, whether adapted to tropical, temperate or polar seas, whether immature, prebreeder, breeder or failed breeder and whether an annual, biennial or non-seasonal nester.

Annual breeding species that only disperse postnuptially, moult at sea at that time, for example *Pterodroma macroptera* which undertakes a complete moult in its 3 months away (Imber, 1976c). Mainly sedentary birds such as *Daption capense, Pachyptila vittata* and *Pagodroma nivea* appear to moult during the postbreeding exodus before their autumn reoccupation of their colonies. However, trans-equatorial migrants delay wing and tail moult until they have reached their 'wintering' quarters and can grow new feathers while in rich feeding grounds. This was first documented for *Puffinus tenuirostris* by Marshall and Serventy (1956a) and is also true of *P. griseus, P. gravis, Oceanites oceanicus* and *Pelagodroma marina*. Delaying flight feather moult until after migration also enables the full-winged birds to feed efficiently in advance of their long return haul across the Equator. The migratory *H. pelagicus* is unusual in starting wing and tail moult during incubation but suspending secondary moult until in winter quarters in the South Atlantic (Scott, 1970 in Ginn & Melville, 1983). In contrast, *Puffinus p. mauretanicus*, not a long-distance migrant, takes about three months to renew its flight feathers, the secondaries starting nearly 2 months after the primaries.

Immature and prebreeding petrels moult earlier than the established breeders. Presumably successive moults begin progressively later until the birds reach non-breeder status and are physiologically capable of reproducing. Advanced pre-breeders and non-breeders are those commonly seen flying during incubation and chick rearing with gaps in their wings as a result of the early incidence of moult. Failed breeders too tend to start shedding soon after abandoning nuptial activity. The successful breeders are generally the last to moult.

Data are not available on the energetics of petrel moulting, but the costs seem likely to be considerable, particularly as feather replacement must be done while recuperating from breeding, laying down reserves for the next breeding episode, and perhaps a long return migration. That all the flight feathers are often not replaced suggests a resource limitation of time, food, nutrients, etc. And, although there is little direct evidence for impairment of feeding when not full-winged, birds in wing moult, albatrosses especially, are common among beach-washed debris. Kinsky (1968) reported that in a major 'wreck' in April that year involving 187 albatrosses of five species, all 45 *D. c. cauta* were in heavy moult with body feathers only half renewed. No *D. c. salvini* were found then, yet after a less severe storm in 1947, about 40 were collected but no *D. c. cauta*, a difference that Kinsky attributed to some, perhaps all, of the 1947 birds being in moult, the *D. c. cauta*, not having started, escaping.

Major studies of tubenose moult include those of Mayaud (1931, 1932b, 1941, 1949–50), Bierman and Voous (1950), Marshall and Serventy (1956a), Stresemann and Stresemann (1966, 1970), Beck (1970), Mougin (1975), Ainley *et al.* (1976), Hunter (1984a), Furness (1988b), Brown (1988b), Weimerskirch (1991), Prince *et al.* (1993), Prince and Rodwell (1994) and Langston and Rohwer (1995). Summaries of moulting patterns for many species are given by Cramp and Simmons (1977) and Marchant and Higgins (1990).

B *Procellariidae, Hydrobatidae and Pelecanoididae*

1 *Wings*

Data for 48 species show that these undergo complete wing moults annually, replacement usually extending over several months. The primary feathers are shed outwards from the wrist, from P1 to P10; that is, in descendent sequence. Several may be lost in rapid succession initially, creating a gap between the primaries and secondaries. Up to eight feathers have been missing from a wing of *F. glacialis*, but more usually only two to four primaries are lost at one time. Thereafter replacement tends to slow down, the longest feathers being the last replaced. The innermost, being shorter, take less time to regrow, for example in *Oceanodroma homochroa* P9 took *c.* 77 days, P1 and P2 only about 32 days (Ainley *et al.*, 1976). The possibility of sexual differences was raised by the 12 *O. leucorhoa* found 'wrecked' in 1952. The six males were all in new feather, the six females all in old and worn plumage (Hazelwood & Gorton, 1954).

Transequatorial migrants shed their feathers rapidly once they have completed their journeys and there are many reports of seas and beaches being strewn with flight feathers, for example British Columbian shores littered with those from *Puffinus griseus*, and primaries and rectrices from *Oceanites oceanicus* along a wash-line in Long Island, New York.

The wing moult of *P. gravis* on the Grand Banks described by Brown (1988a), peaked in June when 92% of 450 birds counted were in 'heavy moult'. He pointed out that the relatively few *P. griseus* overwintering in the North Atlantic do not moult there, in contrast to those going to the North Pacific, and suggested that the North Atlantic birds are non-breeders which have moulted early before flying north.

Ogi *et al.* (1981) recorded wing lengths of *P. griseus* taken at sea in the subarctic North Pacific as 295 ± 12 mm ($n = 51$) in June, 284 ± 19 mm ($n = 50$) in July, 277 ± 22 mm ($n = 35$) in August, 295 ($n = 3$) in September and 299 ± 8 ($n = 34$) in October. Evidently in August the outermost primaries had been shed to be replaced by September, the moult completed by October. These workers saw no evidence of secondary moult and they supposed that it might have occurred immediately following their arrival. Feather loss did not cause flightlessness. Further south, off

California, Chu (1984) followed their plumage changes from the May mixture of old, new, and growing feathers to almost complete regrowth of the remiges by September.

North Sea *F. glacialis* were in wing moult from May to October with few in the north and northwestern areas during the breeding season, but a high proportion moulting in June–September in the east and southeast (Tasker *et al.*, 1987). These authors thought that the latter were non-breeders enjoying the plentiful food locally available. Brown (1988a) reported a similar timetable for non-breeding light and dark morphs in the North Atlantic, where wing moult of breeders seemed complete by April.

Hunter (1984a) estimated that breeding male and female *M. halli* took 183 and 168 days respectively to complete primary moult, *M. giganteus* 228 and 189 to do that, so it is not surprising that such birds started shedding early and increased the rate of moult towards the end of chick rearing as their feeding rate slackened. That such a long moult can be fitted into annual breeding Hunter ascribed to a plenitude of energy-rich food available during the austral summer.

Less is known of secondary moult. The secondaries are replaced concurrently with the primaries but are shed from several centres, some proceeding descendently, others ascendently. For example, in *Oceanodroma homochroa* there were four such loci (Ainley *et al.*, 1976). Similarly, in *P. puffinus* and *P. gravis* three loci of ascendent and one of descendent moult have been described. In *P. griseus* secondary moult has been reported as proceeding in both directions from S4 to S6, in *Oceanites oceanicus* from S1 to S4 inwards and from S5 to S9 outwards. In *P. p. mauretanicus*, which has 19–22 secondaries, Mayaud (1931) found four groups: S1–S4, shed ascendently, were separated at the diastataxis from S5–S13 also shed ascendently; S19–S13 shed descendently; and, finally, three close to the body, S19–S21, were replaced ascendently.

Moult of the remiges of sedentary petrels may start during nesting, for example some *Macronectes* and *Daption* lose primaries in late incubation and during chick rearing (Warham, 1962; Hunter, 1984a). Hunter showed that for *Macronectes* the overlap in moulting and breeding is exceptional, with males starting first. Yet, despite the 6-week gap between the laying of *M. giganteus* and *M. halli*, at South Georgia the males of both began wing moult on about the same date. In *M. giganteus* males were dropping primaries around egg laying, their partners starting at least 60 days later. Most male *M. halli* delayed this until the hatch, their females until the chicks were some 3 weeks old. These differences may reflect the more southerly habitat and shorter summer season of *M. giganteus* so that timing is critical to fit a complete moult into a yearly cycle and, being large, the time to grow the new feathers is also greater than in smaller fulmars. But with these too, successful breeders tend to begin shedding remiges in the late chick stage so may also be up against a time constraint. Wing coverts appear to be moulted with the remiges, but few data are available.

Among petrels of tropical seas the complete wing moult seems to take much longer. In some *Oceanodroma castro* primary moult starts when the chicks are well grown, a full replacement evidently requiring the whole of the postnuptial period at sea. Indeed, in this, and in some *O. homochroa*, early arrivals at the colonies may still be growing new primaries (Harris, 1969b; Ainley *et al.*, 1976).

2 *Rectrices*

Tail moult is poorly documented. There may be no clear pattern, and asymmetrical regrowth is common, for example in *P. p. mauretanicus* (Mayaud, 1931). The fulmars *Daption capense, T. antarctica* and *F. glacialoides* may become virtually tail-less, and, according to Beck (1969), in *D. capense* the loss is partly compensated for by the long tail coverts that preserve the width of the tail and which are renewed later.

With Ashy and Leach's Storm Petrels, Ainley *et al.* (1976) found that the commonest sequence (the feathers being numbered from the central pair outwards) was 6.4.2.1.3.5; that is, the outer pair tended to be shed first, then every other feather from the outside towards the centre, and then from the centre outwards. In *H. pelagicus*, however, the feathers are moulted in pairs from the centre outwards (Scott, 1970 in Ainley *et al.*, 1976). In *O. castro* and *O. leucorhoa*, Harris (1969b) and Allan (1962) discerned no orderly replacement.

In some petrels, perhaps in most, rectrices tend to be shed when primary moult is nearly complete, for example in *P. pacificus, Procellaria aequinoctialis, Pachyptila salvini, Pagodroma nivea, F. glacialoides* and *Oceanites oceanicus*. However, with *Oceanodroma leucorhoa* and *O. homochroa* moult of the tail began during chick rearing before that of the wings (Ainley *et al.*, 1975).

3 *Contour feathers*

These are replaced gradually over several months, perhaps even continuously, and not all are necessarily changed each year, so that there may be several generations clothing the body—data are few. As the mass of the body plumage in seabirds is 3–7% of the body mass (Furness & Burger, 1988), it would make sense to spread the energetic cost of replacement over time without losing effective insulation.

Preening hastens the removal of loosening feathers. The order of the shedding has not been precisely determined but the belly feathers tend to show moult first, at least in some burrowing species, perhaps due to abrasion from moving through the access tunnels. In *O. homochroa*, the feathers of the face were the last to be renewed. In *P. p. mauretanicus*, birds of unknown status were in full moult on the back in July, the contour feathers being replaced over the posterior part first (Mayaud, 1931). Transequatorial migrants replace these feathers in their 'winter' quarters, *P. griseus* off California not finishing body moult until after that of the wings (Chu, 1984).

In non-migratory species the contour feathers are shed mainly during the postbreeding dispersal. With sedentary ones, and those such as fulmars that visit their colonies when not nesting, some body feathers are moulted while breeding, often as early as incubation. Shed feathers accumulate in nests of giant petrels, shearwaters, gadfly and storm petrels, and are often added to the linings by the chicks.

4 *Brood patch*

The timing of brood patch shedding and refeathering has seldom been followed. Figure 10.28 shows the sequence in *O. leucorhoa*. As in some other storm petrels, the

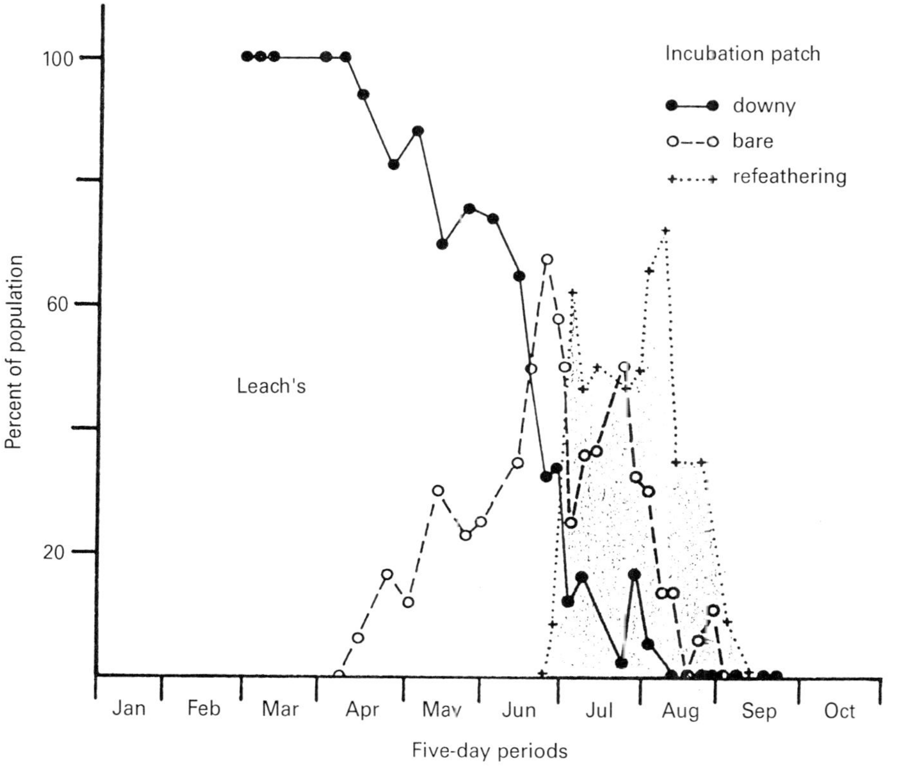

Figure 10.28 *Changes in the development of the incubation patch in* Oceanodroma leucorhoa *during the prebreeding and breeding seasons. From Ainley* et al. *(1975).*

down starts to be shed about a month before the egg appears and refeathering within a week before the hatch. In *O. castro*, Allan (1962) indicated moult of the brood patch down taking only 4 days with vascularization complete 5 days before the lay. The pattern of the feather loss and replacement is unstudied. With *Pterodroma inexpectata* some incubating birds retain a narrow central bar of down which creates the appearance of two brood patches (Warham *et al.*, 1977).

In failed breeders (*M. giganteus, Fregetta tropica, Oceanites oceanicus* and *Fulmarus glacialis*) the patch is refeathered more quickly than in breeding birds according to Conroy (1972), who suggested that the presence of an egg inhibits refeathering in incubating birds. Immatures retain feathered brood patches throughout the year but non-breeders tend to shed down despite lacking eggs: the ages at which this begins have not been determined.

C *Diomedeidae*

Stone (1900) reported the unusual nature of the moult in these birds including the asymmetry in the ages of the flight feathers on the different wings of *Diomedea nigripes* and the mode of replacement, later termed 'wave' moult, 'Staffelmauser' (Stresemann & Stresemann, 1966) or 'serially descendent' (Cramp & Simmons, 1977). Because renewal of the remiges in albatrosses (and perhaps other petrels) may involve simultaneous ascendent and descendent modes, this is not strictly wave or Staffelmauser, as the waves do not travel along the wing in the same direction.

Apart from a few *D. exulans* that shed some primaries towards the end of chick-rearing (Weimerskirch, 1991), all wing, body and tail moult in albatrosses has been at sea between nesting episodes, and breeders and non-breeders are full winged at landfall.

1 *Mollymawks*

Prince *et al.* (1993) followed the development of moult of the primaries and tail in *D. m. melanophrys* (an annual breeder) and *D. chrysostoma* (a biennial one) in relation to age and breeding success. From this, and data on bill and head colour, Prince and Rodwell (1994) described a method for ageing birds up to 6 years old.

The outstanding feature of the moult in *D. melanophrys* is that the outer primaries P8–P10 are shed biennially (Fig. 10.29). The cycle begins during the second winter at sea, when three outer feathers are replaced descendently—Phase 1 in Prince's terminology. In the third winter P8–P10 are kept while P2–P7 are replaced ascendently, starting at P6 or P7. Thereafter, this two-phase pattern between odd and even ages persists into adulthood at 7–13 years, more inner primaries and fewer outer ones being changed with increasing years, and P1 being renewed in either phase. The pattern for *D. chrysostoma* is similar except that by their third year (Phase 2) the four innermost primaries remain intact, so that such birds have only changed 30% of their feathers, *D. melanophrys* 70% by then. Some *D. chrysostoma* keep inner primaries until their sixth year. Both species have changed some 40% of their primaries by about 8 years old.

Breeding *D. melanophrys* continue a 2-year cycle with a descendent regime in Phase 1 and an ascendent one in Phase 2. Breeding Grey-heads stick less closely to this regime. Those breeding 2 years after their last successful episode moult an average of nine primaries the previous year instead of up to five per annum in a normal phase moult. Often P7, P6, P4 and P5 change in the two consecutive years. On average, over the 16–17 months between nestings, 11 primaries are renewed, the 2-year cycle being modified so that in the 12 months before nesting the successful birds had changed eight primaries with 95% of them P8–P10 (as in Phase 1), their quills now being in good condition. Successful *D. melanophrys*, however, only replace three to four of them. Gough Island *D. chlororhynchos* also change more primaries (5.2) when they fail than if they were successful (4.2) (Furness, 1988b).

The pattern of secondary moult in relation to primary moult in non-breeding *D. melanophrys* is shown in Fig. 10.30. At 2 years, most of the wing was of juvenile feathers; only the outermost primaries and the innermost secondaries had been replaced. In the next 3 years, increasing numbers of secondaries were moulted but not until 5 years had all the original secondaries gone and the second generation begun to be replaced.

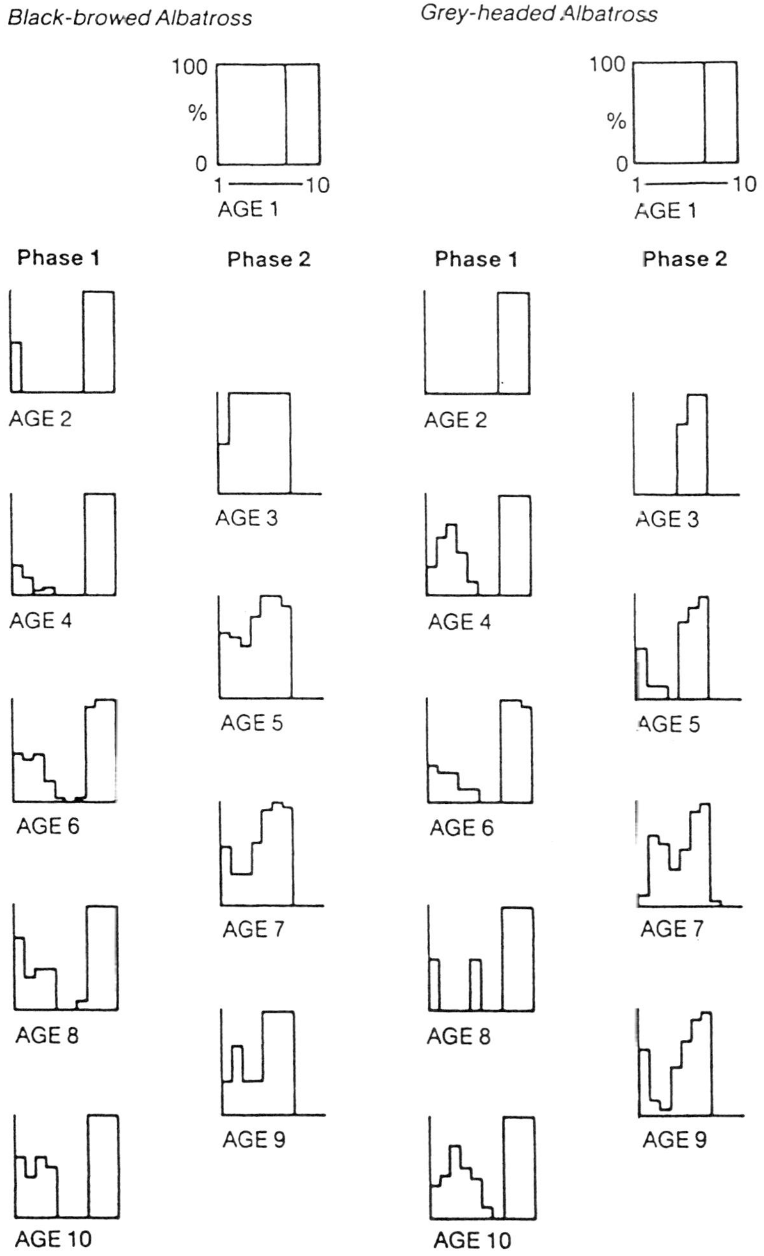

Figure 10.29 *Primary moult in non-breeding* Diomedea melanophrys *and* D. chrysostoma *aged 1–10 years. Primaries are numbered along the 'x' axis, the vertical line separating P8–P10 from the rest. The 'y' axis is the % of birds replacing each primary in the last year From Prince (1993).*

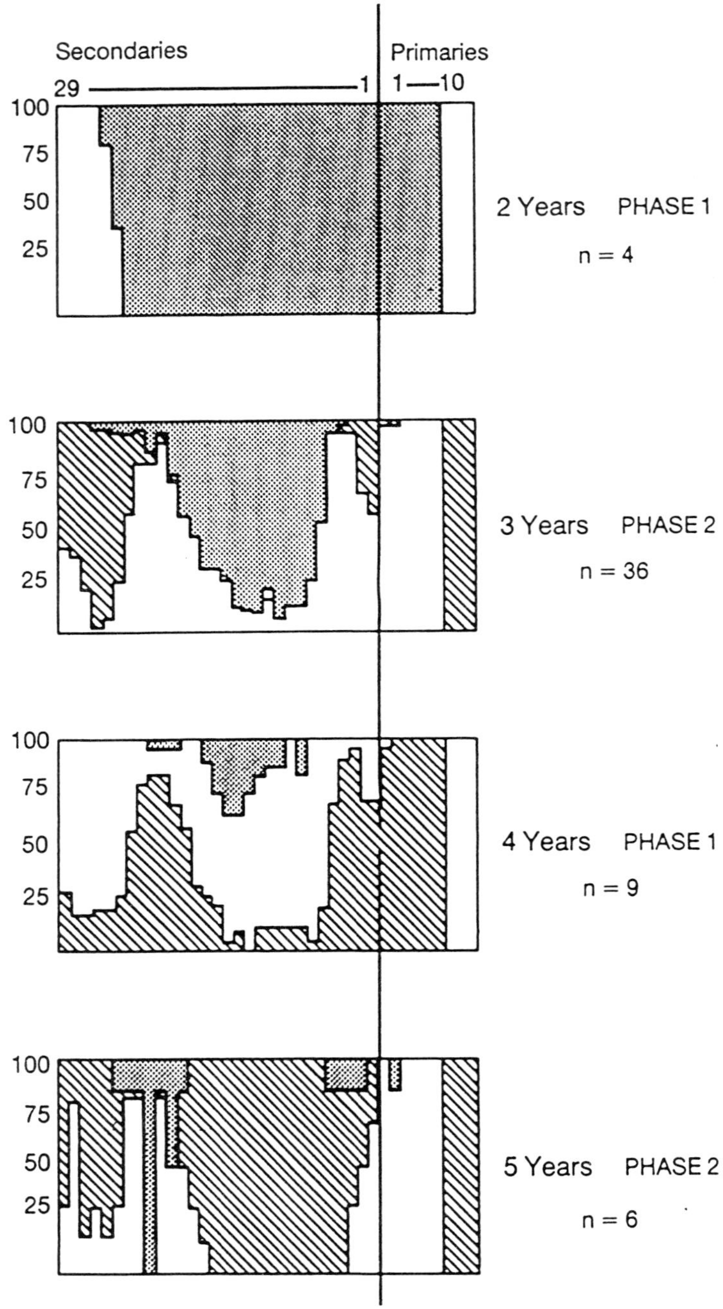

Figure 10.30 *Secondary moult in relation to primary moult in* Diomedea melanophrys *aged 2–5 years. Each box represents one complete wing, with the percentage of feathers of various ages indicated on the 'y' axis. Unshaded areas are recently moulted feathers, hatched areas 1-year-old feathers, stippled areas 2-year-old feathers. From Prince (1993).*

2 North Pacific albatrosses

Langston and Rohwer (1995) found that *D. immutabilis* and *D. nigripes* get their definitive plumage after 34 months at sea. They scored the ages of individual remiges of specimens taken from drift nets and allocated the birds to age classes based on the state of their bursas (Broughton, 1994). As in mollymawks, the primaries were moulted in two series, descendently from P6 and ascendently from P5. Unlike the mollymawks the outer three were renewed annually, birds moulting for the first time replacing only these. Adults might shed all their primaries, only the outer three, or these and all or part of the inner series. When the inner ones were not entirely replaced the next moult began where the previous year's ended. This 'wraparound' system lessens the accumulation of old inner primaries—the older ones being replaced first. An average of 11–12 of the 37 secondaries were also replaced yearly in a complex sequence.

3 Sexual differences in albatross moult

No sexual differences in the patterns of moults in the mollymawks or North Pacific species have been found, but male *D. exulans* checked on the nesting grounds by Weimerskirch (1991) had significantly more renewed primaries than the females—8.3 ± 1.7 ($n = 273$) and 7.3 ± 2.2 ($n = 309$) for breeders and 5.4 ± 2.5 ($n = 116$) and 4.8 ± 2.1 ($n = 125$) for non-breeders. There were three generations of primary feathers which were replaced by descendent wave moult from one to two loci, slightly more in breeding females than in the males. First-time breeding males renewed fewer feathers than established ones but there was no such difference with the females.

4 Rectrices and other feathers

Tail moult in albatrosses takes place simultaneously with that of the wing. Prince *et al.* (1993) found both successful and failed mollymawks renewing their rectrices annually. In *D. exulans* moult has been stated as starting at two loci, R2 and R5, and spreading from the centre outwards. In this species and in four mollymawks Brooke (1981) reported that the 12 rectrices are renewed in a normally symmetrical manner based on three loci on each side of the tail.

The four quills of the alula are also renewed but the mode of renewal is unclear (Brooke and Furness, 1982).

D Ecological factors affecting moult

Harris (1973) was perhaps the first to relate the incompleteness of albatross moult to a shortage of time. Not only do the large remiges take a long time to grow, but the very long breeding period leaves less time for feather replacement. The system does ensure that nesting birds are full-winged and has some flexibility, the extent of the moult being related to bodily condition during the non-breeding period (Weimerskirch, 1991). Langston and Rohwer (1995) found that the outer primaries of the North Pacific birds got heavily abraded and believed this to be a major factor behind their annual replacement.

The work of Harris (1973) shows that *D. irrorata* renews its primary feathers faster than Prince's mollymawks; that is, at a rate of 2.37 per month in successful breeders *vs.* 0.68 and 0.67 per month for *D. chrysostoma* and *D. melanophrys* respectively. Failed *D. irrorata* replaced 1.10 per month as against 0.52 and 0.50 for failed mollymawks. Harris concluded that successful birds, tied to the colonies until late in the season, lacked the 3–4 months needed for complete plumage renewal, so replaced P6–P10 while retaining some of the inner ones. Most wings held three generations of feathers. Prince *et al.* (1993) showed that successful *D. chrysostoma, D. melanophrys, D. chlororhynchos* and *D. exulans* (but not Crozet Island *D. exulans*) also renewed fewer feathers between breeding episodes than failed birds and suggested that the subantarctic species, constrained by the long breeding cycle, must moult when surface productivity is declining whereas, although *D. irrorata* takes even longer to rear its chick because food levels are lower in local seas, it can still renew more feathers because food availability does not fall seasonally as in the south.

On the other hand, successful breeding *D. exulans* of both sexes renewed more primaries than failed ones, having 12 months off-duty to make the change, failed birds only 5–11 months to do that, depending on at what stage failure had occurred. In both categories of females the extent of the moult was linked with the length of the non-breeding season, but this was not so for the males (Weimerskirch, 1991).

XII Embryology

The physiology of petrel embryos was dealt with in Chapter 7.III. None of the classical embryologists studied their ontogeny but Schauinsland (1903) followed the development of those of *D. immutabilis* and *Puffinus pacificus*. In addition to whole embryos (Fig. 10.31), he also showed sections through the blastoderm at gastrulation and later stages.

An earlier study by Studer (1889) mainly dealt with older embryos and especially the changes in the nasal tubes. He examined *Pelecanoides urinatrix, L. brevirostris, Pterodroma lessonii, Pachyptila desolata* and *Fregetta tropica*, again mostly of imprecise ages. Figure 10.24 shows the later formation of the naricorns and their fusion on the top of the bill from their initial lateral position in *L. brevirostris*. In hatchling shearwaters the naricorns are similarly separated.

Recently the organ growth has been followed in embryos of *Puffinus pacificus* by Zhang and Whittow (1992), the only study of this kind on a tubenose and the sole examination of pipped eggs of any bird. In sum, all the main organs increased in size in a linear manner but at varying rates during the pre-pipping stage. After that, however, the stomach, pectoral muscles and eyes grew little while growth of the intestines, liver, heart and leg muscles accelerated; presumably this is one reason why O_2 demand suddenly increases at pipping.

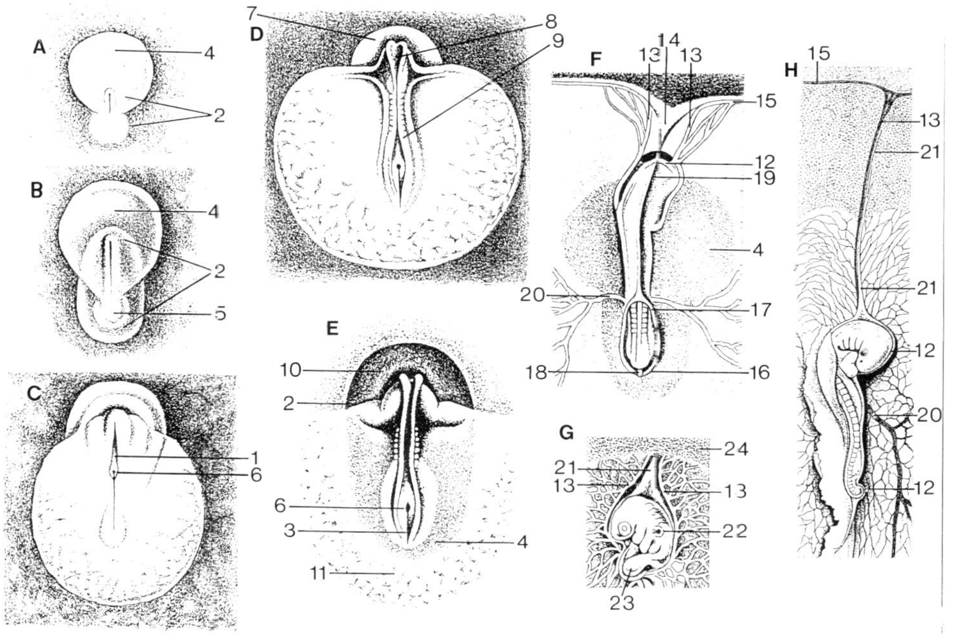

Figure 10.31 *Development of embryos of* Diomedea immutabilis, *precise ages unstated, magnifications various. (A) Complete early embryo with primitive streak forming in mesoderm. (B) Early embryo with further developed primitive streak and anterior end of primitive groove formed. (C) Embryo with anlage of notochord pressing out from forebrain bulge at the anterior end of the primitive streak. (D) An older embryo with somites forming on either side of the notochord, neural folds prominent and primitive streak shortening. (E) Another older embryo with well-developed neural fold. (F) Still older example viewed from below. The primitive streak, now a knot, is almost gone, the amnion folds formed and the vitelline veins anteriorly and arteries posteriorly, well advanced. (G) A well-developed embryo embedded in a network of vitelline vessels, with four visceral clefts and otic and optic capsules. (H) An older embryo viewed from above and embraced by the amnion. 1, Anlage of notochord; 2, mesoblast; 3, primitive streak; 4, area pellucida; 5, area opaca; 6, Hensen's node; 7, proamnion; 8, forebrain with neuropore above; 9, neural fold; 10, blastoderm; 11, anlage for blood and vessels; 12, amnion; 13, vitelline vein; 14, midline of embryo; 15, sinus terminalis; 16, rear of amnion fold; 17, front of amnion fold; 18, primitive knot; 19, mesodermal mesentery; 20, omphalomesenteric artery; 21, anterior end of amnion; 22, otic vesicle; 23, heart tube; 24, mesoderm. From Schauinsland (1903).*

CHAPTER 11
Evolution and Radiation

Contents

I Introduction

The adaptive radiation of the Procellariiformes is one of the most extensive and successful among vertebrates. Even today, with many ranges modified by man, the niches occupied remain very varied. Some, such as *Procellaria westlandica*, form part of the fauna of temperate rainforests (and the subfossil record shows that others did so in recent times), while *Thalassoica antarctica* and *Pagodroma nivea* often breed on the nunataks of Antarctica far from the sea. Man has also provided some new niches, for example the ship-following one for scavengers.

Among a range of proposals concerning tubenose evolution there were two general points of agreement. One is that all arose from a single ancestor. The tubular nostrils of similar design throughout, septal olfactory conchae and peculiar form of the gut emphasize this monophyly. The other is that petrels and penguins are closely related—as first proposed by Furbringer (1888). Characters shared include schizognathous palates, two comparable nestling downs, holorhinal nares, palmate feet, compound rhamphothecae, similar vertebrae, aspects of the shape of the humeri, low body temperatures, considerable fasting ability and vigorous inhalatory and exhalatory calling. A possible link is the presence of tubular nostrils in small chicks of the penguin *Eudyptula minor* (Kinsky, 1960; Warham, pers. obs.). Simpson (1946) drew attention to features of the Miocene skull of *Paraptenodytes* that are more like those of modern tubenoses than to those of modern penguins, for example the shape and relationships of the pterygoid and quadrate bones. However, penguins lack highly developed olfactory bulbs, sense of smell being inoperable under water. There are also similarities in the display repertoires of *Eudyptula* and *Eudyptes* to those of diomedeids and procellariids, and also with sulids and even phalacrocoracids: a comparative analysis of these displays by modern techniques is long overdue.

II The fossil evidence

Tubenoses seem rather unlikely birds to become fossilized, but all four extant families are represented in the fossil record (Appendix I). Named species include seven albatrosses, two or three fulmars, two gadfly petrels, one prion, one *Bulweria*, at least 20 shearwaters, three storm petrels and a diving petrel. There is also a range of unnamed albatrosses, gadfly petrels and shearwaters.

The albatrosses are particularly interesting because six of the named species, including two *Plotornis*, occurred in either the eastern Pacific or in the North Atlantic (*The Petrels*, fig. 2.10), with at least four more in Lower Pliocene deposits at Lee Creek, North Carolina (Olson, 1985c, p. 209). One unnamed *Diomedea* has been identified from the Pliocene of Japan (Ono, 1980). Albatrosses must therefore have been prominent components of the seabird faunas of northern seas during the Miocene and Pliocene. It is possible, of course, that some of the Californian and Carolinan species were ancestral to today's birds, but all the northern ones in the Atlantic seem to have died out during the Pleistocene.

None of these extinct species reached the sizes of today's 'great' albatrosses. However, *Diomedea californica* and *D. anglica* were as big or bigger than today's *D. albatrus*. Others, such as *D. milleri*, were slightly below mollymawk size while the *Plotornis* spp. seem to have been smaller still, helping to close the distribution gap between the smallest albatrosses and the largest procellariids (*The Petrels*, fig. 1.5).

Until recently, the earliest recognized tubenose was the Oligocene *raemondonckii* from Belgium, tentatively placed in *Puffinus* by Olson (1985c). He also reported Upper Oligocene bones from South Carolina that appeared to be procellariid, one rather like *Bulweria*. However, even earlier material, probably procellariiform, includes late Cretaceous *Tytthostonyx glauconiticus* described from a humerus found in New Jersey (Olson and Parris, 1987). These authors tentatively hypothesized this as being similar to the humerus of the stock that gave rise to the Procellariiformes and noted resemblances to an Eocene frigate bird. Nessov (1986) described *Eopuffinus kazachstanensis* from the Upper Palaeocene, *c.* 55 million years ago (MYA) on the basis of a single frontal bone. It was in yellow sandstone near Tashkent laid down in shallow seas along with fish, turtles, primitive notoungulates and sirenians, the discoverer hypothesizing its origin either from the Indian Ocean or from the already opened Atlantic. Also fragmentary are eight fossils from the Lower Eocene of Kent, UK, identified as from four tubenoses by Harrison and Walker (1977). These are *Marinavis longirostris* from a bird about twice as big as *Puffinus tenuirostris*, a very large storm petrel *Primodroma bournei* and *Neptuniavis miranda* and *N. minor*, the former probably larger than *D. exulans* and both assigned to the Procellariidae. If all are correctly diagnosed they imply a wide procellariiform radiation some 50 MYA. Also early in the record is *Hydrotherkornis oregonus* from the Upper Eocene which Miller (1931) thought an alcid but which Chandler (1990) identified as procellariiform. The specimen is a tibiotarsus which Chandler believed very *Daption*-like.

Although the fossil collections are growing and old material remains undescribed, none of the specimens has been definitely identified as 'prototubinare'. Those specimens that are complete or near-complete, such as *P. diatomicus*, seem morphologically similar to modern species, and Olson (1985c) considered that most subgenera of *Puffinus* existed by the Middle Miocene. Among the albatrosses, however, he did note that some skulls were less ossified than those of today's birds, particularly about the nostrils.

Some of today's genera formerly had different distributions. The small gadfly petrel *Pterodroma kurodai* from Aldabra Atoll provides an example, as no '*Cookilaria*'-sized gadfly petrels are found in the Indian Ocean today. Similarly *P. pacificus* or its allies are missing from the Atlantic whereas in the Pleistocene a related species, *P. pacificoides*, occurred on St Helena (Olson, 1975) and an earlier, unnamed relative comes from the Pliocene of North Carolina (Olson, 1985c).

III The past and present radiation

The radiation of the tubenoses seems to have been well in hand by Miocene times and included small and large forms, although, being large, albatross bones may be over-represented in the collections. Both the Tertiary radiation, and more clearly that of today, exemplify Cope's Rule—the tendency to evolve towards larger body size (*The Petrels*, p. 8). In explaining this statistical generalization, Stanley (1973) pointed out that the fossil record of animals shows that in the history of many taxa, size change was associated with speciation events. New taxa have tended to arise at relatively small body size, producing an early concentration of small species both in the higher taxon and in the subgroups within it, thus increasing the tendency for overall size increases with speciation as illustrated by Fig. 11.1. Thus at any particular time the size polygon within a taxon tends to be negatively skewed as it is in the tubenoses today (*The Petrels*, fig. 1.5). Stanley emphasized that species toward the extremes of body size, limited by physical or physiological constraints, tend to be specialized and such species mostly vanish when the niches to which they are peculiarly adapted change or disappear.

Stanley's model of evolution may fit that of the Procellariiformes in which large forms like the North Atlantic albatrosses and the prion *Pachyptila salax* became extinct, and today's size range is even wider than that revealed by the fossil record. The smaller species, breeding faster, would have speciated more rapidly than large forms while those like the 'great' albatrosses, specialized for slope and dynamic soaring, would be among the first to disappear should climatic change lead to persistent calms. Also, such large species, with their smaller populations, are more vulnerable to man-made alterations.

Once established, the procellariiform radiation may have proceeded rather rapidly, perhaps aided by reduced competition at sea and by the new feeding niches made available with the decline of the nekton-eating reptiles.

IV Phylogenies

A *Relationships with other higher taxa*

Partly because of the similarities between fossil and modern species and their perceived links with the penguins whose ancestry was long known to stretch back to

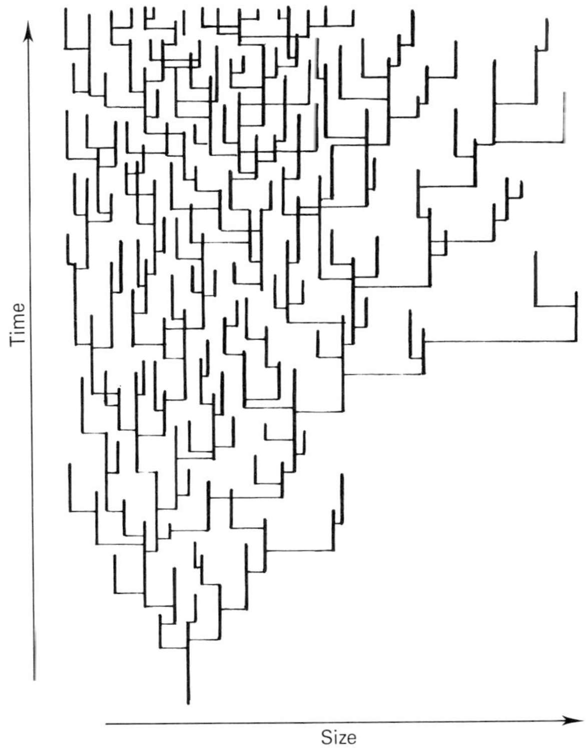

Figure 11.1 *Postulated pattern of size diversification for a typical taxon. Vertical lines represent central, morphologically stable, populations of species, horizontal lines speciation events in which most evolutionary change occurs. After Stanley (1973).*

the Eocene, the Procellariiformes were judged by the earlier morphologists to be an ancient group, perhaps originating in the Cretaceous. Forbes (1882) held this view and (p. 63) listed the anatomical features he thought possessed by the ancestral form. Many of the early theories of their history were reviewed by Sibley and Ahlquist (1990, pp. 528–534) and a belief in their antiquity has also been held in recent times, for example by Cracraft (1981) who considered that a close relationship between the gaviomorphs, penguins, procellariiforms and pelecaniforms suggested a lineage of aquatic species established very early in the radiation of birds, presumably in the early Cretaceous or late Jurassic. From a study of the electrophoretic mobilities of mitochondrial malate dehydrogenase of a wide range of birds, Kuroda *et al.* (1982) placed the tubenoses near the foot of the radiation after the rheas and tinamous. Harper (1978, p. 545) pointed out that his data showed that several major protein groups, for example transferrins, are monomorphic throughout the Procellariiformes, whereas in still rapidly evolving groups such as terrestrial birds, most are polymorphic at this locus, suggesting that the petrels have been adapted to their environments for a very long time.

Animals may retain evidence of an ancient ancestry. With petrels two characteristics may be suggested: their well-developed olfactory sense and their low body temperatures. It seems at least as likely that the large olfactory bulbs are derived from a proavian ancestor which kept an efficient reptilian sense of smell, perhaps to perfect it for detecting food at sea (Wenzel, 1986), rather than that the prototubinare reacquired that sense after it had been mostly lost in the early evolution of birds, confounding Dollo's 'Law'. A low body temperature can also be viewed as a preadaptation to a marine lifestyle. Other possible retentions from a reptilian past include the white egg resistant to chilling, energy saving by torpidity in the young, delayed maturity, long development cycles, long lifespans and a shared ability to hydrolyse wax esters.

The possible homologies of the petrel rhamphotheca with the scales on the snout of modern diapsids were examined over 90 years ago (Lonnberg, 1904). He traced the modified reptilian scales in the compound rhamphothecae of ratites, tinamous, penguins, in several *Diomedea,* and in *Macronectes, Puffinus* and *Fulmarus glacialis.* In *Diomedea* he homologized the single maxillary unguis with the single reptilian rostrale, the culminicorn with the internasal and, below that, on either side, a long narrow shield extending from the nostril ending in a tooth-like projection towards the rostral hook that he believed represented the first labial fused to the second labial to produce the latericorns. Lonnberg equated the naricorns to the reptilian nasals that lie between the labial and internasal shields. He linked the single inter-ramicorn of the albatross with the reptilian mentale that lies between the infralabials which, fused together, form the mandibular unguis. Lonnberg considered the ramicorns to be made from two plates separated by the sulcus (as in *Phoebetria*), the upper a second infralabial, the lower a submandibular. He also stated that the fusion of these two parts in *Diomedea* could be traced to their ontogenetic development. In procellariids, *Pelecanoides* and *Phoebetria* these lower plates tend to separate when digested with alkali, but remain fused in *Diomedea* (see Fig. 10.1). Summarizing a series of papers on the tubenose rhamphotheca, von Boetticher (1955), also used the terminology of reptilian scalation in describing the bill plates, but was evidently unaware of Lonnberg's earlier work.

Another feature often thought to be primitive in birds is dark plumage (e.g. Hutton, 1903; von Boetticher, 1932; Kuroda, 1954; Whittow, 1993), so that *Phoebetria* has been regarded as the albatross with the most primitive characters, particularly in view of its deep sulcus and despite its advanced flight style. Von Boetticher (1932) reviewed plumage colour throughout the order, attempting to relate the predominant heavily pigmented state, primitive in his view, with the complexity of the rhamphothecae and foot scalation.

It may be significant that dark plumage can be accompanied by limited display and vocalization repertoires, for example in *Phoebetria vs. Diomedea* and *Bulweria* with no aerial display and very simple calls *vs.* other procellariids. That the prefrontals are also unfused to the nasals and mesethmoids may also be a primitive character in *Bulweria* (Bourne, 1975). Indeed, one could speculate on the hypothetical prototubinare resembling *Bulweria;* that is, being rather small (*Bulweria* is the smallest procellariid), dark plumaged, with few vocalizations and displays and, as Jouanin *et al.* (1979) pointed out, nesting in natural cavities rather than digging its own burrow. This bird's habit of excavating a nest depression beneath its protective boulder suggests how burrowing may have begun. Such an ancestor would also be expected to have a low body temperature (see Table 7.2) and well-developed olfactory bulbs.

Their development in *Bulweria* awaits examination but the species is said to have a peculiar body odour and possibly olfactory communication. This could be a retained character predating the evolution of a more advanced vocal system.

The Procellariiformes have been linked with a range of orders other than penguins, most particularly the Gaviiformes or divers. These share with the tubinares pale white eggs, schizognathous palates, palmate feet, two coats of nestling down, similarities of feather structure (see Fig. 10.26) and quite large olfactory bulbs (Cobb, 1959/60, fig. 2). Kuroda (1954, p. 11) noted considerable similarities in the development of the vertebral hypapophyses and the big development of the gaviid cnemial crest. Olson (1985c, p. 215) figured a late Eocene penguin that he thought quite loon-like, reported other loon-like material from the Seymour Island penguin deposits, and suggested a common ancestry of penguins and loons.

Links with Charadriiformes have been mooted. From his extensive work on mallophagan parasites, Timmermann (1965) deduced that *Phaethon* and *Fregata*, not the penguins, were nearest the tubenoses. However, Paterson (pers. comm.) points out that penguin feathers form a very uniform habitat unlikely to support a varied louse fauna, so that comparisons with petrels may be invalid. Lonnberg (1904, p. 505) compared the rhamphothecae of skuas with that of *Fulmarus* and found the arrangement of the plates very similar, even proposing that the skua's cere was the homologue of the petrel's nasal tubes.

The unravelling of the relationships of the petrels to other major groups took a new turn with the development of biochemical methods such as Jacob's work on the chemical composition of preen gland waxes (Jacob, 1976). However, the wax of *D. melanophrys* was atypical, suggesting a closer link to the flamingoes than to other petrels, illustrating that chemical taxonomy has its pitfalls (Harper, 1978). Methods for comparing proteins and genetic material and of cladistic analyses of data from which hypothetical dichotomous phylogenetic 'trees' based on character matrices can be constructed, represent a considerable advance.

Cracraft (1981, 1982, 1985) included tubenoses in his cladistic analyses of skeletal morphology. He grouped them together with the Sphenisciformes, Gaviiformes and Pelecaniformes, quoting similarities between diomedeids (which he considered primitive) and pelecaniforms in cranial and postcranial characters.

Prager and Wilson (1980) produced the preliminary phylogeny of Fig. 11.2 from immunological comparisons of related monomeric proteins from a wide range of birds: note the grouping of penguins, loons, albatrosses and herons.

The dendrograms of Kuroda *et al.* (1990) derived by Unpaired Group Mean Averaging (UPGMA) from protein electrophoresis of 15 species gave the storm petrels as the sister group of the rest with the separation of the albatrosses *D. immutabilis* and *D. nigripes* following that of the Procellariidae.

Livezey (1989) included *Pelecanoides urinatrix* in a morphometric analysis of recent and living penguins because of hypotheses that both evolved from a similar flying bird. The substantial differences uncovered were not thought enough to invalidate the hypothesis. A factor preventing *Pelecanoides* from improving its underwater propulsion by becoming flightless may be that such birds would be too small and vulnerable when travelling from the sea to their nests. Significantly, the smallest penguins are burrowers and land after dark. A cladistic analysis briefly described by O'Hara (1989) gave *D. exulans* and *Gavia immer* as sister taxa to the penguins.

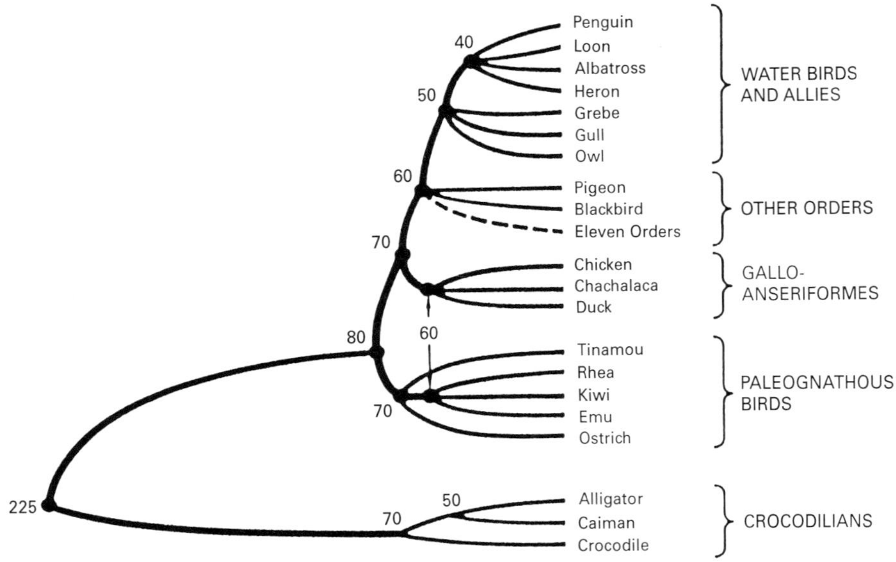

Figure 11.2 *A tree offered as a guide to further research based mainly on immunological comparisons of transferrin, albumin, penalbumin and ovalbumin. More than 1200 pair-wise comparisons were made, including over 500 transferrin immunological distances. From Prager and Wilson (1980).*

Another study based on morphological characters is that of McKitrick (1991b) who included 35 tubenoses in her phylogenetic analysis based on 66 characters of the hind limb. Part of one of her trees is shown in Fig. 11.3. In a different approach, Starck (1993) analysed hatchling characters of a wide range of species from Palaeognathae to the Passeriformes. The resulting cladogram shows the Procellariiformes and Pelecaniformes branching from a common lineage that also gave rise to the Sphenisciformes and Gaviiformes.

McKitrick (1992) married hindlimb and behavioural data on nesting systems with parental care. She included five petrels. The branching patterns in the new trees differed somewhat from the earlier ones, for example in showing *Oceanodroma* diverging before *Oceanites*, but, as in Fig. 11.3, with the penguins as the sister group to the tubenoses.

Comparing the average DNAs of species using the DNA hybridization technique developed for birds by Sibley and Ahlquist (1990) has provided many leads on phylogenetic relationships, particularly of the higher taxa. They measured the degrees of differences in the DNAs of some 1700 species by hybridizing single-stranded, unlabelled DNA with a wide range of single-stranded radioactively labelled DNA. The hybrid material was heated, gradually breaking the hydrogen bonds, and at intervals the released single-stranded DNA was radioassayed, giving a measure of the hybrid molecules melted at each temperature and of the number of their unmatched bases. The more similar the DNA strands, the higher the melting point. For example, in Fig. 11.4, the *L. brevirostris* curve is a 'homoduplex' (a hybrid of labelled and unlabelled DNA from the same bird) and not until *c.* 85°C was 50% of double-stranded DNA lost, whereas the heteroduplexes from *Pelecanoides urinatrix*

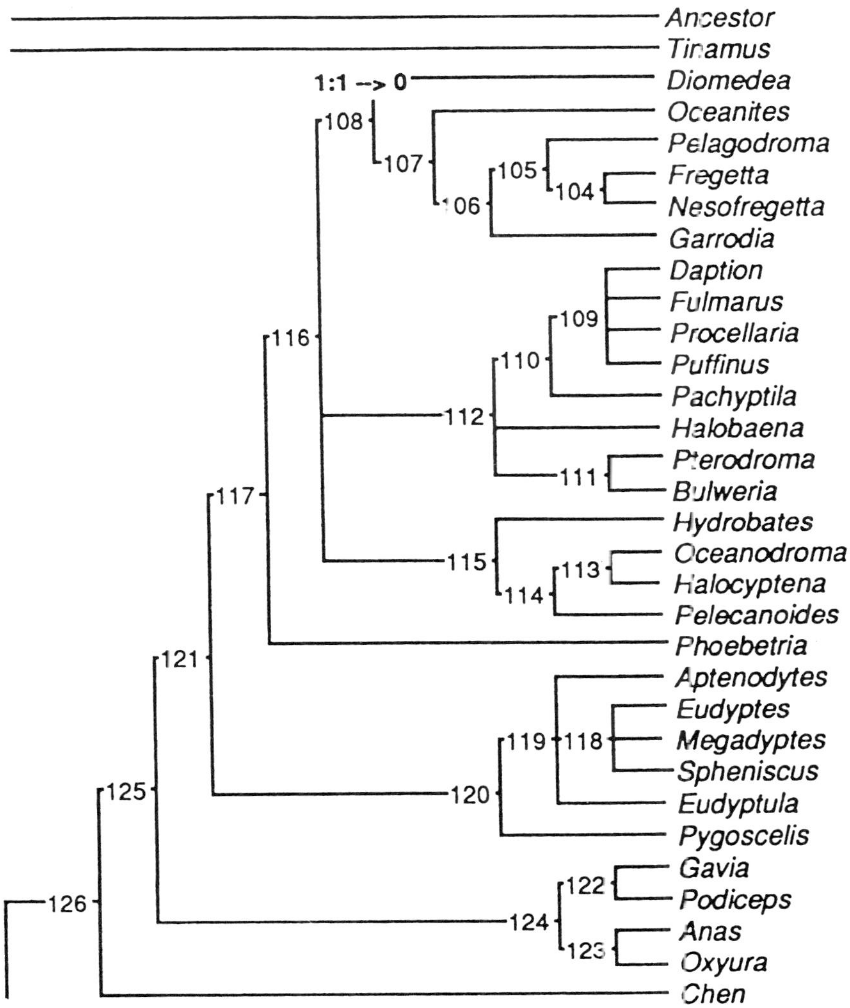

Figure 11.3 *Part of a tree based on hind limb musculature. The nodes are numbered. Note the special position of* Phoebetria *and separation of* Oceanites *from* Garrodia. *From McKitrick (1991b).*

and *Puffinus puffinus* reached that stage at around 81°C and the *Oceanodroma furcata* hybrid at 77.5°C. The plots for the diver *G. stellata*, frigate bird *F. ariel*, penguin *P. papua*, and cormorant *P. pelagicus* show that their hybrid DNAs melted at even lower temperatures.

The hybridization data generated a series of branching trees like those of Fig. 11.5. As with earlier ones, these place the loons and penguins as the groups nearest the Tubinares, more distant relatives being the frigate birds and the Pelecaniformes.

Sibley and Ahlquist (1990) set their DNA clock by reference to dated geological events affecting bird distributions and estimated a median reduction in melting point of 1°C as equal to a span of 4.5 million years. They used this figure to estimate

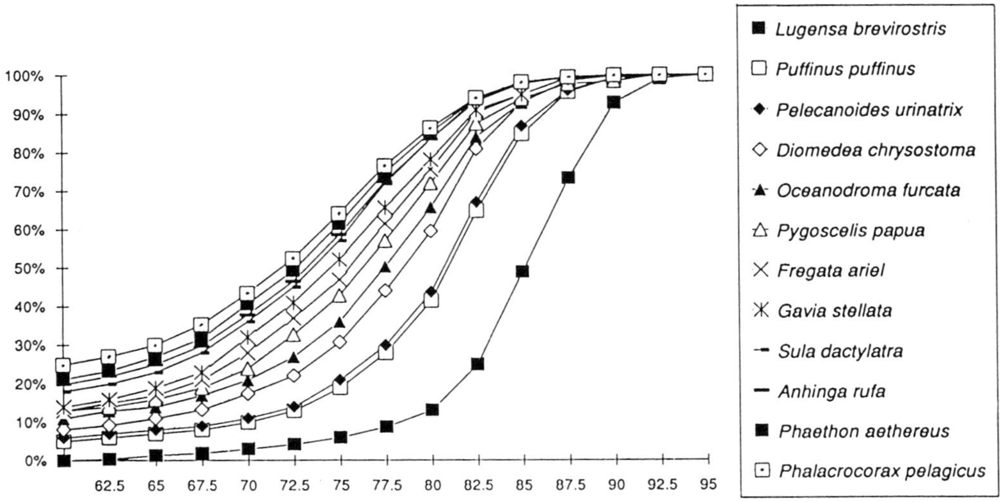

Figure 11.4 *Melting curves for hybrid DNAs with* Lugensa brevirostris × *Manx Shearwater, Common Diving Petrel, Grey-headed Albatross, Fork-tailed Storm Petrel, Gentoo Penguin, Lesser Frigatebird, Red-throated Loon, Masked Booby, African Darter, Red-billed Tropicbird and Pelagic Cormorant. From Sibley and Ahlquist (1990).*

approximate divergence dates, for example in Fig. 11.5 the petrels and gaviids branch off at 45 MYA, the storm petrels from the other tubenoses at 36 MYA.

Sibley and Ahlquist's work, and the bird classification based upon it have been criticized on various counts, but despite these they have clarified a number of problems about relationships and, as they state (p. 536) 'we have not claimed perfection, just progress'.

Particular attention has been paid to southern petrels and penguins in a series of studies by Paterson *et al.* (1993, 1995, in press); Paterson & Gray (in press) and Wallis *et al.* (pers. comm.) which mark an important step forward. Their phylogenies are based on cladistic analyses of data derived from sequencing the mitochondrial 12S ribosomal RNA gene, from isozyme electrophoresis, from behavioural and lifestyle characters, and from the presence or absence of phthirapteran (mallophagan) ectoparasites.

The idea that the evolution of parasites parallels that of their hosts has been around a long time, and Harrison (1916) related six groups of '*Lipeurus*' mallophaga to their procellariiform hosts. These parasites sorted into seven groups associated with: (i) *Garrodia, Oceanites* and *Oceanodroma*; (ii) *Hydrobates* and *Pelagodroma*; (iii) *Daption, Pagodroma* and *Thalassoica*; (iv) *Pterodroma, Puffinus, Procellaria, Bulweria* with *Pelecanoides* as an outlier; (v) *Fulmarus*; (vi) *Macronectes, Diomedea* and *Phoebetria*; and (vii) *Pachyptila* and *Halobaena*. Despite his placement of *Pelagodroma* with *Hydrobates* and his suggestion that *Macronectes* is an albatross, most of his deductions fall into line with more detailed recent results.

Timmermann (1965) set out to check the current systematics of the Procellariiformes against the distribution of the 124 mallophaga to which they were hosts. In general, his system is quite congruent with those from other approaches. He did not erect a petrel phylogeny but rejected the idea that *Pelecanoides* was an auk, and he related *Halobaena* to *Pachyptila* and *Bulweria* to *Pterodroma*.

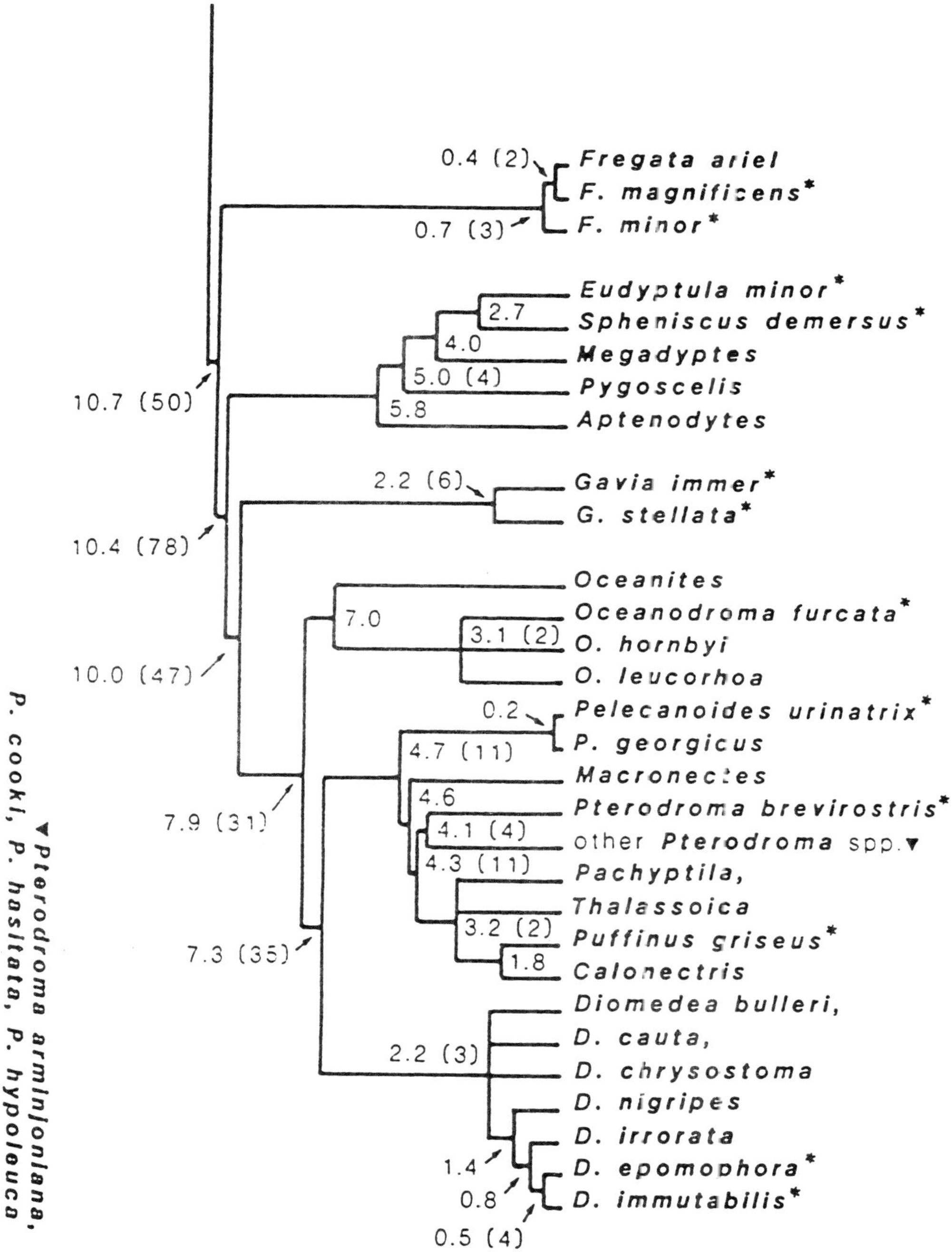

Figure 11.5 *Part of a tree generated by UPGMA analyses of average linkage clustering from DNA/DNA hybrids. Numbers at nodes are differences in °C between hybrids and homoduplexes, numbers in brackets are such values averaged. Askerisks indicate species used as tracers. UPGMA assumes a uniform rate of DNA evolution. From Sibley and Ahlquist (1990).*

Paterson *et al.* (1993) discuss possible causes of error in attempting to reconcile host and parasite phylogenies including host-switching. They thought that this was unlikely to be important as the lice cannot live long off the birds nor on hosts not their own, and that there is little physical contact between different species. However, there are many instances today where two species of petrel share a burrow, or nest cheek by jowl and fight for possession of nest sites, for example *Puffinus griseus* and *Pterodroma inexpectata, Pelagodroma marina* and *Pachyptila vittata,* with one species evicting eggs, chicks or even adults of the other, as the aggressive *Pterodroma nigripennis* is doing to *P. neglecta* and *P. solandri* on Philip Island, Norfolk Island (O. Evans, pers. comm.). So there must have been ample opportunity for switching over the long period of petrel evolution and the possibility cannot be dismissed.

The phylogenies of Paterson *et al.* (1993) used representatives of the Charadriiformes and Pelecaniformes as out-groups. The host phylogeny, based on the presence or absence of 16 louse genera on 18 bird species, is shown in Fig. 11.6A.

Tests for congruency between the parasite and host trees were rather equivocal, although implying that cospeciation had occurred. The authors predicted that three louse genera share the same pattern of descent as their hosts (*Puffinus griseus, P. huttoni, Procellaria westlandica, Pterodroma inexpectata, P. cookii*), and that species parasitizing *Puffinus* are more closely related to the lice of *Procellaria* than to those of *Pterodroma*. The patterns of cospeciation were considered to be masked mainly by relict distributions and Paterson *et al.* (1993) speculated that much of the radiation of the lice predated that of the birds: their rates of evolution may also differ.

Paterson *et al.* (1993) used foraging methods and reproductive, agonistic and life-history patterns for their behavioural data, the resulting trees being compared with those from isozyme electrophoretic and 12SrRNA data. The behavioural/life-history trees proved to have significant cladistic structure and, using special tests for congruence (Page, 1993), were found to match those from protein electrophoresic and ribosomal DNA data significantly better than chance. Figure 11.6B shows a cladogram derived from the combined information, Fig. 11.7 a reconciliation of some louse and host phylogenies.

From his analysis of the sequences of 1143 base pairs of the mitochondrial cytochrome-b gene of 50 species using *Aptenodytes patagonicus* as the out-group, Nunn (1994) confirmed the monophyly of the traditional four families. On his data the earliest to diverge, the Diomedeidae, form the sister group to the rest, the hydrobatids and procellariids being the next to separate. A similar separation of the albatrosses was postulated by earlier work (e.g. Figs 11.5 and 11.6), and from analyses of skeletal characters (Meredith, 1985).

The position of the diving petrels is unclear; should these be given family rank or placed with the Procellariidae as Sibley and Ahlquist's (1990) and Meredith's (1985) analyses imply? The various arrangements from recent studies are contradictory—as a sister group to *Pelagodroma marina* (Fig. 11.6A), to *Daption* (Fig. 11.6B), to *Oceanodroma* and *Halocyptena* (Fig. 11.3), to *Puffinus griseus* by Kuroda *et al.* (1990, fig. 2), and to the Procellariidae *in toto* (Figs 11.5 and 11.9). Thus, at present, the retention of these birds in a separate family seems desirable.

Clearly, the methods so far discussed need refining, perhaps in more careful choice of features examined. For example, in morphometric analyses to use

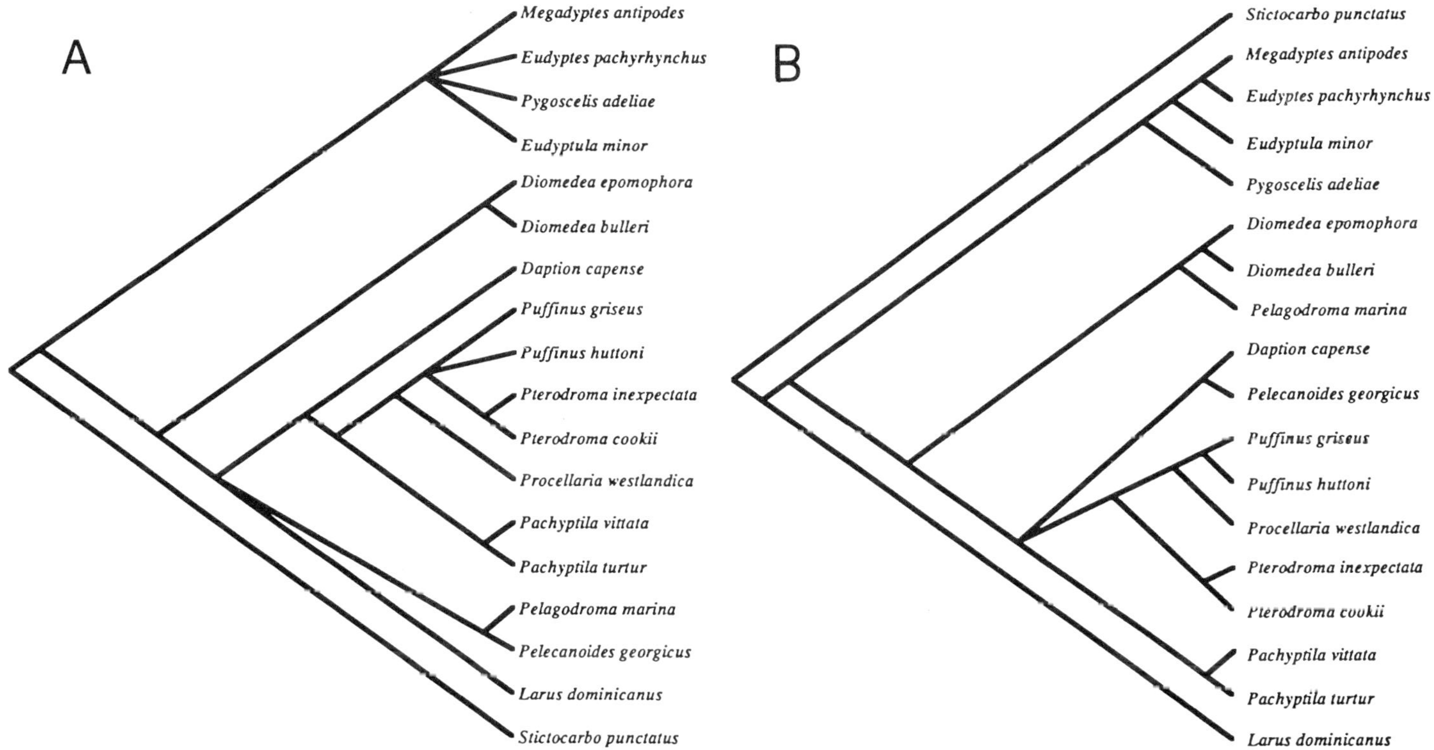

Figure 11.6 *(A) A host cladogram generated from the presence or absence of louse genera. (B) 'Bootstrap' tree (1000 replications) of combined data from feather lice, mt 12S rRNA, isozyme electrolysis and behavioural/life history information. From Paterson* et al. *(1993).*

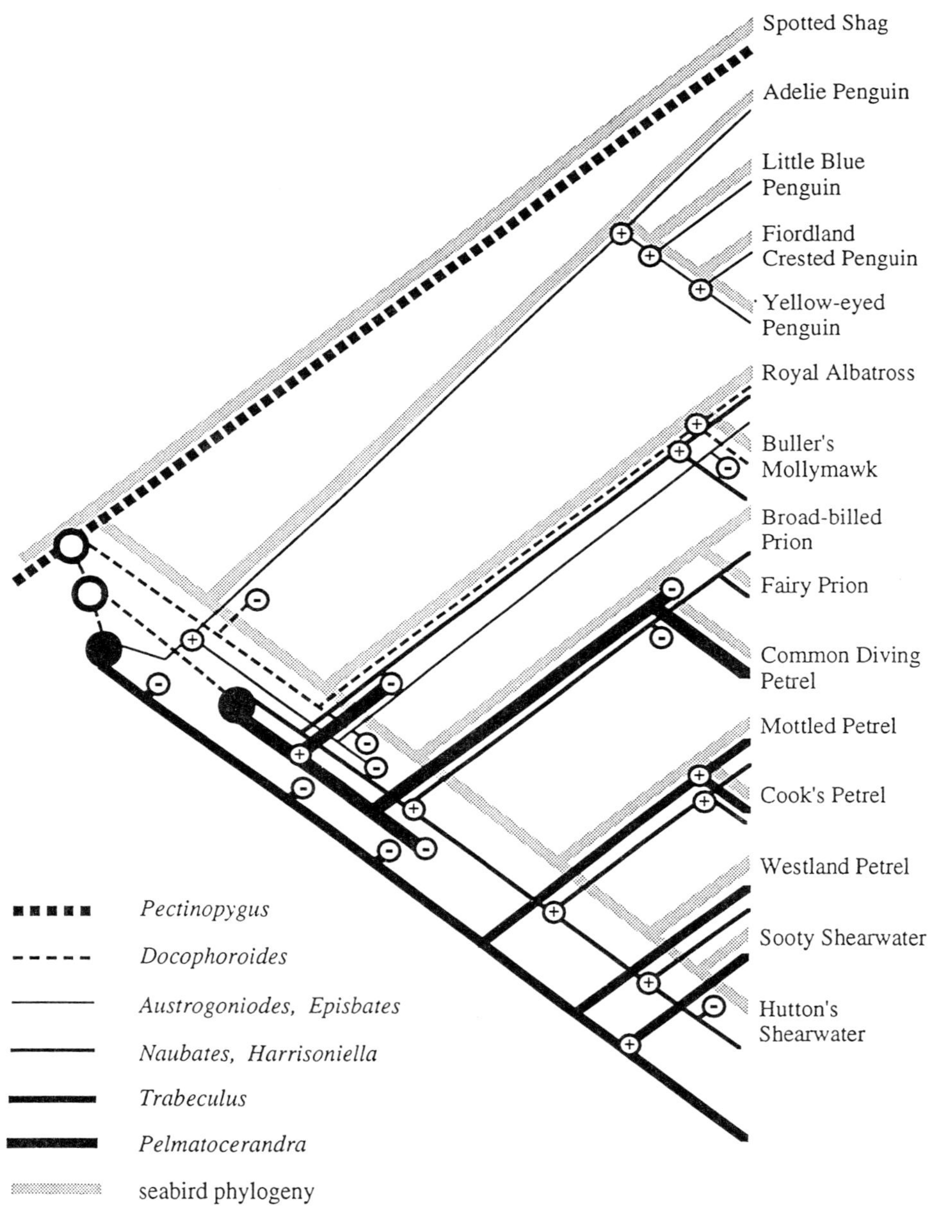

Figure 11.7 *Reconciliation of seabird and lice phylogenies with louse relationships mapped on to the seabird phylogeny. Postulated evolutionary events are shown: cospeciation (+), sorting (−), that is, extinction or missed in sampling, intrahost speciation (open circles), and incongruent duplication (filled circles, potential host-switching). From Paterson & Gray (in press).*

dimensions of several variables all of which scale with body size (*The Petrels*, fig. 1.6) is to use body size repeatedly as if separate variables.

B *Intergeneric and interspecific relationships*

In the past, relationships between and within genera have usually been deduced from similarities of structure, plumage, distribution, behaviour and vocalizations, and have seldom been based on genetical evidence. For example, only two studies of petrel karyotypes appear to have been published. Oguma (1927) determined that *Oceanodroma leucorhoa* had 37 haploid chromosomes shaped more like those of *Lacerta* than of a bird, and Yamashina and Udagawa (1954) figured those of *Calonectris leucomelas* whose haploid number was 42.

Sibley and Ahlquist's (1990) data on seven *Diomedea* spp. gave a tentative phylogeny for that genus (Fig. 11.5). This suggests that the ancestral form first gave rise to the mollymawks with the northern Pacific species and *D. epomophora* evolving later. Mitochondrial 12rRNA and parasite data separated the mollymawks from the great albatrosses and would support the reintroduction of *Thalassarche* (Wallis *et al.*, pers. comm.). However, they only compared *D. epomophora* and *D. bulleri*.

Nunn *et al.* (1994) and Nunn *et al.* (in press) analysed the cytochrome-b gene (1143 base pairs) from 14 albatross species (Fig. 11.8). Their cladistic tree shows the great albatrosses as the sister group to the North Pacific and Galapagos species with the mollymawks radiating from a common ancestor with *Phoebetria*. They suggest *D. chlororhynchos* as exemplifying the archetypical ancestor to these southern forms.

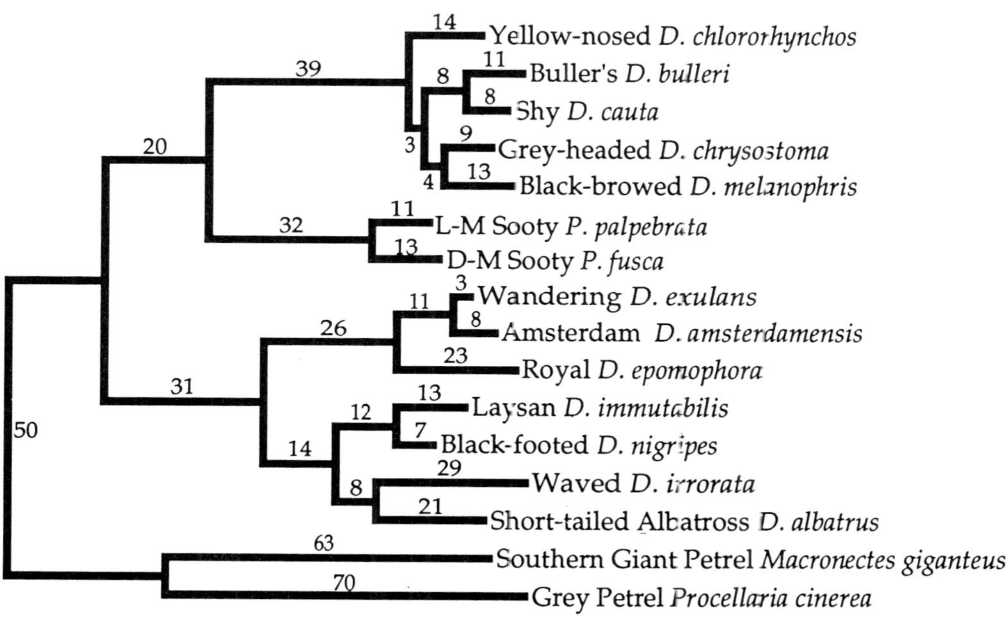

Figure 11.8 *Provisional parsimony analysis of the cytochrome-b gene of 14 albatross species. Numbers indicate synaptomorphic substitutions in the DNA sequences. From* Nunn et al. *(in press).*

Their findings indicate the need for a systematic reappraisal of this family with reversion to some of the earlier names based on morphological characters as previously envisaged (*The Petrels*, p. 30).

These studies also revealed significant genetic differences between the various subspecies of *D. cauta, D. chlororhynchos, D. melanophrys* and *D. exulans* (including *D. e. amsterdamensis*), that would support their separation as full species. The subspecies of *D. epomophora*, which regularly produce fertile hybrids (Robertson, 1993b), would not be affected.

Palma and Pilgrim (1984) found that, alone of the northern hemisphere albatrosses, *D. irrorata* is host to the louse *Harrisoniella ferox*, a species otherwise known only from the southern mollymawks, the other northern albatrosses having different members of this louse genus. The four other kinds of phthiraptera restricted to *D. irrorata* emphasize its special position and possible derivation from ancestral mollymawk stock (cf. Fig. 11.5), an idea not supported by the cytochrome-b data of Fig. 11.8.

From his mitochondrial DNA work, Nunn (1994) also determined an early origin for the fulmars, which may account for the great variations in their size and plumage, and *Fulmarus miocaenus* and *F. hammeri* of the Miocene also suggest that the fulmarine radiation began quite early, perhaps late in the Oligocene.

An early use of electrophoresis was by Shaughnessy (1970a) in comparing the mobilities of transferrin, albumin and haem-binding proteins in the blood of the two giant petrels. Their identical mobilities suggest that these species had only recently evolved. In further studies, Shaughnessy and Conroy (1977) used family data to show that plumage colour in *Macronectes giganteus* is controlled by two autosomal allelic genes with white phase dominant to dark. They also proposed that pure white chicks with pale feet (but dark eyes) such as occur at Macquarie Island (Warham, 1962), are homozygous white, the commoner white birds with dark flecks, heterozygous.

Palma and Pilgrim (1987) described a phthirapteran *Perineus macronecti*, restricted to *Macronectes* which is host also to *Docophoroides, Paraclistus* and *Perineus circumfasciatus*, all mainly parasitic on *Diomedea*. *P. circumfasciatus* is close to *P. macronecti*, their similarity implying that the latter is a relatively recent development. Palma and Pilgrim speculated that *Macronectes* acquired their ancestral infestations from diomedeid hosts, or that the sharing of these genera indicates a closer relationship with the albatrosses than is currently admitted.

From his work on plasma proteins, Harper (1978) deduced that the prions were related to the fulmars and Nunn (1994) also placed *Halobaena* and *Pachyptila* with them, as well as *L. brevirostris*, which he considered a specialized, burrow-nesting fulmar.

For gadfly petrels, Imber (1985a) used a mixture of anatomical, skeletal, behavioural and parasitological data and assumed that the untwisted intestines represented the original condition, the helicoidal state of the 29 species he left in *Pterodroma*, an advanced one. Meredith (1985), in unpublished cladistic analyses of skeletal characters that he believed were derived, produced a phylogeny for 23 species based on principal component analyses. This differs considerably from that of Imber (cf. *The Petrels*, pp. 89, 90), and Meredith conceived the twisted intestines of most *Pterodroma* as a shared derived condition that has been secondarily reduced, for

Table 11.1 Feather Lice (Phthiraptera) of the genera* *Halipeurus, Trabeculus,* and *Naubates* for which gadfly petrels are hosts. From Imber (1985a) and Zonfrillo (1993)

Phthiraptera	*Pterodroma* hosts
Halipeurus: marquesanus group†	
H. noctivagus	*c. cervicalis, p. phaeopygia*
H. heraldicus	*alba, heraldica*
Halipeurus: procellariae group	
H. theresae	*axillaris, hypoleuca, caribbaea, hasitata, feae*
H. leucophryna	*pycrofti, longirostris*
H. turtur	*cookii*
H. consimilis	*inexpectata*
H. kermadecensis	*n. neglecta, arminjoniana, externa*
H. procellariae	*macroptera* subspp., *lessonii, incerta, magentae, m. mollis*
Halipeurus sp. nov.	*madeira, cahow*
Trabeculus hexakon	*axillaris, nigripennis, pycrofti, longirostris, leucoptera* subspp., *cookii, externa*
T. fuscoclypeatus	*n. neglecta, arminjoniana, hasitata*
T. schillingi (small form)	*inexpectata*
T. schillingi	*macroptera* subspp., *lessonii, caribbaea, incerta, m. mollis, madeira, feae*
Naubates damma	*leucoptera* subsp., *cookii*
N. thieli	*solandri*
N. pterodromi	*mollis*
Naubates spp.	*inexpectata, ultima, macroptera* subspp., *lessonii, incerta*

*None of these genera is hosted by *Lugensa*.
†*Pterodroma rostrata* is host to *H. marquesanus*.

example in *P. nigripennis, P. externa cervicalis, P. hypoleuca* and *P. ultima*, not advanced as postulated by Imber.

Meredith agreed with Imber on the distinctiveness of *Pseudobulweria* and *Lugensa*. Compared to the *Pterodroma, Pseudobulweria* had a small cranium, long, narrow pelvis, higher *crista cnemialis*, laterally compressed tarsometatarsus and a narrow articular area of the lower mandible, in addition to the features noted by Imber. *Lugensa* differed in the very wide frontal between the nostrils, short synsacrum, compressed culmen and relatively narrow tibiotarsus and tarsometatarsus.

Timmermann (1965) was uncertain about the relationships of *L. brevirostris* and Imber (1985a) used the existence of its unique feather louse *Beafordiella*, the presence of *Saemundssonia pterodromae* and the absence of *Halipeurus, Naubates* and *Trabeculus*, as evidence of this bird's distinctiveness and its closeness to the fulmars. Imber also saw evidence of secondary infestations (host switching), for example *Naubates thieli*, which is found only on *Pterodroma solandri* (Table 11.1), is related to a group with shearwaters as its host—winter-breeding *P. solandri* may share burrows with summer-breeding *Puffinus pacificus*. Paterson *et al.* (1993) considered one case of incongruence due to a possible secondary infestation, for example an ancestor of the *Harrisonella* group had switched from procellariid petrels to *Diomedea*.

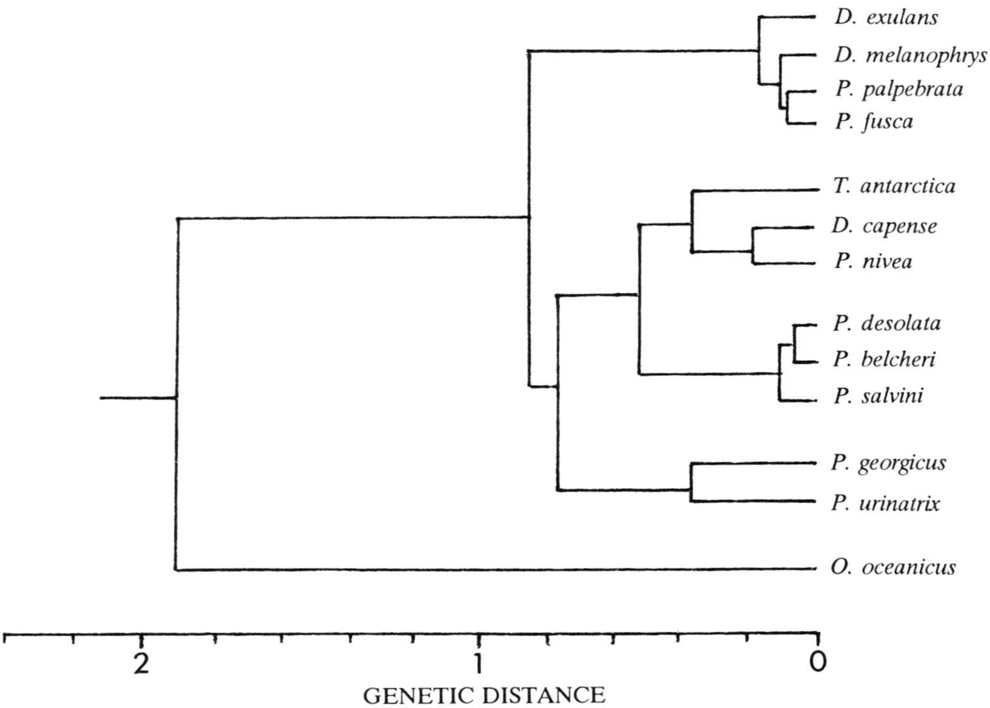

Figure 11.9 *Dendrogram based on Nei's distances and derived by UPGMA for 13 southern petrels. Dates of nodes in MYA calculated as 26.3 × 'd' values. From Viot* et al. *(1993).*

In a smaller study, Zonfrillo (1993) used feather lice to examine the relationships of the *Pterodroma* breeding in the North Atlantic. His analysis of the *Halipeurus* fauna separates *P. madeira* from *P. feae* despite their current geographical proximity, *P. madeira* and *P. cahow* having a common ancestor with *caribbea* and *P. feae*, both having descended from a common stock with *P. hasitata*. The separation of *P. madeira* from *P. feae* was supported by data on measurements and plumage. This arrangement differs from the results of a multivariate analysis of morphological and vocalization characters by Bretagnolle (1995) of all the *P. mollis* 'complex'. He concluded that only *P. mollis* and *P. feae*, each with three subspecies, should be recognized.

Some genetic analyses such as that of Wink *et al.* (1993a) and Nunn (1994) associate *Calonectris* with *Puffinus*; the latter study also identified *Procellaria* as closer to the fulmars than to the shearwaters.

Figure 11.9 is derived from electrophoretic analyses of blood proteins at 29 loci by Viot *et al.* (1993). Like Fig. 11.8 it shows *D. exulans* arising earlier than *D. melanophrys*. Their data fell into five distinct groups corresponding to the customary families and the fulmars. The earliest group, the storm petrels, although only represented by *Oceanites oceanicus*, is estimated to have diverged about 50 MYA. The prions *Pachyptila desolata* and *P. belcheri* were very close, more so than either was to *P. salvini*.

Viot *et al.* also compared mtDNA data for the prions and *Halobaena caerulea*, which they considered related but distinct, with a mean divergence time of 3.7 MYA. The two diving petrels were highly differentiated, their estimated divergence at 9.5 MYA.

Despite Shaughnessy's demonstration of the similarities of some plasma proteins of the giant petrels, their specific distinctiveness was hardly in doubt. However, the existence of very dark breeding *M. halli* on the Chatham Islands, and of Falkland and Gough Island birds with green-tipped bills, usually classified as *M. giganteus*, but showing intermediate characters, was not then appreciated. A comprehensive examination of the genetic constitutions of all the populations (including the small group of *M. halli* still breeding off Stewart Island) could clarify the situation.

Among gadfly petrels, the status of the polychromatic *Pterodroma neglecta* and *P. arminjoniana* breeding sympatrically at the Pitcairns, whether hybridization occurs and so on, might be resolved by discrete sampling of blood for DNA analysis by polymerase chain reaction (PCR). A start at clarification of this situation has been made by Brooke and Rowe (in press). They found dark morph *P. arminjoniana* mating only with dark morphs and nesting away from the light morphs. By combining some behavioural information with sequencing data of the mt cytochrome-b gene, they concluded that two species were involved, naming the dark morph *P. atrata*, the light-phase birds remaining as *P. heraldica*. A similar approach might help determine whether *P. l. pycrofti* deserves full specific status or is best regarded as a subspecies of *P. longirostris*. Meredith (1985), noted that the sterna and rostra of *P. pycrofti* and *P. longirostris* differ, supporting their separation.

One of many uncertainties about *Puffinus* shearwaters is the status of *P. carneipes* vis-à-vis *P. creatopus*: are they but one species? A molecular approach might also resolve the question of whether *Garrodia* should be merged with *Oceanites* as some palaeontologists have proposed, despite a great lack of knowledge of the comparative biology of these genera, for example *Garrodia* is very unusual among small petrels in nesting above ground (see also Fig. 11.3).

The cytochrome-b genes of *Oceanodroma*, and particularly of the unidentified dark specimens that have been turning up recently in the North Atlantic, were analysed using the PCR technique by Dawson *et al.* (1995). The gene sequence of the mystery birds proved identical with those from *O. monorhis* from the Far East, supporting their identification as that species and the view that there is a breeding population in the Atlantic, probably at the Salvage Islands, where a bird with a well-vascularized brood patch was examined in 1994 (Jouanin, pers. comm.).

C *Relationships within species and populations*

Recent investigators of the status of races have tended to use advanced methods of analysis of the data, and some have employed molecular biological techniques such as restriction enzyme analysis of mtDNAs.

An example of morphological character analysis is Power and Ainley's (1986) clarification of the relationships of the populations of *O. leucorhoa* breeding along the Pacific and Atlantic coasts of North America. The Pacific birds range from the

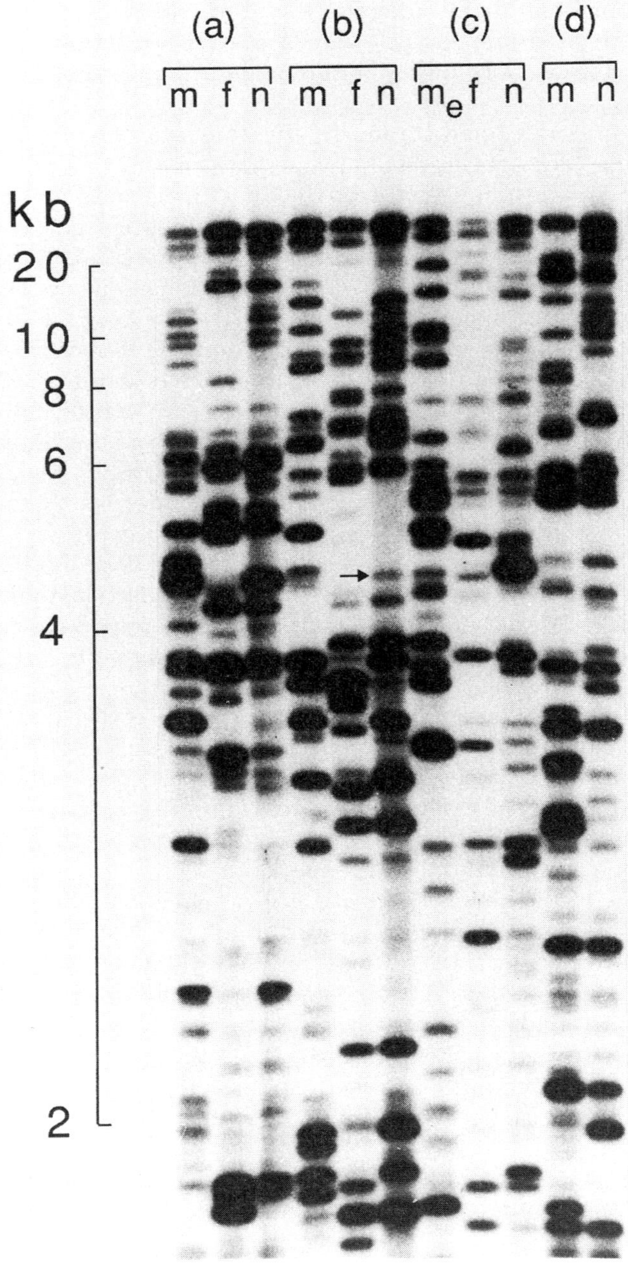

Figure 11.10 *DNA fingerprints for four families of* Fulmarus glacialis *in which the female (f) had at least one EPC. (a) and (b) two families whose nestlings (n) were confirmed by band-sharing analysis to be the offspring of the pair males (m), all nestling bands except one (arrowed) being present in one or both parents. (c) Extra-pair male with a female and her chick: band-sharing shows that he cannot be the father. (d) Pair male and nestling: banding shows that he is the father. From Hunter* et al. *(1992).*

Aleutians down to Mexico, have acquired a variety of subspecific names, and the pattern is complicated by Guadalupe Island having two morphologically distinct summer- and winter-breeding populations. Birds from these various groups vary in size and in the amount of white on the rump. Power and Ainley subjected eight mensural characters and indices of rump colour from 707 specimens to canonical variate analysis and calculated Mahalanobis distances between each pair of locality samples. They detected a clinal pattern of variation with northern birds tending to be longer winged, bigger billed, and with longer tarsi; the southern ones being the smallest in these characters, following Bergmann's Rule. Populations at Les Coronados and San Benitos Islands, however, formed a distinct subset, while the Guadalupe summer birds were smaller than the winter ones. The authors considered these two populations so different as to have experienced different selection pressures from the rest. How much, if any, interbreeding occurs at Guadalupe, was not determined and in Power and Ainley's tentative classification these populations were separated into subspecies *O. l. socorroensis* and *O. l. cheimomnestes*, an arrangement followed in the present and previous volumes.

Some species' problems are less complex. Randi *et al.* (1989), for example, examined the genetic variability of *Calonectris d. diomedea* and *C. d. borealis* by electrophoretic analyses of blood proteins. The birds from the three colonies examined proved not to have significantly different values of heterozygosity or percentages of polymorphism at the six loci sampled and dendrograms based on the genetic distances separated the Mediterranean populations from those of the Azores, supporting their subspecific status. However, the analyses suggested that there is some movement between these populations, with perhaps 4–19 birds being exchanged per generation. Randi *et al.* considered that it is unlikely that the Mediterranean populations of this shearwater are in equilibrium either demographically or genetically, and they suggested that only some 1667 generations have been completed since colonization of the Mediterranean after the Pleistocene glaciations about 10 000 years BP and that only if the effective size was less than 10 000 birds would the effects of gene flow become clearly detectable. Wink *et al.* (1993a) also found that the cytochrome-b sequences of the Mediterranean and Atlantic populations only differed slightly as expected of subspecies. By contrast, comparison of the partially sequenced cytochrome-b genes (300 base pairs) of *Puffinus puffinus puffinus* and *P. p. yelkouan* by Wink *et al.* (1993b) indicated 8.5% nucleotide exchanges, supporting their full specific status.

An example of 'fingerprinting' at the individual level is the study by Swatschek *et al.* (1994) on DNA restriction fragments from the blood of parents and chicks of *C. d. diomedea*. These provided no evidence for extra-pair copulations (EPCs) despite the frequent opportunities. Similarly, Hunter *et al.* (1992), although recording the frequency of EPC among *Fulmarus glacialis*, found from the fingerprints that no extra-pair paternity had actually occurred (Fig. 11.10).

1 Philopatry

The effects of a high level of philopatry (from visual, olfactory and/or auditory imprinting on characteristics of their breeding place in chicks) on speciation in tubenoses must be profound, but it has long been clear that not all birds return to breed at their natal colonies or how could species such as *Pachyptila turtur* have become circumpolar breeders? Furthermore, without some exploratory migration, a population confined to a single location is at risk from catastrophic events such as tsunamis or vulcanism, both likely hazards during the Miocene, when the value of a 'strategic reserve' of prebreeders may have been greater than today. Direct evidence for shifts between colonies is also available, for example for female *Puffinus puffinus* detected by Brooke (1978b).

Proving philopatry has previously involved the recovery of marked birds of known provenance, is usually expensive in time and energy, and only possible in long-term studies. Ovenden *et al.* (1991), however, tested the extent of philopatry using restriction enzyme analyses of mtDNA from 21 *Pachyptila turtur* of mixed sexes from Albatross Island in Bass Strait. All proved to have identical mtDNA haplotypes and, as mtDNA is maternally inherited, Ovenden *et al.* also discussed other means by which the genome homogeneity could have arisen, since Albatross Island would not have existed during the last glaciation at least up to 8000 years BP. These authors hypothesized that the mtDNA state of the colony could have resulted from a founder effect, perhaps from no more than four females and their mates, from which only a dozen generations could have produced the current population of *c.* 10 000 breeding pairs. A similarly small initial number would have been needed if a 'bottleneck' event had left a few birds within the original range of a population otherwise extinct. Either process could have accounted for the lack of mtDNA sequence variation. Ovenden *et al.* also looked at the mtDNA from *P. desolata* and found its genetic divergence from *P. turtur* low. They suggested that the phylogeny of the genus could be investigated by cladistic and phenetic analyses of the presence or absence of restriction sites in the species' mtDNAs.

A different situation seems to hold with the *Puffinus tenuirostris* colonies in the same general area as the *Pachyptila turtur* ones, most of which would also have been unsuitable breeding habitat during the last glaciation unless the birds flew far inland. The mtDNA sequences at 16 colonies were examined by Austin *et al.* (1994). There was no detectable population structure, with two closely related groups of mtDNA haplotypes widespread. This, and other factors, suggested gene flow and a considerable exchange of birds between colonies, despite no records of long-distance shifts by this species from the large numbers banded as chicks recovered. Also, contrary to findings with *P. turtur*, the genetic data for several colonies known to have been established in the last 35 years, suggest they were begun by a large group of birds followed by further substantial immigration.

Among *C. d. diomedea*, Swatschek *et al.* (1994) found from DNA 'finger-printing' that band-sharing coefficients between partners and between apparently unrelated birds averaged 0.362 ± 0.013 *vs.* 0.28 ± 0.049 ($P < 0.05$), suggesting that genetically related clusters develop, perhaps because of first-time nesters settling near to their birthplaces; some males were even in their natal burrow.

It would be instructive to be able to follow the genetic changes in other new or expanding populations, such as the Penikese Island *Puffinus puffinus* or the *D. immutabilis* colonizing islands off Mexico and California, using the new techniques requiring only small amounts of blood.

2 Relicts

The fossil record suggests that extinctions have been the norm during the past history of the Tubinares, and natural, as opposed to man-induced, extinctions presumably continue today. A number of species now reduced to very small numbers may be relicts of former much wider distributions and bigger populations. In many cases numerical relicts result from human predation or actions, for example *Pterodroma cahow* and *P. magentae*. The Chatham Island endemic *P. axillaris* may perhaps be destined to die out naturally and this may be true also of Macgillivray's Petrel *P. macgillivrayi* and Elliot's Storm Petrel *Oceanites gracilis*, all probably numerical relicts with no known history of significant human persecution nor evidence for a former wider range. Indeed, *P. macgillivrayi* and other extant *Pseudobulweria—P. aterrima* and *P. rostrata*—may also be relicts and information on their biochemistry, genetics and behaviour is badly needed. New Zealand seems to be the centre for relict tubenoses today. Subfossil material suggests that *P. bulleri*, *Procellaria westlandica* and *P. parkinsoni*, now restricted to one or two breeding places, were quite widespread on the mainland until recently.

The small *Pterodroma cookii*, breeding only at two places far apart, provides an example of a disjunct geographical relict (Udvardy, 1976). Here again, its disappearance from the intervening mainland where subfossil remains show that it once bred (Millener, 1990), was probably caused by man, by habitat destruction and/or the Maori Rat *R. exulans*. *P. l. pycrofti* provides a parallel case. Now rare and breeding only on a few islands off northern New Zealand, it was formerly part of the avifauna of Norfolk Island and probably also of Lord Howe (Meredith, 1990). Bones of this bird from refuse dumps of Norfolk Island's early settlers suggest that they ate it as well as the better documented Providence Petrel *P. solandri*.

3 Hybridization

What effect hybridization following recolonization of previous breeding ranges, as with *Macronectes*, has had on the present range of species, is conjectural. Current overt cross-breeding is well known only in large surface nesters such as the giant petrels and the albatrosses of the Hawaiian northwestern chain, but the possibilities for hybridization seem high when congeners breed in close proximity in large numbers. For example, although the four kinds of prion at Ile de l'Est tend to use different habitats, cross-breeding between such birds with similar voices would not be surprising. However, despite an attempt by Cox (1980) to explain intergradation of morphological characters in prions by hybridization, no clear examples have come to light and will be hard to detect among such nocturnally active birds.

Some specimens have been suggested as hybrids, for example a possible *Puffinus tenuirostris* × *P. griseus* by Kuroda (1967b) and a possible *P. bulleri* × *P. griseus* (Bartle, pers. comm.), and there are examples of subspecies interbreeding, for example *C. d. diomedea* and *C. d. borealis*, *D. e. epomophora* and *D. e. sanfordi* and *D. m. melanophrys* and *D. m. impavida* (Sanchez, pers. comm.; Robertson, 1993b; G. A. Taylor, pers. comm.). Royal Albatross crosses are of southern females taking northern partners at the Taiaroa Head colony. Their hybrid offspring are of growing importance and to 1992 had produced 32 eggs which appear about 11 days later than those of *D. e. sanfordi* pairs, so extending the laying period for the colony as a whole (Robertson, 1993b).

Overt attempts at courtship between distinct species are probably quite common judging from personal observations of displaying between *D. melanophrys* and *D. chrysostoma* and of *D. albatrus* making advances to *D. nigripes*. Vagrant tubenoses that take up residence far from their normal ranges tend to associate with other species of similar size, colour or behaviour, for example the solitary *D. melanophrys* associating with the Shetland Island gannets. Such imprinting can be very precise so that lone birds may be unaware of other wanderers of their own kin close at hand, as when two *D. albatrus* on Midway Atoll evidently failed to meet.

V Vocalizations as taxonomic characters

If vocalizations act as isolating mechanisms, they may also provide useful taxonomic characters, the more so for those many species that communicate vocally after dark. The calls of congeneric petrels nesting sympatrically have not been compared quantitatively, but it is difficult to find examples where the vocalizations of congeners breeding in the same place are easily confused to our ears. Those of *D. melanophrys* and *D. chrysostoma* and of *D. immutabilis* and *D. nigripes* in mixed or adjacent colonies are easily recognizable (Warham, pers. obs.; Sparling, 1977), as are the shearwaters *P. pacificus* and *P. nativitatis* in similar situations (Shallenberger, 1973).

When using vocalizations as taxonomic characters like has to be compared with like, for example homologous calls from the same class and sex, complete repertoires with complete repertoires, etc. The use of geographic variation in the calls of *Oceanodroma leucorhoa* by Ainley (1980) without allowing for the existence of sexual dimorphism in the voice, provides a case in point (James and Robertson, 1986).

These authors discussed the value of vocalizations in petrel systematics, comparing male calls of *P. puffinus* and *P. assimilis*, of *O. castro* and *Hydrobates pelagicus*. They examined a range of acoustic variables with male body weights and lengths of tarsi, bills and wings using ANOVA and multivariate discriminate function analysis. Although the taxa could be separated solely on their calls, variables linked with body size gave even better separations. The work suggests that vocalizations could be useful in supplementing other characters when examining 'difficult' species.

More recent examples of taxonomic work include analyses of the calls of *Halobaena caerulea* and *Oceanites oceanicus* by Bretagnolle (1989a, 1990a). He used the calls of male and female birds and other behavioural data to support the retention of the genus *Halobaena* and its association with *Pachyptila*, rather than with fulmars such as *Pagodroma*.

If, as seems probable, vocalizations are at least partly genetically based, analyses should prove congruent with those drawn from other data. And, being readily quantifiable in terms of frequencies, harmonic structure, internote duration and the like, may prove more useful behavioural characters than displays or postures less amenable to quantification.

VI Where did the tubenose radiation begin?

Most past students of this group, such as Shuntov (1974) and Harper *et al.* (1990), have concluded that the tubenose radiation took place in the southern hemisphere, partly because of the abundance of species there today. For example, Murphy (1936, p. 773) considered that the birthplace of the diving petrels was probably near Cape Horn where all four extant species can be found, pointing out 'In any group of organisms, we should look to the point of origin for advanced present-day types rather than for primitive present-day types, and, upon this criterion, southern South America fulfils the requirements.'

However, a vast period of time is involved in which islands have arisen and disappeared, straits have opened and closed, wind patterns and climates changed and continental drift continued, so that it could be argued that the tubenoses arose elsewhere and present distributions offer few clues to their birthplace. Indeed, if most speciation occurs at the margins of a species' range on small populations and under different conditions, 'primitive' forms would be expected at the point of origin (Paterson, pers. comm.). Added to which, the majority of fossil petrels have been found in northern deposits, the earliest possible procellariiform. *Tytthostonyx glauconitis*, coming from late Cretaceous strata in New Jersey. Indeed, Kuroda (1954, p. 46) speculated that the tubenoses evolved in the mid-Atlantic/mid-Pacific, spreading eastwards to the Indo-Pacific when the Mediterranean and Indian Oceans were one and then south to form a secondary centre of radiation in the Southern Ocean.

Against such arguments, one can point to the much smaller land masses in the south and the reduced availability there of fossiliferous deposits and palaeontological activity. Nor are the southern fossils negligible, including as they do at least four albatrosses and a range of smaller species (Section IX below). Note too, that there are currently only three genera—*Calonectris, Hydrobates* and *Halocyptena*—endemic to the northern hemisphere, but many restricted to the south.

Another factor arguing for a southern origin for the 'prototubinare' is the total restriction of penguin fossils to the south. These have preserved well, partly because of their heavy, unpneumatized bones. For example, from deposits in mainland New Zealand (mostly from South Island), 17 penguin specimens have been unearthed dating from Late Paleocene/Early Eocene to Pliocene times (Fordyce and Jones, 1990). Today three species still breed there with another six on islands offshore. Thus, the New Zealand region has the greatest concentration of penguin species today and may have done so in Tertiary times. Likewise, it also has the greatest concentration of tubenoses, 36 species breeding on the mainland or on offshore islands; that is, *c.* 35% of the total present radiation. Unfortunately, in contrast to the penguin record, there are no New Zealand tubinare fossils of significant age with the possible exception of Oligocene *Manu antiquus* Marples 1946 which may not be procellariiform (Olson, 1985c). Analyses of some behaviours of modern penguins by Paterson *et al.* (in press) also suggest that the proto-penguin arose in the New Zealand region.

Thus, if the Procellariiformes and Sphenisciformes arose from a common ancestor, a reasonable hypothesis is that this took place in the south, the flighted tubenoses spreading to the tropics and to northern seas, no doubt with some reverse colonizations, while the penguins remained unable to extend northwards except along the cold Humboldt Current.

VII When did the tubenose radiation begin?

Sibley and Ahlquist's (1990) 'molecular clock' placed the divergence of the penguins at about 47 MYA, tying in with the Eocene fossils at 45–50 MYA, the Procellariiformes branching off at about 36 MYA. However, we have no reason to suppose that the earliest fossils so far unearthed represent the earliest of their kin, rather the contrary, for most early fossil penguins and petrels seem rather similar to modern forms and would appear to have needed quite a long period to have evolved to that stage. And although these authors pointed out the probable effects of delayed maturity in procellariiforms causing a slower than average rate of genetic evolution, and tried to correct for this, the effect is likely to be greater than they thought, because breeding in several albatrosses and some other petrels is not only delayed but biennial or erratic (Chastel *et al.*, 1993), while in a range of species established breeders routinely take 'sabbatical' years off. Their forebears may well have behaved similarly. Furthermore, if fossils such as *Tytthostonyx, Eopuffinus* and *Hydrotherikornis* are truly procellariiform, the ancestor must have arisen more than 36 MYA.

VIII What was the prototubinare lifestyle?

We can only speculate on how the early petrels bred, whether colonially or solitarily, whether on the surface or under cover, whether visiting by day or by night. It seems unlikely that there were no potential predators then to favour selection for large size or nesting underground. But as neither small reptiles, carnivorous placentals or marsupials seem likely to be deterred from entering burrows, then, as now, the smaller petrels probably bred on islands or other refuges. It has been suggested that the early tubenoses laid below ground and only evolved into surface nesting when they developed the ability to spit defensive stomach oil. Such a shift would probably have been followed by increases in size—large birds do not burrow and bigger size reduces predation by smaller animals. The longer guard stage then needed would be helped by the bigger birds' capacity for fasting.

Kuroda (1954) suggested that the southern transequatorial migratory shearwaters in the Pacific might have arisen in the north from stock akin to *Puffinus conradi*, and that their migrations today involve a retracing of ancestral pathways, a hypothesis that could be extended to postulate a southern homeland for *P. puffinus* and *Hydrobates pelagicus*. Or the early birds could have discovered rich but distant feeding grounds (perhaps not as distant then as now), during their prebreeding explorations and continued these visits after breeding. However, whether of southern or northern origin, *Fulmarus* spp. show few signs of retracing ancestral tracks nor do the North Pacific albatrosses visit their assumed southern homes. Tubenoses today seem to have occupied all the available niches and, significantly, despite the vast numbers 'wintering' in the opposite hemispheres, including many prebreeders, there is little sign of attempts to colonize the contranuptial areas—just a few lone albatrosses taking up residence in the North Atlantic and *Oceanodroma leucorhoa* in the south (*The Petrels*, p. 189).

IX Procellariiform radiation and past geological and climatic change

The evolution of the tubenoses during Eocene to Pliocene times took place while Australia and New Zealand were moving into lower latitudes and Antarctica was shifting south. In the north, the Tethys Sea gave continued access from the expanding Atlantic through to the Indo-Pacific, the North and South American land masses being far apart. The Tethys evidently did not close until about 10 MYA while the final closing of the Strait of Panama was not until about 3–5 MYA.

The record shows that as far back as the early Oligocene, *c.* 38 MYA, albatrosses and shearwaters such as *Puffinus raemondonckii* inhabited Atlantic seas, and by the Miocene there appears to have been a substantial radiation of petrels there that would have had access to the Indo-Pacific via the Tethys Sea. The marine fauna then included at least four albatrosses whose bones are in the Lower Pliocene deposits of North Carolina (Olson, 1985c) and it seems highly unlikely that seas supporting albatrosses would not also carry a range of smaller petrels, and particularly the shallow and apparently rich waters of the Tethys. Nessov's *Eopuffinus* could have been part of the fauna of that sea. Olson (1985c) believed that most modern subgenera of *Puffinus* had evolved by the Middle Miocene with very little change thereafter; for example he found that *P. conradi* differed in only minor skeletal details from *P. gravis* (Olson, 1985c, p. 211).

In the eastern Pacific during Miocene times there was a fauna of large (*D. californica*) to small albatrosses (*D. milleri*), at least one storm petrel (*Oceanodroma hubbsi*), and doubtless other medium-sized species.

Although the fossil record for the southern hemisphere is naturally sparse, an unnamed albatross was reported by Tambussi and Tonni (1988) from Late Eocene deposits on Seymour Island, Antarctica, the site of Eocene penguin material, another unnamed albatross is from the Miocene of Argentina (Olson, 1984) and *D. thyridata* from the Late Miocene of Victoria (Wilkinson, 1969). Late Miocene/Early Pliocene South African deposits have revealed a small albatross, species of *Procellaria, Calonectris, Puffinus*, a fulmarine, two storm petrels, three species of *Pachyptila*, and at least four kinds of penguin (Olson, 1983, 1985a,b). While some may have been beach-wrecked birds, numerous bones of the large prion *Pachyptila salax* and of *Oceanites zaloscarthmus* indicate breeding, and Olson suggested where offshore islands nearby could have provided suitable nesting places in the Early Pliocene.

The history of the penguins, which is wholly in the south, is better documented and the early evolution of the Tubinares in that region there side by side with the penguins seems a reasonable hypothesis. Harper (1978) suggested that the environmental trigger helping initiation of both groups in the Pacific was the global cooling that began in the Eocene/Oligocene boundary about 38 MYA. This was followed by a prolonged Antarctic glaciation as that continent shifted to the south, the ocean between there and Australia expanding. The glaciation in turn led to extensive erosion and fed 'rock flour'-containing nutrients into the sea encouraging phytoplankton production particularly in the shallow waters over the extensive Campbell Plateau. With Drake Passage already open, important events were the development of the Circumpolar Antarctic Current about 25–30 MYA, initally strong, as it was channelled through the narrow gap between Antarctica and Australia, but later circulating much as at present, as well as the creation of the Antarctic Convergence *c.* 22 MYA.

Thus it is in this area of the southwest Pacific that the main radiation of the penguins, the petrels, and also the mystocete whales (Fordyce, 1977) may have taken place, all taking advantage of the rich planktonic food supply and the demise of the large marine reptiles. During the Oligocene and particularly in the Miocene, vulcanism led to repeated inundations and regressions around the New Zealand region providing a wealth of islands and potential breeding places presumably free of predatory quadrupeds. However, as both saurischian and ornithischian dinosaurs occupied New Zealand in the late Cretaceous, some smaller predatory forms could still have been present when the tubenose radiation began. Aerial predators such as skuas are known from the early Miocene in the North Atlantic and could have also been in the south, so that the protective burrowing habit could have begun early. Vulcanism extended into the Pliocene and would again have affected island availability, some islands perhaps having shorter lives than those of the species using them. Others, such as the Three Kings, Poor Knights and Snares Islands have long histories, perhaps back to the Miocene, and Campbell Island, an eroded shield volcano from the late Miocene, also seems to have been isolated since then (Fleming, 1979). To the north, Norfolk Island had not emerged and the Norfolk Ridge was then probably an archipelago.

Udvardy (1976) drew attention to the fact that the Hawaiian Chain, which has seen islands born, built up, crumbled, denuded and submerged in the last 10 million years, and which has a varied and abundant marine avifauna with tubenoses dominant, strangely lacks endemic seabirds. Such conditions would be expected to encourage endemism, as it seems to have done in the Galapagos and New Zealand.

The Pleistocene seems to have seen the demise of the Atlantic albatrosses, *Puffinus pacificoides* at St Helena and *P. holeae* from the Canary Islands. An eggshell associated with adults and nestlings of *P. holeae* was dated at 31 100 ± 1100 years BP (Walker *et al.*, 1990). Olson's analysis of the Late Miocene/Early Pliocene marine fossils of Cape Province, South Africa, testifies to the wide radiation of petrels and penguins in that region about 5 MYA. The occurrence of *Pelecanoides cymatotrypetes* in this assemblage is the only Tertiary fossil described from this family and its similarity to the bones of modern species shows that diving petrels originated well before Pliocene times, supporting the hypotheses of Fig. 11.9 rather than that of Fig. 11.5. *Oceanites zalocarthmus* was so like modern forms that Olson (1985a) suggested that this and *Pelecanoides cymatotrypetes* might be ancestral to *Pelagodroma (O) marina* and *Pelecanoides urinatrix* respectively. Olson (1985a) pointed out that this marine fauna, with its subantarctic elements of prions and diving petrels, indicated a cooler marine environment in the early Pliocene than at present and that as the Benguela Current did not develop until the late Miocene this fauna probably lived at higher latitudes until then and moved north to South African seas when the cooler conditions developed.

The history of today's North Pacific albatrosses is conjectural. Partly because all of them breed in the northern winter, as if carrying their original breeding timetable from the south, many have thought them derived from a southern mollymawk, *Diomedea albatrus* perhaps from an ancestral form of *D. exulans* whereas Nunn's results (Fig. 11.8) point to all being descended from a common ancestor with the great albatrosses. This includes *D. irrorata*, with its strange physiognomy, rather dark, perhaps neotenous plumage, egg-carrying behaviour, distinctive displays, and flecked down in the chick, as in *D. immutabilis* and *D. nigripes*. Is it (or all three) relicts from Tertiary radiation in the eastern Pacific? Fleming (1932) pointed out that one could imagine a time when Drake Passage was so restricted that the Humboldt Current was strengthened to span the tropics into northern waters and provided a highway for southern cool-water forms. Such an invasion, perhaps repeated at intervals, could have produced birds such as *D. californica*, leaving the present species as remnants of that radiation.

With its glacial and interglacial periods and rising and falling sea levels the Pleistocene must have provided conditions for isolating sections of populations to adapt independently to local conditions so that on the barrier's being relaxed significant interbreeding did not occur, partly because their breeding seasons now differed significantly. Some species, however, may have only evolved in post-Pleistocene times, for example *Fulmarus* and *Macronectes*: the members of these genera appear so similar in their anatomy and behaviour that their separation seems to have been recent. Voous (1949) also invoked Pleistocene glaciations to explain the history of *Fulmarus*. He suggested that when the polar seas used by seabirds shifted to lower latitudes during heavy glacial times, *Fulmarus glacialoides* colonized the North Pacific where the immigrants evolved into *F. glacialis* when they became isolated during interglacials, later spreading through Arctic seas to the North Atlantic. During the last glacial period he envisaged the Pacific and Atlantic stocks becoming separated by the advancing ice, as they still are (*The Petrels*, fig. 3.9). Today, *F. glacialoides* is still more familiar as a wanderer into low latitudes (particularly along the Humboldt Current), than its northern counterpart. On the other hand, bearing in mind the Miocene *Fulmarus* from California, a northern origin is possible (Bourne (1984)).

How the original *Macronectes* stock came to be divided into warmer water, early-laying *M. halli* and colder water, late-laying *M. giganteus*, is problematical. As with other fulmars the time taken from egg laying to chick fledging is short in comparison with other similar-sized petrels, a necessary condition for success in short polar summers, and as *M. halli* has this characteristic despite nesting in milder climates, the ancestral bird may have been a southern one, *M. halli* developing from a population isolated from the rest by laying early and coming to rely more on seals than *M. giganteus*, which became more dependent on penguins—as suggested by Hunter (1987). The present sympatry and hybridization would be a consequence of a secondary expansion of the southern bird, with its different diet and phenology, into the range of the northern one.

The extreme size variation and high genetic variability of *Pagodroma* in Adelie Land, explained by Jouventin and Viot (1985) as the result of separate colonizations of two subspecies after the post-Pleistocene climatic amelioration and their subsequent hybridization was discussed in *The Petrels*, pp. 59, 60.

Geographical or oceanographical changes able to divide populations that eventually speciate would be expected to affect more than one taxon. Are there other

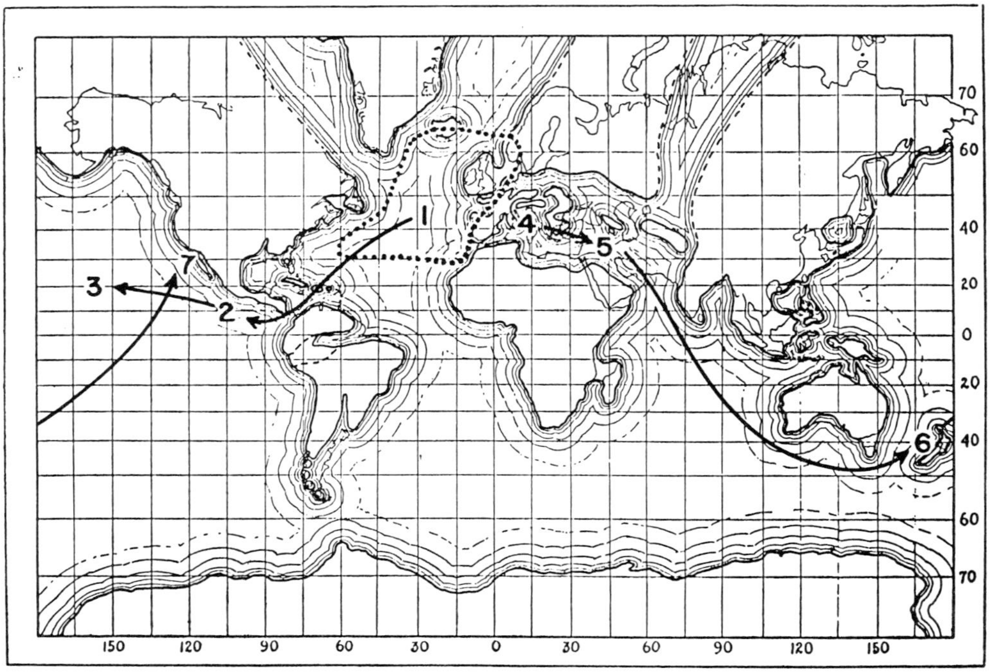

Figure 11.11 *Murphy's hypothesis on the origins of the* Puffinus puffinus *complex. Present-day distributions shown on a map of the Oligocene Period. 1,* Puffinus p. puffinus*; 2,* P. a. auricularis*; 3,* P. a. newelli*; 4,* P. p. mauretanicus*; 5,* P. p. yelkouan*; 6,* P. gavia*; 7,* P. opisthomelas*. (The specific identity and breeding place of* P. huttoni *was then unknown.) From Murphy (1952).*

species-pairs that could have arisen through the same event and simultaneously with *Macronectes*? Possible examples are the *Phoebetria* albatrosses, *P. fusca* feeding in warm seas, *P. palpebrata* in Antarctic ones; and *Pterodroma macroptera* and *P. lessonii*, again warm- and cold-water birds. Where sympatric, the members of these pairs breed at different times. Could the oscillations and vicissitudes of the convergence boundaries about the breeding islands be sufficient to cut off parts of their original stocks in the absence of land or ice barriers? The giant petrels of Gough Island, showing characters of both species, could be close to the original stock as Hunter (1987) pointed out, if not the result of a recent breakdown of reproductive isolation. Clearly, the past history of such sister species might be clarified with further evidence from biochemical and DNA studies.

The Tethys Sea was suggested as the ancestral home of the worldwide *Puffinus puffinus* group of shearwaters by Murphy (1952) (Fig. 11.11). Bourne *et al.* (1988) modified this concept, pointing out that these birds are even now more widespread in northern latitudes than in the south and can be divided into brown-backed (*P. yelkouan, P. mauretanicus, P. opisthomelas* and *P. gavia*) and black-backed forms (*P. puffinus, P. newelli* and *P. huttoni*). They envisaged the small brown-backed *P. lherminieri* and black-backed *P. assimilis* as early derivatives isolated in a warm Indian

Ocean *c.* 15 MYA, thence dispersing throughout warm seas. They hypothesized that the ancestral stock of brown-backed birds was cut off into one of the land-locked seas formed as the Tethys broke up and by the time that the Mediterranean, Black Sea and Atlantic were again connected, the brown-backed birds were distinct enough in phenology and ecology to remain separate from the black-backed ones of the Tethys Ocean. These authors were undecided whether the Pacific Ocean species were most likely to have been derived from Mediterranean brown-backed stock before the formation of the Isthmus of Panama *c.* 2 MYA, or whether they evolved independently in the Gulf of California (*P. auricularis, P. opisthomelas*) and New Zealand (*P. huttoni, P. gavia*). Voous (1976) modified Murphy's hypothesis differently, emphasizing the repeated Miocene dessications of the Mediterranean that left only saline lakes and would have forced seabirds such as *P. assimilis* and *P. lherminieri* into the Indo-Pacific. Voous thought that today's Mediterranean forms of *P. puffinus* resulted from rather recent immigrations, perhaps through the intermittently open Straits of Gibraltar leading to *P. yelkouan*, with another from the Red Sea giving rise to *P. mauretanicus*. Here, as elsewhere, the need is for a multifaceted analysis leading to matching phylogenies and including other, newly extinct members of this group such as *P. holeae, P. nestori, P. olsoni* and *P. spelaeus* (see Appendix I).

An example of recent distributional changes, apparently in response to changing climatic conditions, is provided by bones of *F. glacialis* excavated at archaeological sites in northern Norway (Montevecchi and Hufthammer, 1990). They dated many from about 30 000 to 400 years BP, mainly between 2000–1000 years BP. In modern times this species did not breed in Norway until the beginning of this century and evidence from Icelandic sagas was invoked to suggest that the population decline coincided with severe ice conditions along the coast from the 13th to 18th centuries.

At the other end of the world the dating of the basal layers of 29 samples of frozen stomach oil (probably mostly from *Pagodroma nivea*) from nunataks in east Antarctica by Verkulich and Hiller (1994) indicates an initial colonization *c.* 10 000 years BP, with a continuous expansion as the ice retreated. An even earlier date of 35 000 years BP was gained from a deposit elsewhere in east Antarctica by Hiller *et al.* (1988) (Fig. 11.12).

X Conclusions

Despite the sometimes divergent results available so far, the use of DNA 'finger-printing', PCR, and other techniques of molecular genetics, promises to help resolve many of the outstanding problems of relationships involving the tubenoses, their history and demography, particularly if a reliable 'molecular clock' can be devised and consistent and acceptable criteria established for separating higher taxa. The ability to use even small quantities of blood with PCR or other techniques such as DNA from feather pulp, which pose little danger to the bird, will help in tackling problems with rare species or those like *Puffinus lherminieri* with many races as it will make much more feasible collecting material from all the populations rather than from a few readily available ones.

No doubt such methods also have their limitations, but they promise better data than DNA hybridization or the cleaving of specific sequences using 'restriction

Figure 11.12 *Section of a stomach oil deposit from east Antarctica whose lowest level was dated at 35 000 years* BP. *Photo: Achim Hiller.*

enzymes'. Presumably it may be possible eventually to tie specific sequences to specific characters in the whole animal, and the data derived from such approaches need testing against behavioural, morphological and other evidence based on the living birds which reflect the work of all the active genes. Although morphology may change faster than the DNA does (Parkin, pers. comm.), the greater the congruence between the different approaches, the greater the confidence in the results.

These techniques attempt to construct histories on similarities or differences between extant species. Except for *Pterodroma kurodai*, no fossil species has been included in a 'tree' and the branching patterns seem likely to be rather different if characters of the true ancestral forms could be included. Even now there are some near-complete fossils that could be incorporated in phylogenetic analyses (cf. the fine detail available of the much older *Protoavis* (Chatterjee, 1991), not to mention that known for *Archaeopteryx*).

Thomson (1976) drew attention to the way that diversification of genera has been relatively rapid during evolutionary time but followed almost immediately by rapid extinctions, maximum diversity being maintained for relatively short periods. Tubenose evolution may have proceeded like this. The two-dimensional representation of phylogenies based on extant species must be a considerable simplification compared with the network of twigs needed to represent the complete history.

Until the dust from the many studies of the relationships of the Procellariiformes and related groups now in train has settled, it seems foolish to discard familiar classifications such as that used here, in favour of the latest version from a particular school of thought. This is as true of ornithology today, when increasing numbers of active and knowledgeable non-professionals need a reliable and stable nomenclature, as it was 30 years ago (Alexander *et al.*, 1965). So far, molecular studies suggest that no major readjustments to current taxonomy are needed. Perhaps the main surprises are the association of the sooty albatrosses with the mollymawks and indications that the whole tubenose radiation began quite late in the evolution of birds. Hopefully, in due course it will be possible to consolidate the evidence from all the various approaches to create a classification and a phylogeny commanding general agreement.

CHAPTER 12
Petrels and Man

Contents

I Introduction

The earliest references to petrels seem to be in the classical literature. Aristotle's 'cepphus' (Historia Animalium, Bk.IX 35) has been thought to refer to a shearwater. The bird so called and figured in the *Ornithiaca* of Dionysius (*The Petrels*, p. 150) is *Calonectris d. diomedea* and even if this work does not date from the 1st century BC, as hypothesized by Kadar (1978), the text and illustration provide the first account of a tubenosed bird. The excellent representations of birds in that codex could conceivably have been copied from the figures that Aristotle used in his teaching.

This bird still breeds on the Diomede or Tremiti Islands off Apulia, and Aldrovandi (1657), illustrated a specimen he got from the Augustine friars there (Fig. 12.1).

The famous legend of the metamorphosis of the companions of Diomedes of the

Figure 12.1 *The 'Bird of Diomede',* Calonectris d. diomedea. *From Aldrovandi (1657).*

Figure 12.2 *A Cape Petrel* Daption capense *from Peter Mundy's voyage in 1638. From Temple (1919).*

Odyssey into birds was put into an ornithological perspective by D'Arcy Thompson (1918, 1966), and, after a visit to the Tremiti Islands, by Winthrope (1973). D'Arcy Thompson (1966) listed 11 names from Greek classical authors that he thought might refer to shearwaters, and such birds are mentioned in Pliny's *Natural History* from which much of the information in later works such as those of Turner (1544), Belon (1555) and Gesner (1555), as well as that of Aldrovandi came. These provide a melange of fact and fancy, but do refer to the burrowing habit, the sharp pointed beak and the lachrymose night-time cries. Gesner also includes 'De Puffino Anglico' (*Puffinus puffinus*), as well as *C. diomedea*. Lachmund (1674), with what he thought was a 'Bird of Diomede' before him, was more critical, noting that the beak was made of separate pieces, the nostrils in tubes greatly developed for the sense of smell, and that the skin had a musty odour.

The first ornithological text not following the Aristotelian tradition was the *Ornithology of Francis Willoughby* by John Ray (1678). This figured both *C. diomedea* and *P. puffinus*, the latter under two heads: 'The Puffin of the Isle of Man' *P. anglorum* (p. 333) and the 'Shear-water' (p. 334). None of these writers seems to have known of the Dionysius *Ornithiaca*.

Later came direct reports from seafarers such as Dampier on storm petrels and Lefroy on the Cahow (Butler, 1882). Peter Mundy (Temple, 1919) provided a very early illustration of a familiar bird to sailors in southern seas—*Daption capense* (Fig. 12.2). Petrels could be quite important and many were caught to eke out the rations.

Where tubenoses have been significant to their economy, coastal and island people have given the species names. These are often onomatopoeic such as the Maori 'parara' for *Pachyptila vittata*, others descriptive, such as 'tangi kerikeri' (crier in burrows) for *Pterodroma alba* in the Tuamotus (Emory, 1947).

Many myths, superstitions and taboos evolved around them. Superstitions about petrels and albatrosses as birds of ill omen were widespread. In an early reference to albatrosses, Cosmas (1909) recounts his horror when they followed his ship during a terrifying storm on a 6th century voyage in the southern Indian Ocean. However, sailors still ate them, and, as one wit put it:

A dead loss was the albatross
The 'Ancient Mariner' slew;
The modern Tar, acuter far,
Makes him into stew. (Horton, 1860).

According to Popplewell (1918), finding a true albino *Puffinus griseus* ('jimmy bird') or a white one with dark eyes ('queen bird') was an omen to the Maori muttonbirders of a calamity within a year. A family that had caught and released such a bird lost two children off the rocks the next birding season. Even today, the weird nocturnal cries of *Pterodroma hasitata* form the basis for stories of witches in Cuban folklore (Lee & Vina, 1993).

Such cries and their sombre plumage may have given rise to the Moslem belief that the *yelkouan* shearwaters streaming through the Bosphorus (Fig. 2.10) were ghost birds tenanted by the souls of the damned and the Turks did not allow them to be killed (Stanley, 1902, p. 79). Bourne *et al.* (1988) state that *P. yelkouan* is still known as 'halcyon' in the eastern Mediterranean, while 'yelkouan' is Turkish for the souls of the departed. Winthrope (1973) speculated that rafts of these Bosphorus shearwaters, never found breeding locally, gave rise to the fable of the halcyons that nested on the sea. Armstrong (1958) pointed out that the legend of the birds' earlier human existence could have arisen when sailors landing near breeding colonies tried to explain the subterranean night-time wailings, and that the Tremiti Islands may have been thought haunted long before the reputed landing of Diomedes.

II Petrels as food

Doubtless, tubenoses have been eaten by people since antiquity whenever they were available. For some human populations seabirds, including petrels, were a mainstay of their economy.

The whalers and sealers ate petrels caught at sea or on land—seal colonies often being cheek by jowl with those of tubenoses. Birds caught on lines (*The Petrels*, p. 20) made a welcome addition to the diets of navigators such as Cook and Flinders. Their explorations led to settlements on previously uninhabited places where new species were added to the menus, for example *Diomedea exulans, D. chlororhynchos* and *Puffinus gravis* to those of the Tristan da Cunha settlers, *Pterodroma cahow* to sailors stranded on Bermuda, and *P. solandri* to the convicts on Norfolk Island. Nowadays, in Tasmania *Puffinus tenuirostris* harvesting on land has involved taking well-developed chicks by hand, but formerly adults were caught with dogs, nets and vocal lures. New Zealand Maoris took various species by lighting fires on ridges below flight lines. Phillipps (1958) records their using green vegetation to create smoke that overcame overfliers—probably *Pterodroma macroptera, P. inexpectata* and *Procellaria parkinsoni*—and Buck (1949) described how adult petrels were lured by fires into nets set on coastal cliffs on misty nights.

The eggs of many species were also taken, for example large numbers of those of *Puffinus tenuirostris* were used by Melbourne biscuit makers.

A *Albatrosses*

Until recent times, albatrosses only seem to have been harvested on a regular basis in the North Pacific. In Alaska the Eskimos took fat, easily caught *D. albatrus* among the pack-ice near St Lawrence Island (Murie, 1959), and their bones are abundant in ancient Aleutian village middens and in Indian village mounds in California. Yesner (1976) concluded that *D. albatrus* was an important, if seasonal, resource to the Aleutians in prehistoric times, probably hunted in the inter-island passes. Some *D. nigripes* and *D. immutabilis* were also eaten. Medway (in press) reviews the widespread mortality of smaller albatrosses taken at sea by sailors and their passenger culture last century.

The taking of *D. epomophora sanfordi* and *D. cauta eremita* at the Chatham Islands began with the arrival of the Polynesian Moriori about 400–500 years BP. The expeditions to and from the steep stacks—the Pyramid, Sisters and Forty-fours—were hazardous, and invested with precautions, tapus and invocations to the spirits of wind, tide and sea (details in Skinner & Baucke, 1928). The Morioris travelled in 'wash-through' punt-like vessels made from wood and saplings buoyed by flower stalks of the local *Phormium* flax (Fig. 12.3). The fat squabs were killed by a stroke of a club on the neck, which ensured that the valuable stomach oil was retained. Skinner and Baucke described the subsequent plucking, ripening, cooking in the oil and their own fat, and their storage in bags made from fronds of bull-kelp *Durvillea antarctica*. Having slaughtered the Morioris, the Maori victors continued the harvest, but now, using sailing vessels and whale boats, the practice became less hazardous and more regular.

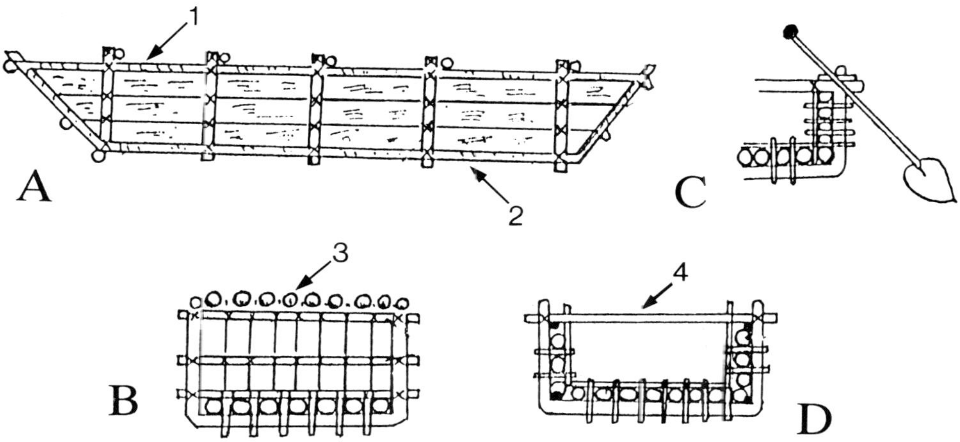

Figure 12.3 *Moriori 'wash-through' canoe for bird hunting, 2.4 m wide, 8.5 m long, 0.9 m deep, on a framework of saplings lashed together with vines. A, Side view; B, end view; C, cross-section of half canoe showing placement of paddle; D, cross-section. 1 & 2, Upper and lower saplings; 3, rolled flax cylinders for flotation; 4, cross-tie. From Skinner and Baucke (1928).*

A trade in preserved albatrosses and other birds developed from about 1841 between The Chathams and the New Zealand mainland, but records of harvest sizes seem non-existent. What is known is summarized by Robertson (1991). A Chatham Islander recalls that in the 1920s they had albatrosses every year, the fishing people living on them for months. The birds were cooked on the beach near her home and preserved in 4-gallon kerosene cans (E. Melluish, pers. comm., 1993). Chicks continued to be taken even after 1953 when New Zealand wildlife was legally protected. For example, in 1990 67–100 birds were taken from Little Sister at the end of September (Robertson, 1991).

After the extinction of *D. exulans* on their main island, the Tristan da Cunhans turned to *D. chlororhynchos* on Nightingale Island, taking eggs and chicks. Methods of taking and storing were described by Munch (1946). Numbers decreased and egg collecting was stopped in 1975.

B *Fulmars*

Some *Macronectes halli* are taken at the Chatham Islands during the albatross hunts, and *Fulmarus glacialis* was eaten at least 30 000 years BP in Norway (Montevecchi and Hufthammer, 1990). In the Pribiloffs and Kurile Islands, *F. g. rodgersi* was eaten back in 1741 (Steller in Krashnaminnikov, 1764), but the main harvesting of fulmars was in Ireland, Greenland, Spitzbergen, Fair Isle and the Faroes. On St Kilda the birds were taken as far back as the 9th century. Fisher (1952a, p. 440) gave catches at the Icelandic colonies from 1897 to 1941. The maximum take was 59 930 birds. Today the Faroes' annual harvest of *c.* 3000 *P. puffinus* chicks and 8–10 000 *F. glacialis* eggs and *c.* 40 000 chicks appears to be sustainable (Kampp, pers. comm.). The methods used were described by Williamson (1945) and Fisher (1952a).

C *Gadfly petrels*

Into historical times the Hawaiians ate the 'Ua'U, *Pterodroma p. sandwichensis*, the 'Ao', *P. a. newelli*, and doubtless the extinct *Pterodroma jugabilis* (Olson & James, 1991). When the Polynesians arrived *c.* 1400 BP, the 'Ua'U probably bred abundantly on all the islands (Olson & James, 1982, 1984), and large-scale use of both adults and chicks followed. Bryan (1908) hunted for 'Ua'U on Molokai in heavy forest at about 1100 m. Dogs barked on finding birds which were dug out of their burrows and killed by a finger forced down the throat and twisted to tear heart and lungs. Some were brought down by imitating their call, particularly early in the season (cf. Chapter 4 XVII). All these were evidently adult birds. However, chicks were preferred but reserved for the chiefs. They were extracted with a stick split at the end and twisted into the down. Some adults were caught in nets stretched over burrow mouths. Where chicks could be reached through a hole sunk over the nest chamber a tree fern stem plug allowed nestlings to be easily extracted in later years. The combined effects of taking adults of a K-selected species and the depredations of cats and mongooses, explain the present endangered status of this subspecies, the records

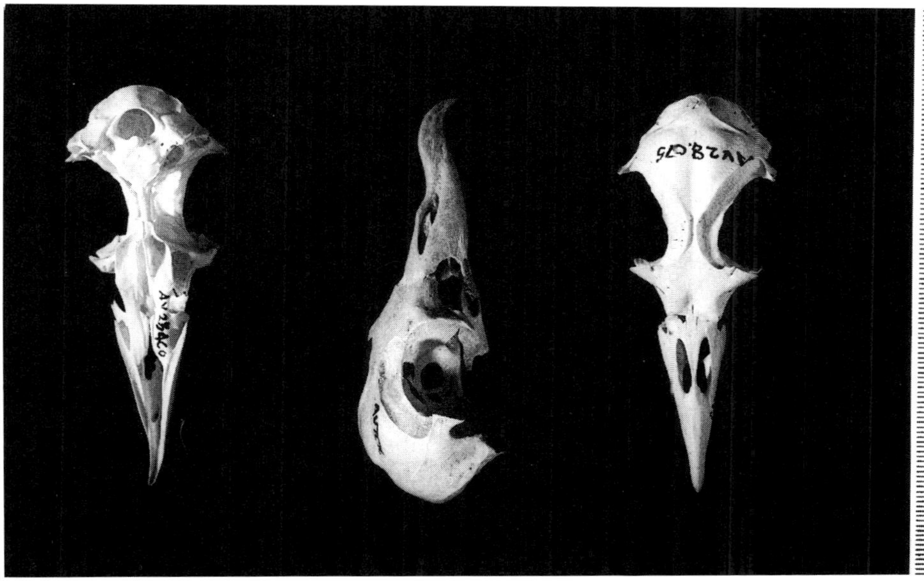

Figure 12.4 *Skulls reputedly of Chatham Island Taiko* Pterodroma magentae, *unearthed from dune deposits on the main island of The Chathams.*

and specimens of which were listed by Banko (1980). The lowland birds on Oahu were evidently wiped out before the arrival of Europeans through overcropping and the depredations of the Polynesian dogs, pigs and rats.

In another investigation on Molokai, Weisler and Gargett (1993) found that the commonest bird remains in rock shelters and sand dunes were those of *Bulweria bulwerii* and *P. hypoleuca*. These authors concluded that many of the species extinct there were used as food, but that this did not necessarily explain the demise of *P. hypoleuca*, a bird also once harvested in the Ogasawara Islands, Japan (Oka, 1994b).

The principal 'muttonbird' of northern New Zealand has been *P. macroptera gouldi*, the only petrel legally taken there today. Permits are issued annually by local Maori committees to descendants of the tribe for that region, the 'Iwi' committees. No other bird may be killed, nor adults, eggs or fledglings that can fly. Digging out of burrows, the use of dogs or fire are prohibited nor may birds be sold. A record of the take is supposed to be kept, but this seems seldom done. One estimate gives the annual catch at about 1700 birds.

On the Chatham Islands the Taiko *P. magentae*, which had large colonies on forested hills of the main island, was last harvested in 1902 and 1903 when 3000 and 600 chicks were killed respectively. Originally, these petrels must have been abundant and their plentiful remains in middens and dune deposits (Fig. 12.4) testify to the species' importance to the islanders' economy (Sutton, 1979). Today only something over 100 birds are thought to exist on The Chathams, 42 having been seen or caught from 1 January 1978 to January 1992 (Crockett, 1995).

In southern New Zealand the common gadfly petrel eaten as a muttonbird was *P. inexpectata* which was (is?) taken along with *Puffinus griseus*. It was highly prized and the boned carcasses were often stored in bull kelp bags. The outside layers of the fronds were separated by a hand to break the internal cell walls, the ends tied off, and the whole inflated (usually with a vacuum cleaner nowadays). Fat was poured in submerging the birds completely, the ends tied up and the whole wrapped in totara bark or newspaper, and the resulting 'poha' protected in a plaited basket.

The gadfly petrel *Pterodroma neglecta* formed an important resource for European settlers on Raoul, the main island of the Kermadecs (*The Petrels*, p. 107). The young birds were plucked, cleaned, degutted, rubbed with dry salt, soaked in brine, put in a smoke house for a day and then pressed down in casks until their oil covered them. At Norfolk Island the use of the 'Bird of Providence' (Hunter, 1793) *P. solandri* to succour starving seamen and convicts has been well documented (Iredale, 1929; Whitley, 1934; Dunn, 1988). Thanks to the daily tallies kept by Lt Ralph Clark, RN and listed by Whitley, more is known about the take there in 1790 than of the New Zealand muttonbird harvests of the 1990s.

Recent examinations of bones from archaeological sites on Pacific islands provide insights into man's use of petrels for food. At uninhabited Henderson Island, the commonest bird remains are those of *P. alba*, which still nests there (Steadman & Olson, 1985). The bones came from cultural strata with ^{14}C dates of 800 and 500 years BP. However, the absence of remains of *P. ultima*, *P. arminjoniana* and *Puffinus pacificus*, which also breed there today, suggested to these researchers that all were exterminated and have recolonized the island since it was abandoned by the Polynesians.

Gadfly petrels have also been eaten in the Caribbean—fires set on the mountains above the breeding places catching *Pterodroma hasitata* (*The Petrels*, p. 93). Labat (1722) described hunting for 'diablotins' (either *P. hasitata* or *P. (h.) caribbea*) on Guadeloupe (translation in Bent, 1922). Here too dogs were used to help find them, the birds dragged out with hooked poles. The plucked and eviscerated bodies were cooked on spits and proved so delicious that, as the season for taking the chicks fell in Lent, the local missionaries, after due deliberation (and tastings!), declared them vegetable food that could be eaten at all times.... According to Porter (1930) the French on Dominica formerly exported 'great quantities' of these birds to Martinique and other local French possessions.

D *Shearwaters*

The two *Procellaria* breeding on the main New Zealand islands were both taken by the Maori but both are currently protected although there may have been some poaching of *P. westlandica* chicks at times.

1 *Calonectris*

There appear to be no accounts in the classical literature of the regular harvesting of *C. d. diomedea* in the Mediterranean. In recent years there are accounts of its use as food, for example by Vaughan (1980), who estimated several hundred eggs being collected from the main colony on Linosa. Archaeological excavations on Lavessi Island, Corsica show that this bird was taken there from the 15th century (Vigne *et*

Figure 12.5 *Cape Verde Island fisherman with a catch of* Calonectris d. edwardsi *chicks. Photo: H.-H. Schleich, from Hazevoet (1991).*

al., 1991). At the Balearic Islands eggs and chicks were taken, the former making better pastry than hen's eggs. These practices appear to have ceased as the birds' numbers declined (Serra, 1986) but on the Chaffarines Islands, Spaniards and Moroccans were killing both young and old in the late 1970s (da Juana *et al.*, 1980).

At Grand Salvage Island taking of *C. d. borealis* began before 1841 and a detailed account of the harvest comes from Zino (1985). Something like 22 000 shearwaters were taken over about 20 days. The birds were sold in Madeira, the feathers went to England for eiderdowns, the depot fat melted down for fishing bait, but what happened to the stomach oil is not clear—probably it was burnt in lamps. Further south on the Canary Islands the chicks were dragged from the burrows and crevices with a hooked rod (Bannerman, 1914). At the Cape Verdes, chicks of *C. d. edwardsi* are still taken (Hazevoet, 1991 and Fig. 12.5).

In Japan, chicks of *C. leucomelas* were gathered for their meat and fat from the Izu Islands for at least 250 years. On Mikura-jima, harvesting peaked at about 30 000 chicks annually between 1951 and 1977. It has since fallen as the islanders no longer depend solely on their own resources. Kurimoto (1937) stated that the catching was supervised by the village master. The fat was used by the inhabitants, the meat salted and exported to the larger island Miyake-jima. The birds are now legally protected. Oka (1994b) reviews the careful harvesting of these birds and their traditional cooking methods. The position of those breeding in Korea seems unknown, but as late as 1972 chicks and eggs were on sale in Taipei markets.

2 *Puffinus*

For the use of *Puffinus puffinus* as food see *The Petrels*, p. 158. The species is seldom taken nowadays, although Alexander (1954) thought that chicks were sometimes still eaten on the Blaskets (Co. Kerry). In the Faroes, the hunt for *P. puffinus* was celebrated on 26 August with versification and a festival (Williamson, 1945). Collins (1884) described the use of *P. gravis* by the Grand Banks fishermen and seeing 200–500 birds suspended in rigging where they could be kept for several days if eviscerated. There are many other references to past use of shearwaters for food, for example chicks of *P. lherminieri* sold in Grenada (Lawrence, 1889), but often the birds no longer remain to be exploited. This shearwater was one of the species present in archaeological sites on Antigua dated at 4300–2500 years BP at a time when the island was first colonized—another case of man-made extinction (Steadman *et al.*, 1984).

Elsewhere, *P. pacificus* chicks are still sold on the Seychelles and taken from the Mascerene Islands, as is *P. assimilis* on the Cape Verdes and *P. gravis* from Nightingale Island—61 000 and 98 000 in 1973 and 1974 according to Richardson (1984).

However, the main shearwater harvests today involve the Tasmanian and New Zealand 'muttonbirds', *P. tenuirostris* and *P. griseus*. The origin of 'muttonbird' is disputed, and may have first been used on Norfolk Island for *Pterodroma solandri* (Dunn, 1988, p. 124). Some aver that it refers to the woolly texture of the chick's down, others from their tasting like mutton (they don't), from their fat which was like the tallow on sheep carcasses, and so on. These shearwaters form the basis for local cottage industries, that of the Tasmanian species being carefully regulated, the New Zealand one much less so.

Coastal aborigines ate *Puffinus tenuirostris* washed on to the beaches, probably during the northwards migration (for example Lampert, 1966, 1971), but the Tasmanians had access to large breeding colonies. Vanderwal and Horton (1984) examined 6858 bones of *P. tenuirostris* (as well as of *Pachyptila turtur* and *D. cauta*) from Maatsuyker Island, identified adult and juvenile material and suggested that broken radii and ulnae showed that marrows were sometimes extracted.

Although aborigines ate the birds long before the Europeans arrived, the sealers' and whalers' boats allowed previously inaccessible colonies to be exploited. Eggs, fat, oil and feathers were mostly used at first, the meat discarded. Eggs were popular and sold to cake manufacturers but also eaten locally. Davies (1846) reported an aboriginal women consuming 52 in 1 day!

Much of today's Tasmanian harvest is exported to Auckland. The processing sheds on the islands are licensed by health inspectors and talleys kept of the birds killed on the different colonies. The 1994 commercial harvest was of some 219 700 chicks, 2740 l of oil, and 252 kg of feathers. The New Zealand catch is also supposed to be counted but in practice the data, if any, have not been provided to the government and although part of the catch is sold to the public, no health inspections are made.

Details of current harvesting methods in Tasmania have been well detailed by Serventy (1958), Hill *et al.* (1981); Skira (1987, 1990) and Callister (1991). The bird is still important to the descendants of the Tasmanian aborigines on Cape Barren Island in Bass Strait, but their harvest has fallen over the years, although they are pressing for more islands to be made available on cultural and historical grounds.

The practice of eating petrels no doubt came with the first Polynesian settlers to New Zealand from the North Pacific about 1000 years ago where petrels were an important food resource. The new arrivals would have found similar birds nesting abundantly around headlands, in coastal forests and on islands offshore. The North Island Maoris apparently did not usually preserve their muttonbirds but ate them as they were collected. However, according to Phillipps (1958), Waikanae Maoris late in the 19th century preserved Kapiti Island *Puffinus griseus* in casks or in kelp containers. Beattie (1954) described how Maoris from Cape Egmont rowed all the way to The Bluff to trade for muttonbirds for their winter food.

In the South Island, the harvest of *P. griseus* comes mainly from islands in Foveaux Strait and off the western and southwestern coasts of Stewart Island. Of many accounts of muttonbirding there, perhaps the most informative are those of Porter (1935) and the more detailed one of Richdale [1948?] based on observations in the 1940s.

About half of these southern islands have Maori owners who control the food-gathering regimes there. Only chicks are taken from 1 April to 31 May, as laid down in the Titi (Muttonbird) Regulations 1978. These stipulate that no adult birds may be taken, that where holes are dug to reach birds these must be refilled and stopped, that dogs must be fully controlled, no cats taken ashore, no fires lit except for domestic purposes or emergencies, and all necessary precautions taken to exclude rats, mice, stoats and weasels. No firearm shall be taken.

None the less, the muttonbird industry in New Zealand is only loosely supervised. Some families are said to harvest up to 5000 birds in a season and the question of overcropping has been raised. Richdale's figure of an annual harvest of 250 000 birds agrees with that estimated by B. D. Bell (pers. comm.). Some small islands are not cropped regularly but may be harvested occasionally—usually by a day visit or evening 'torching'.

Most of the Sooty Shearwaters taken are eaten by the Maori owners and their families, and Davidson (1984) suggested, from archaeological evidence, that the large-scale trade in muttonbirds was a post-European development, spurred on by the acquisition of whale boats. The birds may be sold in New Zealand but may not be exported. Some reach city shops around New Zealand, sometimes from beneficiaries who live in these places. In 1935 the muttonbirders got 5½d per bird; in 1991 each was fetching $6 in the Christchurch and about $12 in Melbourne, so was not cheap, as two birds would be needed for a good meal.

Harvesting at Pohowaitai Island (38 ha) as described by Te Maire R. Tau (pers. comm.) is probably quite typical of the procedure elsewhere. Allocation of working areas between families is determined by drawing lots. The easiest areas are given to the old people, but the most densely burrowed grounds are left untouched until later in the season. At Pohowaitai lines are put down late on the best areas and each family works a strip perhaps 15 m by 100 m and must adhere strictly to these defined limits. The late chicks have lost weight due to the burning off of their fat deposits, their meat is better, and they are highly prized.

The squabs are mostly extracted with a wire probe with a loop at the end. This is hooked under a wing and the chick dragged out. Elsewhere the probe may be used to locate the nesting chamber, a shaft dug to extract the chick and a tree fern plug used to seal the hole in the traditional way. The extracted chicks are mostly killed immediately by pressing their soft skulls against the birder's knuckles: a few of the older Maoris still kill them by biting off their heads. Killing must be done gently so as not to bruise the flesh, the birds carried down to the processing sheds five on each side of a rope with a slip-knot at each end.

Towards the end of the season, when the chicks are feathering rapidly and emerging after dark to explore their surroundings, 'torching' is employed. Battery-operated hand torches are used to reveal the birds on the ground so that they can be picked up. Formerly, carbide acetylene lamps were used, which, in turn, had replaced torches made from bark and fat.

The stomach oil is carefully squeezed out to avoid fouling the down and, after plucking, the carcasses are dipped into near-boiling water and the down rubbed off by hand. Alternatively, rough-plucked chicks are dipped into melted paraffin wax. They are then hung up, the wax hardens, and when broken off takes the remaining down with it.

Before preservation, the head, wings, legs and viscera are discarded. If to be preserved in their fat, the birds are 'tried out' in pots over a fire until the bones soften. The fat comes from the subdermal and visceral stores which are refined by heating and by skimming before the birds are immersed. After cooking, they are allowed to cool and then packed loosely into tins or buckets and completely covered with warm fat to exclude all air. Nowadays, most are preserved in salt. The flesh is rubbed with salt and the birds packed in barrels or plastic buckets with ample salt between each layer in a style similar to that used with the smaller Bass Strait birds. A few are deep frozen.

Formerly, preparations for the 'birding' season took several months and involved special rituals, chants, and lifting of tapus during the opening ceremonies and on the return of the birding parties to their villages (Skinner & Baucke, 1928; Best, 1942). In those days, the islands were reached by canoe. Nowadays most birders go by helicopter, their food and equipment for the 3 months' stay by boat, often being uplifted by helicopter on reaching the islands. The 200–300 'birders' involved take many tonnes of stores each year. However, by the time that all expenses have been met, little profit may remain, so that for many muttonbirding is more of a traditional relaxation rather than a commercial enterprise.

Currently, the regulations are again under review, some Maori tribes having applied under the Treaty of Waitangi to obtain control of Crown Islands and the right to restart harvesting of petrels. This would allow them to take species currently totally protected. For example, one application seeks to take not only *P. griseus* but also *P. carneipes* and *P. gavia* from islands in Cook Strait. As a breeding species, *P. gavia* is restricted to New Zealand, its world population probably quite small by shearwater standards. A proposal to start taking *P. bulleri* from its sole breeding place, the Poor Knights Islands, is unexpected. The Maori living there ate the birds and traded preserved ones with mainland Maori until, in 1823, the islanders were massacred, the place abandoned and declared tapu, since when no birding has taken place.

Chatham Island Polynesians are also pressing for the conservation regulations to be lifted so that they can resume harvesting *P. griseus* from outlying islands. These include South-East Island where illegal muttonbirding on a small scale has been going on for years. The shearwater population here is small, but this is one of the most important islands in Australasia for endemic species, with the world populations of *P. axillaris*, the oystercatcher *Haematopus chathamensis*, and the plover *Thinornis novaeseelandiae*, not to mention a recently established (or re-established?) population of a very rare passerine, *Petroica traversi*.

E *Storm and diving petrels*

Their small bones are not usually numerous in midden deposits, but in New Zealand and the Izu Islands they were eaten and *Oceanodroma castro* appears to have been part of the early Hawaiian's diet (Ziegler in Banko *et al.*, 1991). In the State of Washington, Indians ate *O. leucorhoa*, New South Wales aborigines *Pelagodroma marina* and Peruvians *P. garnotii*.

F *Palatability*

Once the musty smell is ignored, all reports indicate that fresh petrel eggs make good eating, for example Cott (1953) placed those of *F. glacialis* as close to those of the domestic fowl. The chicks too can be tasty, depending on their cooking and treatment. The young 'Manks Puffins' were 'much sought after by the inhabitants, killed, salted, and eaten with potatoes or cabbage' (Fleming, 1828). Opinions vary, however, on the palatability of the adult birds, although starving sailors or sealers couldn't afford to be 'choosey', and, as Fisher (1952a, p. 489a) says, for Greenlanders 'after a season with nothing but walrus and bear meat, any change is for the better'. Bligh's sailors force-fed albatrosses and 'pintados' on board ship with cracked corn to fatten them and rid them of their fishy flavour.

In general, adult fulmars and shearwaters taste unpleasant—the reason given by Maoris for not harvesting adult *Puffinus griseus*. According to Bigelow (1902), this bird was the only seafowl that was 'totally inedible' yet Templeman (1945) found it tastier than *P. gravis*. Adult *C. leucomelas* are also unpalatable (Kobayashi & Cho, 1981), as the Greeks also noted of *C. diomedea*, and gadfly petrel chicks seem generally much preferred to the adults. Hunter (1793) referred to the fishy taste of *Pterodroma solandri* but King (in Hunter, 1793) wrote that they tasted not the least fishy—possibly King was eating chicks, Hunter adults. Fresh muttonbirds are best roasted with breadcrumb stuffing or mixed with mashed potato and deep fried as croquettes, or just eaten with bread and butter, but not boiled—though this is necessary with salt-preserved birds. The latter need boiling in water twice before stuffing and roasting. Nowadays, they are often served as an entree.

Cott (1971) quoted Harrison Matthews that young *D. exulans* are excellent, flesh pale, tender, juicy and well flavoured. Swinburne (1886) found that *D. melanophrys* tasted like hare, and this bird, caught at sea, was once sold in Cape Town fish markets.

III As baits or lures for fishermen

The use of petrels as fishing bait no doubt extends back to prehistoric times but the birds are seldom so used today. Collins (1884) detailed how the Grand Banks fishermen caught 'hagdons' (mainly *Puffinus gravis* with fewer *P. griseus* and *F. glacialis*) on hooked lines baited with codfish livers (Fig. 12.6). Collins also described how *Oceanites oceanicus* were 'tolled up' for bait with a large codfish liver, and when a group had assembled a score or more could be killed with a single stroke of a whip. Albatrosses and other petrels were often used as bait by visitors to the southern islands such as Amsterdam (Jouventin, 1994b) and until recently chicks of *P. p. mauretanicus* also provided bait for Balearic fishermen (Serra, 1986).

Plenge *et al.* (1989) reported the killing of 23 *D. chrysostoma* by Peruvian fishermen. They use the breast feathers for fishing lures and Maltese fishermen still employ feathers of *C. d. diomedea* in this way (Zammit & Borg, 1986/87).

Coastal aborigines in New South Wales fashioned hooks and spikes from split bones of albatrosses and *P. tenuirostris* (Lampert, 1966), the spikes being used for spear fishing.

HAG FISHING.

Figure 12.6 *Fishing for 'hagdons'* (Puffinus gravis *and* P. griseus) *on the Grand Banks. The birds were used as bait by trawl-line cod fishermen. From Collins (1884).*

IV Miscellaneous uses

The use of bird feathers by the millinery trade has been well documented, the main procellariiform victims being *D. immutabilis, D. albatrus* and *D. nigripes* (Rice & Kenyon, 1962). Southern birds suffered less, although the *D. cauta* colony on Albatross Island, from which Surgeon Bass collected birds for fresh meat in December 1798 (Flinders, 1814), was later repeatedly raided by sealers. Three birds yielded a pound of feathers which raised 9d to 1/- in Launceston (Plomley, 1966). Feathers from *P. tenuirostris* were also sold to upholsterers and down from the chicks is still used in sleeping bags. That from *P. griseus* evidently cannot be deodorized satisfactorily and is no longer used. The St Kildans and others used fulmar down for their beds, and feathers of *C. diomedea* from The Salvages were sold to England for eiderdowns and in Madeira for making artifical flowers (Zino, 1985).

Stomach oil has also been used extensively; that for the Northern Fulmar, for instance, was important to the economy of the St Kildans, supplying oil for their lamps and 'medecine for their distempers'. It was stored in containers made from the stomachs of gannets. The use of the oily bodies of storm petrels as candles by drawing wicks through their mouths and cloacae was practised in Scotland and Ireland and by the southern sealers.

The stomach oil of the Tasmanian Muttonbird was sold at 4s to 4s 6d a gallon for use in lamps (Elwes, 1859). It was not only expressed from the adults but also from fledglings that were too small for normal processing. On average, 100 birds produced a gallon of oil (Murray-Smith in Skira, 1990). Nowadays, it is fed to racehorses and used for grooming to impart a shiny coat. Until recently, the oil of *C. diomedea* was used for cleaning brass fittings on boats.

Stomach oil was taken for a variety of ills by the Faroese and St Kildans, and tested for the treatment of tuberculosis in Maoris by Purdy (1900). The fat of male *Pterodroma feae* at the Cape Verde islands was (?is) used for relief of arthritic joints, and de Naurois (1969) feared for the future of that bird because of this attention.

In Tasmania whole *Puffinus tenuirostris* were rendered down—'fatting'. The product was very effective for greasing the skips in coalmines and the skids in sawmills. Some body fat was sold to dairy farmers as a food supplement for calves and some was used in soap making.

The Kamchatkans caught *D. albatrus* with baited hooks and made needles and combs from their wing bones and, according to Krashnaminnikov (1764), made floats for their nets from the carcasses. The Aleut tied spear heads to their shafts with sinews from albatross wings and the St Lawrence Island Eskimos used their beaks with part of the skull attached to bail out their tiny leather canoes. Maoris used hollow albatross bones for making 'pipe' music.

There are several reports of live albatrosses being used as buoys by drowning men, for example the seaman lost overboard at 42°S, 90°E on 24 October 1881, who, despite being encumbered with sea boots, was recovered a hour later clinging to an albatross (Green, 1887).

Wrecked people sometimes tied messages to albatrosses and giant petrels in the hope of rescue. Serventy and Whittell (1962) instance the discovery on 18 September 1887 of a mollymawk (possibly *D. chlororhynchos*) on a beach near Perth wearing a tin neck collar with a punched message that 13 sailors were shipwrecked on the Crozets

4 August that year. The information was telegraphed to London and on to Paris, but by 2 December when a French warship reached the islands there was no sign of the men, just a letter from the captain that they had been wrecked on 9 March and now, on 30 September, were setting off for Possession Island.

O'Reilly (1818) referred to *F. glacialis* as an indicator of whales, and congregations of tubenoses are still useful to sport and commercial fishermen in signalling the presence of upwellings, concentrations of krill, the presence of tuna and marlin, etc. Streams of petrels all heading one way indicate the direction of land, and McGlone *et al.* (1994) suggest that the then vast tide of transequatorial migrant shearwaters speeding southwards regularly each year, might have guided the Polynesian explorers to New Zealand.

To some extent muttonbirding was and is regarded as a sport, although shearwaters have hardly been considered game birds. However, according to Sultana (1986), Maltese 'sportsmen', using motorized rubber dinghies, shoot many *C. diomedea* and other seabirds off the breeding colonies and, although this was made illegal in 1986, it still continues. And in 1992 the rare Barau's Petrel *Pterodroma baraui* was being used for target practice when flying low over the coast to its colonies on Reunion, despite legal protection (Bretagnolle & Attie, 1993).

V Problem petrels

In the early days of Tasmanian settlement some conflicts arose between muttonbirders and graziers. The latter killed birds when burning the *Poa* grassland to encourage new growth, and cattle impacted the soil making burrowing difficult; sheep seem to be less of a problem (Norman, 1970b).

The soils of petrel colonies, rich in nutrients and of good structure, have sometimes been removed as fertilizer or sought after for growing crops, for example the best land on Santa Cruz in the Galapagos is where *P. phaeopygia* nests or nested (Harris, 1970). However, those ploughing up burrows soon discover that this does not deter further activity as established breeders reopen their sites, and there have been minor conflicts of this kind, for example between gardeners and *Puffinus carneipes* and *P. tenuirostris* on Lord Howe and Tasman Islands respectively. On New Island, The Falklands, burrowing *Pachyptila belcheri* were considered pests and pigs used to root them out (Strange, 1980).

More serious has been the problem of aircraft strikes, particularly at Midway Atoll when it was the base for nuclear-armed air patrols. Between 1958 and 1964 there were 300–400 strikes each year. Some 54 000 *D. immutabilis* were killed during control programmes to minimize these collisions. The problem seems to have been less the danger of crashing, than that flights had to be aborted with very expensive, time-consuming dumping of big fuel loads before landing for repairs. However, between November 1961 and 1 January 1965 only 58 out of 19 000 departures involved aborted take-offs or fuel dumping (Fisher, 1966). He concluded that the control measures had not actually reduced strike incidence. A contrary view was that of Robbins (1966).

On Kauai, in the main Hawaiian Group, the expanding population of *D. immutabilis* is again conflicting with man. The birds have nested or attempted to nest for years on the Pacific Missile Range despite harassment programmes. Recently, chicks hatched there have been cross-fostered to pairs at the Kilauea Point colony, others raised by hand. Also troublesome are those trying to nest at Dillinghan Airfield on Oahu where numbers increase each year despite severe disturbance and non-lethal harassment—fireworks, chasing, capture and release elsewhere, destruction of nests, and the removal of chicks for hand raising: these appear to be on-going programmes (Pyle, 1991).

There seems to be no new evidence of a role for tubenoses as vectors for influenza viruses or ornithosis to man (*The Petrels*, pp. 48 and 169). In a study by Austin and Hinshaw (1984) none of 100 *Puffinus griseus* tested showed any signs of type A viruses. Nor were any tick-borne viruses found on *Ixodes uriae* from Icelandic fulmars (Moss *et al.*, 1986). And, despite the many bites that field workers collect, there are few reports of illness resulting.

Albatrosses and giant petrels have been implicated in attacks on people in the sea. Cleland (1956) described how an injured naval rating washed overboard was bitten and scarred by an albatross which, none the less, saved his life by attracting the attention of a searching destroyer, and sailors in life-rafts have viewed approaches of albatrosses and giant petrels with alarm: their sharp bills could easily pierce the rubber fabric.

VI Captivity

Although wild petrels are sometimes very tame, presumably when hungry (see Fig. 4.1), birds of such wide horizons are difficult to keep in captivity. The smaller species appear to be the least able to cope with constraint, but chicks of several have been hand reared, for example *Pterodroma cahow* (Wingate, 1972); *P. feae* (Metcalf in Zino & Zino, 1986); *P. phaeopygia* by Pratt (1971); *Puffinus lherminieri* (Reid, 1877), etc. As mentioned above, chicks of *D. immutabilis* are regularly hand raised in Hawaii and, following the successful rearing of a chick of *D. epomophora* (Robertson & Wright, 1973), this has become a common procedure for abandoned chicks at the Taiaroa Heads colony in New Zealand.

Adults are harder to manage, though some, such as *Puffinus tenuirostris* and *P. lherminieri*, soon learn to feed from the hand and will gorge themselves excessively if supplied *ad libitum*. Zino (1985) recorded a completely tame albino *C. diomedea*. Captive storm petrels seem to die within a week or 10 days (Mathew, 1881); Ticehurst, 1912), although better results might be gained with modern methods. Oil-spitting fulmars cannot be caged with other seabirds because of their readiness to spit deadly stomach oil (Kritzler, 1948; Swennen, 1974).

Albatrosses and giant petrels have been kept for years, for example *D. irrorata*, *D. immutabilis* and *D. nigripes*; the latter was fed for 7 years on squid. C. A. Nicholls, who has restored many beach-wrecked petrels and even had *P. tenuirostris* laying in burrows in her Perth garden, developed techniques for their rehabilitation. She reports (in litt.) that cleanliness is very important. The food must be fresh and not offal. Initially, captives are fed as demanded, three to four times daily for weakened birds. Once or twice a week, half the contents of a multivitamin capsule are given inserted in a slot in the food and one halibut oil capsule administered weekly in the same way. The birds are not allowed to rest on leaves or fabric, which get wet and harbour fungi; layers of newspapers changed daily serve well. Seawater for drinking is given in a large dish and changed frequently. It should be the real thing, not water dosed with salt tablets. After a meal, salt excretion should begin from the bill within a short time and the birds are hosed down to remove any food trapped in the feathers and to encourage preening. Recovered birds need only one feed or less daily. Birds held captive for long periods can be given fresh whole fowls, rabbits or chunks of sheep meat which they can tear at in a natural manner and ingest fur and feathers. Healthy birds can be freed when they have become active, their feathers waterproof, and they are able to beat their wings strongly. Released birds often indulge in an orgy of bathing and drinking, but most seem to recover.

Giant petrels are fed chopped beef or gutted fish, for example herring. For those newly caught, the food is put in a small dish with seawater. They may have to have their beaks dipped into this before they will drink voluntarily. Giant petrels can become very aggressive so may need keeping apart from other birds.

Albatrosses are different and may become very tame. Most need force feeding initially on clean gutted fish prewetted in seawater, and force feeding seawater from a squeeze-bottle may be necessary before they will come to drink themselves. They like a 30–60 cm high mound of sand or clean earth on which to sit and from which they will defecate in the usual radial manner. Their feet are tender, so no standing on concrete floors or wire should be allowed.

VII Conservation

The maintenance of a petrel population in an entire state—its conservation—involves many approaches which are common to those for other animal groups such as the preservation of essential needs (food, safety, suitable breeding places, etc.), but others are specific to tubenoses. Most of the pressures threatening these populations today are man-made and require human actions for their mitigation.

Foods and food availability must have changed greatly during the radiation of these birds but today's big changes mostly derive from increased commercial fisheries although there are little data suggesting that these compete significantly with tubenoses for shared resources. Fisheries can benefit them too—witness the extent of scavenging by *F. glacialis*, *D. melanophrys*, *Procellaria westlandica* and *P. aequinoctialis*. Their populations could well decline if fishing techniques changed, fishing grounds were abandoned, or offal became unavailable. Conservationists can seldom affect food availability but the conservation of fisheries may help associated seabirds too (Bourne, 1981, p. 119).

Maintenance of suitable habitat is essential. Breeding habitats become degraded through excessive rain or other natural causes, but most changes are from unnatural ones. Habitat maintenance often calls for direct action. For example, at Cabbage Tree Island removal of the native *Pisonia* trees has reduced deaths of the small population of *Pterodroma leucoptera* from their adhesive seeds. Erosion, perhaps by the birds themselves, destroys burrows, but more often destruction of binding vegetation by stock, rabbits or fire provides the initial trigger. Rehabilitation may involve replanting, sowing and fertilizing while removing the cause of the degradation. For instance, at Macquarie Island and The Kermadecs overgrazing by rabbits brought erosion which reduced the breeding habitat for birds such as *Pachyptila desolata*, *Pelecanoides urinatrix* and *Pterodroma neglecta*. At the former location the great reduction in rabbits after the introduction of myxomatosis has seen the vegetation recover rapidly giving hope that the fewer rabbits will also result in fewer cats and skuas with benefits to all the tubenoses there (Rounsevell & Brothers, 1984). Ungulates can cause habitat loss through soil impaction and burrow breakdown. The cattle herd on Amsterdam Island, implicated in the decline of *D. e. amsterdamensis*, has been halved and fenced out of the nesting area, with 1600 native *Phylica* seedlings planted in the cattle-free zone (Jouventin, 1994b).

Changes in vegetation, creating dense interlocking ground cover preventing burrowing, may also remove habitat. Rehabilitation may require the use of herbicides or the temporary introduction of goats. Elimination of unwanted plant species, however, provides no guarantee that the vegetation will revert back to its original state. At Taiaroa Heads the coastal slopes used by *D. epomophora* are clothed in tall introduced grasses that are regularly mowed to facilitate the birds' movements. Lack of vegetative cover may also be detrimental. On Torishima Island the improvement to the nesting habitat from planting grass by H. Hasegawa has been partly offset by large-scale erosion. However, major soil conservation work has been done and the *D. albatrus* population continues to grow at about 7% per annum with 146 pairs on eggs in the 1994/95 season (Hasegawa, pers. comm.). At Midway and Kure Atolls the addition of shiploads of soil increased the habitat suited to albatrosses, but the growth of the exotic *Casuarina litorea* and plants such as *Brassica campestris* and *Verbesina enceloides* has rendered areas unsuitable for nesting, and *Casuarina* increases mortality due to collisions on landing.

Creating new habitat is seldom possible. Perhaps occasionally a dredging operation might be adapted to create an island colonizable by birds such as storm petrels, using suitable lures. On a grander scale, a supertanker, decks covered with soil seeded appropriately, could create an artificial island which anchored say, in the subantarctic Pacific, could well be colonized by tubenoses.

Competition on land usually involves alien species competing for space, as when rabbits evict petrels from their burrows, for example *Pterodroma lessonii* (Warham, 1967a) and *C. diomedea* (Bayle & Fernandez, 1992). Sometimes competition is a natural event, for example expanding seal populations destroying nesting habitat (Croxall *et al.*, 1990b). The well-known case of the tropic bird *Phaethon lepturus* evicting *P. cahow* from its nesting sites provides another example. Development of a baffle allowing the petrel to enter but denying access to its larger competitor, as well as the quite complex programme involved due to the philopatry of the tropic birds, was detailed by Wingate (1977).

A *Endangered species*

Birds most at risk appear to be those with small populations confined to very few breeding places. Species such as *Procellaria westlandica*, although nesting only in a very limited area, seem to be doing quite well at present, but would be much safer were another population established, preferably offshore. Others such as *D. irrorata* are carefully monitored, while *D. c. eremita* nesting only on Pyramid Rock, has about

3200 pairs (Robertson, 1991) and seems reasonably safe despite illegal kills by local Maori.

Among other albatrosses, only *D. (exulans) amsterdamensis* is seriously endangered, with but 70 birds thought to be extant (Jouventin, 1994b), and at least one killed on a Japanese longliner. Most of the endangered birds, however, are gadfly petrels. The rarest is probably *Pterodroma macgillivrayi* with but one adult and one fledgling seen alive. Others include *P. magentae* (population perhaps greater than 100 birds (Crockett, 1995), with three chicks fledged that year); *P. madeira* (only eight known burrows); *P. feae* (perhaps 150–200 breeding pairs on The Desertas and a few on The Cape Verdes), and *P. cahow* with 49 nesting pairs in 1995 (Wingate, pers. comm.). *P. aterrima*, with its breeding site undiscovered and with only three examples known in recent years although some have been seen at sea, may be as rare as *P. macgillivrayi*. The Chatham Island Petrel *P. axillaris*, which may number little more than 200 birds (West, 1995), is also endangered. Among storm petrels, *Oceanodroma markhami* and *O. hornbyi* are of indeterminate status and evidently rare.

B *Predator control*

Petrels evolved with predators that killed them on land and at sea, perhaps including marine reptiles as well as extant taxa such as sharks. However, they appear never to have experienced serious predation by large placental mammals, lacked adequate defensive mechanisms against these, and resorted to placental-free habitats for breeding.

Predation by placental carnivores is often influenced by the presence of other alien species, such as rabbits. Elimination of rabbits or rats where these supply a major food resource for cats or mustelids may mean increased pressure on the remaining tubenoses, particularly those that breed in the winter. Recently it has been demonstrated that some of these pests can be controlled or eliminated. Research in progress may also lead to new repellants for rabbits and highly attractive baits for cats and mustelids.

In large areas where complete eradication is impracticable, it may be possible to maintain a *cheval de frise* of traps or poison stations around a colony during the critical period, as has been done with *P. magentae*, for example (Imber *et al*., 1994).

1 *Cats*

Until recently, cat control involved continual effort or numbers soon bounced back. On Kerguelen, for example, 2500 were shot over 5 years but as rabbits were uncontrolled the cats spread and the effort was wasted (Pascal, 1980). He estimated that cats were killing a million petrels there at the end of the 1970s, since when they have extended their range throughout the main island. Fortunately, many of the smaller islands are cat-free. Even under the best conditions, however, control involves many man-hours of work.

Veitch (1985) detailed the approaches used to eradicate cats from eight islands around New Zealand and gave notes on home ranges, methods to lessen the capture of non-target species, field signs of cats, and so on. On Little Barrier Island eradication aimed at protecting *Pterodroma cookii* and *Procellaria parkinsoni* which faced extinction there. This is heavily forested (2817 ha), with little level ground and steep gullies. Elimination took 4 years, involved 128 people (many unpaid) in 3880 man-days of work. Up to 950 Lanes Ace gin traps and 27 000 pieces of fresh fish poisoned with 1080 were used. The poison baits were spread along tracks, the traps set on a grid pattern for ease of checking. As expected, it was difficult to catch the last animals. At least 151 cats were killed. Veitch emphasized the need to ensure that financial support is forthcoming over the whole period required to gain control or elimination, and for getting the acceptance of the project from affected persons such as landowners.

The introduction of the viral disease feline enteritis (feline panleucopaenia) to islands may be a useful tool where it is possible to infect live cats. On Marion Island it reduced a population of some 3400 animals down to a few hundred, when intensive trapping and shooting removed still more. The results on *Pterodroma macroptera*, one of their favoured foods, were dramatic. Its breeding success varied from 0 to 20.5% from 1979 to 1984—cats ate the chicks—but in 1990 rose to 60% with no loss of chicks in study plots (Cooper & Fourie, 1991).

2 *Rats*

The naive belief that cats could control rats has been responsible for many of today's problems. Warfarin in conjuction with a-chlorohydrin succeeded in eliminating *Rattus norvegicus* from small islands around New Zealand, Nonsuch Island, Bermuda, and elsewhere, but extermination became much more feasible with the introduction of brodifacoum baits (e.g. ICI 'Talon' in pellet form; Moors, 1985). This is effective at 0.26 mg kg^{-1} body mass. A single dose is lethal and few non-target animals are affected. The bait is laid inside coloured plastic tubes that protect it from rain and, while allowing the rats ready access, hinder birds from reaching it. The baits are placed in a grid after initial feeding with non-poisoned ones. The poison baits are inspected daily, the pellets replaced as necessary, and when none is taken checks are continued over several months to ensure the last animal has been killed. While such work is labour intensive, the availability of volunteers has made possible the eradication of rats from a number of New Zealand islands, for example *R. norvegicus* from The Noises (21.8 and 9.5 ha) and *R. exulans* from Motuara (59 ha) in Queen Charlotte Sound, breeding place of *Puffinus gavia* and a small population of *P. griseus*.

Brodifacoum has helped control *R. rattus* which attacks chicks of *Pterodroma madeira* at 1500 m on inland cliffs with very difficult access. The rodenticide was incorporated in wax blocks to withstand the wind and rain of the very exposed ledges near the nesting holes (Buckle & Zino, 1989).

Cruz and Cruz (1987a) found that coumatetrayl reduced a *R. rattus* population more quickly than brodifacoum, and used it effectively for bait cordons around selected colonies of *P. phaeopygia* at the Galapagos Islands. No secondary poisoning was seen. The same compound seems to have eradicated *R. norvegicus* from Toro Islets (2.6 ha) near Corsica (Thibault, 1992).

3 Other alien mammals

In many of the placental mammal problems, only one is involved. On The Galapagos, however, conservationists are faced with scattered colonies of *P. phaeopygia* on large islands subject to a range of threats—burros, goats and cattle destroy habitat and trample nests, dogs and pigs dig them up, and cats and rats eat chicks and adults, the mix varying from island to island. Goats breaking through burrows in volcanic scoria caused much erosion (Tomkins, 1985; Cruz & Cruz, 1987b). Control was effected on Floreana by hunting cats, pigs, goats, burros and cattle throughout the breeding season of *P. phaeopygia*. Rodents were controlled as above, cats hunted with dogs and shot and their control helped with 1080 baits which also helped reduce pigs on Santiago. Some small colonies have been fenced to reduce access by dogs and pigs. On Floreana at least, the result has been high reproductive success of the petrels, but The Galapagos exemplify the problems where the birds are on large islands and pest elimination impracticable—control has to be on-going and is expensive, requiring cash input from outside, from tourists and foundations, together with volunteer help in the field.

On small islands, large animals such as pigs and goats can be shot out. This was done in 1936 on Aorangi Island and allowed its natural recolonization by *Puffinus bulleri*. Pigs and goats can be stopped from digging out burrows by covering the ground with heavy-gauge wire netting pegged down to follow ground irregularities. On a limited scale and on flat ground the steel mesh used by builders for reinforcement will defeat any pig, but with both materials it will probably be necessary to open the mesh at some burrow entrances.

Mustelids are seldom associated with petrel colonies except in Hawaii. Control is difficult and may involve a cordon of traps set in drain pipes or other suitable tubes.

Mice are seldom more than a nuisance but are potentially dangerous for storm petrels. They can be eliminated with rodenticides and a lot of labour. Mana Island (217 ha) in Cook Strait, New Zealand was completely rid of the 'millions' present using 5500 bait stations in a 25-m grid pattern.

C Disturbance

Colonial nesting petrels can be intolerant of disturbance when breeding. A classic example is that of giant petrels following the establishment of research stations on southern islands (*The Petrels*, p. 73). Much disturbance arises from the proliferation of weekend recreational boat owners who land on formerly undisturbed islands whose seabirds were thereby automatically protected. Vandalism leading to nesting failure and mortalities is common where petrels breed near to large human populations, for example around the Hawaiian islands and on inshore islands around Australasia. Such unauthorized landings entail the risk of alien introductions, of which there are many examples. The difficulties in monitoring such events may be lessened once satellite surveillance becomes feasible.

At nature reserves, visitors have to be dissuaded from too close approaches with suitable barriers and, with burrowing species, made to keep to marked tracks. Researchers nowadays are more aware of 'observer effects', some degree of which may be impossible to avoid. A useful, if cumbersone, device to reduce damage when crossing densely burrowed, friable ground, is to wear boots with plywood attached to spread the body weight.

D *Pollution*

It is very difficult to protect tubenoses from marine pollutants, and the recent discovery of high levels of DDE and DDT in adult *D. nigripes* and *D. immutabilis* and evidence of eggshell thinning, is ominous.

Petrels are very prone to internal contamination by plastic materials floating on the world's oceans, plastic loads that may be increasing. Small pellets in particular are picked up by dippers and patterers such as prions and storm petrels and larger items by surface seizers such as *Puffinus gravis* and albatrosses (*The Petrels*, p. 45). Plastics have not been found in diving petrels. For reviews see Azzarello and Van Vleet (1987), Ryan (1987a, 1988), Sileo *et al.* (1990b), Moser & Lee (1992) and Spear *et al.* (1995a).

Gizzards may be clogged with plastics. Breeders regurgitate these with the food to the young and the higher incidence of plastic in immatures may be because they lack the means for getting rid of the stuff.

Plastics may stay in the gizzard for months—experiments with *Procellaria aequinoctialis* suggested a half-life of at least 1 year (Ryan & Jackson, 1987). Yet there is no evidence of alimentary canal blockage, of interference with digestion or of stomach ulcers, nor is body condition overly affected (Ryan, 1987b). The material appears not to enter the intestine except as eroded fragments that are passed with the faeces.

Sievert and Sileo (1993) checked the growth and survival of chicks of *D. immutabilis* and *D. nigripes*; endoscopes showed that 27 and 16% respectively had $>22\,cm^3$ of plastic in their proventriculi. Before fledging, the chicks vomited castings containing plastic and other indigestibles. Although those with big loads fledged at lower weights than chicks with small loads, their development was the same and dehydration, not internal mechanical lesions, the cause of death. Postfledging survival was unknown.

E *Enticement, rehabilitation and rescue*

1 *Decoys*

These are increasingly being used both for the renewal of current colonies and to establish new ones. Wandering non-breeders can be induced to land by playbacks of tape recordings when models of their kin in display (for diurnally active birds) or artificial burrows (for nocturnal ones) may induce them to remain, pair up, and eventually breed.

This technique as applied to seabirds seems to have been pioneered by the National Audubon Society to attract terns and Puffins *Fratecula arctica*. Later, Podolsky and Kress (1989) demonstrated the effectiveness of the Fraser Darling effect by inducing *O. leucorhoa* to nest on islands in Muscongus Bay, Maine, on most of which these birds no longer bred, and one, Ross Island, with no record of breeding. Experiments were run using different combinations of calls and burrows (Fig. 12.7). At all four islands the birds eventually occupied preprepared burrows, including 12 out of 40 at Ross Island. The birds tended to choose nests close to the loudspeakers. Although sound stimulation was subsequently discontinued, the petrels in due course raised chicks.

A similar project to attract roaming *D. immutabilis* to breed within a refuge at Kilauea Point, Kauai (*The Petrels*, p. 46) using playbacks and models (Podolsky, 1990) has been very effective with >100 nests started in 1995 (R. L. Pyle, pers. comm.). On Torishima Island a similar effort is being made to start a new *D. albatrus* colony away from the present unsuitable site (Hasegawa, pers. comm.).

Podolsky (1990) found that three-dimensional models in 'sky-call' attitudes were the most attractive, and in a further development to entice more wanderers to colonize Kaohikaipu Island about 1 km off the coast of Oahu, sets of 'sky-calling' decoys, some with model chicks or ceramic eggs have been established (Podolsky & Kress, 1994). One group had no chick or egg decoys and only adult models plus colony and courtship sounds, the other had chick and egg decoys with chick calls superimposed on the colony sounds. This second arrangement attracted 71% of landings.

The birds performed 'sky-call' displays as well as sleeping, self- and allopreening, walking, loafing and mutual billing, 'dancing' with one another and billing with the decoys. 'Sky-calling' decoys with chicks were particularly effective. If these stimuli can be maintained it seems likely that breeding will start, which should relieve some of the conflicts arising when the birds try to nest on inappropriate places such as airports.

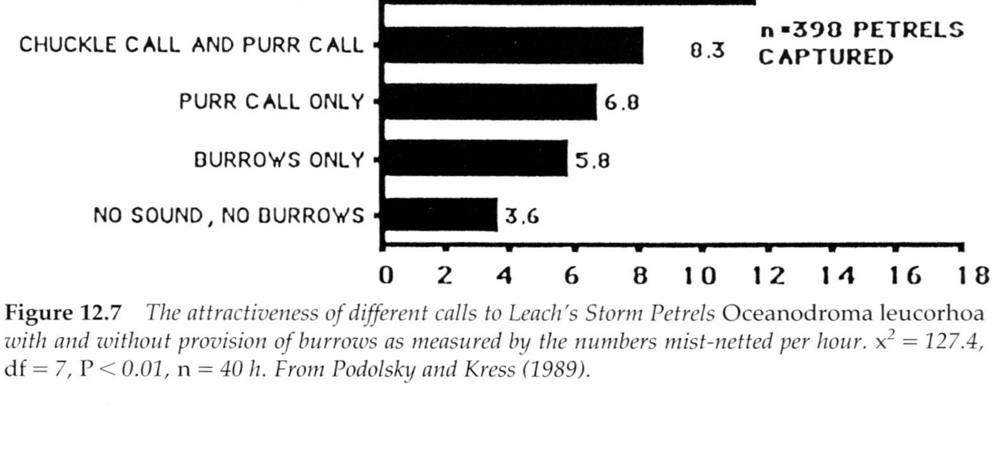

Figure 12.7 *The attractiveness of different calls to Leach's Storm Petrels* Oceanodroma leucorhoa *with and without provision of burrows as measured by the numbers mist-netted per hour.* $x^2 = 127.4$, df = 7, $P < 0.01$, n = 40 *h. From Podolsky and Kress (1989).*

Podolsky and Kress (1992) also experimented with playbacks at colonies of *Pterodroma phaeopygia* on Santa Cruz, Galapagos, using calls at different intensities. Double intensity ones proved the best. Many birds were attracted, most with downy brood patches and judged as non-breeders. In one area without nests, 80 artificial burrows began to be occupied within 5 days of playback being started, with 16 of them being occupied during the first season. One egg was laid in 1988 and by 1991 six pairs had fledged young. This experiment points to the possibility of getting the species to nest in compact groups more easily protected from predators.

2 *Translocation*

Serventy (1967) found that eggs of *Puffinus tenuirostris* cross-fostered to birds on another island produced chicks that returned there, and not to the island of their rightful parents. However, his hypothesis that nestlings are imprinted on their natal colony during their last exploratory nights ashore was not supported by the results from 157 nestlings shifted after their last meals to empty burrows 1–3 km from their natal ones. By 20 years later, a lower proportion of these transferees had come back to their new island than that of chicks born there (Serventy *et al.*, 1989). Either the shifted birds had already become attached to their natal homes before being moved, perhaps while still underground, or they had taken much higher postfledging losses than the locally born birds. This may help explain other translocation failures, such as that reported for *P. puffinus* (Brooke, 1990, p. 40).

More success seems likely with a current attempt to establish a new colony of *Puffinus gavia* on Maud Island, New Zealand. Chicks have been shifted from nests 35 km away and fed twice daily in artificial burrows to fledging. Feeding is via a stomach tube on a blended mixture of 400 g of fresh fish, 60 ml of treacle, 20 ml of olive oil plus vitamin additives. Most birds soon accepted the tube, which facilitated feeding. Later the fish and oil content was increased: 50 g night and morning were needed for a consistent weight gain. A solar-powered playback of the shearwater calls runs from dusk to dawn in the hope of attracting returning adults or, perhaps more important, of imprinting the chicks with their colony dialect.

These birds do not have the highly synchronized laying period of a trans-equatorial migrant shearwater, so the transferred chicks have been of a wide spread of ages and this may mean that few have already become attached to their natal colony. Up to April 1995, one bird from the 1990/91 season has reappeared in the 1994/95 one (B. D. Bell, pers. comm.).

Translocation and care of chicks is extremely labour intensive, plans often frustrated by bad weather, and it would seem that if a putative colony is within the normal range of the species, first attempts should use amplified vocalizations in conjunction with preprepared models and/or burrows.

3 *Artificial burrows*

Quite apart from modifying existing burrows for observation (*The Petrels*, p. 212), tubenoses take to artificial burrows, probably the more so where nest densities are

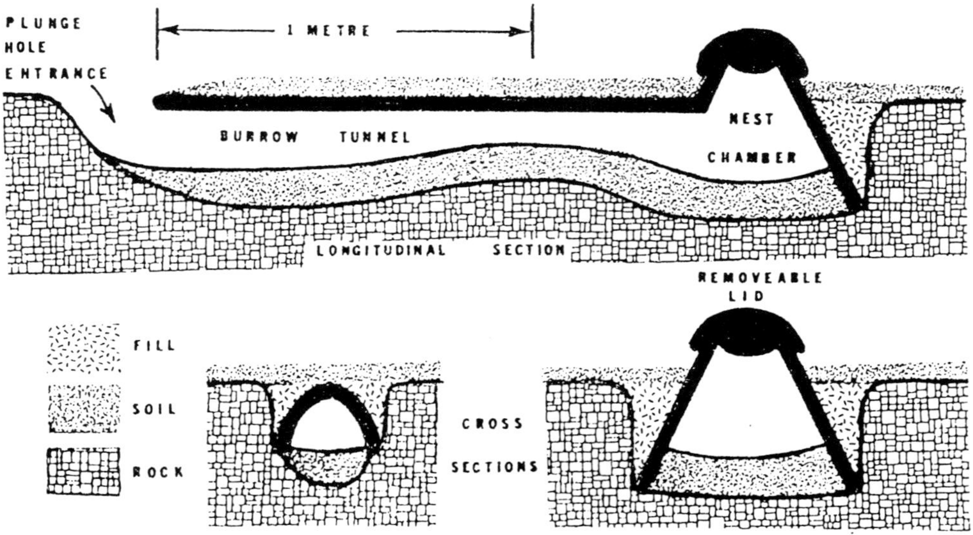

Figure 12.8 *Artificial 'plunge-hole' burrow for* Pterodroma cahow. *Sited on the level tops of islands by trenching along the rock, excavating a nest chamber with a removable lid, and roofing the whole with concrete poured over a wire form. From Wingate (1977).*

high and surface eggs plentiful. Dimensions of the chamber and its tunnel should be like the natural ones, the tunnel in cross-section matching the size and shape of the bird's body and should incorporate a bend if possible. Tunnels larger than necessary may encourage bigger species to enlarge them and compete with the target species.

Nest boxes may be tailored to fit particular situations. Prions used a set slotted into a rock crevice, and Tomkins (1985) designed suites of boxes placed in rockpiles for *Pterodroma phaeopygia*. Wingate (1977) made special 'plunge-hole' burrows for *P. cahow* in situations devoid of soil (Fig. 12.8). Those for *Puffinus gavia* on Maud Island are made from buried 112-mm flexible tubing as used in garden drainage, with a section cut out to produce a sharp bend, and have a soil floor.

As the nests will usually require inspection, the seal of the chamber lid needs to be carefully designed so that rain cannot drain through it nor soil fall on to chick or egg. A properly hinged lid covered with a well-impacted sod or a block of polystyrene helps maintain the natural insulation.

3 *Fledgling rescue*

Strandings of young birds on their maiden flights are particularly common in Hawaii where many pass over the bright lights of hotels and tennis courts to reach the sea (Telfer *et al.*, 1987). Well-published campaigns (Fig. 12.9) and the provision of collecting points cleared daily in local stores and car parks, have enabled many to be rescued instead of being killed by cats and dogs. In 1991, for example, 20 *Pterodroma phaeopygia* and 1811 *Puffinus a. newelli* were turned in on Kauai, 92% of which were banded and released successfully.

Figure 12.9 *Posters seeking the rescue of 'Ua'U grounded by lights, fog, etc. ask for birds to be placed in the cages sited at convenient 'Petrel Rescue Stations' from which they will be collected and, if fit, banded and released.*

4 *Cross-fostering*

The switching of the eggs or young of one species to be raised by another must be a technique almost as old as the domestication of birds, and experimental egg exchanges have shown that some petrels can raise others' chicks. Fisher (1972) found that the fledging successes of *D. nigripes* raised from the egg by *D. immutabilis* (and vice versa) were as good as those of chicks raised by their rightful parents.

The technique could help in conserving a rare species subject to heavy predation by removing eggs for rearing by a related form breeding in safety, or, perhaps, in attempts to establish a new colony in safer habitat. However, the trial transfer of eggs of *P. a. newelli* for hatching by *P. pacificus* by Byrd *et al.* (1984), has apparently not succeeded in establishing new sites. And whether fostered birds will take on behaviours of their foster species and make them unacceptable to their own is unknown.

VIII Fisheries 'by-catch'

No doubt petrels have been killed in the course of fishing since fishing began, and until recently losses from commercial fisheries were insignificant, but with the development of large-scale driftnets, 'mother ships', and lines of baited hooks 100 km long, the picture has changed drastically. Substantial numbers of seabirds get enmeshed, hooked and drowned, although the public became aware of the problem mainly through learning of the losses of sea mammals and turtles. Aspects of these conservation problems were outlined in the various chapters of *The Petrels* and the conservation of albatrosses worldwide was reviewed in detail by Gales (1993).

The scale of the problem was not fully realized until scientists were stationed on ships to monitor these 'by-catches', which were understated in the fishermen's own reports. Observers' figures still tend to be conservative as birds drop from nets or hooks and are missed or are discarded before being counted, while some take bait, hook and all and are found dead or dying on their nests. Losses are most severe where the fishing grounds and those of the birds coincide. Nets lost or discarded at sea also cause deaths as their monofilament material is almost indestructible and continues 'ghost-fishing' for unknown periods. The introduction of exclusive economic zones (EEZs) for 200 miles offshore made an important impact as the responsible countries could control the fishing there through licensing.

The main victims may at times be yearlings, at others, breeders, and either sex or one sex may predominate. For example, of 47 *Procellaria cinerea* drowned on longlines, 43 proved to be females in breeding condition, all probably with nests on Antipodes Island (Murray *et al.*, 1993). The loss of a breeder during the nesting season means that breeding fails. The death of a partner, even if in the non-breeding season, means that at least 1 year's breeding is lost with probably a reduction in success even after a new pair-bond is fashioned.

A *Drift-net fisheries*

One of the successes in the conservation of marine animals in general, and of tubenoses in particular, has been the cessation of drift-net fishing in response to United Nations resolutions calling for a global moratorium on this kind of large-scale pelagic fishing. The main nations involved, Japan, Taiwan and South Korea, seem to have complied.

At their height, these fisheries deployed a formidable distance of nets and eight distinct types of activity were involved; for example, the Japanese mothership fishery for salmon in 1959 had 16 motherships and 460 catcherboats mainly in the Bering Sea and around the Aleutians. The losses of seabirds in this and other pelagic fisheries for tuna, marlin, swordfish, albacore, neon flying squid and smaller coastal operations, as reported by international observers, were summarized by Johnson *et al.* (1993), Ogi *et al.* (1993), DeGange *et al.* (1993), and the species' associations by McKinnell and Waddell (1993). The Japanese mothership operation used nets averaging 16.5 km long and set nearly a million kilometres of them. The salmon gill nets produced a curtain from the surface down to 6–10 m. Most shearwaters were trapped several metres down when diving as the nets were being set or retrieved, whereas albatrosses and fulmars mostly became entangled at the surface when they tried to eat fish caught in the meshes.

In 1990, Ogi *et al.* (1993) estimated that driftnets killed 500 000 seabirds of 36 species, 356 000 being *Puffinus griseus*, 40 000 *P. tenuirostris*—80% of the casualties came from the Japanese squid industry. From an estimated 40% of the catch of *P. griseus* being breeders (banded birds taken and returned are few and of unknown status) and other assumptions, Ogi *et al.* suggested that a <0.2% annual decrease in the breeding stock would result. As the only population estimate for this species dates from 1971 (Warham & Wilson, 1982), it is impossible to cross-check this on the ground.

Other birds caught in fair numbers were *O. furcata*, *P. bulleri*, *P. carneipes*, *F. glacialis* and especially *D. immutabilis* and *D. nigripes*.

B Set net fisheries

Being on a much smaller scale than the drift net ones, these take a smaller toll of seabirds and they are seldom placed near petrel colonies or in deep water. Off Newfoundland, however, many *P. gravis* on their migration are trapped as the gill-nets of the cod fishery are set or retrieved—an estimated 2200 each year between 1981–1984 (Piatt & Nettleship, 1987). Of 388 taken at known depths, 218 were at 0–10 m, 153 at 10–20 m and 17 between 20 and 30 m. The 22 *P. griseus* also killed were similarly distributed.

C Subantarctic squid trawl fishery

About 50 Soviet trawlers fish for arrow squid *Notodarus sloanii* in shelf and slope waters around the subantarctic islands off New Zealand. A considerable mortality of tubenoses occurred from collisions with a netsonde monitoring cable running from the aft gantry out to about 20 m away. The cable is thick, grey and whips about as the boat pitches. Many birds get wrapped around it, especially in overcast, calm and foggy weather, some losing wings or sustaining other damage. The fishery operates round the Auckland Islands and the main victims are the local *D. cauta*. Bartle (1991) autopsied 36 birds, 19 males, 16 females, all breeding adults. These albatrosses have long fed behind trawlers and some become enmeshed as the nets are hauled, but the netsonde cable is the major hazard. More modern ships use hull- or aquaplane-mounted transducers which pose few hazards. Vessels with exposed netsonde cables are no longer licensed to fish in this EEZ.

D Longline fisheries

Declines in populations of *D. exulans* on South Georgia, Macquarie Island and the French Indian Ocean islands aroused suspicions that some 'unnatural' influence was at work. That the longline commercial fisheries were involved was confirmed when scientists recovered 'by-catch' birds with leg bands, 88% of the 16 recoveries of South Georgian birds since 1984 being from longliners, 64% of these being adults (Croxall *et al.*, 1990a). At a rate of 0.41 birds per 1000 hooks, Brothers (1991) calculated a loss of 44 000 albatrosses—9625 *D. exulans*, 19 250 *D. melanophrys*, 1375 *D. chrysostoma*, 19 250 *D. cauta* and 4125 *Phoebetria palpebrata*.

Some are taken while breeding and both the South Georgian and Crozet Islands catch includes more females than males since these favour subtropical waters where the longliners work (Fig. 12.10). Ringed South Georgian mollymawk recoveries have increasingly come from these fisheries. Juveniles drown more than adults, which

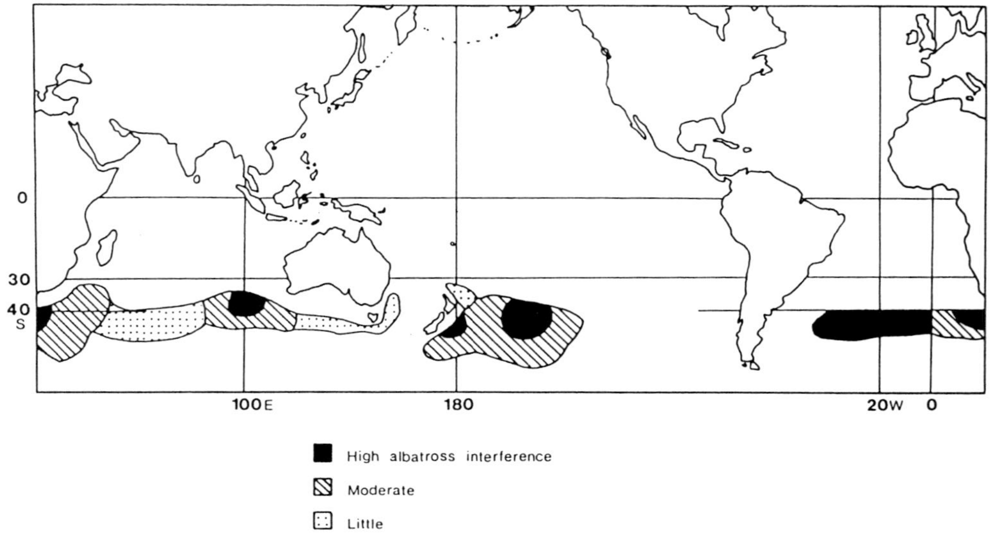

Figure 12.10 *Distributions of regions of albatross interference with longline fishing operations in the Southern Ocean. From Brothers (1991).*

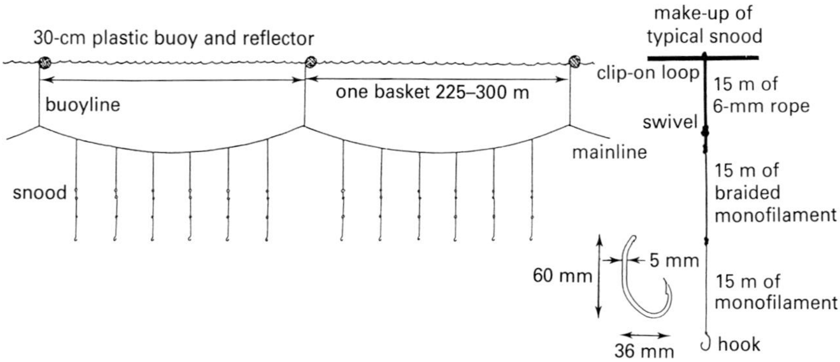

Figure 12.11 *Design of a typical Japanese longline made of 5–15 snoods arranged into 'baskets' supported by buoys carrying lights and radio-beacons and looped together for 100–130 km. The squid-baited hooks are 60–180 m down.*

may help explain the low recruitment rates in recent years (Prince *et al.*, 1994b). Declines in the populations of *D. chrysostoma* at Campbell Island have also occurred, and again, longlining is suspected.

Although the populations of *D. exulans* on The Crozets and Kerguelen have increased recently, apparently because the longliners have moved further east, this fishery now appears to pose the greatest danger to non-target marine vertebrates. In the Southern Ocean, the Japanese tuna fishery deployed 108 million hooks yearly between 1981 and 1986 (Fig. 12.11). A longline carries about 3000 hooks and is fed out over one side of the stern usually in the evening at about 400 m min^{-1} with the ship steaming at *c.* 10 knots. About every 6 s a prebaited 'snood' is clipped on to the mainline and thrown out. The whole takes about 5–6 h to set, fishes for 4–5 h and is hauled in over about 12 h. Birds are mostly caught when they grab and swallow the baits as these hit the water. If hook or bait is swallowed or snagged in the mouth the petrel is dragged under and drowned (Fig. 12.12).

Imber (1994) and Brothers (1991) found that when the more agile *Daption capense, Pterodroma macroptera* or *Procellaria cinerea* grabbed or dived for the bait, others closing in caused a commotion which attracted mollymawks. These often scattered the smaller birds and took the bait unless *D. exulans* displaced them. Most baits were taken without the birds being harmed, mollymawks tending to nibble at them rather than gulp them down.

A range of species is caught. In New Zealand waters these are mainly adult *Diomedea exulans, D. bulleri* and *P. cinerea*. Except for the last species (Bartle, 1990) the sex ratios of the by-catch tend to be equal but the predominance of adults taken is serious for a K-selected species. The effects on the breeding populations in New Zealand cannot be directly checked as too few of these have been censused accurately, although this is being rectified.

The position of the Wandering Albatrosses at Macquarie Island is better known. The breeding population increased from 17 in 1956 to a maximum of 97 in 1966 and then fell at 8.1% annually. It is now in danger of extinction. De la Mare and Kerry (1994) show how this decline has coincided with the increase in long-line hooks set in the tuna fisheries (Fig. 12.13). Moloney *et al.* (1994) have devised a model for assessing the impact of longline fishing on these populations.

Vaske (1991) revealed that tuna fishers off Brazil take a significant number of petrels other than albatrosses. During 52 fishing days observers recorded a 'by-catch' of 58 *Procellaria a. aequinoctialis*, six *P. a. conspicillata*, four *D. exulans*, two *D. melanophrys* and one *F. glacialis*. At this rate, with 10 vessels and 270 days fishing, some 2650 birds would be killed annually including the rare *P. a. conspicillata* race of the Shoemaker with only *c.* 1000 pairs on Inaccessible Island. And this is an expanding fishery. A newly established South Africa hake long-line fishery has been estimated as killing about 8000 *P. aequinoctialis* annually (K. N. Barnes, pers. comm.).

Fishermen do not care to have their baits stolen by birds, nor to have their activities restricted for environmental reasons, and bait loss costs the bluefin tuna industry an estimated $Aus7 million (Brothers, 1991).

One device to reduce 'by-catch' deploys a bird-scare line with coloured streamers to hang like a curtain from the stern to discourage birds from sweeping close enough to reach the baits before they disappear (Fig. 12.14). In trials this cut bait-taking quite dramatically (Murray *et al.*, 1993). The long-term efectiveness of these 'tori-lines' has been questioned, and discarding offal on the opposite side of the boat to draw birds away from the hooks also works for but a short time.

The quicker the baits are submerged, the less time there is for them to be taken, so weighted bait lines would help. Brothers (1991) found that where the baits were thrown less than 5 m from the ship the bird catch increased, as these baits, trapped in the turbulence, were near the surface longer than those thrown to clear the wake. An automatic bait-throwing device recently deployed on >100 boats may relieve the problem. Bait condition is another factor. Those still frozen tend to float, thawed ones sink, so less of the latter are taken.

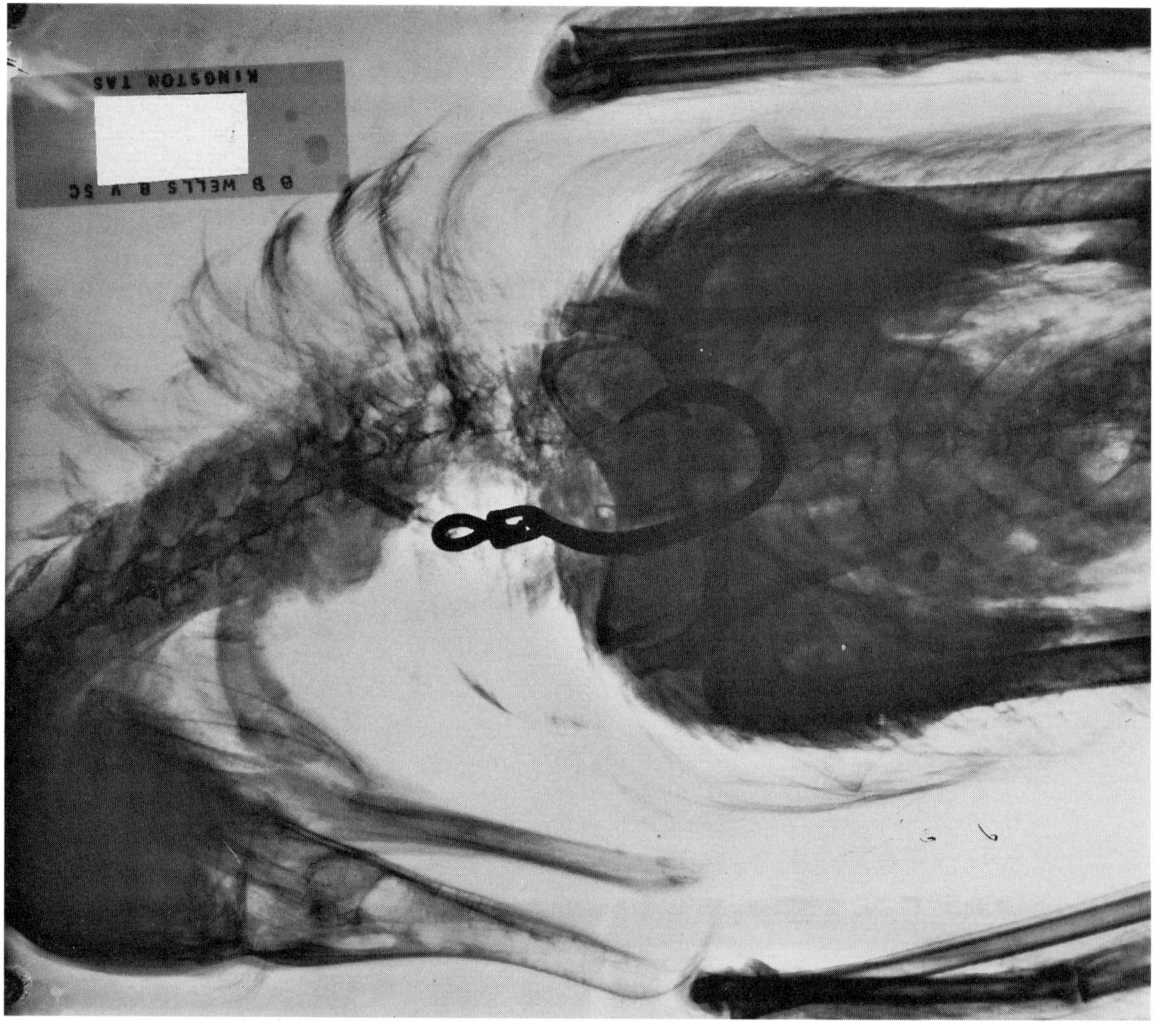

Figure 12.12 *Radiograph of Wandering Albatross that swallowed bait and hook. Photo: Courtsey N. Brothers.*

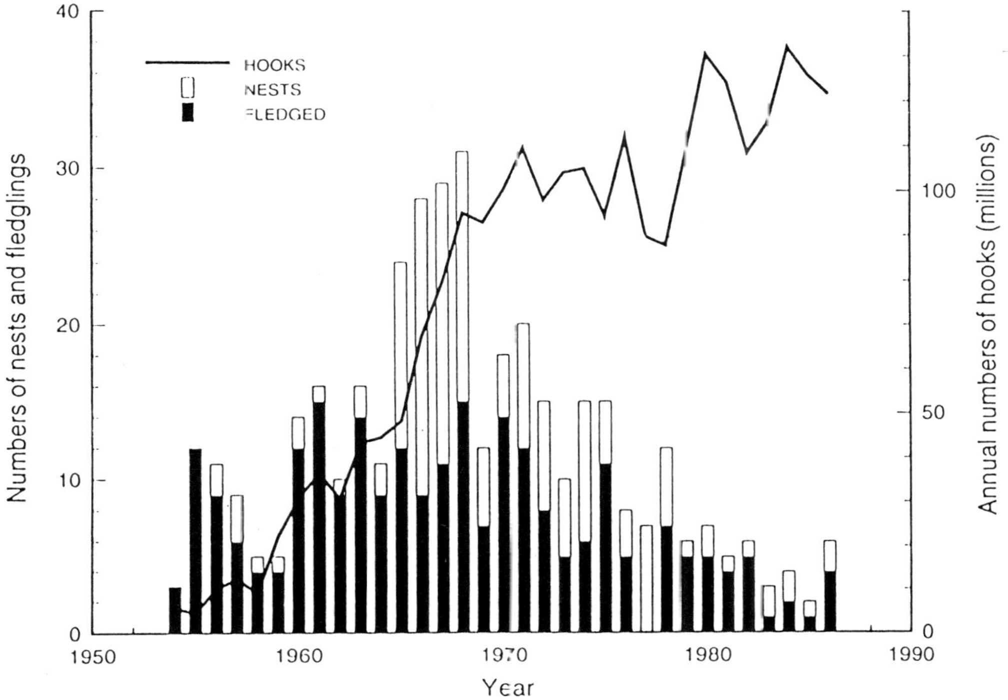

Figure 12.13 *Numbers of nests with eggs and fledglings of Wandering Albatrosses* Diomedea exulans *at Macquarie Island from 1954 to 1987 and the total number of longline hooks (in millions) deployed in the southern tuna fisheries. From de la Mare and Kerry (1994).*

Finally, as albatrosses feed mainly by day, setting the lines after dark may mitigate the damage (though not to nocturnal feeders such as *Procellaria cinerea*) and since 1992 Japanese long-liners in New Zealand waters have been required either to fit a streamer line or to adopt night setting.

Other problems are more localized. In the lobster fishery at Tristan da Cunha the catch is processed at night under powerful arc lamps on board ship close to Inaccessible, Nightingale and Gough Islands that have some of the world's largest tubenose populations. Ryan (1991) found that dazzled birds crashed into the superstructures, landed on decks to become contaminated with oil, and many stranded in the scuppers and lifeboats were later killed by gulls and skuas. The

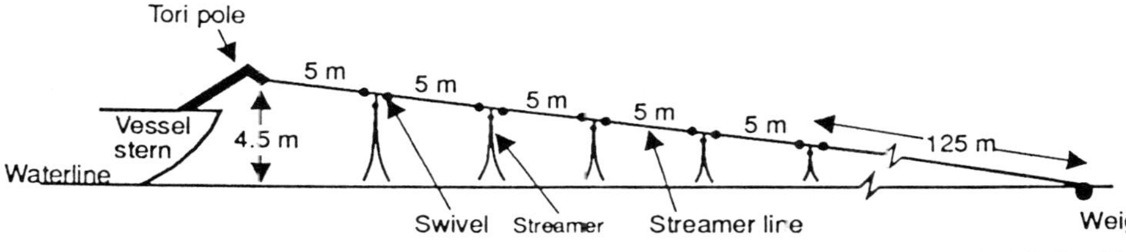

Figure 12.14 *Diagram of a streamer scare or 'tori' line to Conservation of Antarctic Marine Living Resources specifications (CCAMLR, 1990).*

Genus *Eopuffinus* Nessov, 1986
- *E. kazachstanensis* Nessov, 1986. Upper Palaeocene, Central Asia

Genus *Puffinus* Brisson, 1760
- *P. raemondonckii* (van Beneden, 1871). Lower Oligocene, Belgium
- *P. micraulax* Brodkorb, 1963. Lower Miocene, South Carolina
- *P. aquitanicus* (Milne-Edwards, 1874). Middle Miocene, France
- *P. antiquus* (Milne-Edwards, 1874). Middle Miocene, France
- *P. conradi* Marsh, 1870. Middle Miocene, Maryland
- *P. inceptor* Wetmore, 1930. Middle Miocene, California
- *P. diatomicus* Miller, 1925. Upper Miocene, California
- *P. calhouni* Howard, 1968. Upper Miocene, California
- *P. barnesi* Howard, 1978. Upper Miocene, California
- *P. tedfordi* Howard, 1971. Lower Pliocene, Mexico
- *P. felthami* Howard, 1949. Lower Pliocene, California
- *P. kanakoffi* Howard, 1949. Upper Pliocene, California
- *P. priscus* L. H. Miller, 1961. Upper Miocene, California
- *P. mitchelli* L. H. Miller, 1961. Upper Miocene, California
- *P. holeae* Walker, Wragg and Harrison, 1990. Upper Pleistocene, Canary Is
- *P. nestori* Alcover, 1989. Upper Pleistocene, Pityusae Is
- *P. olsoni* McMinn, Jaune and Alcover, 1990. Recent, Canary Is
- *P. spelaeus* Holdaway and Worthy, 1994. Recent, South I., New Zealand
- *P. gilmorei* Chandler, 1990. Upper Pliocene, California
- *P. pacificoides* Olson, 1975. Pleistocene, St Helena

Genus *Neptuniavis* Harrison and Walker, 1977
- *N. miranda* Harrison and Walker, 1977. Lower Eocene, England
- *N. minor* Harrison and Walker, 1977. Lower Eocene, England

FAM. MARINAVIDAE

Genus *Marinavis* Harrison and Walker, 1977
- *M. longirostris* Harrison and Walker, 1977. Lower Eocene, England

FAM. HYDROBATIDAE

Genus *Oceanites* Keyserling and Blasius, 1840
- *O. zaloscarthmus* Olson, 1985a. Lower Pliocene, South Africa

Genus *Oceanodroma* Reichenbach, 1853
- *O. hubbsi* L. H. Miller, 1951. Upper Miocene, California

Genus *Primodroma* Harrison and Walker, 1977
- *P. bournei* Harrison and Walker, 1977. Lower Eocene, England

FAM. PELECANOIDIDAE

Genus *Pelecanoides* Lacepede, 1799
- *P. cymatotrypetes* Olson, 1985a. Lower Pliocene, South Africa

*Assignment as a tubenose is debatable

APPENDIX II
Some Key References to the Parasites of Petrels

Trematodes

Yamaguti, S. (1958). *Systema Helminthum. Vol.1. The Digenetic Trematodes of Vertebrates, part 1.* Interscience, New York.

Cestodes

Szpotanska, J. (1925). Etude sur les Tetrabothriides des Procellariiformes. *Bull Acad Polonaise Sci Lettres Ser b Sci Natur* **3**, 673–727.

Yamaguti, S. (1959). *Systema Helminthum. Vol. 2. The Cestodes of Vertebrates.* Interscience, New York.

Murav'eva, S.I. (1969). [A study of the Tetrabothrids of tubenosed Antarctic birds]. *Mater Nauch Confer Vses Obshch Gel'mint* **1**, 170–178.

Nematodes

Johnston, T.F. and Mawson, P.M. (1942). Nematodes from Australian albatrosses and petrels. *Trans R Soc S Aust* **66**, 66–70.

Yamaguti, S. (1961). *Systema Helminthum, Vol. 3. The Nematodes of Vertebrates, part 1.* Interscience, New York.

Fleas

Smit, F.G.A.M. (1979). The Fleas of New Zealand (Siphonaptera). *J R Soc NZ* **9**, 143–232.

Pilgrim, R.L.C. (1980). The New Zealand Flea Fauna. *Proc Inter Confr Fleas, Astonwold, Peterborough, U.K.* (R. Traub and H. Starke, eds) pp. 173–184. Balkema, Rotterdam.

Feather Lice

Edwards, R.L. (1961). Studies of the Philopteridae (Mallophaga) from birds of the Order Procellariiformes. 1. The Genus *Halipeurus* Thompson. *J Parasitol* **47**, 125–157.

Clay, T. and Moreby, C. (1970). Mallophaga and Anoplura of subAntarctic Islands. *Pacif Insects Monogr* **23**, 216–220.

Horning, D.S., Palma, R.L. and Pilgrim, R.L.C. (1980). The Lice (Insecta: Phthiraptera) from the Snares Islands, New Zealand. *Natn Mus NZ Misc Ser* **3**, 1–17.

Furness, R.W. and Palma, R.L. (1992). Phthiraptera of petrels and skuas from Gough Island, South Atlantic Ocean. *Seabird* **14**, 33–42.

Mites

Dubinin, M.N. (1949). [The fauna of the Feather Mites on birds of the Order Procellariiformes]. *Parazitologicheskii Sbornik—Zoologicheski Inst Akad Nauk SSSR* **11**, 201–228.

Mironov, S.V. (1989). [A new subgenus and three new species of the Feather Mite genus *Zachvatkinia* from Procellariiformes]. *Parazitologiya (Leningr)* **23**, 309–319.

Black-browed and Grey-headed Albatrosses *Diomedea melanophris* and *D. chrysostoma* at Bird-Island, South Georgia. *Ibis* **127**, 523–529.

Atkin, K. (1979). Flight characters of Leach's Petrel. *Br Birds* **72**, 334–336.

Atkinson, I.A.E. (1985). The spread of commensal species of *Rattus* to oceanic islands and their effects on island avifaunas. In *Conservation of Island Birds* (P.J. Moors, ed.) pp. 35–81. ICBP Techn. Publ. **3**, Cambridge.

Au, D.W. (1991). Polyspecific nature of tuna schools: shark, dolphin and seabird associates. *Fishery Bull US* **89**, 343–354.

Au, D.W.K. & Pitman, R.L. (1986). Seabird interactions with dolphins and tuna in the eastern tropical Pacific. *Condor* **88**, 304–317.

Au, D.W. & Pitman, R.L. (1988). Seabird relationships with tropical tunas and dolphins. In *Seabirds and Other Marine Vertebrates: Competition, Predation and Other Interactions* (J. Burger, ed.) pp. 174–212. Columbia University Press, New York.

Audubon, J.J.L. (1835–39). *Ornithological Biography; or, An Account of the Habits of the Birds of the United States of America*, etc. Vols 3, 4 & 5. Black, Edinburgh.

Austin, F.J. & Hinshaw, V.S. (1984). The isolation of influenza A viruses and paramyxoviruses from feral ducks in New Zealand. *Aust J Exp Biol Med Sci* **62**, 355–360.

Austin, J.J., Carter, E. & Parkin, D.T. (1993). Genetic evidence for extra-pair fertilisation in socially monogamous Short-tailed Shearwater *Puffinus tenuirostris* (Procellariiformes, Procellariidae) using DNA-fingerprinting. *Aust J Zool* 41, 11–11.

Austin, J.J., White, R.W.G. & Ovenden, J.R. (1994). Population-genetic structure of a philopatric, colonially nesting seabird, the Short-tailed Shearwater (*Puffinus tenuirostris*). *Auk* **111**, 70–79.

Azzarello, M.Y. & Van Vleet, E.S. (1987). Marine birds and plastic pollution. *Mar Ecol Progr Ser* **37**, 295–303.

Bailey, A.M. (1956). Birds of Midway and Laysan Islands. *Denver Mus Nat Hist Pict* **12**, 1–130.

Bailey, A.M. & Sorensen, J.H. (1962). *SubAntarctic Campbell Island. Denver Mus Nat Hist Proc* **10**, 1–305.

Bailey, R. (1966). The sea-birds of the southeast coast of Arabia. *Ibis* **108**, 224–264.

Bailey, R.S. (1968). The pelagic distribution of sea-birds in the western Indian Ocean. *Ibis* **110**, 493–519.

Bailey, R.S., Pocklington, R. & Willis, P.R. (1968). Storm-petrels *Oceanodroma* spp. in the Indian Ocean. *Ibis* **110**, 27–34.

Bailey, S.F., Pyle, P. & Spear, L.B. (1989). Dark *Pterodroma* petrels in the North Pacific: identification, status, and North American occurrence. *Am Birds* **43**, 400–415.

Baines, A.C. (1889). The sailing flight of the albatross. *Nature, Lond* **40**, 9–10.

Baltz, D.M. & Morejohn, G.V. (1977). Food habits and niche overlap of seabirds wintering on Monterey Bay, California. *Auk* **94**, 526–543.

Bang, B.G. (1960). Anatomical evidence for olfactory function in some species of birds. *Nature, Lond* **188**, 547–549.

Bang, B.G. (1965). Anatomical adaptations for olfaction in the Snow Petrel. *Nature, Lond* **205**, 513–514.

Bang, B.G. (1966). The olfactory apparatus of tubenosed birds (Procellariiformes). *Acta Anat* **65**, 391–415.

Bang, B.G. (1971). Functional anatomy of the olfactory system in 23 orders of birds. *Acta Anat Suppl* **58**, 1–76.

Bang, B.G. & Cobb, S. (1968). The size of the olfactory bulb in 108 species of birds. *Auk* **85**, 55–61.

Bang, B.G. & Wenzel, B.M. (1985). Nasal cavity and olfactory system. In *Form and Function in Birds*. Vol. 3 (A.S. King & J. McLelland, eds) pp. 195–225. Academic Press, London.

Banko, W.E. (1980). History of endemic Hawaiian birds. Sea birds: Hawaiian Dark-rumped Petrel ('Ua'u). *CPSU/UH Avian history Rep 026/10*. University of Hawaii, Manoa.

Banko, W.E., Banko, P.C. & David, R.E. (1991). Specimens and probable breeding activity of the Band-rumped Storm-Petrel on Hawaii. *Wilson Bull* **103**, 650–655.

Bannerman, D.A. (1914). The distribution and nidification of the Tubinares in the North Atlantic Islands. *Ibis (10th Ser)* **2**, 438–494.

Barrat, A. & Mougin, J.L. (1974). Donnees numeriques sur la zoogeographie de l'avifaune Antarctique et Subantarctique. *Comm Natn Fr Rech Antarct* **33**, 1–18.

Barritt, M.K. (1992). Some observations in winter in the tropical South Atlantic. *Sea Swallow* **41**, 4–10.

Bartholomew, G.A. & Howell, T.R. (1964). Experiments on nesting behaviour of Laysan and Black-footed Albatrosses. *Anim Behav* **12**, 549–559.

Bartle, J.A. (1968). Observations on the breeding habits of Pycroft's Petrel. *Notornis* **15**, 70–99.

Bartle, J.A. (1990). Sexual segregation of foraging zones in procellariiform birds: implications of accidental capture on commercial fishery longlines of Grey Petrels (*Procellaria cinerea*). *Notornis* **37**, 146–150.

Bartle, J.A. (1991). Incidental capture of seabirds in the New Zealand sub-Antarctic squid trawl fishery, 1990. *Bird Conserv Inter* **1**, 351–359.

Bartle, J.A. (Sandy), Hu, D., Stahl, J.-C., Pyle, P., Simons, T.R. & Woodby, D. (1993). Status and ecology of gadfly petrels in the temperate North Pacific. In *The Status, Ecology and Conservation of Marine Birds of the North Pacific* (J. Vermeer, K.T. Briggs, K.H. Morgan & D. Siegel-Causey, eds) pp. 101–111. Can Wildl Serv Spec Publ, Ottawa.

Bartle, J.A. & Stahl, J.-C. (in press). Distribution of transequatorial migrant gadfly petrels in the Pacific in relation to time of breeding and moult. *Rec Natn Mus NZ*.

Bayle, P. & Fernandez, O. (1992). Protection of Cory's Shearwater *Calonectris diomedea* by limitation of population of feral Rabbits *Oryctolagus* on the Frioul Archipelago (Marseilles, France). *Avocetta* **16**, 67–70.

Beattie, H. (1954). *Our Southernmost Maoris*. Author, Dunedin.

Bech, C., Brent, R., Pedersen, P.F., Rasmussen, J.G. & Johansen, K. (1982). Temperature regulation in chicks of the Manx Shearwater *Puffinus puffinus*. *Ornis Scand* **13**, 206–210.

Bech, C., Mehlum, F. & Haftorn, S. (1988). Development of chicks during extreme cold conditions: the Antarctic Petrel (*Thalassoica antarctica*). *Acta XIX Congr Int Orn* 2, 1447–1456.

Bech, C., Mehlum, F. & Haftorn, S. (1991). The thermoregulatory abilities in chicks of the Antarctic Petrel (*Thalassoica antarctica*). *Polar Biol* **11**, 233–238.

Beck, J.R. (1969). Food, moult and age of first breeding in the Cape Pigeon, *Daption capensis* Linnaeus. *Br Antarct Surv Bull* **21**, 33–44.

Beck, J.R. (1970). Breeding seasons and moult in some smaller Antarctic petrels. In *Antarctic Ecology*, Vol. 11 (M.W. Holdgate, ed.) pp. 542–550. Academic Press, London.

Beck, J.R. & Brown, D.W. (1971). The breeding biology of the Black-bellied Storm-Petrel *Fregetta tropica*. *Ibis* **113**, 73–90.

Beck, J.R. & Brown, D.W. (1972). The biology of Wilson's Storm Petrel, *Oceanites oceanicus* (Kuhl), at Signy Island, South Orkney Islands. *Br Antarct Surv Scient Rep* **69**, 1–54.

Beddard, F.E. (1898). *The Structure and Classification of Birds*. Longmans, Green, London.

Beebe, W. (1926). *The Arcturus Adventure*. Putnam, New York.

Beebe, W. (1935). Rediscovery of the Bermuda Cahow. *Bull N Y Zool Soc* **38**, 187–190.

Belon, P. (1555). *L'Histoire de la Nature des Oyseaux, avec Leurs Descriptions, et Naifs Portraicts Retirez du Naturel*. Cauellat, Paris.

Bennett, A.G. (1927). The adaptability of sub-Antarctic and Antarctic birds to local conditions. *Emu* **26**, 259–263.

Bretagnolle, V. (1989a). Calls of Wilson's Storm Petrel: functions, individual and sexual recognitions, and geographic variation. *Behaviour* **111**, 98–112.

Bretagnolle, V. (1989b). Temporal progression of the Giant-Petrel courtship. *Ethology* **80**, 245–254.

Bretagnolle, V. (1990a). Behavioural affinities of the Blue Petrel *Halobaena caerulea*. *Ibis* **132**, 102–105.

Bretagnolle, V. (1990b). Effet de la lune sur l'activite des petrels (classe Aves) aux iles Salvages (Portugal). *Can J Zool* **68**, 1404–1409.

Bretagnolle, V. (1993). Adaptive significance of seabird coloration: the case of the Procellariiformes. *Am Nat* **142**, 141–173.

Bretagnolle, V. (1995). Systematics of the Soft-plumaged Petrel *Pterodroma mollis* (*Procellariidae*): new insight from the study of vocalizations. *Ibis* **137**, 207–218.

Bretagnolle, V. (1995a). Acoustic communication in a group of non-passerine birds, the petrels. In *Ecology and Evolution of Acoustic Communication in Birds* (D.E. Kroodsma & E.H. Miller, eds). Cornell University Press, New York.

Bretagnolle, V. (in pressb). Information exchange, colonial breeding and long-term monogamy: constraints acting on acoustic communication in waterbirds. In *Colonial breeding in waterbirds* (F. Cezilly, H. Haffner & D.N. Nettleship, eds). Oxford University Press.

Bretagnolle, V. & Attie, C. (1991). Status of Barau's Petrel (*Pterodroma baraui*): colony sites, breeding population and taxonomic affinities. *Colonial Waterbirds* **14**, 25–33.

Bretagnolle, V. & Attie, C. (1993). Massacre d'une espece protegee sur le territoire francais: le petrel de Barau. *Courrier Nature* **138**, 40.

Bretagnolle, V. & Lequette, B. (1990). Structural variation in the call of the Cory's Shearwater (*Calonectris diomedea*) Aves Procellariidae. *Ethology* **85**, 313–323.

Bretagnolle, V. & Robisson, P. (1991). Species-specific recognition in birds: an experimental investigation of Wilson's Storm-Petrel (*Procellariiformes, Hydrobatidae*) by means of digitalized signals. *Can J Zool* **69**, 1669–1673.

Bretagnolle, V. & Thibault, J.-C. (in press). A method for sexing fledglings in Cory's Shearwater, a comment on sex ratio variations over a four year period. *Auk*.

Bretagnolle, V. & Thomas, T. (1990). Seabird distribution between Tasmania and Adelie Land (Antarctica) and comparison with nearby Antarctic Sectors. *Emu* **90**, 97–107.

Bretagnolle, V., Zotier, R. & Jouventin, P. (1990). Comparative population biology of four prions (Genus *Pachyptila*) from the Indian Ocean and consequences for their taxonomic status. *Auk* **107**, 305–316.

Briggs, K.T., Tyler, W.B. & Lewis, D.B. (1985a). Comparison of ship and aerial surveys of birds at sea. *J Wildl Manage* **49**, 405–411.

Briggs, K.T., Tyler, W.B. & Lewis, D.B. (1985b). Aerial surveys for seabirds: methodological experiments. *J Wildl Manage* **49**, 412–417.

Briggs, K.T., Tyler, W.B., Lewis, D.B. & Carlson, D.R. (1987). Bird communities at sea off California: 1975–1983. *Studies in Avian Biol* **11**, 1–74.

Broady, P.A. (1979). Quantitative studies on the terrestrial algae of Signy Island, South Orkney Islands. *Br Antarct Surv Bull* **47**, 31–41.

Broady, P.A. (1989). Survey of algae and other terrestrial biota at Edward VIII Peninsula, Marie Byrd Land. *Antarct Sci* **1**, 215–224.

Broady, P.A., Adams, C.J., Cleary, P.J. & Weaver, S.D. (1989). Ornithological observations at Edward VII Peninsula, Antarctica, in 1987–88. *Notornis* **36**, 53–61.

Brooke, M. (1990). *The Manx Shearwater*. Poyser, London.

Brooke, M. de L. (1978a). Some factors affecting the laying date, incubation and breeding success of the Manx Shearwater, *Puffinus puffinus*. *J Anim Ecol* **47**, 477–495.

Brooke, M. de L. (1978b). The dispersal of female Manx Shearwaters. *Ibis* **120**, 546–551.

Brooke, M. de L. (1978c). A test for visual location of the burrow by Manx Shearwaters *Puffinus puffinus*. *Ibis* **120**, 347–349.

Brooke, M. de L. (1978d). Weights and measurements of the Manx Shearwater, *Puffinus puffinus*. *J Zool Lond* **186**, 359–374.

Brooke, M. de L. (1978e). Sexual differences in the voice and individual vocal recognition in the Manx Shearwater (*Puffinus puffinus*). *Anim Behav* **26**, 622–629.

Brooke, M. de L. (1986). The vocal systems of two nocturnal petrels, the White-chinned *Procellaria aequinoctialis* and the Grey *P. cinerea*. *Ibis* **128**, 502–512.

Brooke, M. de L. (1988). Sexual dimorphism in the voice of the Greater Shearwater. *Wilson Bull* **100**, 319–323.

Brooke, M. de L. (1989). Determination of the absolute visual threshold of a nocturnal seabird, the Common Diving Petrel *Pelecanoides urinatrix*. *Ibis* **131**, 290–300.

Brooke, M. de L. & Klages, N. (1986). Squid beaks regurgitated by Greyheaded and Yellownosed Albatrosses, *Diomedea chrysostoma* and *D. chlororhynchos* at the Prince Edward Islands. *Ostrich* **57**, 203–206.

Brooke, M. de L. & Prince, P.A. (1991). Nocturnality in Seabirds. *Acta XX Congr Int Orn* **2**, 1113–1121.

Brooke, R.K. (1981). Modes of moult of flight feathers in albatrosses. *Cormorant* **9**, 13–18.

Brooke, R.K. & Furness, B.L. (1982). Reversed modes of moult of flight feathers in the Black-browed Albatross *Diomedea melanophris*. *Cormorant* **10**, 27–30.

Brooke, M. de L. & Rowe, G. (in press). Behavioural and molecular evidence for granting specific status to light and dark morphs of the Herald Petrel *Pterodroma heraldica*. *Ibis*.

Brooks, A. (1937). The patagial fan in the Tubinares. *Condor* **39**, 82–83.

Brosset, A. (1963). La reproduction des oiseaux de mer des Iles Galapagos en 1962. *Alauda* **31**, 81–109.

Brothers, N.P. (1984). Breeding, distribution and status of burrow nesting petrels at Macquarie Is. *Aust Wildl Res* **11**, 113–131.

Brothers, N. (1991). Albatross mortality and associated bait loss in the Japanese Longline Fishery in the Southern Ocean. *Biol Conserv* **55**, 255–268.

Brothers, N.P. & Brown, M.J. (1987). The potential use of Fairy Prions (*Pachyptila turtur*) as monitors of heavy metal levels in Tasmanian waters. *Mar Pollut Bull* **18**, 132–134.

Broughton, J.M. (1994). Size of Bursa of Fabricius in relation to gonad size and age in Laysan and Black-footed Albatrosses. *Condor* **96**, 203–207.

Brown, C.R. (1988). Energy expenditure during incubation in four species of sub-Antarctic burrowing petrels. *Ostrich* **59**, 67–70.

Brown, C.R. & Adams, N.J. (1984). Basal metabolic rate and energy expenditure during incubation in the Wandering Albatross (*Diomedea exulans*). *Condor* **86**, 187–191.

Brown, C.R. & Adams, N.J. (1988). Egg temperature, embryonic metabolism, and water loss from the eggs of sub-Antarctic Procellariiformes. *Physiol Zool* **61**, 126–136.

Brown, C.R. & Prys-Jones, R.P. (1988). Development of homeothermy in chicks of sub-Antarctic burrowing petrels. *S Afr J Zool* **23**, 288–294.

Brown, D.A. (1966). Breeding Biology of the Snow Petrel *Pagodroma nivea* (Forster). *Aust Natn Antarct Res Exped Sci Rep B (1) Zool* **89**, 1–63.

Brown, R.G.B. (1970). Fulmar distribution: a Canadian perspective. *Ibis* **112**, 44–51.

Brown, R.G.B. (1980a). Seabirds as marine animals. In *Behavior of Marine Animals, Vol. 4* (J. Burger, B.L. Olla & H.E. Winn, eds) pp. 1–39. Plenum, New York.

Brown, R.G.B. (1980b). The pelagic ecology of seabirds. *Trans Linn Soc NY* **9**, 15–21.

Brown, R.G.B. (1986). *Revised Atlas of Eastern Canadian Seabirds*. 1. *Shipboard Surveys*. Can Wildl Serv, Dartmouth.

Brown, R.G.B. (1988a). The wing-moult of fulmars and shearwaters (*Procellariidae*) in Canadian Arctic waters. *Can Field-Nat* **102**, 203–208.

Clarke, A. (1977). Contamination of Peregrine Falcons *(Falco peregrinus)* with fulmar stomach oil. *J Zool Lond* **181**, 11–20.

Clarke, A. & Prince, P.A. (1976). The origin of stomach oil in marine birds: analyses of the stomach oil from six species of subAntarctic procellariiform birds. *J Exp Mar Biol Ecol* **23**, 15–30.

Clausen, G., Sanson, R. & Storesund, A. (1971). The HbO_2 dissociation curve of the Fulmar and the Herring Gull. *Respir Physiol* **12**, 66–70.

Cleland, J.B. (1956). Birds as they affect the health and well-being of man in Australia. *Emu* **56**, 183–193.

Cline, D.R., Siniff, D.B. & Erickson, A.W. (1969). Summer birds of the pack ice in the Weddell Sea, Antarctica. *Auk* **86**, 701–716.

Coates, B.J. (1985). *The birds of Papua New Guinea. Including the Bismark Archipelago* and *Bougainville*. Dove Publ, Alderley, Queensland.

Cobb, S. (1959/60). A note on the size of the avian olfactory bulb. *Epilepsia* **1**, 394–402.

Cobb, S. (1960). Observations on the comparative anatomy of the avian brain. *Perspectives in Biology & Medecine* **3**, 383–408.

Collins, C.T. & Tikasingh, E.S. (1974). Status of the Great Shearwater in Trinidad, West Indies. *Bull Br Orn Club* **94**, 96–99.

Collins, J.W. (1884). Notes on the habits and methods of capture of various species of sea birds that occur on the fishing banks off the eastern coast of North America, and which are used as bait for catching codfish by New England fishermen. *Rep US Commissioner Fish and Fisheries for 1882*, 311–335.

Condon, H.T. (1939). The cranial osteology of certain Tubinares. *Trans R Soc S Aust* **63**, 311–328.

Cone, C.D. (1964). A mathematical analysis of the dynamic soaring flight of the albatross with ecological interpretations. *Virginia Inst Mar Sci Spec Scient Rep* **50**, 1–104.

Conroy, J.W.H. (1972). Ecological aspects of the histology of the Giant Petrel, *Macronectes giganteus* (Guelin), in the maritime Antarctic. *Br Antarct Surv Sci Rep* **75**, 1–74.

Conroy, J.W.H. & French, M.C. (1974). Organochlorine levels in two species of Antarctic birds. *Br Antarct Surv Bull* **38**, 43–47.

Cooper, J. (1974). Albatross displays off the south-west coast of South Africa. *Notornis* **21**, 234–238.

Cooper, J. & Fourie, A. (1991). Improved breeding success of Great-winged Petrels *Pterodroma macroptera* following control of feral cats *Felis catus* at sub-Antarctic Marion Island. *Bird Conserv Inter* **1**, 171–175.

Cooper, J., Underhill, L.G. & Avery, G. (1991). Primary molt and transequatorial migration of the Sooty Shearwater. *Condor* **93**, 724–730.

Cooper, J., Wilson, R.P. & Adams, N.J. (1993). Timing of foraging by the Wandering Albatross *Diomedea exulans. Proc NIPR Symp Polar Biol* **6**, 55–61.

Cooper, J., Henley, S.R. & Klages, N.T.W. (1992). The diet of the Wandering Albatross *Diomedea exulans* at SubAntarctic Marion Island. *Polar Biol* **12**, 477–484.

Copestake, P.G., Croxall, J.P. & Prince, P.A. (1988). Use of cloacal sexing techniques in mark–recapture estimates of breeding population size in Wilson's Storm Petrel *Oceanites oceanicus* at South Georgia. *Polar Biol* **8**, 271–279.

Cosmas, I. (1909). *The Christian Topography of Cosmas, an Egyptian Monk* (Trnsl J.W. McCrindle). Franklin, New York.

Costa, D.P. & Prince, P.A. (1987). Foraging energetics of Grey-headed Albatrosses *Diomedea chrysostoma* at Bird Island, South Georgia. *Ibis* **129**, 149–158.

Cott, H.B. (1953). The exploitation of wild birds for their eggs. *Ibis* **95**, 409–449.

Cott, H.B. (1971). The edibility of birds, etc. *Proc Zool Soc Lond* **116**, 371–524.

Coues, E. (1864–66). A critical review of the family Procellariidae. *Proc Acad Nat Sci Philad* **1864**, 72–91, 116–144; 1866, 25–33, 134–197.

Coulson, J.C. & Wooller, R.D. (1976). Differential survival rates among breeding Kiltiwake Gulls *Rissa Tridactyla* (L.). *J Anim Ecol* **45**, 205–214.

Coulter, M.C. & Risebrough, R.W. (1973). Shell-thinning in eggs of the Ashy Petrel (*Oceanodroma homochroa*) from the Farallon Islands. *Condor* **75**, 254–255.

Cowan, A.N. (1979). Ornithological studies at Casey, Antarctica, 1977–1978. *Aust Bird Watcher* **8**, 69–90.

Coward, T.A. (1895). Manx Shearwater breeding on the coast of Carnarvonshire. *Zoologist*, **3rd Ser 19**, 72.

Cox, J.B. (1980). Some remarks on the breeding distribution and taxonomy of the prions (Procellariidae: *Pachyptila*). *Rec S Aust Mus* **18**, 91–121.

Cracraft, J. (1968). The lacrimal-ectethmoid bone complex in birds: a single character analysis. *Am Midland Nat* **80**, 316–359.

Cracraft, J. (1981). Toward a phylogenetic classification of Recent birds of the World (Class Aves). *Auk* **98**, 681–714.

Cracraft, J. (1982). Phylogenetic relationships and monophyly of Loons, Grebes and Hesperornithiform birds, with comments on the early history of birds. *Syst Zool* **31**, 35–56.

Cracraft, J. (1985). Monophyly and phylogenetic relationships of the Pelecaniformes: a numerical cladistic analysis. *Auk* **102**, 834–853.

Cramp, S. & Simmons, K.E.L. (eds) (1977, 1983). *The Birds of the Western Palearctic*. Vols I & III. Oxford University Press, Oxford.

Cramp, S., Bourne, W.R.P. & Saunders, D. (1974). *The Seabirds of Britain and Ireland*. Collins, London.

Crockett, D.E. (1995). Rediscovery of the Chatham Island Taiko *Pterodroma magentae*. *Notornis* **41 (Suppl.)**, 49–60.

Crossin, R.S. (1974). The storm petrels (Hydrobatidae). *Smithson Contr Zool* **158**, 154–205.

Croxall, J.P. (1982a). Aspects of the population demography of Antarctic and Sub-antarctic seabirds. *Com Natn Fr Rech Antarct*. **51**, 479–488.

Croxall, J.P. (1982b). Energy costs of incubation and moult in petrels and penguins. *J Anim Ecol* **51**, 177–194.

Croxall, J.P. (1991). Constraints on reproduction in albatrosses. *Acta XX Congr Int Orn* **1**, 281–302.

Croxall, J.P. & Gaston, A.J. (1988). Patterns of reproduction in high-latitude Northern- and Southern-Hemisphere Seabirds. *Acta XIX Congr Int Orn* **1**, 1176–1194.

Croxall, J.P. & North, A.W. (1988). Fish prey of Wilson's Storm Petrel *Oceanites oceanicus* at South Georgia. *Br Antarct Surv Bull* **78**, 37–42.

Croxall, J.P. & Prince, P.A. (1980). Food, feeding ecology and biological segregation of seabirds at South Georgia. *Biol J Linn Soc*. **14**, 103–131.

Croxall, J.P. & Prince, P.A. (1994). Dead or alive, night or day: how do albatrosses catch squid? *Antarct Sci* **6**, 155–162.

Croxall, J.P. & Rothery, P. (1991). Population regulation in seabirds: implications of their demography for conservation. In *Bird Population Studies: Relevance to Conservation and Management* (C.M. Perrins, J.D. Le Breton & G.M. Hirons, eds) pp. 272–296. Oxford University Press, Oxford.

Croxall, J.P., Ricketts. C. & Prince, P.A. (1984a). Impact of seabirds on marine resources, especially krill, of South Georgia waters. In *Seabird Enegetics* (G.C. Whittow & H. Rahn, eds) pp. 285–317. Plenum, New York.

Croxall, J.P., Prince, P.A., Hunter, I., McInnes, S.J. & Copestake, P.G. (1984b). The seabirds of the Antarctic Peninsula, Islands of the Scotia Sea, and Antarctic Continents between 80° W and 20° W: their status and conservation. In *Status and Conservation of the World's Seabirds* (J.P. Croxall, P.G.H. Evans & R.W. Schreiber. eds) pp. 637–666. ICBP, Cambridge.

Dzerzhinsky, F.Y. & Yudin, K.A. (1979). [Homology of jaw muscles of the Tuatara and birds]. *Ornitologiya* **14**, 14–34.
Eakin, R.R., Dearborn, J.H. & Townsend, W.C. (1986). Observations of marine birds in the South Atlantic Ocean in the late austral Autumn. *Antarct Res Ser* **44**, 69–86.
Ealey, E.H.M. (1954). Analysis of stomach contents of some Heard Island birds. *Emu* **54**, 204–210.
Elliott, H.F.I. (1957). A contribution to the Ornithology of the Tristan da Cunha Group. *Ibis* **99**, 545–586.
Elliott, J.E. & Noble, D.G. (1993). Chlorinated hydrocarbon contaminants in marine birds of the temperate North Pacific. In *The Status, Ecology, and Conservation of Marine Birds of the North Pacific* (K. Vermeer, K.T. Briggs, K.H. Morgan & D. Siegel-Causey, eds) pp. 241–253. Can Wildl Serv, Ottawa.
Elliott, J.E., Scheuhammer, A.M., Leighton, F.A. & Pearce, P.A. (1992). Heavy metal and metallothionen concentrations in Atlantic Canadian seabirds. *Arch Environ Contam Toxiocol* **22**, 63–73.
Ellis, H.I. (1984). Energetics of free-ranging seabirds. In *Seabird Energetics* (G.C. Whittow & H. Rahn, eds) pp. 203–234. Plenum, New York.
Elton, C.S. (1958). *The Ecology of Invasions by Animals and Plants*. Methuen, London.
Elwes, R. (1859). Note on the breeding and mode of capture of the Short-tailed Petrel, or Mutton-bird (*Puffinus obscurus*), in the islands of Bass's Straits. *Ibis* **1st Ser 1**, 397–399.
Ely, C.A. & Clapp, R.B. (1973). The natural history of Laysan Island, Northwestern Hawaiian Islands. *Atoll Res Bull* **171**, 1–361.
Emory, K.P. (1947). Tuamotuan bird names. *J Polynesian Soc* **56**, 188–196.
Engelskjon, T. (1986). Botany of two Antarctic mountain ranges: Gjelsvikfjella and Muhlig-Hofmannfjella, Dronning Maud Land. *Polar Res* **4**, 205–224.
Enquist, M., Plane, E. & Roed, J. (1985). Aggressive communication in Fulmars (*Fulmarus glacialis*). *Anim Behav* **33**, 1007–1020.
Enticott, J.W. (1986). Associations between seabirds and cetaceans in the African Sector of the Southern Ocean. *S Afr J Antarct Res* **16**, 25–28.
Erickson, A.W., Gilbert, J.R., Petrides, G.A., Oehlenschlager, R.J., Sinha, A.S. & Otis, J. (1972). Population of seals, whales and birds in the Bellingshausen and Amundsen Seas. *Antarct J US* **7**, 70–72.
Erickson, J.G. (1955). Flight behavior of the Procellariiformes. *Auk* **72**, 415–420.
Erickson, R.A., Bailey, S.F. & Yee, D.G. (1989). The spring season March 1–May 31, 1989. Middle Pacific Coast Region. *Am Birds* **43,** 531–535.
Erikstad, K.E. (1990). Winter diets of four seabird species in the Barents Sea after a crash in Capelin stock. *Polar Biol* **10**, 619–627.
Evans, P.G.H. (1982). Associations between seabirds and cetaceaus: a review. *Mammal Rev* **12**, 187–206.
Evers, N. & Foster, H.J. (1920). Note on Mutton-bird oil. *Pharmaceut J Pharmacist* **4th Ser 51**, 100.
Eydoux, F. & Souleyet, A. (1838). Sur la temperature de l'homme et des oiseaux. *C R Hebd Acad Sci Paris* **6**, 456–457.
Eyton, T.C. (1867). *Osteologia Avium; or a Sketch of the Osteology of Birds*. Hobson, Wellington (Salop).
Faber, F. (1822). *Prodromus der Islandischen ornithologie oder geschichte der Vogel Islands*. Riopping, Kobenhaven.
Falla, R.A. (1922). Notes on petrels washed ashore, West Coast, Auckland Province, N.Z. *Emu* **21**, 206–207.
Falla, R.A. (1924). Discovery of a breeding place of Buller's Shearwater, Poor Knights Island, N.Z. *Emu* **24**, 37–43.
Falla, R.A. (1937). Birds. *BANZ Antarct Res Exped 1929–1931. Rep Ser B* **2**, 1–288.

Falla, R.A. (1960). Oceanic birds as dispersal agents. *Proc R Soc Lond* **152B**, 655–659.

Farner, D.S. (1956). Body temperature of the Fairy Prion (*Pachyptila turtur*) in flight and at rest. *J Appl Physiol* **8**, 546–548.

Farner, D.S. & Serventy, D.L. (1959). Body temperature and the ontogeny of thermoregulation in the Slender-billed Shearwater. *Condor* **61**, 426–433.

Fefer, S.I., Harrison, C.S., Naughton, M.B. & Shallenberger, R.J. (1984). Synopsis of results of recent seabird research conducted in the Northwestern Hawaiian Islands. In *Proceedings 2nd Symposium on Resource Investigations in the Northwestern Hawaiian Islands* (R.W. Grigg & K.Y. Tanoue, eds) Vol 1, pp. 9–76. University of Hawaii Sea Grant Coll. Honolulu.

Ferreira, R.E.C. & Wormell, P. (1971). Fertiliser response of vegetation on ultrabasic terraces on Rhum. *Trans Bot Soc Edinb* **41**, 149–154.

Fineran, B.A. (1973). A botanical survey of seven Mutton-bird islands, South-west Stewart Island. *J R Soc NZ* **3**, 475–526.

Fisher, H.I. (1961). Weights and measurements of organs of Bonin Island Petrels, *Pterodroma leucoptera hypoleuca*. *Auk* **78**, 269–271.

Fisher, H.I. (1966). Airplane–albatross collisions on Midway Atoll. *Condor* **68**, 229–242.

Fisher, H.I. (1971). The Laysan Albatross: its incubation, hatching and associated behaviors. *Living Bird* **10**, 19–78.

Fisher, H.I. (1972). Sympatry of Laysan and Black-footed Albatrosses. *Auk* **89**, 381–402.

Fisher, H.I. (1975a). Longevity of the Laysan Albatross, *Diomedea immutabilis*. *Bird-banding* **46**, 1–6.

Fisher, H.I. (1975b). Mortality and survival in the Laysan Albatross, *Diomedea immutabilis*. *Pacific Sci* **29**, 279–300.

Fisher, H.I. (1976). Some dynamics of a breeding colony of Laysan Albatrosses. *Wilson Bull* **88**, 121–142.

Fisher, H.I. & Fisher, J.R. (1972). The oceanic distribution of the Laysan Albatross *Diomedea immutabilis*. *Wilson Bull* **84**, 7–27.

Fisher, H.I. & Fisher, M.L. (1969). The visits of Laysan Albatrosses to the breeding colony. *Micronesica* **5**, 173–221.

Fisher, J. (1952a). *The Fulmar*. Collins, London.

Fisher, J. (1952b). The changes in the distribution of the Fulmar (*Fulmarus glacialis*). *Proc 10th Int Orn Congr*, 449–462.

Fisher, J. (1954). Evolution and bird sociality. In *Evolution as a Process* (J. Huxley, A.C. Harding & E.B. Ford, eds) pp. 71–83. Allen & Unwin, London

Fisher, J. (1966). The Fulmar population of Britain and Ireland, 1959. *Bird Study* **13**, 5–76.

Fisher, J. & Lockley, R.M. (1954). *Sea-birds*. Collins, London.

Fisher, W.K. (1904). The Albatross dance at Sea. *Condor* **6**, 78.

Fleming, C.A. (1941). Notes on the Neozelanic forms of the subgenus *Cookilaria*. *Emu* **41**, 69–80.

Fleming, C.A. (1979). *The Geological History of New Zealand and its Life*. Auckland University Press, Auckland.

Fleming, C.A. (1982). *George Edward Lodge. The unpublished New Zealand bird paintings*. Nova Pacifica, Wellington.

Fleming, J. (1828). *History of British Animals, etc*. Bell & Bradfute, Edinburgh.

Flinders, M. (1814). *A voyage to Terra Australis, Vol 1*. Nicol, London.

Flint, E.A. & Fineran, B.A. (1969). Observations on the climate, peats and terrestrial algae of the Snares Islands. *NZ J Sci* **12**, 286–301.

Forbes, W.A. (1882). Report on the anatomy of the petrels (*Tubinares*), collected during the voyage of HMS *Challenger*. *Zool Challenger Exped* **4 (11)**, 1–64.

Fordyce, R.E. (1977). The development of the Circum-Antarctic Current and the evolution of the Mysticeti (Mammalia: Cetacea). *Palaeogeogr Palaeoclimatol Palaeoecol* **21**, 265–271.

Gillham, M.E. (1961b). Plants and seabirds of granite islands in south-east Victoria. *Proc R Soc Victoria* **74**, 21–35.
Gillham, M.E. (1961c). Modification of sub-Antarctic flora on Macquarie Island by sea birds and sea elephants. *Proc R Soc Victoria* **74**, 1–12.
Gillham, M.E. (1962). Granite islands of South-east Victoria as a seabird habitat. *Proc R Soc Victoria* **75**, 45–63.
Gillham, M.E. (1963). Breeding habitats of the White-faced Storm Petrel (*Pelagodroma marina*) in Eastern Bass Strait. *Proc R Soc Tasmania* **97**, 33–41.
Gillham, M.E. & Thomson, J.A. (1961). Old and new storm petrel rookeries in Port Phillip Bay. *Proc R Soc Victoria* **74**, 37–46.
Ginn, H.B. & Melville, D.S. (1983). *Moult in Birds*. Guide 19. Br Trust Orn, Tring.
Glauert, L. (1946). The Little Shearwater's year. *Emu* **46**, 187–192.
Glenny, F.H. (1955). Modifications of pattern in the aortic arch system of birds and their phylogenetic significance. *Proc US Natn Mus* **104**, 535–621.
Gobeil, R.E. (1969). Arteries in the heart region of the Greater Shearwater. *Auk* **86**, 343–345.
Godley, E.J. (1989). The flora of Antipodes Island. *NZ J Bot* **27**, 531–563.
Godman, F. du Cane (1907–1910). *A Monograph of the Petrels.* Witherby, London.
Gould, J. (1865). *Handbook to the Birds of Australia. Vol. 2.* Author, London.
Gould, P. & King, W.B. (1967). Records of four species of *Pterodroma* from the Central Pacific Ocean. *Auk* **84**, 591–594.
Gould, P.J. (1983). Seabirds between Alaska and Hawaii. *Condor* **85**, 286–291.
Gould, P.J. & Forsell, D.J. (1989). Techniques for shipboard surveys of marine birds. *US Fish & Wildl Techn Rep* **25**, 1–22.
Gould, P.J. & Hobbs, R. (1993). Population dynamics of the Laysan and other albatrosses in the North Pacific. *N Pacif Comm Bull* **53**, 485–497.
Gould, P.J., Forsell, D.J. & Lensink, C.J. (1982). *Pelagic Distribution and Abundance of Seabirds in the Gulf of Alaska and Eastern Bering Sea.* US Fish Wildl Serv, Washington.
Grafe, F. (1973). Verbreitung des Grossen Sturmtauchers (*Puffinus gravis*) vor der SE-Kuste Gronlands in August 1966. *Vogelwelt* **94**, 175–182.
Graham, H.D. (1890). *The birds of Iona and Mull.* David Douglas, Edinburgh.
Grant, G.S. (1984). Energy cost of incubation to the parent. Seabird in *Seabird Energetics* (G.C. Whittow & H. Rahn, eds) pp. 59–71. Plenum Press, NY.
Grant, G.S. (1985). Rete mirabile ophthalicum and intercarotid anastomosis in Procellarii-formes taken off the Carolina coast. *Brimleyana* **11**, 81–86.
Grant, G.S. & Whittow, G.C. (1983). Metabolic cost of incubation in the Laysan Albatross and Bonin Petrel. *Comp Biochem Physiol* **74**A, 77–82.
Grant, G.S. & Whittow, G.C. (1984). Metabolic rate of Laysan Albatross and Bonin Petrel chicks on Midway Atoll. *Pacif Sci* **38**, 170–176.
Grant, G.S., Pettit, T.N. & Whittow, G.C. (1981). Rat predation on Bonin Petrel eggs on Midway Atoll. *J Fld Orn* **52**, 336–338.
Grant, G.S., Pettit, T.N., Rahn, H., Whittow, G.C. & Paganelli, C.V. (1982a). Regulation of water loss from Bonin Petrel (*Pterodroma hypoleuca*) eggs. *Auk* **99**, 236–242.
Grant, G.S., Pettit, T.N., Rahn, H., Whittow, G.C. & Paganelli, C.V. (1982b). Water loss from Laysan and Black-footed Albatross eggs. *Physiol Zool* **55**, 405–414.
Grant, G.S., Warham, J., Pettit, T.N. & Whittow, G.C. (1983). Reproductive behavior and vocalizations of the Bonin Petrel. *Wilson Bull* **95**, 522–539.
Gray, R. (1871). *The Birds of the West of Scotland Including the Outer Hebrides.* Murray, Glasgow.
Grebmeier, J. & Harrison, N.M. (1992). Seabird feeding on benthic amphipods facilitated by Gray Whale activity in the northern Bering Sea. *Mar Ecol Prog Ser* **80**, 125–133.
Green, B. & Brothers, N. (1989). Water and sodium turnover and estimated food consumption rates in free-living Fairy Prions (*Pachyptila turtur*) and Common Diving Petrels (*Pelecanoides urinatrix*). *Physiol Zool* **62**, 702–715.

Green, J.F. (1887). *Ocean Birds*. Porter, London.

Green, K. (1986a). Food of the Cape Pigeon (*Daption capense*) from Princess Elizabeth Land, East Antarctica. *Notornis* **33**, 151–154.

Green, K. (1986b). Observations on the food of the Southern Giant Petrel near Davis, Antarctica. *Notornis* **33**, 90–94.

Green, R.H. (1974). Albatross Island. *Rec Queen Victoria Mus* **51**, 1–17.

Griffiths, A.M. (1981). Biases in censuses of pelagic seabirds at sea in the Southern Ocean. In *Proceedings of the Symposium on Birds of the Sea and Shore* (J. Cooper, ed.) pp. 189–196. African Seabird Grp, Cape Town.

Griffiths, A.M. (1983). Factors affecting the distribution of the Snow Petrel (*Pagodroma nivea*) and the Antarctic Petrel (*Thalassoica antarctica*). *Ardea* **71**, 145–150.

Griffiths, A.M., Seigfried, W.R. & Abrams, R.W. (1982). Ecological structure of a pelagic seabird community in the Southern Ocean. *Polar Biol* 1, 39–46.

Groscolas, R., Schreiber, L. & Morin, F. (1991). The use of tritriated water to determine protein and lipid utilization in fasting birds: a validation study in incubating Great-winged Petrels, *Pterodroma macroptera*. *Physiol Zool* **64**, 1217–1233.

Gross, W.A.O. (1935). The life history cycle of Leach's Petrel (*Oceanodroma leucorhoa leucorhoa*) on the outer sea islands of the Bay of Fundy. *Auk* **52**, 382–399.

Grubb, T.C. (1972). Smell and foraging in shearwaters and petrels. *Nature Lond* **237**, 404–405.

Grubb, T.C. (1973). Colony location by Leach's Petrel. *Auk* **90**, 78–82.

Grubb, T.C. (1974). Olfactory navigation to the nesting burrow in Leach's Petrel (*Oceanodroma leucorrhoa*). *Anim Behav* **22**, 192–202.

Grubb, T.C. (1979). Olfactory guidance of Leach's Storm Petrel to the breeding island. *Wilson Bull* **91**, 141–143.

Guillotin, M. & Jouventin, P. (1980). Le Petrel des Neiges a Pointe Geologie. *Gerfaut* **70**, 51–72.

Gurney, J.H. (1922). On the sense of smell in birds. *Ibis* **11th Ser 4**, 225–253.

Guzman, J.R. & Myres, M.T. (1983). The occurrence of Shearwaters (*Puffinus* spp.) off the west coast of Canada. *Can J Zool* **61**, 2064–2077.

Gynn, E. (1984). Dead Shearwaters on Skokholm. *Skomer & Skokholm Bull* **7**, 10–11.

Hagen, Y. (1952). Birds of Tristan da Cunha. *Results Norw Scient Exped Tristan da Cunha 1937–1938*, **20.** Kommisjon Hos Jacob Dybwad. Oslo.

Hagerup, O. (1926). Communities of birds. Oecological Studies. 1. The communities of birds of the Northern Atlantic Ocean. *Vidensk Medd Dansk naturhist Foren* **82**, 127–151.

Hall, A.J., Tasker, M.L. & Webb, A. (1987). The marine distribution of Sooty Shearwater, Manx Shearwater, Storm Petrel and Leach's Petrel in the North Sea. *Seabird* **10**, 60–70.

Hall, K.J. & Williams, A.J. (1981). Animals as agents of erosion at sub-Antarctic Marion Island. *S Afr J Antarct Res* **10/11**, 18–24.

Halse, S.A. & Halse, N.J. (1988). Seabirds and shorebirds at Ningaloo in winter, with comments on Hutton's Shearwater. *West Aust Nat* **17**, 97–106.

Hamanaka, T. (1984). [Heavy metals accumulation in seabirds]. *Mar Sci Monthly* **16**, 245–251.

Hamlet, M.P. & Fisher, H.I. (1967). Air sacs of respiratory origin in some procellariiform birds. *Condor* **69**, 586–595.

Haney, J.C. (1985). Band-rumped Storm-petrel occurrences in relation to upwelling off the coast of the South-eastern United States. *Wilson Bull* **97**, 543–547.

Haney, J.C. (1987a). Aspects of the pelagic ecology and behavior of the Black-capped Petrel (*Pterodroma hasitata*). *Wilson Bull* **99**, 153–168.

Haney, J.C. (1987b). Effects of a Gulf Stream warm-core ring on summer seabird distribution in the Northwest Atlantic Ocean. *Limnol Oceanogr* **32**, 665–673.

Haney, J.C. (1989a). Remote characterization of marine bird habitats with satellite imagery. *Colonial Waterbirds* **12**, 67–77.

Haney, J.C. (1989b). Iterative techniques for characterizing marine bird habitats with time-series of satellite images. *Colonial Waterbirds* **12**, 78–89.

Hector, J.A.L. (1988). Reproductive endocrinology of albatrosses. *Acta XIX Congr Int Orn* **2**, 1702–1709.

Hector, J.A.L. & Goldsmith, A.R. (1985). The role of prolactin during incubation: comparative studies of three *Diomedea* albatrosses. *Gen Comp Endocrinol* **60**, 236–243.

Hector, J.A.L. & Harvey, S. (1986). Corticosterone secretion through long incubation shifts in *Diomedea* albatrosses. *Gen Comp Endocr* **62**, 349–352.

Hector, J.A.L., Croxall, J.P. & Follett, B.K. (1986a). Reproductive endocrinology of the Wandering Albatross *Diomedea exulans* in relation to biennial breeding and deferred sexual maturity. *Ibis* **128**, 9–22.

Hector, J.A.L., Follett, B.K. & Prince, P.A. (1986b). Reproductive endocrinology of the Black-browed Albatross *Diomedea melanophris* and the Grey-headed Albatross *D. chrysostoma*. *J Zool Lond* **208**, 237–253.

Hector, J.A.L., Pickering, S.P.C., Croxall, J.P. & Follett, B.K. (1990). The endocrine basis of deferred sexual maturity in the Wandering Albatross, *Diomedea exulans* L. *Funct Ecol* **4**, 59–66.

Heinemann, D. (1981). A rangefinder for pelagic bird censusing. *J Wildl Manage* **45**, 489–493.

Heinemann, D., Hunt, G. & Everson, I. (1989). Relationships between the distributions of marine avian predators and their prey, *Euphausia superba*, in Bransfield Strait and Southern Drake Passage, Antarctica. *Mar Ecol Progr Ser* **58**, 3–16.

Helbig, A. (1983). Mass occurrence of Manx Shearwaters (*Puffinus puffinus*) at the coast of southern California coincides with exceptionally warm water temperatures. *Ardea* **71**, 161–162.

Hemery, G. & Jouanin, C. (1988). Statut et origine geographique des populations de Petrels Culblanc (*Oceanodroma leucorhoa leucorhoa*) presentes dans le Golfe de Gascogne. *Alauda* **56**, 238–245.

Hemery, G., d'Elbee, E. & Terrasse, J.-F. (1986). Regulation d'une population de Petrels-tempete *Hydrobates pelagicus* par reproduction intermittente. *C R Acad Sci Paris* **303 Ser III**, 353–356.

Hepburn, I.R. & Randall, R.E. (1975). Nest site distribution of the fulmar (*Fulmarus glacialis*) within the Monach Isles National Nature Reserve, Outer Hebrides, U.K. *J Biogeogr* **2**, 223–228.

Hill, A.H., Thompson, C. & McCuaig, M.A. (1981). Tasmanian Muttonbird harvesting and the future of the species. In *Exploited and Endangered Wildlife* (R. Jones, ed.) pp. 76–146. Centre for Environmental Studies, Hobart.

Hill, G.J.E. & Barnes, A. (1989). Census and distribution of Wedge-tailed Shearwater *Puffinus pacificus* burrows on Heron Island, Great Barrier Reef Marine Park. *Emu* **89**, 135–139.

Hiller, A., Wand, U., Kampf, H. & Stakerbrandt, W. (1988). Occupation of Antarctic Continent by petrels during the past 35 000 years: inferences from a C^{14} study of stomach oil deposits. *Polar Biol* **9**, 69–77.

Hills, S. & Fiscus, C.H. (1988). Cephalopod beaks from the stomachs of Northern Fulmars (*Fulmarus glacialis*) found dead on the Washington coast. *Murrelet* **69**, 15–20.

Hindwood, K.A. (1940). The birds of Lord Howe Island. *Emu* **40**, 1–86.

Hofer, H. (1950). Zur morphologie der Kiefermuskulatur der Vogel. *Zool Jahrb Abt Anat* **70**, 428–555.

Hoffman, W., Heinemann, D. & Wiens, J.A. (1981). The ecology of seabird feeding flocks in Alaska. *Auk* **98**, 437–456.

Holdgate, M.W. & Wace, N.M. (1961). The influence of man on the floras and faunas of southern islands. *Polar Rec* **10**, 475–493.

Holgersen, H. (1957). Ornithology of the "Brategg" Expedition. *Scient Results "Brategg" Exped* 1947–1948. **No 4**. Grieggs, Bergen.

Holt, D.W. (1987). Short-eared Owl, *Asio flammens*, predation of Leach's Storm-petrels, *Oceanodroma leucorhoa* in Massachusetts. *Can Fld-Nat* **101**, 448–450.

Honda, K., Marcoveccio, J.E., Kan, S., Tatsukawa, R. & Ogi, H. (1990). Metal concentrations in pelagic seabirds from the North Pacific Ocean. *Arch Environ Contam Toxicol* **19**, 704–711.

Horgan, I.E. & Barrett, J.A. (1985). The use of lipid profiles in comparing the diet of seabirds. In *Antarctic Nutrient Cycles and Food Webs* (W.R. Siegfried, P.R. Condy & R.M. Laws, eds) pp. 493–497. Springer-Verlag, Berlin.

Horning, D.S. & Horning, C.J. (1974). Bird records of the 1971–1973 Snares Islands, New Zealand, Expedition. *Notornis* **21**, 13–24.

Horton, E. (1860). Use of the albatross. *Zoologist* **18**, 6981.

Hosie, G.W. & Ritz, D.A. (1980). Relationship between lipids of mutton birds (*Puffinus tenuirostris*) and that of their food, Krill (*Nyctiphanes australis*). *Aust Mar Sci Bull* **71**, 16–17.

Howell, A.B. (1917). Birds of the islands off the coast of Southern California. *Pacif Coast Avifauna* **12**, 1–127.

Howell, S.N.G. & Webb, S. (1990). The seabirds of Las Islas Revillagigedo, Mexico. *Wilson Bull* **102**, 140–146.

Howell, T.R. & Bartholomew, G.A. (1961a). Temperature regulation in Laysan and Black-footed Albatrosses. *Condor* **63**, 185–197.

Howell, T.R. & Bartholomew, G.A. (1961b). Temperature regulation in nesting Bonin Petrels, Wedge-tailed Shearwaters and Christmas Island Shearwaters. *Auk* **78**, 343–354.

Huber, L.N. (1971). Notes on the migration of the Wilson's Storm Petrel *Oceanites oceanicus* near Eniwetok Atoll, Western Pacific Ocean. *Notornis* **18**, 38–42.

Hudson, A.V. & Furness, R.W. (1988). Utilization of discarded fish by scavenging seabirds behind whitefish trawlers in Shetland. *J Zool Lond* **215**, 151–166.

Hudson, R. (1966). Adult survival estimates for two Antarctic petrels. *Br Antarct Surv Bull* **8**, 63–73.

Huin, N. (1995). Diving depths of White-chinned Petrels. *Condor* **96**, 1111–1113.

Hunt, G.L. Jr (1990). The distribution of seabirds at sea: physical and biological aspects of their marine environment. In *Current Topics in Avian Biology* (R. van den Elzen, K.-L. Schuchmann & K. Schmidt-Koenig, eds) pp. 167–171. Deutschen Orn Gesellschaft, Berlin.

Hunt, G.L. Jr (1991a). Marine ecology of seabirds in polar oceans. *Am Zool* **31**, 131–142.

Hunt, G.L. Jr (1991b). Marine birds and ice-influenced environments of polar oceans. *J Mar System* **2**, 233–240.

Hunt, G.L. Jr & Nettleship, D.N. (1988). Seabirds of high-latitude northern and southern environments. *Proc XIX Congr Int Orn* **1**, 1143–1155.

Hunt, G.L. Jr & Schneider, D.C. (1987). Scale-dependent processes in the physical and biological environment of marine birds. In *Seabirds: Feeding Biology and Role in Marine Ecosystems* (J.P. Croxall, ed.) pp. 7–41. Cambridge University Press, Cambridge.

Hunt, G.L. Jr & Veit, R.R. (1983). Marine bird distribution in Antarctic waters. *Antarct J US* **18**, 167–179.

Hunt, G.L. Jr, Gould, P.J., Forsell, D.J. & Peterson, H. (1981). Pelagic distribution of marine birds in the eastern Bering Sea. In *The Eastern Bering Sea Shelf: Oceanography and Resources*, Vol. 2 (D.W. Hood & S.A. Calder, eds) pp. 689–718. University of Washington Press, Seattle.

Hunter, F.M., Burke, T. & Watts, S.E. (1992). Frequent copulation as a method of paternity assurance in the Northern Fulmar. *Anim Behav* **44**, 149–156.

Hunter, I., Croxall, J.P. & Prince, P.A. (1982). The distribution and abundance of burrowing

Jacob, J. (1982). Stomach oils. In *Avian Biology* **VI** (D.S. Farner & J.R. King, eds) pp. 325–340. Academic Press, New York.

Jacob, J. & Hoerschelmann, H. (1982). Chemotaxonomische untersuchungen zur systematic der Rohrennasen (Procellariiformes). *J Orn* **123**, 63–84.

Jacquinot, H. & Pucheran, J. (1853). Zoologie. Mammiferes et oiseaux. In *Voyage au Pol Sud et dans l'Oceanie sur les corvettes L'Astrolabe et La Zeelee*, etc. Vol. 3. Gide & Baudry, Paris.

James, P.C. (1984). Sexual dimorphism in the voice of the British Storm Petrel *Hydrobates pelagicus. Ibis* **126**, 89–92.

James, P.C. (1985a). The vocal behaviour of the Manx Shearwater *Puffinus puffinus. Z Tierpsychol* **67**, 269–283.

James, P.C. (1985b). Geographical and temporal variation in the calls of the Manx Shearwater *Puffinus puffinus* and the British Storm Petrel *Hydrobates pelagicus. J Zool Lond* **207A**, 331–344.

James, P.C. (1986a). How do Manx Shearwaters *Puffinus puffinus* find their burrows? *Ethology* **71**, 287–294.

James, P.C. (1986b). The filoplumes of the Manx Shearwater *Puffinus puffinus. Bird Study* **33**, 117–120.

James, P.C. & Robertson, H.A. (1985a). The call of Bulwer's Petrel (*Bulweria bulwerii*), and the relationship between intersexual call divergence and aerial calling in the nocturnal Procellariiformes. *Auk* **102**, 878–882.

James, P.C. & Robertson, H.A. (1985b). The calls of male and female Madeiran Storm-Petrels (*Oceanodroma castro*). *Auk* **102**, 391–393.

James, P.C. & Robertson, H.A. (1985c). Sexual dimorphism in the voice of the Little Shearwater *Puffinus assimilis. Ibis* **127**, 388–390.

James, P.C. & Robertson, H.A. (1985d). First record of Swinhoe's Storm Petrel *Oceanodroma monorhis* in the Atlantic Ocean. *Ardea* **73**, 105–106.

James, P.C. & Robertson, H.A. (1985e). The use of playback recordings to detect and census burrowing seabirds. *Seabird* **8**, 18–20.

James, P.C. & Robertson, H.A. (1986). How useful are vocalizations in petrel systematics? *Emu* **86**, 186–189.

Jansen, F.H. (1981). [The migration of the Sooty Shearwater *Puffinus griseus* along the Netherlands Coast]. *Limosa* **54**, 117–126.

Jansen, F.H. (1983). De Windfactor. *Trekvogels* **0**, 31–40.

Jehl, J.R. Jr (1973a). Late autumn observations of pelagic birds off southern California. *West Birds* **4**, 45–52.

Jehl, J.R. Jr (1973b). The distribution of marine birds in Chilean waters in winter. *Auk* **90**, 114–135.

Jehl, J.R. Jr (1974a). The near-shore avifauna of the middle American west coast. *Auk* **91**, 681–699.

Jehl, J.R. Jr (1974b). The distribution and ecology of marine birds over the continental shelf of Argentina in winter. *Trans San Diego Soc Nat Hist* **17**, 217–234.

Jehl, J.R. Jr (1982). The biology and taxonomy of Townsend's Shearwater. *Gerfaut* **72**, 121–135.

Jenkins, J. (1970). Black-capped and other petrels near the Kermadecs. *Notornis* **17**, 130–131.

Jenkins, J.A.F. (1979). Observations on the Wedge-tailed Shearwater in the south-west Pacific. *Notornis* **26**, 331–348.

Jenkins, J.A.F. (1980). Seabird records from Tonga—an account based on the literature and recent observations. *Notornis* **27**, 205–234.

Jenkins, J.A.F. (1988). The distribution of Buller's Shearwater (*Puffinus bulleri*) in New Zealand coastal waters and in the Tasman Sea. *Notornis* **35**, 203–215.

Jennings, P. (1983). The recent status of the Manx Shearwater on The Calf of Man. *Peregrine* **5**, 225.

Jespersen, P. (1930). Ornithological observations in the north Atlantic Ocean. *Oceanogr Rep Danish 'Dana' Exped 1920–22* **7**, 1–36.

Johansen, K. & Millard, R.W. (1973). Vascular responses to temperature in the foot of the Giant Fulmar, *Macronectes giganteus*. *J Comp Physiol* **85**, 47–64.

Johansen, K. & Millard, R.W. (1974). Cold-induced neurogenic vasodilation in skin of the Giant Fulmar *Macronectes giganteus*. *Am J Physiol* **227**, 1232–1235.

Johnson, D.W., Shaffer, T.L. & Gould, P.J. (1993). Incidental catch of marine birds in the North Pacific High Seas Driftnet Fisheries in 1990. *North Pacif Commission Bull* **53 (III)**, 473–483.

Johnson, O.W. (1968). Some morphological features of avian kidneys. *Auk* **85**, 216–228.

Johnson, P.N. (1975). Vegetation and flora of the Solander Islands, southern New Zealand. *NZ J Bot* **13**, 189–213.

Johnson, P.N. (1982). Botanical notes on some southern New Zealand islands. *NZ J Bot* **20**, 121–130.

Johnstone, G.W. (1974). Field characters and behaviour at sea of giant petrels in relation to their oceanic distribution. *Emu* **74**, 209–218.

Johnstone, G.W. (1977). Comparative feeding ecology of the giant petrels *Macronectes giganteus* (Gmelin) and *M. halli* (Mathews). In *Adaptations Within Antarctic Ecosystems* (G. Llano, ed.) pp. 647–668. Gulf Publ. Houston.

Johnstone, G.W. (1982). Zoology. In *Expedition to the Australian Territory of Heard and McDonald Islands 1980* (C. Veenstra & J. Manning, eds) pp. 33–39. Techn Rep **31**, Dept Natn Development, Canberra.

Johnstone, G.W. (1985). Threats to birds on subantarctic islands. In *Conservation of Island Birds* (P.J. Moors, ed.) pp. 101–121. Int Coun Bird Preserv Tech Publ **3**, Cambridge.

Joly, Y., Frenot, Y. & Vernon, P. (1987). Environmental modifications of a subantarctic peat-bog by the Wandering Albatross (*Diomedea exulans*): a preliminary study. *Polar Biol* **8**, 61–72.

Jones, E. (1977). Ecology of the Feral Cat, *Felis catus* (L.), (Carnivora: Felidae) on Macquarie Island. *Aust Wildl Res* **4**, 249–262.

Jones, E. (1980). A survey of burrow-nesting petrels at Macquarie Island based upon remains left by predators. *Notornis* **27**, 11–20.

Jones, F.W. (1937). The breeding of the White-faced Storm Petrel (*Pelagodroma marina*) on South Australian Islands. *S Aust Orn* **14**, 35–41.

Jones, J.M. (1859). *The Naturalist in Bermuda*. Reeves & Turner, London.

Jouanin, C. & Gill, F.B. (1967). Recherche du Petrel de Barau, *Pterodroma baraui*. *Oiseau Revue Fr Orn* **37**, 1–19.

Jouanin, C., Roux, F. & Zino, A. (1969). Visites aux lieux de nidification de *Pterodroma mollis 'deserta'*. *Oiseau Revue Fr Orn* **39**, 161–175.

Jouanin, C., Mongin, J.-L., Roux, F. & Zino, A. (1979). Le Petrel de Bulwer *Bulweria bulwerii* dans l'archipel de Madere et aux iles Selvagens. *Oiseau Revue Fr Orn* **49**, 165–184.

Jouventin, P. (1977). Olfaction in Snow Petrels. *Condor* **79**, 198–199.

Jouventin, P. (1990). Shy Albatrosses *Diomedea cauta salvini* breeding on Penguin Island, in the Crozet Archipelago, Indian Ocean. *Ibis* **132**, 126–127.

Jouventin, P. (1994a). Les populations d'oiseaux marins des T.A.A.F.: resume de 20 annees de recherche. *Alauda* **62**, 44–47.

Jouventin, P. (1994b). Past, present and future of Amsterdam Island (Indian Ocean) and its avifauna. *Birdlife Conserv Ser* **1**, 122–132.

Jouventin, P. & Lequette, B. (1990). The dance of the Wandering Albatross. *Emu* **90**, 122–131.

Jouventin, P. & Mougin, J.-L. (1981). Les strategies adaptatives des Oiseaux de mer. *Revue Ecol (Terre et Vie)* **35**, 217–272.

Jouventin, P. & Robin, J.P. (1984). Olfactory experiments on some Antarctic birds. *Emu* **84**, 46–48.

Jouventin, P. & Viot, C-R. (1985). Morphological and genetic variability of Snow Petrels *Pagodroma nivea*. *Ibis* **127**, 430–441.

Jouventin, P. & Weimerskirch, H. (1984). L'Albatross Fuligineux a Dos Sombre *Phoebetria fusca*, exemple de strategie d'adaptation extreme a la vie pelagique. *Rev Ecol* (*Terre Vie*) **39**, 401–429.

Jouventin, P. & Weimerskirch, H. (1988). Demographic strategies of southern albatrosses. *Acta XIX Int Congr Orn* **1**, 857–865.

Jouventin, P. & Weimerskirch, H. (1990a). Satellite tracking of Wandering Albatrosses. *Nature Lond* **343**, 746–748.

Jouventin, P. & Weimerskirch, H. (1990b). Long-term changes in seabird and seal populations in the Southern Ocean. In *Antarctic Ecosystems. Ecological Change and Conservation* (K.R. Kerry & G. Hempel, eds) pp. 208–213. Springer, Berlin.

Jouventin, P. & Weimerckirch, H. (1991). Changes in the population size and demography of southern seabirds: management implications. In *Bird Population Studies. Relevance to Conservation and Management* (C.M. Perrins, J.-D. Lebreton & G.J.M. Hirons, eds) pp. 297–314. Oxford University Press, Oxford.

Jouventin, P., Monicault, G. De & Blosseville, J.M. (1981). La danse de l'Albatross, *Phoebetria fusca*. *Behaviour* **78**, 43–80.

Jouventin, P., Mougin, J.-L., Stahl, J.-C. & Weimerskirch, H. (1982). La segregation ecologique entres les oiseaux des Iles Crozet. Donnees preliminaires. *Com Natn Fr Rech Antarct* **51**, 457–467.

Jouventin, P., Ridoux, V., Stahl, J.-C. & Weimerskirch, H. (1988). La segregation ecologique des petrels des Iles Crozet. *Rev Ecol (Terre Vie)* **43**, 357–366.

Jungius, H. & Hirsch, U. (1979). Herzfrequenz anderungen bei Brutvogeln in Galapagos als Folge von Storungen durch Besucher. *J Orn* **120**, 299–310.

Kadar, Z. (1978). *Survivals of Greek Zoological Illuminations in Byzantine Manuscripts*. Akad Kiado, Budapest.

Kay, G.T. (1953). The Fulmar's bill. *Scott Nat* **65**, 125–127.

Kendeigh, S.C., Dol'Nik, V.R. & Gavrilov, V.M. (1977). Avian Energetics. In *Granivorous Birds in Ecosystems* (J. Pinowski & S.C. Kendeigh, eds) pp. 127–204. Cambridge University Press, Cambridge.

Kennedy, J.N. (1917). Notes on birds observed in the North Sea and North Atlantic Ocean during the autumn and winter of 1914. *Ibis* **10th Ser 5**, 30–41.

Kennedy, P.C. (1978). Vegetation and soils of North Island, Foveaux Strait, New Zealand. *NZ J Bot* **16**, 419–434.

Kenyon, K.W. & Rice, D.W. (1958). Homing of Laysan Albatrosses. *Condor* **60**, 3–6.

Kerry, K.R., Horne, R.S.C. & Dorward, D.F. (1983). Record of the Short-tailed Shearwater *Puffinus tenuirostris* in Antarctic waters. *Emu* **83**, 35–36.

Kharitonov, S.P. & Tikhonov, A.V. (1982). [Some sound signals of the Fulmar (*Fulmarus glacialis*) and Fork-tailed Petrel (*Oceanodroma leucorhoa*) during the nest period]. *Zool Zh* **61**, 1097–1099.

King, W.B. (1970). The trade wind zone oceanography pilot study Part VII: observations of sea birds March 1964 to June 1965. *US Fish Wildl Serv Spec Scient Rep Fish* **586**. Washington.

King, W.B. (1974). Pelagic studies of seabirds in the Central and Eastern Pacific Ocean. *Smithson Contr Zool* **158**, 1–277.

Kinsky, F.C. (1958). 8th Annual Report of the Ornithological Society of New Zealand Ringing Committee for the year ending 31 March 1957. *Notornis* **8 (Suppl)**, 30 pp.

Kinsky, F.C. (1960). The yearly cycle of the Northern Blue Penguin (*Eudyptula minor noaehollandiae*) in the Wellington Harbour area. *Rec Dominion Mus* **3**, 145–218.

Kinsky, F.C. (1968). An unusual seabird mortality in the southern North Island (New Zealand) April, 1968. *Notornis* **15**, 143–155.

Kirkwood, J.K., Cunningham, A.A., Hawkey, C., Howlett, J. & Perrins, C.M. (1995). Hematology of fledgling Manx Shearwaters (*Puffinus puffinus*) with and without 'Puffinosis'. *J Wildl Dis* **31**, 96–98.

Klages, N., Gales, R. & Pemberton, D. (1990). The stomach contents of Antarctic Petrels *Thalassoica antarctica* feeding young chicks at Scullin Monolith, Mawson Coast, Antarctica. *Polar Biol* **10**, 545–547.

Klassen, M. & Drent, R. (1991). An analysis of hatchling resting metabolism: in search of ecological correlates that explain deviations from allometric relations. *Condor* **93**, 612–629.

Klemm, R.D. (1969). Comparative myology of the hind limb of procellariiform birds. *South Illinois Univ Monogr Sci Ser* **2**, 1–269.

Klinckowstrom, A. von (1890). Les lobes olfactives du *Fulmarus glacialis*. *Biol Foren Stockholm* **3**, 10–11.

Knight, G.C., Walker C.H., Cabot, D.C. & Harris, M.P. (1981). The activity of two hepatic microsomal enzymes in sea birds. *Comp Biochem Physiol* **68C**, 127–132.

Knight, G.C. & Walker, C.H. (1982). A study of the hepatic microsomal monogenase of sea birds and its relationships to organochlorine pollutants. *Comp Biochem Physiol* **73C**, 211–221.

Kobayashi, K. & Cho, H. (1981). [*Birds of Taiwan*]. Maeda Graphic Arts, Kyoto.

Kostelecka-Myrcha, A. & Myrcha, A. (1989). Changes of the red blood picture during nestling development of Wilson's Storm Petrel (*Oceanites oceanicus* Kuhl). *Pol Polar Res* **10**, 151–162.

Krashnaminnikov, S.P. (1764). *The History of Kamschatka, and the Kurilski Islands, with the Countries Adjacent* (Transl. J. Grieve). Jeffrys, London.

Krasnow, L.D. & Sanger, G.A. (1982). Feeding ecology of marine birds in the nearshore waters of Kodiak Island. In *Final Report to the Outer Continental Shelf Environmental Assessment Program*, pp. 1–113. US Fish Wildl Serv, Anchorage.

Kritzler, H. (1948). Observations on behavior in captive fulmars. *Condor* **50**, 5–15.

Kurimoto, S. (1937). [Muttonbirding of *Puffinus leucomelas* on Mikura Island]. *Yancho* **4**, 27–31.

Kuroda, N. (1953). On the skeletons of *Puffinus nativitatis* and *Pagodroma nivea*. *Tori* **13**, 50–67.

Kuroda, N. (1954). *On the Classification and Phylogeny of the Order Tubinares, Particularly the Shearwaters* (*Puffinus*). Author, Tokyo.

Kuroda, N. (1955a). On the osteology of the Gadfly-Petrels, *Pterodroma*. *Annot Zool Japonenses* **28**, 171–177.

Kuroda, N. (1955b). Osteological notes on the Buller's Shearwater, *Puffinus bulleri* Salvin. *Annot Zool Japonenses* **28**, 167–170.

Kuroda, N. (1957a). A brief note on the pelagic migration of the Tubinares. *Misc Rep Yamashina Inst Orn Zool* **11**, 436–449.

Kuroda, N. (1957b). [On the histological change of the proventriculus in the Tubinares (Avis)]. *Dobutsugaku Zasshi* [*Zoological Magazine*] **66,** 27–31.

Kuroda, N. (1959). Osteological notes on *Pterodroma hypoleuca* (Aves). *Annot Zool Japonenses* **32**, 147–151.

Kuroda, N. (1960a). Analysis of sea bird distribution in the northwest Pacific Ocean. *Pacif Sci* **14**, 55–67.

Kuroda, N. (1960b). [Histological observation of the proventriculus of *Diomedea*]. *Misc Rep Yamashina's Inst Orn Zool* **2**, 139–142.

Kuroda, N. (1960c). [On the pectoral muscles of birds]. *Misc Rep Yamashina Inst Orn Zool* **14**, 50–59.

Kuroda, N. (1962). On the cervical muscles of birds. *Misc Rep Yamashina Inst Orn* **3**, 189–211.

Kuroda, N. (1967a). Morpho-anatomical analysis of parallel evolution between the Diving Petrel and Ancient Auk, with comparative osteological data on other species. *Misc Rep Yamashina Inst Orn Zool* **5**, 111–137.

Kuroda, N. (1967b). Note on the whitish underparts of *Puffinus tenuirostris* and a supposed hybrid between *P. griseus*. *Misc Rep Yamashina Inst Orn Zool* **5**, 194–197.

Kuroda, N. (1983). Some osteological notes on Procellariiformes. *Tori* **32**, 41–61.

Kuroda, N. (1986). On the intestinal twistings in Gadfly-Petrels and comparative notes on the digestive tract in Procellariiformes. *Jap J Orn* **35**, 1–14.

Kuroda, N. (1988). A distributional analysis of *Diomedea immutabilis* and *D. nigripes* in the North Pacific. *J Yamashina Inst Orn* **20**, 1–20.

Kuroda, N. (1991). Distributional patterns and seasonal movements of Procellariiformes in the North Pacific. *J Yamashina Inst Orn* **23**, 23–84.

Kuroda, N., Kakizawa, R., Hori, H., Osaka, Y., Usuda, N. & Utida, S. (1982). Evolution of mitochondrial malate dehydrogenase in birds. *J Yamashina Inst Orn* **14**, 1–15.

Kuroda, N., Kakizawa, R. & Watada, M. (1990). Genetic divergence and relationships in fifteen species of Procellariiforms. *J Yamashina Inst Orn* **22**, 114–123.

Kuroda, N.H. (1963). A fragmental observation on the avian kidney. *Misc Rep Yamashina Inst Orn Zool* **3**, 274–286.

Labat, J.-B. (1722). *Noveau Voyage aux Isles de l'Amerique Contenant l'Histoire Naturelle de ces Pays, etc.* Vol. 1. Paris.

Lacan, F. & Mougin, J.-L. (1974). Les oiseaux des Iles Gambier et de quelques atolls orientaux de l'Archipel des Tuamotu (Ocean Pacifique). *Oiseau Revue Fr Orn* **44**, 192–280.

Lack, D. (1968). *Ecological Adaptations for Breeding in Birds*. Methuen, London.

Lack, D. & Lockley, R.M. (1938). Skokholm Bird Observatory homing experiments. *Br Birds* **31**, 242–248.

Lachmund, F. (1674). *De ave Diomedea Dissertatio cum Vera Ejus Effigie Ari Incisa*, etc. Andream Frisium, Amsterdam.

Lambert, K. (1984). Der Kerguelensvogel, *Pterodroma brevirostris* Lesson, 1831, im Sudatlantik. *Beitr Vogelkd Jena* **30**, 191–202.

Lampert, R.J. (1966). An excavation at Durras North, New South Wales. *Archaeol Physical Anthropol Oceania* **1**, 83–118.

Lampert, R.J. (1971). Burrill Lake and Currarong. Coastal sites in southern New South Wales. *Terra Australis* **1**, 1–86.

Langston, N.E. & Rohwer, S. (1995). Unusual patterns of primary molt in Laysan and Black-footed Albatrosses. *Condor* **97**, 1–19.

Laurent, U.B.G. (1981). Hyaluronate in aqueous humour. *Exp Eyes Res* **33**, 147–155.

Lavauden, L. & Mourgue, M. (1918). Contribution a l'etude du Thalassidrome Tempete dans la Mediterranee. *Rev fr Orn* **10**, 305–309.

Lawrence, G.N. (1889). An account of the breeding habits of *Puffinus auduboni* in the island of Grenada, West Indies, with a note on *Zenaida rubripes*. *Auk* **6**, 19–21.

Lawrie, R.A. (1952). Biochemical differences between red and white muscle. *Nature Lond* **170**, 122–123.

Lee, D.S. (1986). Seasonal distribution of marine birds in North Carolina waters, 1975–1986. *Am Birds* **40**, 409–412.

Lee, D.S. (1987). First record of Wilson's Storm-Petrels on a Christmas bird count. *Am Birds* **41**, 1331–1333.

Lee, D.S. & Vina, N. (1993). A re-evaluation of the status of the endangered Black-capped Petrel, *Pterodroma hasitata*, in Cuba. *Ornitologia Neotropical* **4**, 99–101.

Lee, K.-S. & Won, P.-O. (1988). [Breeding biology of Swinhoe's Petrel *Oceanodroma monorhis* (Swinhoe) on Chibal Islet, Korea]. *Bull Inst Orn Kyung Hee Univ* **2**, 39–60.

Le Grand, H.E. Jr (1981). The autumn migration. August 1–November 30, 1980. Southern Atlantic Coast Region. *Am Birds* **35**, 170–172.

Le Mao, P. & Yesou, P. (1993). The annual cycle of Balearic Shearwaters and Western Yellow-legged Gulls: some ecological considerations. In *Status and Conservation of Seabirds, Ecogeography and Mediterranean Action Plan* (J.S. Aguilar, X. Monbailliu & A.M. Paterson, eds) pp. 135–145. Soc Espan Orn, Madrid.

Lensink, C.J. (1984). The status and conservation of seabirds in Alaska. In *Status and Conservation of the World's Seabirds* (J.P. Croxall, P.G.H. Evans & R.W. Schreiber, eds) pp. 13–27. ICBP, Cambridge.

Lequette, B. & Jouventin, P. (1991a). The dance of the Wandering Albatross II: Acoustic Signals. *Emu* **91**, 172–178.

Lequette, B. & Jouventin, P. (1991b). Comparison of visual and vocal signals of Great Albatrosses. *Ardea* **79**, 383–393.

Lequette, B., Verheyden, C. & Jouventin, P. (1989). Olfaction in subAntarctic seabirds: its phylogenetic and ecological significance. *Condor* **91**, 732–735.

Lesson, R.P. & Garnot, P. (1828). Zoology. In *Voyage Antour du Monde, Execute par Ordre du Roi, sur la Corvette de sa Majestie, La Coquille, etc* (L.I. Duperrey, ed.) Vol. 1(2), pp. 361–743. Bertrand, Paris.

Lewis, R.W. (1966). Studies of the glyceryl ethers of the Stomach Oil of Leach's Petrel *Oceanodroma leucorhoa* (Viellot). *Comp Biochem Physiol* **19**, 363–377.

Lewis, R.W. (1969a). Studies on the Stomach Oils of marine animals – II. Oils of some procellariiform birds. *Comp Biochem Physiol* **31**, 725–731.

Lewis, R.W. (1969b). Distribution of dietary squalene in a Wandering Albatross, *Diomedea exulans. Nature Lond* **224**, 1220–1221.

Lill, A. & Baldwin, J. (1983). Weight changes and the mode of depot fat accumulation in migratory Short-tailed Shearwaters. *Aust J Zool* **31**, 891–902.

Lipinski, M.R. & Jackson, S. (1989). Surface-feeding on cephalopods by procellariiform seabirds in the southern Benguela region, South Africa *J Zool Lond* **218**, 549–563.

Littlepage, J.L. (1964). Seasonal variation in lipid content of two Antarctic marine crustacea. In *Biologie Antarctique* (R. Carrick, M. Holdgate & J. Prevost, eds) pp. 463–470. Hermann, Paris.

Liversidge, R. & LeGras, G.M. (1981). Observations of seabirds off the eastern Cape, South Africa, 1958–1963. In *Proc Symp Birds of the Sea and Shore* (J. Cooper, ed.) pp. 149–167. Afr Seabird Grp, Cape Town.

Livezey, B.C. (1989). Morphometric patterns in Recent and fossil penguins (Aves, Sphenisciformes). *J Zool Lond* **219**, 269–307.

Lock, J.W., Thompson, D.R., Furness, R.W. & Bartle, J.A. (1992). Metal concentrations in seabirds of the New Zealand region. *Environ Pollut* **75**, 289–300

Lockie, J.D. (1952). A comparison of some aspects of the retinae of the Manx Shearwater, Fulmar Petrel, and House Sparrow. *Q J Micr Sci* **93**, 347–356.

Lockley, R.M. (1931). Further notes on the breeding habits of the Manx Shearwater. *Br Birds* **24**, 202–207.

Lockley, R.M. (1932). On the breeding habits of the Storm-Petrel, with special reference to its incubation and fledging-periods. *Br Birds* **25**, 206–211.

Lockley, R.M. (1942). *Shearwaters*. Dent, London.

Lockley, R.M. (1952). Notes on the birds of the Berlengas (Portugal), the Desertas and Baixo (Madeira) and the Salvages. *Ibis* **94**, 144–158.

Lockley, R.M. (1953). On the movements of the Manx Shearwater at sea during the breeding season. *Br Birds* **46 (Suppl)**, 1–48.

Longstaff, T.G. (1924). Notes from Spitzbergen, 1923. *Ibis* **11th Ser 6**, 480–495.

Lonnberg, E. (1904). On the homologies of the different pieces of the compound rhamphotheca of birds. *Arkiv Zool* **1**, 473–512.

Loomis, L.M. (1918). A review of the albatrosses, petrels, and diving petrels. *Proc Calif Acad Sci 4th Ser* **2 (2)**, 1–187.

Loomis, L.M. (1921). Remarks on the migration of southern hemisphere albatrosses and petrels. *Auk* **38**, 527–531.

Love, J.A. (1978). Leach's and Storm petrels on North Rona: 1971–1974. *Ringing & Migration* **2**, 15–19.

Lovenskjold, H.L. (1954). Studies on the Avifauna of Spitzbergen. *Norsk Polarinst Skr* **103**, 1–131.

Lowe, P.R. (1925). On the classification of the Tubinares or Petrels. *Proc Zool Soc Lond* **1925**, 1433–1443.

Lucas, A.M. & Stettenheim, P.R. (1972). *Avian Anatomy Integument.* US Dept Agriculture Handbook **362**. Washington, DC.

Luders, D.J. (1977). Behaviour of Antarctic Petrels and Antarctic Fulmars before laying. *Emu* **77**, 208–214.

Ludwig, J.P., Summer, C.L., Jones, P., Giesy, J.P., Auman, H.J., Rolland, R.M. & Colborn, T.L. (in press). Polychlorinated diaromatic hydrocarbons (PCDDs, PCDFs, PCBs, PCNs, DDTs) and other xenobiotics in North Pacific albatrosses: concentrations of planar compounds exceed levels for reproductive effects. In *Proc 1st Inter Conf on the Biology and Conservation of Albatrosses* (G. Robertson ed.). Surrey Beatty, Sydney.

Luke, B.G., Johnstone, G.W. & Woehler, E.J. (1989). Organochlorine pesticides, PCBs and mercury in Antarctic and subAntarctic seabirds. *Chemosphere* **19**, 2007–2021.

Lukowski, A.B. (1983). DDT and its metabolites in the tissues and eggs of migrating Antarctic seabirds from the regions of the South Shetland Islands. *Polish Polar Res* **4**, 135–141.

Lukowski, A.B., Karolewski, M.A. & Gorski, T. (1987). Polychlorinated biphenyls in the tissues of Antarctic marine migratory birds and penguins from the breeding colony on King George Island (South Shetland Islands). *Polish Polar Res* **8**, 179–187.

M'Cormick, R. (1842). A sketch of the Antarctic regions, embracing a few passing remarks, geological and ornithological. *Tasm J Nat Sci* **1**, 241–247.

Macdonald, D. & Green, R.H. (1963). Albatross Island. *Emu* **63,** 23–31.

Macdonald, M.A. (1977). An analysis of the recoveries of British-ringed Fulmars. *Bird Study* **24**, 208–214.

Mactavish, B. (1992). Atlantic Provinces Region. *Am Birds* **46**, 1114–1116.

Maesako, Y. (1985). Community structure of *Machilus thunbergii* forests disturbed by birds (*Calonectris leucomelas*: Streaked Shear-water) on Kanmurijima Island, Kyoto Prefecture, Japan. *Jap J Ecol* **35**, 387–400.

Magnan, A. (1922). Les characteristiques des oiseaux suivant le mode de Vol. *Ann Sci Nat Zool* **(10)5**, 125–330.

Magnan, A. (1925). *Le Vol a Voile*. Roche d'Estrez, Paris.

Mahoney, S.A. & Jehl, J.R. Jr (1984). Body water content in marine birds. *Condor* **86**, 208–209.

Manikovski, S. (1971). The influence of meterological factors on the behaviour of sea birds. *Acta Zool Cracoviensia* **16**, 581–668.

Marchant, S. (1977). A seawatch on the southern coast of New South Wales. *Emu* **77**, 9–18.

Marchant, S. & Higgins, P.J. (eds) (1990). *Handbook of Australian, New Zealand & Antarctic Birds,* Vol. 1A. *Ratites to Petrels*. Oxford University Press, Melbourne.

Marks, J.S. & Hall, C.S. (1992). Tool use by Bristle-thighed Curlews feeding on albatross eggs. *Condor* **94**, 1032–1034.

Marks, J.S. & Leasure, S.M. (1992). Breeding biology of Tristram's Storm-Petrel on Laysan Island. *Wilson Bull* **104**, 719–731.

Marshall, A.J. (1949). On the function of the interstitium of the testis. The sexual cycle of a wild bird, *Fulmarus glacialis* (L.). *Quart J Micr Sci* **90**, 265–280.

Marshall, A.J. & Serventy, D.L. (1956a). Moult adaptation in relation to long-distance migration in petrels. *Nature Lond* **177**, 943.

Marshall, A.J. & Serventy, D.L. (1956b). The breeding cycle of the Short-tailed Shearwater, *Puffinus tenuirostris* (Temminck), in relation to trans-equatorial migration and its environment. *Proc Zool Soc Lond* **127**, 489–510.

Marshall, A.J. & Serventy, D.L. (1957). On the post-nuptial rehabilitation of the avian testis tunic. *Emu* **57**, 59–63.

Martin, A.R. (1986). Feeding association between dolphins and shearwaters around the Azores Islands. *Can J Zool* **64**, 1372–1374.

Martin, D.W. & Myres, M.T. (1969). Observations on the distribution and migration of some seabirds off the outer coasts of British Columbia and Washington State, 1946–1949. *Syesis* **2**, 241–256.

Martin, G. (1990). Designer eyes for seabirds of the night. *New Scient* **128**, 46–48.

Martin, G.R. & Brooke, M. de L. (1991). The eye of a procellariiform seabird, the Manx Shearwater, *Puffinus puffinus*: visual fields and optical structure. *Brain Behav Evol* **37**, 65–78.

Maruyama, N., Oka, N., Watabe, Y., Kuroda, N., Skira, I.J. & Kono, H. (1986). [Re-examination of the migration routes of the Short-tailed Shearwater by the trans-Pacific Survey]. In *Synthetic Study of the Mass Mortality of Short-tailed Shearwater* Puffinus tenuirostris (N. Kuroda, ed.) pp. 36–42. Study Grp of Short-tailed Shearwater, Abiko.

Mathew, M.A. (1881). Habits of the Storm Petrel in captivity. *Zoologist* **3rd Ser 5**, 489–490.

Mathews, G.M. (1928). *The Birds of Norfolk and Lord Howe Islands and the Australian South Polar Quadrant*. Witherby, London.

Mathews, G.M. (1935). Some remarks on the patagial fan of the three British breeding petrels. *Br Birds* **29**, 201–202.

Mathews, G.M. (1936). The ossification of certain tendons in the patagial fan of Tubinares. *Bull Br Orn Club* **56**, 45–50.

Mathews, G.M. & Iredale, T. (1921). *A Manual of the Birds of Australia*. Witherby, London.

Mathiu, P.M., Whittow, G.C. & Dawson, W.R. (1992). Hatching and establishment of thermoregulation in the Wedge-tailed Shearwater (*Puffinus pacificus*) *Physiol Zool* **65**, 583–603.

Matochik, J.A., Reems, C.N. & Wenzel, B.M. (1991). A brain atlas of the Northern Fulmar (*Fulmarus glacialis*) in stereotaxic coordinates. *Brain Behavior Evol* **37**, 215–244.

Matthews, G.V.T. (1953). Navigation in the Manx Shearwater. *J Exp Biol* **30**, 370–396.

Matthews, G.V.T. (1964). Individual experience as a factor in the navigation of Manx Shearwaters. *Auk* **81**, 132–146.

Matthews, G.V.T. (1968). *Bird Navigation*. Cambridge University Press, Cambridge.

Matthews, L.H. (1929). The birds of South Georgia. *Discovery Rep* **1**, 563–592.

Matthews, L.H. (1949). The origin of stomach oil in the petrels, with comparative observations on the avian proventriculus. *Ibis* **91**, 373–392.

Mayaud, N. (1931). Contribution a l'etude de la mue des puffins. *Alauda*, **2nd Ser 2**, 230–249.

Mayaud, N. (1932a). Considerations sur la morphologie et la systematique de quelques puffins. *Alauda* **4**, 41–78.

Mayaud, N. (1932b). Sur la mue des Puffins. *Alauda* **2nd Ser 4**, 110–112.

Mayaud, N. (1934). Considerations sur la morphologie et la systematique de quelques puffins (addendum). *Alauda* **6**, 87–95.

Mayaud, N. (1941). Etudes sur les plumages et les mues. I. La distinction des ages chez certains Hydrobatides. *Alauda* **11** (**Spec No**), xliv-xlvi.

Mayaud, N. (1949–50). Nouvelles precisions sur la mue des procellariens. *Alauda* **17/18**, 144–155, 222–233.

McCanch, N.V. (1981). Predation on Manx Shearwaters by Grey Seals. *Br Birds* **74**, 348.

McClatchie, S., Hutchinson, D. & Nordin, K. (1989). Aggregation of avian predators and zooplankton prey in Otago shelf waters. *J Plankton Res* **11**, 361–374.

McDermond, D.K. & Morgan, K.H. (1993). Status and conservation of North Pacific albatrosses. In *The Status, Ecology, and Conservation of Marine Birds of the North Pacific* (K. Vermeer, K.T. Briggs, K.H. Morgan & D. Siegel-Causey, eds) pp. 70–81. Can Wildl Serv Spec Publ, Ottawa.

McDonald, D.B. & Caswell, H. (1993). Matrix methods for avian demography. *Current Orn* **10**, 139–185.

McGlone, M.S., Anderson, A.J. & Holdaway, R.N. (1994). An ecological approach to the Polynesian settlement of New Zealand. In *The Origins of the First New Zealanders* (D.G. Sutton, ed.) pp. 136–163. Auckland University Press, Auckland.

McKinnell, S. & Waddell, B. (1993). Associations of species caught in the Japanese Large Scale Pelagic Squid Driftnet Fishery in the Central North Pacific Ocean: 1988–1990. *North Pacif Commission Bull* **53(III)**, 91–109.

McKitrick, M.C. (1991a). Forelimb mylogy of loons (Gaviiformes), with comments on the relationship of loons and tubenoses (Procellariiformes). *Zool J Linn Soc* **102**, 115–152.

McKitrick, M.C. (1991b). Phylogenetic analysis of avian hindlimb musculature. *Misc Publ Mus Zool Univ Michigan* **179**, 1–85.

McKitrick, M.C. (1992). Phylogenetic analysis of avian parental care. *Auk* **109**, 828–846.

McLelland, J. (1979). Digestive system. In *Form and Function in Birds*, Vol. 1 (A.S. King & J. McLelland, eds) pp. 69–181. Academic Press, London.

Mead, C. & Clark, J. (1990). Report on bird ringing for Britain and Ireland for 1989. *Ringing & Migration* **11**, 137–176.

Medway, D.G. (in press). Human induced mortality of southern albatrosses at sea in the 19th century – a brief historical review. In *Proc 1st Inter Conf on the Biology and Conservation of Albatrosses* (G. Robertson ed.). Surrey Beatty, Sydney.

Mees, G.F. (1976). Mass mortality of *Puffinus gravis* (O'Reilly) on the coast of Suriname (Aves, Procellariidae). *Zool Meded Leiden* **49**, 269–271.

Meeth, P. & Meeth, K. (1983). Seabird observations from six Pacific Ocean crossings. *Sea Swallow* **32**, 58–65.

Mehlum, F. (1990). Seabird distribution in the northern Barents Sea marginal ice-zone during late summer. *Polar Res* **8**, 61–65.

Mercier, G. (1987). Le Fulmar (*Fulmarus glacialis*) estivant et nicheur en Picardie. *Avocette* **11**, 15–40.

Meredith, C. (1985). The vertebrate fossil fauna of Norfolk island, and the phylogeny of the genus *Pterodroma*. Unpubl PhD thesis, Monash University.

Meredith, C. (1990). Vertebrate fossil faunas from islands in Australasia and the Southwest Pacific. In *Vertebrate Palaeontology of Australasia* (P. Vickers-Rich, J.M. Monaghan, R.F. Baird & T.H. Rich, eds) pp. 1345–1382. Chapman & Hall, London.

Meseth, E.H. (1968). The behavior of the Laysan Albatross, *Diomedea immutabilis* on its breeding ground. Unpubl PhD thesis, Southern Illinois University.

Meseth, E.H. (1975). The dance of the Laysan Albatross *Diomedea immutabilis. Behaviour* **54**, 217–257.

Midtgard, U. (1988). Comparative morphology of the avian circulatory system. *Acta XIX Congr Int Orn* **2**, 2445–2454.

Miles, J.A.R. & Shrivastav, J.B. (1951). Ornithosis in certain sea-birds. *J Anim Ecol* **20**, 195–200.

Miller, A.H. (1931). An auklet from the Eocene of Oregon. *Univ Calif Publ Geol Sci* **20**, 23–26.

Miller, L. (1936). Some maritime birds observed off San Diego, California. *Condor* **38**, 9–16.

Millener, P.R. (1990). Evolution, extinction and the subfossil record of New Zealand's avifauna. In *A Flying Start* (B.J. Gill & B.D. Heather, eds) pp. 93–100. Random Century & Orn Soc NZ, Auckland.

Milne-Edwards, A. (1867–68). *Recherches Anatomiques et Paleontologiques pour servir a l'Histoire des Oiseaux Fossiles de la France*, Vol.1. Masson, Paris.

Milne-Edwards, A. & Grandidier, A. (1882). *Histoire Physique, Naturelle, et Politique de Madagascar*. Grandidier, Paris.

Milsom, W.K., Johansen, K. & Millard, R.W. (1973). Blood respiratory properties in some Antarctic birds. *Condor* **75**, 472–474.

Minami, H., Minagawa, M. & Ogi, H. (1995). Changes in stable carbon and nitrogen isotope ratios in Sooty and Short-tailed Shearwaters during their northward migration. *Condor* **97**, 565–574.

Mitchell, P.C. (1901). On the intestinal tract in birds; with remarks on the valuation and nomenclature of zoological characters *Trans Linn Soc Lond* **2nd Ser Zool 8**, 173–275.

Mitchell, P.C. (1913). The peroneal muscles in birds. *Proc Zool Soc Lond* **1913**, 1039–1072.

Mizutani, H., Hasegawa, H. & Wada, E. (1986). High nitrogen isotope ratios for soils of seabird rookeries. *Biogeochemistry* **2**, 221–247.

Mobberley, D.H. (1974). Concentration of seabirds in North Pacific, 26th and 27th April 1973. *Sea Swallow* **23**, 72–73.

Mochizuki, H. & Kasuga, I. (1985). Seabirds in the Australian sector of the Southern Ocean, January and February, 1984. *Trans Tokyo Univ Fisheries* **6**, 155–165.

Moloney, C.L., Cooper, J., Ryan, P.G. & Siegfried, W.R. (1994). Use of a population model to assess the impact of longline fishing on Wandering Albatross *Diomedea exulans* populations. *Biol Conserv* **70**, 195–203.

Montague, T.L. (1984). The food of Antarctic Petrels (*Thalassoica antarctica*). *Emu* **84**, 244–245.

Montague, T.L. (1988). Birds of Prydz Bay, Antarctica: distribution and abundance. *Hydrobiologia* **165**, 227–237.

Montague, T.L., Cullen, J.M. & Fitzherbert, K. (1986). The diet of the Short-tailed Shearwater *Puffinus tenuirostris* during its breeding season. *Emu* **86**, 207–213.

Monteiro, L.R., Furness, R.W. & del Nevo, A.J. (1995). Mercury levels in seabirds from the Azores, mid-North Atlantic Ocean. *Arch Environ Contam Toxicol* **28**, 304–309.

Montevecchi, W.A. & Hufthammer, A.K. (1990). Zooarchaeological implications for prehistoric distributions of seabirds along the Norwegian coast. *Arctic* **43**, 110–114.

Montevecchi, W.A., Blundon, E., Coombes, G., Porter, J. & Rice, P. (1978). Northern Fulmar breeding range extended to Baccalieu Island, Newfoundland. *Can Fld-Nat* **92**, 80–82.

Montevecchi, W.A., Birt-Friesen, V.L. & Cairns, D.K. (1992). Reproductive energetics and prey harvest of Leach's Storm-petrels in the Northwest Atlantic. *Ecology* **73**, 823–832.

Montgomery, H.H. (1892). Some account of the mutton birds, or Sooty Petrels (*Nectris brevicaudus*) etc. *Pap Proc R Soc Tasm* **1891**, 1–9.

Moors, P.J. (1980). Southern Great Skuas on Antipodes Island, New Zealand: observations on food, breeding, and growth of chicks. *Notornis* **27**, 133–146.

Moors, P.J. (1985). Eradication campaigns against *Rattus norvegicus* on the Noises Islands, New Zealand, using Brodifacoum and 1080. In *Conservation of Island Birds* (P.J. Moors, ed.) pp. 143–155. *Int Coun Bird Preserv Techn Publ* **3**, Cambridge.

Moors, P.J. & Atkinson, I.A.E. (1984). Predation on seabirds by introduced mammals, and factors affecting its severity. In *Status and Conservation of the World's Seabirds* (J.P. Croxall, P.G.H. Evans & R.W. Schreiber, eds) pp. 667–690. *Int Coun Bird Preserv Tech Publ* **2**, Cambridge.

Morgan, K.H., Vermeer, K. & McKelvey, R.W. (1991). *Atlas of Pelagic Birds of Western Canada*. Occas Pap **72**, Can Wildl Serv, Ottawa.

Morgan, K.R., Chappell, M.A. & Bucher, T.L. (1992). Ventilatory oxygen extraction in relation to ambient temperature in four Antarctic seabirds. *Physiol Zool* **65**, 1092–1113.

Morgan, W.L. (1982). Feeding methods of the Short-tailed Shearwater *Puffinus tenuirostris. Emu* **82**, 226–227.

Morgan, W.L. & Ritz, D.A. (1982). Comparison of the feeding apparatus in the mutton-bird, *Puffinus tenuirostris* (Temminck) and the Fairy Prion, *Pachyptila turtur* (Kuhl) in relation to the capture of Krill, *Nyctiphanes australis. J Exp Mar Biol Ecol* **59**, 6–76.

Morzer Bruyns, W.F.J. (1965). Birds seen during west to east trans-Pacific crossing along Equatorial Counter-Current around latitude 7°N in the autumn of 1960. *Sea Swallow* **17**, 57–66.

Moser, M.L. & Lee, D.S. (1992). A fourteen-year survey of plastic ingestion by western North Atlantic seabirds. *Colonial Waterbirds* **15**, 83–94.

Moss, S.R., Petersen, A. & Nuttall, P.A. (1986). Tick-borne viruses in Icelandic seabird colonies. *Acta Naturalia Islandica* **32**, 1–19.

Mougin, J.-L. (1967). Etude ecologique des deux especes de fulmars le Fulmar Atlantique (F. *glacialis*) et le Fulmar Antarctique (F. *glacialoides*). *Oiseau Revue Fr Orn* **37**, 57–103.

Mougin, J.-L. (1970a). Les albatross fuligineux *Phoebetria palpebrata* et *P. fusca* de l'Ile de la Possession (Archipel Crozet). *Oiseau Revue Fr Orn* **40, No Spec,** 37–61.

Mougin, J.-L. (1970b). Le petrel a Menton Blanc *Procellaria aequinoctialis* de l'Ile de la Possession (Archipel Crozet). *Oiseau Revue Fr Orn* **40, No Spec,** 62–96.

Mougin, J.-L. (1970c). Observations ecologiques sur les Grands Albatross (*Diomedea exulans*) de l'Ile de la Possession (Archipel Crozet) en 1968. *Oiseau Revue Fr Orn* **40**, 16–36.

Mougin, J.-L. (1975). Ecologie comparee des procellariidae Antarctiques et Subantarctiques. *Com natn Fr rech antarct* **36**, 1–195.

Mougin, J.-L. (1989). Donnees preliminaires sur la structure et la dynamique de la population de Petrels de Bulwer *Bulweria bulwerii* de l'Ile Selvagem Grande (30°09′N, 15°52′W). *C R Acad Sci Paris* **308, Ser III**, 103–106.

Mougin, J.-L. & Prevost, J. (1980). Evolution annuelle des effectifs et des biomasses des oiseaux antarctiques. *Rev Ecol (Terre Vie)* **34**, 101–133.

Mougin, J.-L. & Roux, F. (1988). La stabilite des effectifs des Puffins Cendres *Calonectris diomedea borealis* de l'Ile Selvagem Grande (30°09′N, 15°52′W) de 1983 a 1986. *Bocagiana* **116**, 1–6.

Mougin, J.-L., Jouanin, C. & Roux, F. (1987). Structure et dynamique de la population du Puffins Cendres *Calonectris diomedea borealis* de l'Ile Salvagem Grande (30°09′N, 15°52′W). *Oiseau Revue Fr Orn* **57**, 201–225.

Mougin, J.-L., Jouanin, C. & Roux, F. (1988). Les migrations du Puffin Cendre *Calonectris diomedea. Oiseau Revue Fr Orn* **58**, 303–318.

Muirhead, S.J. & Furness, R.W. (1988). Heavy metal concentrations in the tissues of seabirds from Gough Island, South Atlantic Ocean. *Mar Pollut Bull* **19**, 278–283.

Munch, P.A. (1946). Sociology of Tristan da Cunha. *Results Norw Scient Exped Tristan da Cunha 1937–1938* **13**, 1–330.

Munday, B.L. (1966). "Limey disease" of mutton birds (*Puffinus tenuirostris*). In *Diseases of Tasmania's Free-living Animals* (N.A. Staley, ed.) pp. 12–16. *Tasmania Dept Agr Res Bull* **5**.

Munday, B.L., Mason, R.W., Wells, R.J.H. & Arundel, J.H. (1971). Further studies on 'Limey-Disease' of Tasmanian Mutton Birds (*Puffinus tenuirostris*). *J Wildl Diseases* **7**, 126–129.

Munro, G.C. (1947). Effect of a tidal wave on some seabird nesting islands. *Elepaio* **7**, 43–44, 51–52.

Murie, O.J. (1959). The fauna of the Aleutian Islands and Alaska Peninsula. *North Am Fauna* **61**, 1–406.

Murphy, R.C. (1918). A study of the Atlantic *Oceanites. Bull Am Mus Nat Hist* **38**, 117–146.

Murphy, R.C. (1924a). Birds collected during the Whitney South Sea Expedition. II. *Am Mus Novit* **124**, 1–13.

Murphy, R.C. (1924b). The marine ornithology of the Cape Verde Islands, with a list of all the birds of the Archipelago. *Bull Am Mus Nat Hist* **50**, 211–278.

Murphy, R.C. (1925). *Bird Islands of Peru*. Putnam, New York & London.

Murphy, R.C. (1936). *Oceanic Birds of South America*. Macmillan & Am Mus Nat Hist, New York.

Murphy, R.C. (1952). The Manx Shearwater, *Puffinus puffinus*, as a species of world-wide distribution. *Am Mus Novit* **1586**, 1–21.

Murphy, R.C. (1958). The vertebrates of Scope November 7–December 16, 1956. *US Dept Int Spec Scient Rep Fish* **279**, 101–111.

Murphy, R.C. (1964). Systematics and distribution of Antarctic petrels. In *Biologie Antarctique* (R. Carrick, M. Holdgate & J. Prevost, eds) pp. 349–358. Hermann, Paris.

Murphy, R.C. & Harper, F. (1921). A review of the Diving Petrels. *Bull Am Mus Nat Hist* **44**, 495–554.

Murphy, R.D. and Irving, S. (1951). A review of the Frigate-petrels (*Pelagodroma*). *Am Mus Novit* **1506**, 1–17.

Murphy, R.C. & Pennoyer, J.M. (1952). Larger petrels of the genus *Pterodroma*. *Am Mus Novit* **1580**, 1–43.

Murray, T.E., Bartle, J.A., Kalish, S.R. & Taylor, P.R. (1993). Incidental capture of seabirds by Japanese Southern Bluefin Tuna longline vessels in New Zealand waters, 1988–1992. *Bird Conserv Inter* **3**, 181–210.

Murrish, D.E. & Guard, C.L. (1977). Cardiovascular adaptations of the giant petrel, *Macronectes giganteus*, to the Antarctic environment. In *Adaptations Within Antarctic Ecosystems* (G. Llano, ed.) pp. 511–530. Smithson Inst, Washington.

Murrish, D.E. & Tirrell, P.C. (1981). Respiratory heat and water exchange in three species of Antarctic birds. *Antarct J US* **16**, 148–150.

Mushin, R. & Ashburner, M. (1952). Gastrointestinal microflora of mutton birds (*Puffinus tenuirostris*) healthy and affected with "Limy" disease. *J Bact* **33**, 1260–1267.

Myers, M.T. (1979). Long-term climatic and oceanographic cycles regulating seabird distributions and numbers. In *Conservation of Marine Birds of Northern North America* (J.C. Bartonek & D.N. Nettleship, eds) pp. 3–7. *US Fish Wildl Serv Rep* **11**, Washington.

Mykytowycz, R. (1963). "Limy-bird Disease" in chicks of the Tasmanian Mutton-bird (*Puffinus tenuirostris*). *Avian Diseases* **7**, 67–79.

Mykytowycz, R., Dane, D.S. & Beech, M. (1955). Ornithosis in the petrel, *Puffinus tenuirostris* (Temminck). *Austral J Exp Biol* **33**, 629–636.

Myrcha, A. & Kostelecka-Myrcha, A. (1979). Blood picture in some species of Antarctic birds. *Bull Acad Polon Sci Ser Sci Biol* **Cl 11, 27**, 911–915.

Naarding, J.A. (1980). Study of the Short-tailed Shearwater *Puffinus tenuirostris* in Tasmania. *Natn Parks Wildl Serv Tasm Rep*, 78 pp.

Nagler, C.T. & Smith, P.C. (1992). Vocalizations of nestling Leach's Storm-Petrels. *Condor* **94**, 1002–1006.

Nagy, K.A. (1989). Doubly-labeled water studies of vertebrate physiological ecology. In *Stable Isotopes in Ecological Research* (P.W. Rundel, J.R. Ehleringer & K.A. Nagy, eds) pp. 270–287. Springer-Verlag, Berlin.

Naito, Y., Nasu, K. & Suzuki, H. (1979). Distribution of the Antarctic sea-birds in the outer margin of the summer pack-ice area. *Antarct Rec (Tokyo)* **66**, 50–63.

Nakamura, K. (1982). Distribution of the gadfly petrels of the genus *Pterodroma* in the Antarctic and subAntarctic regions of the Australian Sector, austral summer 1981. *Trans Tokyo Univ Fish* **5**, 203–211.

Nakamura, K. & Hasegawa, M. (1979). A brief note on distribution of Buller's Shearwater, *Puffinus bulleri*, in Japan and the adjacent seas. *J Yamashina Inst Orn* **11**, 123–127.

Nakamura, K. & Tanaka, Y. (1977). Distribution and migration of two species of the genus *Pterodroma* in the North Pacific. *Misc Rep Yamashina Inst Orn* **9**, 112–120.

Nelson, B. (1968). *Galapagos: Islands of Birds*. Longmans, London.

Nelson, J.B. (1966). Flighting behaviour of Galapagos Storm Petrels. *Ibis* **108**, 430–432.

Nessov, L.A. (1986). [The first finding of the Late Cretaceous bird *Ichthyornis* in the old world and some other bird bones from the Cretaceous and Palaeogene of Soviet Middle Asia]. *Proc Zool Inst Leningrad* **147**, 31–38.

Nettleship, D.N. & Montgomerie, R.D. (1974). The Northern Fulmar, *Fulmarus glacialis*, breeding in Newfoundland. *Am Birds* **28**, 16.

Newcomer, M.W. & Silber, G.K. (1989). Sightings of the Laysan Albatross in the northern Gulf of California, Mexico. *West Birds* **20**, 134–135.

Newell, R.G. (1968). Influx of Great Shearwaters in autumn 1965. *Br Birds* **61**, 145–159.

Newman, D.G. (1987). Burrow use and population densities of Tuatara (*Sphenodon punctatus*) and how they are influenced by Fairy Prions (*Pachyptila turtur*) on Stephens Island, New Zealand. *Herpetofauna* **43**, 336–344.

Nicholls, D., Murray, D., Battam, H., Robertson, G., Moors, P., Butcher, E. & Hildebrandt, M. (1995). Satellite tracking of the Wandering Albatross *Diomedea exulans* around Australia and in the Indian Ocean. *Emu* **95**, 223–230.

Nicholson, J.K. & Kendall, M.D. (1983). The fine structure of dark or intercalated cells from the distal and collecting tubules of avian kidneys. *J Anat* **136**, 145–156.

Nicholson, J.K. & Osborn, D. (1983). Kidney lesions in pelagic seabirds with high tissue levels of cadmium and mercury. *J Zool Lond* **200**, 99–118.

Nishigai, M., Saeki, R., Ishitani, R., Sugimori, F., Ishibashi, Y., Oka, N. & Nakama, A. (1981). [Pathological investigations on the cause of death of Slender-billed Shearwaters collected in the Johga-shima area]. *J Yamashina Inst Orn* **13**, 82–88.

Nitzsch, C.L. (1867). *Nitzsch's Pterylography* (Transl P.L. Sclater). Royal Society, London.

Noble, D.G. & Elliott, J.E. (1986). Environmental contaminants in Canadian seabirds 1968–1985: trends and effects. *Techn Rep* **Ser 13**, Canadian Wildlife Service.

Norheim, G. (1987). Levels and interactions of heavy metals in sea birds from Svalbard and the Antarctic. *Environ Pollut* **47**, 83–94.

Norheim, G. & Kjos-Hanssen, B. (1984). Persistent chlorinated hydrocarbons and mercury in birds caught off the west coast of Spitzbergen. *Environ Pollut* **33A**, 143–152.

Norheim, G., Somme, L. & Holt, G. (1982). Mercury and persistent chlorinated hydrocarbons in Antarctic birds from Bouvetoya and Dronning Maud Land. *Environ Pollut* **28A**, 233–240.

Norman, F.I. (1967). The interactions of plants and animals on Rabbit Island, Wilson's Promontory, Victoria. *Proc R Soc Victoria* **80**, 193–200.

Norman, F.I. (1969). A note on the copulation of the Tasmanian muttonbird. *Emu* **69**, 47–49.

Norman, F.I. (1970a). Ecological effects of rabbit reduction on Rabbit Island, Wilson's Promontory, Victoria. *Proc R Soc Victoria* **83**, 235–252.

Norman, F.I. (1970b). The effects of sheep on the breeding success and habitat of the Short-tailed Shearwater, *Puffinus tenuirostris* (Temminck). *Aust J Zool* **18**, 215–229.

Norman, F.I. (1970c). Food preferences of an insular population of *Rattus rattus*. *J Zool Lond* **162**, 493–503.

Norman, F.I. (1971). Predation by the fox (*Vulpes vulpes* L.) on colonies of the Short-tailed Shearwater (*Puffinus tenuirostris* (Temminck)) in Victoria, Australia. *J Appl Ecol* **8**, 21–32.

Norman, F.I. (1975). The murine rodents *Rattus rattus, exulans,* and *norvegicus* as avian predators. *Atoll Res Bull* **182**, 1–13.

Norman, F.I. & Ward, S.J. (1992). Food and aspects of growth in the Antarctic Petrel and Southern Fulmar breeding at Hop Island, Rauer Group. *Emu* **92**, 207–222.

Norman, F.I., Whitehead, M.D., Ward, S. J. & Arnould, P.J.Y. (1992). Aspects of the breeding biology of Antarctic Petrels and Southern Fulmars in the Rauer Group, East Antarctica. *Emu* **92**, 193–206.

Nunn, G.B. (1994). A mitochondrial DNA phylogeny of the petrels Procellariiformes. *J Orn* **135**, 34.

Nunn, G.B., Cooper, J., Jouventin, P., Robertson, C.J.R. & Robertson G. (in press). Evolutionary relationships among extant albatrosses (Procellariiformes: Diomedeidae) established from complete cytochrome-*b* gene sequences. *Auk*.

Nunn, G.B., Arctander, P., Bretagnolle, V., Cooper, J., Robertson, C. & Robertson, G. (1994). Molecular phylogeny of the Albatrosses. *J Orn* **135**, 352.

O'Day, K. (1940). The fundus and fovea centralis of the albatross (*Diomedea cauta cauta* (Gould)). *Br J Opthal* **24**, 201–207.

O'Hara, (1989). Phylogeny of living penguins. Abstract 49. *Am Zool* **29**. 11A.

O'Reilly, B. (1818). *Greenland, and the Adjacent Seas, and the North-west Passage to the Pacific Ocean*, etc. Baldwin, Craddock & Joy, London.

Oatley, T.B. (1979). Underwater swimming by albatrosses. *Cormorant* **7**, 31.

Oberbauer, T.A., Cibit, C. & Lichtwardt, E. (1989). Notes from Isla Guadalupe. *West Birds* **20**, 89–90.

Obst, B.S. (1985). Densities of Antarctic seabirds at sea and the presence of the krill *Euphausia superba*. *Auk* **102**, 540–549.

Obst, B.S. (1986). Wax digestion in Wilson's Storm-petrel. *Wilson Bull* **98**, 189–195.

Obst, B.S. & Hunt, G.L. Jr (1990). Marine birds feed at Gray Whale mud plumes in the Bering Sea. *Auk* **107**, 678–688.

Obst, B.S. & Nagy, K.A. (1992). Field energy expenditure of the Southern Giant-Petrel. *Condor* **94**, 801–810.

Obst, B.S. & Nagy, K.A. (1993). Stomach oil and the energy budget of Wilson's Storm-Petrel nestlings. *Condor* **95**, 792–805.

Obst, B.S., Nagy, K.A. & Ricklefs, R.E. (1987). Energy utilization by Wilson's Storm-petrel (*Oceanites oceanicus*). *Physiol Zool* **60**, 200–210.

Ogi, H. (1984). Feeding ecology of the Sooty Shearwater in the western subarctic North Pacific Ocean. In *Marine Birds: Their Feeding Ecology and Commercial Fisheries Relationships* (D.N. Nettleship, G.A. Sanger & P.F. Springer, eds) pp. 78–84. Min Supply & Services, Canada.

Ogi, H., Kubodera, T. & Nakamura, K. (1980). The pelagic feeding ecology of the Short-tailed Shearwater (*Puffinus tenuirostris*) in the subarctic Pacific region. *J Yamashina Inst Orn* **12**, 157–182.

Ogi, H., Shimazaki, K. & Nakamura, K. (1981). Sooty Shearwaters in the subArctic North Pacific: seasonal changes in body weight and molt. *Res Inst N Pacif Fish Hokkaido Univ* **Spec Vol**, 207–215.

Ogi, H., Tanaka, H. & Yoshida, H. (1986). The occurrence of Mottled Petrels in the Bering Sea. *Mem Natn Inst Polar Res (Tokyo)* **Spec Issue 44**, 153–159.

Ogi, H., Yatsu, A., Hatanaka, H. & Nitta, A. (1993). The mortality of seabirds by Driftnet Fisheries in the North Pacific. *North Pacif Commission Bull* **53 (III)**, 499–518.

Oguma, K. (1937). Chromosome number of sea-birds new to cytology. *J Fac Sci Hokkaido Imp Univ* **Ser 6**, 265–282.

Ohlendorf, H.M. (1993). Marine birds and trace elements in the temperate North Pacific. In *The Status, Ecology, and Conservation of Marine Birds of the North Pacific* (D. Vermeer,

K.T. Briggs, K.H. Morgan & D. Siegel-Causey, eds) pp. 232–240. Can Wildl Serv, Ottawa.

Ohlendorf, H.M. & Harrison, C.S. (1986). Mercury, selenium, cadmium and organochlorines in eggs of three Hawaiian seabird species. *Environ Pollut* **11B**, 169–191.

Ohyama, Y., Naito, Y. & Nakamura, K. (1981). Visual observations of sea birds in the Southern Ocean in the 1979–1980 summer. *Antarct Rec (Tokyo)* **73**, 37–57.

Oka, N. (1994a). Underwater feeding of three shearwaters: Pale-footed (*Puffinus carneipes*), Sooty (*Puffinus griseus*) and Streaked (*Calonectris leucomelas*) Shearwaters. *J Yamashina Inst Orn* **26**, 81–84.

Oka, N. (1994b). Sustainable exploitation of Streaked Shearwaters *Calonectris leucomelas* on Mikura Island, off the Izu Peninsula, Japan. *J Yamashina Inst Orn* **26**, 99–108.

Oka, N. & Maruyama, N. (1985). Visual evaluation of tibiotarsus and femur marrows as a method of estimating nutritive conditions of Short-tailed Shearwaters. *J Yamashina Inst Orn* **17**, 57–65.

Oka, N. & Maruyama, N. (1986). Mass mortality of Short-tailed Shearwaters along the Japanese coast. *Tori* **34**, 97–104.

Okamoto, B. (1972). [*Calonectris leucomelas* on Kanmurijima Island]. Komine Shoten, Tokyo.

Okazaki, M., Oshida, Y., Maloney, R. & Warham, J. (1993). Effects of Sooty Shearwaters *Puffinus griseus* on surface soils on Motuara Island, New Zealand. *J Yamashina Inst Orn* **25**, 137–143.

Oliver, W.R.B. (1945). Avian evolution in New Zealand and Australia. *Emu* **45**, 55–72 & 119–152.

Ollason, J.C. & Dunnet, G.M. (1978). Age, experience and other factors affecting the breeding success of the fulmar, *Fulmarus glacialis*, in Orkney. *J Anim Ecol* **47**, 961–976.

Ollason, J.C. & Dunnet, G.M. (1983). Modelling annual changes in numbers of breeding Fulmars, *Fulmarus glacialis*, at a colony in Orkney. *J Anim Ecol* **52**, 185–198.

Ollason, J.C. & Dunnet, G.M. (1988). Variation in breeding success in fulmars. In *Reproductive Success* (T.H. Clutton-Brock, ed.) pp. 268–278. University of Chicago Press, Chicago.

Olson, S.L. (1975). Paleornithology of St Helena Island, South Atlantic Ocean. *Smithson Contr Paleobiol* **23**, 1–49.

Olson, S.L. (1983). Fossil seabirds and changing marine environments in the late Tertiary of South Africa. *S Afr J Sci* **72**, 399–402.

Olson, S.L. (1984). Evidence of a large albatross in the Miocene of Argentina (Aves: Diomedeidae). *Proc Biol Soc Wash* **97**, 741–743.

Olson, S.L. (1985a). Early Pliocene Procellariiformes (Aves) from Langebaanweg, South-western Cape Province, South Africa. *Ann S Afr Mus* **95**, 123–145.

Olson, S.L. (1985b). An early Pliocene marine avifauna from Duinefontein, Cape Province, South Africa. *Ann S Afr Mus* **95**, 147–164.

Olson, S.L. (1985c). The fossil record of birds. In *Avian Biology* Vol. VIII. (D.S. Farner & J.R. King, eds) pp. 79–252. Academic Press, London.

Olson, S.L. (1990). A very early observation of helicoidal intestines in the Procellariidae. *Notornis* **37**, 24–25.

Olson, S.L. & James, H.F. (1982). Prodromus of fossil avifauna of the Hawaiian Islands. *Smithson Contr Zool* **365**, 1–59.

Olson, S.L. & James, H.F. (1984). The role of the Polynesians in the extinction of the avifauna of the Hawaiian Islands. In *Quaternary Extinctions: A Prehistoric Revolution* (P.S. Martin & R.G. Klein, eds) pp. 768–780. University of Arizona Press, Tucson.

Olson, S.L. & James, H.F. (1991). Descriptions of thirty-two new species of birds from the Hawaiian Islands: Part I Non-Passeriformes. *Orn Monogr* **45**. Am Orn Union, Washington.

Olson, S.L. & Parris, D.C. (1987). The Cretaceous birds of New Jersey. *Smithson Contr Paleobiol* **63**, 1–22.

Olsthoorn, J.C.M. & Nelson, J.B. (1990). The availability of breeding sites for some British seabirds. *Bird Study* **37**, 145–164.

Ono, K. (1980). [Pliocene tubinare bird from Kakegawa, Shizuoka Prefecture, Japan]. *Mem Natn Sci Mus Tokyo* **13**, 29–34.

Orians, G.H. (1958). A capture–recapture analysis of a shearwater population. *J Anim Ecol* **27**, 71–86.

Orton, M.N. (1968). Notes on Antarctic Petrels. *Emu* **67**, 225–229.

Osborn, D., Harris, M.P. & Nicholson, J.K. (1979). Comparative tissue distribution of mercury, cadmium and zinc in three species of pelagic seabirds. *Comp Biochem Physiol* **64C**, 61–67.

Osborn, D., Harris, M.P. & Young, W.J. (1987). Relations between tissue contaminant concentrations in a small sample of seabirds. *Comp Biochem Physiol* **87C**, 415–420.

Osburn, P.I. (1911). Collecting Socorro and Black Petrels in Lower California. *Condor* **13**, 31–34.

Ovenden, J.R., Wust-Saucy, A., Bywater, R., Brothers, N. & White, R W.G. (1991). Genetic evidence for philopatry in a colonially nesting seabird, the Fairy Prion (*Pachyptila turtur*). *Auk* **108**, 688–694.

Ozawa, K. (1967). Distribution of Sea Birds in Austral Summer Season in the Southern Ocean. *Antarct Rec (Toyko)* **29**, 1–36.

Ozawa, K., Mimura, K., Egoshi, H. & Nagano, K. (1964). Observations of sea birds of the Southern Ocean (II). *J Tokyo Univ Fish Spec Ed* **7**, 1–42.

Ozawa, K., Yamada, T., Kira, M. & Shimizu, T. (1968). Observations of sea-birds in the Southern Ocean (III). *J Tokyo Univ Fish Spec Ed* **9**, 51–100.

Paganelli, C.V. (1980). The physics of gas exchange across the avian eggshell. *Am Zool* **20**, 329–338.

Page, R.D.M. (1993). *Component User Manual (Release 2.2.)* Nat Hist Mus, London.

Palma, R.L. & Pilgrim, R.L.C. (1983). The genus *Bedfordiella* (Mallophaga: Philopteridae) and a note on the lice from the Kerguelen Petrel (*Pterodroma brevirostris*). *Rec Natn Mus NZ* **2**, 145–150.

Palma, R.L. & Pilgrim, R.L.C. (1984). A revision of the genus *Harrisoniella* (Mallophaga: Philopteridae). *NZ J Zool* **11**, 145–166.

Palma, R.L. & Pilgrim, R.L.C. (1987). A revision of the genus *Perineus* (Phthiraptera: Philopteridae). *NZ J Zool* **14**, 563–586.

Palmer, R.S. (ed.) (1962). *Handbook of North American Birds*. Vol. 1. Yale University Press, New Haven and London.

Paris, P. (1913). Recherches sur la glaude uropygienne des oiseaux. *Arch Zool Exp Gen* **53**, 139–276.

Pascal, M. (1980). Structure et dynamique de la population de chats harets de l'archipel de Kerguelen. *Mammalia* **44**, 161–182.

Paterson, A.M., Gray, R.D. & Wallis, G.P. (1993). Parasites, petrels and penguins: does louse presence reflect seabird phylogeny? *Int J Parasitology* **23**, 515–526.

Paterson, A.M., Wallis, G.P. & Gray, R.D. (1995). Petrels, Penguins and Parsimony: does cladistic analysis of behavior reflect seabird phylogeny? *Evolution* **49**, 974–989.

Paterson, A.M. & Gray, R.D. (in press). Host–parasite cospeciation, host switching and missing the boat. In *Host–parasite coevolution: general principles and avian models* (D.H. Clayton & J. Moore, eds). Oxford University Press.

Paterson, A.M., Gray, R.D. & Wallis, G.P. (in press). Behavioural evolution in penguins: phylogenetic implications, homology and character evolution. *Syst Biol.*

Paulian, P. (1953). Pinnepedes, Cetaces, Oiseaux des Iles Kerguelen et Amsterdam. *Mem Inst Scient Madagascar* **Ser A (8)**, 111–234.

Paulian, P. (1959). Observations sur l'aptitude a la plongee chez les Procellariiformes. *Oiseau Revue Fr Orn* **29**, 128–130.

Payne, M.R. & Prince, P.A. (1979). Identification and breeding biology of the diving petrels *Pelecanoides georgicus* and *P. urinatrix exsul* at South Georgia. *NZ J Zool* **6**, 299–318.

Peakall, D.B., Tremblay, J., Kinter, W.B. & Miller, D.S. (1981). Endocrine dysfunction in seabirds caused by ingested oil. *Environ Res* **24**, 6–14.

Peakall, D.B., Jeffrey, D.A. & Boersma, D. (1987). Mixed-function oxidase activity in seabirds and its relationship to oil pollution. *Comp Biochem Physiol* **88C**, 151–154.

Pearce, P.A., Elliott, J.E., Peakall, D.B. & Norstrom, R.J. (1989). Organochlorine contaminants in eggs of seabirds in the Northwest Atlantic, 1968–1984. *Environ Pollut* **56**, 217–235.

Pearson, J.G. (1899). Notes on some birds of eastern North Carolina. *Auk* **16**, 246–250.

Penny, M. (1974). *The Birds of the Seychelles and Outlying Islands*. Collins, London.

Pennycuick, C.J. (1960). Gliding flight of the Fulmar Petrel. *J Exp Biol* **37**, 330–338.

Pennycuick, C.J. (1978). Fifteen testable predictions about bird flight. *Oikos* **30**, 165–176.

Pennycuick, C.J. (1982). The flight of petrels and albatrosses (Procellariiformes), observed in South Georgia and its vicinity. *Phil Trans R Soc Lond* **B300**, 75–106.

Pennycuick, C.J. (1987a). Flight of auks (Alcidae) and other northern seabirds compared with southern Procellariiformes: ornithodolite observations. *J Exp Biol* **128**, 335–347.

Pennycuick, C.J. (1987b). Flight of seabirds. In *Seabirds: Feeding Biology and Role in Marine Ecosystems* (J. P. Croxall, ed.) pp. 43–62. Cambridge University Press, Cambridge.

Pennycuick, C.J. (1989). *Bird Flight Performance. A Practical Calculation Manual*. Oxford University Press, Oxford.

Pennycuick, C.J. & Webbe, D. (1959). Observations on the Fulmar in Spitzbergen. *Br Birds* **52**, 321–332.

Pennycuick, C.J., Croxall, J.P. & Prince, P.A. (1984). Scaling of foraging radius and growth rate in petrels and albatrosses (Procellariiformes). *Ornis Scand* **15**, 145–154.

Percival, E. (1942). The juvenile plumage of some birds and an interpretation of its nature. *Trans R Soc NZ* **72**, 6–20.

Perkins, R.C.L. (1913). Vertebrata. In *Fauna Hawaiiensis*, Vol. 1 (D. Sharp, ed.) pp. 365–466. Cambridge University Press, Cambridge.

Pernkopf, E. (1937). Die Vergleichung der verschiedenen Formtypen des Vorderdarmes der Kranioten. In *Handbuch der Vergleichenden Anatomie der Wirbeltiere* (L. Bolk, E. Goppert, E. Kallius & W. Lubosch, eds) Vol. III, pp. 477–562. Urban and Schwarzenburg, Berlin.

Perrins, C.M., Harris, M.P. & Britton, C.K. (1973). Survival of Manx Shearwaters *Puffinus puffinus*. *Ibis* **115**, 535–548.

Perry, R. (1946). *A Naturalist on Lindisfarne*. Lindsay Drummond, London.

Pettit, T.N. & Whittow, G.C. (1982a). The initiation of pulmonary respiration in a bird embryo: blood and air cell gas tensions. *Resp Physiol* **48**, 199–208.

Pettit, T.N. & Whittow, G.C. (1982b). The initiation of pulmonary respiration in a bird embryo: tidal volume and frequency. *Resp Physiol* **48**, 209–218.

Pettit, T.N., Whittow, G.C. & Grant, G.S. (1981). Rete mirabile ophthalicum in Hawaiian seabirds. *Auk* **98**, 844–846.

Pettit, T.N., Grant, G.S., Whittow, G.C., Rahn, H. & Paganelli, C. (1982a). Embryonic oxygen consumption and growth of Laysan and Black-footed Albatrosses. *Am J Physiol* **242**, R121-R128.

Pettit, T.N., Grant, G.S., Whittow, G.C., Rahn, H. & Paganelli, C. (1982b). Respiratory gas exchange and growth of Bonin Petrel embryos. *Physiol Zool* **55**, 162–170.

Pettit, T.N., Whittow, G.C. & Ellis, H.I. (1984). Food and energetic requirements of seabirds at French Frigate Shoals, Hawaii. In *Resource Investigations in the Northwestern Hawaiian Islands* Vol. 2 (R.W. Grigg & K.Y. Tanoue, eds) pp. 265–282. University of Hawaii Sea Grant College Program, Honolulu.

Pettit, T.N., Ellis, H.I. & Whittow, G.C. (1985). Basal metabolic rate in tropical seabirds. *Auk* **102**, 172–174.

Pettit, T.N., Nagy, K.A., Ellis, H.I. & Whittow, G.C. (1988). Incubation energetics of the Laysan Albatross. *Oecologia (Berlin)* **74**, 546–550.

Phillipps, W.J. (1958). Some notes on muttonbirding in New Zealand. *Notornis* **7**, 189–191.

Phillips, J.H. (1963). The pelagic distribution of the Sooty Shearwater. *Ibis* **105**, 340–353.

Phillips, W.W.A. (1954). Petrels, shearwaters and other oceanic birds in the North Indian Ocean. *J Bombay Nat Hist Soc* **52**, 1–15.

Phillips, W.W.A. (1955). Wilson Petrel [*Oceanites oceanicus* (Kuhl)] in Indo-Ceylon waters, with special reference to the 1954 southward migration. *J Bombay Nat Hist Soc* **53**, 132–133.

Piatt, J.F. & Nettleship, D.N. (1987). Incidental catch of marine birds and mammals in fishing nets off Newfoundland, Canada. *Mar Pollut Bull* **18**, 344–349.

Pickering, S.P.C. (1989). Attendance patterns and behaviour in relation to experience and pair-bond formation in the Wandering Albatross *Diomedea exulans* at South Georgia. *Ibis* **131**, 183–195.

Pierce, R.J. & Parrish, G.R. (1993). Birds of Motuopao Island, Northland, New Zealand. *Tane* **34**, 59–67.

Pierotti, R. (1988). Associations between marine birds and mammals in the Northwest Atlantic Ocean. In *Seabirds and Other Marine Vertebrates: Competition, Predation and Other Interactions* (J. Burger, ed.) pp. 31–58. Columbia University Press, New York.

Pinder, R. (1966). The Cape Pigeon, *Daption capensis* Linnaeus, at Signy Island, South Orkney Islands. *Br Antarct Surv Bull* **8**, 19–47.

Pitman, R.L. (1985). The marine birds of Alijos Rocks, Mexico. *West Birds* **16**, 81–92.

Pitman, R.L. (1986). *Atlas of Seabird Distribution and Relative Abundance in the Eastern Tropical Pacific*. Administrative Rep. LT-86–02C, Southwest Fisheries Center, La Jolla.

Pitman, R.L. & Ballance, L.T. (1990). Daytime feeding by Leach's Storm-petrel on a midwater fish in the eastern Tropical Pacific. *Condor* **92**, 524–527.

Pitman, R.L. & Ballance, L.T. (1992). Parkinson's Petrel distribution and foraging ecology in the eastern Pacific: aspects of an exclusive feeding relationship with dolphins. *Condor* **94**, 825–835.

Place, A.R. (1991). The avian digestive system—an optimally designed plug-flow chemical reactor with recycle? *Acta XX Congr Int Orn* **2**, 913–919.

Place, A.R. (1992a). Comparative aspects of lipid digestion and absorption: physiological correlates of wax ester digestion. *Am J Physiol* **263**, R464-R471.

Place, A.R. (1992b). Bile is essential for lipid assimilation in Leach's Storm Petrel, *Oceanodroma leucorhoa*. *Am J Physiol* **263**, R389-R399.

Place, A.R. & Butler, R.G. (1986). The importance of bile salts to wax ester assimilation in Leach's Storm-petrel *Oceanodroma leucorhoa*. *Bull Mt Desert Island Biol Lab* **26**, 79–83.

Place, A.R. & Roby, D.D. (1986). Assimilation and deposition of dietary fatty alcohols in Leach's Storm-petrel *Oceanodroma leucorhoa*. *J Exp Zool* **240**, 149–161.

Place, A.R. & Stiles, E.W. (1992). Living off the wax of the land: bayberries and Yellow-rumped Warblers. *Auk* **109**, 334–345.

Place, A.R., Stoyan, N.C., Ricklefs, R.E. & Butler, R.G. (1989). Physiological basis of stomach oil formation in Leach's Storm-petrel (*Oceanodroma leucorhoa*). *Auk* **106**, 687–699.

Place, A.R., Sievert, P., Butler, R.G. (1991). The volume of stomach oils increases during prefledging weight loss in Leach's Storm-petrel (*Oceanodroma leucorhoa*) chicks. *Auk* **108**, 709–733.

Plant, A.R. (1989). Occurrence of filoplumes in Storm Petrels. *Seabird* **12**, 32–34.

Platania, S.P., Grant, G.S. & Lee, D. (1986). Core temperatures of non-nesting Western Atlantic Seabirds. *Brimleyana* **12**, 13–18.

Plenge, M.A., Parker, T.A., Hughes, R.A. & O'Neill, J.P. (1989). Additional notes on the distribution of birds in west Central Peru. *Gerfaut* **79**, 55–68.

Plomley, N.J.B. (1966). *Friendly Mission. The Tasmanian Journals and Papers of George Augustus Robinson 1829–1834.* Tasmanian Historical Res Assn, Hobart.

Plotz, J., Weidel, H. & Bersch, M. (1991). Winter aggregations of marine mammals and birds in the north-eastern Weddell Sea pack ice. *Polar Biol* **11**, 303–309.

Pocklington, R. (1979). An oceanographic interpretation of seabird distributions in the Indian Ocean. *Mar Biol* **51**, 9–21.

Podolsky, R.H. (1990). Effectiveness of social stimuli in attracting Laysan Albatross to new potential nesting sites. *Auk* **107**, 119–125.

Podolsky, R.H. & Kress, S.W. (1989). Factors affecting colony formation in Leach's Storm-Petrel. *Auk* **106**, 332–336.

Podolsky, R. & Kress, S.W. (1992). Attraction of the endangered Dark-rumped Petrel to recorded vocalizations in the Galapagos Islands. *Condor* **94**, 448–453.

Podolsky, R. & Kress, S. (1994). Final report summarizing results of the 1993–1994 Field Season. Attraction of Laysan Albatross to Kaohikaipu Island, Oahu. Natn Audubon Soc, Ithaca.

Poncy, R. (1926). Manx Shearwater on the Lake of Geneva. *Ibis* **12th Ser 2**, 238.

Poole, A. & Gill, F. (1993). *The Birds of North America.* Acad Nat Sci, Philadelphia & Am Orn Union, Washington.

Popova, M.F. (1972). [On morpho-functional adaptations of the neck in swimming and diving birds]. *Vestn Zool* **6**, 54–60.

Popplewell, D.L. (1918). On a partially white form of *Puffinus griseus* Gmelin. *Trans NZ Inst* **50**, 144–145.

Porter, S. (1930). Notes on the birds of Dominica. *Avicultural Mag* **4**, 146–158.

Porter, S. (1935). Notes on New Zealand birds. The Sooty Shearwater or 'Mutton Bird' (*Puffinus griseus*). *Avicult Mag* **4th Ser 13**, 34–43.

Power, D.M. & Ainley, D.G. (1986). Seabird geographic variation: similarity among populations of Leach's Storm-petrel. *Auk* **103**, 575–585.

Powers, K.D. (1982). A comparison of two methods of counting birds at sea. *J Field Orn* **53**, 209–222.

Powers, K.D. (1983). *Pelagic Distributions off the northeastern United States.* NOAA, Washington.

Powers, K.D. & Van Os, J.A. (1979). A concentration of Greater Shearwaters in the western North Atlantic. *Am Birds* **33**, 253.

Powlesland, R.G. (1985). Seabirds found dead on New Zealand beaches in 1983 and a review of albatross recoveries since 1960. *Notornis* **32**, 23–41.

Powlesland, R.G. (1986). Seabirds found dead on New Zealand beaches in 1984 and a review of fulmar recoveries since 1960. *Notornis* **33**, 171–184.

Prager, E.M. & Wilson, A.C. (1980). Phylogenetic relationships and rates of evolution in birds. *Acta XVII Congr Int Orn* **2**, 1209–1214.

Pratt, J.J. (1971). The survival of a petrel. *Elepaio* **31**, 66–67.

Prevost, (1953). Notes sur la reproduction du Fulmar Antarctique *Fulmarus glacialoides* A. Smith. *Alauda* **21**, 157–164.

Prince, P.A. (1980a). The food and feeding ecology of the Blue Petrel (*Halobaena caerulea*) and Dove Prion (*Pachyptila desolata*). *J Zool Lond* **190**, 59–76.

Prince, P.A. (1980b). The food and feeding ecology of the Grey-headed Albatross *Diomedea chrysostoma* and Black-browed Albatross *D. melanophris*. *Ibis* **122**, 476–488.

Prince, P.A. (1993). Moult in Black-browed and Grey-headed Albatrosses *Diomedea melanophris* and *D. chrysostoma*. *Ibis* **135**, 121–131.

Prince, P.A. & Croxall, J.P. (1983). Birds of South Georgia: new records and re-evaluation of status. *Br Antarct Surv Bull* **59**, 15–27.

Prince, P.A. & Francis, M.D. (1984). Activity budgets of foraging Grey-headed Albatrosses. *Condor* **86**, 297–300.

Prince, P.A. & Jones, M. (1992). Maximum dive depths attained by South Georgia diving petrel *Pelecanoides georgicus* at Bird Island, South Georgia. *Antarct Sci* **4**, 433–434.

Prince, P.A. & Morgan, R.A. (1987). Diet and feeding ecology of Procellariiformes. In *Seabirds: Feeding Biology and Role in Marine Ecosystems* (J.P. Croxall, ed.) pp. 135–171. Cambridge University Press, Cambridge.

Prince, P.A. & Rodwell, S.P. (1994). Ageing immature Black-browed and Grey-headed Albatrosses using moult, bill and plumage characteristics. *Emu* **94**, 246–254.

Prince, P.A., Wood, A.G., Barton, T. & Croxall, J.P. (1992). Satellite tracking of Wandering Albatrosses (*Diomedea exulans*) in the South Atlantic. *Antarct Sci* **4**, 31–36.

Prince, P.A., Rodwell, S., Jones, M. & Rothery, P. (1993). Moult in Black-browed and Grey-headed Albatrosses *Diomedea melanophris* and *D. chrysostoma*. *Ibis* **135**, 121–131.

Prince, P.A., Huin, N. & Weimerskirch, H. (1994a). Diving depths of albatrosses. *Antarct Sci* **6**, 353–354.

Prince, P.A., Rothery, P., Croxall, J.P. & Wood, A.G. (1994b). Population dynamics of Black-browed and Grey-headed Albatrosses *Diomedea melanophris* and *D.chrysostoma* at Bird Island, South Georgia. *Ibis* **136**, 50–71.

Pryor, M.A. (1964). [Silver-grey Fulmar and Antarctic Petrel of Haswell Islet.] *Soviet Antarct Exped Inform Bull* **50**, 38–41.

Pryor, M.E. (1968). The avifauna of Haswell Island, Antarctica. *Antarct Res Ser* **12**, 57–82.

Pulich, W. (1982). Documentation and status of Cory's Shearwater in the western Gulf of Mexico. *Wilson Bull* **94**, 381–385.

Purdy, J.S. (1900). Tuberculosis and feeding with mutton birds. *Australas Med Gazette* **1900**, 469–470.

Pycraft, W.P. (1899). Contributions to the osteology of birds. Part III. *Tubinares. Proc Zool Soc Lond* **1899**, 381–411.

Pycraft, W.P. (1907). On the systematic position of the petrels. In *A Monograph of the Petrels* (F. Du Cane Godman, ed.) pp. xv–xxxi. Witherby, London.

Pyle, R.L. (1991). The winter season. Hawaiian Islands Region. *Am Birds* **45**, 324–325.

Quay, W.B. (1972). Infrequency of pineal atrophy among birds and its relation to nocturnality. *Condor* **74**, 33–45.

Quay, W.B. (1987). Comparative study of the anal glands of birds. *Auk* **84**, 379–389.

Quinlan, S.E. (1983). Avian and River Otter predation in a storm-petrel colony. *J Wildl Manage* **47**, 1036–1043.

Rahn, H., Ackermann, R.A. & Paganelli, C.V. (1984). Eggs, yolk and embryonic growth rate. In *Seabird Energetics* (G.C. Whittow & H. Rahn, eds) pp. 89–112. Plenum, New York.

Rahn, H. & Huntington, C.E. (1988). Eggs of Leach's Storm Petrel: O_2 uptake, water loss and microclimate of the nest. *Comp Biochem Physiol* **91A**, 519–521.

Rahn, H. & Whittow, G.C. (1988). Adaptations to a pelagic life: eggs of the albatross, shearwater and petrel. *Comp Biochem Physiol* **91A**, 415–423.

Rainbow, P.S. (1989). Copper, cadmium and zinc concentrations in oceanic amphipod and euphausid crustaceans, as a source of heavy metals to pelagic seabirds. *Mar Biol* **103**, 513–518.

Randall, R.M. & Randall, B.M. (1986). The seasonal occurrence of Leach's Storm Petrel *Oceanodroma leucorhoa* at St Croix Island, South Africa. *Ostrich* **57**, 157–161.

Randi, E., Spina, F. & Massa, B. (1989). Genetic variability in Cory's Shearwater (*Calonectris diomedea*). *Auk* **106**, 411–417.

Ranft, R. & Slater, P.J.B. (1987). Absence of ultrasonic calls from night-flying storm petrels *Hydrobates pelagicus*. *Bird Study* **34**, 92–93.

Rankin, M.N. & Duffey, E.A.G. (1948). A study of the bird life of the North Atlantic. *Br Birds* **41 (Suppl)**, 1–42.

Rasmussen, E.V. (1985). [The occurrence of the Sooty Shearwater *Puffinus griseus* in Denmark]. *Dansk Orn Foren Tidsskr* **79**, 1–9.

Rau, G.H., Ainley, D.G., Bengtson, J.L., Torres, J.J. & Hopkins, T.L. (1992). $^{15}N/^{14}N$ and $^{13}C/^{12}$ in Weddell Sea birds, seals and fish: implications for diet and trophic structure. *Mar Ecol Progr Ser* **84**, 1–8.

Ray, J. (1678). *The Ornithology of Francis Willoughby*, etc. Ray Soc, London.

Rayner, J.M.V. (1983). Form and function in avian flight. *Current Ornithology* **5**, 1–66.

Rechten, C. (1985). The Waved Albatross in 1983—El Nino leads to complete breeding failure. In *El Nino en las Galapagos: El Evento de 1982–1983* (G. Robinson & E.M. del Pino, eds) pp. 227–237. Fundacion Charles Darwin par las Islas Galapagos, Quito.

Rees, E.I.S. (1961). Notes on the food of the Greater Shearwater. *Sea Swallow* **14**, 54–55.

Rees, E.I.S. (1963). Marine birds in the Gulf of St Lawrence and Strait of Belle Isle during November. *Can Fld-Nat* **77**, 98–107.

Reid, S.G. (1877). Birds of the Bermudas. *Zoologist* **3rd Ser 1**, 473–493.

Reinhardt, J. (1871). Om en hidtil ubekjendt knogle i hovedskallen hos Turakoerne (*Musophagides*, Sundev.), med nogle bemaerkniger om de lignende knogler hos andre fuglefamilier. *Vidensk Medd Danske Naturh Foren Kbn* **1871**, 326–349.

Reinhardt, T.J. (1874). Sur la structure anatomique des ailes dans la famille des petrels. *Gervais' J Zool* **3**, 139–144.

Renzoni, A., Focardi, S., Fossi, C., Leonzio, C. & Mayol, J. (1986). Comparison between concentrations of mercury and other contaminants in eggs and tissues of Cory's Shearwater *Calonectris diomedea* collected on Atlantic and Mediterranean Islands. *Environ Pollut* **40A**, 17–35.

Ribic, C.A. & Ainley, D.G. (1988/89). Constancy of seabird species assemblages: an exploratory look. *Biol Oceangr* **6**, 175–202.

Ribic, C.A., Ainley, D.G. & Spear, L.B. (1992). Effects of El Nino and La Nina on seabird assemblages in the Equatorial Pacific. *Mar Ecol Progr Ser* **80**, 109–124.

Rice, D.W. & Kenyon, K.W. (1962). Breeding cycles and behavior of Laysan and Black-footed Albatrosses. *Auk* **79**, 517–567.

Richardson, M.E. (1984). Aspects of the ornithology of the Tristan da Cuna Group and Gough Island, 1972–1974. *Cormorant* **12**, 123–201.

Richdale, L.E. (1939). A Royal Albatross nesting on the Otago Peninsula, New Zealand. *Emu* **38**, 467–488.

Richdale, L.E. (1943). The Kuaka, or diving petrel, *Pelecanoides urinatrix* (Gmelin). *Emu* **43**, 24–48.

Richdale, L.E. (1944a). The Sooty Shearwater in New Zealand. *Condor* **46**, 93–107.

Richdale, L.E. (1944b). The Titi Wainui or Fairy Prion, *Pachyptila turtur* (Kuhl). *Trans R Soc NZ* **74**, 32–48; 165–181.

Richdale, L.E. (1945). Supplementary notes on the Diving Petrel. *Trans R Soc NZ* **75**, 42–53.

Richdale, L.E. [1948]. *Maori and Mutton-bird*. Author, Dunedin.

Richdale, L.E. (1949). The pre-egg stage in Buller's Mollymawk. *Biol Mongr* **2**. Author, Dunedin.

Richdale, L.E. (1950). The pre-egg stage in the albatross family. *Biol Mongr* **3**. Author, Dunedin.

Richdale, L.E. (1952). Post-egg period in albatrosses. *Biol Mongr* **4**. Author, Dunedin.

Richdale, L.E. (1963). Biology of the Sooty Shearwater *Puffinus griseus*. *Proc Zool Soc Lond* **141**, 1–117.

Richdale, L.E. (1965a). Biology of the birds of Whero Island, New Zealand, with special reference to the diving petrel and the White-faced Storm Petrel. *Trans Zool Soc Lond* **31**, 1–86.

Richdale, L.E. (1965b). Breeding behaviour of the Narrow-billed and the Broad-billed Prion on Whero Island, New Zealand. *Trans Zool Soc Lond* **31**, 87–155.

Richdale, L.E. & Warham, J. (1973). Survival, pair bond retention and nest-site tenacity in Buller's Mollymawk. *Ibis* **115**, 257–263.

Richter, R. (1937). Eineges uber die Lebenweise des Eissturmvogels (*Fulmarus glacialis* L.). *J Orn* **85**, 187–200.

Ricklefs, R.E. (1974). Energetics of reproduction in birds. In *Avian Energetics* (R.A. Paynter, ed.) pp. 152–292. Nuttall Orn Club No. 15.

Ricklefs, R.E. (1989). Adaptations to cold in bird chicks. In *Physiology of Cold Adaptation in Birds* (C. Bech & R.E. Reinertsen, eds) pp. 329–338. Plenum, New York.

Ricklefs, R.E. & Matthew, K.K. (1983). Rates of oxygen consumption in four species of seabird at Palmer Station, Antarctic Peninsula. *Comp Biochem Physiol* **74A**, 885–888.

Ricklefs, R.E. & Roby, D.D. (1983). Development of homeothermy in the diving petrels *Pelecanoides urinatrix exsul* and *P. georgicus* and the Antarctic Prion *Pachyptila desolata*. *Comp Biochem Physiol* **75A**, 307–311.

Ricklefs, R.E., Roby, D.D. & Williams, J.B. (1986). Daily energy expenditure by adult Leach's Storm-petrels during the nesting cycles. *Physiol Zool* **59**, 649–660.

Ricklefs, R.E., White, S.C. & Cullen, J. (1980). Energetics of postnatal growth in Leach's Storm-petrel. *Auk* **97**, 566–575.

Ridoux, V. (1987). Feeding association between seabirds and Killer Whales, *Orcinus orca*, around the subantarctic Crozet Islands. *Can J Zool* **65**, 2113–2115.

Ridoux, V. (1989). Impact des oiseaux de mer sur les ressources marines autour des Iles Crozet: Estimation preliminaire. In *Actes Colloque Rech Fr Terres Australes, Strasbourg, 1987* (L. Laubier, S. Gautier & I. Martin, eds) pp. 85–92.

Ridoux, V. (1994). The diets and dietary segregation of seabirds at the subAntarctic Crozet Islands. *Mar Orn* **22**, 1–192.

Ridoux, V. & Offredo, C. (1989). The diets of five summer breeding seabirds in Adelie Land, Antarctica. *Polar Biol* **9**, 137–145.

Rijke, A.M. (1970). Wettability and phylogenetic development of feather structure in water birds. *J Exp Biol* **52**, 469–479.

Risebrough, R.W. (1974). Transfer of organochlorine pollutants to Antarctica. In *Adaptations within Antarctic Ecosystems* (G. Llano, ed.) pp. 1203–1210. Smithson Inst, Houston.

Risebrough, R.W. & Carmignani, G.M. (1972). Chlorinated hydrocarbons in Antarctic birds. In *Proc Colloquium on Conservation Problems in Antarctica, 10–12 Sept 1971, Blacksburg, Virginia, USA* (B.C. Parker, ed.) pp. 63–78. Allen Press, Lawrence, Kansas.

Ristow, D. & Wink, M. (1980). Sexual dimorphism of Cory's Shearwater. *Il-Merrill* **21**, 9–12.

Robbins, C.S. (1966). *Birds and aircraft on Midway Islands 1959–63 Investigations. Special Scient Rep Wildl* **85**, pp. 63. Dept Interior, Washington.

Robbins, C.S. & Rice, D.W. (1974). Recoveries of banded Laysan Albatrosses (*Diomedea immutabilis*) and Black-footed Albatrosses (*D. nigripes*). *Smithson Contr Zool* **158**, 232–277.

Robel, D. & Konigstedt, D. (1976). Das Vorkommen des Schwarzschnabel-Sturmtauchers (*Puffinus puffinus*) an der Westkuste des Schwarzen Meeres. *Faunist Abhandlungen Staatliches Mus Tierkunde Dresden* **6**, 1–15.

Roberson, D. & Bailey, S.F. (1991). *Cookilaria* petrels in the eastern Pacific Ocean: identification and distribution, Pts I & II. *Am Birds* **45**, 399–402, 1067–1081.

Roberts, B. (1940). The life cycle of Wilson's Petrel *Oceanites oceanicus* (Kuhl). *Br Graham Land Exped Scient Rep* **1(2)**, 141–194.

Robertson, C.J.R. (1975a). Report on the distribution status and breeding biology of the Royal Albatross, Wandering Albatross and White-capped Mollymawk on the Auckland

Islands. In *Preliminary Results of the Auckland Islands Expedition 1972–73* (J.C. Yaldwyn, ed.) pp. 143–151. Dept Lands & Survey, Wellington.

Robertson, C.J.R. (1975b). Yellow-nosed Mollymawk (*Diomedea chlororhynchos*) recorded in the Chatham Islands. *Notornis* **22**, 342–344.

Robertson, C.J.R. (1985). Article "Royal Albatross" p. 60 in *Complete Book of New Zealand Birds*. Reader's Digest, Sydney.

Robertson, C.J.R. (1991). Question on the harvesting of Toroa in the Chatham Islands. *Sci Res Ser* **35**, Dept Conservation, Wellington.

Robertson, C.J.R. (1993a). Survival and longevity of the Northern Royal Albatross *Diomedea epomophora sanfordi* at Taiaroa Head 1937–93. *Emu* **93**, 269–276.

Robertson, C.J.R. (1993b). Timing of egg laying in the Royal Albatross (*Diomedea epomophora*) at Taiaroa Head 1937–1992. *Conserv Advisory Sci Notes* **50**. Dept Conservation, Wellington.

Robertson, C.J.R. (1994). Albatross research on (Motuhara) Forty-Fours Islands: 6–15 December 1993. *Conservation Advisory Sci Notes* **70**. Dept Conservation, Wellington.

Robertson, C.J.R. & Bell, B.D. (1984). Seabird status and conservation in the New Zealand region. In *Status and Conservation of the World's Seabirds* (J. Croxall, P.G.H. Evans and R.W. Schreiber, eds) pp. 573–586. *Int Coun Bird Preserv Techn Publ* **2**, Cambridge.

Robertson, C.J.R. & Kinsky, F.C. (1972). The dispersal movements of the Royal Albatross (*Diomedea epomophora*). *Notornis* **19**, 289–301.

Robertson, C.J.R. & Wright, A. (1973). Successful hand-rearing of an abandoned Royal Albatross chick. *Notornis* **20**, 49–58.

Roby, D.D. (1991). Diet and postnatal energetics in convergent taxa of plankton-feeding seabirds. *Auk* **108**, 131–146.

Roby, D.D. & Ricklefs, R.E. (1986). Energy expenditure in adult Least Auklets and diving petrels during the chick-rearing period. *Physiol Zool* **59**, 661–678.

Roby, D.D., Place, A.R. & Ricklefs, R.E. (1986). Assimilation and deposition of wax esters in planktivorous seabirds. *J Exp Zool* **238**, 29–41.

Roby, D.D. Brink, K.L. & Place, A.R. (1989). Relative passage rates of lipid and aequeous digesta in the formation of stomach oils. *Auk* **106**, 303–313.

Rodhouse, P.G., Clarke, M.R. & Murray, A.W.A. (1987). Cephalopod prey of the Wandering Albatross *Diomedea exulans*. *Mar Biol* **96**, 1–10.

Rodhouse, P.G., Prince, P.A., Clarke, M.R. & Murray, A.W.A. (1990). Cephalopod prey of the Grey-headed Albatross *Diomedea chrysostoma*. *Mar Biol* **104**, 353–362.

Rodhouse, P.G. & Prince, P.A. (1993). Cephalopod prey of the Black-browed Albatross *Diomedea melanophris* at South Georgia. *Polar Biol* **13**, 373–376.

Rogers, A.E.F. (1980). Seabirds observed between Sydney and Buenos Aires. *Notornis* **27**, 69–78.

Rosenheim, O. & Webster, T.A. (1927). The stomach oil of the Fulmar Petrel (*Fulmarus glacialis*). *Biochem J* **21**, 111–117.

Rothery, P. & Prince, P.A. (1990). Survival and breeding frequency in albatrosses. *Ring* **13**, 61–74.

Rothschild, W. & Hartert, E. (1899). A review of the ornithology of the Galapagos Islands. *Novit Zool* **6**, 85–205.

Rounsevell, D.E. & Brothers, N.P. (1984). The status and conservation of seabirds at Macquarie Island. In *Status and Conservation of the World's Seabirds* (J.P. Croxall, P.G.H. Evans & R.W. Schreiber, eds) pp. 587–592. ICBP, Cambridge.

Routh, M. (1949). Ornithological observations in the Antarctic seas. *Ibis* **91**, 577–606.

Rowan, M.K. (1951). The Yellow-nosed Albatross *Diomedea chlororhynchos* Gmelin, at its breeding grounds in the Tristan da Cunha Group. *Ostrich* **22**, 139–155.

Rowan, M.K. (1952). The Greater Shearwater *Puffinus gravis* at its breeding grounds. *Ibis* **94**, 97–121.

Rowan, M.K. (1965). Regulation of sea-bird numbers. *Ibis* **107**, 54–59.

Rueppell, G. (1977). The course of the upper-side flow on a wing model of the Fulmar (*Fulmarus glacialis*) in slow flight. *Fortsch Zool* **24**, 287–295.

Ryan, P.G. (1987a). The incidence and characteristics of plastic particles ingested by seabirds. *Mar Environ Res* **23**, 175–206.

Ryan, P.G. (1987b). The effects of ingested plastic on seabirds: correlations between plastic load and body condition. *Environ Pollut* **46**, 119–125.

Ryan, P.G. (1988). Intra-specific variation in plastic ingestion by seabirds and the flux of plastic through seabird populations. *Condor* **90**, 446–452.

Ryan, P.G. (1991). The impact of the commercial lobster fishery on seabirds at the Tristan da Cunha Islands, South Atlantic Ocean. *Biol Conserv* **57**, 339–350.

Ryan, P.G. & Cooper, J. (1989). The distribution and abundance of aerial seabirds in relation to Antarctic Krill in the Prydz Bay region, Antarctica, during the late Summer. *Polar Biol* **10**, 199–209.

Ryan, P.G. & Jackson, S. (1987). The lifespan of ingested plastic particles in seabirds and their effect on digestive efficiency. *Mar Pollut Bull* **18**, 217–219.

Ryan, P.G. & Moloney, C.L. (1988). Effect of trawling on bird and seal distributions in the Southern Benguela region. *Mar Ecol Progr Ser* **45**, 1–11.

Ryan, P.G. & Moloney, C.L. (1991a). Prey selection and temporal variation in the diet of Subantarctic Skuas at Inaccessible Island, Tristan da Cunha. *Ostrich* **62**, 52–58.

Ryan, P.G. & Moloney, C.L. (1991b). Tristan Thrushes kill adult White-bellied Storm-Petrels. *Wilson Bull* **103**, 130–132.

Ryan, P.G. & Rose, B. (1989). Migrant Seabirds. In *Oceans of Life off Southern Africa* (A.I.L. Payne & R.J.M. Crawford, eds) pp. 274–287. Vlaeberg, Cape Town.

Ryan, P.G. & Watkins, B.P. (1989). The influence of physical factors and ornithogenic products on plant and arthropod abundance at an inland nunatak group in Antarctica. *Polar Biol* **10**, 151–160.

Ryan, P.G., Avery, G., Rose, B., Ross, G.J.B., Sinclair, J.C. & Vernon, C.J. (1989). The Southern Ocean seabird irruption to South African waters during winter 1984. *Cormorant* **17**, 41–55.

Sachs, G. (1993). Minimaler Windbedarf fur den dynamischen Segelflug der Albatrosse. *J Orn* **134**, 435–445.

Sagar, P.M. & Warham, J. (in press). Breeding biology of the Southern Buller's Mollymawks *Diomedea bulleri bulleri*. In *Proc 1st Inter Conf on the Biology and Conservation of Albatrosses* (G. Robertson, ed.). Surrey Beatty, Sydney.

Sagar, P.M. & Warham, J. (1993). A long-lived Southern Buller's Mollymawk (*Diomedea bulleri bulleri*) with a small egg. *Notornis* **40**, 303–304.

Sagar, P.M., Molloy, J., Tennyson, A.J.D. & Butler, D. (1994). Numbers of Buller's Mollymawk breeding at the Snares Islands. *Notornis* **41**, 85–92.

Saiff, E.I. (1974). The middle ear of the skull of birds. The Procellariiformes. *Zool J Linn Soc* **54**, 213–240.

Salamolard, M. & Weimerskirch, H. (1993). Relationship between foraging effort and energy requirement throughout the breeding season in the Wandering Albatross. *Funct Ecol* **7**, 643–652.

Salomonsen, F. (1935). Aves. In *Zoology of the Faroes*. Vol. 3(2) (R. Sparck, A.S. Jensen, W. Lundbeck & H. Mortensen, eds) pp. 1–269. Host, Kobenhavn.

Salomonsen, F. (1965). The geographical variation of the fulmar (*Fulmarus glacialis*) and the zones of marine environment in the North Atlantic. *Auk* **82**, 327–355.

Salvin, O. (1896). Tubinares (petrels and albatrosses). *Catalogue of the Birds of the British Museum* **25**, 340–455.

Sanger, G.A. (1974a). Laysan Albatross (*Diomedea immutabilis*). *Smithson Contr Zool* **158**, 129–153.

Sanger, G.A. (1974b). Black-footed Albatross (*Diomedea nigripes*). *Smithson Contr Zool* **158**, 96–128.

Sanger, G.A. (1983). Diets and food web relationships of seabirds in the Gulf of Alaska and adjacent marine regions. In *Final Report to the Outer Shelf Environmental Assessment Program*. US Fish Wildl Serv, Anchorage.

Sanger, G.A. (1987). Trophic levels and trophic relationships of seabirds in the Gulf of Alaska. In *Seabirds: Feeding Biology and Role in Marine Ecosystems* (J.P. Croxall, ed.) pp. 229–257. Cambridge University Press, Cambridge.

Schauinsland, H. (1903). Beitrage zur Entwicklungsgeschichte und Anatomie der Wirbeltiere. I, II, III. *Zoologica Stuttgart* **16**, 1–168.

Schmidt-Nielsen, K. (1960). The salt-secreting gland of marine birds. *Circulation* **21**, 955–967.

Schmidt-Nielsen, K., Jorgensen, C.B. & Osaki, H. (1958). Extrarenal salt excretion in birds. *Am J Physiol* **193**, 101–107.

Schmitz, E. (1894). Die Puffinen-Jagd auf den Selvagens. *Orn Jb Palaarkt Faunegeb* **5**, 19–20.

Schneider, D.C. & Duffy, D.C. (1985). Scale-dependent variability in seabird abundance. *Mar Ecol Progr Ser* **25**, 211–218.

Schneider, D.C., Hunt, G.L. Jr & Harrison, N.M. (1986). Mass and energy transfer to seabirds in the southeastern Bering Sea. *Continental Shelf Res* **5**, 241–257.

Schramm, M. (1983). Predation by subantarctic skuas *Catharacta antarctica* on burrowing petrels at Marion Island. *S Afr J Antarct Res* **13**, 41–44.

Schramm, M. (1986). The diet of chicks of Greatwinged, Kerguelen and Softplumaged Petrels at the Prince Edward Islands. *Ostrich* **57**, 9–15.

Scofield, P. (1990). Responses of Cook's and Black Petrels to artificial sounds. *Notornis* **37**, 129–130.

Scorer, R.S. (1958). *Natural Aerodynamics*. Pergamon Press, London.

Scouler, J. (1826). A voyage to Madeira, Brazil, Juan Fernandez, and the Galapagos Islands performed in 1824 and 1825, etc. *Edinb J Sci* **5**, 195–214.

Sekora, P.C., Byrd, G.V. & Gibson, D.D. (1979). Breeding distribution and status of marine birds in the Aleutian Islands, Alaska. *US Fish Wildl Serv Res Rep* **11**, 33–46.

Selby, P.J. (1833). *Illustrations of British Ornithology*. Lizars, Edinburgh.

Selous, E. (1927). *Realities of bird life*. Constable, London.

Serra, J.M. (1986). Human impact on seabirds in the Balearic Islands. *NATO ASI Ser* **G12**, 379–396.

Serventy, D.L. (1952). Movements of the Wilson Storm-petrel in Australian Seas. *Emu* **52**, 105–116.

Serventy, D.L. (1953). Movements of pelagic sea-birds in the Indo-Pacific region. *Proc 7th Pacif Sci Congr* **4**, 394–407.

Serventy, D.L. (1958). Mutton-birding. *Aust Encyclopedia* **6**, 233–234.

Serventy, D.L. (1967). Aspects of the population ecology of the Short-tailed Shearwater *Puffinus tenuirostris*. *Proc XIV Int Orn Congr* 165–190.

Serventy, D.L. & Curry, P.J. (1984). Observations on colony size, breeding success, recruitment and inter-colony dispersal in a Tasmanian colony of Short-tailed Shearwaters *Puffinus tenuirostris* over a 30-year period. *Emu* **84**, 71–79.

Serventy, D.L. & Whittell, H.M. (1962). *Birds of Western Australia*. 3rd edn. Paterson Brokensha, Perth.

Serventy, D.L., Gunn, B.M., Skira, I.J., Bradley, J.S. & Wooller, R.D. (1989). Fledgling translocation and philopatry in a seabird. *Oecologia* **81**, 428–429.

Serventy, D.L., Serventy, V.N. & Warham, J. (1971). *The Handbook of Australian Sea-birds*. Reed, Sydney.

Shallenberger, R.J. (1973). Breeding biology, homing behavior and communication patterns of the Wedge-tailed Shearwater *Puffinus pacificus chlororhynchus*. PhD thesis, University of California.

Shallenberger, R.J. (1975). Olfactory use in the Wedge-tailed Shearwater (*Puffinus pacificus*) on Manana Is. In *Olfaction & Taste V* (A. Denton & J.P. Coughlan, eds) pp. 355–359. Academic Press, London & New York.

Shaughnessy, P.D. (1970a). Serum protein of two sibling species of giant petrel (*Macronectes* spp.). *Comp Biochem Physiol* **33**, 721–723.

Shaughnessy, P.D. (1970b). The genetics of plumage phase dimorphism of the Southern Giant Petrel *Macronectes giganteus*. *Heredity* **25**, 501–506.

Shaughnessy, P.D. & Conroy, J.W.H. (1977). Further data on the inheritance of plumage phases of the Southern Giant Petrel *Macronectes giganteus*. *Br Antarct Surv Bull* **45**, 25–28.

Shaughnessy, P.D. & Fairall, N. (1976). Notes on seabirds at Gough Island. *S Afr J Antarct Res* **6**, 23–25.

Shaughnessy, P.D. & Voisin, J.-F. (1981). Observations of giant petrels *Macronectes* spp. along the Atlantic coast of Southern Africa. In *Proc Symp Birds Sea Shore* (J. Cooper, ed.) pp. 199–213. Afr Seabird Grp, Cape Town.

Shibuya, T. & Tonosaki, K. (1972). Electrical responses of single olfactory receptor cells in some vertebrates. In *Olfaction and Taste IV* (D. Schneider, ed.) pp. 102–108. Wissenschaftliche Verlag, Stuttgart.

Shiomi, K. & Ogi, H. (1992). Feeding ecology and body size dependence on diet of the Sooty Shearwater *Puffinus griseus*, in the North Pacific. *Proc NIPR Symp Polar Biol* **5**, 105–113.

Shirihai, H. (1987). Shearwaters and other tubenoses at Eilat. *Dutch Birding* **9**, 152–157.

Shufeldt, R.W. (1887). On a collection of birds' sterna and skulls, collected by Dr Thomas H. Streets, US Navy. *Proc US Natn Mus* **10**, 376–387.

Shufeldt, R.W. (1888). Observations upon the osteology of the Order Tubinares and Steganopodes. *Proc US Natn Mus* **2**, 235–315.

Shufeldt, R.W. (1907). On the osteology of the Tubinares. *Am Nat* **41**, 109–124.

Shufeldt, R.W. (1922). A comparative study of some subfossil remains of birds from Bermuda, including the 'Cahow'. *Ann Carnegie Mus Nat Hist* **13**, 333–418.

Shuntov, V.P. (1974). *Sea Birds and the Biological Structure of the Ocean*. Translation. US Dept Commerce, Springfield.

Sibley, C.G. & Ahlquist, J.E. (1990). *Phylogeny and Classification of Birds. A Study in Molecular Evolution*. Yale University Press, New Haven.

Siegel-Causey, D. & Kharitonov, S.P. (1990). The evolution of coloniality. *Current Orn* **7**, 285–330.

Siegfried, W.R. (1982). The roles of birds in ecological processes affecting the functioning of the terrestrial ecosystem at Sub-Antarctic Marion Island. *CNFRA* **51**, 493–499.

Siegfried, W.R., Williams, A.J., Burger, A.E. & Berruti, A. (1978). Mineral and energy contributions of eggs of selected species of seabirds to the Marion Island terrestrial ecosystem. *S Afr J Antarct Res* **8**, 75–87.

Sievert, P.R. & Sileo, L. (1993). The effects of ingested plastic on growth and survival of albatross chicks. In *The Status, Ecology and Conservation of Marine Birds in the North Pacific* (K. Vermeer, K.T. Briggs, K.H. Morgan & D. Siegel-Causey, eds) pp. 212–217. Canadian Wildl Serv Spec Publ, Ottawa.

Sievert, P.R., Butler, R. & Place, A.R. (1990). Total body water and its turnover in normal and salt-loaded nestling Leach's Storm-Petrels, *Oceanodroma leucorhoa*. *Bul Mt Desert Island Biol Lab* **29**, 143–146.

Sileo, L. & Fefer, S.I. (1987). Paint chip poisoning of Laysan Albatross at Midway Atoll. *J Wildl Diseases* **23**, 432–437.
Sileo, L., Sievert, P.R. & Samuel, M.D. (1990a). Causes of mortality of albatross chicks at Midway Atoll. *J Wildl Diseases* **26**, 329–338.
Sileo, L., Sievert, P.R., Samuel, M.D. & Fefer, S.I. (1990b). Prevalence and characteristics of plastic ingested by Hawaiian Seabirds. In *Proc 2nd Int Conf Marine Debris, 2–7 April 1989* (R.S. Shomura & M.L. Godfrey, eds) pp. 665–681. NOAA Tech Memo TM NHFS SWFSC-154.
Simmons, K.E.L. (1972). Some adaptive features of seabird plumage types. *Br Birds* **65**, 465–479.
Simons, T.R. (1981). Behavior and attendance patterns of the Fork-tailed Storm-Petrel. *Auk* **98**, 145–158.
Simons, T.R. (1984). A population model of the endangered Hawaiian Dark-rumped Petrel. *J Wildl Mgmt* **48**, 1065–1076.
Simons, T.R. (1985). Biology and behavior of the endangered Hawaiian Dark-rumped Petrel. *Condor* **87**, 229–245.
Simons, T.R. & Whittow, G.C. (1984). Energetics of breeding Dark-rumped Petrels. In *Seabird Energetics* (G.C. Whittow & H. Rahn, eds) pp. 159–181. Plenum, New York.
Simpson, G.G. (1946). Fossil Penguins. *Bull Am Mus Nat Hist* **87**, 1–99.
Sinclair, J.C. (1980). Subantarctic Skua *Catharacta antarctica* predation techniques on land and sea. *Cormorant* **8**, 3–6.
Skinner, H.D. & Baucke, W. (1928). The Morioris. *Mem Bernice P Bishop Mus* **9**, 343–384.
Skira, I.J. (1979). Underwater feeding by Short-tailed Shearwaters. *Emu* **79**, 43.
Skira, I.J. (1986). Food of the Short-tailed Shearwater, *Puffinus tenuirostris,* in Tasmania. *Aust Wildl Res* **13**, 481–488.
Skira, I.J. (1987). Socio-economic aspects of muttonbirding in Tasmania, Australia. In *The Value of Birds* (A.W. Diamond & F.L. Filion, eds) pp. 63–75. ICBP Techn Publ **6**, Cambridge.
Skira, I.J. (1990). Human exploitation of the Short-tailed Shearwater (*Puffinus tenuirostris). Pap Proc R Soc Tasmania* **124**, 77–90.
Skira, I.J., Wapstra, J.E., Towney, G.N. & Naarding, J.A. (1985). Conservation of the Short-tailed Shearwater *Puffinus tenuirostris* in Tasmania, Australia. *Biol Conserv* **37**, 225–236.
Sklepkovych, B.O. & Montevecchi, W.A. (1989). The world's largest known nesting colony of Leach's Storm Petrels on Baccalieu Island, Newfoundland. *Am Birds* **43**, 38–42.
Sleptsov, M.M. (1960). [Buller's Shearwater in USSR waters]. *Ornithologia* **3**, 410–412.
Smith, G.T., Wingfield, J.C. & Veit, R.R. (1994). Adrenocortical response to stress in the Common Diving Petrel, *Pelecanoides urinatrix. Physiol Zool* **67**, 526–537.
Smith, R.I.L. (1990). Signy Island as a paradigm of biological and environmental change in Antarctic ecosystems. In *Antarctic Ecosystems. Ecological Change and Conservation* (K.R. Kerry & G. Hempel, eds) pp. 32–50. Springer, Berlin.
Smith, V.R. (1976). The effect of burrowing species of Procellariidae on the nutrient status of inland tussock grasslands on Marion Island. *J S Afr Bot* **42**, 265–272.
Smith, V.R. (1977). The chemical composition of Marion Island soils, plants and vegetation. *S Afr J Antarct Res* **7**, 28–39.
Smith, V.R. (1978). Animal–plant–soil nutrient relationships on Marion Island (Subantarctic). *Oecologia, Berlin* **32**, 239–253.
Smith, V.R. (1979). The influence of seabird manuring on the phosphorus status of Marion Island (Subantarctic) soils. *Oecologia, Berlin* **41**, 123–126.
Sorensen, J.H. (1950). The Light-mantled Sooty Albatross at Campbell Island. *Cape Exp Ser Bull* **8**, 1–30. DSIR, Wellington.
Sparling, D.W. (1977). Sounds of Laysan and Black-footed Albatrosses. *Auk* **94**, 256–269.

Spear, L. & Ainley, D.G. (1993). Kleptoparasitism by Kermadec Petrels, jaegers and skuas in the eastern Tropical Pacific: evidence of mimicry by two species of *Pterodroma*. *Auk* **110**, 222–233.

Spear, L.B., Ainley, D.G. & Ribic, C.A. (1995a). Incidence of plastic in seabirds from the Tropical Pacific, 1984–91: relation with distribution of species, sex, age, season, year and body weight. *Mar Environ Res* **40**, 123–146.

Spear, L., Nur, N. & Ainley, D.G. (1992a). Estimating absolute densities of flying seabirds using analyses of relative movement. *Auk* **109**, 385–389.

Spear, L.B., Howell, S.N.G. & Ainley, D.G. (1992b). Notes on the at-sea identification of some Pacific Gadfly Petrels (Genus: *Pterodroma*). *Colonial Waterbirds* **15**, 202–218.

Spear, L.B., Ainley, D.G., Nur, N. & Howell, S.N.G. (1995b). Population size and factors affecting at-sea distributions of four endangered Procellariids in the Tropical Pacific. *Condor* **97**, 613–618.

Speich, S.M. & Pitman, R.L. (1984). River otter occurrence and predation on nesting marine birds in the Washington Islands Wilderness. *Murrelet* **65**, 25–27.

Stahl, J.-C. (1987). Distribution des oiseaux marins dans le sud-ouest de l'Ocean Indien: donnees preliminaires de la compagne Apsara II—Antiprod III. In *Les Rapports des Campagnes a la Mer MD38/APSARA II—ANTIPROD III a Bord du 'Marion Dufresne' 16 Janvier–22 Fevrier 1984* (M. Fontugne & M. Fiala, eds) pp. 175–190. Terre Austr Antarct Fr Publ Rech **84**.

Stahl, J.-C., Jouventin, P., Mougin, J.-L., Roux, J.P. & Weimerskirch, H. (1985). The foraging zones of seabirds in the Crozet Islands sector of the Southern Ocean. In *Antarctic Nutrient Cycles and Food Webs* (W.R. Siegfried, P.R. Condy & R.M. Laws, eds) pp. 478–485. Springer, Berlin.

Stallcup, R.W. (1976). Pelagic birds of Monterey Bay (revised 1981). *West Birds* **7**, 113–136.

Stanley, E. (1902). *A Familiar History of Birds*. Longmans, Green, London.

Stanley, S.M. (1973). An explanation for Cope's Rule. *Evolution* **27**, 1–26.

Starck, J.M. (1993). Evolution of avian ontogenies. *Current Orn* **10**, 275–366.

Steadman, D.W. & Olson, S.L. (1985). Bird remains from an archaeological site on Henderson Island, South Pacific: man-caused extinctions on an "uninhabited" island. *Proc Natl Acad Sci USA* **82**, 6191–6195.

Steadman, D.W., Pregill, G.K. & Olson, S.L. (1984). Fossil vertebrates from Antigua, Lesser Antilles: evidence for late Holocene human-caused extinctions in the West Indies. *Proc Natn Acad Sci USA* **81**, 4448–4451.

Steele, W.K. & Klages, N.T. (1986). Diet of the Blue Petrel at Sub-Antarctic Marion Island. *S Afr J Zool* **21**, 253–256.

Stone, C.J., Webb, A., Barton, C., Ratcliffe, N., Reed, T.C., Tasker, M.L., Camphuysen, C.J. & Pienkowski, M.W. (1995). *An Atlas of Seabird Distribution in North-west European Waters*. Joint Nat Conserv Coun, Peterborough.

Stone, W. (1900). Report on the birds and mammals collected by the McIlhenny Expedition to Pt. Barrow, Alaska. *Proc Acad Nat Sci Philadelphia* **1900**, 4–49.

Stonehouse, B. (1964). A wreck of juvenile Sooty Shearwaters (*Puffinus griseus*) in South Canterbury. *Notornis* **11**, 46–48.

Storey, A.E. (1984). Function of Manx Shearwater calls in mate attraction. *Behaviour* **89**, 73–89.

Storey, A.E. & Grimmer, B.L. (1986). Effect of illumination on the nocturnal activities of Manx Shearwaters: colony avoidance or inconspicuous behaviour. *Bird Behav* **6**, 85–89.

Storey, A.E. & Lien, J. (1985). Development of the first North American colony of Manx Shearwaters. *Auk* **102**, 395–401.

Strange, I.J. (1980). The Thin-billed Prion, *Pachyptila belcheri* at New Island, Falkland Islands. *Gerfaut* **70**, 411–445.

Stresemann, E. & Stresemann, V. (1966). Die Mauser der Vogel. *J Orn* **107 (Spec Part)**, 1–445.

Stresemann, E. & Stresemann, V. (1970). Uber Mauser und Zug von *Puffinus gravis*. *J Orn* **111**, 378–393.

Strikwerda, T.E., Fuller, M.R., Seegar, W.S., Howey, P.W. & Black, H.D. (1986). Bird-borne satellite transmitter and location program. *Johns Hopkins APL Techn Digest* **7**, 203–208.

Strong, R.M. (1911). On the olfactory organs and the sense of smell in birds. *J Morph* **22**, 619–660.

Studer, T. (1889). *Zoologie und Geologie. Die Forschungsreise S.M.S. 'Gazelle' in den Jahren 1874 bis 1876, etc.* Vol. III. Mittler, Berlin.

Sudbury, A.W., Tomkins, R.J. & Gibson, J.D. (1985). Recurring pairs of Wandering Albatrosses at a New South Wales winter feeding ground? *Emu* **85**, 195–197.

Sugimori, F., Oka, N. & Ishibashi, Y. (1985). The degree of skull ossification as a means of aging Short-tailed Shearwaters. *J Yamashina Inst Orn* **17**, 159–165.

Sultana, J. (1986). Seabird conservation problems in the Maltese Islands. In *Mediterranean Marine Avifauna* (Medmaravis & X. Monbailliu, eds) pp. 324–429. *NATO ASI* **Ser Vol G12**.

Summerhayes, C.P., Hofmeyr, P.K. & Rioux, R.H. (1974). Seabirds off the southwestern coast of Africa. *Ostrich* **45**, 83–109.

Summerhayes, V.S. & Elton, C.S. (1923). Contributions to the ecology of Spitzbergen and Bear Island. *J Ecol* **11**, 214–286.

Sundin, B. (1985). Sallsamt havsvalespel pa Skomvaer, Rost. [Storm petrels attracted by lighthouse diesel generator]. *Var Fugelfauna* **8**, 41.

Sutherland, F.L. (1965). Dispersal of pumice, supposedly from the 1962 South Sandwich Islands eruption, on southern Australian shores. *Nature Lond* **207**, 1332–1335.

Sutton, D.G. (1979). Island and coastal fowling strategies of the prehistoric Moriori. In *Birds of a Feather* (A. Anderson, ed.) pp. 123–139. *NZ Archaeol Assn Monogr U.*

Suzuki, A., Kanzaki, N., Maruyama, N. & Oka, N. (1986). [Re-examination of the migratory routes of the Short-tailed Shearwater from a simulated analysis of energy expenditure]. In *Synthetic Study of the Mass-Mortality of the Short-tailed Shearwater* Puffinus tenuirostris. pp. 43–62. Study Group Short-tailed Shearwater, Abiko.

Swales, M.P. (1965). The Sea-birds of Gough Island. *Ibis* **107**, 17–42, 215–229.

Swanson, N.M. & Merritt, F.D. (1974). The breeding cycle of the Wedge-tailed Shearwater on Mutton Bird Island, NSW. *Aust Bird Bander* **12**, 3–9.

Swatschek, I., Ristow, D. & Wink, M. (1994). Mate fidelity and parentage in Cory's Shearwater (*Calonectris diomedea*)—field studies and DNA-fingerprinting. *J Molecular Ecol* **3**, 259–262.

Swennen, C. (1974). Observations on the effect of ejection of stomach oil by the fulmar *Fulmarus glacialis* on other birds. *Ardea* **62**, 111–117.

Swennen, C. & Duiven, P. (1982). [On the behaviour of a Manx Shearwater *Puffinus P. puffinus*]. *Limosa* **55**, 140–141.

Swinburne, S. (1886). Notes on birds observed on various voyages between England and the Cape of Good Hope. *Proc R Physical Soc Edinb* **9**, 193–201.

Tambussi, C.P. & Tonni, E.P. (1988). Un Diomedeidae (Aves: Procellariiformes) del Eoceno Tardio de Antartida. *Jornados Argentino Paleontologia Vertebrados* **5**, 34.

Tanaka, Y. (1986). [Distribution and migration of the Solander's Petrel *Pterodroma solandri* in the North Pacific in relation to sea surface temperatures]. *J Yamashina Inst Orn* **18**, 55–62.

Tanaka, Y. & Inaba, F. (1981). [The distribution and migration of the White-necked Petrel *Pterodroma externa cervicalis*, in the west area of the North Pacific Ocean and the Japanese waters]. *J Yamashina Inst Orn* **13**, 61–68.

Tanaka, Y. & Kaneko, Y. (1983). [Distribution and migration of the Bonin Petrel *Pterodroma hypoleuca* in the Northwest Pacific in relation to sea surface temperatures]. *Tori* **32**, 119–127.

Tanaka, Y., Kaneko, Y. & Sato, S. (1985). Distribution and migration of smaller petrels of the genus *Pterodroma* in the Northwest Pacific. *J Yamashina Inst Orn* **17**, 23–31.

Taoka, M. & Okumura, H. (1988). Playback experiments of synthesised sounds in Leach's Storm-Petrels (*Oceanodroma leucorhoa*). *Jap Women's Univ J* **35**, 129–133.

Taoka, M. & Okumura, H. (1989). Individuality of Chatter-call and selective response to the bird's own call in Leach's Storm-petrels *Oceanodroma leucorhoa*. *Jap Women's Univ J* **36**, 107–112.

Taoka, M. & Okumura, H. (1990). Sexual differences in flight calls and the cue for vocal sex recognition of Swinhoe's Storm-Petrels. *Condor* **92**, 571–575.

Taoka, M., Sato, T. & Okumura, H. (1987). [Sexual difference and vocal sex recognition in the Purr Call of Leach's Storm-Petrel (*Oceanodroma leucorhoa*)]. *Jap Women's Univ J* **34**, 113–117.

Taoka, M., Sato, T., Kamada, T. & Okumura, H. (1988). Situation-specifics of vocalizations in Leach's Storm-Petrel *Oceanodroma leucorhoa*. *J Yamashina Inst Orn* **20**, 82–90.

Taoka, M., Sato, T., Kamada, T. & Okumura, H. (1989a). Sexual dimorphism of chatter-calls and vocal sex recognition in Leach's Storm-Petrels (*Oceanodroma leucorhoa*). *Auk* **106**, 498–501.

Taoka, M., Won, P. & Okumura, H. (1989b). Vocal behavior of Swinhoe's Storm-Petrel (*Oceanodroma monorhis*). *Auk* **106**, 471–474.

Taoka, M., Sato, T., Kamada, T. & Okumura, H. (1989c). Heterosexual response to playback calls of the Leach's Storm-Petrel *Oceanodroma leucorhoa* *J Yamashina Inst Orn* **21**, 84–89.

Tasker, M.L., Jones, P.H., Dixon, T. & Blake, B.F. (1984). Counting seabirds at sea from ships: a review of methods employed and a suggestion for a standardized approach. *Auk* **101**, 567–577.

Tasker, M.L., Webb, A., Hall, A.J., Pienowski, M.K. & Langslow, D.R. (1987). *Seabirds in the North Sea*. Nat Conserv Coun, Aberdeen.

Tazawa, H. & Whittow, G.C. (1994). Embryonic heart rate and oxygen pulse in two procellariiform seabirds, *Diomedea immutabilis* and *Puffinus pacificus*. *J Comp Physiol* **B163**, 642–648.

Telfer, T.C., Sincock, J.L., Byrd, G.V. & Reed, J.R. (1987). Attraction of Hawaiian seabirds to lights: conservation efforts and effects of moon phase. *Wildl Soc Bull* **15**, 405–413.

Temple, R.C. (1919). *The Travels of Peter Mundy, in Europe and Asia, 1608–1667*. Vol. III, Part II. Hakluyt Soc, London.

Templeman, (1945). Observations on some Newfoundland sea-birds. *Can Fld-Nat* **59**, 136–138.

Tennyson, A. & Taylor, G. (1990a). Curtis Island. *OSNZ News* **57**, 10.

Tennyson, A.J.D. & Taylor, G.A. (1990b). Behaviour of *Pterodroma* petrels in response to "war-whoops". *Notornis* **37**, 121–128.

Thibault, J.-C. (1973). Notes ornithologiques Polynesiennes. I.—Les Iles Gambier. *Alauda* **41**, 111–119.

Thibault, J.-C. (1992). Eradication of the Brown Rat from the Toro Islets (Corsica): remarks about an unwanted colonizer. *Avocetta* **16**, 114–117.

Thibault, J.-C. (1994). Nest-site tenacity and mate fidelity in relation to breeding success in Cory's Shearwater *Calonectris diomedea*. *Bird Study* **41**, 25–28.

Thibault, J.-C. & Holyoak, D.T. (1978). Vocal and olfactory displays in the genera *Bulweria* and *Pterodroma*. *Ardea* **66**, 53–56.

Thomas, D.H. (1988). Osmoregulation in wild birds: retrospect and prospects. *Acta XIX Congr Int Orn* **2**, 1511–1522.

Thomas, D.H. & Wink, M. (1985). Osmoregulation in pelagic marine birds (Procellariidae). [Abstract]. *Acta 18th Int Congr Orn* **2**, 1182.

Thomas, G. (1982). The food and feeding cology of the Light-mantled Sooty Albatross at South Georgia. *Emu* **82**, 92–100.

Thompson, d'Arcy, W. (1918). The Birds of Diomede. *Classical Rev* **32**, 92–96.

Thompson, d'Arcy, W. (1966). *A Glossary of Greek Birds*. George Olms, Hildesheim.

Thompson, D.B.A. & Thompson, P.S. (1980). Breeding Manx Shearwater *Puffinus puffinus*. *Hebridean Nat* **4**, 54–65.

Thompson, D.R. (1990). Metal levels in marine vertebrates. In *Heavy Metals in the Marine Environment* (R.W. Furness & P.S. Rainbow, eds) pp. 143–182. CRC Press, Baton Rouge.

Thompson, D.R. & Furness, R.W. (1989a). Comparison of the levels of total and organic mercury in seabird feathers. *Mar Poll Bull* **20**, 577–579.

Thompson, D.R. & Furness, R.W. (1989b). The chemical form of mercury stored in South Atlantic seabirds. *Environ Pollut* **60**, 305–317.

Thompson, D.R., Stewart, F.M. & Furness, R.W. (1990). Using seabirds to monitor mercury in marine environments. The validity of conversion ratios for tissue comparisons. *Mar Pollut Bull* **21**, 339–342.

Thompson, D.R., Furness, R.W. & Walsh, P.M. (1992). Historical changes in mercury concentrations in the marine ecosystem of the north and north-east Atlantic Ocean as indicated by seabird feathers. *J Appl Ecol* **29**, 79–84.

Thompson, D.R., Furness, R.W. & Lewis, S.A. (1993). Temporal and spatial variation in mercury concentrations in some albatrosses and petrels from the sub-Antarctic. *Polar Biol* **13**, 239–244.

Thompson, K.R. (1989). An assessment of the potential for competition between seabirds and fisheries in the Falkland Islands. *Rep Falkland Is Found, Brighton*. 94 pp.

Thompson, K.R. (1992). Quantitative analysis of the use of discards from squid trawlers by Black-browed Albatrosses *Diomedea melanophris* in the vicinity of the Falkland Islands. *Ibis* **134**, 11–21.

Thompson, K.R. & Riddy, M.D. (1995). Utilization of offal and discards from 'finfish' trawlers around the Falkland Islands by the Black-browed Albatross *Diomedea melanophris*. *Ibis* **137**, 198–206.

Thompson, K.R. & Rothery, P. (1991). A census of the Black-browed Albatross *Diomedea melanophrys* on Steeple Jason Island, Falkland Islands. *Biol Conserv* **56**, 39–48.

Thomson, A.L. (1965). The transequatorial migration of the Manx Shearwater (Puffin des Anglais). *Oiseau Revue fr Orn* **35 (No Spec)**, 130–140.

Thomson, K.S. (1976). Explanation of large scale extinctions of lower vertebrates. *Nature Lond* **261**, 578–580.

Thoresen, A.C. (1969). Observations on the breeding behaviour of the diving petrel *Pelecanoides u. urinatrix* (Gmelin). *Notornis* **16**, 241–260.

Ticehurst, C.B. (1912). Notes on a Storm Petrel in captivity. *Avicultural Mag* **3rd Ser 3**, 111–113.

Tickell, W.L.N. (1962). The Dove Prion, *Pachyptila desolata* Gmelin. *Falkland Is Depend Surv Scient Rep* **33**, 1–55.

Tickell, W.L.N. (1964). Feeding preferences of the albatrosses *Diomedea melanophris* and *D. chrysostoma* at South Georgia. In *Biologie Antarctique* (R. Carrick, M.W. Holdgate & J. Prevost, eds) pp. 383–387. Hermann, Paris.

Tickell, W.L.N. (1966). Movements of Black-browed and Grey-headed Albatrosses in the South Atlantic. *Emu* **66**, 357–367.

Tickell, W.L.N. (1968). The biology of the great albatrosses, *Diomedea exulans* and *Diomedea epomophora*. *Antarct Res Ser* **12**, 1–55.

Tickell, W.L.N. (1984). Behaviour of Blackbrowed and Greyheaded Albatrosses at Bird Island, South Georgia. *Ostrich* **56**, 64–85.

Tickell, W.L.N. (1993). Atlas of Southern Hemisphere Albatrosses. *Bull Pacific Seabird Grp* **20**, 22–38.

Tickell, W.L.N. & Pinder, R. (1972). Chick recognition by albatrosses. *Ibis* **114**, 543–548.

Timmermann, G. (1965). Die Federlingsfauna der Sturmvogel und die Phylogenese des procellariiformen Vogelstammes. *Abh Verh Naturw Ver Hamburg, NF* **Bd8 Suppl**, 1–249.

Tirrell, P.C. & Murrish, D.E. (1979). Cutaneous blood flow in the giant petrel. *Antarct J US* **14**, 169–171.

Tomkins, R.J. (1985). Breeding success and mortality of Dark-rumped Petrels in the Galapagos, and control of their predators. In *Conservation of Island Birds* (P.J. Moors, ed.) pp. 159–175. Int Coun Bird Preserv, Cambridge.

Tomkins, R.J. & Milne, B.J. (1991). Differences among Dark-rumped Petrel (*Pterodroma phaeopygia*) populations within the Galapagos Archipelago. *Notornis* **38**, 1–35.

Trivelpiece, W.Z., Butler, R.D.G., Miller, D.S. & Peakall, D.B. (1984). Reduced survival of chicks of oil-dosed adult Leach's Storm-Petrels. *Condor* **86**, 81–82.

Tulloch, R.J. (1977). Behaviour of Manx Shearwaters. *Br Birds* **70**, 35.

Turner, W. (1544). *Avium Praecipaurum, quarum apud Plinium et Aristotelem mentio est, brevisis et succinta Historia*. Gymnicus, Cologne.

Udvardy, M.D.F. (1976). Did Santa Rosalia like sea-birds? *Proc 16th Int Orn Congr*, 739–746.

Ulrich, F. (1904). Zur Kenntnis der Luftsacke bei *Diomedea exulans* und *Diomedea fuliginosa*. *Wiss Ergeb Deutsch Tiefsee-Exped "Valdivia"* **7**, 319–342.

van Aarde, R.J. (1980). The diet and feeding behaviour of feral cats *Felis catus* at Marion Island. *S Afr J Wildl Res* **10**, 123–128.

Vanderwal, R.L. & Horton, D. (1984). Coastal southwest Tasmania. *Terra Australis* **9**, 1–137.

van Franeker, J.A. (1994). A comparison of methods for counting seabirds at sea in the Southern Ocean. *J Field Orn* **65**, 96–108.

van Oordt, G.J. & Kruijt, J.P. (1953). On the pelagic distribution of some Procellariiformes in the Atlantic and Southern Oceans. *Ibis* **95**, 615–637.

van Rensburg, P.J.J. & Bester, M.N. (1988). The effect of cat *Felis catus* predation on three breeding Procellariidae species on Marion Island. *S Afr J Zool* **23**, 301–305.

van Ryzin, M.T. & Fisher, H.I. (1976). The age of Laysan Albatrosses, *Diomedea immutabilis*, at first breeding. *Condor* **78**, 1–9.

van Zinderen Bakker, E.M. Jr (1971). Comparative avian ecology. In *Marion and Prince Edward Islands* (E.M. van Zinderen Bakker Sr. J.M. Winterbottom & R.A. Dyer, eds) pp. 161–172. Balkema, Cape Town.

Vaske, T. Jnr. (1991). Seabirds mortality on longline fishing for tuna in Southern Brazil. *Ciencia Cultura* **43**, 388–390.

Vaughan, R. (1980). Notes on Cory's Shearwater (*Calonectris diomedea*) and some other birds on Linosa, Pelagic Isles. *Riv ital Orn* **50**, 143–154.

Veit, R.R. & Hunt, G.L. Jr (1991). Broadscale density and aggregation of pelagic birds from a circumnavigational survey of the Antarctic Ocean. *Auk* **108**, 790–800.

Veit, R.R. & Hunt, G.L. Jr (1992). The spatial distribution of seabirds near the South Orkney Islands and the Weddell–Scotia confluence. *Polar Biol* **11**, 637–641.

Veit, R.R., Silverman, E.D. & Everson, I. (1993). Aggregation patterns of pelagic predators and their principal prey, Antarctic Krill, near South Georgia. *J Anim Ecol* **62**, 551–564.

Veitch, C.R. (1985). Methods of eradicating cats from offshore islands in New Zealand. In *Conservation of Island Birds* (P.J. Moors, ed.) pp. 125–141. ICBP, Cambridge.

Verheyden, C. & Jouventin, P. (1994). Olfactory behavior of foraging Procellariformes. *Auk* **111**, 285–291.

Verkulich, S. & Hiller, A. (1994). Holocene deglaciation of the Bunger Hills revealed by ^{14}C

measurements on stomach oil deposits in Snow Petrel colonies. *Antarct Sci* **6**, 395–399.

Vermeer, K. & Devito, K. (1988). The importance of *Paracallisoma coecus* and myctophid fishes to nesting Fork-tailed and Leach's Storm-petrels in the Queen Charlotte Islands, British Columbia. *J Plankton Res* **10**, 63–75.

Verwey, J. (1927). A specimen of *Fulmarus glacialis* (L.) with feathered legs. *Ardea* **16**, 11–19.

Vigne, J.-D., Lefevre, C., Thibault, J.-C. & Guyot, I. (1991). Contribution archeozoologiques a l'histoire recente des oiseaux marins de l'Ile Lavezzi (Corse-XIVe–XXe Siecles). *Alauda* **59**, 11–21.

Vinson, J.-M. (1976). Notes sur les Procellariens de l'Ile Ronde. *Oiseau Revue fr Orn* **46**, 1–24.

Viot, C.R., Jouventin, P. & Bried, J. (1993). Population genetics of southern seabirds. *Mar Orn* **21**, 1–25.

Vleck, C.M. & Kenagy, G.J. (1980). Embryonic metabolism of the Fork-tailed Storm-petrel: physiological patterns during prolonged and interrupted incubation. *Physiol Zool* **53**, 32–42.

Voisin, J.-F. (1968). Les petrels geants (*Macronectes halli* et *Macronectes giganteus*) de l'Ile de la Possession. *Oiseau Revue fr Orn* **38**, 95–122.

Voisin, J.-F. (1978). Observations sur le comportement des petrels geants de l'Archipel Crozet. *Alauda* **46**, 209–234.

Voisin, J.-F. (1990). Movements of giant petrels *Macronectes* spp. banded as chicks at Iles Crozet and Kerguelen. *Mar Orn* **18**, 27–36.

Voisin, J.-F. (1991). Sur le regime et l'ecologie alimentaires des Petrels geants *Macronectes halli* et *M.giganteus* de l'archipel Crozet. *Oiseau Revue fr Orn* **61**, 39–49.

von Boetticher, H. (1930). Beitrage zur Kenntnis der Nasenrohren der Sturmvogel (Tubinares). *Senckenbergiana* **12**, 91–100.

von Boetticher, H. (1932). Die Gefiederfarbung der Sturmvogel (*Tubinares*) und ihre Beziehungen zu der allegemeinen phyletischen Entwicklung dieser Vogelgruppe. *Jenaische Z Naturwiss* **67**, 107–123.

von Boetticher, H. (1934). Der hornige Schnabeluberzug (die Rhamphotheke) der Albatrosse. *Kocsag* **7**, 40–47.

von Boetticher, H. (1955). *Albatrosse und andere Sturmvogel*. Ziemsen Verlag, Wittenberg Lutherstadt.

von Mangold, O. (1948). Die Nase der segelnden Vogel ein Organ des Stromungssinnes? *Naturwissenschaften* **33**, 19–23.

Voous, K.H. (1949). The morphological, anatomical, and distributional relationship of the Arctic and Antarctic Fulmars (Aves, Procellariidae). *Ardea* **37**, 113–122.

Voous, K.H. (1965). White-faced Storm Petrels in the Indian Ocean: Correction and additions. *Ardea* **53**, 237.

Voous, K.H. (1976). The birds of the tropical 'Middle Seas', past and present. *Proc 16th Int Orn Congr*, 697–704.

Voous, K.H. & Wattel, J. (1963). Distribution and migration of the Greater Shearwater. *Ardea* **51**, 143–157.

Wagner, R. (1840). Anatomie characteristik der Gattung *Puffinus*. In *Naturgeschichte der Vogel Deuthschlands* (J. J. Naumann, ed.) pp 614–617. Fleischer, Liepzig.

Wahl, T.R. (1978). Seabirds in the Northwestern Pacific Ocean and South Central Bering Sea in June 1975. *West Birds* **9**, 45–66.

Wahl, T.R. (1984). Observations on the diving behavior of the Northern Fulmar. *West Birds* **15**, 131–133.

Wahl, T.R. (1985). The distribution of Buller's Shearwater (*Puffinus bulleri*) in the North Pacific Ocean. *Notornis* **32**, 109–117.

Wahl, T.R., Ainley, D.G., Benedict, A.H. & DeGange, A.R. (1989). Associations between seabirds and water-masses in the northern Pacific Ocean in summer. *Mar Biol* **103**, 1–11.

Walker, C.A., Wragg, G.M., Harrison, C.J.O. (1990). A new shearwater from the Pleistocene of the Canary Islands and its bearing on the evolution of certain *Puffinus* shearwaters. *Historical Biol* **3**, 203–224.

Walker, C.H., Knight, G.C., Chipman, J.K. & Ronis, M.J.J. (1984). Hepatic microsomal monooxygenases of sea birds. *Mar Environ Res* **14**, 416–419.

Walker, K., Elliott, G., Nicholls, D., Murray, D. & Dilks, P. (1995). Satellite tracking of Wandering Albatross (*Diomedea exulans*) from the Auckland Islands: preliminary results. *Notornis* **42**, 127–137.

Walton, D.W.H. & Smith, R.I.L. (1979). The chemical composition of South Georgian vegetation. *Br Antarct Surv Bull* **49**, 117–135.

Warham, J. (1952). A family of shearwaters. *Country Life* **111**, 1250–1251

Warham, J. (1955). Observations on the Little Shearwater at the nest. *West Aust Nat* **5**, 31–39.

Warham, J. (1956). The breeding of the Great-winged Petrel, *Pterodroma macroptera*. *Ibis* **98**, 171–185.

Warham, J. (1958a). The nesting of the Little Penguin *Endyptula minor*. *Ibis* **100**, 605–616.

Warham, J. (1958b). The nesting of the Shearwater *Puffinus carneipes*. *Auk* **75**, 1–14.

Warham, J. (1960). Some aspects of breeding behaviour in the Short-tailed Shearwater. *Emu* **60**, 75–87.

Warham, J. (1962). The biology of the Giant Petrel *Macronectes giganteus*. *Auk* **79**, 139–160.

Warham, J. (1964). Marked Sooty Shearwaters (*Puffinus griseus*) in the Northern Hemisphere. *Ibis* **106**, 390–391.

Warham, J. (1967a). The White-headed Petrel *Pterodroma lessoni* at Macquarie Island. *Emu* **67**, 1–22.

Warham, J. (1967b). Snares Island birds. *Notornis* **14**, 122–139.

Warham, J. (1969). Notes on some Macquarie Island birds. *Notornis* **16**, 190–197.

Warham, J. (1971). Body temperatures of petrels. *Condor* **73**, 214–219.

Warham, J. (1972). Aspects of the biology of the Erect-crested Penguin *Eudyptes sclateri*. *Ardea* **60**, 145–184.

Warham, J. (1975). Fulmar behaviour. *Scott Birds* **8**, 319–321.

Warham, J. (1976). Aerial displays by large petrels. *Notornis* **23**, 255–357.

Warham, J. (1977a). Wing loadings, wing shapes, and flight capabilities of Procellariiformes. *NZ J Zool* **4**, 73–83.

Warham, J. (1977b). The incidence, functions and ecological significance of petrel stomach oils. *Proc NZ Ecol Soc* **24**, 84–93.

Warham, J. (1979). The voice of the Soft-plumaged Petrel (*Pterodroma mollis*). *Notornis* **26**, 357–360.

Warham, J. (1981). Does Hutton's Shearwater circumnavigate Australia? *Emu* **81**, 44.

Warham, J. (1988a). Responses of *Pterodroma* petrels to man-made sounds. *Emu* **88**, 109–111.

Warham, J. (1988b). Vocalisations of *Procellaria* petrels. *Notornis* **35**, 169–183.

Warham, J. & Bell, B.D. (1979). The birds of Antipodes Island, New Zealand. *Notornis* **26**, 121–169.

Warham, J. & Bennington, S.L. (1983). A census of Buller's Albatross *Diomedea bulleri* at the Snares Islands, New Zealand. *Emu* **83**, 112–114.

Warham, J. & Fitzsimons, C.H. (1987). The vocalisations of Buller's Mollymawk, *Diomedea bulleri* (Aves: Diomedeidae), with some comparative data on other albatrosses. *NZ J Zool* **14**, 65–79.

Warham, J. & Wilson, G.J. (1982). The size of the Sooty Shearwater population at the Snares Islands, New Zealand. *Notornis* **29**, 23–30.

Warham, J., Watts, R. & Dainty, R.J. (1976). The composition, energy content and function of the stomach oils of petrels (Order Procellariiformes). *J Exp Mar Biol Ecol* **23**, 1–13.

Warham, J., Keeley, B.R. & Wilson, G.J. (1977). Breeding of the Mottled Petrel. *Auk* **94**, 1–17.

Warman, S.R. & Warman, C.E. (1985). The food of Buzzards on Skokholm, Pembrokeshire. *Skomer Skokholm Bull* **9**, 8–9.

Wasilewski, A. (1986). Ecological aspects of the breeding cycle in the Wilson's Storm Petrel, *Oceanites oceanicus* (Kuhl), at King George Island (South Shetland Islands, Antarctica). *Polish Polar Res* **7**, 173–216.

Watabe, Y., Oka, N. & Maruyama, N. (1987). Season appearances of Short-tailed (*Puffinus tenuirostris*) and Sooty (*Puffinus griseus*) Shearwaters on the Tokyo-Kushiro Line, Japan. *J Yamashina Inst Orn* **19**, 117–124.

Watanuki, Y. (1985). Food of breeding Leach's Storm-petrels (*Oceanodroma leucorhoa*). *Auk* **102**, 884–886.

Watanuki, Y. (1986). Moonlight avoidance behavior in Leach's Storm-petrels as a defence against Slaty-backed Gulls. *Auk* **103**, 14–22.

Waters, W.E. (1964). Observations on small petrels at St Kilda, 1961–62. *Scott Birds* **3**, 73–81.

Watson, G.E. (1968). Synchronous wing and tail molt in diving petrels. *Condor* **70**, 182–183.

Watson, G.E. & Divorky, G.J. (1971). Identification of *Diomedea leptorhyncha* Coues 1866, an albatross with remarkably small salt glands. *Condor* **73**, 487–489.

Watson, G.E., Angle, J.P., Harper, P.C., Bridge, M.A., Schlatter, R.P., Tickell, W.L.N., Boyd, J.C. & Boyd, M.M. (1971). *Birds of the Antarctic and Subantarctic*. Am Geogr Soc, New York.

Watson, G.E., Lee, D.S. & Backus, E.S. (1986). Status and subspecific identity of White-faced Storm-petrels in the Western North Atlantic Ocean. *Am Birds* **40**, 401–407.

Watters, J.J. (1853). *The Natural History of the Birds of Ireland*. McGlashan, Dublin.

Watts, R. & Warham, J. (1976). Structure of the intact lipids of petrel stomach oils. *Lipids* **11**, 423–429.

Webb, A., Harrison, N.M., Leaper, G.M., Steele, R.D., Tasker, M.L. & Pienkowski, M.W. (1990). *Seabird Distribution West of Britain*. Nature Conservancy Council, Aberdeen.

Weimerskirch, H. (1982). La strategie de reproduction de l'Albatross Fuligineux a dos Sombre. *Comm Natn fr Rech Antarct* **51**, 437–447.

Weimerskirch, H. (1991). Sex-specific differences in molt strategy in relation to breeding in the Wandering Albatross. *Condor* **93**, 731–737.

Weimerskirch, H. (1992). Reproductive effort in long-lived birds: age-specific patterns of condition, reproduction and survival in the Wandering Albatross. *Oikos* **64**, 464–473.

Weimerskirch, H. & Jouventin, P. (1987). Population dynamics of the Wandering albatross, *Diomedea exulans*, of the Crozet Islands: causes and consequences of the population decline. *Oikos* **49**, 315–322.

Weimerskirch, H. & Robertson, G. (1994). Satellite tracking of Light-mantled Sooty Albatrosses. *Polar Biol* **14**, 643–652.

Weimerskirch, H. & Wilson, R.P. (1992). When do Wandering Albatrosses *Diomedea exulans* forage? *Mar Ecol Progr Ser* **86**, 297–300.

Weimerskirch, H., Jouventin, P., Mougin, J.L., Stahl, J.C. & van Beveren, M. (1985). Banding recoveries and the dispersal of seabirds breeding in French austral and Antarctic Territories. *Emu* **85**, 22–33.

Weimerskirch, H., Jouventin, P. & Stahl, J.C. (1986). Comparative ecology of six albatross species breeding on the Crozet Islands. *Ibis* **128**, 195–213.

Weimerskirch, H., Clobert, J. & Jouventin, P. (1987). Survival in five southern albatrosses and its relationship to their life history. *J Anim Ecol* **56**, 1043–1055.

Weimerskirch, H., Bartle, J.A., Jouventin, P. & Stahl, J.C. (1988). Foraging ranges and partitioning of feeding zones in three species of Southern albatross. *Condor* **90**, 214–219.

Weimerskirch, H., Zotier, R. & Jouventin, P. (1989a). The avifauna of the Kerguelen Islands. *Emu* **89**, 15–29.

Weimerskirch, H., Lequette, B. & Jouventin, P. (1989b). Development and maturation of plumage in the Wandering Albatross *Diomedea exulans. J Zool Lond* **219**, 411–421.

Weimerskirch, H., Salamolard, M. & Jouventin, P. (1992). Satellite telemetry of foraging movements in the Wandering Albatross. In *Wildlife Telemetry: Remote Monitoring and Tracking of Animals* (I.G. Priede & S.M. Swift, eds) pp. 185–198. Ellis Holwood, Chichester.

Weimerskirch, H., Salamolard, M., Sarrazin, F. & Jouventin, P. (1993). Foraging strategy of Wandering Albatrosses through the breeding season: a study using satellite telemetry. *Auk* **110**, 325–342.

Weimerskirch, H., Chastel, O., Ackermann, L., Chaurand, T., Cuenot-Chaillet, F., Hindermeyer, X. & Judas, J. (1994a). Alternate long and short foraging trips in pelagic seabird parents. *Anim Behav* **47**, 472–476.

Weimerskirch, H., Doncaster, C.P. & Cuenot-Chaillett, F. (1994b). Pelagic seabirds and the marine environment: foraging patterns of Wandering Albatrosses in relation to prey availability and distribution. *Proc R Soc Lond* **255B**, 91–97.

Weimerskirch, H., Chastel, O. & Ackermann, L. (1995). Adjustment of parental effort to manipulated foraging ability in a pelagic seabird, the Thin-billed Prion *Pachyptila belcheri. Behav Ecol Sociobiol* **36**, 11–16.

Weimerskirch, H. & Sagar, P.M. (in press). Diving depths of shearwaters. *Ibis*.

Weisler, M.I. & Gargett, R.H. (1993). Pacific island avian extinctions: the taphonomy of human predation. *Archaeol Oceania* **28**, 85–93.

Wenzel, B.M. (1967). Olfactory perception in birds. In *Olfaction and Taste 2* (T. Hayashi, ed.) pp. 203–217. Pergamon Press, Oxford.

Wenzel, B.M. (1980). Chemoreception in Seabirds. In *Behavior of Marine Animals 4* (J. Burger, B.L. Olla & H.E. Winn, eds) pp. 41–67. Plenum, New York.

Wenzel, B.M. (1986). The ecological and evolutionary challenges of procellariiform olfaction. In *Chemical Signals in Vertebrates 4* (D. Duvall, D. Muller-Schwarze & R.M. Silverstein, eds) pp. 357–368. Plenum, New York.

Wenzel, B.M. (1987). The olfactory and related systems in birds. In *The Terminal Nerve* (Nervus Terminalis). *Ann NY Acad Sci* **519**, 137–149.

Wenzel, B.M. (1991). Olfactory abilities of birds. *Acta XX Congr Int Orn* **3**, 1820–1829.

Wenzel, B.M. & Meisami, E. (1987). Number, Size, and Density of Mitral Cells in the olfactory bulbs of the Northern Fulmar and Rock Dove. In *Olfaction and Taste 9*, pp. 700–701. NY Acad Sci, New York.

Wenzel, B.M. & Meisami, E. (1990). Quantitative characteristics of the olfactory system of the Northern Fulmar (*Fulmarus glacialis*): a pattern for sensitive odor detection? In *10th Int Symp Olfaction & Taste, Oslo* (K.B. Doving, ed.) p. 317. GCS A/S, Oslo.

Wenzel, B.M. & Sieck, M.H. (1972). Olfactory perception and bulbar electrical activity in several avian species. *Physiol Behav* **9**, 287–293.

West, J.A. (1995). Chatham Petrel (*Pterodroma axillaris*)—an overview. *Notornis* **41 (Suppl)**, 19–26.

West, J.A. & Imber, M.J. (1986). Some foods of Buller's Mollymawk *Diomedea bulleri. NZ J Zool* **13**, 169–174.

West, J.A. & Nilsson, R.J. (1995). Habitat use and burrow densities of burrow-nesting seabirds on South East Island, Chatham Islands, New Zealand. *Notornis* **41 (Suppl)**, 27–37.

Westerskov, K. (1963). Ecological factors affecting the distribution of a nesting Royal Albatross population. *Proc 13th Int Orn Congr* **2**, 795–811.

Whitehead, M.D. (1991). Food resource utilization by seabirds breeding in Prydz Bay, Antarctica. *Acta XX Congr Int Orn* **3**, 1384–1392.

Whitehouse, M.J. & Veit, R.R. (1994). Distribution and abundance of seabirds and fur seals near the Antarctic Peninsula during the austral winter, 1986. *Polar Biol* **14**, 325–330.

Whitley, G. (1934). The doom of the Bird of Providence, *Pterodroma melanopus* (Gmelin). *Aust Zool* **8**, 42–49.

Whitlock, F.L. (1927). Ocean derelicts. *Emu* **26**, 154–157.

Whittow, G.C. (1984). Physiological ecology of incubation in tropical seabirds. *Studies Avian Biol* **8**, 42–72.

Whittow, G.C. (1993). Black-footed Albatross *Diomedea nigripes*. Laysan Albatross *Diomedea immutabilis*. In *The birds of North America* (A. Poole & F. Gill, eds) Nos 65 & 66. Acad Nat Sci, Philadelphia.

Whittow, G.C., Pettit, T.N., Ackerman, R.A. & Paganelli, C.V. (1987). Temperature regulation in a burrow-nesting tropical seabird, the Wedge-tailed Shearwater (*Puffinus pacificus*). *J Comp Physiol* **B157**, 607–614.

Whittow, G.C., Simons, T.R. & Pettit, T.N. (1984). Water loss from the eggs of a tropical sea bird (*Pterodroma phaeopygia*) at high altitude. *Comp Biochem Physiol* **78A**, 537–540.

Wiens, J.A. (1984). Modelling the energy requirements of seabird populations. In *Seabird Energetics* (G.C. Whittow & H. Rahn, eds) pp. 255–283. Plenum, New York.

Wilkinson, H.E. (1969). Description of an Upper Miocene albatross from Beaumaris, Victoria, Australia, and a review of fossil Diomedeidae. *Mem Natn Mus Victoria* **29**, 41–51.

Willett, G. (1919). Notes on the nesting of two little-known species of petrel. *Condor* **21**, 60–61.

Williams, A.J. (1978). Mineral and energy contributions of petrels (Procellariiformes) killed by cats, to the Marion Island terrestrial ecosystem. *S Afr J Antarct Res* **8**, 49–53.

Williams, A.J., Burger, A.E. & Berruti, A. (1978). Mineral and energy contributions of selected species of seabirds to the Marion Island terrestrial ecosystem. *S Afr J Antarct Res* **8**, 53–59.

Williams, G.R. (1960). The birds of the Pitcairn Islands, Central South Pacific Ocean. *Ibis* **102**, 58–70.

Williams, J.B. & Ricklefs, R.E. (1984). Egg temperature and embryo metabolism in some high-latitude procellariiform birds. *Physiol Zool* **57**, 118–127.

Williamson, K. (1945). The economic importance of sea-fowl in the Faeroe Islands. *Ibis* **87**, 249–269.

Willis, G.S. (1954). The Wandering Albatross. Some notes on its capture. *Sea Swallow*, **7**, 20–22.

Wilson, C.A., Cahn, R.D. & Kaplan, N.O. (1963). Functions of the two forms of lactic dehydrogenase in the breast muscles of birds. *Nature Lond* **197**, 331–334.

Wilson, J.A. (1975). Sweeping flight and soaring by albatrosses. *Nature Lond* **257**, 307–308.

Wilson, R.P. (1984). An improved stomach pump for penguins and other seabirds. *J Field Ornithol* **55**, 109–111.

Wilson, R.P., Hustler, K., Ryan, P.G., Burger, A.E. & Noldeke, E.C. (1992). Diving birds in cold water: do Archimedes and Boyle determine energetic costs? *Am Nat* **140**, 179–200.

Wingate, D.B. (1964). Discovery of breeding Black-capped Petrels on Hispaniola. *Auk* **81**, 147–159.

Wingate, D.B. (1972). First successful hand-rearing of an abandoned Bermuda Petrel chick. *Ibis* **114**, 97–101.

Wingate, D.B. (1977). Excluding competitors from Bermuda Petrel nesting burrows. In *Proc Symp on Management Techniques for Preserving Endangered Birds* (S.A. Temple, ed.) pp. 93–102. University of Wisconsin Press & Croom Helm, Madison.

Wink, M., Wink, C. & Ristow, D. (1982). Brutbiologie mediterraner Gelbschnabelsturmtaucher (*Calonectris diomedea diomedea*). *Seevogel* **Spec No**, 127–135.

Wink, M., Scharlau, W. & Ristow, D. (1987). Niedrige Ei- und Korpertemperatur ('Hypothermie') bein brutenden Gelbschnabel sturmtauchern (*Calonectris diomedea*). *J Orn* **128**, 334–338.

Wink, M., Heidrich, P., Kahl, U., Swatschek, I., Witt, H.-H. & Ristow, D. (1993a). Inter- and intraspecific variation of the nucleotide sequence of Cytochrome *b* gene in Cory's (*Calonectris diomedea*), Manx Shearwater (*Puffinus puffinus*) and the Fulmar (*Fulmarus glacialis*). *Z Naturforsch* **48c**, 504–509.

Wink, M., Heidrich, P. & Ristow, D. (1993b). Genetic evidence for speciation of the Manx Shearwater (*Puffinus puffinus*) and Mediterranean Shearwater (*Puffinus yelkouan*). *Vogelwelt* **114**, 226–232.

Winthrope, I. (1973). The shearwaters of Diomedes. *Seabird Rep* **3**, 37–40.

Witherby, H.F., Jourdain, F.C.R., Ticehurst, N.F. & Tucker, B.W. (1944). *The Handbook of British Birds. Vol. 4*. Witherby, London.

Withers, P.C. (1979). Aerodynamics and hydrodynamics of the 'hovering' flight of Wilson's Storm Petrel. *J Exp Biol* **80**, 83–91.

Wittenberger, J.F. & Hunt, G.L. Jr (1985). The adaptive significance of coloniality in birds. *Avian Biol* **8**, 2–78.

Wodzicki, K. & Laird, M. (1970). Birds and bird lore in the Tokelau Islands. *Notornis* **17**, 247–276.

Woehler, E.J. & Johnstone, G.W. (1988). Banding studies of giant petrels *Macronectes* spp., at Macquarie Island. *Pap Proc R Soc Tasm* **122**, 143–152.

Wood, C.A. (1917). *The Fundus Oculi of Birds especially as viewed by the Ophthalmoscope*. Lakeside Press, Chicago.

Wood, C.J. (1973). The flight of albatrosses (A computer simulation). *Ibis* **115**, 244–256.

Wood, K.A. (1990a). Temporal and zonal patterns of abundance of shearwaters (*Puffinus*) off central New South Wales. *Aust Wildl Res* **17**, 453–466.

Wood, K.A. (1990b). Seasonal abundance and marine habitats of *Procellaria* fulmarine and gadfly petrels off central New South Wales. *Notornis* **37**, 81–105.

Wood, K.A. (1990c). Seasonal abundance and marine habitat of storm-petrels (Oceanitidae) off central New South Wales. *Corella* **14**, 37–43.

Wood, K.A. (1992). Seasonal abundance and spatial distribution of albatrosses off central New South Wales. *Aust Bird Watcher* **14**, 207–225.

Wood, K.A. (1993). Feeding behaviour, offal preferences and tarsus shape of *Puffinus* shearwaters off Central New South Wales. *Notornis* **40**, 123–127.

Wood-Jones, F. (1937). The olfactory organ of the Tubinares. *Emu* **36**, 281–286; **37**, 10–13 & 128–131.

Woods, R.W. (1970). The avian ecology of a tussock island in the Falkland Islands. *Ibis* **112**, 15–24.

Woodward, P.W. (1972). The natural history of Kure Atoll, Northwestern Hawaiian Islands. *Atoll Res Bull* **164**, 318 pp.

Wooller, R.D., Bradley, J.S., Serventy, D.L. & Skira, I.J. (1988). Factors contributing to reproductive success in Short-tailed Shearwaters (*Puffinus tenuirostris*). *Acta XIX Congr Int Orn* **1**, 848–856.

Wooller, R.D., Bradley, J.S., Skira, I.J. & Serventy, D.L. (1989). Short-tailed Shearwater. In *Lifetime Reproduction in Birds* (I. Newton, ed.) pp. 405–417. Academic Press, London.

Wooller, R.D., Bradley, J.S., Skira, I.J. & Serventy, D.L. (1990). The reproductive success of Short-tailed Shearwaters *Puffinus tenuirostris* in relation to their age and breeding experience. *J Anim Ecol* **59**, 161–170.

Wooller, R.D., Bradley, J.S. & Croxall, J.P. (1992). Long-term population studies of seabirds. *Tree* **7**, 111–114.

Wormell, P. (1969). Red Deer (*Cervus elaphus*) as predator on Manx Shearwater (*Procellaria puffinus*). *Deer* **1**, 289.

Wormell, P. (1976). The Manx Shearwaters of Rhum. *Scott Birds* **9**, 103–118.

Wynne-Edwards, V.C. (1935). On the habits and distribution of birds in the North Atlantic. *Proc Boston Soc Nat Hist* **40**, 233–346.

Wynne-Edwards, V.C. (1939). Intermittent breeding of the Fulmar (*Fulmarus glacialis* (L.)), with some general observations on non-breeding in sea-birds. *Proc Zool Soc Lond* **109A**, 127–132.

Wynne-Edwards, V.C. (1962). *Animal Dispersion in Relation to Social Behaviour*. Oliver & Boyd, Edinburgh.

Yamashina, Y. & Udagawa, T. (1954). The chromosomes of the Streaked Shearwater *Puffinus leucomelas*. *Misc Rep Yamashina Inst Orn Zool* **5**, 220–221.

Yesner, D.R. (1976). Aleutian Island albatrosses: a population history. *Auk* **93**, 263–280.

Yesou, P. (1982). A propos de la presence remarquable du Puffin Cendre *Calonectris diomedea* pres des cotes du golfe de Gascogne et de la mer Celtique en 1980. *Oiseau Revue fr Orn* **52**, 197–217.

Yocom, C. (1947). Notes on behavior and abundance of the Black-footed Albatross in the Pacific water off the continental North American shores. *Auk* **64**, 507–523.

Yoshida, H. & Ogi, H. (1987). [Morphological and ecological differences between Short-tailed Shearwaters and Sooty Shearwaters]. *Aquabiol Tokyo* **9**, 274–279.

Yoshida, N. (1981). [*Seabirds that Climb Trees*.] Ubunsha, Tokyo.

Yudin, K.A. (1957). [Certain adaptive peculiarities of the wing in the birds of the order Tubinares]. *Zool Zh* **36**, 1859–1873.

Yudin, K.A. (1961). [On the mechanisms of the jaw in Charadriiformes, Procellariiformes and some other birds]. *Trudy Zool Inst Leningrad* **29**, 257–302.

Zammit, R.C. & Borg, J. (1986–87). Notes on the breeding of the Cory's Shearwater in the Maltese Islands. *Il-Merill* **24**, 1–9.

Zhang, Q. & Whittow, G.C. (1992). Embryonic oxygen consumption and organ growth in the Wedge-tailed Shearwater. *Growth Development Aging* **56**, 205–214.

Zink, R.M. (1978). Birds of the Weddell Sea. *Antarct J US* **13**, 142–145.

Zink, R.M. (1981). Observations of seabirds during a cruise from Ross Island to Anvers Island, Antarctica. *Wilson Bull* **93**, 1–20.

Zino, P.A. (1971). The breeding of Cory's Shearwater *Calonectris diomedea* on the Salvage Islands. *Ibis* **113**, 212–217.

Zino, P.A. (1985). A short history of the Shearwater hunt on the Great Salvage and recent developments on this island. *Bocagiana* **84**, 1–9.

Zino, P.A. & Zino, F. (1986). Contribution to the study of the petrels of the genus *Pterodroma* in the Archipelago of Madeira. *Bol Mus Municipal Funchal* **38**, 141–165.

Zonfrillo, B. (1982a). The response of Storm Petrels to calls of other species. *Scott Birds* **12**, 85–86.

Zonfrillo, B. (1982b). Dark eye-patch of Fulmar. *Br Birds* **75**, 225–226.

Zonfrillo, B. (1993). Relationships of the *Pterodroma* petrels from the Madeira Archipelago inferred from their feather lice. *Bol Mus Mun Funchal* **Sup 2**, 325–331.

Species Index

Subject Index